Advances in Experimental Medicine and Biology

Volume 1210

Series Editors
Wim E. Crusio, *CNRS and University of Bordeaux UMR 5287, Institut de Neurosciences Cognitives et Intégratives d'Aquitaine, Pessac Cedex, France*
John D. Lambris, *University of Pennsylvania, Philadelphia, PA, USA*
Nima Rezaei, *Children's Medical Center Hospital, Tehran University of Medical Sciences, Tehran, Iran*

More information about this series at http://www.springer.com/series/5584

Scott M. Dehm • Donald J. Tindall
Editors

Prostate Cancer

Cellular and Genetic Mechanisms
of Disease Development
and Progression

Second Edition

Editors
Scott M. Dehm
Department of Laboratory Medicine
and Pathology
Department of Urology
Masonic Cancer Center
University of Minnesota
Minneapolis, MN, USA

Donald J. Tindall
Departments of Urology, and
Biochemistry & Molecular Biology
Mayo Clinic College of Medicine
Rochester, MN, USA

ISSN 0065-2598 ISSN 2214-8019 (electronic)
Advances in Experimental Medicine and Biology
ISBN 978-3-030-32658-6 ISBN 978-3-030-32656-2 (eBook)
https://doi.org/10.1007/978-3-030-32656-2

© Springer Nature Switzerland AG 2013, 2019
This work is subject to copyright. All rights are reserved by the Publisher, whether the whole or part of the material is concerned, specifically the rights of translation, reprinting, reuse of illustrations, recitation, broadcasting, reproduction on microfilms or in any other physical way, and transmission or information storage and retrieval, electronic adaptation, computer software, or by similar or dissimilar methodology now known or hereafter developed.
The use of general descriptive names, registered names, trademarks, service marks, etc. in this publication does not imply, even in the absence of a specific statement, that such names are exempt from the relevant protective laws and regulations and therefore free for general use.
The publisher, the authors, and the editors are safe to assume that the advice and information in this book are believed to be true and accurate at the date of publication. Neither the publisher nor the authors or the editors give a warranty, express or implied, with respect to the material contained herein or for any errors or omissions that may have been made. The publisher remains neutral with regard to jurisdictional claims in published maps and institutional affiliations.

This Springer imprint is published by the registered company Springer Nature Switzerland AG
The registered company address is: Gewerbestrasse 11, 6330 Cham, Switzerland

Preface

Prostate cancer is the most frequently diagnosed non-cutaneous cancer in men and the second leading cause of male cancer deaths in the United States. The two main risk factors for developing prostate cancer are being male and aging, with additional contributions from genetic and environmental factors. These genetic and environmental factors may also account for disparities in prostate cancer among certain populations, such as men of African ancestry. The development of new technologies for characterization of prostate cancer specimens at the DNA, RNA, protein, microenvironment, and metabolite level has identified multiple ways the disease can be categorized into specific subtypes. These subtypes may be reflective of the original, single cell within a prostate gland that accumulates sufficient genetic damage—mutations, copy number alterations, and chromosomal rearrangements—to progress to prostate cancer. Identifying and characterizing the prostate cancer cell of origin could illuminate strategies to prevent carcinogenesis, cellular transformation, and ultimately prevent prostate cancer. Additionally, defining prostate cancer subtypes based on genomic, transcriptomic, or other features has clinical relevance because these subtypes have been shown to provide prognostic information about disease course or predictive information about therapeutic responses. Finally, investigating how the various cell types and metabolites within a localized or metastatic prostate tumor interact and communicate could provide opportunities for developing new targeted therapies.

The purpose of this book is to provide a contemporary overview of the causes and consequences of prostate cancer within these cellular and genetic frameworks. This book will provide an overview on key cellular and genetic aspects of prostate cancer, written by experts in the fields of epidemiology, toxicology, cell biology, genetics, genomics, cell–cell interactions, cell signaling, hormone signaling, and transcriptional regulation. The following subjects will be reviewed:

- The interplay of genetics, environment, lifestyle, and diet in the development and progression of prostate cancer.
- The cell types in the normal prostate gland that are susceptible to mutagenesis and transformation to become prostate cancer, and how they contribute to different subtypes of prostate cancer.
- Heterogeneity of localized and metastatic prostate cancer and subclassification schemes based on genomic and transcriptomic profiles that are being used to facilitate patient stratification.

- The various cell types that comprise a prostate cancer when the disease is localized to the prostate, or has metastasized to distant sites.
- The metabolic milieu and energetic requirements of prostate cancer cells.

Endocrine and cell intrinsic signal transduction and transcriptional regulatory pathways that control prostate cancer cell proliferation, survival, invasion, and metastasis.

Overall, these chapters should provide the reader with a thorough understanding of the origin, development, and progression of prostate cancer from a cellular and genetic perspective. This book will be distinguished from other books on prostate cancer because of this focus on cellular and genetic mechanisms as opposed to clinical diagnosis and management. As a result, this book will be of broad interest to basic and translational scientists with familiarity of these topics, as well as trainees at earlier stages of their research careers.

Minneapolis, MN, USA   Scott M. Dehm
Rochester, MN, USA   Donald J. Tindall

Contents

Diet and Lifestyle in Prostate Cancer   1
Kathryn M. Wilson and Lorelei A. Mucci

Dietary Carcinogens and DNA Adducts in Prostate Cancer   29
Medjda Bellamri and Robert J. Turesky

**Genetic, Environmental, and Nuclear Factors Governing
Genomic Rearrangements**   57
Susmita G. Ramanand and Ram S. Mani

Cells of Origin for Prostate Cancer   67
Li Xin

Prostate Cancer Genomic Subtypes   87
Michael Fraser and Alexandre Rouette

Prostate Cancer Transcriptomic Subtypes   111
Daniel E. Spratt

**Immunological Complexity of the Prostate Cancer
Microenvironment Influences the Response to Immunotherapy** . . .   121
Nataliya Prokhnevska, Dana A. Emerson, Haydn T. Kissick, and
William L. Redmond

The Tumor Microenvironments of Lethal Prostate Cancer   149
William L. Harryman, Noel A. Warfel, Raymond B. Nagle, and
Anne E. Cress

The Bone Microenvironment in Prostate Cancer Metastasis   171
Anthony DiNatale and Alessandro Fatatis

Prostate Cancer Energetics and Biosynthesis   185
Chenchu Lin, Travis C. Salzillo, David A. Bader, Sandi R.
Wilkenfeld, Dominik Awad, Thomas L. Pulliam, Prasanta Dutta,
Shivanand Pudakalakatti, Mark Titus, Sean E. McGuire,
Pratip K. Bhattacharya, and Daniel E. Frigo

**Canonical and Noncanonical Androgen Metabolism
and Activity**   239
Karl-Heinz Storbeck and Elahe A. Mostaghel

vii

Germline and Somatic Defects in DNA Repair Pathways in Prostate Cancer 279
Sara Arce, Alejandro Athie, Colin C. Pritchard, and Joaquin Mateo

The Role of RB in Prostate Cancer Progression 301
Deborah L. Burkhart, Katherine L. Morel, Anjali V. Sheahan,
Zachary A. Richards, and Leigh Ellis

Interplay Among PI3K/AKT, PTEN/FOXO and AR Signaling in Prostate Cancer 319
Yuqian Yan and Haojie Huang

Androgen Receptor Dependence 333
Aashi P. Chaturvedi and Scott M. Dehm

Wnt/Beta-Catenin Signaling and Prostate Cancer Therapy Resistance 351
Yunshin Yeh, Qiaozhi Guo, Zachary Connelly, Siyuan Cheng,
Shu Yang, Nestor Prieto-Dominguez, and Xiuping Yu

Epigenetic Regulation of Chromatin in Prostate Cancer 379
Ramakrishnan Natesan, Shweta Aras, Samuel Sander Effron,
and Irfan A. Asangani

Oncogenic ETS Factors in Prostate Cancer 409
Taylor R. Nicholas, Brady G. Strittmatter, and Peter C. Hollenhorst

Neural Transcription Factors in Disease Progression 437
Daksh Thaper, Sepideh Vahid, and Amina Zoubeidi

Index 463

Contributors

Shweta Aras Department of Cancer Biology, Abramson Family Cancer Research Institute, Epigenetics Institute, Perelman School of Medicine, University of Pennsylvania, Philadelphia, PA, USA

Sara Arce Vall d'Hebron Institute of Oncology, Barcelona, Spain

Irfan A. Asangani Department of Cancer Biology, Abramson Family Cancer Research Institute, Epigenetics Institute, Perelman School of Medicine, University of Pennsylvania, Philadelphia, PA, USA

Alejandro Athie Vall d'Hebron Institute of Oncology, Barcelona, Spain

Dominik Awad Department of Cancer Systems Imaging, The University of Texas MD Anderson Cancer Center, Houston, TX, USA

The University of Texas MD Anderson Cancer Center, UTHealth Graduate School of Biomedical Sciences, Houston, TX, USA

David A. Bader Department of Molecular and Cellular Biology, Baylor College of Medicine, Houston, TX, USA

Medjda Bellamri Department of Medicinal Chemistry, Cancer and Cardiovascular Research Building, University of Minnesota, Minneapolis, MN, USA

Masonic Cancer Center, University of Minnesota, Minneapolis, MN, USA

Pratip K. Bhattacharya Department of Cancer Systems Imaging, The University of Texas MD Anderson Cancer Center, Houston, TX, USA

The University of Texas Health Science Center at Houston, Houston, TX, USA

Deborah L. Burkhart Department of Oncologic Pathology, Dana-Farber Cancer Institute, Boston, MA, USA

Aashi P. Chaturvedi Masonic Cancer Center, University of Minnesota, Minneapolis, MN, USA

Siyuan Cheng Department of Biochemistry & Molecular Biology, LSU Health Sciences Center, Shreveport, LA, USA

Zachary Connelly Department of Biochemistry & Molecular Biology, LSU Health Sciences Center, Shreveport, LA, USA

Anne E. Cress University of Arizona Cancer Center, Tucson, AZ, USA

Scott M. Dehm Masonic Cancer Center, University of Minnesota, Minneapolis, MN, USA

Department of Laboratory Medicine and Pathology, University of Minnesota, Minneapolis, MN, USA

Department of Urology, University of Minnesota, Minneapolis, MN, USA

Anthony DiNatale Department of Pharmacology and Physiology, Drexel University College of Medicine, Philadelphia, PA, USA

Program in Prostate Cancer, Sidney Kimmel Cancer Center, Thomas Jefferson University, Philadelphia, PA, USA

Prasanta Dutta Department of Cancer Systems Imaging, The University of Texas MD Anderson Cancer Center, Houston, TX, USA

Samuel Sander Effron Department of Cancer Biology, Abramson Family Cancer Research Institute, Epigenetics Institute, Perelman School of Medicine, University of Pennsylvania, Philadelphia, PA, USA

Leigh Ellis Department of Oncologic Pathology, Dana-Farber Cancer Institute, Boston, MA, USA

Department of Pathology, Brigham and Women's Hospital, Boston, MA, USA

The Broad Institute, Cambridge, MA, USA

Dana A. Emerson Molecular Microbiology and Immunology, Oregon Health and Science University, Portland, OR, USA

Earle A. Chiles Research Institute, Providence Cancer Institute, Portland, OR, USA

Alessandro Fatatis Department of Pharmacology and Physiology, Drexel University College of Medicine, Philadelphia, PA, USA

Program in Prostate Cancer, Sidney Kimmel Cancer Center, Thomas Jefferson University, Philadelphia, PA, USA

Michael Fraser Computational Biology Program, Ontario Institute for Cancer Research, Toronto, ON, Canada

Daniel E. Frigo Department of Cancer Systems Imaging, The University of Texas MD Anderson Cancer Center, Houston, TX, USA

Center for Nuclear Receptors and Cell Signaling, University of Houston, Houston, TX, USA

Department of Biology and Biochemistry, University of Houston, Houston, TX, USA

Department of Genitourinary Medical Oncology, The University of Texas MD Anderson Cancer Center, Houston, TX, USA

Molecular Medicine Program, The Houston Methodist Research Institute, Houston, TX, USA

Qiaozhi Guo Merton College, University of Oxford, Oxford, UK

William L. Harryman University of Arizona Cancer Center, Tucson, AZ, USA

Peter C. Hollenhorst Medical Sciences, Indiana University School of Medicine, Bloomington, IN, USA

Haojie Huang Department of Biochemistry and Molecular Biology, Mayo Clinic College of Medicine and Science, Rochester, MN, USA

Department of Urology, Mayo Clinic College of Medicine and Science, Rochester, MN, USA

Mayo Clinic Cancer Center, Mayo Clinic College of Medicine and Science, Rochester, MN, USA

Haydn T. Kissick Department of Urology, Emory University, Atlanta, GA, USA

Chenchu Lin Department of Cancer Systems Imaging, The University of Texas MD Anderson Cancer Center, Houston, TX, USA

The University of Texas MD Anderson Cancer Center, UTHealth Graduate School of Biomedical Sciences, Houston, TX, USA

Ram S. Mani Department of Pathology, UT Southwestern Medical Center, Dallas, TX, USA

Department of Urology, UT Southwestern Medical Center, Dallas, TX, USA

Harold C. Simmons Comprehensive Cancer Center, UT Southwestern Medical Center, Dallas, TX, USA

Joaquin Mateo Vall d'Hebron Institute of Oncology, Barcelona, Spain

Sean E. McGuire Department of Molecular and Cellular Biology, Baylor College of Medicine, Houston, TX, USA

Department of Radiation Oncology, The University of Texas MD Anderson Cancer Center, Houston, TX, USA

Katherine L. Morel Department of Oncologic Pathology, Dana-Farber Cancer Institute, Boston, MA, USA

Elahe A. Mostaghel Clinical Research Division, Fred Hutchinson Cancer Research Center, Seattle, WA, USA

Department of Medicine, University of Washington, Seattle, WA, USA

Geriatric Research, Education and Clinical Center, VA Puget Sound Health Care System, Seattle, WA, USA

Lorelei A. Mucci Channing Division of Network Medicine, Harvard Medical School, Brigham and Women's Hospital, Boston, MA, USA

Raymond B. Nagle Department of Pathology, University of Arizona Cancer Center, Tucson, AZ, USA

Ramakrishnan Natesan Department of Cancer Biology, Abramson Family Cancer Research Institute, Epigenetics Institute, Perelman School of Medicine, University of Pennsylvania, Philadelphia, PA, USA

Taylor R. Nicholas Department of Biology, Indiana University, Bloomington, IN, USA

Nestor Prieto-Dominguez Department of Biochemistry & Molecular Biology, LSU Health Sciences Center, Shreveport, LA, USA

Colin C. Pritchard University of Washington, Seattle, WA, USA

Nataliya Prokhnevska Department of Urology, Emory University, Atlanta, GA, USA

Shivanand Pudakalakatti Department of Cancer Systems Imaging, The University of Texas MD Anderson Cancer Center, Houston, TX, USA

Thomas L. Pulliam Department of Cancer Systems Imaging, The University of Texas MD Anderson Cancer Center, Houston, TX, USA

Center for Nuclear Receptors and Cell Signaling, University of Houston, Houston, TX, USA

Department of Biology and Biochemistry, University of Houston, Houston, TX, USA

Susmita G. Ramanand Department of Pathology, UT Southwestern Medical Center, Dallas, TX, USA

William L. Redmond Earle A. Chiles Research Institute, Providence Cancer Institute, Portland, OR, USA

Zachary A. Richards Department of Oncologic Pathology, Dana-Farber Cancer Institute, Boston, MA, USA

Alexandre Rouette Computational Biology Program, Ontario Institute for Cancer Research, Toronto, ON, Canada

Travis C. Salzillo Department of Cancer Systems Imaging, The University of Texas MD Anderson Cancer Center, Houston, TX, USA

The University of Texas MD Anderson Cancer Center, UTHealth Graduate School of Biomedical Sciences, Houston, TX, USA

Anjali V. Sheahan Department of Oncologic Pathology, Dana-Farber Cancer Institute, Boston, MA, USA

Daniel E. Spratt Department of Radiation Oncology, University of Michigan Medical Center, Ann Arbor, MI, USA

Karl-Heinz Storbeck Department of Biochemistry, Stellenbosch University, Stellenbosch, South Africa

Brady G. Strittmatter Department of Molecular and Cellular Biochemistry, Indiana University, Bloomington, IN, USA

Daksh Thaper Department of Urologic Sciences, Faculty of Medicine, University of British Columbia, Vancouver, BC, Canada

Vancouver Prostate Centre, Vancouver, BC, Canada

Mark Titus Department of Genitourinary Medical Oncology, The University of Texas MD Anderson Cancer Center, Houston, TX, USA

Robert J. Turesky Department of Medicinal Chemistry, Cancer and Cardiovascular Research Building, University of Minnesota, Minneapolis, MN, USA

Masonic Cancer Center, University of Minnesota, Minneapolis, MN, USA

Sepideh Vahid Vancouver Prostate Centre, Vancouver, BC, Canada

Noel A. Warfel University of Arizona Cancer Center, Tucson, AZ, USA

Sandi R. Wilkenfeld Department of Cancer Systems Imaging, The University of Texas MD Anderson Cancer Center, Houston, TX, USA

The University of Texas MD Anderson Cancer Center, UTHealth Graduate School of Biomedical Sciences, Houston, TX, USA

Kathryn M. Wilson Channing Division of Network Medicine, Harvard Medical School, Brigham and Women's Hospital, Boston, MA, USA

Li Xin Department of Urology, University of Washington, Seattle, WA, USA

Yuqian Yan Department of Biochemistry and Molecular Biology, Mayo Clinic College of Medicine and Science, Rochester, MN, USA

Shu Yang Department of Biochemistry & Molecular Biology, LSU Health Sciences Center, Shreveport, LA, USA

Yunshin Yeh Department of Pathology, Overton Brooks Medical Center, Shreveport, LA, USA

Xiuping Yu Department of Biochemistry & Molecular Biology, LSU Health Sciences Center, Shreveport, LA, USA

Department of Urology, LSU Health Sciences Center, Shreveport, LA, USA

Amina Zoubeidi Department of Urologic Sciences, Faculty of Medicine, University of British Columbia, Vancouver, BC, Canada

Vancouver Prostate Centre, Vancouver, BC, Canada

Diet and Lifestyle in Prostate Cancer

Kathryn M. Wilson and Lorelei A. Mucci

Prostate cancer is the most commonly diagnosed non-skin cancer among men in the United States (US). In addition, three million men in the US are prostate cancer survivors. Given the large public health burden of prostate cancer, the identification of the factors associated with prostate cancer prevention could improve health and outcomes for men.

A variety of diet and lifestyle factors have been studied with respect to prostate cancer risk in large, prospective cohort studies. More recently, researchers have begun to study the association of diet and lifestyle with prostate cancer survival after diagnosis. Cohort studies are generally considered to be a higher level of evidence than case-control studies, which are susceptible to recall bias and selection bias. For this reason, we focus on results from prospective cohort studies when possible. The major cohort studies with results discussed in this chapter are summarized in Table 1.

In spite of this work, few modifiable risk factors have been firmly established as playing a role in prostate cancer. Among modifiable risk factors, smoking and obesity are consistently associated with higher risk specifically of advanced prostate cancer. There is also consider-able evidence for a positive association between dairy intake and overall prostate cancer risk, and an inverse association between cooked tomato/lycopene intake and risk of advanced disease. Several other dietary factors consistently associated with risk in observational studies, including selenium and vitamin E, have been cast into doubt by results from clinical trials. Results for other well-studied dietary factors, including fat intake, red meat, fish, vitamin D, soy and phytoestrogens are mixed.

Migrant studies have found that moving from countries with low prostate cancer incidence to countries with high incidence increases the risk of prostate cancer over time. Among Japanese [1, 2] and Korean [3] immigrants to the US, Chinese immigrants to the US and Canada [4], and European immigrants to Australia [5] prostate cancer risk is much higher than that of their native counterparts, but still below that of white men born in the US, Canada, and Australia. This suggests that there are important environmental contributors to prostate cancer risk in addition to strong genetic factors.

The lack of well-established modifiable risk factors for prostate cancer compared to other common cancers is likely due to several possibilities. First, prostate cancer has among the highest heritability of all common cancers [6]; second, early life exposures may play an important role in risk, rather than mid- and later-life exposures assessed in most epidemiological

K. M. Wilson (✉) · L. A. Mucci
Channing Division of Network Medicine, Harvard Medical School, Brigham and Women's Hospital, Boston, MA, USA
e-mail: kwilson@hsph.harvard.edu

© Springer Nature Switzerland AG 2019
S. M. Dehm, D. J. Tindall (eds.), *Prostate Cancer*, Advances in Experimental Medicine and Biology 1210, https://doi.org/10.1007/978-3-030-32656-2_1

Table 1 Overview of major prospective studies discussed in the chapter

Study name	Primary institution	Location	Design	Start–End dates	N male participants at baseline	Age at baseline; other entry criteria
Health Professionals Follow-up Study (HPFS)	Harvard University	U.S.	Observational cohort study	1986 to ongoing follow-up	51,529	40–75 years
Physicians' Health Study I and II (PHS)	Harvard University/ Brigham & Women's Hospital	U.S.	Randomized trials of cardiovascular disease and cancer prevention (PHS I: aspirin, beta-carotene; PHS II: Vitamins C, E, beta-carotene, multivitamin)	PHS I: 1982–1995 PHS II: 1997–2007 (Follow-up for prostate cancer through 2012)	PHS I: 22,071 PHS II: 14,642	PHS I: 40–84 PHS II: ≥50 years
NIH-AARP Diet and Health Study	National Cancer Institute (US)	U.S. (6 states, 2 metro areas)	Observational cohort study	1995/6 to ongoing follow-up	322,363	50–71 years
Prostate Cancer Prevention Trial (PCPT)	Southwest Oncology Group (SWOG)	U.S.	Randomized trial of prostate cancer prevention with finasteride	1993–2003	18,882	≥55 years; Normal DRE and PSA < 3 ng/ml at baseline
Selenium and Vitamin E Cancer Prevention Trial (SELECT)	Southwest Oncology Group (SWOG)	U.S.	Randomized trial of prostate cancer prevention with selenium and Vitamin E	2001/4–2008	35,533	≥50 years
European Prospective Investigation into Cancer and Nutrition (EPIC)	International Agency for Research on Cancer (IARC)-World Health Organization (WHO)	10 Western European countries	Observational cohort study	1992/9 to ongoing follow-up	153,427	35–74 years at most centers
VITamin D and OmegA-3 TriaL (VITAL)	Harvard University/ Brigham & Women's Hospital	U.S.	Randomized trial of cardiovascular disease and cancer prevention with vitamin D and omega-3 fatty acids	2010–2017	12,786	≥50 years

Diet and Lifestyle in Prostate Cancer

Study	Institution	Country	Study type	Years	N	Age
Alpha-Tocopherol, Beta-Carotene Cancer Prevention Study (ATBC)	National Cancer Institute (US) and National Institute for Health and Welfare of Finland	Finland	Randomized trial of lung cancer prevention with alpha-tocopherol and beta-carotene	1985–1993	29,133	50–69 years; current smokers
Cancer Prevention Study II Nutrition Study (CPS II)	American Cancer Society	U.S. (21 states)	Observational cohort study	1992/3 to ongoing follow-up	86,402	50–74 years, approximately
Multi-ethnic Cohort Study (MEC)	University of Hawaii and University of Southern California	U.S. (Hawaii & Los Angeles)	Observational cohort study	1993/6 to ongoing follow-up	96,810	45–75 years
VITamins and Lifestyle Study (VITAL)	University of Washington/ Fred Hutchinson Cancer Research Center	U.S. (Washington State)	Observational cohort study	2000/2 to ongoing follow-up	35,242	50–76 years

All studies were funded by U.S. NIH institutes, health/cancer agencies in other countries, and foundations with the following exceptions: PHS I study agents and packaging were provided by drug/supplement companies; the study was otherwise funded by NIH. PHS II was partially funded by BASF Corporation, and study agents and packing were provided by supplement companies; the study was otherwise funded by NIH. PCPT was partially funded by Merck in addition to NIH/NCI. SELECT study agents and packaging were provided by the supplement companies; the study was otherwise funded by NIH. VITAL trial study agents and packaging were provided by supplement companies and Quest Diagnostics provided blood measurements free of charge; the study was otherwise funded by NIH studies. Finally, prostate-specific antigen (PSA) screening plays a critical role in prostate cancer detection and incidence rates, which has important implications for epidemiological studies. It is important to understand the impact of PSA screening on prostate cancer epidemiology, as screening must be considered in the interpretation of risk factor studies. We will briefly discuss this below, and we will then review current evidence on dietary and other lifestyle factors, including tobacco use, obesity, and physical activity. For each of these potential risk factors, we will discuss findings for both incidence and survival, if available. A summary of risk factors for prostate cancer is provided in Table 2.

The Impact of PSA Screening on Epidemiological Studies of Prostate Cancer

Autopsy studies have shown that latent prostate cancer is quite common. A review of 19 studies of prostate cancer prevalence upon autopsy found that 29% of white men aged 60–69 and 36% aged 70–79 had undiagnosed prostate cancer at the time of death, with even higher prevalence among black men [7]. Screening by prostate specific antigen (PSA) allows for the detection of these lesions, many of which would never come to light clinically in the absence of screening. Thus, PSA screening has a great impact on incidence rates and the clinical presentation of prostate cancer in populations where it has been introduced.

The mix of indolent and aggressive disease in a given study population will depend on the time period of the study and the PSA screening practices in the population. This makes is difficult to compare relative risks for "total" prostate cancer across studies, as relative risks will reflect a weighted average of indolent and aggressive disease specific to the time and place in which the study was conducted. To deal with this, it is helpful to assess associa-

Table 2 Summary of epidemiological evidence on diet and lifestyle in prostate cancer

Risk factor	Direction of effect[a]
Well-confirmed risk factors	
Height	↑
Probable relationship exists, based on substantial data	
Insulin-like growth factor 1	↑
Smoking	↑ (advanced disease)
Obesity/Body mass index	↑ (advanced disease)
Obesity/Body mass index	↓ (localized disease)
Physical activity (vigorous)	↓ (advanced disease)
Dairy intake	↑
Fish intake	↓ (advanced disease)
Lycopene/tomato intake	↓
Weak, if any, relationship exists, based on substantial data	
Total fat intake	–
Alpha-tocopherol supplements	–
Selenium supplements	–
Childhood/young adult body size	–
Inconsistent findings or limited study to date	
Long-chain marine fatty acid intake	–
Alpha-linolenic acid intake	↑
Red & processed meat intake	–
Calcium intake	↑
Vitamin D	↓ (advanced disease)
Soy/phytoestrogen intake	↓
Dietary vitamin E intake	↓ (advanced disease)
Dietary selenium intake	↓

[a]Arrows indicate direction of relationship: ↑: increase in risk; ↓: decrease in risk; —: no association

tions separately for fatal or advanced stage disease and localized disease, and/or to stratify results by pre-PSA and PSA time periods. Some studies also look at associations for high-grade and low-grade disease separately; however, changes in grading over time, along with between-pathologist differences in grading, introduce substantial misclassification into grade-based categorization in large cohort studies [8].

Dietary Factors and Prostate Cancer Risk and Survival

A western dietary pattern has long been suspected to contribute to prostate cancer risk based on ecologic studies that compared prostate cancer mortality rates around the world and migrant studies as discussed above. Early ecologic studies demonstrated the striking disparity in animal product and fat consumption between high-risk (US, Sweden) and low-risk (Japan, China) countries [9]. Such studies are quite limited, however, as they do not assess individual-level behaviors and subsequent disease outcomes and are unable to account for confounding elements.

Western dietary patterns are also implicated based on the strong evidence of positive associations of adult height and circulating insulin-like growth factor 1 (IGF-1) levels with prostate cancer risk. Both height and circulating IGF-1 levels reflect, in part, nutritional status.

Height. A systematic review found 22 out of 25 studies of height and prostate cancer incidence reported positive associations [10]. The dose-response meta-analysis found a significant 4% increase in risk of prostate cancer overall per 5 cm increase in height (95% CI 1.03–1.05). The same review found that four of five studies on height and risk of prostate cancer mortality reported positive associations. Results were consistent for non-advanced, advanced, and fatal disease. Two large pooled analyses [11, 12] found very similar results for associations of height with prostate cancer incidence as well as mortality.

Adult height partially reflects nutritional status in early life [13–15], which impact circulating growth factors and other hormones during childhood and puberty [16]. Although genetics plays a major role in determining height, [17], dietary factors are essential to reach the genetic potential. Total energy intake, protein, and dairy intake in childhood are all associated with greater attained height [18–20].

IGF-1. Total energy intake and protein intake are, in turn, positively associated with circulating IGF-1 in children and adults in observational studies [21, 22] and in feeding trials [23]. In addition, adult height is positively correlated with IGF-1. [24] IGF-1 is a major growth-regulating molecule, which is a potent mitogen that can also inhibit apoptosis. It is secreted mainly by the liver but is also produced in several other tissues, including the prostate, in response to growth hormone.

Adult levels of circulating IGF-1 have consistently been associated with increased risk of prostate cancer. A pooled analysis of data from 3700 cases and 5200 controls in 12 prospective cohort studies [25] found an odds ratio of 1.38 (95% CI 1.19–1.60, p-trend = <0.001) for prostate cancer comparing the top to bottom quintile of serum IGF1. The association was stronger for low-grade than for high-grade disease, but did not vary by stage of disease. Adjustment for various sex hormone levels also measured in 8 of the 12 studies did not affect the IGF-1 results. Results from a meta-analysis of 42 retrospective and prospective studies found a similar significant and positive association for IGF-1 and prostate cancer risk [26].

Evidence supporting associations of height and IGF-1 with prostate cancer is quite consistent across study populations. In combination with results from migrant studies and the large geographical variation in incidence rates of prostate cancer, this suggests that nutritional status plays some role, perhaps early in life, in the development of prostate cancer. A variety of specific dietary factors have been investigated in detail with respect to prostate cancer risk, including fat intake, meat intake, intake of fish and marine-derived long-chain fatty acids, dairy products and calcium, vitamin D, tomato and lycopene, soy and phytoestrogens, vitamin E, and selenium. Each is discussed in detail below.

Fat Intake

Driven by early ecologic studies, dietary fat intake has been of great interest in studies of diet and prostate cancer, as high fat intake, particularly from animal sources, is a major attribute of the western diet. However, a meta-analysis of 14 prospective cohort studies found no association between total fat intake and risk of prostate cancer [27]. Intakes of saturated, polyunsaturated, and monounsaturated fats were also not associated with risk, and fat intakes were not associated with risk of advanced stage disease.

Several prospective studies have evaluated intake of specific fatty acids in relation to prostate cancer. In western diets, alpha-linolenic acid (ALA) is the principal dietary n-3 (or omega-3) fatty acid. Commonly consumed foods rich in ALA include: mayonnaise, vegetable oils, margarine, walnuts, cheese, beef, pork, and lamb. Several meta-analyses of ALA intake and prostate cancer have found no association with overall disease risk, though there appears to be significant heterogeneity between studies [28–30]. The Health Professionals Follow-up Study (HPFS) cohort found that ALA intake was not associated with overall prostate cancer risk but was associated with significantly increased risk of fatal prostate cancer [31]. This positive association between ALA intake and risk of lethal disease was observed among cases diagnosed prior to the advent of PSA screening (~1994), and not for cases diagnosed in the PSA-screening era [32]. Two other prospective studies of ALA intake did not find increased risks for more advanced prostate cancer; however, both were limited by low case numbers [33, 34], while a fourth prospective study did not examine associations specifically with advanced disease [35].

A pooled analysis [36] of individual-level data from seven prospective studies with measured blood levels of ALA found no association with overall prostate cancer risk. There was some suggestion of heterogeneity by stage of disease (p = 0.032), with no significant association for localized disease and a borderline significant *inverse* association for risk of advanced prostate cancer (OR 0.68, 95% CI 0.46–1.00); however, the number of advanced cases was low in the seven included studies. There was no suggestion of an association between blood ALA levels and risk of high- or low-grade disease. Overall it does not appear that ALA is an important risk factor for prostate cancer; however, the evidence in this area is unusually inconsistent across studies.

The role of long-chain n-3 fatty acids in prostate cancer has also been debated; see the section below on fish intake for discussion.

Survival. Fewer studies have examined the association between fat intake and survival among men with prostate cancer, but these studies are consistent in finding improved survival with increased vegetable fats and poorer survival with saturated and animal fats. In the Physicians' Health Study (PHS) men who consumed more saturated fat in place of carbohydrate in the post-diagnosis diet were at increased risk of cancer-specific and all-cause mortality [37]. Increased intake of vegetable fats after diagnosis was associated with lower risk of all-cause, but not cancer-specific, mortality. In HPFS, post-diagnosis vegetable fat intake was associated with significantly lower risk of both cancer-specific and all-cause mortality. Higher intakes of saturated and trans fats were positively associated with all-cause mortality [38]. A study of men diagnosed with prostate cancer in Sweden found that those reporting a higher intake of saturated fat at the time of diagnosis were at significantly greater risk of prostate cancer mortality [39]. To date, the literature has been consistent in showing that higher intakes of vegetable fats and lower intake of animal and saturated fats after diagnosis are associated with improved cancer-specific and overall survival.

Meat Intake

Red meat and processed meat have both been intensively studied as possible risk factors for prostate cancer, as both are notable components of the western diet. However, meat intake does not appear to be associated with prostate cancer risk.

A meta-analysis of 11 prospective studies found a combined relative risk for extreme categories of red meat intake of 0.98 (95% CI 0.93–1.04) for total prostate cancer and 1.01 (95% CI 0.94–1.09) for advanced prostate cancer (8 studies) [40]. A recent pooled analysis of individual data from 15 cohort studies also found no association between red meat intake and risk of total or fatal prostate cancer [41].

For processed meat, the meta-analysis of 11 prospective studies found a relative risk for extreme categories of intake of 1.05 (95% CI 0.99–1.12) for total prostate cancer and 1.10 (95% CI 0.95–1.27) for advanced cancer (8 studies) [40]. There was evidence of publication bias for processed meat studies, and risk estimates were weaker in more recent studies that adjusted for more potential confounders. The pooled analysis of 15 cohorts found a suggestion of a slight increase in risk of total prostate cancer for the highest category of processed meat intake, though there was no significant trend across categories (HR 1.04, 95% CI 1.01–1.08, p-trend = 0.29); there was no association for fatal disease [41]. Similarly, several more recent studies found no associations for red or processed meat. One paper found no evidence that red or processed meat was associated with risk total or advanced prostate cancer among African-Americans in the NIH-AARP Diet and Health Study [42]. A study [43] focused on the PSA screening era in the HPFS cohort also found no associations of meat intake with lethal prostate cancer. Finally, a study in the Netherlands found no association between low- or no meat consumption and risk of prostate cancer [44].

One possible mechanism by which red meat could raise the risk of cancer is through heterocyclic amines (HCA) formed during cooking [45–48]. HCAs are mutagenic compounds formed during cooking of muscle of meat and fish at high temperatures. Preference for doneness of red meat and calculated intakes of common HCAs have been studied with respect to prostate cancer in several prospective studies with mixed results. Three found no clear associations between doneness or HCA intake and risk of prostate cancer [47–49]. Three found positive associations between well done red meat [42, 45, 46, 50], as well as HCA intake [50], and risk of prostate cancer, including advanced disease. Overall, intakes of red and processed meat do not appear related to prostate cancer risk; however, well-done red meat and the associated carcinogens may play some role.

Survival. In a study of post-diagnosis meat intake and survival among men diagnosed with apparently localized prostate cancer, one study found suggestive but not statistically significant associations between intake of red meat and poultry and risk of lethal prostate cancer [43]. Another study found that a "Western dietary pattern", characterized by higher intakes of red and processed meats, high-fat dairy, and refined grains, was associated with increased risk of prostate cancer-specific and all-cause mortality [51]. Finally, a study [52] among men surgically treated for localized cancer found that lower intakes of red meat, particularly well-done red meat, and higher intakes of poultry and fish were associated with lower risk of PSA recurrence, independent of stage and grade of disease. Overall, it appears that lower intakes of red meat may be associated with improved survival, which is consistent with the findings on post-diagnosis fat intake and survival discussed above.

Fish Intake and Marine Fatty Acids

Populations with a high consumption of fish, for example in Japan and among Alaskan natives, have lower rates of prostate cancer than populations with western dietary patterns, where fish intake is generally lower [53–55]. Fish contain long-chain marine n-3 fatty acids (eicosapentaenoic acid, EPA, [20:5n-3] and docosahexaenoic acid, DHA, [22:6n-3]), which can modify inflammatory pathways and may therefore affect prostate cancer risk and progression [56]. Indeed, a study among men without cancer in the Prostate Cancer Prevention Trial (PCPT, a randomized trial of finasteride for prostate cancer prevention) found that men with higher serum levels of n-3 fatty acids had lower levels of prostatic inflammation [57].

However, the role of fish and long-chain fatty acids in prostate cancer has been debated due to reports from PCPT and The Selenium and Vitamin E Cancer Prevention Trial (SELECT) of significant *positive* associations between higher concentrations of serum long-chain n-3 fatty acids and risk of high-grade disease [58]. Neither trial had enough advanced or fatal cases to study those outcomes separately.

The European Prospective Investigation into Cancer and Nutrition (EPIC) cohort found a positive association between serum EPA and risk of high-grade disease, consistent with PCPT and SELECT. However, they observed no association with advanced or fatal disease [59]. The PHS cohort [60] assessed whole blood fatty acid content and also found no association with advanced disease, and a significant inverse association with localized disease.

A pooled analysis of seven prospective studies [36], including all four discussed above, found a significantly increased risk of total prostate cancer with higher serum levels of both EPA and DHA. Risk was approximately 15% higher for men in the highest quintile of either fatty acid compared to men in the lowest quintile of that fatty acid. However, there was significant heterogeneity between studies (p = 0.02 for EPA, p < 0.001 for DHA). Thus there may be modest positive associations between blood levels of marine long-chain fatty acids and risk of total prostate cancer; however, it is unclear if these associations are causal, and the reason for heterogeneity across studies is unclear. Differences in PSA screening may explain some of the heterogeneity. The positive associations were stronger for cases diagnosed after 2000 than those diagnosed earlier. In addition, the vast majority of cases from the SELECT and PCPT trials were screen-detected, whereas the PHS, which found inverse associations, contained many cases diagnosed between 1982 and 1995, prior to the onset of widespread PSA screening in the US.

Multiple studies have examined questionnaire-assessed intake of fish and fish-derived long-chain fatty acids. A 2010 meta-analysis of fish intake found no association between fish consumption and incidence of total prostate cancer; for the highest versus lowest category of intake across 12 cohort studies the relative risk was 1.01 (95% CI 0.90–1.14) [61]. However, in four cohort studies of prostate cancer-specific mortality, fish intake was associated with a significantly lower risk (RR 0.37, 95% CI 0.18–0.74). A recent systematic review [62] of long-chain n-3 fatty acids

found similar results, with no association for risk of overall prostate cancer, and an inverse association for prostate cancer mortality. Of seven studies investigating associations between long-chain n-3 fatty acid intake and prostate cancer-specific mortality, five found significant inverse associations and two found non-significant inverse associations. A study in Iceland, where there is a tradition of fish-oil consumption, found that fish oil consumption in later life was associated with a lower risk of advanced prostate cancer [63].

The recently completed VITAL trial [64] tested marine n-3 fatty acids (at a dose of 1 g per day, equal to about 4 servings/week of fish) in the primary prevention of cardiovascular disease and cancer among 12,786 men 50 years of age or older. Mean follow-up was 5.3 years. There was no difference in the incidence of prostate cancer (a predefined secondary outcome) between groups. (N = 411 total cases, RR = 1.15, 95% CI 0.94–1.39). However, low power and a relatively short follow-up time limit the conclusions that can be drawn from this null result.

Overall, current evidence is quite mixed regarding associations of serum fatty acid levels and fish intake with risk of disease by both grade and stage. Additional studies in cohorts with long-term follow-up are needed to draw conclusions about the role of fish, fish oil supplements, and specific long-chain omega-3 fatty acids in prostate cancer risk and progression.

Dairy Products and Calcium

Dairy products, in addition to containing a substantial amount of animal fat, are the most common dietary sources of calcium and vitamin D, all of which have been implicated in prostate cancer risk. The strong correlation between dairy foods and these nutrients create challenges in trying to disentangle their independent effects. A meta-analysis conducted as a part of the AICR/WCRF Continuous Update Project found a statistically significant increased risk of total prostate cancer with higher intakes of dairy products and dietary calcium (i.e. from foods, not supplements) [10]. The combined estimate across 15 cohort studies found a 7% increased risk per 400 g of dairy products per day (95% CI 2–12%) and 5% increased risk per 400 mg of dietary calcium (95% CI 2–9%). Intakes of milk, cheese, and total calcium (i.e. foods plus supplements) were also positively associated with risk. There was evidence of non-linear associations for calcium (from foods alone and total intake), with positive associations more pronounced at very high intakes (>1500 mg/day).

Associations according to stage of disease were less clear, with significant positive associations for non-advanced disease but not for advanced disease for both dairy and dietary calcium. However, across five studies of fatal cancer, the association with dairy was an 11% increased risk per 400 g per day (95% CI −8% to 33%), quite similar to the risk estimate for overall prostate cancer, but with lower power and a wide confidence interval.

Estimates for total calcium were somewhat weaker than for dietary calcium, suggesting that some other component of dairy, rather than calcium itself, is driving the dietary calcium estimates. However, interpretation of this, too, is complicated, as the only two studies that examined total calcium and fatal prostate cancer found a significantly increased risk (RR 1.11, 95% CI 1.02–1.21). In addition, there was significant heterogeneity for the total calcium estimate based on study follow-up time, with a non-significant association with total prostate cancer in six studies with less than 10 years of follow-up, but a significant positive association in three studies with 10 or more years of follow-up. This heterogeneity by follow-up time is supported by a report from the HPFS [65], which found a significant association between total calcium intake 12–16 years prior to diagnosis of advanced prostate cancer, but not for shorter time periods between intake and diagnosis. This suggests that calcium may play a role early in the carcinogenesis process.

Possible mechanisms linking dairy or calcium and prostate cancer risk include the down-regulating effect of high calcium intake on vitamin D levels [66] and the positive association between dairy and IGF-1 levels [67]. The positive association observed for low-fat or skim milk argues against dairy fat playing an important role in the association.

Interestingly, the HPFS cohort also found a positive association between phosphorus intake and risk of total, lethal, and high-grade prostate cancer, independent of the association with calcium [65]. In contrast to the pattern observed for calcium, the phosphorus association was strongest for intakes shortly before the time of diagnosis (0–4 years). Phosphorus, like calcium, is concentrated in dairy, but is more widespread in other foods than is calcium. Fewer studies have examined phosphorus than calcium, particularly with respect to advanced or fatal disease, but this should be explored in other studies. High phosphorus intake increases parathyroid hormone, which promotes bone remodeling [68]. Prostate cancer preferentially metastasizes to bone and is more likely to spread to bone with higher remodeling activity [69, 70].

Overall, there is substantial evidence that dairy intake is associated with increased prostate cancer risk; however, the role of calcium, phosphorus, or other specific components is less clear.

Survival. There have been three studies of post-diagnosis dairy intake and prostate cancer survival among men diagnosed with apparently localized disease. HPFS and a Swedish study both found that higher post-diagnosis intake of whole milk was associated with worse survival [71, 72], while higher intake of low-fat dairy was associated with improved survival that was statistically significant in HPFS and suggestive in the Swedish population. On the other hand, PHS reported that intake of total dairy, including both high- and low-fat dairy foods, was associated with increased risk of all-cause and prostate cancer-specific mortality [73]. Thus, evidence has consistently shown that high-fat dairy after diagnosis is associated with worse survival, whereas the role of low-fat dairy is uncertain.

Vitamin D

Vitamin D, which is an important regulator of calcium homeostasis, has also been considered as a prostate cancer risk factor. The main source of vitamin D is endogenous production in the skin resulting from sun exposure, and diet is a secondary source. Dihydroxyvitamin D [1,25(OH)$_2$D] is a steroid hormone involved in regulating differentiation and proliferation of many cell types, including prostate epithelial cells, which express functional vitamin D receptors. 1,25(OH)$_2$D is the most biologically active form, whereas hydroxyvitamin D [25(OH)D] is found in much higher concentrations in blood and better reflects sun and dietary exposure, as its levels are less strictly regulated by the body.

A meta-analysis of circulating 25(OH)D levels in 14 prospective nested case-control studies found no association with total prostate cancer risk (OR 1.04, 95% CI 0.99–1.10) [74]. In six studies of aggressive prostate cancer, defined as a mix of high grade and advanced stage, there was also no association (OR 0.98, 95% CI 0.84, 1.15); however, there was evidence of heterogeneity between studies. A similar null association was found in another meta-analysis [75]. A consortium of cohort studies with 518 fatal cases and 2986 controls similarly found no association between 25(OH)D and risk of fatal prostate cancer. However, there was evidence that the 25(OH)D association may be modified by genetic variation in several vitamin D-related genes [76].

Five large studies published after these meta-analyses have had mixed results. The PCPT [77] found no association overall, but a significant inverse association between 25(OH)D and high-grade cancer. Conversely, a Swedish study [78] found a suggestive positive association with total prostate cancer, and the ATBC study [79], a cohort of Finnish smokers, found a significantly increased risk of total and aggressive (stage 3 or 4 or Gleason grade 8+) disease. Finally, the SELECT [80] trial reported a U-shaped relationship between circulating Vitamin D and risk of total prostate cancer, with significantly lower risk in the middle quintile relative to the lowest quintile and no difference in risk between the highest and lowest quintiles.

The recently completed VITAL trial [81] tested vitamin D3 (cholecalciferol) at a dose of 2000 IU per day (together with fish oil) in the primary prevention of cardiovascular disease and cancer among 12,786 men 50 years of age or older with mean follow-up of 5.3 years. There was no association between vitamin D supplementation and prostate cancer incidence (N = 411 total events, RR 0.88 (0.72–1.07)). However, as with the VITAL results for fish oil supplements, the limited power and short follow-up time of the trial limit how informative this null result is.

1,25(OH)$_2$D has been less studied than 25(OH)D. However, a meta-analysis of seven prospective studies of 1,25(OH)$_2$D found no association with total prostate cancer (OR 1.00, 95% CI 0.87–1.14). Only two studies have looked at 1,25(OH)$_2$D and risk of aggressive disease (both based on high grade, or advanced stage, or prostate cancer death) with a suggestive combined odds ratio of 0.86 (95% CI 0.72–1.02) [82].

Survival. In spite of the null findings for associations between vitamin D and incidence of total or aggressive disease, there is some evidence that vitamin D plays a role in prostate cancer progression. Several studies have found inverse associations between 25(OH)D and survival among prostate cancer patients [83–86], though others have not [79, 87]. In addition, genetic variants in the vitamin D pathway are associated with risk of recurrence or progression and prostate cancer-specific mortality [76, 85, 88]. Genetic variants in the vitamin D receptor were associated with Gleason score in some studies [89, 90], and high expression of the vitamin D receptor protein in prostate cancer tissue was associated with lower risk of prostate cancer mortality among men with prostate cancer in the HPFS and PHS cohorts, with adjustment for PSA at diagnosis, Gleason grade, and stage [91].

Thus, while vitamin D exposure does not seem to be associated with lower risk of prostate cancer incidence, several lines of evidence suggest that the vitamin D pathway may play a role in prostate cancer progression.

Lycopene and Tomatoes

Tomatoes and lycopene, a carotenoid consumed mainly from tomato products, have been the focus of many studies due to early reports of a significant inverse association between intake and risk of prostate cancer [92]. A meta-analysis [93] found significant inverse associations between both questionnaire-assessed lycopene intake and circulating lycopene levels and risk of prostate cancer. Across 16 case-control and 9 prospective studies, men with the highest lycopene intake had a 12% lower risk of prostate cancer compared to men with the lowest intakes (95% CI 0.78–0.98, p = 0.02).

A pooled analysis of data from 15 prospective studies found no association between circulating lycopene levels and risk of overall prostate cancer (OR 0.97, 95% CI 0.86–1.08 for top versus bottom quintile). However, there was significant heterogeneity by stage, with higher lycopene associated with lower risk of advanced, but not localized disease. Men in the highest quintile of circulating lycopene had a HR of 0.65 (95% CI 0.46–0.91, p-trend = 0.03) for advanced stage or fatal prostate cancer compared to men in the lowest quintile.

These findings are in agreement with those from the HPFS cohort [94] based on lycopene intake assessed by questionnaire, which found a stronger inverse association with lethal cancer (death or distant metastatic disease) than for total prostate cancer. The HR for lethal cancer was 0.72 (95% CI 0.56–0.94, p-trend = 0.04) for the highest versus lowest quintile. This inverse association with lethal cancer was stronger among a sub-cohort of men who received PSA tests (HR 0.47, 95% CI 0.29–0.75, p-trend = 0.009), suggesting the association is not related to differences in screening or detection.

Cooked or processed tomato products, such as tomato sauce, tomato soup, and ketchup, offer more readily bioavailable sources of lycopene than fresh tomatoes [95]. Accordingly, some epidemiologic studies have found significant inverse effects for tomato sauce while reporting weaker results for raw tomato intake and no significant influence for tomato juice [96]. The correlation between dietary estimates of lycopene based on food frequency questionnaires and circulating levels measured in blood are relatively low, ranging from 0 to 0.47 [96]. A clinical trial found that men assigned to consume one serving per day of either tomato sauce, tomato juice, or tomato soup for at least 2 weeks had significant increases in both plasma and prostatic lycopene levels [97].

Experimental studies suggest that lycopene can inhibit angiogenesis, perhaps through regulation of vascular endothelial growth factor (VEGF) and the PI3K-Akt and ERK/p38 signaling pathways [98–101]. Interestingly, three measures of tumor angiogenesis—microvessel diameter and area and irregularity of the vessel lumen—were all associated with lycopene intake such that those with higher intakes had more favorable angiogenesis markers [94]. These angiogenesis markers are associated with risk of lethal disease independent of grade [102]. Overall, there is fairly consistent evidence that lycopene is associated with lower risk of advanced or fatal prostate cancers, and experimental evidence supports this observation.

Survival. Among men diagnosed with aggressive prostate cancers in the American Cancer Society's Cancer Prevention Study II Nutrition Cohort (CPS II), high lycopene intake before and after diagnosis was consistently associated with improved survival; however, there was no association between lycopene intake and survival when all prostate cancer cases were included [103].

Soy/Phytoestrogens

Traditional Asian diets are notably high in phytoestrogens, chiefly from soy-based foods. Intakes of soy and phytoestrogens are low in the typical western diet. Because there are stark differences in incidence of prostate cancer between Asian and western countries, these foods and compounds have been of interest with respect to prostate cancer risk.

Dietary phytoestrogens, naturally occurring constituents of plants, are divided into two main categories: lignans and isoflavonoids. Lignans occur in whole-grain bread, seeds, berries, vegetables, and tea, while the main source of isoflavonoids is soy beans and soy products. The primary isoflavones in soy are genistein and daidzein.

Animal studies have suggested that phytoestrogens may play a role in prostate cancer initiation and progression through estrogenic effects, inhibition of angiogenesis, antioxidant activity, stimulation of apoptosis, and inhibition of cell growth [104–108].

A meta-analysis [109] of 16 studies of soy intake and total prostate cancer risk found a significant inverse association, with a relative risk of 0.70 (95% CI 0.58–0.85) for the highest versus lowest intakes. The association was stronger for unfermented soy foods (including soy milk, tofu, and soybeans) and was not significant for fermented soy foods (miso and natto). In addition, the inverse association was more pronounced among the nine case-control studies than in the seven prospective cohort studies. In four prospective studies that examined risk of advanced disease specifically there was no association with soy intake. The meta-analysis also included nine studies of circulating genistein and seven of circulating daidzein and found no association between these isoflavones and prostate cancer risk.

Another meta-analysis of questionnaire-assessed phytoestrogen intakes [110] found a significant inverse association with total prostate cancer risk in 18 case-control studies (RR for highest versus lowest category 0.69, 95% CI 0.57–0.81) and a borderline significant inverse association in 11 cohorts studies (RR 0.87, 95% CI 0.89–1.00). However, there was a suggestion of publication bias. In addition, the borderline significant association in prospective studies suggests that selection and/or recall bias may explain some of the results seen in the case-control studies.

The Multiethnic Cohort Study (MEC) [111], conducted among men in Hawaii and California, was the largest study included in both meta-analyses. There was a suggestion of an inverse association between soy foods and overall prostate cancer risk (HR 0.90, 95% CI 0.80–1.01, p-trend = 0.20 for the highest versus lowest tertile), and a borderline significant association for high-grade or nonlocalized prostate (HR 0.78, 95% CI 0.62–0.98, p-trend = 0.05). There was no inverse association between soy food and prostate cancer risk among the Japanese-American men in the study, who had higher soy intakes than the white, Latino, and African-American men. Two cohort studies in Japan that involved study populations with much higher soy intake than in MEC found suggestive, but not statistically significant, inverse associations with prostate cancer [112, 113].

Overall, weak inverse associations have been observed between soy and isoflavone intake; however, the lack of association for circulating isoflavone levels, along with the weaker results among prospective studies and among populations with higher soy intake suggest that the observed associations may be due to bias (selection bias, recall bias, confounding) rather than an underlying causal association. Additional prospective cohort studies are needed in populations with high soy intake to determine whether there is, in fact, an inverse association with disease risk.

Survival. One clinical trial [114] of soy protein supplementation and biochemical recurrence enrolled 177 men at high risk of recurrence following radical prostatectomy and randomized them to a daily soy protein isolate supplement versus placebo. Treatment lasted up to two years. The trial was stopped early due to a lack of treatment effect. It should be noted that adding soy protein isolate to an overall western dietary pattern has different nutritional effects than substituting soy foods for other foods in the diet, which is what was studied in the epidemiological studies of prostate cancer incidence discussed above. The association between soy food or isoflavone intake and long-term survival among men with prostate cancer has not been studied.

Vitamin E and Alpha-tocopherol

Vitamin E refers to a group of fat-soluble compounds, including tocopherols and tocotrienols, which have antioxidant and pro-immune properties. Gamma-tocopherol is the most common tocopherol in the US diet, but plasma levels of alpha-tocopherol are higher than those of gamma-tocopherol [115]. Alpha-tocopherol is the biologically most active form, and current dietary recommendations for vitamin E in the US are based on alpha-tocopherol alone. Possible anti-carcinogenic actions of vitamin E include its ability to reduce DNA damage and inhibit malignant cellular transformation [116, 117]. In experimental models, derivatives of vitamin E inhibit growth, induce apoptosis, and enhance therapeutic effects in human prostate cancer cells [118, 119].

Interest in Vitamin E with respect to prostate cancer was driven by secondary results of the Alpha-Tocopherol Beta-Carotene Cancer Prevention (ATBC) Study [120]. ATBC was a randomized trial of lung cancer prevention among male smokers in Finland. While alpha-tocopherol supplementation had no effect on lung cancer risk, men given alpha-tocopherol had a 32% reduction in prostate cancer risk compared to placebo [121]. Several years earlier, a large trial of a multi-nutrient supplement in Linxian, China found that vitamin E (in combination with selenium and beta-carotene) reduced overall cancer mortality [122]. These results, along with laboratory evidence and some epidemiologic support, motivated two trials of vitamin E supplementation on the risk of prostate cancer, SELECT and the Physicians' Health Study II (PHS II).

SELECT was a trial of selenium and vitamin E supplementation and prostate cancer risk, conducted among 35,533 men from the US, Canada, and Puerto Rico. The study was planned for 7–12 years but was stopped early due to a lack of efficacy for risk reduction [123]. The initial report, based on an average of 5.5 years of treatment, found a non-significant suggestion of *increased* prostate cancer risk among men receiving 400 IU/day of alpha-tocopherol. With additional follow-up time, the vitamin E group was found to have a significantly increased risk of prostate cancer (RR 1.17, 99% CI 1.004–1.36, p = 0.008, among 1149 cases) [124]. Interestingly, there was not a statistically significant increased risk of prostate cancer in the vitamin E and selenium combination group (HR 1.05, 95% CI 0.89–1.22), suggesting the two may interact. In fact, SELECT was designed as a four-group trial rather than a factorial trial based on the hypothesis that the two agents, both of which have anti-oxidant activity, may interact [125].

PHS II, conducted contemporaneously with SELECT, was a randomized trial of vitamin E and vitamin C supplement use and prostate cancer risk among 14,642 US physicians. With a median of 8 years of follow-up, there was no effect of 400 IU of vitamin E taken every other day on incidence of prostate cancer (HR 0.97, 95% CI 0.85–1.09) [126]. PHS II was a factorial design, so the vitamin E estimate was made across groups of vitamin C supplement use; however, there was no suggestion of an interaction between vitamin E and vitamin C supplementation.

Together, the SELECT and PHS II results suggest that vitamin E supplement use is at best ineffective and possibly harmful with respect to prostate cancer risk. This is in contrast to the ATBC findings that spurred these trials. Of note, all men in the ATBC trial were smokers, and the prostate cancers were diagnosed prior to the advent of PSA screening, and thus were generally aggressive. The SELECT and PHS II trials were done in the PSA screening era and could not specifically study advanced or fatal prostate cancers. Of 2279 prostate cancers diagnosed in SELECT through July 2011, only nine were diagnosed with stage T3 disease, three with N1 disease, and 13 with metastatic disease. Even the ability to study high-grade cancer was limited, with 613 (27%) cases grade 7 and above, only 134 of which were grade 8–10 [124]. In addition, only 8% of men in SELECT and 4% of men in PHS II were current smokers, so neither trial could address the effect of vitamin E specifically among smokers.

Interestingly, epidemiological studies of vitamin E and prostate cancer risk have tended to support the ATBC results, with generally null associations for overall prostate cancer, but inverse associations for advanced disease and among smokers. In the VITamins And Lifestyle (VITAL) study, a cohort study in Washington state designed specifically to examine supplement use and cancer risk, intake of supplemental vitamin E over 10 years was not associated with overall prostate cancer risk, but was associated with a reduced risk of advanced prostate cancer (n = 123; HR 0.43, 95% CI 0.19–1.0 for average intake ≥400 IU/day vs. none) [127]. Other epidemiological studies have similarly found a protective association limited to ever smokers, including a prospective study of dietary vitamin E intake [128], and a study of vitamin E supplementation and lethal prostate cancer risk [129].

A pooled analysis of 13 prospective studies of blood alpha-tocopherol and prostate cancer risk found significant inverse associations overall and for advanced prostate cancer (regionally invasive, distant metastatic, or fatal cancer), with an odds ratio for advanced disease of 0.74 (95% CI 0.59–0.92; p-trend = 0.001) for the highest versus lowest quintile, based on 1226 advanced cases [130]. There was significant heterogeneity by disease aggressiveness, with no association for non-advanced disease. In addition, there was no association among never smokers (OR 0.99, 95% CI 0.82–1.18), and significant inverse associations for former and current smokers (OR 0.84, 95% CI 0.72–0.97 for former; OR 0.82, 95% CI 0.73–0.93 for current), although the p-value for heterogeneity by smoking was not statistically significant.

Overall, use of vitamin E supplements for prostate cancer prevention is not supported; however, diets higher in alpha-tocopherol appear to be associated with lower risk of advanced disease, particularly among smokers. The underlying mechanisms for this association among smokers are unclear.

Selenium

The trace element selenium is not itself an antioxidant, but it is an essential element for the antioxidant enzyme glutathione peroxidase. It is also required for the function of other selenoproteins involved in exerting anti-tumor effects, including apoptosis and inhibition of cellular proliferation [131–133]. Dietary intake of selenium depends on the selenium content of soil in which foods are grown, which varies greatly by geography. Ecologic studies have suggested an inverse association between selenium soil content and prostate cancer incidence [134]. Because selenium contents in specific foods vary based on the selenium content of the soil, epidemiological studies of selenium must be based on biological sampling, primarily measuring selenium levels in blood or toenails, rather than questionnaire-based diet assessments. Since the activity of some selenoenzymes plateau with higher selenium level [135], the chemopreventive effect of supplemental selenium is expected to be greatest in populations with low baseline selenium exposure, with little marginal effect among selenium-replete populations [136].

Like vitamin E, selenium was tested in the SELECT trial based on secondary results of other randomized trials. The Nutritional Prevention of Cancer Trial, designed to study the effect of selenium supplementation on non-melanoma skin cancer recurrence, found a 63% reduction in prostate cancer risk among men taking selenium supplements [137]. With additional follow-up time, the protective effect was seen only among men with low baseline levels of PSA or selenium [136]. Another trial of selenium (with vitamin E and beta-carotene) in Linxian, China found a reduction in total cancer mortality in China [122].

As discussed above, the SELECT trial was stopped early due to lack of efficacy of the supplements [123]. With additional follow-up, there was still no association between selenium and prostate cancer risk (RR 1.09, 99% CI 0.93–1.27) [124]. In addition, baseline selenium status (measured in toenails) was not associated with prostate cancer risk among men in the trial, and baseline status did not modify the association between selenium supplementation and risk [138]. As with vitamin E, conclusions about selenium drawn from SELECT are limited by the small number of advanced and high-grade cases.

A recent Mendelian randomization study [139] among over 70,000 men in the PRACTICAL consortium used a gene score based on 11 SNPs that predict circulating selenium levels as a non-confounded proxy for selenium status to investigate whether selenium might be causally related to prostate cancer risk. The results were similar to SELECT, with no association with overall prostate cancer risk. There was a non-significant suggestion of increased risk of aggressive disease (OR = 1.21, 95% CI 0.98–1.49). However, the genetic instrument, while very significantly associated with circulating selenium levels (p < 5 × 10⁻⁸), explained only 2.5–5% of variation in these levels, limiting how informative this study is for shedding light on the true association between circulating selenium and prostate cancer risk.

In contrast to the SELECT results, observational studies of selenium and prostate cancer risk have been quite consistent in finding inverse associations. A recent pooled analysis [140] of individual-level data from 15 prospective studies found that nail selenium levels were associated with lower risk of total and aggressive prostate cancer, while blood levels were associated with lower risk of aggressive disease. For aggressive prostate cancer, the OR for men in the highest versus lowest quintile of nail selenium was 0.18 (95% CI 0.11–0.31), and for blood selenium was 0.43 (95% CI 0.21–0.87). A recent report from a Danish cohort [141] also found an inverse association between plasma selenium and risk of high-grade prostate cancer (HR 0.77, 95% CI 0.64–0.94, p-trend = 0.009) but no association with total or advanced stage disease. Two recent meta-analyses of blood selenium [142] and toenail selenium [143] also found significant inverse associations.

These results are unusually consistent and strong among studies of dietary factors and prostate cancer risk. Because selenium status depends on the geographical source of foods in the diet rather than the selection of specific foods, it is difficult to imagine how confounding by other aspects of a healthy diet or lifestyle could explain the magnitude of the results from observational studies. The results of the SELECT trial do not support the use of selenium supplements for the prevention of prostate cancer in middle-aged and older men. However, the association between selenium and prostate cancer risk and survival is still not completely clear.

Survival. A study in HPFS found that use of selenium supplements of 140 mcg/day was associated with significantly increased risk of prostate cancer mortality among men diagnosed with localized prostate cancer. The association was independent of pre-diagnosis supplement use, use of other supplements, and stage and grade of disease at diagnosis [144]. The authors suggest the possibility of a U-shaped relationship between selenium status and cancer incidence and progression, with adverse effects at very low and very high levels.

Other Lifestyle Factors

Tobacco

Although strongly linked to a number of cancers, cigarette smoking does not appear to be associated with overall prostate cancer incidence. A meta-analysis [145] of 15 studies prior to 1995 (i.e., the pre-PSA era), found a pooled relative risk for current smoking and risk of prostate cancer of 1.06 (95% CI 0.98–1.15). For 18 studies completed after 1995 (the PSA screening era), there was a significant inverse association with current smoking (RR 0.84, 95% CI 0.79–0.89). This likely reflects the fact that current smokers are less likely to undergo PSA screening and are therefore not as likely as non-smokers to be diagnosed with prostate cancer. This heterogeneity by time period highlights the importance of accounting for PSA screening in studies of prostate cancer incidence. A previous meta-analysis [146] similarly found no association between current smoking and prostate cancer incidence; however, it did show a positive association with risk among the heaviest smokers measured by cigarettes per day or pack-years.

In contrast to the lack of association for overall prostate cancer, a positive association between smoking and prostate cancer mortality has been documented consistently, as noted by the Surgeon General's 2014 report [147]. A meta-analysis of 21 prospective cohort studies of smoking and prostate cancer mortality found that current cigarette smoking was associated with a 24% increased risk of fatal disease (95% CI 18–31%), with little evidence of heterogeneity between studies [145]. There was a significant dose-response relationship between number of cigarettes smoked per day and mortality. There was a suggestion of increased risk for former smoking and subsequent prostate cancer mortality, with a 6% increase in risk (95% CI 0–13%). In the HPFS cohort smokers who had quit less than 10 years previously were at increased risk of fatal prostate cancer (HR 1.73, 95% CI 1.00–3.01), but that longer-term former smokers were not at significantly increased risk (HR 1.04, 95% CI 0.66–1.64). Thus smoking is consistently observed to be associated with risk of advanced or fatal prostate cancer and appears to play a role in disease progression, in spite of its lack of association with overall incidence.

Survival. In line with findings on incidence of lethal disease, studies of smoking and survival among prostate cancer patients suggests that smoking is associated with increased prostate cancer-specific mortality as well as total mortality [148–154]. A pooled analysis of five prospective cohort studies found that current smoking was associated with a 40% higher risk of prostate cancer mortality (95% CI 20–70%) among prostate cancer patients [155]. A meta-analysis of 28 studies including both population-based and clinically-based study populations with varying treatments found that current smokers at treatment have worse overall mortality (HR 1.96, 95% CI 1.69–2.28), prostate cancer-specific mortality (HR 1.79, 95% CI 1.47–2.20), and recurrence-free survival (HR 1.48, 95% CI 1.28–1.72) than never smokers. Virtually all of the included studies adjusted for age at diagnosis, stage, and grade, and associations were similar across studies judged at high or low risk of bias. Another meta-analysis [156] among patients with localized prostate cancer undergoing primary radical prostatectomy or radiotherapy found very similar results.

Obesity

Because obesity can influence endogenous levels of sex hormones [157, 158], as well as the insulin/IGF axis, both of which are relevant to prostate cancer, it has been studied in many epidemiologic studies. Body mass index (BMI), measured as height (m)/weight $(kg)^2$ is the most commonly used measure of obesity in these studies. At the population level, BMI is highly correlated with other measures of adiposity and is uncorrelated with height [159, 160]; it is strongly predictive of mortality [161]. However, it does not perform well in the very elderly, when high BMI may begin to reflect lean body mass rather than adiposity [162].

The association between BMI and total prostate cancer incidence is somewhat inconsistent. A meta-analysis of 27 studies found a borderline significant combined relative risk of 1.03 (95% CI 1.00–1.07, p = 0.11) per 5 unit increase in BMI [163]. However, BMI is consistently associated with a lower risk of localized disease but an increased risk of advanced disease. Because of this heterogeneity by stage, the association between BMI and total prostate cancer varies across populations depending on PSA screening and the case mix found in that time and place. A meta-analysis of 13 prospective studies [164] found a relative risk per 5 unit increase in BMI of 0.94 (95% CI 0.91–0.97) for localized prostate cancer, and 1.09 (95% CI 1.02–1.16) for advanced prostate cancer. (The definitions of localized and advanced were a mix of advanced stage and high-grade, depending on the original studies.)

The AICR/WCRF Continuous Update Project report on prostate cancer [10] concluded that greater body fatness is a "probable" cause of advanced prostate cancer. Their meta-analysis of 23 studies of advanced cancer found a relative risk per 5 unit increase in BMI of 1.08 (95% CI 1.04–1.12). For 12 studies of prostate cancer mortality, the combined relative risk per 5 unit increase in BMI was 1.11 (95% CI 1.06–1.17). These results are consistent with more recently published results from the large European EPIC cohort [165], which found a hazard ratio for fatal prostate cancer of 1.14 (95% CI 1.02–1.27) per five unit increase in BMI. Two recent meta-analyses [164, 166] also found similar magnitudes of association with prostate cancer mortality, as did a pooled analysis of 57 prospective studies [167] from Europe, Japan, and the USA, comprising 1242 prostate cancer deaths.

The NIH-AARP cohort [168], which includes over 150,000 U.S. men, studied BMI trajectories from early adulthood onward. The BMI trajectories were not associated with total prostate cancer incidence. However, among never-smokers, BMI trajectories that resulted in obesity during adulthood were associated with a twofold increased risk of fatal prostate cancer compared to men who maintained a healthy BMI. These results highlight the importance of accounting for smoking in studies of obesity due to the strong inverse association between smoking and body weight and the positive association between smoking and prostate cancer survival.

Survival. Higher BMI is fairly consistently associated with poorer outcomes among men diagnosed with prostate cancer. In a meta-analysis of six studies of survival after prostate cancer diagnosis, the relative risk of prostate cancer mortality was 1.20 (95% CI 0.99–1.46) for a five unit increase in BMI around the time of diagnosis or treatment [164]. There was significant heterogeneity in this estimate due to the inclusion of the largest study, which found a non-significant inverse association with mortality based on 4 years of follow-up. Two studies among men with prostate cancer in Sweden published after the meta-analysis found significantly increased risks of prostate cancer-specific mortality with higher BMI [169, 170]. In 16 studies of biochemical recurrence after treatment, a five unit increase in BMI was associated with a relative risk of 1.21 (95% CI 1.11–1.31) [164].

Body Size in Early Life

Childhood obesity is inconsistently associated with adult prostate cancer risk. Four studies have examined pre-puberty body size (8–10 years) [171–174], with two reporting inverse associations, including for advanced disease [172, 173], while two others found no associations [171, 174]. One of these studies was from HPFS [174], which was an update of a previous report from this cohort [175], which found significant inverse associations between obesity at age 10 and risk of advanced and metastatic disease. However, this association was no longer observed with 16 additional years of follow-up. It is possible that childhood body size influences risk of prostate cancer less among older men. One study of body size at the time of puberty also found no association with prostate cancer risk [176].

The HPFS found an inverse association between BMI at age 21 and risk of advanced and lethal (death or distant metastasis) prostate cancer, independent of later life and earlier life body size [174]. Two other studies found similar inverse associations with advanced [172] or fatal [177] disease; however, other studies have found no associations [178–181]. A review of studies on total prostate cancer incidence suggested no relationship or a weak positive relationship [182].

Adiposity is known to increase estrogen and decrease androgen serum concentrations in men [157]. Hence, a childhood or early adult hormonal milieu characterized by low exposure to the stimulating effect of androgens on the prostate might protect against the disease. However, overall there is no consistent association between childhood and young adult body size and prostate cancer risk.

Weight Change

A meta-analysis of adult weight gain, from around age 18 to 25 until study entry in mid or late life and risk of prostate cancer found no clear association with overall risk [183]. Among eight prospective studies, the combined relative risk for the highest versus lowest weight gain category was 0.98 (95% CI 0.91–1.06). A dose-response meta-analysis of four studies also found no association, but there was a suggestion of an inverse association for localized disease (RR 0.96 for 5 kg weight gain, 95% CI 0.92–1.00) and a suggestion of a positive association for advanced disease at diagnosis (RR 1.04 for 5 kg weight gain, 95% CI 0.99–1.09). In line with these suggestive findings, several cohort studies have found significant positive associations between adult weight gain and prostate cancer mortality [181, 184, 185].

Given that obesity itself is associated with increased risk of advanced and fatal prostate cancer, it is difficult to separate the effect of weight gain during adulthood from the effect of obesity. The large NIH-AARP study [168] discussed above found a similar increase in risk of fatal prostate cancer among never smokers for all weight change trajectories ending in obesity, regardless of the specific timing of the weight gain. A study [186] among men diagnosed with localized prostate cancer in the HPFS cohort found that weight gain from age 21 to the time of diagnosis was associated with worse survival among never smoking men, whereas BMI itself at the time of diagnosis was not associated with survival.

Four studies have examined short-term weight gain around the time of prostate cancer diagnosis. Two studies used mortality as the outcome. One [169] found a significantly increased risk of prostate cancer-specific mortality for weight gain of >5% compared to stable weight in the 5–10 years after diagnosis of localized prostate cancer, and a significantly increased risk of total mortality for weight loss of >5%. Another [186] found no association between weight change in the 4 or 8 years prior to diagnosis and prostate cancer mortality among men diagnosed with localized disease. Two other studies among men treated with prostatectomy reported that weight gain in the year before surgery [187] or from 1 year before to 5 years after surgery [188] were associated with increased risk of biochemical recurrence. Additional studies of the role of weight changes before and after diagnosis and prostate cancer survival are needed.

Physical Activity

Physical activity is associated with reduced risk of several types of cancer. Multiple biological mechanisms for this have been proposed, including enhanced immune system function [189], changes in the endogenous hormonal milieu [190–192], reduction in inflammation [193–196], and reduced obesity [197]. Both obesity and metabolic syndrome have been associated with increased risk of advanced prostate cancer and worse prostate cancer-specific survival and response to hormonal therapy [198], so the positive systemic effects of exercise may impact prostate cancer risk and survival.

Physical activity has not been associated with overall prostate cancer risk. A meta-analysis [199] of 27 cohort studies and 23 case-control studies found a summary relative risk of 0.99 (95% CI 0.94–1.04) comparing the highest versus lowest categories of activity. Interestingly, a population-based Norwegian cohort study [200], which also found no association between higher levels of activity and risk of overall prostate cancer, did report a positive association between sitting time and risk. Men who reported sitting for 8 or more hours per day had a 22% (95% CI 5–42%) increased risk of prostate cancer compared to those who reported less than 8 h/day of sitting time.

Results on the association between physical activity and risk of advanced or fatal disease are mixed. Two prospective cohort studies, HPFS [201] and the CPS II [202], found inverse associations between higher levels of recreational physical activity and the risk of advanced or fatal disease, independent of BMI. However, four other cohorts, EPIC [203], the NIH-AARP Diet and Health Study [204], the Swedish National March Cohort [205], and PHS [206] found no associations between greater activity and risk of disease by stage or grade. Overall, a meta-analysis [199] of 10 cohort studies found no association between pre-diagnosis physical activity and prostate cancer mortality, with a relative risk of 0.93 (95% CI 0.81–1.08) for the highest versus lowest categories of activity.

The assessment of long-term physical activity levels is challenging. Study participants are often asked to report on the type, intensity, and duration of their average physical activity, both currently and in the past. The resulting misclassification may be responsible for the weak and often nonsignificant findings. Subgroups less prone to measurement error, such as, men who engage in a consistent program of vigorous activity, may offer the best chance of detecting a relationship between exercise and prostate cancer if one exists. The HPFS analysis is unique in that it was based on repeated prospective assessments of physical activity every 2 years over 14 years of follow-up. It found significantly lower risks of advanced and fatal disease for high levels of vigorous activity, but not for more moderate activity [201]. It is possible that repeated assessments of physical activity over time and study populations with a wide range of activity, including very active participants, is required for an inverse association between physical activity and advanced prostate cancer risk to emerge.

Survival. Physical activity may improve prostate cancer survival and may also ameliorate some of the adverse effects of therapy [207].

The few observational studies of activity after diagnosis and prostate cancer progression have reported beneficial associations. A meta-analysis [199] of four cohort studies of physical activity after diagnosis and prostate cancer mortality found a significant inverse association, with a relative risk of 0.69 (95% CI 0.55–0.85) comparing the highest to lowest activity categories. In the HPFS cohort, both moderate activity ($\geq$2.5 h/week) and vigorous activity ($\geq$1.25 h/week) were both associated with significantly improved overall and prostate cancer-specific survival [208, 209]. The authors estimated that 13–16% of deaths among men diagnosed with non-metastatic prostate cancer in the study population would have been prevented over 10 years if all men had engaged in 1.25 h/week or more of vigorous activity, and 5–10% of deaths could have been avoided with engagement in moderative activity.

A study of PSA recurrence in prostate cancer patients found similar decreases in risk with higher activity levels [210]. This lends support to the results for mortality because PSA recurrence is less susceptible to bias due to reverse causation (i.e., decreasing activity levels in response to disease progression) than prostate cancer mortality is.

Overall, while evidence on physical activity and prostate cancer incidence is mixed, it appears that activity is beneficial among men with prostate cancer.

Summary and Future Directions

Although an inherited genetic component may be larger for prostate cancer than for most other malignancies, evidence that lifestyle factors are important is also overwhelming; the substantial geographic variation and changing incidence among migrants demonstrate this as well. A summary of the evidence for diet and other lifestyle factors and prostate cancer risk is provided in Table 2. Substantial data supports that smoking and obesity/higher BMI are associated with increased risk of advanced prostate cancer, while obesity it inversely associated with risk of localized disease. In addition, an inverse association between vigorous activity and risk of advanced disease seems likely. Dietary factors associated with prostate cancer risk and survival are less well established. Of those studied, it seems probable that dairy intake is associated with increased risk, while fish intake and lycopene/tomato intake are associated with lower risk. However, even these dietary factors remain somewhat controversial within the research community.

Aside from these three dietary factors, most of the evidence on diet and prostate cancer is inconclusive. The role of calcium, vitamin D, and soy/phytoestrogen intake remains to be clarified. And the SELECT trial has complicated the interpretation of the data on Vitamin E and selenium.

SELECT and PHS II established that use of Vitamin E supplements in middle age and later are at best not protective, and possibly harmful, with respect to prostate cancer risk. However, circulating levels of vitamin E are very consistently associated with lower risk in observational studies, with no clear sources of confounding or other biases that might explain these results. The role of dietary and supplemental Vitamin E thus remains uncertain.

SELECT also found that use of selenium supplements in middle age and beyond are not protective for prostate cancer. However, observational studies are quite consistent in finding a substantially lower risk of prostate cancer among men with higher toenail or blood levels of selenium, and again, there are no clear sources of confounding or other bias that seem to explain these results. Thus the role of selenium in prostate cancer remains unclear and controversial.

Finally, the clear positive associations between height and circulating IGF-1 and prostate cancer risk, along with the long natural history of prostate cancer, suggest that dietary factors in childhood and adolescence likely impact prostate cancer risk; however, specific relationships have yet to be established, as studying early life exposures presents methodological challenges.

In practical terms, men concerned with prostate cancer risk should be encouraged to stop smoking, be as physically active as possible, and achieve or maintain a healthy weight. These recommendations also have the advantage of having a positive impact on risk of type 2 diabetes, cardiovascular disease, and other chronic diseases. Reducing dairy intake while increasing consumption of fish and tomato products is also reasonable advice. Finally, men should be counseled against taking Vitamin E or selenium supplements at levels higher than those found in multivitamins. (This is particularly true for Vitamin E given that meta-analysis of randomized trials find that high-dose Vitamin E supplements increase total mortality [211, 212].) Further research is needed to support more specific dietary recommendations for prostate cancer prevention and for preventing recurrence and progression in prostate cancer patients.

References

1. H. Shimizu, R.K. Ross, L. Bernstein, R. Yatani, B.E. Henderson, T.M. Mack, Cancers of the prostate and breast among Japanese and white immigrants in Los Angeles County. Br. J. Cancer **63**(6), 963–966 (1991)
2. G. Maskarinec, J.J. Noh, The effect of migration on cancer incidence among Japanese in Hawaii. Ethn. Dis. **14**(3), 431–439 (2004)
3. J. Lee, K. Demissie, S.E. Lu, G.G. Rhoads, Cancer incidence among Korean-American immigrants in the United States and native Koreans in South Korea. Cancer Control **14**(1), 78–85 (2007)
4. A.J. Hanley, B.C. Choi, E.J. Holowaty, Cancer mortality among Chinese migrants: a review. Int. J. Epidemiol. **24**(2), 255–265 (1995)
5. M. McCredie, S. Williams, M. Coates, Cancer mortality in migrants from the British Isles and continental Europe to New South Wales, Australia, 1975-1995. Int. J. Cancer **83**(2), 179–185 (1999)
6. L.A. Mucci, J.B. Hjelmborg, J.R. Harris, et al., Familial risk and heritability of cancer among twins in nordic countries. JAMA **315**(1), 68–76 (2016)
7. J.L. Jahn, E.L. Giovannucci, M.J. Stampfer, The high prevalence of undiagnosed prostate cancer at autopsy: implications for epidemiology and treatment of prostate cancer in the Prostate-specific Antigen-era. Int. J. Cancer **137**(12), 2795–2802 (2015)
8. J.R. Stark, S. Perner, M.J. Stampfer, et al., Gleason score and lethal prostate cancer: does 3 + 4 = 4 + 3? J. Clin. Oncol. **27**(21), 3459–3464 (2009)
9. D.P. Rose, A.P. Boyar, E.L. Wynder, International comparisons of mortality rates for cancer of the breast, ovary, prostate, and colon, and per capita food consumption. Cancer **58**(11), 2363–2371 (1986)
10. D. Aune, D.A. Navarro Rosenblatt, D.S. Chan, et al., Dairy products, calcium, and prostate cancer risk: a systematic review and meta-analysis of cohort studies. Am. J. Clin. Nutr. **101**(1), 87–117 (2015)
11. G.D. Batty, F. Barzi, M. Woodward, et al., Adult height and cancer mortality in Asia: the Asia Pacific Cohort Studies Collaboration. Ann. Oncol. **21**(3), 646–654 (2010)
12. Emerging Risk Factors C, Adult height and the risk of cause-specific death and vascular morbidity in 1 million people: individual participant meta-analysis. Int. J. Epidemiol. **41**(5), 1419–1433 (2012)
13. L.A. Proos, Anthropometry in adolescence—secular trends, adoption, ethnic and environmental differences. Horm. Res. **39**(Suppl 3), 18–24 (1993)
14. T.J. Cole, Secular trends in growth. Proc. Nutr. Soc. **59**(2), 317–324 (2000)
15. J. Fudvoye, A.S. Parent, Secular trends in growth. Ann. Endocrinol. (Paris) **78**(2), 88–91 (2017)
16. A. Juul, P. Bang, N.T. Hertel, et al., Serum insulin-like growth factor-I in 1030 healthy children, adolescents, and adults: relation to age, sex, stage of puberty, testicular size, and body mass index. J. Clin. Endocrinol. Metab. **78**(3), 744–752 (1994)
17. G. Beunen, M. Thomis, H.H. Maes, et al., Genetic variance of adolescent growth in stature. Ann. Hum. Biol. **27**(2), 173–186 (2000)
18. A.S. Wiley, Does milk make children grow? Relationships between milk consumption and height in NHANES 1999-2002. Am. J. Hum. Biol. **17**(4), 425–441 (2005)
19. S.J. Whiting, H. Vatanparast, A. Baxter-Jones, R.A. Faulkner, R. Mirwald, D.A. Bailey, Factors that affect bone mineral accrual in the adolescent growth spurt. J. Nutr. **134**(3), 696S–700S (2004)
20. A. Alimujiang, G.A. Colditz, J.D. Gardner, Y. Park, C.S. Berkey, S. Sutcliffe, Childhood diet and growth in boys in relation to timing of puberty and adult height: the Longitudinal Studies of Child Health and Development. Cancer Causes Control **29**(10), 915–926 (2018)

21. L.E. Underwood, Nutritional regulation of IGF-I and IGFBPs. J. Pediatr. Endocrinol. Metab. **9**(Suppl 3), 303–312 (1996)
22. E. Giovannucci, M. Pollak, Y. Liu, et al., Nutritional predictors of insulin-like growth factor I and their relationships to cancer in men. Cancer Epidemiol. Biomark. Prev. **12**(2), 84–89 (2003)
23. W.J. Smith, L.E. Underwood, D.R. Clemmons, Effects of caloric or protein restriction on insulin-like growth factor-I (IGF-I) and IGF-binding proteins in children and adults. J. Clin. Endocrinol. Metab. **80**(2), 443–449 (1995)
24. L.B. Signorello, H. Kuper, P. Lagiou, et al., Lifestyle factors and insulin-like growth factor 1 levels among elderly men. Eur. J. Cancer Prev. **9**(3), 173–178 (2000)
25. A.W. Roddam, N.E. Allen, P. Appleby, et al., Insulin-like growth factors, their binding proteins, and prostate cancer risk: analysis of individual patient data from 12 prospective studies. Ann. Intern. Med. **149**(7), 461–471 (2008)., W483-468
26. M.A. Rowlands, D. Gunnell, R. Harris, L.J. Vatten, J.M. Holly, R.M. Martin, Circulating insulin-like growth factor peptides and prostate cancer risk: a systematic review and meta-analysis. Int. J. Cancer **124**(10), 2416–2429 (2009)
27. C. Xu, F.F. Han, X.T. Zeng, T.Z. Liu, S. Li, Z.Y. Gao, I. Fat Intake, Not linked to prostate cancer: a systematic review and dose-response meta-analysis. PLoS One **10**(7), e0131747 (2015)
28. J.A. Simon, Y.H. Chen, S. Bent, The relation of alpha-linolenic acid to the risk of prostate cancer: a systematic review and meta-analysis. Am. J. Clin. Nutr. **89**(5), 1558S–1564S (2009)
29. M. Carayol, P. Grosclaude, C. Delpierre, Prospective studies of dietary alpha-linolenic acid intake and prostate cancer risk: a meta-analysis. Cancer Causes Control **21**(3), 347–355 (2010)
30. A.J. Carleton, J.L. Sievenpiper, R. de Souza, G. McKeown-Eyssen, D.J. Jenkins, Case-control and prospective studies of dietary alpha-linolenic acid intake and prostate cancer risk: a meta-analysis. BMJ Open **3**(5) (2013)
31. E. Giovannucci, Y. Liu, E.A. Platz, M.J. Stampfer, W.C. Willett, Risk factors for prostate cancer incidence and progression in the health professionals follow-up study. Int. J. Cancer **121**(7), 1571–1578 (2007)
32. J. Wu, K.M. Wilson, M.J. Stampfer, W.C. Willett, E.L. Giovannucci, A 24-year prospective study of dietary alpha-linolenic acid and lethal prostate cancer. Int. J. Cancer **142**(11), 2207–2214 (2018)
33. D.O. Koralek, U. Peters, G. Andriole, et al., A prospective study of dietary alpha-linolenic acid and the risk of prostate cancer (United States). Cancer Causes Control **17**(6), 783–791 (2006)
34. A.G. Schuurman, P.A. van den Brandt, E. Dorant, H.A. Brants, R.A. Goldbohm, Association of energy and fat intake with prostate carcinoma risk: results from The Netherlands Cohort Study. Cancer **86**(6), 1019–1027 (1999)
35. S.Y. Park, S.P. Murphy, L.R. Wilkens, B.E. Henderson, L.N. Kolonel, Fat and meat intake and prostate cancer risk: the multiethnic cohort study. Int. J. Cancer **121**(6), 1339–1345 (2007)
36. F.L. Crowe, P.N. Appleby, R.C. Travis, et al., Circulating fatty acids and prostate cancer risk: individual participant meta-analysis of prospective studies. J. Natl. Cancer Inst. **106**(9) (2014)
37. E.L. Van Blarigan, S.A. Kenfield, M. Yang, et al., Fat intake after prostate cancer diagnosis and mortality in the Physicians' Health Study. Cancer Causes Control **26**(8), 1117–1126 (2015)
38. E.L. Richman, S.A. Kenfield, J.E. Chavarro, et al., Fat intake after diagnosis and risk of lethal prostate cancer and all-cause mortality. JAMA Intern. Med. **173**(14), 1318–1326 (2013)
39. M.M. Epstein, J.L. Kasperzyk, L.A. Mucci, et al., Dietary fatty acid intake and prostate cancer survival in Orebro County, Sweden. Am. J. Epidemiol. **176**(3), 240–252 (2012)
40. D.D. Alexander, P.J. Mink, C.A. Cushing, B. Sceurman, A review and meta-analysis of prospective studies of red and processed meat intake and prostate cancer. Nutr. J. **9**, 50 (2010)
41. K. Wu, D. Spiegelman, T. Hou, et al., Associations between unprocessed red and processed meat, poultry, seafood and egg intake and the risk of prostate cancer: a pooled analysis of 15 prospective cohort studies. Int. J. Cancer **138**(10), 2368–2382 (2016)
42. J.M. Major, A.J. Cross, J.L. Watters, A.R. Hollenbeck, B.I. Graubard, R. Sinha, Patterns of meat intake and risk of prostate cancer among African-Americans in a large prospective study. Cancer Causes Control **22**(12), 1691–1698 (2011)
43. E.L. Richman, S.A. Kenfield, M.J. Stampfer, E.L. Giovannucci, J.M. Chan, Egg, red meat, and poultry intake and risk of lethal prostate cancer in the prostate-specific antigen-era: incidence and survival. Cancer Prev. Res. (Phila.) **4**(12), 2110–2121 (2011)
44. A.M. Gilsing, M.P. Weijenberg, R.A. Goldbohm, P.C. Dagnelie, P.A. van den Brandt, L.J. Schouten, Vegetarianism, low meat consumption and the risk of lung, postmenopausal breast and prostate cancer in a population-based cohort study. Eur. J. Clin. Nutr. **70**(6), 723–729 (2016)
45. R. Sinha, Y. Park, B.I. Graubard, et al., Meat and meat-related compounds and risk of prostate cancer in a large prospective cohort study in the United States. Am. J. Epidemiol. **170**(9), 1165–1177 (2009)
46. S. Koutros, A.J. Cross, D.P. Sandler, et al., Meat and meat mutagens and risk of prostate cancer in the Agricultural Health Study. Cancer Epidemiol. Biomark. Prev. **17**(1), 80–87 (2008)
47. S. Sharma, X. Cao, L.R. Wilkens, et al., Well-done meat consumption, NAT1 and NAT2 acetylator genotypes and prostate cancer risk: the multiethnic

cohort study. Cancer Epidemiol. Biomark. Prev. **19**(7), 1866–1870 (2010)

48. A. Sander, J. Linseisen, S. Rohrmann, Intake of heterocyclic aromatic amines and the risk of prostate cancer in the EPIC-Heidelberg cohort. Cancer Causes Control **22**(1), 109–114 (2011)

49. S. Rohrmann, K. Nimptsch, R. Sinha, et al., Intake of meat mutagens and risk of prostate cancer in a cohort of U.S. health professionals. Cancer Epidemiol. Biomark. Prev. **24**(10), 1557–1563 (2015)

50. A.J. Cross, U. Peters, V.A. Kirsh, et al., A prospective study of meat and meat mutagens and prostate cancer risk. Cancer Res. **65**(24), 11779–11784 (2005)

51. M. Yang, S.A. Kenfield, E.L. Van Blarigan, et al., Dietary patterns after prostate cancer diagnosis in relation to disease-specific and total mortality. Cancer Prev. Res. (Phila.) **8**(6), 545–551 (2015)

52. K.M. Wilson, L.A. Mucci, B.F. Drake, et al., Meat, fish, poultry, and egg intake at diagnosis and risk of prostate cancer progression. Cancer Prev. Res. (Phila.) **9**(12), 933–941 (2016)

53. P.A. Nutting, W.L. Freeman, D.R. Risser, et al., Cancer incidence among American Indians and Alaska Natives, 1980 through 1987. Am. J. Public Health **83**(11), 1589–1598 (1993)

54. J. Zhang, S. Sasaki, K. Amano, H. Kesteloot, Fish consumption and mortality from all causes, ischemic heart disease, and stroke: an ecological study. Prev. Med. **28**(5), 520–529 (1999)

55. E. Dewailly, G. Mulvad, H. Sloth Pedersen, J.C. Hansen, N. Behrendt, J.P. Hart Hansen, Inuit are protected against prostate cancer. Cancer Epidemiol. Biomark. Prev. **12**(9), 926–927 (2003)

56. J.M. Chan, P.H. Gann, E.L. Giovannucci, Role of diet in prostate cancer development and progression. J. Clin. Oncol. **23**(32), 8152–8160 (2005)

57. S.H. Nash, J.M. Schenk, A.R. Kristal, et al., Association between serum phospholipid fatty acids and intraprostatic inflammation in the placebo arm of the prostate cancer prevention trial. Cancer Prev. Res. (Phila.) **8**(7), 590–596 (2015)

58. T.M. Brasky, C. Till, E. White, et al., Serum phospholipid fatty acids and prostate cancer risk: results from the prostate cancer prevention trial. Am. J. Epidemiol. **173**(12), 1429–1439 (2011)

59. F.L. Crowe, N.E. Allen, P.N. Appleby, et al., Fatty acid composition of plasma phospholipids and risk of prostate cancer in a case-control analysis nested within the European Prospective Investigation into Cancer and Nutrition. Am. J. Clin. Nutr. **88**(5), 1353–1363 (2008)

60. J.E. Chavarro, M.J. Stampfer, H. Li, H. Campos, T. Kurth, J. Ma, A prospective study of polyunsaturated fatty acid levels in blood and prostate cancer risk. Cancer Epidemiol. Biomark. Prev. **16**(7), 1364–1370 (2007)

61. K.M. Szymanski, D.C. Wheeler, L.A. Mucci, Fish consumption and prostate cancer risk: a review and meta-analysis. Am. J. Clin. Nutr. **92**(5), 1223–1233 (2010)

62. M. Aucoin, K. Cooley, C. Knee, et al., Fish-derived omega-3 fatty acids and prostate cancer: a systematic review. Integr. Cancer Ther. **16**(1), 32–62 (2017)

63. J.E. Torfadottir, U.A. Valdimarsdottir, L.A. Mucci, et al., Consumption of fish products across the lifespan and prostate cancer risk. PLoS One **8**(4), e59799 (2013)

64. J.E. Manson, N.R. Cook, I.M. Lee, et al., Marine n-3 fatty acids and prevention of cardiovascular disease and cancer. N. Engl. J. Med. **380**(1), 23–32 (2019)

65. K.M. Wilson, I.M. Shui, L.A. Mucci, E. Giovannucci, Calcium and phosphorus intake and prostate cancer risk: a 24-y follow-up study. Am. J. Clin. Nutr. **101**(1), 173–183 (2015)

66. E. Giovannucci, Dietary influences of 1,25(OH)2 vitamin D in relation to prostate cancer: a hypothesis. Cancer Causes Control **9**(6), 567–582 (1998)

67. J. Ma, E. Giovannucci, M. Pollak, et al., Milk intake, circulating levels of insulin-like growth factor-I, and risk of colorectal cancer in men. J. Natl. Cancer Inst. **93**(17), 1330–1336 (2001)

68. A. Berruti, L. Dogliotti, G. Gorzegno, et al., Differential patterns of bone turnover in relation to bone pain and disease extent in bone in cancer patients with skeletal metastases. Clin. Chem. **45**(8 Pt 1), 1240–1247 (1999)

69. A. Schneider, L.M. Kalikin, A.C. Mattos, et al., Bone turnover mediates preferential localization of prostate cancer in the skeleton. Endocrinology **146**(4), 1727–1736 (2005)

70. J. Sturge, M.P. Caley, J. Waxman, Bone metastasis in prostate cancer: emerging therapeutic strategies. Nat. Rev. Clin. Oncol. **8**(6), 357–368 (2011)

71. A. Pettersson, J.L. Kasperzyk, S.A. Kenfield, et al., Milk and dairy consumption among men with prostate cancer and risk of metastases and prostate cancer death. Cancer Epidemiol. Biomark. Prev. **21**(3), 428–436 (2012)

72. M.K. Downer, J.L. Batista, L.A. Mucci, et al., Dairy intake in relation to prostate cancer survival. Int. J. Cancer **140**(9), 2060–2069 (2017)

73. M. Yang, S.A. Kenfield, E.L. Van Blarigan, et al., Dairy intake after prostate cancer diagnosis in relation to disease-specific and total mortality. Int. J. Cancer **137**(10), 2462–2469 (2015)

74. R. Gilbert, C. Metcalfe, W.D. Fraser, et al., Associations of circulating 25-hydroxyvitamin D with prostate cancer diagnosis, stage and grade. Int. J. Cancer **131**(5), 1187–1196 (2012)

75. S. Gandini, M. Boniol, J. Haukka, et al., Meta-analysis of observational studies of serum 25-hydroxyvitamin D levels and colorectal, breast and prostate cancer and colorectal adenoma. Int. J. Cancer **128**(6), 1414–1424 (2011)

76. I.M. Shui, A.M. Mondul, S. Lindstrom, et al., Circulating vitamin D, vitamin D-related genetic variation, and risk of fatal prostate cancer in the

National Cancer Institute Breast and Prostate Cancer Cohort Consortium. Cancer **121**(12), 1949–1956 (2015)

77. J.M. Schenk, C.A. Till, C.M. Tangen, et al., Serum 25-hydroxyvitamin D concentrations and risk of prostate cancer: results from the Prostate Cancer Prevention Trial. Cancer Epidemiol. Biomark. Prev. **23**(8), 1484–1493 (2014)

78. J. Brandstedt, M. Almquist, J. Manjer, J. Malm, D. Vitamin, PTH, and calcium and the risk of prostate cancer: a prospective nested case-control study. Cancer Causes Control **23**(8), 1377–1385 (2012)

79. D. Albanes, A.M. Mondul, K. Yu, et al., Serum 25-hydroxy vitamin D and prostate cancer risk in a large nested case-control study. Cancer Epidemiol. Biomark. Prev. **20**(9), 1850–1860 (2011)

80. A.R. Kristal, C. Till, X. Song, et al., Plasma vitamin D and prostate cancer risk: results from the Selenium and Vitamin E Cancer Prevention Trial. Cancer Epidemiol. Biomark. Prev. **23**(8), 1494–1504 (2014)

81. J.E. Manson, N.R. Cook, I.M. Lee, et al., Vitamin D supplements and prevention of cancer and cardiovascular disease. N. Engl. J. Med. **380**(1), 33–44 (2019)

82. R. Gilbert, R.M. Martin, R. Beynon, et al., Associations of circulating and dietary vitamin D with prostate cancer risk: a systematic review and dose-response meta-analysis. Cancer Causes Control **22**(3), 319–340 (2011)

83. S. Tretli, E. Hernes, J.P. Berg, U.E. Hestvik, T.E. Robsahm, Association between serum 25(OH) D and death from prostate cancer. Br. J. Cancer **100**(3), 450–454 (2009)

84. F. Fang, J.L. Kasperzyk, I. Shui, et al., Prediagnostic plasma vitamin D metabolites and mortality among patients with prostate cancer. PLoS One **6**(4), e18625 (2011)

85. I.M. Shui, L.A. Mucci, P. Kraft, et al., Vitamin D-related genetic variation, plasma vitamin D, and risk of lethal prostate cancer: a prospective nested case-control study. J. Natl. Cancer Inst. **104**(9), 690–699 (2012)

86. J. Brandstedt, M. Almquist, J. Manjer, J. Malm, D. Vitamin, PTH, and calcium in relation to survival following prostate cancer. Cancer Causes Control **27**(5), 669–677 (2016)

87. S.K. Holt, S. Kolb, R. Fu, R. Horst, Z. Feng, J.L. Stanford, Circulating levels of 25-hydroxyvitamin D and prostate cancer prognosis. Cancer Epidemiol. **37**(5), 666–670 (2013)

88. S.K. Holt, E.M. Kwon, J.S. Koopmeiners, et al., Vitamin D pathway gene variants and prostate cancer prognosis. Prostate **70**(13), 1448–1460 (2010)

89. L. Chen, G. Davey Smith, D.M. Evans, et al., Genetic variants in the vitamin D receptor are associated with advanced prostate cancer at diagnosis: findings from the prostate testing for cancer and treatment study and a systematic review. Cancer Epidemiol. Biomark. Prev. **18**(11), 2874–2881 (2009)

90. S. Gandini, P. Gnagnarella, D. Serrano, E. Pasquali, S. Raimondi, Vitamin D receptor polymorphisms and cancer. Adv. Exp. Med. Biol. **810**, 69–105 (2014)

91. W.K. Hendrickson, R. Flavin, J.L. Kasperzyk, et al., Vitamin D receptor protein expression in tumor tissue and prostate cancer progression. J. Clin. Oncol. **29**(17), 2378–2385 (2011)

92. E. Giovannucci, A. Ascherio, E.B. Rimm, M.J. Stampfer, G.A. Colditz, W.C. Willett, Intake of carotenoids and retinol in relation to risk of prostate cancer. J. Natl. Cancer Inst. **87**(23), 1767–1776 (1995)

93. J.L. Rowles 3rd, K.M. Ranard, J.W. Smith, R. An, J.W. Erdman Jr., Increased dietary and circulating lycopene are associated with reduced prostate cancer risk: a systematic review and meta-analysis. Prostate Cancer Prostatic Dis. **20**(4), 361–377 (2017)

94. K. Zu, L. Mucci, B.A. Rosner, et al., Dietary lycopene, angiogenesis, and prostate cancer: a prospective study in the prostate-specific antigen era. J. Natl. Cancer Inst. **106**(2), djt430 (2014)

95. C. Gartner, W. Stahl, H. Sies, Lycopene is more bioavailable from tomato paste than from fresh tomatoes. Am. J. Clin. Nutr. **66**(1), 116–122 (1997)

96. E. Giovannucci, E.B. Rimm, Y. Liu, M.J. Stampfer, W.C. Willett, A prospective study of tomato products, lycopene, and prostate cancer risk. J. Natl. Cancer Inst. **94**(5), 391–398 (2002)

97. E.M. Grainger, C.W. Hadley, N.E. Moran, et al., A comparison of plasma and prostate lycopene in response to typical servings of tomato soup, sauce or juice in men before prostatectomy. Br. J. Nutr. **114**(4), 596–607 (2015)

98. C.M. Yang, Y.T. Yen, C.S. Huang, M.L. Hu, Growth inhibitory efficacy of lycopene and beta-carotene against androgen-independent prostate tumor cells xenografted in nude mice. Mol. Nutr. Food Res. **55**(4), 606–612 (2011)

99. S. Elgass, A. Cooper, M. Chopra, Lycopene inhibits angiogenesis in human umbilical vein endothelial cells and rat aortic rings. Br. J. Nutr. **108**(3), 431–439 (2012)

100. M.L. Chen, Y.H. Lin, C.M. Yang, M.L. Hu, Lycopene inhibits angiogenesis both in vitro and in vivo by inhibiting MMP-2/uPA system through VEGFR2-mediated PI3K-Akt and ERK/p38 signaling pathways. Mol. Nutr. Food Res. **56**(6), 889–899 (2012)

101. C.S. Huang, C.H. Chuang, T.F. Lo, M.L. Hu, Antiangiogenic effects of lycopene through immunomodualtion of cytokine secretion in human peripheral blood mononuclear cells. J. Nutr. Biochem. **24**(2), 428–434 (2013)

102. L.A. Mucci, A. Powolny, E. Giovannucci, et al., Prospective study of prostate tumor angiogenesis and cancer-specific mortality in the health professionals follow-up study. J. Clin. Oncol. **27**(33), 5627–5633 (2009)

103. Y. Wang, E.J. Jacobs, C.C. Newton, M.L. McCullough, Lycopene, tomato products and prostate cancer-specific mortality among men

diagnosed with nonmetastatic prostate cancer in the Cancer Prevention Study II Nutrition Cohort. Int. J. Cancer **138**(12), 2846–2855 (2016)

104. M.J. Messina, V. Persky, K.D. Setchell, S. Barnes, Soy intake and cancer risk: a review of the in vitro and in vivo data. Nutr. Cancer **21**(2), 113–131 (1994)

105. D.M. Tham, C.D. Gardner, W.L. Haskell, Clinical review 97: potential health benefits of dietary phytoestrogens: a review of the clinical, epidemiological, and mechanistic evidence. J. Clin. Endocrinol. Metab. **83**(7), 2223–2235 (1998)

106. H. Adlercreutz, W. Mazur, Phyto-oestrogens and Western diseases. Ann. Med. **29**(2), 95–120 (1997)

107. A. Bylund, J.X. Zhang, A. Bergh, et al., Rye bran and soy protein delay growth and increase apoptosis of human LNCaP prostate adenocarcinoma in nude mice. Prostate **42**(4), 304–314 (2000)

108. E. Kyle, L. Neckers, C. Takimoto, G. Curt, R. Bergan, Genistein-induced apoptosis of prostate cancer cells is preceded by a specific decrease in focal adhesion kinase activity. Mol. Pharmacol. **51**(2), 193–200 (1997)

109. C.C. Applegate, J.L. Rowles, K.M. Ranard, S. Jeon, J.W. Erdman, Soy consumption and the risk of prostate cancer: an updated systematic review and meta-analysis. Nutrients **10**(1) (2018)

110. M. Zhang, K. Wang, L. Chen, B. Yin, Y. Song, Is phytoestrogen intake associated with decreased risk of prostate cancer? A systematic review of epidemiological studies based on 17,546 cases. Andrology **4**(4), 745–756 (2016)

111. S.Y. Park, S.P. Murphy, L.R. Wilkens, B.E. Henderson, L.N. Kolonel, S. Multiethnic Cohort, Legume and isoflavone intake and prostate cancer risk: the Multiethnic Cohort Study. Int. J. Cancer **123**(4), 927–932 (2008)

112. N.E. Allen, C. Sauvaget, A.W. Roddam, et al., A prospective study of diet and prostate cancer in Japanese men. Cancer Causes Control **15**(9), 911–920 (2004)

113. N. Kurahashi, M. Iwasaki, S. Sasazuki, et al., Soy product and isoflavone consumption in relation to prostate cancer in Japanese men. Cancer Epidemiol. Biomark. Prev. **16**(3), 538–545 (2007)

114. M.C. Bosland, I. Kato, A. Zeleniuch-Jacquotte, et al., Effect of soy protein isolate supplementation on biochemical recurrence of prostate cancer after radical prostatectomy: a randomized trial. JAMA **310**(2), 170–178 (2013)

115. Q. Jiang, Natural forms of vitamin E: metabolism, antioxidant, and anti-inflammatory activities and their role in disease prevention and therapy. Free Radic. Biol. Med. **72**, 76–90 (2014)

116. M. Meydani, Vitamin E. Lancet **345**(8943), 170–175 (1995)

117. S.N. Meydani, M.G. Hayek, Vitamin E and aging immune response. Clin. Geriatr. Med. **11**(4), 567–576 (1995)

118. E.A. Ripoll, B.N. Rama, M.M. Webber, Vitamin E enhances the chemotherapeutic effects of adria-

mycin on human prostatic carcinoma cells in vitro. J. Urol. **136**(2), 529–531 (1986)

119. K. Gunawardena, D.K. Murray, A.W. Meikle, Vitamin E and other antioxidants inhibit human prostate cancer cells through apoptosis. Prostate **44**(4), 287–295 (2000)

120. Alpha-Tocopherol BCCPSG, The effect of vitamin E and beta carotene on the incidence of lung cancer and other cancers in male smokers. N. Engl. J. Med. **330**(15), 1029–1035 (1994)

121. O.P. Heinonen, D. Albanes, J. Virtamo, et al., Prostate cancer and supplementation with alpha-tocopherol and beta-carotene: incidence and mortality in a controlled trial. J. Natl. Cancer Inst. **90**(6), 440–446 (1998)

122. W.J. Blot, J.Y. Li, P.R. Taylor, et al., Nutrition intervention trials in Linxian, China: supplementation with specific vitamin/mineral combinations, cancer incidence, and disease-specific mortality in the general population. J. Natl. Cancer Inst. **85**(18), 1483–1492 (1993)

123. S.M. Lippman, E.A. Klein, P.J. Goodman, et al., Effect of selenium and vitamin E on risk of prostate cancer and other cancers: the Selenium and Vitamin E Cancer Prevention Trial (SELECT). JAMA **301**(1), 39–51 (2009)

124. E.A. Klein, I.M. Thompson Jr., C.M. Tangen, et al., Vitamin E and the risk of prostate cancer: the Selenium and Vitamin E Cancer Prevention Trial (SELECT). JAMA **306**(14), 1549–1556 (2011)

125. S.M. Lippman, P.J. Goodman, E.A. Klein, et al., Designing the Selenium and Vitamin E Cancer Prevention Trial (SELECT). J. Natl. Cancer Inst. **97**(2), 94–102 (2005)

126. J.M. Gaziano, R.J. Glynn, W.G. Christen, et al., Vitamins E and C in the prevention of prostate and total cancer in men: the Physicians' Health Study II randomized controlled trial. JAMA **301**(1), 52–62 (2009)

127. U. Peters, A.J. Littman, A.R. Kristal, R.E. Patterson, J.D. Potter, E. White, E. Vitamin, selenium supplementation and risk of prostate cancer in the Vitamins and lifestyle (VITAL) study cohort. Cancer Causes Control **19**(1), 75–87 (2008)

128. V.A. Kirsh, R.B. Hayes, S.T. Mayne, et al., Supplemental and dietary vitamin E, beta-carotene, and vitamin C intakes and prostate cancer risk. J. Natl. Cancer Inst. **98**(4), 245–254 (2006)

129. J.M. Chan, M.J. Stampfer, J. Ma, E.B. Rimm, W.C. Willett, E.L. Giovannucci, Supplemental vitamin E intake and prostate cancer risk in a large cohort of men in the United States. Cancer Epidemiol. Biomark. Prev. **8**(10), 893–899 (1999)

130. T.J. Key, P.N. Appleby, R.C. Travis, et al., Carotenoids, retinol, tocopherols, and prostate cancer risk: pooled analysis of 15 studies. Am. J. Clin. Nutr. **102**(5), 1142–1157 (2015)

131. G.F. Combs Jr., S.B. Combs, The nutritional biochemistry of selenium. Annu. Rev. Nutr. **4**, 257–280 (1984)

132. C. Redman, J.A. Scott, A.T. Baines, et al., Inhibitory effect of selenomethionine on the growth of three selected human tumor cell lines. Cancer Lett. **125**(1–2), 103–110 (1998)

133. D.G. Menter, A.L. Sabichi, S.M. Lippman, Selenium effects on prostate cell growth. Cancer Epidemiol. Biomark. Prev. **9**(11), 1171–1182 (2000)

134. M.P. Rayman, The importance of selenium to human health. Lancet **356**(9225), 233–241 (2000)

135. J. Neve, Human selenium supplementation as assessed by changes in blood selenium concentration and glutathione peroxidase activity. J. Trace Elem. Med. Biol. **9**(2), 65–73 (1995)

136. A.J. Duffield-Lillico, B.L. Dalkin, M.E. Reid, et al., Selenium supplementation, baseline plasma selenium status and incidence of prostate cancer: an analysis of the complete treatment period of the Nutritional Prevention of Cancer Trial. BJU Int. **91**(7), 608–612 (2003)

137. L.C. Clark, G.F. Combs Jr., B.W. Turnbull, et al., Effects of selenium supplementation for cancer prevention in patients with carcinoma of the skin. A randomized controlled trial. Nutritional Prevention of Cancer Study Group. JAMA **276**(24), 1957–1963 (1996)

138. A.R. Kristal, A.K. Darke, J.S. Morris, et al., Baseline selenium status and effects of selenium and vitamin E supplementation on prostate cancer risk. J. Natl. Cancer Inst. **106**(3), djt456 (2014)

139. J. Yarmolinsky, C. Bonilla, P.C. Haycock, et al., Circulating selenium and prostate cancer risk: a mendelian randomization analysis. J. Natl. Cancer Inst. **110**(9), 1035–1038 (2018)

140. N.E. Allen, R.C. Travis, P.N. Appleby, et al., Selenium and prostate cancer: analysis of individual participant data from fifteen prospective studies. J. Natl. Cancer Inst. **108**(11) (2016)

141. M. Outzen, A. Tjonneland, E.H. Larsen, et al., Selenium status and risk of prostate cancer in a Danish population. Br. J. Nutr. **115**(9), 1669–1677 (2016)

142. Z. Cui, D. Liu, C. Liu, G. Liu, Serum selenium levels and prostate cancer risk: a MOOSE-compliant meta-analysis. Medicine (Baltimore) **96**(5), e5944 (2017)

143. K. Sayehmiri, M. Azami, Y. Mohammadi, A. Soleymani, Z. Tardeh, The association between Selenium and Prostate Cancer: a systematic review and meta-analysis. Asian Pac. J. Cancer Prev. **19**(6), 1431–1437 (2018)

144. S.A. Kenfield, E.L. Van Blarigan, N. DuPre, M.J. Stampfer, L.G. E, J.M. Chan, Selenium supplementation and prostate cancer mortality. J. Natl. Cancer Inst. **107**(1), 360 (2015)

145. F. Islami, D.M. Moreira, P. Boffetta, S.J. Freedland, A systematic review and meta-analysis of tobacco use and prostate cancer mortality and incidence in prospective cohort studies. Eur. Urol. **66**(6), 1054–1064 (2014)

146. M. Huncharek, K.S. Haddock, R. Reid, B. Kupelnick, Smoking as a risk factor for prostate cancer: a meta-analysis of 24 prospective cohort studies. Am. J. Public Health **100**(4), 693–701 (2010)

147. Services USDoHaH, *The Health Consequences of Smoking-50 Years of Progress: A Report of the Surgeon General* (U.S. Department of Health and Human Services, Centers for Disease Control and Prevention, National Center for Chronic Disease Prevention and Health Promotion, Office on Smoking and Health, Atlanta, 2014)

148. M.G. Oefelein, M.I. Resnick, Association of tobacco use with hormone refractory disease and survival of patients with prostate cancer. J. Urol. **171**(6 Pt 1), 2281–2284 (2004)

149. T. Pickles, M. Liu, E. Berthelet, et al., The effect of smoking on outcome following external radiation for localized prostate cancer. J. Urol. **171**(4), 1543–1546 (2004)

150. J. Pantarotto, S. Malone, S. Dahrouge, V. Gallant, L. Eapen, Smoking is associated with worse outcomes in patients with prostate cancer treated by radical radiotherapy. BJU Int. **99**(3), 564–569 (2007)

151. D.M. Moreira, J.A. Antonelli, J.C. Presti Jr., et al., Association of cigarette smoking with interval to biochemical recurrence after radical prostatectomy: results from the SEARCH database. Urology **76**(5), 1218–1223 (2010)

152. C.E. Joshu, A.M. Mondul, C.L. Meinhold, et al., Cigarette smoking and prostate cancer recurrence after prostatectomy. J. Natl. Cancer Inst. **103**(10), 835–838 (2011)

153. S.A. Kenfield, M.J. Stampfer, J.M. Chan, E. Giovannucci, Smoking and prostate cancer survival and recurrence. JAMA **305**(24), 2548–2555 (2011)

154. M. Rieken, S.F. Shariat, L.A. Kluth, et al., Association of cigarette smoking and smoking cessation with biochemical recurrence of prostate cancer in patients treated with radical prostatectomy. Eur. Urol. **68**(6), 949–956 (2015)

155. B.D. Carter, N.D. Freedman, E.J. Jacobs, Smoking and mortality—beyond established causes. N. Engl. J. Med. **372**(22), 2170 (2015)

156. B. Foerster, C. Pozo, M. Abufaraj, et al., Association of smoking status with recurrence, metastasis, and mortality among patients with localized prostate cancer undergoing prostatectomy or radiotherapy: a systematic review and meta-analysis. JAMA Oncol. **4**(7), 953–961 (2018)

157. R. Pasquali, F. Casimirri, S. Cantobelli, et al., Effect of obesity and body fat distribution on sex hormones and insulin in men. Metabolism **40**(1), 101–104 (1991)

158. C.S. Mantzoros, E.I. Georgiadis, Body mass and physical activity are important predictors of serum androgen concentrations in young healthy men. Epidemiology **6**(4), 432–435 (1995)

159. J.C. Seidell, K.M. Flegal, Assessing obesity: classification and epidemiology. Br. Med. Bull. **53**(2), 238–252 (1997)

160. R.W. Taylor, D. Keil, E.J. Gold, S.M. Williams, A. Goulding, Body mass index, waist girth, and waist-to-hip ratio as indexes of total and regional adiposity in women: evaluation using receiver operating characteristic curves. Am. J. Clin. Nutr. **67**(1), 44–49 (1998)

161. Global BMIMC, E. Di Angelantonio, N. Bhupathiraju Sh, et al., Body-mass index and all-cause mortality: individual-participant-data meta-analysis of 239 prospective studies in four continents. Lancet **388**(10046), 776–786 (2016)

162. J.C. Seidell, T.L. Visscher, Body weight and weight change and their health implications for the elderly. Eur. J. Clin. Nutr. **54**(Suppl 3), S33–S39 (2000)

163. A.G. Renehan, M. Tyson, M. Egger, R.F. Heller, M. Zwahlen, Body-mass index and incidence of cancer: a systematic review and meta-analysis of prospective observational studies. Lancet **371**(9612), 569–578 (2008)

164. A. Discacciati, N. Orsini, A. Wolk, Body mass index and incidence of localized and advanced prostate cancer—a dose-response meta-analysis of prospective studies. Ann. Oncol. **23**(7), 1665–1671 (2012)

165. A. Perez-Cornago, P.N. Appleby, T. Pischon, et al., Tall height and obesity are associated with an increased risk of aggressive prostate cancer: results from the EPIC cohort study. BMC Med. **15**(1), 115 (2017)

166. X. Zhang, G. Zhou, B. Sun, et al., Impact of obesity upon prostate cancer-associated mortality: a meta-analysis of 17 cohort studies. Oncol. Lett. **9**(3), 1307–1312 (2015)

167. Prospective Studies C, G. Whitlock, S. Lewington, et al., Body-mass index and cause-specific mortality in 900 000 adults: collaborative analyses of 57 prospective studies. Lancet **373**(9669), 1083–1096 (2009)

168. S.P. Kelly, H. Lennon, M. Sperrin, et al., Body mass index trajectories across adulthood and smoking in relation to prostate cancer risks: the NIH-AARP diet and health study. Int. J. Epidemiol. **48**(2), 464–473 (2019)

169. S.E. Bonn, F. Wiklund, A. Sjolander, et al., Body mass index and weight change in men with prostate cancer: progression and mortality. Cancer Causes Control **25**(8), 933–943 (2014)

170. A. Cantarutti, S.E. Bonn, H.O. Adami, H. Gronberg, R. Bellocco, K. Balter, Body mass index and mortality in men with prostate cancer. Prostate **75**(11), 1129–1136 (2015)

171. A.W. Hsing, J. Deng, I.A. Sesterhenn, et al., Body size and prostate cancer: a population-based case-control study in China. Cancer Epidemiol. Biomark. Prev. **9**(12), 1335–1341 (2000)

172. W.R. Robinson, J. Stevens, M.D. Gammon, E.M. John, Obesity before age 30 years and risk of advanced prostate cancer. Am. J. Epidemiol. **161**(12), 1107–1114 (2005)

173. M. Barba, I. Terrenato, H.J. Schunemann, et al., Indicators of sexual and somatic development and

adolescent body size in relation to prostate cancer risk: results from a case-control study. Urology **72**(1), 183–187 (2008)

174. E. Moller, K.M. Wilson, J.L. Batista, L.A. Mucci, K. Balter, E. Giovannucci, Body size across the life course and prostate cancer in the health professionals follow-up study. Int. J. Cancer **138**(4), 853–865 (2016)

175. E. Giovannucci, E.B. Rimm, M.J. Stampfer, G.A. Colditz, W.C. Willett, Height, body weight, and risk of prostate cancer. Cancer Epidemiol. Biomark. Prev. **6**(8), 557–563 (1997)

176. S.O. Andersson, J. Baron, R. Bergstrom, C. Lindgren, A. Wolk, H.O. Adami, Lifestyle factors and prostate cancer risk: a case-control study in Sweden. Cancer Epidemiol. Biomark. Prev. **5**(7), 509–513 (1996)

177. A. Discacciati, N. Orsini, S.O. Andersson, O. Andren, J.E. Johansson, A. Wolk, Body mass index in early and middle-late adulthood and risk of localised, advanced and fatal prostate cancer: a population-based prospective study. Br. J. Cancer **105**(7), 1061–1068 (2011)

178. A.G. Schuurman, R.A. Goldbohm, E. Dorant, P.A. van den Brandt, Anthropometry in relation to prostate cancer risk in the Netherlands cohort study. Am. J. Epidemiol. **151**(6), 541–549 (2000)

179. G.G. Giles, G. Severi, D.R. English, et al., Early growth, adult body size and prostate cancer risk. Int. J. Cancer **103**(2), 241–245 (2003)

180. A.J. Littman, E. White, A.R. Kristal, Anthropometrics and prostate cancer risk. Am. J. Epidemiol. **165**(11), 1271–1279 (2007)

181. M.E. Wright, S.C. Chang, A. Schatzkin, et al., Prospective study of adiposity and weight change in relation to prostate cancer incidence and mortality. Cancer **109**(4), 675–684 (2007)

182. W.R. Robinson, C. Poole, P.A. Godley, Systematic review of prostate cancer's association with body size in childhood and young adulthood. Cancer Causes Control **19**(8), 793–803 (2008)

183. N. Keum, D.C. Greenwood, D.H. Lee, et al., Adult weight gain and adiposity-related cancers: a dose-response meta-analysis of prospective observational studies. J. Natl. Cancer Inst. **107**(2) (2015)

184. C. Rodriguez, S.J. Freedland, A. Deka, et al., Body mass index, weight change, and risk of prostate cancer in the cancer prevention study II nutrition cohort. Cancer Epidemiol. Biomark. Prev. **16**(1), 63–69 (2007)

185. J.K. Bassett, G. Severi, L. Baglietto, et al., Weight change and prostate cancer incidence and mortality. Int. J. Cancer **131**(7), 1711–1719 (2012)

186. B.A. Dickerman, T.U. Ahearn, E. Giovannucci, et al., Weight change, obesity and risk of prostate cancer progression among men with clinically localized prostate cancer. Int. J. Cancer **141**(5), 933–944 (2017)

187. B.M. Whitley, D.M. Moreira, J.A. Thomas, et al., Preoperative weight change and risk of adverse outcome following radical prostatectomy: results from

the Shared Equal Access Regional Cancer Hospital database. Prostate Cancer Prostatic Dis. **14**(4), 361–366 (2011)

188. C.E. Joshu, A.M. Mondul, A. Menke, et al., Weight gain is associated with an increased risk of prostate cancer recurrence after prostatectomy in the PSA era. Cancer Prev. Res. (Phila.) **4**(4), 544–551 (2011)

189. I.M. Lee, Exercise and physical health: cancer and immune function. Res. Q. Exerc. Sport **66**(4), 286–291 (1995)

190. J.L. Durstine, W.L. Haskell, Effects of exercise training on plasma lipids and lipoproteins. Exerc. Sport Sci. Rev. **22**, 477–521 (1994)

191. P.F. Kokkinos, B. Fernhall, Physical activity and high density lipoprotein cholesterol levels: what is the relationship? Sports Med. **28**(5), 307–314 (1999)

192. P. Kokkinos, J. Myers, Exercise and physical activity: clinical outcomes and applications. Circulation **122**(16), 1637–1648 (2010)

193. S.P. Helmrich, D.R. Ragland, R.W. Leung, R.S. Paffenbarger Jr., Physical activity and reduced occurrence of non-insulin-dependent diabetes mellitus. N. Engl. J. Med. **325**(3), 147–152 (1991)

194. J.E. Manson, D.M. Nathan, A.S. Krolewski, M.J. Stampfer, W.C. Willett, C.H. Hennekens, A prospective study of exercise and incidence of diabetes among US male physicians. JAMA **268**(1), 63–67 (1992)

195. J.L. Abramson, V. Vaccarino, Relationship between physical activity and inflammation among apparently healthy middle-aged and older US adults. Arch. Intern. Med. **162**(11), 1286–1292 (2002)

196. E.S. Ford, Does exercise reduce inflammation? Physical activity and C-reactive protein among U.S. adults. Epidemiology **13**(5), 561–568 (2002)

197. A.P. Simopoulos, Energy imbalance and cancer of the breast, colon and prostate. Med. Oncol. Tumor Pharmacother. **7**(2–3), 109–120 (1990)

198. J. Flanagan, P.K. Gray, N. Hahn, et al., Presence of the metabolic syndrome is associated with shorter time to castration-resistant prostate cancer. Ann. Oncol. **22**(4), 801–807 (2011)

199. I.N. Benke, M.F. Leitzmann, G. Behrens, D. Schmid, Physical activity in relation to risk of prostate cancer: a systematic review and meta-analysis. Ann. Oncol. **29**(5), 1154–1179 (2018)

200. V. Rangul, E.R. Sund, P.J. Mork, O.D. Roe, A. Bauman, The associations of sitting time and physical activity on total and site-specific cancer incidence: results from the HUNT study, Norway. PLoS One **13**(10), e0206015 (2018)

201. E.L. Giovannucci, Y. Liu, M.F. Leitzmann, M.J. Stampfer, W.C. Willett, A prospective study of physical activity and incident and fatal prostate cancer. Arch. Intern. Med. **165**(9), 1005–1010 (2005)

202. A.V. Patel, C. Rodriguez, E.J. Jacobs, L. Solomon, M.J. Thun, E.E. Calle, Recreational physical activity and risk of prostate cancer in a large cohort of U.S. men. Cancer Epidemiol. Biomark. Prev. **14**(1), 275–279 (2005)

203. N.F. Johnsen, A. Tjonneland, B.L. Thomsen, et al., Physical activity and risk of prostate cancer in the European Prospective Investigation into Cancer and Nutrition (EPIC) cohort. Int. J. Cancer **125**(4), 902–908 (2009)

204. S.C. Moore, T.M. Peters, J. Ahn, et al., Physical activity in relation to total, advanced, and fatal prostate cancer. Cancer Epidemiol. Biomark. Prev. **17**(9), 2458–2466 (2008)

205. A. Grotta, M. Bottai, H.O. Adami, et al., Physical activity and body mass index as predictors of prostate cancer risk. World J. Urol. **33**(10), 1495–1502 (2015)

206. S.A. Kenfield, J.L. Batista, J.L. Jahn, et al., Development and application of a lifestyle score for prevention of lethal prostate cancer. J. Natl. Cancer Inst. **108**(3) (2016)

207. L. Bourke, D. Smith, L. Steed, et al., Exercise for men with prostate cancer: a systematic review and meta-analysis. Eur. Urol. **69**(4), 693–703 (2016)

208. S.A. Kenfield, M.J. Stampfer, E. Giovannucci, J.M. Chan, Physical activity and survival after prostate cancer diagnosis in the health professionals follow-up study. J. Clin. Oncol. **29**(6), 726–732 (2011)

209. B.A. Dickerman, E. Giovannucci, C.H. Pernar, L.A. Mucci, M.A. Hernan, Guideline-based physical activity and survival among US men with nonmetastatic prostate cancer. Am. J. Epidemiol. **188**(3), 579–586 (2019)

210. E.L. Richman, S.A. Kenfield, M.J. Stampfer, A. Paciorek, P.R. Carroll, J.M. Chan, Physical activity after diagnosis and risk of prostate cancer progression: data from the cancer of the prostate strategic urologic research endeavor. Cancer Res. **71**(11), 3889–3895 (2011)

211. E.R. Miller 3rd, R. Pastor-Barriuso, D. Dalal, R.A. Riemersma, L.J. Appel, E. Guallar, Meta-analysis: high-dosage vitamin E supplementation may increase all-cause mortality. Ann. Intern. Med. **142**(1), 37–46 (2005)

212. G. Bjelakovic, D. Nikolova, L.L. Gluud, R.G. Simonetti, C. Gluud, Mortality in randomized trials of antioxidant supplements for primary and secondary prevention: systematic review and meta-analysis. JAMA **297**(8), 842–857 (2007)

Dietary Carcinogens and DNA Adducts in Prostate Cancer

Medjda Bellamri and Robert J. Turesky

Introduction

The World Health Organization (WHO) has reported that prostate cancer (PC) is the second most common cancer in men worldwide, with an estimated 1.1 million incident cases and 0.3 million deaths occurring in 2012 [1]. PC is more commonly diagnosed in economically developed countries, which may be attributed to more extensive PC screening programs. The major risk factors identified for PC include aging, family history, and ethnicity, with African-American men having a two-time higher risk compared to Caucasians [2]. The occurrence of PC varies widely worldwide. Many studies of migrant populations show a significant increase in the incidence of PC and mortality rates in migrants from regions of the world with a low prevalence of PC, following their relocation to countries with high risk for PC, suggesting that environmental or dietary factors influence the risk factors for PC

development [3]. The frequent consumption of high-fat diets, dairy products, red meats, and alcohol are implicated as risk factors for PC [4]. However, the precise role of dietary factors and specific chemicals in the diet and mechanisms involved in the development of this malignancy remain unclear.

The Diet as a Risk Factor for Human PC

High-Fat Diet

Dietary fat and several fatty acids are postulated to play a role in PC etiology and tumor progression, although the findings of epidemiologic studies are inconsistent. Some studies found a strong positive association between fat consumption, PC incidence and mortality [5–10], whereas other investigations have not detected a correlation [11–14].

Several studies conducted *in vivo* in animal models and *in vitro* have shown a role for a high-fat diet in the development and progression of PC. Tissue culture medium conditioned with adipose tissue obtained from mice fed with high-fat Western-style foods enhanced cell proliferation, migration, and invasion of human prostate cancer cells *in vitro* [15, 16]. In the transgenic adenocarcinoma mouse prostate (TRAMP) and xenograft models, circulating adipokine and cytokine alter-

M. Bellamri · R. J. Turesky (✉)
Department of Medicinal Chemistry, Cancer and Cardiovascular Research Building, University of Minnesota, Minneapolis, MN, USA

Masonic Cancer Center, University of Minnesota, Minneapolis, MN, USA
e-mail: Rturesky@umn.edu

© Springer Nature Switzerland AG 2019
S. M. Dehm, D. J. Tindall (eds.), *Prostate Cancer*, Advances in Experimental Medicine and Biology 1210, https://doi.org/10.1007/978-3-030-32656-2_2

ations and other factors induced by a high-fat diet contributed to PC progression [15–18]. Although strong evidence supports the effects of a high-fat diet on PC development and progression, the exact mechanism(s) by which a high-fat diet underlines PC etiology remain uncertain. Several hypotheses are proposed including intake of fatty acids, resulting in inflammation, induction of oxidative stress, and cell signaling alteration.

Fatty Acids

Fatty acids, such as *n-3*, and *n-6* polyunsaturated fatty acids, and their metabolites are involved in numerous pathways that can affect PC development and progression. For example, *n-6* fatty acids linoleic acid and arachidonic acid enhance proliferation of human prostate cell lines [19, 20]. Moreover, *n-6* fatty acids are precursors of eicosanoids, which are converted to prostaglandins (PGs). The *n-6* fatty acid arachidonic acid is metabolized by the enzyme cyclooxygenase (Cox-1 and Cox-2) to form prostaglandin E2 (PGE2) [21]. PGE2 is a short-lived hormone-like molecule involved in cell proliferation, cell differentiation and inflammation [21]. Notably, the growth stimulation of PC cells, by treatment with arachidonic acid, is correlated to the induction of *COX2* expression and an increased synthesis of PGE2 [19, 22]. At the molecular level, PGE2 binds to EP4 and EP2 receptors, resulting in the subsequent activation of the protein kinase A (PKA) pathway, which leads to expression of early growth-related response genes including *c-fos* [23]. Arachidonic acid also activates the phosphatidylinositol-4,5-bisphosphate 3-kinase signaling pathway (PI3K/Akt) [20]. The PI3K/Akt cascade is involved in the progression and aggressiveness of PC. In fact, after long-term androgen deprivation therapy, there is constitutive activation of the PI3K/Akt pathway, a mechanism that leads to increased resistance of tumor cells to apoptosis [24, 25]. The activation of the PI3K/Akt pathway by arachidonic acid in human prostate cells also results in the activation of the Nuclear Factor Kappa Beta (NF-κB) cascade [20]. The induction of the NF-κB pathway increases cell resistance to chemotherapy and radiation therapy. Moreover, the activation of NF-κB also stimulates tumor cell growth, the inhibition of apoptosis, and enhances tumor invasion, metastasis, and angiogenesis [26] (Fig. 1).

In contrast to the *n-6* polyunsaturated fatty acids, *n-3* long-chain fatty acids protect against PC development. For example, a significant decrease in the growth of PC xenografts occurs in nude mice fed with a diet containing high levels of eicosapentaenoic and docosahexaenoic acids [27]. These effects have been supported by *in vitro* studies where both eicosapentaenoic acid and docosahexaenoic acid inhibit the proliferation of human PC cell lines [19, 28]. Moreover, eicosapentaenoic acid and docosahexaenoic acid prevent the progression of human prostate cells toward an aggressive androgen-independent phenotype. At the molecular level, eicosapentaenoic and docosahexaenoic acid treatments inhibit the PI3K/Akt signaling pathway and decrease expression of the androgen receptor (AR), a master regulator of prostate cell proliferation and PC development [29].

Inflammation

Inflammation often occurs in the prostates of aging men, and plays a critical role in the development of benign prostatic hyperplasia (BPH) and PC incidence [30–32]. Androgen levels, genetic predisposition, obesity, and a high-fat diet are associated with BPH and PC [33].

The association between a high-fat diet, the induction of inflammation, and PC markers have been reported in several *in vivo* studies. Prostatic inflammation correlates with cell proliferation and an increase in prostate gland size of mice consuming a high-fat diet [34, 35]. Consumption of a high-fat diet also elevates ataxin levels in the adipose tissue, leading to a significant increase in the production of lysophosphatidic acid, which can act directly on the prostate and induce hyperplasia and cell proliferation [36].

While inflammation is associated with an enhancement of PC development, there is not a clear understanding of the mechanisms involved

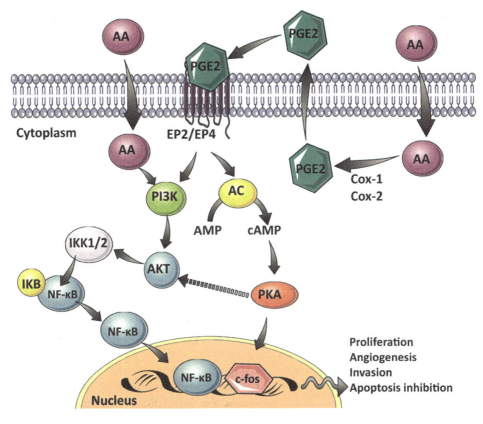

Fig. 1 Effect of Arachidonic acid (AA) and its metabolite prostaglandin E2 (PGE2) on cell signaling in human prostate cells

in this effect. Several studies have reported the involvement of immune cells and the production of pro-inflammatory cytokines. In clinical studies, prostate of patients with BPH contain infiltrates of macrophages, T-lymphocytes, and B-lymphocytes that are chronically activated [37]. These infiltrating cells produce cytokines including IL-2, IFN-γ IL-6, IL-8, IL-17, and TGF-β that maintain a chronic immune response and induce persistent intra-prostatic inflammation and fibromuscular growth by an autocrine or paracrine effect [38, 39]. These pro-inflammatory cytokines can modulate prostate growth in mice fed a high-fat diet [40] and correlate with the production of pro-inflammatory cytokines such as IL-1α, IL-1β, IL-6, IL-17 and TNF-α [41, 42]. These cytokines can induce prostate growth through induction of secondary mediators such as Cox-2. For example, IL-17 serves to stabilize and increase the enzymatic activity of Cox-2 [43]. Of note, the induction of Cox-2 expression in the prostate epithelium is associated with increased cell proliferation and apoptotic resistance [44]. Furthermore, the treatment of human prostate cells *in vitro* with serum obtained from obese mice containing elevated levels of pro-inflammatory cytokines promotes cell proliferation, invasion, migration and, epithelial-mesenchymal transition [45].

Oxidative Stress

A disproportionate generation of reactive oxygen species (ROS) causes tissue injury, DNA damage, and post-translational DNA modifications, which can lead to neoplasia in the human prostate [46, 47]. ROS are generated from the mitochondrial respiratory chain, an uncontrolled arachidonic acid cascade, and NADPH oxidase [33].

The expression of NADPH oxidase subunits such as gp91phox, p47phox, and p22phox is increased in the prostates of mice fed a high-fat diet [34]. Also of note, human PC cells harbor an increased level of ROS compared to normal prostate cells [48]. The ROS activity correlates with dysregulation of the NADPH oxidase system, which is a critical event for the malignant phenotype of human PC cells [47, 49–52].

At the molecular level, continual oxidative stress leads to the activation of two critical signaling pathways: the signal transducer and activator of transcription (STAT-3) and NF-κB pathways [33]. The activation of both cascades leads to the expression of transcription factors required for regulating genes involved in proliferation, survival, angiogenesis, invasion, and inflammation [53] (Fig. 2).

A high-fat diet also leads to increased activation of NF-κB in many organs of mice, including prostate [54]. In humans, there is constitutive activation of NF-κB in prostate adenocarcinoma [55]. Moreover, this constitutive activation is associated with upregulation of pro-survival molecules including Bcl-2, Bcl-XL, and Mcl-1 [56]. Similarly, increased STAT-3 activation and its DNA binding occur in the prostate of mice fed a high-fat diet [57]. In human PC cells, the inhibition of STAT-3 results in the inhibition of cell proliferation and a significant decrease in cell viability [58].

Dairy Products

Several epidemiological studies have reported that frequent intake of high-fat dairy products is associated with an increased risk of developing PC [59–61]; however, other studies failed to observe this association [62, 63]. The role of high-dairy fat intake in PC risk is supported by studies conducted *in vitro* where milk modulated and promoted the proliferation of the human prostate cancer cell lines LNCaP and PC-3 [64, 65]. Saturated fat intake, high-calcium intake, decreased circulating levels of 1,25-dihydroxyvitamin D (the active form of vitamin D), and increasing levels of insulin-like growth factor-1 (IGF-1) are several potential mechanisms by which milk and dairy product intake may impact the incidence and the progression of PC.

Saturated Fat Intake

Saturated fat is another likely factor in dairy products that may influence the development and the aggressiveness of PC. A higher intake of low-fat milk is associated with a greater risk of non-aggressive PC, whereas whole-fat milk intake is frequently associated with a higher incidence of aggressive PC phenotypes [63, 66–69].

High-Calcium Intake

Intake of calcium above the recommended daily doses (~1000 mg/day) is associated with increased risk of developing PC but also with aggressive and highly malignant PC [70–75]. The underlying mechanisms of high calcium intake and the risk of PC are not yet elucidated. Over-activation of the calcium-sensing receptor and calcium-dependent voltage-gated channel expressed in human prostate cells by the high levels of ionized calcium circulating in the bloodstream are two potential mechanisms involved in PC etiology [76–79]. The stimulation of these receptors by extracellular calcium increases PC cell proliferation, apoptosis resistance, and metastatic potential in *vitro* and in *vivo* [80–83].

Vitamin D

Several epidemiological studies have reported an association between low levels of vitamin D and higher risk for PC [84–86]. The modulation of vitamin D metabolism and the decrease of 1,25-dihydroxyvitamin D levels are associated with an increased risk of PC [87]. Once ingested, vitamin D is metabolized to its biologically active form 1,25-dihydroxyvitamin D through a two-step oxidation. The first oxidation reaction catalyzed by CYP2R1 occurs in the liver leading to the formation of 25-hydroxyvitamin D, and the

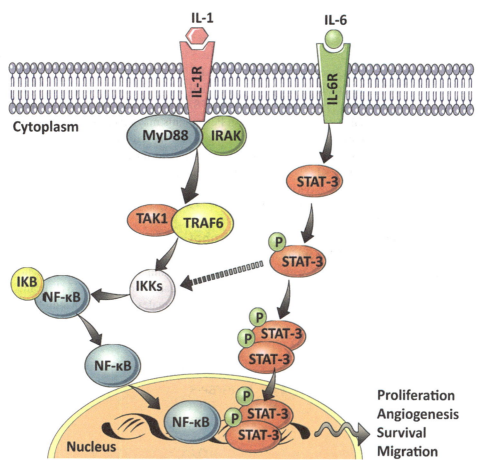

Fig. 2 Effect of pro-inflammatory cytokines such as IL-6 and IL-1 on the activation and the cross talk of STAT-3 and NF-κB pathways

second oxidation catalyzed by CYP27B1 occurs in the kidney, producing the 1,25-dihydroxyvitamin D [88]. 1,25-Dihydroxy vitamin D has antiproliferative effects that are driven through the nuclear Vitamin D receptor (VDR) pathway, leading to the expression of genes involved in cell cycle arrest, cell apoptosis and differentiation [89]. VDR is expressed in both normal and cancer prostate cells [90–92]. In human prostate cells, 1,25-dihydroxyvitamin D produces antiproliferative effects [92–96], reduces oxidative stress [97], and up-regulates pro-apoptotic genes [98]. More than 2000 genes are modulated by 1,25-dihydroxy vitamin D, including genes encoding for androgen metabolism [99]. More detailed studies are required to elucidate the critical roles of vitamin D in PC development.

IGF-1

The IGF system includes three ligands (insulin, IGF-1, IGF-2), their receptors (insulin receptor (INSR), IGF-1 receptor (IGF-1R), the mannose 6-phosphate receptor (M6P/IGF-2R), and six circulating IGF-binding proteins (IGFBP1–6) [100]. In the human prostate, every element of this system is expressed in normal, hyperplastic, and neoplastic prostate tissues, as well as in primary and cancer cell lines [101–110]. The IGF system plays a critical role in normal gland growth and development of the prostate [104, 108, 111, 112]. A higher serum IGF-1 concentration is correlated with an increased risk of PC [113–119]. The biological functions of IGF-I are mediated primarily through the IGF-IR, a tyrosine kinase transmembrane receptor that binds IGF-I with higher affinity than IGF-II [120]. Interestingly, inhibition of IGF-1R is associated with decreased androgen-dependent and androgen-independent growth *in vitro* as well as a suppression of *in vivo* tumor growth and PC cell invasiveness [108, 121–123]. Conversely, activation of IGF-1R by its ligand IGF-1 leads to the activation of several signaling pathways including mitogen-activated protein kinase (MAPK) and PI3K/Akt [124]. The activation of these signaling pathways induces proliferation and migration, and inhibits apoptosis in the PC cell [125–128] (Fig. 3).

The IGFBPs provide an additional, extracellular mechanism to regulate IGF activity. The IGFBPs bind to IGF-1 and IGF-2 with high affinity and thereby diminish their binding to IGF-R, resulting in the inhibition of the IGF signaling pathway [100]. Altered IGFBP plasma levels are found in PC patients, and a decrease in IGFBP-3 is associated with higher risk and progression of PC [113, 114, 129, 130]. IGFBP-3 is a substrate for the serum protease PSA [104]. Therefore, high levels of PSA in PC patients may result in a decrease in circulating levels of IGFBP-3 by proteolytic cleavage, leading to an increase in IGFs including IGF-1, thus facilitating disease progression (Fig. 3).

Alcohol

Alcohol consumption accounts for about 5% of all cancer deaths worldwide [131]. In the USA, 92% of adult males self-report a long-term use of

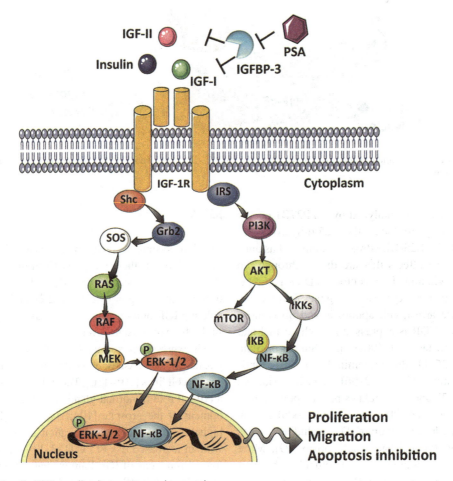

Fig. 3 Insulin/IGF signaling in prostate carcinogenesis

alcohol [132], and up to 3.7% of total cancer deaths are linked to alcohol [133]. Elevated alcohol consumption can contribute to a number of malignancies, including cancer of the oral cavity, pharynx, larynx, esophagus, and liver of both sexes and colorectal cancer in women. However, findings from epidemiological studies on the role of alcohol consumption in PC risk are inconsistent. Several studies found that alcohol consumption is a risk factor for PC [134–136], whereas other studies reported a decreased risk of PC [137]. The compilation of meta-analyses are also inconsistent: several reports found no association between alcohol consumption and PC risk [138–140], whereas others reported a significantly increased risk of PC with alcohol [141–144]. One meta-analysis reported an increased risk for PC for men drinking more than 50 g of alcohol per day, with the risk becoming slightly higher for men who consume more than 100 g per day [141]. Three other meta-analyses also reported a significantly increased risk in PC for light and moderate drinking (one to four drinks per day) [142, 143] or the equivalent of up 24 g of alcohol per day [144]. An association between alcohol intake and the degree of aggressiveness of PC was reported in some studies [145–148], but not other studies [135, 149, 150].

Ethanol is the primary form of alcohol in alcoholic beverages. Ethanol is classified as a human carcinogen [151]. The genotoxic effects and carcinogenicity of ethanol are thought to be driven by its major metabolite, acetaldehyde. Acetaldehyde forms DNA adducts in human cells *in vitro* and *in vitro* [152]. In humans, levels of acetaldehyde DNA adducts present in lymphocytes are seven times higher in alcohol users compared non-users [153]. *In vivo* studies demonstrated that ethanol is efficiently bioactivated into acetaldehyde in rat prostate by different enzymatic pathways involving xanthine oxidoreductase and cytochrome P450 2E1 [154, 155]. Moreover, acetaldehyde formation is linked to an increase in prostate epithelial cell death and ultrastructural alterations in epithelial cells including chromatin condensation around the perinuclear membrane and endoplasmic reticulum dilatation, an ultra-structural marker of endoplasmic reticulum stress [156]. The rat prostate lacks alcohol dehydrogenase and aldehyde dehydrogenase activities, resulting in an accumulation of acetaldehyde and thus, an increase in genomic damage in the prostate of rats exposed to ethanol [156]. Chronic ethanol exposure also leads to oxidative stress and a diminution in the antioxidant defense system in the rat ventral prostate [156, 157].

An association of PC risk with cancer aggressiveness was observed with high intake of beer [145, 146, 148]; while modest protective effects were observed for red wine consumption in some but not all studies [158–160]. The protective effect of wine, especially red wine, is likely attributed to its high contents of polyphenols such as flavonoids and resveratrol [161]. Polyphenolic compounds harbor antioxidant and anti-androgenic activities and therefore are thought to act as anti-carcinogens [162]. *In vitro*, nanomolar concentrations polyphenols inhibit cell growth in a dose and time-dependent manner in both androgen-dependent (LNCaP) and androgen-independent (DU145 and PC-3) human PC cell lines. Treatment of LNCaP and PC-3 cells with flavonoids, including catechin, epicatechin, and quercetin, inhibits cell proliferation, whereas resveratrol is the most potent inhibitor of DU145 cell growth. The proposed mechanism for the antiproliferative effect of polyphenols is through the modulation of NO production [163].

Red and Processed Meat

Many epidemiological studies have focused on the role of red and processed meats in PC risk. Some meta-analyses report an elevated risk for PC with frequent consumption of meats, whereas other studies failed to find an overall effect on PC risk [164, 165]. It is hypothesized that DNA damaging agents, including heme iron, N-nitroso compounds (NOCs) formed in processed meats [166], polycyclic aromatic hydrocarbons (PAHs) formed in smoked meats and meats cooked under flame [167], and heterocyclic aromatic amines (HAAs) formed in well-done grilled meats [168, 169], contribute to PC risk. Of note, the risk of

PC for African American men is ~2-fold greater than for Caucasians [2]. One paradigm proposed for the increased risk of PC in African-American men is based on their preference for frequent consumption of well-done cooked meats containing the HAA, 2-amino-1-methyl-6-phenylimidazo[4,5-*b*]pyridine (PhIP). PhIP is a rodent prostate carcinogen and potential human prostate carcinogen [170, 171], and may explain the higher risk of PC for African-American men compared to white men [172].

Heme Iron

Once ingested, heme-containing proteins such as myoglobin or hemoglobin are hydrolyzed to peptides, amino acids, and heme iron. Feeding ferriheme to rodents induces cytotoxicity, enhances cell proliferation of colonic mucosa, promotes oxidative stress, and thus, may contribute to colorectal cancer [173]. Heme iron is transported through the bloodstream to all organs of the body and can catalyze oxidative reactions, causing DNA, protein and lipid oxidation in multiple organs including the prostate [174]. The overall level of free radical damage induced by heme-catalyzed oxidation is estimated to be comparable to that resulting from ionizing radiation [174]. Two epidemiological studies have examined the role of heme iron in PC development [175, 176]. One reported a positive association between total heme iron intake and advanced PC risk, whereas another reported no associations of the dietary factors with PC risk irrespective of stage or grade. Thus, additional studies are necessary to evaluate any potential associations between heme iron consumption and PC.

N-Nitroso Compounds (NOCs)

Carcinogenic NOCs include two chemical classes, *N*-nitrosamines and *N*-nitrosamides, formed by the reaction of nitrosating agents derived from nitrite with amines and amides respectively [177]. Nitrites added to processed meat serve as anti-bacterial agents as well as cur-

ing agents, and they produce the characteristic red-pink color of cured meats. However, nitrites also react with amines in processed meats to produce dietary sources of NOCs. Moreover, the consumption of processed meat is a significant dietary source of nitrite, secondary amines, and amides, which can undergo nitrosation to form NOCs within the gastrointestinal tract [177–179]. The ingestion of heme contained in red meat can stimulate the endogenous formation of NOCs in the digestive tract [180–182]. More than 300 NOCs have been detected in 39 different animal species, including six species of nonhuman primates. Of these, 85% of *N*-nitrosamines and 92% of *N*-nitrosamides were reported to induce cancer in multiple organs including liver, lung, esophagus, bladder, and pancreas [183]. NOCs or their metabolites alkylate DNA. While *N*-nitrosamides react spontaneously with DNA, *N*-nitrosamines require metabolism by cytochromes, such as by cytochrome P450 2E1, which is expressed in the gastrointestinal tract [178]. Among the different types of DNA adducts formed with NOCs, the alkylation of the O^6-position of guanine is a primary lesion that induces G to A transitions [184–186]. The majority of epidemiological studies have focused on the role of NOCs in gastric, esophageal, and colorectal cancers [187–192]. Two epidemiology studies studied the etiology of NOCs and PC risk: there was no significant association between dietary NOCs and risk of development of PC in either study [176, 192]. Thus, further studies on the role of processed meats and NOCs in PC risk are warranted.

Polycyclic Aromatic Hydrocarbons (PAHs)

PAHs constitute a broad class of compounds that have two or more fused aromatic rings. PAHs arise by the incomplete combustion or high-temperature pyrolysis of organic materials [193]. PAHs are ubiquitous environmental pollutants that occur as complex mixtures but never as individual components [194]. Several PAHs are classified as human carcinogens by the (WHO) [151, 193]. Apart from occupational exposure, such as the case of coke-oven workers [195–197], the general population is exposed primarily to PAHs from dietary sources [167] and cigarette smoke [198]. The preparation of meats, mainly by a direct open flame, results in pyrolysis of the fat drippings, leading to the formation of PAHs, which are deposited through the smoke particulates on the surface of the grilled meats [199, 200]. Various PAHs occur in some charcoal-broiled, grilled, and smoked meats [194, 201–203]. The estimates of the daily dietary intake for the general population are imprecise and range widely. The levels of total daily PAH intake range from 3.7 μg up to 17 μg [167]. The most well-studied PAH is benzo[a]pyrene (B[a]P). B[a]P occurs in some grilled meats at levels up to ~4 ng/g [204, 205].

The carcinogenic properties of PAHs are attributed to their ability to form mutation-prone DNA adducts [193]. PAHs undergo metabolism by cytochrome P450 enzymes to form reactive dihydrodiol epoxides, which react with DNA to form covalent adducts, leading to mutations [206]. B[a]P-DNA adducts are formed in human prostate cells *in vitro* after exposure to B[a]P [207–209]. B[a]P treatment also leads to an increase in DNA double-strand breaks when measured by the comet assay [208, 210]. PAH adducts, including B[a]P adducts, are frequently detected in human prostate tissues by immuno-histochemistry (IHC)-[211–215] with an antibody, which was raised against B[a]P-modified DNA, but also cross-reacted with DNA adducts of at least five other PAHs [216]. Levels of PAH-DNA adducts is higher in adjacent human non-tumor prostate tissue compared with prostate tumor tissue, possibly due a higher cell proliferation rate in the tumor [211, 213, 217]. The occurrence of putative PAH-DNA adducts is associated with a higher risk for PC and cancer recurrence after prostatectomy within 1–2 years after surgery [211]. This risk was prominent in patients younger than 60 years old, patients with advanced-stage disease, and African Americans patients [211]. However, these data should be interpreted with caution since IHC is not a specific method of DNA adduct detection, even for assays performed with monoclonal antibodies,

where possible cross-reactivity of the antibodies with other DNA adducts or endogenous cellular components can lead to false positivity. The occurrence of DNA adducts of B[a]P was not confirmed in one cohort of PC patients when analyzed by liquid chromatography/mass spectrometry (LC/MS), a more specific analytical method than IHC [218]. Thus, there is a critical need to characterize DNA adducts on the same specimens by IHC and LC/MS to determine the validity of the analyses.

Heterocyclic Aromatic Amines (HAAs) and PC

HAA Formation and Sources of Exposure

HAAs are a class of more than 25 genotoxic chemicals known to form in cooked meats, fish, poultry, and tobacco smoke [168, 169, 219]. HAAs are sub-classified into the aminoimida-zoarenes (AIAs) and the high-temperature pyrolytic HAAs (Fig. 4). AIAs contain the *N*-methyl-2-aminoimidazole moiety derived from creatinine in muscle tissue. The AIAs form in meats, fish, and poultry cooked at temperatures above 150 °C and arise through the reaction of pyridine or pyrazines, derived from Strecker reactions, and condensation with creatine [220, 221]. Pyrolytic HAAs form by high-temperature pyrolysis (>250 °C) of protein or amino acids, such as glutamic and tryptophan. Pyrolytic HAAs occur when proteinaceous foods are heated at temperatures generally above 250 °C [168, 222, 223]. Several HAAs are also formed in tobacco smoke [223, 224]. 2-Amino-9*H*-pyrido[2,3-*b*] indole (AαC), a pyrolysis product of tryptophan, is the major carcinogenic HAA formed in combusted tobacco and occurs in mainstream tobacco smoke at levels up to 258 ng per cigarette [225–227]. Unexpectedly, PhIP, an AIA containing the *N*-methyl-2-aminoimidazole moiety of creatine, was detected in tobacco smoke [224]; however, the mechanism of PhIP formation in tobacco smoke has not been determined. The principal sources of exposure to most HAAs occur through

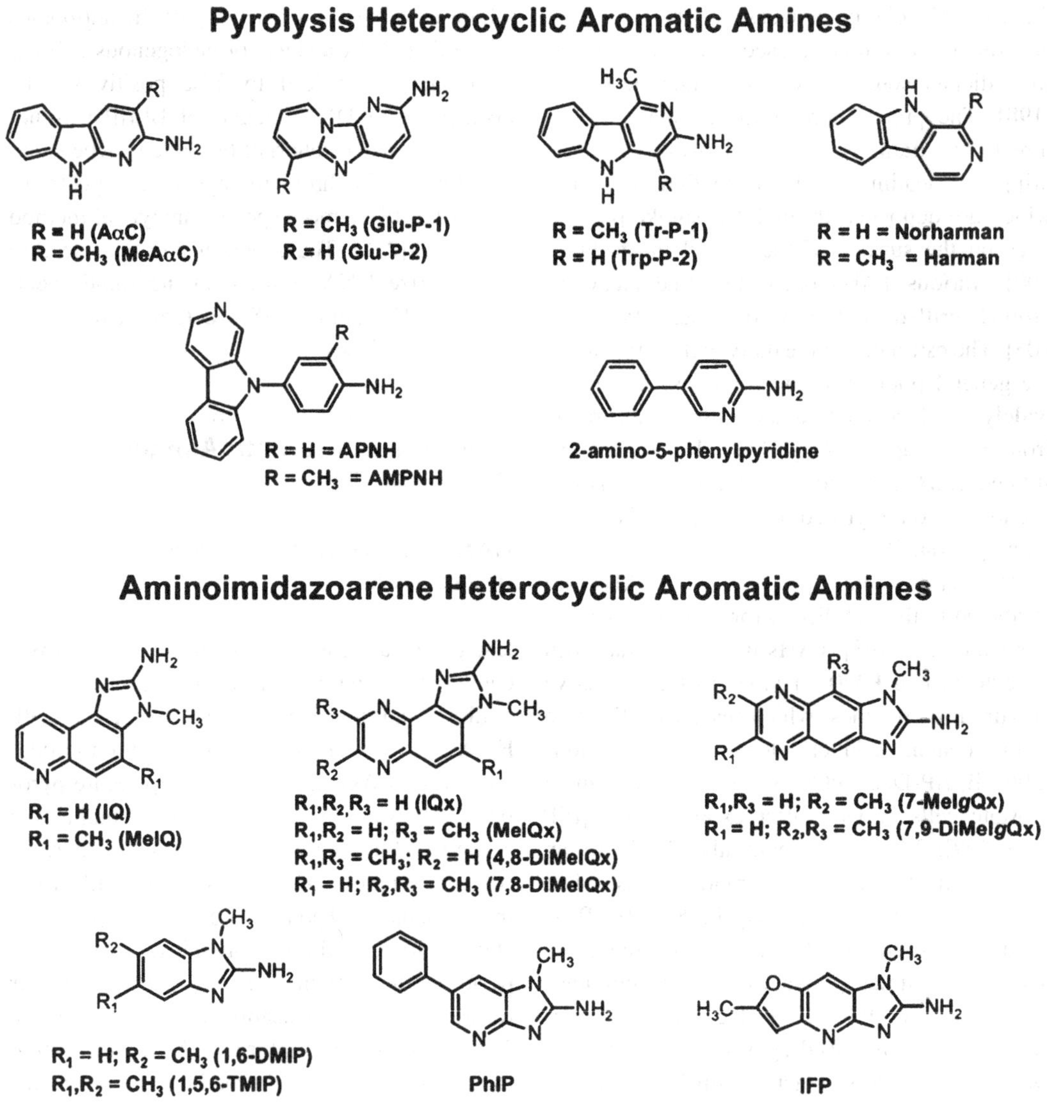

Pyrolysis Heterocyclic Aromatic Amines

R = H (AαC)
R = CH₃ (MeAαC)

R = CH₃ (Glu-P-1)
R = H (Glu-P-2)

R = CH₃ (Tr-P-1)
R = H (Trp-P-2)

R = H = Norharman
R = CH₃ = Harman

R = H = APNH
R = CH₃ = AMPNH

2-amino-5-phenylpyridine

Aminoimidazoarene Heterocyclic Aromatic Amines

R_1 = H (IQ)
R_1 = CH₃ (MeIQ)

R_1,R_2,R_3 = H (IQx)
R_1,R_2 = H; R_3 = CH₃ (MeIQx)
R_1,R_3 = CH₃; R_2 = H (4,8-DiMeIQx)
R_1 = H; R_2,R_3 = CH₃ (7,8-DiMeIQx)

R_1,R_3 = H; R_2 = CH₃ (7-MeIgQx)
R_1 = H; R_2,R_3 = CH₃ (7,9-DiMeIgQx)

R_1 = H; R_2 = CH₃ (1,6-DMIP)
R_1,R_2 = CH₃ (1,5,6-TMIP)

PhIP

IFP

Fig. 4 Chemical structures of prevalent HAAs in cooked meat

the consumption of well-done cooked meats and poultry [168, 228]. HAA formation in cooked meats generally occurs at the low parts-per-billion (ppb) range, but the levels of some HAAs can approach several hundred ppb in well-done cooked meats or poultry [168, 219, 229, 230]. PhIP and 2-amino-3,8-dimethylimidazo[4,5-*f*] quinoxaline (MeIQx) are often the most prevalent HAAs formed in cooked meats and poultry [219, 221, 230–233]. The average dietary HAAs intake ranges from less than 2 to up to 25 ng/kg per day [172, 234].

Bioactivation and Formation of DNA Adducts

HAAs undergo extensive metabolism by hepatic cytochrome P450 1A2 (CYP 1A2)-catalyzed N-oxidation of the exocyclic amine groups, leading to the formation of N-hydroxy-HAAs, as reactive intermediates [228, 235, 236]. CYP 1A1 and CYP 1B1 catalyze this reaction in extrahepatic tissues [228, 235]. The *N*-hydroxy-HAAs are further bioactivated by acetylation or sulfation catalyzed by N-acetyltransferases (NATs) or sulfotransferases (SULTs), respectively, either in the liver or extrahepatic tissues [228]. These unstable esters react with DNA to form DNA adducts [228, 237]. HAA-DNA adducts are mainly formed at C-8 on deoxyguanosine (dG) through the exocyclic amine of the HAAs, to produce dG-C8-HAA adducts as the major DNA adducts [238, 239].

Carcinogenesis of HAAs

HAAs are multisite carcinogens in rodents and induce cancers of the oral cavity, liver, stomach, colon, pancreas, and the mammary gland in females [168, 170]. Notably, PhIP is the only HAA reported to induce PC in rodents [168, 170]. Carcinogenesis studies in rodents used chronic doses of HAAs ranging between 0.1 to 64 mg/kg/day to induce tumors [168, 228]. These doses are more a million-fold higher than the daily intake of HAAs. Thus, we might surmise that the levels of human exposure are too low to contribute to human cancers. However, a linear relationship between HAA dose and HAA-DNA adduct formation occurs in rodent tissues for PhIP, MeIQx, and IQ [240–242], signifying mutation-prone DNA adducts of HAAs can still form in tissues at dosing regimens approaching human exposure levels.

Animal toxicity studies may underestimate the carcinogenic potential of HAAs in humans. For example, the levels of HAA-DNA adducts formed in primary human hepatocytes are significantly higher than those formed in primary rat hepatocytes, under the same doses and times of exposure [243]. Human CYP1A2, which is principally expressed in the liver, is the major CYP involved in the metabolism of many HAAs. Human CYP1A2 is catalytically more efficient than the rat homolog in the bioactivation of PhIP and MeIQx, and perhaps other HAAs [244]. Human CYP1A2 and human liver microsomes preferentially bioactivate HAAs through N-oxidation of the exocyclic amine group. In contrast, rat CYP1A2 and rat liver microsomes preferentially catalyze the detoxication of HAAs by oxidation of the heterocyclic rings [245]. The superior activity (lower K_m and higher k_{cat}) and ability of human CYP1A2 to catalyze N-oxidation can explain the higher levels of HAA adducts formed in human compared to rat hepatocytes. Several HAA-DNA adducts have been detected in human tissues by various techniques, indicating that even at low levels of exposure, HAAs can form DNA adducts in humans [246–256].

PhIP DNA Damage, Mutation, and Carcinogenicity in Prostate

Rodent Studies

PhIP is the only HAA studied thus far that targets the prostate as a principal site for DNA adduct formation and carcinogenesis in rodents [168]. PhIP undergoes metabolism to form high levels of DNA adducts in the prostate of Wistar and Fischer 344 rats [170, 257–260] and induces high levels of mutations in the prostate of the Big Blue *lacI* transgenic rat [260, 261]. PhIP is a prostate carcinogen in the Fischer 344 rat [170]. PhIP also induces prostate tumors in CYP1A-humanized (hCYP1A) mice but not in wild-type mice [262]. This finding reinforces the concept that human CYP1A is superior to the rodent orthologue in the bioactivation of PhIP [244, 263].

Extensive inflammation occurs in the dorsolateral prostate lobe marked by CD45[+] mononuclear leukocyte and CD8[+] T lymphocyte infiltration in PhIP-induced tumors in the prostates of hCYP1A mice [262]. This inflammation is associated with atrophic glands, high-grade prostatic intraepithelial neoplasia, and oxidative stress [262, 264]. In contrast, the prostatic intraepithelial neoplasia lesions are significantly less severe and infrequently associated with inflammation and oxidative stress in the ventral prostate glands [262]. These observations are noteworthy because the dorsolateral prostate is homologous to the human peripheral prostate zone, the most common site of PC development in humans [265]. Similarly, PhIP treatment leads to inflammation with a marked increase in mastocyte and macrophage infiltration and glandular atrophy of the prostate of

Fischer 344 rats [261, 266]. These pathologies induced by PhIP in rodent models of PC are significant because inflammation, oxidative stress, and glandular atrophy are common features in the pathology of human PC [267].

Several key features of cancer biology often reported in human PC occur in PhIP-induced prostate tumors in rodents. For example, treatment of hCYP1A mice with PhIP results in a time-dependent increase in expression of AR protein in prostate tumor epithelial cells [262]. An up-regulation of AR leading to a higher rate of cell proliferation occurs in human PC [268]. Furthermore, PhIP-induced tumors in h-CYP1A mice display significant decreases in levels of E-Cadherin and p63 expression [262]. E-Cadherin is an epithelial cell adhesion molecule involved in the maintenance of normal cell architecture, while the p63 transcription factor has multiple functions in cancer cell biology. PhIP-treatment in Fisher 344 and Big Blue rats also results in significant increases in the levels of Ki-67, a well-established marker of cell proliferation, in the intraepithelial neoplasia regions of the prostate [261, 266]. Dysregulated expression and distribution of these proteins are hallmarks of epithelial malignancies and serve as major diagnostic criteria for human PC [269–271].

PhIP-induced tumors in the h-CYP1A mice also exhibit increased levels of oxidative stress markers, including 8-oxo-dG and nitrotyrosine, markers of oxidative DNA damage and reactive nitrogen species. PhIP treatment results in the up-regulation of COX-2 expression, a cyclooxygenase that catalyzes the formation of pro-inflammatory prostaglandins and a loss of Nuclear factor (erythroid-derived 2)-like 2 (Nrf2) a key transcription factor responsible of the expression of many cytoprotective proteins [264]. Oxidative stress is a key contributor to the development of PC in humans [46, 272]. The PTEN/PI3K/Akt signaling pathway is another critical feature of cell proliferation, cell cycle progression, and survival. Activation of AKT in response to oxidative stress and the loss of PTEN are critical events in human PC progression [273–275]. PhIP-induced prostate tumors in h-CYP1A mice display a significant decrease in PTEN expression and an elevation of phospho-AKT, leading to cell proliferation [264].

These mechanistic data in rodent models reinforce the biological plausibility that PhIP plays an important role in dietary-linked human PC. However, the biological events observed in rodents occurred at very high doses of PhIP treatment—up to 200 mg/kg. Humans are exposed to one million fold or lower daily amounts of PhIP [172], and the capacity of such lower concentrations of PhIP to induce similar biological effects has not been investigated. Therefore, the interpretation of the carcinogenic effects of PhIP in PC of rodents and their extrapolation to human PC should be done with caution.

Human Studies

PhIP is the only HAA reported thus far to form DNA adducts in the prostate of human PC patients [218, 255, 276–278] (Fig. 5). This biomarker data provides support for some of the epidemiological studies that have linked the frequent consumption of well-done cooked red meat containing PhIP with increased risk of PC [279–281]. However, other investigations have failed to find an association between cooked red meat and increased PC risk [175, 282, 283]. The concentrations of PhIP and other HAAs can vary by more than 100-fold in cooked meats [169, 219]. There is a critical need to conduct such epidemiological studies with more precise exposure measurements of HAAs.

The frequency of detection and the levels of PhIP-DNA adducts in human prostate range widely between studies, depending on the analytical method of adduct measurement. For example, using a high-resolution LC/MS method, PhIP-DNA adducts were detected in 13 out of 54 PC patients with levels ranging from 2 to 120 adducts per 10^9 DNA bases [218, 278]. However, PhIP-DNA adducts were detected in a very high percentage of prostate tissues in another cohort, occurring at levels exceeding several adducts per 10^7 DNA bases, when measured by IHC [255, 276, 277]. These discrepancies in adduct measurements may imply a high level of false positivity obtained

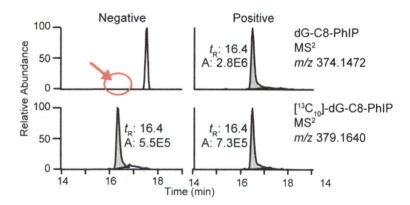

Fig. 5 Representative chromatograms at the MS² scan stage of human prostate samples that were negative and positive for dG-C8-PhIP

by IHC, possibly due to cross-reactivity of the polyclonal antibodies raised against PhIP-modified DNA with other DNA adducts or endogenous cellular components [284].

Cytotoxicity and DNA adduct formation induced by PhIP and other HAAs has been studied in primary and PC cell lines. The parent HAAs, PhIP, MeIQx, IQ and AαC were not toxic at doses up to 10 μM and formed low levels of DNA adducts in LNCaP cells [285]. However, HONH-PhIP, the genotoxic metabolite of PhIP, induced a dose-dependent increase in cytotoxicity, whereas HONH-MeIQx, HONH-IQ, and HONH-AαC were not toxic [285, 286]. Moreover, HONH-PhIP forms DNA adducts at levels that are 20-fold higher than other HONH-HAAs in LNCaP cells [285]. These data suggest that the initial bioactivation step of PhIP to form HONH-PhIP occurs in the liver through CYP 1A2-catalyzed N-oxidation, followed by systemic circulation to reach the prostate, where bioactivation is mediated by Phase II enzymes [285] (Fig. 6). Similar data were reported in primary human prostate epithelial cells, where HONH-PhIP formed 50- to 100-fold higher levels of DNA adducts than IQ, MeIQx, and HONH-MeIQx [207, 287].

HONH-PhIP also induces unscheduled DNA synthesis, and DNA single-strand breaks in primary human prostate epithelial cells at 100-time higher levels than HONH-MeIQx [208, 210]. The higher susceptibility of human prostate cells to the DNA-damaging and genotoxic effect of PhIP compared to other prominent HAAs formed in cooked meats recapitulates the DNA adduct biomarker data in prostate tissues of PC patients and provides support for the possible causal role of PhIP in human PC.

PhIP can also act through non-genotoxic mechanisms via androgenic effects to contribute to the development of PC. PhIP binds to the AR and modulates cell proliferation. Using, *in silico* analysis, binding of PhIP and HONH-PhIP to the AR was found to be comparable to that of the endogenous AR ligand, dihydrotestosterone (DHT), when based on the predicted free energy of binding [288]. Through computational docking studies, both PhIP and HONH-PhIP displayed similar binding modes to DHT and docked with high affinity in the same cavity of the AR ligand binding domain as DHT [288]. Moreover, treatment of the human prostate epithelial cell line LNCaP with PhIP or HONH-PhIP up-regulated AR and PSA expression [288].

PhIP also induces proliferation of human prostate epithelial cells in an AR-independent manner, through the activation of pro-proliferation cell signaling pathways. Low concentrations of PhIP (10^{-12} to 10^{-8} mol/L) increase the proliferation, migration and invasion properties of PC-3, an AR-negative human prostate cell line [289]. Proliferation and migration are mediated through the activation of the ERK signal transduction cascade and a rapid, transient increase in phosphorylation of both MEK1/2 and ERK1/2. Interestingly, mitogenic stimulation with epithelial growth factor (EGF), induces the same pattern of activation [290]. Proliferation, migration, and invasiveness are crucial events in the oncogenic progression of cells [291]. Thus, all these biological phenomena induced by PhIP suggest a

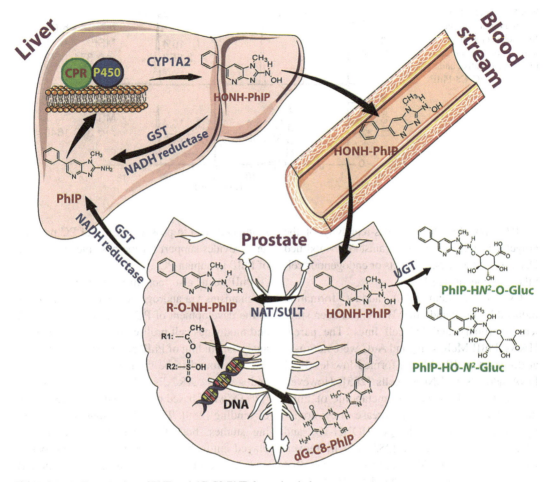

Fig. 6 Metabolic activation of PhIP and dG-C8-PhIP formation in human prostate

carcinogenic potential of PhIP in human prostate. The challenge in risk assessment is to determine if the amounts of PhIP in the diet are sufficient to be a significant risk for PC.

Conclusion

There is growing mechanistic and epidemiological data supporting a role for the diet in the development of PC. Multiple mechanisms and hypotheses have been brought forward for PC risk, ranging from different classes of dietary genotoxicants acting as initiators of PC to different dietary factors involved in tumor promotion. However, the precise roles of specific genotoxicants and nutritional factors in PC remain to be clarified. Prospective epidemiological studies on PC risk with improved assessments of dietary habits, including protective nutrient biomarkers in plasma and urine are needed [292]. The identification of micronutrients that protect against PC [293, 294], and the detection of biomarkers of DNA damage in the prostate, such as DNA adducts, by specific mass spectrometric methods and their linkage to mutations [218, 285], can advance our understanding of the micronutrients and genotoxicants in the diet that impact PC risk.

Acknowledgements A portion of the research conducted in the Turesky laboratory was supported by R01CA122320.

References

1. L.A. Torre, F. Bray, R.L. Siegel, J. Ferlay, J. Lortet-Tieulent, A. Jemal, Global cancer statistics, 2012. CA Cancer J. Clin. **65**, 87–108 (2015)
2. D.G. Bostwick, H.B. Burke, D. Djakiew, S. Euling, S.M. Ho, J. Landolph, H. Morrison, B. Sonawane, T. Shifflett, D.J. Waters, B. Timms, Human prostate cancer risk factors. Cancer **101**, 2371–2490 (2004)
3. P.H. Gann, Risk factors for prostate cancer. Rev. Urol. **4**(Suppl 5), S3–S10 (2002)
4. N.R. Perdana, C.A. Mochtar, R. Umbas, A.R. Hamid, The risk factors of prostate cancer and its prevention: a literature review. Acta Med. Indones. **48**, 228–238 (2016)
5. C. Pelser, A.M. Mondul, A.R. Hollenbeck, Y. Park, Dietary fat, fatty acids, and risk of prostate cancer in the NIH-AARP diet and health study. Cancer Epidemiol. Biomark. Prev. **22**, 697–707 (2013)
6. E. Giovannucci, E.B. Rimm, G.A. Colditz, M.J. Stampfer, A. Ascherio, C.G. Chute, W.C. Willett, A prospective study of dietary fat and risk of prostate cancer. J. Natl. Cancer Inst. **85**, 1571–1579 (1993)
7. D.W. West, M.L. Slattery, L.M. Robison, T.K. French, A.W. Mahoney, Adult dietary intake and prostate cancer risk in Utah: a case-control study with special emphasis on aggressive tumors. Cancer Causes Control **2**, 85–94 (1991)
8. M.L. Slattery, M.C. Schumacher, D.W. West, L.M. Robison, T.K. French, Food-consumption trends between adolescent and adult years and subsequent risk of prostate cancer. Am. J. Clin. Nutr. **52**, 752–757 (1990)
9. A. Lophatananon, J. Archer, D. Easton, R. Pocock, D. Dearnaley, M. Guy, Z. Kote-Jarai, L. O'Brien, R.A. Wilkinson, A.L. Hall, E. Sawyer, E. Page, J.F. Liu, et al., Dietary fat and early-onset prostate cancer risk. Br. J. Nutr. **103**, 1375–1380 (2010)
10. M.M. Lee, R.T. Wang, A.W. Hsing, F.L. Gu, T. Wang, M. Spitz, Case-control study of diet and prostate cancer in China. Cancer Causes Control **9**, 545–552 (1998)
11. F.L. Crowe, T.J. Key, P.N. Appleby, R.C. Travis, K. Overvad, M.U. Jakobsen, N.F. Johnsen, A. Tjonneland, J. Linseisen, S. Rohrmann, H. Boeing, T. Pischon, A. Trichopoulou, et al., Dietary fat intake and risk of prostate cancer in the European prospective investigation into cancer and nutrition. Am. J. Clin. Nutr. **87**, 1405–1413 (2008)
12. S.Y. Park, S.P. Murphy, L.R. Wilkens, B.E. Henderson, L.N. Kolonel, Fat and meat intake and prostate cancer risk: the multiethnic cohort study. Int. J. Cancer **121**, 1339–1345 (2007)
13. T.J. Key, P.B. Silcocks, G.K. Davey, P.N. Appleby, D.T. Bishop, A case-control study of diet and prostate cancer. Br. J. Cancer **76**, 678–687 (1997)
14. P. Ghadirian, A. Lacroix, P. Maisonneuve, C. Perret, G. Drouin, J.P. Perrault, G. Beland, T.E. Rohan, G.R. Howe, Nutritional factors and prostate cancer: a case-control study of French Canadians in Montreal, Canada. Cancer Causes Control **7**, 428–436 (1996)
15. M.B. Hu, H. Xu, W.H. Zhu, P.D. Bai, J.M. Hu, T. Yang, H.W. Jiang, Q. Ding, High-fat diet-induced adipokine and cytokine alterations promote the progression of prostate cancer in vivo and in vitro. Oncol. Lett. **15**, 1607–1615 (2018)
16. H.J. Cho, G.T. Kwon, H. Park, H. Song, K.W. Lee, J.I. Kim, J.H. Park, A high-fat diet containing lard accelerates prostate cancer progression and reduces survival rate in mice: possible contribution of adipose tissue-derived cytokines. Nutrients **7**, 2539–2561 (2015)
17. J.C. Mavropoulos, W.C. Buschemeyer 3rd, A.K. Tewari, D. Rokhfeld, M. Pollak, Y. Zhao, P.G. Febbo, P. Cohen, D. Hwang, G. Devi, W. Demark-Wahnefried, E.C. Westman, B.L. Peterson, et al., The effects of varying dietary carbohydrate and fat content on survival in a murine LNCaP prostate cancer xenograft model. Cancer Prev. Res. (Phila.) **2**, 557–565 (2009)
18. G. Llaverias, C. Danilo, Y. Wang, A.K. Witkiewicz, K. Daumer, M.P. Lisanti, P.G. Frank, A western-type diet accelerates tumor progression in an autochthonous mouse model of prostate cancer. Am. J. Pathol. **177**, 3180–3191 (2010)
19. M. Hughes-Fulford, Y. Chen, R.R. Tjandrawinata, Fatty acid regulates gene expression and growth of human prostate cancer PC-3 cells. Carcinogenesis **22**, 701–707 (2001)
20. M. Hughes-Fulford, C.F. Li, J. Boonyaratanakornkit, S. Sayyah, Arachidonic acid activates phosphatidylinositol 3-kinase signaling and induces gene expression in prostate cancer. Cancer Res. **66**, 1427–1433 (2006)
21. D. Wang, R.N. Dubois, Prostaglandins and cancer. Gut **55**, 115–122 (2006)
22. R.R. Tjandrawinata, R. Dahiya, M. Hughes-Fulford, Induction of cyclo-oxygenase-2 mRNA by prostaglandin E2 in human prostatic carcinoma cells. Br. J. Cancer **75**, 1111–1118 (1997)
23. Y. Chen, M. Hughes-Fulford, Prostaglandin E2 and the protein kinase A pathway mediate arachidonic acid induction of c-fos in human prostate cancer cells. Br. J. Cancer **82**, 2000–2006 (2000)
24. K. Pfeil, I.E. Eder, T. Putz, R. Ramoner, Z. Culig, F. Ueberall, G. Bartsch, H. Klocker, Long-term androgen-ablation causes increased resistance to PI3K/Akt pathway inhibition in prostate cancer cells. Prostate **58**, 259–268 (2004)
25. H. Murillo, H. Huang, L.J. Schmidt, D.I. Smith, D.J. Tindall, Role of PI3K signaling in survival and progression of LNCaP prostate cancer cells to the androgen refractory state. Endocrinology **142**, 4795–4805 (2001)
26. M.S. Mendonca, W.T. Turchan, M.E. Alpuche, C.N. Watson, N.C. Estabrook, H. Chin-Sinex, J.B. Shapiro, I.E. Imasuen-Williams, G. Rangel, D.P. Gilley, N. Huda, P.A. Crooks, R.H. Shapiro,

DMAPT inhibits NF-kappaB activity and increases sensitivity of prostate cancer cells to X-rays in vitro and in tumor xenografts in vivo. Free Radic. Biol. Med. **112**, 318–326 (2017)

27. J.M. Connolly, M. Coleman, D.P. Rose, Effects of dietary fatty acids on DU145 human prostate cancer cell growth in athymic nude mice. Nutr. Cancer **29**, 114–119 (1997)

28. D.P. Rose, Effects of dietary fatty acids on breast and prostate cancers: evidence from in vitro experiments and animal studies. Am. J. Clin. Nutr. **66**, 1513S–1522S (1997)

29. W. Friedrichs, S.B. Ruparel, R.A. Marciniak, L. deGraffenried, Omega-3 fatty acid inhibition of prostate cancer progression to hormone independence is associated with suppression of mTOR signaling and androgen receptor expression. Nutr. Cancer **63**, 771–777 (2011)

30. A. Sciarra, G. Mariotti, S. Salciccia, A. Autran Gomez, S. Monti, V. Toscano, F. Di Silverio, Prostate growth and inflammation. J. Steroid Biochem. Mol. Biol. **108**, 254–260 (2008)

31. B. Gurel, M.S. Lucia, I.M. Thompson Jr., P.J. Goodman, C.M. Tangen, A.R. Kristal, H.L. Parnes, A. Hoque, S.M. Lippman, S. Sutcliffe, S.B. Peskoe, C.G. Drake, W.G. Nelson, et al., Chronic inflammation in benign prostate tissue is associated with high-grade prostate cancer in the placebo arm of the prostate cancer prevention trial. Cancer Epidemiol. Biomark. Prev. **23**, 847–856 (2014)

32. G.T. MacLennan, R. Eisenberg, R.L. Fleshman, J.M. Taylor, P. Fu, M.I. Resnick, S. Gupta, The influence of chronic inflammation in prostatic carcinogenesis: a 5-year followup study. J. Urol. **176**, 1012–1016 (2006)

33. E. Shankar, N. Bhaskaran, G.T. MacLennan, G. Liu, F. Daneshgari, S. Gupta, Inflammatory signaling involved in high-fat diet induced prostate diseases. J. Urol. Res. **2**. (2015)

34. E.V. Vykhovanets, E. Shankar, O.V. Vykhovanets, S. Shukla, S. Gupta, High-fat diet increases NF-kappaB signaling in the prostate of reporter mice. Prostate **71**, 147–156 (2011)

35. X. Cai, R. Haleem, S. Oram, J. Cyriac, F. Jiang, J.T. Grayhack, J.M. Kozlowski, Z. Wang, High fat diet increases the weight of rat ventral prostate. Prostate **49**, 1–8 (2001)

36. P. Kulkarni, R.H. Getzenberg, High-fat diet, obesity and prostate disease: the ATX-LPA axis? Nat. Clin. Pract. Urol. **6**, 128–131 (2009)

37. G. Kramer, G.E. Steiner, A. Handisurya, U. Stix, A. Haitel, B. Knerer, A. Gessl, C. Lee, M. Marberger, Increased expression of lymphocyte-derived cytokines in benign hyperplastic prostate tissue, identification of the producing cell types, and effect of differentially expressed cytokines on stromal cell proliferation. Prostate **52**, 43–58 (2002)

38. G.E. Steiner, U. Stix, A. Handisurya, M. Willheim, A. Haitel, F. Reithmayr, D. Paikl, R.C. Ecker, K. Hrachowitz, G. Kramer, C. Lee, M. Marberger, Cytokine expression pattern in benign prostatic hyperplasia infiltrating T cells and impact of lymphocytic infiltration on cytokine mRNA profile in prostatic tissue. Lab. Investig. **83**, 1131–1146 (2003)

39. J.E. Konig, T. Senge, E.P. Allhoff, W. Konig, Analysis of the inflammatory network in benign prostate hyperplasia and prostate cancer. Prostate **58**, 121–129 (2004)

40. J.L. St Sauver, S.J. Jacobsen, Inflammatory mechanisms associated with prostatic inflammation and lower urinary tract symptoms. Curr. Prostate Rep. **6**, 67–73 (2008)

41. C.T. De Souza, E.P. Araujo, S. Bordin, R. Ashimine, R.L. Zollner, A.C. Boschero, M.J. Saad, L.A. Velloso, Consumption of a fat-rich diet activates a proinflammatory response and induces insulin resistance in the hypothalamus. Endocrinology **146**, 4192–4199 (2005)

42. H. Xu, M.B. Hu, P.D. Bai, W.H. Zhu, S.H. Liu, J.Y. Hou, Z.Q. Xiong, Q. Ding, H.W. Jiang, Proinflammatory cytokines in prostate cancer development and progression promoted by high-fat diet. Biomed. Res. Int. **2015**, 249741 (2015)

43. W.H. Faour, A. Mancini, Q.W. He, J.A. Di Battista, T-cell-derived interleukin-17 regulates the level and stability of cyclooxygenase-2 (COX-2) mRNA through restricted activation of the p38 mitogen-activated protein kinase cascade: role of distal sequences in the 3′-untranslated region of COX-2 mRNA. J. Biol. Chem. **278**, 26897–26907 (2003)

44. W. Wang, A. Bergh, J.E. Damber, Chronic inflammation in benign prostate hyperplasia is associated with focal upregulation of cyclooxygenase-2, Bcl-2, and cell proliferation in the glandular epithelium. Prostate **61**, 60–72 (2004)

45. R.S. Price, D.A. Cavazos, R.E. De Angel, S.D. Hursting, L.A. deGraffenried, Obesity-related systemic factors promote an invasive phenotype in prostate cancer cells. Prostate Cancer Prostatic Dis. **15**, 135–143 (2012)

46. L. Khandrika, B. Kumar, S. Koul, P. Maroni, H.K. Koul, Oxidative stress in prostate cancer. Cancer Lett. **282**, 125–136 (2009)

47. B. Kumar, S. Koul, L. Khandrika, R.B. Meacham, H.K. Koul, Oxidative stress is inherent in prostate cancer cells and is required for aggressive phenotype. Cancer Res. **68**, 1777–1785 (2008)

48. T.P. Szatrowski, C.F. Nathan, Production of large amounts of hydrogen peroxide by human tumor cells. Cancer Res. **51**, 794–798 (1991)

49. S.D. Lim, C. Sun, J.D. Lambeth, F. Marshall, M. Amin, L. Chung, J.A. Petros, R.S. Arnold, Increased Nox1 and hydrogen peroxide in prostate cancer. Prostate **62**, 200–207 (2005)

50. S.S. Brar, Z. Corbin, T.P. Kennedy, R. Hemendinger, L. Thornton, B. Bommarius, R.S. Arnold, A.R. Whorton, A.B. Sturrock, T.P. Huecksteadt, M.T. Quinn, K. Krenitsky, K.G. Ardie, et al., NOX5 NAD(P)H oxidase regulates growth and apoptosis in

DU 145 prostate cancer cells. Am. J. Physiol. Cell Physiol. **285**, C353–C369 (2003)

51. K. Block, J.M. Ricono, D.Y. Lee, B. Bhandari, G.G. Choudhury, H.E. Abboud, Y. Gorin, Arachidonic acid-dependent activation of a p22(phox)-based NAD(P)H oxidase mediates angiotensin II-induced mesangial cell protein synthesis and fibronectin expression via Akt/PKB. Antioxid. Redox Signal. **8**, 1497–1508 (2006)

52. J.L. Arbiser, J. Petros, R. Klafter, B. Govindajaran, E.R. McLaughlin, L.F. Brown, C. Cohen, M. Moses, S. Kilroy, R.S. Arnold, J.D. Lambeth, Reactive oxygen generated by Nox1 triggers the angiogenic switch. Proc. Natl. Acad. Sci. U. S. A. **99**, 715–720 (2002)

53. S.I. Grivennikov, M. Karin, Dangerous liaisons: STAT3 and NF-kappaB collaboration and crosstalk in cancer. Cytokine Growth Factor Rev. **21**, 11–19 (2010)

54. H. Carlsen, F. Haugen, S. Zadelaar, R. Kleemann, T. Kooistra, C.A. Drevon, R. Blomhoff, Diet-induced obesity increases NF-kappaB signaling in reporter mice. Genes Nutr. **4**, 215–222 (2009)

55. S. Shukla, G.T. MacLennan, P. Fu, J. Patel, S.R. Marengo, M.I. Resnick, S. Gupta, Nuclear factor-kappaB/p65 (Rel A) is constitutively activated in human prostate adenocarcinoma and correlates with disease progression. Neoplasia **6**, 390–400 (2004)

56. S.D. Catz, J.L. Johnson, Transcriptional regulation of bcl-2 by nuclear factor kappa B and its significance in prostate cancer. Oncogene **20**, 7342–7351 (2001)

57. E. Shankar, E.V. Vykhovanets, O.V. Vykhovanets, G.T. Maclennan, R. Singh, N. Bhaskaran, S. Shukla, S. Gupta, High-fat diet activates pro-inflammatory response in the prostate through association of Stat-3 and NF-kappaB. Prostate **72**, 233–243 (2012)

58. B.S. Gill, S. Kumar, Navgeet. Evaluating anti-oxidant potential of ganoderic acid A in STAT 3 pathway in prostate cancer. Mol. Biol. Rep. **43**, 1411–22 (2016)

59. D. Aune, D.A. Navarro Rosenblatt, D.S. Chan, A.R. Vieira, R. Vieira, D.C. Greenwood, L.J. Vatten, T. Norat, Dairy products, calcium, and prostate cancer risk: a systematic review and meta-analysis of cohort studies. Am. J. Clin. Nutr. **101**, 87–117 (2015)

60. X. Gao, M.P. LaValley, K.L. Tucker, Prospective studies of dairy product and calcium intakes and prostate cancer risk: a meta-analysis. J. Natl. Cancer Inst. **97**, 1768–1777 (2005)

61. H.L. Newmark, R.P. Heaney, Dairy products and prostate cancer risk. Nutr. Cancer **62**, 297–299 (2010)

62. Y. Park, P.N. Mitrou, V. Kipnis, A. Hollenbeck, A. Schatzkin, M.F. Leitzmann, Calcium, dairy foods, and risk of incident and fatal prostate cancer: the NIH-AARP diet and health study. Am. J. Epidemiol. **166**, 1270–1279 (2007)

63. A. Pettersson, J.L. Kasperzyk, S.A. Kenfield, E.L. Richman, J.M. Chan, W.C. Willett, M.J. Stampfer, L.A. Mucci, E.L. Giovannucci, Milk and dairy consumption among men with prostate cancer and risk of metastases and prostate cancer death. Cancer Epidemiol. Biomark. Prev. **21**, 428–436 (2012)

64. S.W. Park, J.Y. Kim, Y.S. Kim, S.J. Lee, S.D. Lee, M.K. Chung, A milk protein, casein, as a proliferation promoting factor in prostate cancer cells. World J. Mens Health **32**, 76–82 (2014)

65. P.L. Tate, R. Bibb, L.L. Larcom, Milk stimulates growth of prostate cancer cells in culture. Nutr. Cancer **63**, 1361–1366 (2011)

66. Y. Song, J.E. Chavarro, Y. Cao, W. Qiu, L. Mucci, H.D. Sesso, M.J. Stampfer, E. Giovannucci, M. Pollak, S. Liu, J. Ma, Whole milk intake is associated with prostate cancer-specific mortality among U.S. male physicians. J. Nutr. **143**, 189–196 (2013)

67. W. Lu, H. Chen, Y. Niu, H. Wu, D. Xia, Y. Wu, Dairy products intake and cancer mortality risk: a meta-analysis of 11 population-based cohort studies. Nutr. J. **15**, 91 (2016)

68. D. Tat, S.A. Kenfield, J.E. Cowan, J.M. Broering, P.R. Carroll, E.L. Van Blarigan, J.M. Chan, Milk and other dairy foods in relation to prostate cancer recurrence: data from the cancer of the prostate strategic urologic research endeavor (CaPSURE). Prostate **78**, 32–39 (2018)

69. M. Yang, S.A. Kenfield, E.L. Van Blarigan, K.M. Wilson, J.L. Batista, H.D. Sesso, J. Ma, M.J. Stampfer, J.E. Chavarro, Dairy intake after prostate cancer diagnosis in relation to disease-specific and total mortality. Int. J. Cancer **137**, 2462–2469 (2015)

70. J.M. Chan, E. Giovannucci, S.O. Andersson, J. Yuen, H.O. Adami, A. Wolk, Dairy products, calcium, phosphorous, vitamin D, and risk of prostate cancer (Sweden). Cancer Causes Control **9**, 559–566 (1998)

71. J.M. Chan, M.J. Stampfer, J. Ma, P.H. Gann, J.M. Gaziano, E.L. Giovannucci, Dairy products, calcium, and prostate cancer risk in the physicians' health study. Am. J. Clin. Nutr. **74**, 549–554 (2001)

72. E. Giovannucci, E.B. Rimm, A. Wolk, A. Ascherio, M.J. Stampfer, G.A. Colditz, W.C. Willett, Calcium and fructose intake in relation to risk of prostate cancer. Cancer Res. **58**, 442–447 (1998)

73. E. Kesse, S. Bertrais, P. Astorg, A. Jaouen, N. Arnault, P. Galan, S. Hercberg, Dairy products, calcium and phosphorus intake, and the risk of prostate cancer: results of the French prospective SU.VI. MAX (Supplementation en Vitamines et Mineraux Antioxydants) study. Br. J. Nutr. **95**, 539–545 (2006)

74. P.N. Mitrou, D. Albanes, S.J. Weinstein, P. Pietinen, P.R. Taylor, J. Virtamo, M.F. Leitzmann, A prospective study of dietary calcium, dairy products and prostate cancer risk (Finland). Int. J. Cancer **120**, 2466–2473 (2007)

75. K.M. Wilson, I.M. Shui, L.A. Mucci, E. Giovannucci, Calcium and phosphorus intake and prostate cancer risk: a 24-y follow-up study. Am. J. Clin. Nutr. **101**, 173–183 (2015)
76. N. Prevarskaya, R. Skryma, G. Bidaux, M. Flourakis, Y. Shuba, Ion channels in death and differentiation of prostate cancer cells. Cell Death Differ. **14**, 1295–1304 (2007)
77. J.L. Sanders, N. Chattopadhyay, O. Kifor, T. Yamaguchi, E.M. Brown, Ca(2+)-sensing receptor expression and PTHrP secretion in PC-3 human prostate cancer cells. Am. J. Physiol. Endocrinol. Metab. **281**, E1267–E1274 (2001)
78. S. Yano, R.J. Macleod, N. Chattopadhyay, J. Tfelt-Hansen, O. Kifor, R.R. Butters, E.M. Brown, Calcium-sensing receptor activation stimulates parathyroid hormone-related protein secretion in prostate cancer cells: role of epidermal growth factor receptor transactivation. Bone **35**, 664–672 (2004)
79. E.M. Weaver, F.J. Zamora, Y.A. Puplampu-Dove, E. Kiessu, J.L. Hearne, M. Martin-Caraballo, Regulation of T-type calcium channel expression by sodium butyrate in prostate cancer cells. Eur. J. Pharmacol. **749**, 20–31 (2015)
80. J. Liao, A. Schneider, N.S. Datta, L.K. McCauley, Extracellular calcium as a candidate mediator of prostate cancer skeletal metastasis. Cancer Res. **66**, 9065–9073 (2006)
81. K.I. Lin, N. Chattopadhyay, M. Bai, R. Alvarez, C.V. Dang, J.M. Baraban, E.M. Brown, R.R. Ratan, Elevated extracellular calcium can prevent apoptosis via the calcium-sensing receptor. Biochem. Biophys. Res. Commun. **249**, 325–331 (1998)
82. C.V. Vaz, D.B. Rodrigues, S. Socorro, C.J. Maia, Effect of extracellular calcium on regucalcin expression and cell viability in neoplastic and non-neoplastic human prostate cells. Biochim. Biophys. Acta **1853**, 2621–2628 (2015)
83. Y. Sun, S. Selvaraj, A. Varma, S. Derry, A.E. Sahmoun, B.B. Singh, Increase in serum Ca2+/Mg2+ ratio promotes proliferation of prostate cancer cells by activating TRPM7 channels. J. Biol. Chem. **288**, 255–263 (2013)
84. Y. Xu, X. Shao, Y. Yao, L. Xu, L. Chang, Z. Jiang, Z. Lin, Positive association between circulating 25-hydroxyvitamin D levels and prostate cancer risk: new findings from an updated meta-analysis. J. Cancer Res. Clin. Oncol. **140**, 1465–1477 (2014)
85. J.M. Schenk, C.A. Till, C.M. Tangen, P.J. Goodman, X. Song, K.C. Torkko, A.R. Kristal, U. Peters, M.L. Neuhouser, Serum 25-hydroxyvitamin D concentrations and risk of prostate cancer: results from the Prostate Cancer Prevention Trial. Cancer Epidemiol. Biomark. Prev. **23**, 1484–1493 (2014)
86. A.R. Kristal, C. Till, X. Song, C.M. Tangen, P.J. Goodman, M.L. Neuhauser, J.M. Schenk, I.M. Thompson, F.L. Meyskens Jr., G.E. Goodman, L.M. Minasian, H.L. Parnes, E.A. Klein, Plasma vitamin D and prostate cancer risk: results from the selenium and vitamin E cancer prevention trial.

Cancer Epidemiol. Biomark. Prev. **23**, 1494–1504 (2014)
87. J.P. Bonjour, T. Chevalley, P. Fardellone, Calcium intake and vitamin D metabolism and action, in healthy conditions and in prostate cancer. Br. J. Nutr. **97**, 611–616 (2007)
88. D.D. Bikle, Vitamin D metabolism, mechanism of action, and clinical applications. Chem. Biol. **21**, 319–329 (2014)
89. T. Ylikomi, I. Laaksi, Y.R. Lou, P. Martikainen, S. Miettinen, P. Pennanen, S. Purmonen, H. Syvala, A. Vienonen, P. Tuohimaa, Antiproliferative action of vitamin D. Vitam. Horm. **64**, 357–406 (2002)
90. M. Kivineva, M. Blauer, H. Syvala, T. Tammela, P. Tuohimaa, Localization of 1,25-dihydroxyvitamin D3 receptor (VDR) expression in human prostate. J. Steroid Biochem. Mol. Biol. **66**, 121–127 (1998)
91. G.J. Miller, G.E. Stapleton, J.A. Ferrara, M.S. Lucia, S. Pfister, T.E. Hedlund, P. Upadhya, The human prostatic carcinoma cell line LNCaP expresses biologically active, specific receptors for 1 alpha,25-dihydroxyvitamin D3. Cancer Res **52**, 515–520 (1992)
92. D.M. Peehl, R.J. Skowronski, G.K. Leung, S.T. Wong, T.A. Stamey, D. Feldman, Antiproliferative effects of 1,25-dihydroxyvitamin D3 on primary cultures of human prostatic cells. Cancer Res. **54**, 805–810 (1994)
93. G.J. Miller, G.E. Stapleton, T.E. Hedlund, K.A. Moffat, Vitamin D receptor expression, 24-hydroxylase activity, and inhibition of growth by 1alpha,25-dihydroxyvitamin D3 in seven human prostatic carcinoma cell lines. Clin. Cancer Res. **1**, 997–1003 (1995)
94. J. Moreno, A.V. Krishnan, D. Feldman, Molecular mechanisms mediating the anti-proliferative effects of Vitamin D in prostate cancer. J. Steroid Biochem. Mol. Biol. **97**, 31–36 (2005)
95. R.J. Skowronski, D.M. Peehl, D. Feldman, Vitamin D and prostate cancer: 1,25 dihydroxyvitamin D3 receptors and actions in human prostate cancer cell lines. Endocrinology **132**, 1952–1960 (1993)
96. X.Y. Zhao, D.M. Peehl, N.M. Navone, D. Feldman, 1alpha,25-dihydroxyvitamin D3 inhibits prostate cancer cell growth by androgen-dependent and androgen-independent mechanisms. Endocrinology **141**, 2548–2556 (2000)
97. B.Y. Bao, H.J. Ting, J.W. Hsu, Y.F. Lee, Protective role of 1 alpha, 25-dihydroxyvitamin D3 against oxidative stress in nonmalignant human prostate epithelial cells. Int. J. Cancer **122**, 2699–2706 (2008)
98. M. Guzey, S. Kitada, J.C. Reed, Apoptosis induction by 1alpha,25-dihydroxyvitamin D3 in prostate cancer. Mol. Cancer Ther. **1**, 667–677 (2002)
99. D.L. Trump, J.B. Aragon-Ching, Vitamin D in prostate cancer. Asian J. Androl. **20**, 244–252 (2018)
100. M. Pollak, Insulin and insulin-like growth factor signalling in neoplasia. Nat. Rev. Cancer **8**, 915–928 (2008)

101. G. Fiorelli, A. De Bellis, A. Longo, S. Giannini, A. Natali, A. Costantini, G.B. Vannelli, M. Serio, Insulin-like growth factor-I receptors in human hyperplastic prostate tissue: characterization, tissue localization, and their modulation by chronic treatment with a gonadotropin-releasing hormone analog. J. Clin. Endocrinol. Metab. **72**, 740–746 (1991)

102. P. Bonnet, E. Reiter, M. Bruyninx, B. Sente, D. Dombrowicz, J. de Leval, J. Closset, G. Hennen, Benign prostatic hyperplasia and normal prostate aging: differences in types I and II 5 alpha-reductase and steroid hormone receptor messenger ribonucleic acid (mRNA) levels, but not in insulin-like growth factor mRNA levels. J. Clin. Endocrinol. Metab. **77**, 1203–1208 (1993)

103. E. Kaicer, C. Blat, J. Imbenotte, F. Troalen, O. Cussenot, F. Calvo, L. Harel, IGF binding protein-3 secreted by the prostate adenocarcinoma cells (PC-3): differential effect on PC-3 and normal prostate cell growth. Growth Regul. **3**, 180–189 (1993)

104. P. Cohen, D.M. Peehl, B. Baker, F. Liu, R.L. Hintz, R.G. Rosenfeld, Insulin-like growth factor axis abnormalities in prostatic stromal cells from patients with benign prostatic hyperplasia. J. Clin. Endocrinol. Metab. **79**, 1410–1415 (1994)

105. T. Barni, B.G. Vannelli, R. Sadri, C. Pupilli, P. Ghiandi, M. Rizzo, C. Selli, M. Serio, G. Fiorelli, Insulin-like growth factor-I (IGF-I) and its binding protein IGFBP-4 in human prostatic hyperplastic tissue: gene expression and its cellular localization. J. Clin. Endocrinol. Metab. **78**, 778–783 (1994)

106. J.A. Figueroa, A.V. Lee, J.G. Jackson, D. Yee, Proliferation of cultured human prostate cancer cells is inhibited by insulin-like growth factor (IGF) binding protein-1: evidence for an IGF-II autocrine growth loop. J. Clin. Endocrinol. Metab. **80**, 3476–3482 (1995)

107. J.A. Figueroa, S. De Raad, L. Tadlock, V.O. Speights, J.J. Rinehart, Differential expression of insulin-like growth factor binding proteins in high versus low Gleason score prostate cancer. J. Urol. **159**, 1379–1383 (1998)

108. Z. Pietrzkowski, G. Mulholland, L. Gomella, B.A. Jameson, D. Wernicke, R. Baserga, Inhibition of growth of prostatic cancer cell lines by peptide analogues of insulin-like growth factor 1. Cancer Res. **53**, 1102–1106 (1993)

109. C. Boudon, G. Rodier, E. Lechevallier, N. Mottet, B. Barenton, C. Sultan, Secretion of insulin-like growth factors and their binding proteins by human normal and hyperplastic prostatic cells in primary culture. J. Clin. Endocrinol. Metab. **81**, 612–617 (1996)

110. P. Cohen, D.M. Peehl, G. Lamson, R.G. Rosenfeld, Insulin-like growth factors (IGFs), IGF receptors, and IGF-binding proteins in primary cultures of prostate epithelial cells. J. Clin. Endocrinol. Metab. **73**, 401–407 (1991)

111. Z. Pietrzkowski, D. Wernicke, P. Porcu, B.A. Jameson, R. Baserga, Inhibition of cellular proliferation by peptide analogues of insulin-like growth factor 1. Cancer Res. **52**, 6447–6451 (1992)

112. P. Cohen, H.C. Graves, D.M. Peehl, M. Kamarei, L.C. Giudice, R.G. Rosenfeld, Prostate-specific antigen (PSA) is an insulin-like growth factor binding protein-3 protease found in seminal plasma. J. Clin. Endocrinol. Metab. **75**, 1046–1053 (1992)

113. J.M. Chan, M.J. Stampfer, E. Giovannucci, P.H. Gann, J. Ma, P. Wilkinson, C.H. Hennekens, M. Pollak, Plasma insulin-like growth factor-I and prostate cancer risk: a prospective study. Science **279**, 563–566 (1998)

114. A.P. Chokkalingam, M. Pollak, C.M. Fillmore, Y.T. Gao, F.Z. Stanczyk, J. Deng, I.A. Sesterhenn, F.K. Mostofi, T.R. Fears, M.P. Madigan, R.G. Ziegler, J.F. Fraumeni Jr., A.W. Hsing, Insulin-like growth factors and prostate cancer: a population-based case-control study in China. Cancer Epidemiol. Biomark. Prev. **10**, 421–427 (2001)

115. S.M. Harman, E.J. Metter, M.R. Blackman, P.K. Landis, H.B. Carter, Baltimore Longitudinal Study on A, Serum levels of insulin-like growth factor I (IGF-I), IGF-II, IGF-binding protein-3, and prostate-specific antigen as predictors of clinical prostate cancer. J. Clin. Endocrinol. Metab. **85**, 4258–4265 (2000)

116. P. Stattin, A. Bylund, S. Rinaldi, C. Biessy, H. Dechaud, U.H. Stenman, L. Egevad, E. Riboli, G. Hallmans, R. Kaaks, Plasma insulin-like growth factor-I, insulin-like growth factor-binding proteins, and prostate cancer risk: a prospective study. J. Natl. Cancer Inst. **92**, 1910–1917 (2000)

117. A. Wolk, C.S. Mantzoros, S.O. Andersson, R. Bergstrom, L.B. Signorello, P. Lagiou, H.O. Adami, D. Trichopoulos, Insulin-like growth factor 1 and prostate cancer risk: a population-based, case-control study. J. Natl. Cancer Inst. **90**, 911–915 (1998)

118. T. Shaneyfelt, R. Husein, G. Bubley, C.S. Mantzoros, Hormonal predictors of prostate cancer: a meta-analysis. J. Clin. Oncol. **18**, 847–853 (2000)

119. R.C. Travis, P.N. Appleby, R.M. Martin, J.M.P. Holly, D. Albanes, A. Black, H.B.A. Bueno-de-Mesquita, J.M. Chan, C. Chen, M.D. Chirlaque, M.B. Cook, M. Deschasaux, J.L. Donovan, et al., A meta-analysis of individual participant data reveals an association between circulating levels of IGF-I and prostate cancer risk. Cancer Res. **76**, 2288–2300 (2016)

120. J.I. Jones, D.R. Clemmons, Insulin-like growth factors and their binding proteins: biological actions. Endocr. Rev. **16**, 3–34 (1995)

121. M. Grzmil, B. Hemmerlein, P. Thelen, S. Schweyer, P. Burfeind, Blockade of the type I IGF receptor expression in human prostate cancer cells inhibits proliferation and invasion, up-regulates IGF binding protein-3, and suppresses MMP-2 expression. J. Pathol. **202**, 50–59 (2004)

122. P. Burfeind, C.L. Chernicky, F. Rininsland, J. Ilan, J. Ilan, Antisense RNA to the type I insulin-like

growth factor receptor suppresses tumor growth and prevents invasion by rat prostate cancer cells in vivo. Proc. Natl. Acad. Sci. U. S. A. **93**, 7263–7268 (1996)

123. J.D. Wu, A. Odman, L.M. Higgins, K. Haugk, R. Vessella, D.L. Ludwig, S.R. Plymate, In vivo effects of the human type I insulin-like growth factor receptor antibody A12 on androgen-dependent and androgen-independent xenograft human prostate tumors. Clin. Cancer Res. **11**, 3065–3074 (2005)

124. H. Hartog, J. Wesseling, H.M. Boezen, W.T. van der Graaf, The insulin-like growth factor 1 receptor in cancer: old focus, new future. Eur. J. Cancer **43**, 1895–1904 (2007)

125. G. Rodriguez-Berriguete, B. Fraile, P. Martinez-Onsurbe, G. Olmedilla, R. Paniagua, M. Royuela, MAP kinases and prostate cancer. J. Signal Transduct. **2012**, 169170 (2012)

126. R.S. Liao, S. Ma, L. Miao, R. Li, Y. Yin, G.V. Raj, Androgen receptor-mediated non-genomic regulation of prostate cancer cell proliferation. Transl. Androl. Urol. **2**, 187–196 (2013)

127. W. Lim, M. Jeong, F.W. Bazer, G. Song, Coumestrol inhibits proliferation and migration of prostate cancer cells by regulating AKT, ERK1/2, and JNK MAPK cell signaling cascades. J. Cell. Physiol. **232**, 862–871 (2017)

128. P.K. Majumder, W.R. Sellers, Akt-regulated pathways in prostate cancer. Oncogene **24**, 7465–7474 (2005)

129. J.M. Chan, M.J. Stampfer, J. Ma, P. Gann, J.M. Gaziano, M. Pollak, E. Giovannucci, Insulin-like growth factor-I (IGF-I) and IGF binding protein-3 as predictors of advanced-stage prostate cancer. J. Natl. Cancer Inst. **94**, 1099–1106 (2002)

130. L. Li, H. Yu, F. Schumacher, G. Casey, J.S. Witte, Relation of serum insulin-like growth factor-I (IGF-I) and IGF binding protein-3 to risk of prostate cancer (United States). Cancer Causes Control **14**, 721–726 (2003)

131. Y.C. Lee, M. Hashibe, Tobacco, alcohol, and cancer in low and high income countries. Ann. Glob. Health **80**, 378–383 (2014)

132. L. Degenhardt, W.T. Chiu, N. Sampson, R.C. Kessler, J.C. Anthony, M. Angermeyer, R. Bruffaerts, G. de Girolamo, O. Gureje, Y. Huang, A. Karam, S. Kostyuchenko, J.P. Lepine, et al., Toward a global view of alcohol, tobacco, cannabis, and cocaine use: findings from the WHO world mental health surveys. PLoS Med. **5**, e141 (2008)

133. D.E. Nelson, D.W. Jarman, J. Rehm, T.K. Greenfield, G. Rey, W.C. Kerr, P. Miller, K.D. Shield, Y. Ye, T.S. Naimi, Alcohol-attributable cancer deaths and years of potential life lost in the United States. Am. J. Public Health **103**, 641–648 (2013)

134. H.D. Sesso, R.S. Paffenbarger Jr., I.M. Lee, Alcohol consumption and risk of prostate cancer: the Harvard Alumni Health study. Int. J. Epidemiol. **30**, 749–755 (2001)

135. J.L. Watters, Y. Park, A. Hollenbeck, A. Schatzkin, D. Albanes, Alcoholic beverages and prostate cancer in a prospective US cohort study. Am. J. Epidemiol. **172**, 773–780 (2010)

136. R.B. Hayes, L.M. Brown, J.B. Schoenberg, R.S. Greenberg, D.T. Silverman, A.G. Schwartz, G.M. Swanson, J. Benichou, J.M. Liff, R.N. Hoover, L.M. Pottern, Alcohol use and prostate cancer risk in US blacks and whites. Am. J. Epidemiol. **143**, 692–697 (1996)

137. P.C. Dagnelie, A.G. Schuurman, R.A. Goldbohm, P.A. Van den Brandt, Diet, anthropometric measures and prostate cancer risk: a review of prospective cohort and intervention studies. BJU Int. **93**, 1139–1150 (2004)

138. M.P. Longnecker, Alcohol consumption and risk of cancer in humans: an overview. Alcohol **12**, 87–96 (1995)

139. M.S. Morton, K. Griffiths, N. Blacklock, The preventive role of diet in prostatic disease. Br. J. Urol. **77**, 481–493 (1996)

140. L.K. Dennis, Meta-analysis for combining relative risks of alcohol consumption and prostate cancer. Prostate **42**, 56–66 (2000)

141. V. Bagnardi, M. Rota, E. Botteri, I. Tramacere, F. Islami, V. Fedirko, L. Scotti, M. Jenab, F. Turati, E. Pasquali, C. Pelucchi, C. Galeone, R. Bellocco, et al., Alcohol consumption and site-specific cancer risk: a comprehensive dose-response meta-analysis. Br. J. Cancer **112**, 580–593 (2015)

142. K. Middleton Fillmore, T. Chikritzhs, T. Stockwell, A. Bostrom, R. Pascal, Alcohol use and prostate cancer: a meta-analysis. Mol. Nutr. Food Res. **53**, 240–255 (2009)

143. M. Rota, L. Scotti, F. Turati, I. Tramacere, F. Islami, R. Bellocco, E. Negri, G. Corrao, P. Boffetta, C. La Vecchia, V. Bagnardi, Alcohol consumption and prostate cancer risk: a meta-analysis of the dose-risk relation. Eur. J. Cancer Prev. **21**, 350–359 (2012)

144. J. Zhao, T. Stockwell, A. Roemer, T. Chikritzhs, Is alcohol consumption a risk factor for prostate cancer? A systematic review and meta-analysis. BMC Cancer **16**, 845 (2016)

145. N.P. Papa, R.J. MacInnis, H. Jayasekara, D.R. English, D. Bolton, I.D. Davis, N. Lawrentschuk, J.L. Millar, J. Pedersen, G. Severi, M.C. Southey, J.L. Hopper, G.G. Giles, Total and beverage-specific alcohol intake and the risk of aggressive prostate cancer: a case-control study. Prostate Cancer Prostatic Dis. **20**, 305–310 (2017)

146. S.E. McGregor, K.S. Courneya, K.A. Kopciuk, C. Tosevski, C.M. Friedenreich, Case-control study of lifetime alcohol intake and prostate cancer risk. Cancer Causes Control **24**, 451–461 (2013)

147. N. Sawada, M. Inoue, M. Iwasaki, S. Sasazuki, T. Yamaji, T. Shimazu, S. Tsugane, Alcohol and smoking and subsequent risk of prostate cancer in Japanese men: the Japan Public Health Center-based prospective study. Int. J. Cancer **134**, 971–978 (2014)

148. Z. Gong, A.R. Kristal, J.M. Schenk, C.M. Tangen, P.J. Goodman, I.M. Thompson, Alcohol consumption, finasteride, and prostate cancer risk: results from the prostate cancer prevention trial. Cancer **115**, 3661–3669 (2009)

149. E.T. Chang, M. Hedelin, H.O. Adami, H. Gronberg, K.A. Balter, Alcohol drinking and risk of localized versus advanced and sporadic versus familial prostate cancer in Sweden. Cancer Causes Control **16**, 275–284 (2005)

150. J.H. Fowke, L. Howard, G.L. Andriole, S.J. Freedland, Alcohol intake increases high-grade prostate cancer risk among men taking dutasteride in the REDUCE trial. Eur. Urol. **66**, 1133–1138 (2014)

151. Humans IWGotEoCRt, Personal habits and indoor combustions. Volume 100 E. A review of human carcinogens. IARC Monogr. Eval. Carcinog. Risks Hum. **100**, 1–538 (2012)

152. S. Balbo, P.J. Brooks, Implications of acetaldehyde-derived DNA adducts for understanding alcohol-related carcinogenesis. Adv. Exp. Med. Biol. **815**, 71–88 (2015)

153. J.L. Fang, C.E. Vaca, Detection of DNA adducts of acetaldehyde in peripheral white blood cells of alcohol abusers. Carcinogenesis **18**, 627–632 (1997)

154. G.D. Castro, A.M. Delgado de Layno, M.H. Costantini, J.A. Castro, Cytosolic xanthine oxidoreductase mediated bioactivation of ethanol to acetaldehyde and free radicals in rat breast tissue. Its potential role in alcohol-promoted mammary cancer. Toxicology **160**, 11–18 (2001)

155. G.D. Castro, A.M. Delgado de Layno, M.H. Costantini, J.A. Castro, Rat ventral prostate microsomal biotransformation of ethanol to acetaldehyde and 1-hydroxyethyl radicals: its potential contribution to prostate tumor promotion. Teratog. Carcinog. Mutagen. **22**, 335–341 (2002)

156. M.I. Gomez, C.R. de Castro, S.L. Fanelli, L.N. Quintans, M.H. Costantini, J.A. Castro, G.D. Castro, Biochemical and ultrastructural alterations in the rat ventral prostate due to repetitive alcohol drinking. J. Appl. Toxicol. **27**, 391–398 (2007)

157. G.D. Castro, M.H. Costantini, J.A. Castro, Rat ventral prostate xanthine oxidase-mediated metabolism of acetaldehyde to acetyl radical. Hum. Exp. Toxicol. **28**, 203–208 (2009)

158. S. Sutcliffe, E. Giovannucci, M.F. Leitzmann, E.B. Rimm, M.J. Stampfer, W.C. Willett, E.A. Platz, A prospective cohort study of red wine consumption and risk of prostate cancer. Int. J. Cancer **120**, 1529–1535 (2007)

159. C. Chao, R. Haque, S.K. Van Den Eeden, B.J. Caan, K.Y. Poon, V.P. Quinn, Red wine consumption and risk of prostate cancer: the California men's health study. Int. J. Cancer **126**, 171–179 (2010)

160. W.M. Schoonen, C.A. Salinas, L.A. Kiemeney, J.L. Stanford, Alcohol consumption and risk of prostate cancer in middle-aged men. Int. J. Cancer **113**, 133–140 (2005)

161. S.A. Aherne, N.M. O'Brien, Dietary flavonols: chemistry, food content, and metabolism. Nutrition **18**, 75–81 (2002)

162. C.S. Yang, J.M. Landau, M.T. Huang, H.L. Newmark, Inhibition of carcinogenesis by dietary polyphenolic compounds. Annu. Rev. Nutr. **21**, 381–406 (2001)

163. M. Kampa, A. Hatzoglou, G. Notas, A. Damianaki, E. Bakogeorgou, C. Gemetzi, E. Kouroumalis, P.M. Martin, E. Castanas, Wine antioxidant polyphenols inhibit the proliferation of human prostate cancer cell lines. Nutr. Cancer **37**, 223–233 (2000)

164. L.C. Bylsma, D.D. Alexander, A review and meta-analysis of prospective studies of red and processed meat, meat cooking methods, heme iron, heterocyclic amines and prostate cancer. Nutr. J. **14**, 125 (2015)

165. W.G. Gathirua-Mwangi, J. Zhang, Dietary factors and risk for advanced prostate cancer. Eur. J. Cancer Prev. **23**, 96–109 (2014)

166. W. Lijinsky, N-Nitroso compounds in the diet. Mutat. Res. **443**, 129–138 (1999)

167. D.H. Phillips, Polycyclic aromatic hydrocarbons in the diet. Mutat. Res. **443**, 139–147 (1999)

168. T. Sugimura, K. Wakabayashi, H. Nakagama, M. Nagao, Heterocyclic amines: mutagens/carcinogens produced during cooking of meat and fish. Cancer Sci. **95**, 290–299 (2004)

169. J.S. Felton, M. Jagerstad, M.G. Knize, K. Skog, K. Wakabayashi, Contents in foods, beverages and tobacco, in *Food Borne. Carcinogens Heterocyclic Amines*, ed. by M. Nagao, T. Sugimura, (John Wiley & Sons, Chichester, 2000), pp. 31–71

170. T. Shirai, M. Sano, S. Tamano, S. Takahashi, M. Hirose, M. Futakuchi, R. Hasegawa, K. Imaida, K. Matsumoto, K. Wakabayashi, T. Sugimura, N. Ito, The prostate: a target for carcinogenicity of 2-amino-1-methyl-6-phenylimidazo[4,5-b]pyridine (PhIP) derived from cooked foods. Cancer Res. **57**, 195–198 (1997)

171. V. Bouvard, D. Loomis, K.Z. Guyton, Y. Grosse, F.E. Ghissassi, L. Benbrahim-Tallaa, N. Guha, H. Mattock, K. Straif, Carcinogenicity of consumption of red and processed meat. Lancet Oncol. **16**, 1599–1600 (2015)

172. K.T. Bogen, G.A. Keating, U.S. dietary exposures to heterocyclic amines. J. Expo. Anal. Environ. Epidemiol. **11**, 155–168 (2001)

173. A.L. Sesink, D.S. Termont, J.H. Kleibeuker, R. Van der Meer, Red meat and colon cancer: the cytotoxic and hyperproliferative effects of dietary heme. Cancer Res. **59**, 5704–5709 (1999)

174. A. Tappel, Heme of consumed red meat can act as a catalyst of oxidative damage and could initiate colon, breast and prostate cancers, heart disease and other diseases. Med. Hypotheses **68**, 562–564 (2007)

175. R. Sinha, Y. Park, B.I. Graubard, M.F. Leitzmann, A. Hollenbeck, A. Schatzkin, A.J. Cross, Meat and meat-related compounds and risk of prostate can-

cer in a large prospective cohort study in the United States. Am. J. Epidemiol. **170**, 1165–1177 (2009)

176. P.G. Jakszyn, N.E. Allen, L. Lujan-Barroso, C.A. Gonzalez, T.J. Key, A. Fonseca-Nunes, A. Tjonneland, N. Fons-Johnsen, K. Overvad, B. Teucher, K. Li, H. Boeing, A. Trichopoulou, et al., Nitrosamines and heme iron and risk of prostate cancer in the European prospective investigation into cancer and nutrition. Cancer Epidemiol. Biomark. Prev. **21**, 547–551 (2012)

177. M. Dietrich, G. Block, J.M. Pogoda, P. Buffler, S. Hecht, S. Preston-Martin, A review: dietary and endogenously formed N-nitroso compounds and risk of childhood brain tumors. Cancer Causes Control **16**, 619–635 (2005)

178. S.S. Mirvish, Role of N-nitroso compounds (NOC) and N-nitrosation in etiology of gastric, esophageal, nasopharyngeal and bladder cancer and contribution to cancer of known exposures to NOC. Cancer Lett. **93**, 17–48 (1995)

179. P. Mende, B. Spiegelhalder, R. Preussmann, Trace analysis of nitrosated foodstuffs for nitrosamides. Food Chem. Toxicol. **29**, 167–172 (1991)

180. A.J. Cross, J.R. Pollock, S.A. Bingham, Haem, not protein or inorganic iron, is responsible for endogenous intestinal N-nitrosation arising from red meat. Cancer Res. **63**, 2358–2360 (2003)

181. R. Hughes, A.J. Cross, J.R. Pollock, S. Bingham, Dose-dependent effect of dietary meat on endogenous colonic N-nitrosation. Carcinogenesis **22**, 199–202 (2001)

182. J.C. Lunn, G. Kuhnle, V. Mai, C. Frankenfeld, D.E. Shuker, R.C. Glen, J.M. Goodman, J.R. Pollock, S.A. Bingham, The effect of haem in red and processed meat on the endogenous formation of N-nitroso compounds in the upper gastrointestinal tract. Carcinogenesis **28**, 685–690 (2007)

183. G.G. Kuhnle, S.A. Bingham, Dietary meat, endogenous nitrosation and colorectal cancer. Biochem. Soc. Trans. **35**, 1355–1357 (2007)

184. R. Saffhill, G.P. Margison, P.J. O'Connor, Mechanisms of carcinogenesis induced by alkylating agents. Biochim. Biophys. Acta **823**, 111–145 (1985)

185. S.A. Belinsky, T.R. Devereux, R.R. Maronpot, G.D. Stoner, M.W. Anderson, Relationship between the formation of promutagenic adducts and the activation of the K-ras protooncogene in lung tumors from A/J mice treated with nitrosamines. Cancer Res. **49**, 5305–5311 (1989)

186. E.L. Loechler, C.L. Green, J.M. Essigmann, In vivo mutagenesis by O6-methylguanine built into a unique site in a viral genome. Proc. Natl. Acad. Sci. U. S. A. **81**, 6271–6275 (1984)

187. P. Jakszyn, S. Bingham, G. Pera, A. Agudo, R. Luben, A. Welch, H. Boeing, G. Del Giudice, D. Palli, C. Saieva, V. Krogh, C. Sacerdote, R. Tumino, et al., Endogenous versus exogenous exposure to N-nitroso compounds and gastric cancer risk in the European Prospective Investigation

into Cancer and Nutrition (EPIC-EURGAST) study. Carcinogenesis **27**, 1497–1501 (2006)

188. E. De Stefani, P. Boffetta, M. Mendilaharsu, J. Carzoglio, H. Deneo-Pellegrini, Dietary nitrosamines, heterocyclic amines, and risk of gastric cancer: a case-control study in Uruguay. Nutr. Cancer **30**, 158–162 (1998)

189. D. Pobel, E. Riboli, J. Cornee, B. Hemon, M. Guyader, Nitrosamine, nitrate and nitrite in relation to gastric cancer: a case-control study in Marseille, France. Eur. J. Epidemiol. **11**, 67–73 (1995)

190. C. La Vecchia, B. D'Avanzo, L. Airoldi, C. Braga, A. Decarli, Nitrosamine intake and gastric cancer risk. Eur. J. Cancer Prev. **4**, 469–474 (1995)

191. P. Knekt, R. Jarvinen, J. Dich, T. Hakulinen, Risk of colorectal and other gastro-intestinal cancers after exposure to nitrate, nitrite and N-nitroso compounds: a follow-up study. Int. J. Cancer **80**, 852–856 (1999)

192. Y.H. Loh, P. Jakszyn, R.N. Luben, A.A. Mulligan, P.N. Mitrou, K.T. Khaw, N-Nitroso compounds and cancer incidence: the European Prospective Investigation into Cancer and Nutrition (EPIC)-Norfolk study. Am. J. Clin. Nutr. **93**, 1053–1061 (2011)

193. Humans IWGotEoCRt, Some non-heterocyclic polycyclic aromatic hydrocarbons and some related exposures. IARC Monogr. Eval. Carcinog. Risks Hum. **92**, 1–853 (2010)

194. P. Mottier, V. Parisod, R.J. Turesky, Quantitative determination of polycyclic aromatic hydrocarbons in barbecued meat sausages by gas chromatography coupled to mass spectrometry. J. Agric. Food Chem. **48**, 1160–1166 (2000)

195. P. Boffetta, N. Jourenkova, P. Gustavsson, Cancer risk from occupational and environmental exposure to polycyclic aromatic hydrocarbons. Cancer Causes Control **8**, 444–472 (1997)

196. L. Pyy, M. Makela, E. Hakala, K. Kakko, T. Lapinlampi, A. Lisko, E. Yrjanheikki, K. Vahakangas, Ambient and biological monitoring of exposure to polycyclic aromatic hydrocarbons at a coking plant. Sci. Total Environ. **199**, 151–158 (1997)

197. G. Grimmer, G. Dettbarn, J. Jacob, Biomonitoring of polycyclic aromatic hydrocarbons in highly exposed coke plant workers by measurement of urinary phenanthrene and pyrene metabolites (phenols and dihydrodiols). Int. Arch. Occup. Environ. Health **65**, 189–199 (1993)

198. A.T. Vu, K.M. Taylor, M.R. Holman, Y.S. Ding, B. Hearn, C.H. Watson, Polycyclic aromatic hydrocarbons in the mainstream smoke of popular U.S. cigarettes. Chem. Res. Toxicol. **28**, 1616–1626 (2015)

199. M. Jagerstad, K. Skog, Genotoxicity of heat-processed foods. Mutat. Res. **574**, 156–172 (2005)

200. J.G. Lee, S.Y. Kim, J.S. Moon, S.H. Kim, D.H. Kang, H.J. Yoon, Effects of grilling procedures on levels of

polycyclic aromatic hydrocarbons in grilled meats. Food Chem. **199**, 632–638 (2016)

201. O. Viegas, P. Novo, E. Pinto, O. Pinho, I.M. Ferreira, Effect of charcoal types and grilling conditions on formation of heterocyclic aromatic amines (HAs) and polycyclic aromatic hydrocarbons (PAHs) in grilled muscle foods. Food Chem. Toxicol. **50**, 2128–2134 (2012)

202. P. Simko, Factors affecting elimination of polycyclic aromatic hydrocarbons from smoked meat foods and liquid smoke flavorings. Mol. Nutr. Food Res. **49**, 637–647 (2005)

203. S.Y. Chung, R.R. Yettella, J.S. Kim, K. Kwon, M.C. Kim, D.B. Min, Effects of grilling and roasting on the levels of polycyclic aromatic hydrocarbons in beef and pork. Food Chem. **129**, 1420–1426 (2011)

204. N. Kazerouni, R. Sinha, C.H. Hsu, A. Greenberg, N. Rothman, Analysis of 200 food items for benzo[a]pyrene and estimation of its intake in an epidemiologic study. Food Chem. Toxicol. **39**, 423–436 (2001)

205. J.L. Domingo, M. Nadal, Human dietary exposure to polycyclic aromatic hydrocarbons: a review of the scientific literature. Food Chem. Toxicol. **86**, 144–153 (2015)

206. T. Shimada, C.L. Hayes, H. Yamazaki, S. Amin, S.S. Hecht, F.P. Guengerich, T.R. Sutter, Activation of chemically diverse procarcinogens by human cytochrome P-450 1B1. Cancer Res **56**, 2979–2984 (1996)

207. J.A. Williams, F.L. Martin, G.H. Muir, A. Hewer, P.L. Grover, D.H. Phillips, Metabolic activation of carcinogens and expression of various cytochromes P450 in human prostate tissue. Carcinogenesis **21**, 1683–1689 (2000)

208. F.L. Martin, K.J. Cole, G.H. Muir, G.G. Kooiman, J.A. Williams, R.A. Sherwood, P.L. Grover, D.H. Phillips, Primary cultures of prostate cells and their ability to activate carcinogens. Prostate Cancer Prostatic Dis. **5**, 96–104 (2002)

209. E. Hruba, L. Trilecova, S. Marvanova, P. Krcmar, L. Vykopalova, A. Milcova, H. Libalova, J. Topinka, A. Starsichova, K. Soucek, J. Vondracek, M. Machala, Genotoxic polycyclic aromatic hydrocarbons fail to induce the p53-dependent DNA damage response, apoptosis or cell-cycle arrest in human prostate carcinoma LNCaP cells. Toxicol. Lett. **197**, 227–235 (2010)

210. G.G. Kooiman, F.L. Martin, J.A. Williams, P.L. Grover, D.H. Phillips, G.H. Muir, The influence of dietary and environmental factors on prostate cancer risk. Prostate Cancer Prostatic Dis. **3**, 256–258 (2000)

211. B.A. Rybicki, C. Neslund-Dudas, C.H. Bock, A. Rundle, A.T. Savera, J.J. Yang, N.L. Nock, D. Tang, Polycyclic aromatic hydrocarbon—DNA adducts in prostate and biochemical recurrence after prostatectomy. Clin. Cancer Res. **14**, 750–757 (2008)

212. B.A. Rybicki, N.L. Nock, A.T. Savera, D. Tang, A. Rundle, Polycyclic aromatic hydrocarbon-DNA adduct formation in prostate carcinogenesis. Cancer Lett. **239**, 157–167 (2006)

213. B.A. Rybicki, A. Rundle, A.T. Savera, S.S. Sankey, D. Tang, Polycyclic aromatic hydrocarbon-DNA adducts in prostate cancer. Cancer Res. **64**, 8854–8859 (2004)

214. D. Tang, O.N. Kryvenko, Y. Wang, M. Jankowski, S. Trudeau, A. Rundle, B.A. Rybicki, Elevated polycyclic aromatic hydrocarbon-DNA adducts in benign prostate and risk of prostate cancer in African Americans. Carcinogenesis **34**, 113–120 (2013)

215. K. John, N. Ragavan, M.M. Pratt, P.B. Singh, S. Al-Buheissi, S.S. Matanhelia, D.H. Phillips, M.C. Poirier, F.L. Martin, Quantification of phase I/II metabolizing enzyme gene expression and polycyclic aromatic hydrocarbon-DNA adduct levels in human prostate. Prostate **69**, 505–519 (2009)

216. A. Weston, D.K. Manchester, M.C. Poirier, J.S. Choi, G.E. Trivers, D.L. Mann, C.C. Harris, Derivative fluorescence spectral analysis of polycyclic aromatic hydrocarbon-DNA adducts in human placenta. Chem. Res. Toxicol. **2**, 104–108 (1989)

217. N.L. Nock, D. Tang, A. Rundle, C. Neslund-Dudas, A.T. Savera, C.H. Bock, K.G. Monaghan, A. Koprowski, N. Mitrache, J.J. Yang, B.A. Rybicki, Associations between smoking, polymorphisms in polycyclic aromatic hydrocarbon (PAH) metabolism and conjugation genes and PAH-DNA adducts in prostate tumors differ by race. Cancer Epidemiol. Biomark. Prev. **16**, 1236–1245 (2007)

218. S. Xiao, J. Guo, B.H. Yun, P.W. Villalta, S. Krishna, R. Tejpaul, P. Murugan, C.J. Weight, R.J. Turesky, Biomonitoring DNA adducts of cooked meat carcinogens in human prostate by nano liquid chromatography-high resolution tandem mass spectrometry: identification of 2-amino-1-methyl-6-phenylimidazo[4,5-b]pyridine DNA adduct. Anal. Chem. **88**, 12508–12515 (2016)

219. W. Ni, L. McNaughton, D.M. LeMaster, R. Sinha, R.J. Turesky, Quantitation of 13 heterocyclic aromatic amines in cooked beef, pork, and chicken by liquid chromatography-electrospray ionization/tandem mass spectrometry. J. Agric. Food Chem. **56**, 68–78 (2008)

220. M. Jagerstad, K. Skog, S. Grivas, K. Olsson, Formation of heterocyclic amines using model systems. Mutat. Res. **259**, 219–233 (1991)

221. K.I. Skog, M.A. Johansson, M.I. Jagerstad, Carcinogenic heterocyclic amines in model systems and cooked foods: a review on formation, occurrence and intake. Food Chem. Toxicol. **36**, 879–896 (1998)

222. D. Yoshida, T. Matsumoto, R. Yoshimura, T. Matsuzaki, Mutagenicity of amino-alpha-carbolines in pyrolysis products of soybean globulin. Biochem. Biophys. Res. Commun. **83**, 915–920 (1978)

223. T. Matsumoto, D. Yoshida, H. Tomita, Determination of mutagens, amino-alpha-carbolines in grilled foods and cigarette smoke condensate. Cancer Lett. **12**, 105–110 (1981)

224. S. Manabe, K. Tohyama, O. Wada, T. Aramaki, Detection of a carcinogen, 2-amino-1-methyl-6-phenylimidazo[4,5-b]pyridine (PhIP), in cigarette smoke condensate. Carcinogenesis **12**, 1945–1947 (1991)

225. D. Yoshida, T. Matsumoto, Amino-alpha-carbolines as mutagenic agents in cigarette smoke condensate. Cancer Lett. **10**, 141–149 (1980)

226. C.J. Smith, X. Qian, Q. Zha, S.C. Moldoveanu, Analysis of alpha- and beta-carbolines in mainstream smoke of reference cigarettes by gas chromatography-mass spectrometry. J. Chromatogr. A **1046**, 211–216 (2004)

227. L. Zhang, D.L. Ashley, C.H. Watson, Quantitative analysis of six heterocyclic aromatic amines in mainstream cigarette smoke condensate using isotope dilution liquid chromatography-electrospray ionization tandem mass spectrometry. Nicotine Tob. Res. **13**, 120–126 (2011)

228. R.J. Turesky, L. Le Marchand, Metabolism and biomarkers of heterocyclic aromatic amines in molecular epidemiology studies: lessons learned from aromatic amines. Chem. Res. Toxicol. **24**, 1169–1214 (2011)

229. M.G. Knize, J.S. Felton, Formation and human risk of carcinogenic heterocyclic amines formed from natural precursors in meat. Nutr. Rev. **63**, 158–165 (2005)

230. R. Sinha, N. Rothman, E.D. Brown, C.P. Salmon, M.G. Knize, C.A. Swanson, S.C. Rossi, S.D. Mark, O.A. Levander, J.S. Felton, High concentrations of the carcinogen 2-amino-1-methyl-6-phenylimidazo- [4,5-b]pyridine (PhIP) occur in chicken but are dependent on the cooking method. Cancer Res. **55**, 4516–4519 (1995)

231. M.G. Knize, F.A. Dolbeare, K.L. Carroll, D.H. Moore 2nd, J.S. Felton, Effect of cooking time and temperature on the heterocyclic amine content of fried beef patties. Food Chem. Toxicol. **32**, 595–603 (1994)

232. K. Skog, A. Solyakov, P. Arvidsson, M. Jagerstad, Analysis of nonpolar heterocyclic amines in cooked foods and meat extracts using gas chromatography-mass spectrometry. J. Chromatogr. A **803**, 227–233 (1998)

233. R. Sinha, N. Rothman, C.P. Salmon, M.G. Knize, E.D. Brown, C.A. Swanson, D. Rhodes, S. Rossi, J.S. Felton, O.A. Levander, Heterocyclic amine content in beef cooked by different methods to varying degrees of doneness and gravy made from meat drippings. Food Chem. Toxicol. **36**, 279–287 (1998)

234. G.A. Keating, K.T. Bogen, Estimates of heterocyclic amine intake in the US population. J. Chromatogr. B Analyt. Technol. Biomed. Life Sci. **802**, 127–133 (2004)

235. F.G. Crofts, T.R. Sutter, P.T. Strickland, Metabolism of 2-amino-1-methyl-6-phenylimidazo[4,5-b]pyridine by human cytochrome P4501A1, P4501A2 and P4501B1. Carcinogenesis **19**, 1969–1973 (1998)

236. A.R. Boobis, A.M. Lynch, S. Murray, R. de la Torre, A. Solans, M. Farre, J. Segura, N.J. Gooderham, D.S. Davies, CYP1A2-catalyzed conversion of dietary heterocyclic amines to their proximate carcinogens is their major route of metabolism in humans. Cancer Res. **54**, 89–94 (1994)

237. H. Glatt, U. Pabel, W. Meinl, H. Frederiksen, H. Frandsen, E. Muckel, Bioactivation of the heterocyclic aromatic amine 2-amino-3-methyl-9H-pyrido [2,3-b]indole (MeAalphaC) in recombinant test systems expressing human xenobiotic-metabolizing enzymes. Carcinogenesis **25**, 801–807 (2004)

238. H.A. Schut, E.G. Snyderwine, DNA adducts of heterocyclic amine food mutagens: implications for mutagenesis and carcinogenesis. Carcinogenesis **20**, 353–368 (1999)

239. R.J. Turesky, P. Vouros, Formation and analysis of heterocyclic aromatic amine-DNA adducts in vitro and in vivo. J. Chromatogr. B Analyt. Technol. Biomed. Life Sci. **802**, 155–166 (2004)

240. K.W. Turteltaub, J.S. Felton, B.L. Gledhill, J.S. Vogel, J.R. Southon, M.W. Caffee, R.C. Finkel, D.E. Nelson, I.D. Proctor, J.C. Davis, Accelerator mass spectrometry in biomedical dosimetry: relationship between low-level exposure and covalent binding of heterocyclic amine carcinogens to DNA. Proc. Natl. Acad. Sci. U. S. A. **87**, 5288–5292 (1990)

241. R.J. Turesky, R.M. Box, J. Markovic, E. Gremaud, E.G. Snyderwine, Formation and persistence of DNA adducts of 2-amino-3-methylimidazo[4,5-f] quinoline in the rat and nonhuman primates. Mutat. Res. **376**, 235–241 (1997)

242. S. Fukushima, H. Wanibuchi, K. Morimura, S. Iwai, D. Nakae, H. Kishida, H. Tsuda, N. Uehara, K. Imaida, T. Shirai, M. Tatematsu, T. Tsukamoto, M. Hirose, et al., Existence of a threshold for induction of aberrant crypt foci in the rat colon with low doses of 2-amino-1-methyl-6-phenolimidazo[4,5-b] pyridine. Toxicol. Sci. **80**, 109–114 (2004)

243. G. Nauwelaers, E.E. Bessette, D. Gu, Y. Tang, J. Rageul, V. Fessard, J.M. Yuan, M.C. Yu, S. Langouet, R.J. Turesky, DNA adduct formation of 4-aminobiphenyl and heterocyclic aromatic amines in human hepatocytes. Chem. Res. Toxicol. **24**, 913–925 (2011)

244. R.J. Turesky, A. Constable, J. Richoz, N. Varga, J. Markovic, M.V. Martin, F.P. Guengerich, Activation of heterocyclic aromatic amines by rat and human liver microsomes and by purified rat and human cytochrome P450 1A2. Chem. Res. Toxicol. **11**, 925–936 (1998)

245. R.J. Turesky, V. Parisod, T. Huynh-Ba, S. Langouët, F.P. Guengerich, Regioselective differences in C(8)- and N-oxidation of 2-amino-3,8-dimethylimidazo[4,5-*f*]quinoxaline by human

and rat liver microsomes and cytochromes P450 1A2. Chem. Res. Toxicol. **14**, 901–911 (2001)

246. Y. Totsuka, K. Fukutome, M. Takahashi, S. Takashi, A. Tada, T. Sugimura, K. Wakabayashi, Presence of N^2-(deoxyguanosin-8-yl)-2-amino-3,8-dimethylimidazo[4,5-f]quinoxaline (dG-C8-MeIQx) in human tissues. Carcinogenesis **17**, 1029–1034 (1996)

247. K.H. Dingley, K.D. Curtis, S. Nowell, J.S. Felton, N.P. Lang, K.W. Turteltaub, DNA and protein adduct formation in the colon and blood of humans after exposure to a dietary-relevant dose of 2-amino-1-methyl-6-phenylimidazo[4,5-b]pyridine. Cancer Epidemiol. Biomark. Prev. **8**, 507–512 (1999)

248. M.D. Friesen, K. Kaderlik, D. Lin, L. Garren, H. Bartsch, N.P. Lang, F.F. Kadlubar, Analysis of DNA adducts of 2-amino-1-methyl-6-phenylimidazo[4,5-b]pyridine in rat and human tissues by alkaline hydrolysis and gas chromatography/electron capture mass spectrometry: validation by comparison with 32P-postlabeling. Chem. Res. Toxicol. **7**, 733–739 (1994)

249. K.W. Turteltaub, R.J. Mauthe, K.H. Dingley, J.S. Vogel, C.E. Frantz, R.C. Garner, N. Shen, MeIQx-DNA adduct formation in rodent and human tissues at low doses. Mutat. Res. **376**, 243–252 (1997)

250. T.J. Lightfoot, J.M. Coxhead, B.C. Cupid, S. Nicholson, R.C. Garner, Analysis of DNA adducts by accelerator mass spectrometry in human breast tissue after administration of 2-amino-1-methyl-6-phenylimidazo[4,5-b]pyridine and benzo[a]pyrene. Mutat. Res. **472**, 119–127 (2000)

251. J. Zhu, P. Chang, M.L. Bondy, A.A. Sahin, S.E. Singletary, S. Takahashi, T. Shirai, D. Li, Detection of 2-amino-1-methyl-6-phenylimidazo[4,5-b]-pyridine-DNA adducts in normal breast tissues and risk of breast cancer. Cancer Epidemiol. Biomark. Prev. **12**, 830–837 (2003)

252. J. Zhu, A. Rashid, K. Cleary, J.L. Abbruzzese, H. Friess, S. Takahashi, T. Shirai, D. Li, Detection of 2-amino-1-methyl-6-phenylimidazo [4,5-b]-pyridine (PhIP)-DNA adducts in human pancreatic tissues. Biomarkers **11**, 319–328 (2006)

253. C. Magagnotti, R. Pastorelli, S. Pozzi, B. Andreoni, R. Fanelli, L. Airoldi, Genetic polymorphisms and modulation of 2-amino-1-methyl-6-phenylimidazo[4,5-b]pyridine (PhIP)-DNA adducts in human lymphocytes. Int. J. Cancer **107**, 878–884 (2003)

254. M.A. Malfatti, K.H. Dingley, S. Nowell-Kadlubar, E.A. Ubick, N. Mulakken, D. Nelson, N.P. Lang, J.S. Felton, K.W. Turteltaub, The urinary metabolite profile of the dietary carcinogen 2-amino-1-methyl-6-phenylimidazo[4,5-b]pyridine is predictive of colon DNA adducts after a low-dose exposure in humans. Cancer Res. **66**, 10541–10547 (2006)

255. D. Tang, J.J. Liu, A. Rundle, C. Neslund-Dudas, A.T. Savera, C.H. Bock, N.L. Nock, J.J. Yang, B.A. Rybicki, Grilled meat consumption and PhIP-DNA adducts in prostate carcinogenesis. Cancer Epidemiol. Biomark. Prev. **16**, 803–808 (2007)

256. E.E. Bessette, S.D. Spivack, A.K. Goodenough, T. Wang, S. Pinto, F.F. Kadlubar, R.J. Turesky, Identification of carcinogen DNA adducts in human saliva by linear quadrupole ion trap/multistage tandem mass spectrometry. Chem. Res. Toxicol. **23**, 1234–1244 (2010)

257. J.K. Kim, M.A. McCormick, C.M. Gallaher, D.D. Gallaher, S.P. Trudo, Apiaceous vegetables and cruciferous phytochemicals reduced PhIP-DNA adducts in prostate but not in pancreas of Wistar rats. J. Med. Food **21**, 199–202 (2018)

258. K.H. Dingley, E.A. Ubick, M.L. Chiarappa-Zucca, S. Nowell, S. Abel, S.E. Ebeler, A.E. Mitchell, S.A. Burns, F.M. Steinberg, A.J. Clifford, Effect of dietary constituents with chemopreventive potential on adduct formation of a low dose of the heterocyclic amines PhIP and IQ and phase II hepatic enzymes. Nutr. Cancer **46**, 212–221 (2003)

259. C.L. Archer, P. Morse, R.F. Jones, T. Shirai, G.P. Haas, C.Y. Wang, Carcinogenicity of the N-hydroxy derivative of 2-amino-1-methyl-6-phenylimidazo[4,5-b] pyridine, 2-amino-3, 8-dimethyl-imidazo[4,5-f]quinoxaline and 3, 2′-dimethyl-4-aminobiphenyl in the rat. Cancer Lett. **155**, 55–60 (2000)

260. G.R. Stuart, J. Holcroft, J.G. de Boer, B.W. Glickman, Prostate mutations in rats induced by the suspected human carcinogen 2-amino-1-methyl-6-phenylimidazo[4,5-b]pyridine. Cancer Res. **60**, 266–268 (2000)

261. Y. Nakai, W.G. Nelson, A.M. De Marzo, The dietary charred meat carcinogen 2-amino-1-methyl-6-phenylimidazo[4,5-b]pyridine acts as both a tumor initiator and promoter in the rat ventral prostate. Cancer Res. **67**, 1378–1384 (2007)

262. G. Li, H. Wang, A.B. Liu, C. Cheung, K.R. Reuhl, M.C. Bosland, C.S. Yang, Dietary carcinogen 2-amino-1-methyl-6-phenylimidazo[4,5-b]pyridine-induced prostate carcinogenesis in CYP1A-humanized mice. Cancer Prev. Res. **5**, 963–972 (2012)

263. C. Chen, X. Ma, M.A. Malfatti, K.W. Krausz, S. Kimura, J.S. Felton, J.R. Idle, F.J. Gonzalez, A comprehensive investigation of 2-amino-1-methyl-6-phenylimidazo[4,5-b]pyridine (PhIP) metabolism in the mouse using a multivariate data analysis approach. Chem. Res. Toxicol. **20**, 531–542 (2007)

264. J.X. Chen, G. Li, H. Wang, A. Liu, M.J. Lee, K. Reuhl, N. Suh, M.C. Bosland, C.S. Yang, Dietary tocopherols inhibit PhIP-induced prostate carcinogenesis in CYP1A-humanized mice. Cancer Lett. **371**, 71–78 (2016)

265. D.S. Oliveira, S. Dzinic, A.I. Bonfil, A.D. Saliganan, S. Sheng, R.D. Bonfil, The mouse prostate: a basic anatomical and histological guideline. Bosn. J. Basic Med. Sci. **16**, 8–13 (2016)

266. A.D. Borowsky, K.H. Dingley, E. Ubick, K.W. Turteltaub, R.D. Cardiff, R. Devere-White,

Inflammation and atrophy precede prostatic neoplasia in a PhIP-induced rat model. Neoplasia **8**, 708–715 (2006)

267. Y. Nakai, N. Nonomura, Inflammation and prostate carcinogenesis. Int. J. Urol. **20**, 150–160 (2013)

268. C.D. Chen, D.S. Welsbie, C. Tran, S.H. Baek, R. Chen, R. Vessella, M.G. Rosenfeld, C.L. Sawyers, Molecular determinants of resistance to antiandrogen therapy. Nat. Med. **10**, 33–39 (2004)

269. P.J. Richmond, A.J. Karayiannakis, A. Nagafuchi, A.V. Kaisary, M. Pignatelli, Aberrant E-cadherin and alpha-catenin expression in prostate cancer: correlation with patient survival. Cancer Res. **57**, 3189–3193 (1997)

270. S. Signoretti, D. Waltregny, J. Dilks, B. Isaac, D. Lin, L. Garraway, A. Yang, R. Montironi, F. McKeon, M. Loda, p63 is a prostate basal cell marker and is required for prostate development. Am. J. Pathol. **157**, 1769–1775 (2000)

271. J.S. Ross, H.L. Figge, H.X. Bui, A.D. del Rosario, H.A. Fisher, T. Nazeer, T.A. Jennings, R. Ingle, D.N. Kim, E-cadherin expression in prostatic carcinoma biopsies: correlation with tumor grade, DNA content, pathologic stage, and clinical outcome. Mod. Pathol. **7**, 835–841 (1994)

272. G. Gupta-Elera, A.R. Garrett, R.A. Robison, K.L. O'Neill, The role of oxidative stress in prostate cancer. Eur. J. Cancer Prev. **21**, 155–162 (2012)

273. A. Di Cristofano, P.P. Pandolfi, The multiple roles of PTEN in tumor suppression. Cell **100**, 387–390 (2000)

274. X. Wang, K.D. McCullough, T.F. Franke, N.J. Holbrook, Epidermal growth factor receptor-dependent Akt activation by oxidative stress enhances cell survival. J. Biol. Chem. **275**, 14624–14631 (2000)

275. P. Cairns, K. Okami, S. Halachmi, N. Halachmi, M. Esteller, J.G. Herman, J. Jen, W.B. Isaacs, G.S. Bova, D. Sidransky, Frequent inactivation of PTEN/MMAC1 in primary prostate cancer. Cancer Res. **57**, 4997–5000 (1997)

276. D. Tang, J.J. Liu, C.H. Bock, C. Neslund-Dudas, A. Rundle, A.T. Savera, J.J. Yang, N.L. Nock, B.A. Rybicki, Racial differences in clinical and pathological associations with PhIP-DNA adducts in prostate. Int. J. Cancer **121**, 1319–1324 (2007)

277. D. Tang, O.N. Kryvenko, Y. Wang, S. Trudeau, A. Rundle, S. Takahashi, et al., 2-amino-1-methyl-6-phenylimidazo[4,5-b]pyridine (PhIP)-DNA adducts in benign prostate and subsequent risk for prostate cancer. Int. J. Cancer **133**, 961–971 (2013)

278. B.H. Yun, S. Xiao, L. Yao, S. Krishnamachari, T.A. Rosenquist, K.G. Dickman, A.P. Grollman, P. Murugan, C.J. Weight, R.J. Turesky, A rapid throughput method to extract DNA from formalin-fixed paraffin-embedded tissues for biomonitoring carcinogenic DNA adducts. Chem. Res. Toxicol. **30**, 2130–2139 (2017)

279. A.E. Norrish, L.R. Ferguson, M.G. Knize, J.S. Felton, S.J. Sharpe, R.T. Jackson, Heterocyclic amine content of cooked meat and risk of prostate cancer. J. Natl. Cancer Inst. **91**, 2038–2044 (1999)

280. S. Rohrmann, K. Nimptsch, R. Sinha, W.C. Willett, E.L. Giovannucci, E.A. Platz, K. Wu, Intake of meat mutagens and risk of prostate cancer in a cohort of U.S. health professionals. Cancer Epidemiol. Biomark. Prev. **24**, 1557–1563 (2015)

281. A.D. Joshi, R. Corral, C. Catsburg, J.P. Lewinger, J. Koo, E.M. John, S.A. Ingles, M.C. Stern, Red meat and poultry, cooking practices, genetic susceptibility and risk of prostate cancer: results from a multiethnic case-control study. Carcinogenesis **33**, 2108–2118 (2012)

282. S. Sharma, X. Cao, L.R. Wilkens, J. Yamamoto, A. Lum-Jones, B.E. Henderson, L.N. Kolonel, M.L. Le, Well-done meat consumption, NAT1 and NAT2 acetylator genotypes and prostate cancer risk: the multiethnic cohort study. Cancer Epidemiol. Biomark. Prev. **19**, 1866–1870 (2010)

283. A. Sander, J. Linseisen, S. Rohrmann, Intake of heterocyclic aromatic amines and the risk of prostate cancer in the EPIC-Heidelberg cohort. Cancer Causes Control **22**, 109–114 (2011)

284. S. Takahashi, S. Tamano, M. Hirose, N. Kimoto, Y. Ikeda, M. Sakakibara, M. Tada, F.F. Kadlubar, N. Ito, T. Shirai, Immunohistochemical demonstration of carcinogen-DNA adducts in tissues of rats given 2-amino-1-methyl-6-phenylimidazo[4,5-b] pyridine (PhIP): detection in paraffin-embedded sections and tissue distribution. Cancer Res. **58**, 4307–4313 (1998)

285. M. Bellamri, S. Xiao, P. Murugan, C.J. Weight, R.J. Turesky, Metabolic activation of the cooked meat carcinogen 2-amino-1-methyl-6-phenylimidazo[4,5-b]pyridine in human prostate. Toxicol. Sci. **163**, 543–556 (2018)

286. C.P. Nelson, L.C. Kidd, J. Sauvageot, W.B. Isaacs, A.M. De Marzo, J.D. Groopman, W.G. Nelson, T.W. Kensler, Protection against 2-hydroxyamino-1-methyl-6-phenylimidazo[4,5-b]pyridine cytotoxicity and DNA adduct formation in human prostate by glutathione S-transferase P1. Cancer Res. **61**, 103–109 (2001)

287. C.Y. Wang, M. Debiec-Rychter, H.A. Schut, P. Morse, R.F. Jones, C. Archer, C.M. King, G.P. Haas, N-acetyltransferase expression and DNA binding of N-hydroxyheterocyclic amines in human prostate epithelium. Carcinogenesis **20**, 1591–1595 (1999)

288. M. Glass-Holmes, B.J. Aguilar, R.D. Gragg, S. Darling-Reed, C.B. Goodman, Characterization of 2-amino-1-methyl-6-phenylimida zo[4,5b]pyridine at androgen receptor: mechanistic support for its role in prostate cancer. Am. J. Cancer Res. **5**, 191–200 (2015)

289. S.K. Creton, H. Zhu, N.J. Gooderham, The cooked meat carcinogen 2-amino-1-methyl-6-phenylimidazo[4,5-b]pyridine activates the extracellular signal regulated kinase mitogen-activated

protein kinase pathway. Cancer Res. **67**, 11455–11462 (2007)

290. E. Cano, L.C. Mahadevan, Parallel signal processing among mammalian MAPKs. Trends Biochem. Sci. **20**, 117–122 (1995)

291. D. Hanahan, R.A. Weinberg, The hallmarks of cancer. Cell **100**, 57–70 (2000)

292. J. Petimar, K.M. Wilson, K. Wu, M. Wang, D. Albanes, P.A. van den Brandt, M.B. Cook, G.G. Giles, E.L. Giovannucci, G.E. Goodman, P.J. Goodman, N. Hakansson, K. Helzlsouer, et al., A pooled analysis of 15 prospective cohort studies on the association between fruit, vegetable, and mature bean consumption and risk of prostate cancer. Cancer Epidemiol. Biomark. Prev. **26**, 1276–1287 (2017)

293. J.H. Ahn-Jarvis, S.K. Clinton, E.M. Grainger, K.M. Riedl, S.J. Schwartz, M.L. Lee, R. Cruz-Cano, G.S. Young, G.B. Lesinski, Y. Vodovotz, Isoflavone pharmacokinetics and metabolism after consumption of a standardized soy and soy-almond bread in men with asymptomatic prostate cancer. Cancer Prev. Res. (Phila.) **8**, 1045–1054 (2015)

294. E.M. Grainger, C.W. Hadley, N.E. Moran, K.M. Riedl, M.C. Gong, K. Pohar, S.J. Schwartz, S.K. Clinton, A comparison of plasma and prostate lycopene in response to typical servings of tomato soup, sauce or juice in men before prostatectomy. Br. J. Nutr. **114**, 596–607 (2015)

Genetic, Environmental, and Nuclear Factors Governing Genomic Rearrangements

Susmita G. Ramanand and Ram S. Mani

Introduction

Genomic rearrangements are a characteristic of cancer genomes. The field of cancer genomics underwent a paradigm shift in 2005 when recurrent chromosomal rearrangements, previously thought to be largely unique to lymphomas and leukemias, were discovered in prostate cancer (PCa), a solid tumor [1]. Since then, in addition to knowing *TMPRSS2-ERG* gene fusions to be the most common in PCa, seen in over 50% of cases from early stage to metastatic, insights into the gene fusions and their causes have increased tremendously. As a consequence of *TMPRSS2-ERG* gene fusion, the transcription factor ERG is over-expressed in all stages of PCa, including castration-resistant prostate cancer (CRPC). The 5′ fusion partner genes typically represent androgen regulated, transcriptionally active genes (e.g. *TMPRSS2, SLC45A3*), and the 3′ fusion partner genes represent proto-oncogenes—primarily represented by the ETS family of transcription

factors (*ERG, ETV1, ETV4 and ETV5*) [1–4]. Conceptually, the 5′ partner genes provide the promoter-enhancer elements which drive the over-expression of the 3′ partner genes, thereby contributing to oncogenic transformation. While ETS fusions are the most prevalent and observed in ~60% of PCa, non-ETS fusions involving members of the RAF kinase pathway are also observed in a small subset of PCa (~1%) [5]. Although the predominant 3′ partner genes like *ERG* are rearranged in other cancers, the rearrangements of the major 5′ partner genes (e.g. *TMPRSS2, SLC45A3*) are unique to PCa. These gene fusions are causally associated with PCa etiology and serve as diagnostic biomarkers as well as candidates for therapeutic targeting. The ETS gene fusions are early events in PCa development and can thereby serve as markers for clonal heterogeneity. Thus, it is important to understand the origins of recurrent genomic rearrangements that contribute to gene fusion formation in PCa.

Multiple studies have identified diverse factors that underlie the formation of PCa genomic rearrangements, ranging from outside environmental stressors to genetic and nuclear factors [6]. By integrating these studies, we suggest that the factors triggering genomic rearrangements can broadly impact two features—the formation of DNA double strand breaks (DSBs), and misrepair of DNA DSBs (Fig. 1). In this chapter we focus on the individual triggering factors and highlight the interplay between these factors in the context of PCa.

S. G. Ramanand
Department of Pathology, UT Southwestern Medical Center, Dallas, TX, USA

R. S. Mani (✉)
Department of Pathology, UT Southwestern Medical Center, Dallas, TX, USA

Department of Urology, UT Southwestern Medical Center, Dallas, TX, USA

Harold C. Simmons Comprehensive Cancer Center, UT Southwestern Medical Center, Dallas, TX, USA
e-mail: ram.mani@utsouthwestern.edu

© Springer Nature Switzerland AG 2019
S. M. Dehm, D. J. Tindall (eds.), *Prostate Cancer*, Advances in Experimental Medicine and Biology 1210, https://doi.org/10.1007/978-3-030-32656-2_3

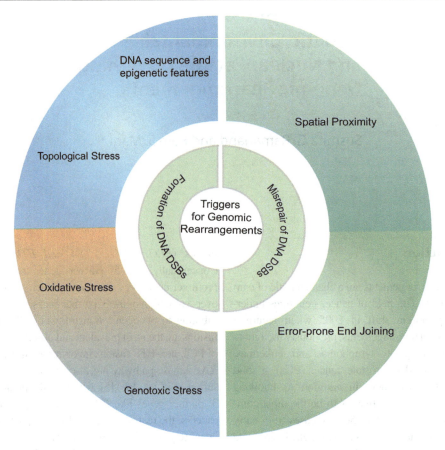

Fig. 1 Triggers for genomic rearrangements. Genomic rearrangements can be triggered by a plethora of factors that are summarized into distinct categories. The onset of genomic rearrangements is preceded by the formation of DNA DSBs and their mis-repair, as depicted in the inner circle. The outer circle shows factors that influence the formation of DNA DSBs or their mis-repair

Formation of DNA Double Strand Breaks (DSBs)

A factor that is critical for gene rearrangements is the formation of DNA DSBs in the genome. These breaks are sensed by the DNA damage response (DDR) signaling pathways, which then facilitate the accurate repair of DSBs by recruiting the appropriate DNA repair mechanisms. Accumulation of DSBs can at times result in their mis-repair, leading to the formation of genomic rearrangements. Thus, it is important to understand the processes and pathways contributing to the formation of recurrent DNA DSBs in PCa development.

Cellular Stressors

Various forms of cellular stress damage the genome and contribute to the formation of DNA DSBs. These cellular stressors include genotoxic stress, oxidative stress, topological stress, replication stress, metabolic stress, among others. A multitude of cellular stressors have been implicated in the formation of DSBs that underlie the formation of recurrent gene fusions in PCa.

Genotoxic Stress

Genotoxic stressors like ionizing radiation (IR) can induce DNA DSBs in cells. The LNCaP and LAPC-4 PCa cell lines are androgen responsive, but are negative for TMPRSS2-ERG gene fusions. These two cell lines are commonly used to study the formation of TMPRSS2-ERG gene fusions. Multiple studies have shown that treatment of androgen responsive cells with the androgen receptor (AR) ligand, dihydrotestosterone (DHT) and IR can result in the robust formation of TMPRSS2-ERG gene fusions [7, 8]. Mechanistically, androgen signaling induces the proximity of TMPRSS2 and ERG genes (to be discussed later), while IR contributes to DNA DSBs—higher IR doses are associated with an increase in the formation of gene fusions. Although IR is a potent extrinsic source of DNA damage, it is not likely to contribute to DNA DSBs involved in PCa gene rearrangements. This is because the prostate is an internal organ and men with PCa are not usually exposed to IR. Thus, alternate genotoxic stressors are likely to contribute to DNA breaks in PCa development.

Oxidative Stress

In addition to direct effects leading to DNA damage, IR treatment also induces the formation of reactive oxygen species (ROS). Increased levels of intracellular ROS contributes to oxidative stress, which in turn promotes DNA damage and the formation of DNA DSBs. ROS can also damage DNA and produce single-strand DNA breaks (SSBs). Clustering of SSBs can result in the spontaneous formation of DSBs. Approximately 1% of single strand lesions are converted to DSBs in normal human cells per cell cycle [9]. A combination of environmental and cellular factors can culminate in oxidative stress. For instance, chronic inflammation is considered a risk factor for PCa initiation [10, 11]. Inflammation induces oxidative stress leading to the formation of DNA DSBs in *TMPRSS2* and *ERG* loci, thereby increasing the likelihood of *TMPRSS2-ERG* gene fusion formation [12]. Also, environment plays an important role in triggering inflammation, as well as factors such as diet, lifestyle, infection, and aging. Thus, oxidative stress is likely to be a physiologically relevant source of DNA DSBs that contributes to PCa-associated genomic rearrangements.

The prostate specific homeobox gene, *NKX3.1*, is required for normal prostate differentiation [13, 14]. Studies employing genetically engineered mouse models and tissue recombination assays indicate that loss of *Nkx3.1* function is a critical event in mouse PCa initiation, with various *Nkx3.1* mutant mouse models recapitulating early stages of prostate carcinogenesis [15–18]. Consistent with its role as a tumor suppressor, decreased NKX3.1 protein expression is associated with focal prostatic atrophy, prostatic intraepithelial neoplasia (PIN), and adenocarcinoma in humans [19, 20]. Mechanistically, loss of *Nkx3.1* in mice contributes to prostate carcinogenesis in mice, in part, by increasing oxidative damage to DNA and proteins [21]. *Nkx3.1* loss associated oxidative stress in mice is possibly due to deregulated expression of several antioxidant and prooxidant enzymes, including glutathione peroxidase 2 and 3 (*GPx2* and *GPx3*), peroxiredoxin 6 (*Prdx6*), and sulfhydryl oxidase Q6 (*Qscn6*). Although inflammation can induce oxidative stress via multiple mechanisms [22–26], the observation that inflammatory cytokines like TNF-α and interleukin-1β can induce ubiquitination and proteasomal degradation of NKX3.1 provides a circumstantial link between inflammation and oxidative stress in the context of the prostate gland [27]. Down-regulation of NKX3.1 enhances the formation of TMPRSS2-ERG gene rearrangements in LNCaP cells upon treatment with various cellular stressors like IR, etoposide and doxorubicin [28]. In addition to NKX3.1 down-regulation, disruption of the Nrf2-antioxidant axis has been linked to increased oxidative stress and DNA damage that may lead to the initiation of cellular transformation in the prostate gland [29]. As ETS gene fusions are early events in PCa etiology, these studies suggest that oxidative stress is a significant source of DNA DSBs in the prostate gland and is a likely contributor to the formation of recurrent gene fusions.

Topological Stress

AR, the central effector of the androgen signaling pathway is a master transcription factor that is necessary for the normal development and differentiation of the prostate gland. Paradoxically, AR also contributes to PCa development and is mis-regulated in every stage of PCa progression from clinically localized PCa to lethal metastatic castration resistant prostate cancer (mCRPC). Upon ligand binding, AR undergoes dimerization and relocates from the cytoplasm to the nucleus where it cooperates with pioneer transcription factors like FOXA1 and binds to enhancers that contain androgen response elements (AREs) [30, 31]. Chromatin bound AR recruits co-regulators and activates a sub-set of these enhancers, which in turn are hypothesized to interact with other enhancers and promoters to regulate the activity of RNA polymerase II (RNA Pol II) [32, 33]. Androgen signaling-associated 3D genome organization and transcriptional regulation contribute to topological stress in the genome. The enzyme DNA topoisomerase II beta (TOP2B) resolves topological stress by the formation of transient DNA DSBs that allow DNA strands to pass through one another, followed by immediate rejoining of the DSBs [34, 35]. TOP2B catalytic activity is required for AR mediated transcriptional regulation [36]. In addition to gene promoters, TOP2B binding is also observed at the boundaries of topologically associating domains (TADs) that are marked by occupancy of structural proteins like CTCF and cohesin [37, 38]. Defects in TOP2B activity can result in the formation of persistent DNA DSBs that can serve as substrates for the formation of genomic rearrangements. The strong co-expression of AR and TOP2B in prostatic intraepithelial neoplasia (PIN) cells, the PCa precursor lesion, highlights the clinical relevance of topological stress in the formation of DNA DSBs [36].

In addition to topological stress via changes in 3D genome organization, elevated transcription can result in the establishment of R loops formed by the hybridization of nascent RNA with the template DNA [39, 40]. During R loop formation, the non-template DNA strand loops out in to a single stranded structure that is vulnerable to DNA damage, leading to DNA DSBs. Multiple cellular mechanisms regulate R loop formation. Active transcription is associated with the accumulation of negative supercoils behind the elongating RNA Pol II. These negative supercoils

promote the opening of double-stranded DNA and facilitate the annealing of nascent RNA to the DNA template strand and subsequent R loop formation by mechanisms that are not completely understood. The DNA topoisomerases, including TOP1 and TOP3B, suppress R loop formation by reducing the negative supercoiling of DNA [41–43]. The Ribonucleases H (RNase H) enzymes resolve R loops by specifically hydrolyzing the RNA in the RNA/DNA hybrids. Thus, defective topoisomerase activity can contribute to topological stress and DNA DSBs in the *TMPRSS2* gene, which is highly expressed in the prostate.

Interplay Between Cellular Stressors

Oxidative stress is a common underlying theme for many cellular stressors. The cellular effects of exposure to IR are partially mediated by oxidative stress via ROS formation [44, 45]. IR-induced ROS formation can occur within minutes and can be sustained for several days [46]. Oxidative stress by ROS can influence both the short-term as well as long-term DNA damaging effects of IR. Treatment of human cells with hydrogen peroxide (H_2O_2), a ROS, triggers topoisomerase II (TOP2) mediated DNA damage [47]. Thus, inflammation-induced oxidative stress can potentially trigger AR induced DNA DSBs via TOP2B poisoning. The transcription-induced, R-loop associated single stranded structures can undergo damage from oxidative stress that can result in DNA DSBs when left unrepaired [48, 49]. In summary, a complex interplay between various cellular stressors create DNA DSBs that when mis-repaired can result in genomic rearrangements.

DNA Sequence and Epigenetic Features

DNA sequence and epigenetic features dictate the formation of DNA DSBs by cellular stressors [6]. Although the transcription-topological stress axis can explain the formation of DNA DSBs in *TMPRSS2*, it is important to note that the *ERG* gene is not expressed in the prostate gland. Gene fusions result in the transcriptional up-regulation of *ERG*. Oxidative stress-induced DNA DSBs in the *ERG* locus are likely to be independent of androgen signaling or transcription [12]. The structure of the *ERG* gene and its chromatin environment influence the formation of *ERG* gene fusions. Intron 3 of *ERG* is the most commonly rearranged intron in the gene. This intron spans ~130 Kb and is one of the longest introns in the human genome. A significant proportion of oxidative stress–induced DNA DSBs in the *ERG* gene overlap with DNase I hypersensitivity sites (DHSs); oxidative stress induced DNA DSBs in the *TMPRSS2* gene do not significantly overlap with DHSs. Thus, the open chromatin sites within the *ERG* locus are vulnerable to the formation of DNA DSBs [12]. More generally, the large intron 3 size in *ERG* affords the enrichment of epigenetic features that influence the formation of DNA DSBs in a transcription-independent manner.

Mis-Repair of DNA DSBs

Human cells are constantly exposed to endogenous and exogenous DNA damaging agents. Estimates suggest that ~10–50 DNA DSBs are formed in normal human cells per cell cycle [9, 50]. Despite the constant threat of DNA DSBs, our cells efficiently fix these lesions by using various DNA repair pathways. Homologous recombination (HR) and the non-homologous end joining (NHEJ) are the two major mechanisms for the repair of DNA DSBs. However, occasionally, DNA DSBs can be mis-repaired to form genomic rearrangements. Multiple mechanisms can account for the mis-repair of DNA DBSs. For example, a significant increase in the formation of DSBs can trump the DNA repair machinery and increase the likelihood of genomic rearrangements. These rearrangements are also influenced by the spatial proximity of DNA DSBs and can be further accentuated by defects in DNA repair.

Spatial Proximity

The proximity of gene fusion partners is a critical factor underlying the formation of recurrent gene fusions. *TMPRSS2-ERG* gene fusions represent-

ing intra-chromosomal gene rearrangements are observed in >50% of PCa—the two genes are located 3 Mb apart on human chromosome 21q22.2. The *TMPRSS2* and *ERG* genes also form inter-chromosomal gene fusions, albeit at much lower frequencies. Additional inter-chromosomal gene fusions include TMPRSS2-ETV1 and SLC45A4-ERG, which are observed in <5% of PCa [2, 51]. The low frequency of inter-chromosomal ETS gene fusions in comparison to intra-chromosomal ETS gene fusions is not due to cancer associated evolutionary selection, rather it reflects the frequency of formation of these oncogenic lesions. This conclusion is supported by the observation that oxidative stress induces *TMPRSS2-ERG* gene fusions at a much higher rate than *TMPRSS2-ETV1* gene fusions in cell-based models [12]. Thus, proximity by location in the chromosomal arm contributes, in part, to the high prevalence of TMPRSS2-ERG gene fusions. In addition to proximity by location, stimulation of LNCaP cells with the AR ligand, DHT, juxtaposes the *TMPRSS2* and *ERG* genes—this effect is not observed in AR negative cells [7]. Multiple independent studies have demonstrated androgen-induced proximity of *TMPRSS2* and *ERG* genes in AR positive cells, although the detailed mechanisms are still not clear [7, 8, 28, 52, 53].

The presence of intrinsically disordered regions (IDRs) in many transcription factors and coactivator proteins can promote dynamic and cooperative interactions among components of the transcriptional machinery via liquid-liquid phase separation [54–59]. In the context of the prostate gland, AR facilitates transcriptional regulation by promoting long-range chromatin interactions between promoters, enhancers and other regulatory elements [33]. Thus, transcriptional regulation by AR can be a significant contributor to the three-dimensional proximity of *TMPRSS2* and *ERG* genes, although additional mechanisms, such as the regulation of replication timing, may also play a role [53, 60]. Mechanistically, the proximity of gene fusion partner loci increases the probability of mis-repair and gene fusion formation when both the loci simultaneously harbor DNA DSBs. These studies explain why stimulation of LNCaP cells with DHT and IR results in the robust formation of *TMPRSS2-ERG* gene fusions.

Error-Prone End-Joining and the Formation of Genomic Rearrangements

NHEJ is the predominant DNA repair pathway involved in the repair of DNA DSBs as it operates during all stages of the cell cycle; HR, a template-dependent pathway, is restricted to S and G2 phases of the cell cycle. As NHEJ is a template-independent pathway, it is favored in the G1 phase for repair of DNA DSBs. The luminal epithelial cells of the prostate express higher levels of AR and TMPRSS2 in comparison to the basal epithelial cells. The ETS gene fusions are hypothesized to originate in the luminal cells, although basal cells can show luminal-like properties when appropriate transcription factors like *NKX3.1* are activated [14]. The prostate luminal cells are post mitotic and prostate basal cells have a very low mitotic index. For these reasons, replication stress and cell division are unlikely to be a major source of recurrent DNA DSBs. Additionally, at any given time, the vast majority of luminal and basal cells in the prostate will be in G1 phase, suggesting that HR is unlikely to be the major DNA repair pathway in pre-malignant prostate epithelial cells. For these reasons, NHEJ is the primary pathway for the repair of DNA DSBs in normal prostate epithelial cells.

Aging, the most significant risk factor for prostate cancer development, is linked to inflammation and oxidative DNA damage. The NHEJ pathway efficiently repairs most of the DNA DSBs in prostate cells. The simultaneous presence of two or more DNA DSBs in close spatial proximity can occasionally result in the mis-repair of these lesions to form genomic rearrangements. The CRISPR-Cas9 system has been employed to engineer *de novo* TMPRSS2-ERG gene fusions in LNCaP cells. Using this system, it was shown that knock-down of NHEJ components such as PRKDC, PAXX, Artemis, KU70, KU80, XRCC4, LIG4, NHEJ1, XPF, 53BP1, and WRN blocked TMPRSS2-ERG gene fusion formation [61]. TMPRSS2-ERG genomic breakpoint analysis of clinical PCa specimens have revealed that most junctions are either blunt or display short microhomologies of 1–4 bp [62].

These studies indicate an essential role for NHEJ in the mis-repair of co-existing DNA DSBs, leading to the formation of gene fusions. Although inherited mutations in HR DNA repair genes were observed in 4.6% of men with localized PCa, it is not clear if these men have a higher incidence of gene fusions [63]. Inherited mutations in NHEJ DNA repair genes are not common in localized PCa. We conclude that the error-prone end-joining ultimately resulting in ETS gene fusions are not due to defects in NHEJ per se, but rather are due to a preponderance of DNA DSBs in the context of 3D genome architecture.

Novel Mediators of ETS Gene Rearrangements

Studies on human PCa specimens and cell-based model systems have identified novel mediators of ETS gene rearrangements. The *CHD1* gene, which encodes an ATP-dependent chromatin-remodeling enzyme, is deleted in ~8% of PCa specimens [64–67]. Independent studies have shown an association between *CHD1* deletions and the absence of *ETS* gene fusions [64, 68]. Inactivation of *CHD1* blocks the formation of *ERG* rearrangements in LNCaP cells treated with the topoisomerase inhibitor, doxorubicin [68]. This is presumably due to the impairment of AR dependent transcription.

BRD2 and BRD4 are members of the bromodomain and extraterminal (BET) family of chromatin readers which bind to acetylated proteins like histones to regulate transcription and promote DNA repair. Treatment with BET inhibitors (BETi), or siRNA-based knock-down of BRD2 and BRD4 blocked NHEJ activity as well as CRISPR-Cas9 mediated TMPRSS2-ERG formation [61]. When compared to either normal prostate or SPOP-mutated PCa, the expression of BRD4 is significantly elevated in ERG fusion positive PCa [61]. In addition to DNA repair, BET proteins play an essential role in AR dependent transcriptional regulation [69]. Thus BET proteins participate in at least two stages in the formation of gene fusions: (a) mediating AR dependent transcription, and (b) facilitating the end-joining steps of DNA repair.

As BRD4 also has an essential role in the B cell class switch recombination (CSR) [70], these studies support the concept that mediators of normal cellular processes are often co-opted and hijacked to promote malignancy.

In PCa specimens, the presence of *TMPRSS2-ERG* gene fusions is highly correlated with lower levels of NKX3.1 protein expression [28]. In addition to its established role in protection against oxidative damage [21], NKX3.1 is an upstream regulator of AR signaling and luminal cell lineage identity [14]. Furthermore, NKX3.1 participates in DNA damage response signaling by accelerating ATM activation [71, 72]. Thus, it is conceivable that loss of NKX3.1 function can influence various steps in the formation of ETS gene fusions.

The Formation of Complex Genomic Rearrangements by Chromoplexy

The advent of Next Generation Sequencing (NGS) has uncovered several aspects of structural variations that are characteristic of PCa. Analysis of whole genome sequencing (WGS) and their copy number profiles in PCa specimens enabled identification of a unique set of events that are oncogenic in nature, i.e. large chains of rearrangements in multiple chromosomes whose effects promote tumorigenesis [65, 73, 74]. Computational modeling of these events or rearrangements from WGS and copy number profiling predicted that their origins were similar to chromothripsis [75], and that the chromosomal disarray was a result of cumulative effects of a handful of events that occurred during the course of tumor development. This phenomenon of complex genome rearrangements, which is seen predominantly in PCa, termed "chromoplexy", originates from the Greek word *pleko*, meaning 'to braid'. Chromoplexy is generally characterized by the formation of simultaneous co-dependent inter-chromosomal and intra-chromosomal rearrangements, resulting in translocations and deletions, computationally predicted to have all originated within a single cell cycle (Fig. 2). A feature of chromoplexy is the presence of 'chains' where the breakpoints of rearrangements involving cancer gene loci map to the reference genome near breakpoints from other rearrangements that have taken place at the same time. Such characteristic breakpoint distributions appear to reflect collections of broken DNA ends that are shuffled and ligated to one another in an aberrant configuration. This gives rise to deletion bridges, which in the context of chromoplexy are not simple deletions that bring cis-adjoining regions from the same chromosome together, rather these represent multiple rearrangements that are ultimately connected as closed chain events.

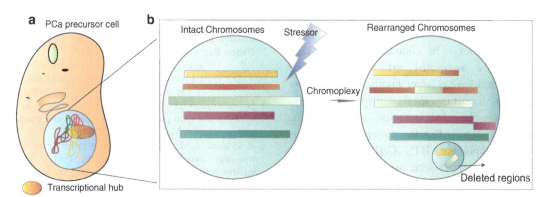

Fig. 2 Chromoplexy. Cancer genomes can evolve in spurts by chromoplexy, a type of genomic rearrangement which is typically associated with transcription-associated 'open-chromatin'. Panel **a** depicts a normal cell in homeostasis where processes like metabolism and transcription are taking place. Panel **b** shows a nucleus, which when exposed to external stressors, swiftly gets rearranged to form several rearrangements involving deletions and translocations

While chromothripsis and chromoplexy are both formed as a result of a catastrophic event, resulting in genomic breakpoints that are typically repaired by NHEJ, in the former, there are hundreds of breakpoints formed in one or two chromosomes, and in the latter, the breakpoints are not clustered and are seen in chains of rearrangements, numbering in tens, and spanning multiple chromosomes. Also, it is interesting to note that chromothripsis is usually seen as a single event that occurs in the earlier stages of cancer, whereas chromoplexy can occur at multiple times during cancer evolution, as evidenced by clonal heterogeneity. Many sequential events involving chromoplectic rearrangements, over a period of time, produce genomic dysregulation in varying degrees.

The origin of chromoplexy is not yet clear, but is thought to occur as a result of DNA damage induced by transcription factor binding. It is likely to be driven by occurrence of DSBs, mediated by transcriptional topological stress in sites that are co-localized with AR, which in the restrictive confines of the nucleus leads to a chain of events preceded by shattering of neighboring chromosomes followed by reassembly, and ultimately multiple translocations, and even deletions.

Chromoplexy can give rise to distinct molecular subtypes. This is well exemplified by the occurrence of TMPRSS2-ERG fusion in ETS$^+$ CHD1^{WT} and ETS$^-$ CHD1^{del} tumors. Characteristically, rearrangements in ETS$^+$ CHD1^{WT} tumors are typically proximal to highly expressed loci, and are inter-chromosomal; display features of chromoplexy. In contrast, ETS$^-$ CHD1^{del} tumors are associated with heterochromatin regions and have a larger number of intra-chromosomal rearrangements displaying chromothripsis-like characteristics. An interesting feature to note is that the pattern of rearrangement breakpoints in ETS$^+$ CHD1^{WT} support the notion of a transcription-associated mechanism for chromoplexy. Chromoplexy, in addition to favoring the rearrangement of growth-promoting genes, can also inactivate multiple tumor suppressor genes coordinately. Thus, chromoplexy can give rise to punctuated tumor evolution in PCa and possibly other malignancies.

Conclusions and Future Directions

Recurrent genomic rearrangements involving the ETS transcription factors are causally associated with PCa etiology. We have outlined the various steps involved in the formation of ETS gene fusions. By compiling the multitude of scientific literature over the past two decades, we have presented the key cellular pathways, molecular players and environmental triggers that regulate the various steps in the formation of ETS gene fusions. Understanding the origins of genomic rearrangements can shed light on tumor etiology and also pave the way to develop strategies to prevent, delay, and treat PCa.

Acknowledgements We thank the members of the Mani laboratory for insights and comments; Peter Ly for discussions and manuscript edits. R.S.M. acknowledges funding support from the NIH Pathway to Independence (PI) Award (R00CA160640), the CPRIT Individual Investigator Research Award (RP190454), and the US Department of Defense Prostate Cancer Research Program (PCRP)—Impact Award (W81XWH-17-1-0675).

References

1. S.A. Tomlins et al., Recurrent fusion of TMPRSS2 and ETS transcription factor genes in prostate cancer. Science **310**(5748), 644–648 (2005)
2. S.A. Tomlins et al., Distinct classes of chromosomal rearrangements create oncogenic ETS gene fusions in prostate cancer. Nature **448**(7153), 595–599 (2007)
3. S.A. Tomlins et al., TMPRSS2:ETV4 gene fusions define a third molecular subtype of prostate cancer. Cancer Res. **66**(7), 3396–3400 (2006)
4. B.E. Helgeson et al., Characterization of TMPRSS2:ETV5 and SLC45A3:ETV5 gene fusions in prostate cancer. Cancer Res. **68**(1), 73–80 (2008)
5. N. Palanisamy et al., Rearrangements of the RAF kinase pathway in prostate cancer, gastric cancer and melanoma. Nat. Med. **16**(7), 793–798 (2010)
6. R.S. Mani, A.M. Chinnaiyan, Triggers for genomic rearrangements: insights into genomic, cellular and environmental influences. Nat. Rev. Genet. **11**(12), 819–829 (2010)
7. R.S. Mani et al., Induced chromosomal proximity and gene fusions in prostate cancer. Science **326**(5957), 1230 (2009)
8. C. Lin et al., Nuclear receptor-induced chromosomal proximity and DNA breaks underlie specific translocations in cancer. Cell **139**(6), 1069–1083 (2009)

9. M.M. Vilenchik, A.G. Knudson, Endogenous DNA double-strand breaks: production, fidelity of repair, and induction of cancer. Proc. Natl. Acad. Sci. U. S. A. **100**(22), 12871–12876 (2003)

10. A.M. De Marzo et al., Inflammation in prostate carcinogenesis. Nat. Rev. Cancer **7**(4), 256–269 (2007)

11. K.S. Sfanos et al., The inflammatory microenvironment and microbiome in prostate cancer development. Nat. Rev. Urol. **15**(1), 11–24 (2018)

12. R.S. Mani et al., Inflammation-induced oxidative stress mediates gene fusion formation in prostate cancer. Cell Rep. **17**(10), 2620–2631 (2016)

13. R. Bhatia-Gaur et al., Roles for Nkx3.1 in prostate development and cancer. Genes Dev. **13**(8), 966–977 (1999)

14. A. Dutta et al., Identification of an NKX3.1-G9a-UTY transcriptional regulatory network that controls prostate differentiation. Science **352**(6293), 1576–1580 (2016)

15. M.M. Shen, C. Abate-Shen, Molecular genetics of prostate cancer: new prospects for old challenges. Genes Dev. **24**(18), 1967–2000 (2010)

16. M.J. Kim et al., Nkx3.1 mutant mice recapitulate early stages of prostate carcinogenesis. Cancer Res. **62**(11), 2999–3004 (2002)

17. M.J. Kim et al., Cooperativity of Nkx3.1 and Pten loss of function in a mouse model of prostate carcinogenesis. Proc. Natl. Acad. Sci. U. S. A. **99**(5), 2884–2889 (2002)

18. S.A. Abdulkadir et al., Conditional loss of Nkx3.1 in adult mice induces prostatic intraepithelial neoplasia. Mol. Cell. Biol. **22**(5), 1495–1503 (2002)

19. C.R. Bethel et al., Decreased NKX3.1 protein expression in focal prostatic atrophy, prostatic intraepithelial neoplasia, and adenocarcinoma: association with gleason score and chromosome 8p deletion. Cancer Res. **66**(22), 10683–10690 (2006)

20. C. Bowen et al., Loss of NKX3.1 expression in human prostate cancers correlates with tumor progression. Cancer Res. **60**(21), 6111–6115 (2000)

21. X. Ouyang et al., Loss-of-function of Nkx3.1 promotes increased oxidative damage in prostate carcinogenesis. Cancer Res. **65**(15), 6773–6779 (2005)

22. H. Blaser et al., TNF and ROS crosstalk in inflammation. Trends Cell Biol. **26**(4), 249–261 (2016)

23. N.S. Chandel, P.T. Schumacker, R.H. Arch, Reactive oxygen species are downstream products of TRAF-mediated signal transduction. J. Biol. Chem. **276**(46), 42728–42736 (2001)

24. Y.S. Kim et al., TNF-induced activation of the Nox1 NADPH oxidase and its role in the induction of necrotic cell death. Mol. Cell **26**(5), 675–687 (2007)

25. J.J. Kim et al., TNF-alpha-induced ROS production triggering apoptosis is directly linked to Romo1 and Bcl-X(L). Cell Death Differ. **17**(9), 1420–1434 (2010)

26. B. Yazdanpanah et al., Riboflavin kinase couples TNF receptor 1 to NADPH oxidase. Nature **460**(7259), 1159–1163 (2009)

27. M.C. Markowski, C. Bowen, E.P. Gelmann, Inflammatory cytokines induce phosphorylation and ubiquitination of prostate suppressor protein NKX3.1. Cancer Res. **68**(17), 6896–6901 (2008)

28. C. Bowen, T. Zheng, E.P. Gelmann, NKX3.1 suppresses TMPRSS2-ERG gene rearrangement and mediates repair of androgen receptor-induced DNA damage. Cancer Res. **75**(13), 2686–2698 (2015)

29. D.A. Frohlich et al., The role of Nrf2 in increased reactive oxygen species and DNA damage in prostate tumorigenesis. Oncogene **27**(31), 4353–4362 (2008)

30. M. Nadal et al., Structure of the homodimeric androgen receptor ligand-binding domain. Nat. Commun. **8**, 14388 (2017)

31. M.E. van Royen et al., Stepwise androgen receptor dimerization. J. Cell Sci. **125**(Pt 8), 1970–1979 (2012)

32. C. Dai, H. Heemers, N. Sharifi, Androgen signaling in prostate cancer. Cold Spring Harb. Perspect. Med. **7**(9) (2017)

33. Z. Zhang et al., An AR-ERG transcriptional signature defined by long-range chromatin interactomes in prostate cancer cells. Genome Res. **29**(2), 223–235 (2019)

34. M.E. Ashour, R. Atteya, S.F. El-Khamisy, Topoisomerase-mediated chromosomal break repair: an emerging player in many games. Nat. Rev. Cancer **15**(3), 137–151 (2015)

35. Y. Pommier et al., Roles of eukaryotic topoisomerases in transcription, replication and genomic stability. Nat. Rev. Mol. Cell Biol. **17**(11), 703–721 (2016)

36. M.C. Haffner et al., Androgen-induced TOP2B-mediated double-strand breaks and prostate cancer gene rearrangements. Nat. Genet. (2010)

37. A. Canela et al., Genome organization drives chromosome fragility. Cell **170**(3), 507–521.e18 (2017)

38. L. Uuskula-Reimand et al., Topoisomerase II beta interacts with cohesin and CTCF at topological domain borders. Genome Biol. **17**(1), 182 (2016)

39. K. Skourti-Stathaki, N.J. Proudfoot, A double-edged sword: R loops as threats to genome integrity and powerful regulators of gene expression. Genes Dev. **28**(13), 1384–1396 (2014)

40. A. Aguilera, B. Gomez-Gonzalez, DNA-RNA hybrids: the risks of DNA breakage during transcription. Nat. Struct. Mol. Biol. **24**(5), 439–443 (2017)

41. A. El Hage et al., Loss of Topoisomerase I leads to R-loop-mediated transcriptional blocks during ribosomal RNA synthesis. Genes Dev. **24**(14), 1546–1558 (2010)

42. S. Tuduri et al., Topoisomerase I suppresses genomic instability by preventing interference between replication and transcription. Nat. Cell Biol. **11**(11), 1315–1324 (2009)

43. Y. Yang et al., Arginine methylation facilitates the recruitment of TOP3B to chromatin to prevent R loop accumulation. Mol. Cell **53**(3), 484–497 (2014)

44. P.A. Riley, Free radicals in biology: oxidative stress and the effects of ionizing radiation. Int. J. Radiat. Biol. **65**(1), 27–33 (1994)

45. E.I. Azzam, J.P. Jay-Gerin, D. Pain, Ionizing radiation-induced metabolic oxidative stress and prolonged cell injury. Cancer Lett. **327**(1–2), 48–60 (2012)
46. R. Ameziane-El-Hassani et al., NADPH oxidase DUOX1 promotes long-term persistence of oxidative stress after an exposure to irradiation. Proc. Natl. Acad. Sci. U. S. A. **112**(16), 5051–5056 (2015)
47. T.K. Li et al., Activation of topoisomerase II-mediated excision of chromosomal DNA loops during oxidative stress. Genes Dev. **13**(12), 1553–1560 (1999)
48. N.R. Pannunzio, M.R. Lieber, AID and reactive oxygen species can induce DNA breaks within human chromosomal translocation fragile zones. Mol. Cell **68**(5), 901–912.e3 (2017)
49. N.R. Pannunzio, M.R. Lieber, Concept of DNA lesion longevity and chromosomal translocations. Trends Biochem. Sci. **43**(7), 490–498 (2018)
50. M.R. Lieber, The mechanism of double-strand DNA break repair by the nonhomologous DNA end-joining pathway. Annu. Rev. Biochem. **79**, 181–211 (2010)
51. M.A. Rubin, C.A. Maher, A.M. Chinnaiyan, Common gene rearrangements in prostate cancer. J. Clin. Oncol. **29**(27), 3659–3668 (2011)
52. N.C. Bastus et al., Androgen-induced TMPRSS2:ERG fusion in nonmalignant prostate epithelial cells. Cancer Res. **70**(23), 9544–9548 (2010)
53. N. Coll-Bastus et al., DNA replication-dependent induction of gene proximity by androgen. Hum. Mol. Genet. **24**(4), 963–971 (2015)
54. H. Lu et al., Phase-separation mechanism for C-terminal hyperphosphorylation of RNA polymerase II. Nature **558**(7709), 318–323 (2018)
55. B.R. Sabari et al., Coactivator condensation at super-enhancers links phase separation and gene control. Science **361**(6400) (2018)
56. S. Chong et al., Imaging dynamic and selective low-complexity domain interactions that control gene transcription. Science **361**(6400) (2018)
57. W.K. Cho et al., Mediator and RNA polymerase II clusters associate in transcription-dependent condensates. Science **361**(6400), 412–415 (2018)
58. D. Hnisz et al., A phase separation model for transcriptional control. Cell **169**(1), 13–23 (2017)
59. A. Boija et al., Transcription factors activate genes through the phase-separation capacity of their activation domains. Cell **175**(7), 1842–1855.e16 (2018)
60. Q. Du et al., Replication timing and epigenome remodelling are associated with the nature of chromosomal rearrangements in cancer. Nat. Commun. **10**(1), 416 (2019)
61. X. Li et al., BRD4 promotes DNA repair and mediates the formation of TMPRSS2-ERG gene rearrangements in prostate cancer. Cell Rep. **22**(3), 796–808 (2018)
62. C. Weier et al., Nucleotide resolution analysis of TMPRSS2 and ERG rearrangements in prostate cancer. J. Pathol. **230**(2), 174–183 (2013)
63. C.C. Pritchard et al., Inherited DNA-repair gene mutations in men with metastatic prostate cancer. N. Engl. J. Med. **375**(5), 443–453 (2016)
64. C.S. Grasso et al., The mutational landscape of lethal castration-resistant prostate cancer. Nature **487**(7406), 239–243 (2012)
65. M.F. Berger et al., The genomic complexity of primary human prostate cancer. Nature **470**(7333), 214–220 (2011)
66. D. Robinson et al., Integrative clinical genomics of advanced prostate cancer. Cell **161**(5), 1215–1228 (2015)
67. Cancer Genome Atlas Research Network, The molecular taxonomy of primary prostate cancer. Cell **163**(4), 1011–1025 (2015)
68. L. Burkhardt et al., CHD1 is a 5q21 tumor suppressor required for ERG rearrangement in prostate cancer. Cancer Res. **73**(9), 2795–2805 (2013)
69. I.A. Asangani et al., Therapeutic targeting of BET bromodomain proteins in castration-resistant prostate cancer. Nature **510**(7504), 278–282 (2014)
70. A. Stanlie et al., Chromatin reader Brd4 functions in Ig class switching as a repair complex adaptor of nonhomologous end-joining. Mol. Cell **55**(1), 97–110 (2014)
71. C. Bowen, E.P. Gelmann, NKX3.1 activates cellular response to DNA damage. Cancer Res. **70**(8), 3089–3097 (2010)
72. C. Bowen et al., Functional activation of ATM by the prostate cancer suppressor NKX3.1. Cell Rep. **4**(3), 516–529 (2013)
73. S.C. Baca et al., Punctuated evolution of prostate cancer genomes. Cell **153**(3), 666–677 (2013)
74. M.M. Shen, Chromoplexy: a new category of complex rearrangements in the cancer genome. Cancer Cell **23**(5), 567–569 (2013)
75. P.J. Stephens et al., Massive genomic rearrangement acquired in a single catastrophic event during cancer development. Cell **144**(1), 27–40 (2011)

Cells of Origin for Prostate Cancer

Li Xin

Definition of Cells of Origin for Cancer

The cells-of-origin for cancer is generally defined as the cell within a tissue from which cancer originates [1, 2]. Despite the simple definition, the term can be confusing under certain circumstances. For example, the initial genetic or epigenetic alterations may occur in one type of cell in a tissue, which then differentiates into another type of cell. The latter cell type may then respond to the preexisting oncogenic signaling, undergo uncontrolled proliferation, and initiate the primary tumor. In this scenario, the identity of the cell-of-origin can be ambiguous [3]. Therefore, it has been suggested that the former cell type be termed as the cell-of-mutation and the latter cell type as the cell-of-origin for cancer [4]. This helps clarify some ambiguity, but this is only clinically meaningful when the tumors originating from these two types of cells display distinct clinical features when given the same oncogenic signal. To simplify the discussion in this review, I will refer to the cells of origin for prostate cancer as the cells in which the initial oncogenic signaling occurs. The topic of the cells of origin for prostate cancer has been reviewed previously [4–7] and you

may refer to those for more comprehensive opinions.

Significance of Investigating Cells of Origin for Cancer

The identity of the cells of origin for various types of cancers has been studied extensively. It has been demonstrated that both stem cells and committed progenitors can serve as the cells of origin for cancer [8]. Different cellular entities are differentially susceptible to distinct oncogenic signaling [9, 10]. The same genomic alteration in different cell types can lead to distinct tumor subtypes [11, 12]. Thus, investigating the cells of origin for cancer is not merely a curiosity, but can provide insights into how tumors initiate and how tumor heterogeneity evolves. In addition, by investigating the identity of the cells of origin for cancer, we can learn more about their features such as unique antigens that they express, specific signaling pathways that are required for their survival, and mechanisms they employ to fight stress signaling, etc. This information may help us identify novel markers for early cancer detection, discover therapeutic vulnerability of the tumors, predict potential mechanisms for therapeutic resistance, and develop novel therapeutic strategies.

L. Xin (✉)
Department of Urology, University of Washington, Seattle, WA, USA
e-mail: xin18@uw.edu

© Springer Nature Switzerland AG 2019
S. M. Dehm, D. J. Tindall (eds.), *Prostate Cancer*, Advances in Experimental Medicine and Biology 1210, https://doi.org/10.1007/978-3-030-32656-2_4

Epithelial Cell Lineages in Normal Human and Rodent Prostates

A prerequisite to investigate the identity of the cells of origin for prostate cancer is a comprehensive understanding of the normal prostate epithelial cell lineages. Three types of prostate epithelial cells have been defined based on their histological appearance and the expression of specific antigens: the luminal, basal, and neuroendocrine cells. The columnar luminal epithelial cells form a single layer surrounding the lumen. These cells carry out the secretory function of the prostate gland and express various markers such as cytokeratin 8 and 18, androgen receptor (AR), prostate specific antigen (PSA), Nkx3.1, etc. The luminal cells are surrounded by a continuous layer of cuboidal basal cells that rest on the basement membrane. Basal cells also expressed specific markers such as p63, and cytokeratin 5 and 14, etc. The function of basal cells remains unclear. Electron microscopic analysis shows the presence of junction-like structures between the basal and luminal cells as well as between the basal cells [13]. This suggests that basal cells may regulate luminal cell biology via inter-cell cargo transport and may also serve as a barrier to maintain prostate tissue homeostasis. Neuroendocrine (NE) cells are rare and sporadically distributed among the glands. These cells have long dendritic processes with nerve-like varicosities [14, 15]. NE cells express synaptophysin and many neuropeptides such as chromogranin A, somatostatin, etc. Some NE cells can reach to the glandular lumen (open type) whereas others only have dendritic cytoplasmic processes migrating into surrounding epithelial cells (closed type). Based on their secretory proteins and morphology, NE cells have been suspected of carrying out both growth and exocrine secretory activities.

Rodent and human prostates are anatomically different. The human prostate displays a lobular structure that can be stratified into four different anatomical zones [16, 17]. The peripheral zone (PZ) includes the dorsolateral and apical parts of the gland. The transition zone (TZ) consists of two lobes and is located between the proximal prostatic urethra and lateral parts of the PZ. The central zone surrounds the ejaculatory ducts and is located at the base of the prostate between the peripheral and transitional zones. Finally, the anterior fibromuscular stroma forms the convexity of the anterior external surface and lacks glandular tissues. The rodent prostate has a tubular structure that is composed of four different lobes: anterior, ventral, dorsal and lateral lobes. An additional difference is that there is a higher fraction of stromal cells in the human prostate than in the rodent prostate.

Despite these differences, the human and mouse prostates share many histological similarities. Mouse prostates also contain the three types of epithelial cells. In contrast to the 1:1 ratio of basal/luminal cells in human prostates, there are more luminal cells in the rodent prostate (ratio of basal/luminal cell number about 1:4) [13]. Therefore, the basal cells in rodent prostates are extended much more than the human basal cells and form a discontinuous layer in contrast to the continuous basal cell layer in the human prostate. Because of this, the rodent prostate epithelium is classified as pseudostratified columnar.

Developmental Origin of Prostate Epithelial Cells

The prostate basal and luminal cells are both derived from the urogenital sinus epithelial cells of the endodermal origin [18]. This has also been demonstrated experimentally by a lineage tracing study in mice, which revealed that the p63[+] urogenital sinus epithelial cells give rise to both the basal and luminal cells in adult mice [19]. However, whether the NE cells also share the same developmental origin remains a question. In a prostate regeneration assay, it has been shown that mouse prostate basal epithelial cells can generate neuroendocrine cells [20], but whether this also happens at the developmental stage in the native prostate microenvironment remains a question. At present, there is no indisputable report in a lineage tracing study demonstrating that NE cells can be generated from cells of the urogenital sinus origin. On the other hand, there is evidence that neuroendocrine cells are derivatives of neuroendocrine progenitors migrating from the neural crest [21]. Immunostaining of the NE cell marker chromogranin A in the human embryonic urogenital sinus revealed that NE cells are absent in the epithelium before the 18th gestation week but show up afterwards. This observation supports the migration theory but cannot exclude the possibility that the NE cell lineage arises from epithelial stem cells at a later developmental stage. On the other hand, a lineage tracing study using Wnt1-Cre showed that some NE cells are derived from the Wnt1-expressing neural crest cells [22]. However, unpublished observations from my group showed that the Wnt1-Cre and Sox10-Cre lines label a significant amount of the prostate basal and luminal epithelial cells, suggesting that the specificity of these neural crest driver transgenic mouse lines are not as tight as expected. In summary, the exact embryonic origin of the neuroendocrine cells remains to be defined.

Inter-lineage Hierarchy in Normal Prostate Epithelium in Adults

The mouse and human prostatic epithelial buds are solid initially and then undergo canalization to form lumen. During this period of development, most cells express the makers of both basal and luminal cells including cytokeratins 5, 8, 14, 18, 19 and P63 [23, 24]. Definitive differentiation of basal and luminal cells starts at the end of the 15th gestation week in humans [23]. In mice, definitive lineage specification is not completed until 5–14 days postnatally in the different prostate lobes [24]. The relationship between the mouse basal and luminal cells at the postnatal developmental stage has been investigated: using a lineage tracing strategy, it was demonstrated that the luminal cells undergo lineage commitment and become unipotent after 2 weeks postnatally, whereas some basal epithelial cells possess bipotent differentiation capability and can generate both basal and luminal

cells at that time [25]. This was corroborated by a careful analysis of spindle orientation of dividing cells in conjunction with the identity of daughter cells [26]. During postnatal development, mouse luminal cells usually undergo symmetric division to generate two luminal daughter cells. However, basal cells can undergo both symmetric and asymmetric division to generate both basal and luminal cells.

At adulthood, prostate epithelial cells turn over very slowly compared to those in the skin or small intestine. However, prostate epithelial cells possess extensive regenerative capability, as was demonstrated in the rodent: When rodents were castrated, their prostates shrunk drastically. This is due to apoptosis of the luminal epithelial cells since many of these cells rely on androgen signaling for survival [27–29]. However, when androgen was replaced, prostate tissues grew quickly back to their normal size. This cycle of involution and regeneration can be repeated up to 30 times [30]. This observation implies the existence of stem cells or progenitor cells.

Much effort has been made to investigate the inter-lineage hierarchical structure in the prostate epithelium, particularly the relationship between the basal and luminal cells. Historically, there has been extensive debates regarding different theories. Initially, much evidence supported the concept that basal cells lie at a higher level in the hierarchy among the other lineages (see discussion below). Subsequently, there were also accumulating reports indicating that the luminal cells also contribute to the maintenance of the other cell lineages in the prostate (see discussion below). During the debates over the years, it has been appreciated that the controversies reflect the functional plasticity of both the basal and luminal cells. They both can display facultative stem cell activity under experimental or pathological conditions. More recently, with the application of lineage-tracing technology in mice, it has been shown that basal and luminal cells are largely independently sustained. This is at least true in adult mice, although the question remains open in the human prostate.

Basal Stem/Progenitor Cell Theory

Since basal cells do not rely on androgen signaling for their survival and express many stem cell-associated antigens such as P63, telomerase, Bcl-2 etc., it was hypothesized that prostate stem cells reside in the basal cell compartment. Those cells can generate both basal cells and luminal cells or even neuroendocrine cells via a transit-amplifying cell population [31]. This hypothesis is supported by immunostaining of human specimens with antibodies against various types of cytokeratins. Intermediate cell types that are K19+, K5++/K18+, or K5+/K18++ have been identified in both human and rat prostate tissues and were suggested to represent snapshots of transition from basal cells to luminal-like cells [32–35]. A recent study using comparative bioinformatics analyses, which supports this hypothesis, proposed a two-step regulatory mechanism of human prostate basal stem cell differentiation, in which retinoic acid induces expression of a set of genes that preferentially respond to androgens during terminal prostate epithelial differentiation [36].

Inspired by research of skin stem cells, early investigators tried to use an *in vitro* culture system to grow putative mouse and human prostate stem cells that can generate holoclones. In those studies, the cell population that expanded in culture always displayed a more basal cell-like phenotype because they expressed high levels of the basal cell markers such as Keratin 14 and 5. However, upon confluence, cells spontaneously adapted a more luminal cell-like phenotype as evidenced by increased expression of the luminal cell markers Keratin 8 and 18 [37–40]. This basal-to-luminal differentiation can be programed *in vitro* by defined growth factors [41]. Subsequently, in a type of 3-dimension sphere assay, where dissociated prostate epithelial cells are cultured either in suspension in low attachment plates or immobilized inside Matrigel with or without stimulation with the urogenital sinus mesenchymal cells, it was also demonstrated that only the prostate basal cells can grow into spheroids [42, 43]. These spheroids contained cells that display a more basal cell-like phenotype. These cells could undergo some degree of differentiation though they could not fully differentiate into mature luminal cells [43]. However, when transplanted *in vivo*, these cells can generate glandular structures containing both the basal and luminal cells [42]. Although there is no evidence that neuroendocrine cells can be generated from basal cells *in vitro*, these studies further support the concept that stem cells/progenitors may reside in the basal cell compartment, depending on the stringency of the definition for stem cells.

This hypothesis is further supported by *ex vivo* studies after a tissue recombination assay was modified and adapted to monitor stem cell activity in the prostate [44, 45]. In these studies, mouse and human prostate basal cells are mixed with rat or mouse urogenital sinus mesenchymal cells and transplanted under the kidney capsules of immunodeficient male mice supplemented with androgen. In all these studies, basal cells can generate glandular structures that contain at least basal and luminal cells [20, 46–52]. Of note, one of the studies showed formation of neuroendocrine cells from basal cells [20]. In contrast, the basal cell-depleted cells in intact adult mice rarely displayed the same capability in the same assay.

Collectively, these *in vitro* and *ex vivo* functional studies support the concept that prostate stem cells or progenitor cells reside in the basal cell compartment. However, there is a caveat. The stem cell assays described above were all performed under experimental conditions that are different from normal physiological conditions. In the prostate regeneration assay, prostate basal cells were dissociated from their native environment and stimulated with embryonic urogenital sinus mesenchymal cells that provide signaling, which is absent in the adult mouse prostate. Therefore, the stem cell activity that the basal cells displayed in these assays may not contribute to the maintenance of the prostate epithelial homeostasis *in vivo*.

To overcome the limitations of the previous studies, several independent groups have employed a lineage tracing approach to investigate the relationship between basal and luminal cells. With this approach, the basal cell lineage in adult mice is labeled with fluorescent proteins, and the identity of their progeny is traced and verified by immunostaining or flow cytometric analyses. In these studies, the investigators employed transgenic mice expressing a CreER^{T2} transgene driven by the basal cell-specific promoters (Keratin 5, 14, P63) to specifically label the basal cells [26, 49, 53–55]. The labeling was performed in adult mice to specifically investigate how adult mouse prostate epithelial homeostasis is maintained. A surprising consensus from four of the studies is that basal cells generate mostly basal cells [49, 53, 54]. The efficiency of color-labeling basal cells was low-to-moderate in these studies. In two of these studies, it was reported that a very small percentage of the basal cells can give rise to luminal cells [26, 49, 54]. This rare event supports the concept that some basal cells possess bipotent differentiation potential in their native microenvironment but may also reflect the imperfect specificity of the promoters in the CreER^{T2} driver line.

Interestingly, the result from another lineage tracing study using a P63-CreER^{T2} line is quite different. This study showed that the p63-expressing basal cells can generate both luminal and neuroendocrine cells effectively. The reason for this discrepancy is unclear. Instead of the fluorescent proteins used in the other three studies, β-galactosidase activity was employed in this study to trace cells, which is less effective in defining the cell boundary at high resolution. In addition, in contrast to the transgenic approach of the other two CreER^{T2} drivers (Keratin5 and 14), CreER^{T2} was knocked in and replaced one endogenous allele of p63 in this model. p63 is a critical transcription factor regulating specification and proliferative potential of basal cells [56–58]. It is possible that p63 haploinsufficiency affects basal cell biology. Therefore, whether the conclusion from this study reflects that of the wild type mouse prostate remains uncertain.

More recently, a meticulous effort was made to attempt a lineage tracing in the human prostate. This approach took advantage of a sporadic mutation of mitochondrial DNA that leads to cytochrome C oxidase deficiency [59, 60]. The mutation therefore can be easily detected by an enzymatic histochemistry. Using quantitative clonal mapping with 3D reconstruction, the authors showed streams of cells containing both basal and luminal cells, and even neuroendocrine cells, with the mutation dispersed from juxtaurethral ducts throughout the entire glandular network. The authors proposed that this reflects direct long-ranging epithelial flows from multipotent basal stem cells. However, it is unknown when the initial mutation occurred. If it happened at an early developmental stage, then the result suggests that these cell lineages have a similar developmental origin.

In summary, the studies over the past several decades have demonstrated clearly that both mouse and human basal cells possess the potential to differentiate into luminal cells and perhaps neuroendocrine cells. However, under physiological conditions, and even during induced artificial regeneration in adult mice, basal cells seem to contribute very marginally to the other cell lineages. Whether this applies to the human prostate remains a question. Single cell trajectory analysis [61] may provide additional insights, but the conclusion will still be suggestive and needs to be further validated by novel technologies in the future.

Luminal Stem/Progenitor Cell Theory

Luminal cells are considered to be terminally differentiated because they do not manifest stem cell features in various *in vitro* and *ex vivo* assays. However, in all those assays, dissociated single cells were used as the research focus. Luminal cells are particularly susceptible to anoikis [62]. Once isolated as single cells they tend to die under these in vitro or *ex vivo* conditions before their stem cell potential can be evaluated. Therefore, those studies cannot exclude the bipotent potential of the luminal cells.

It has long been argued that basal cells and luminal cells are self-replicating cell types. Both cell types actively divide and undergo extensive proliferation in castrated mice when androgen is replaced. An early study demonstrated the existence of label-retaining cells within the luminal cell lineage, which implies the existence of stem cells in the luminal cell lineage [63]. The first compelling *in vivo* evidence for a luminal-like stem/progenitor cell came from a study of castration-resistant Nkx3.1-expressing (CARN) cells [64]. CARN cells are identified in castrated adult mice using an NKX3.1-CreER[T2] knock-in mouse model. CARN cells display a luminal phenotype. The authors labeled these cells with green fluorescence protein in castrated mice using a lineage tracing approach and replaced androgen to induce prostate regeneration. CARN cells can generate both basal and luminal cells. CARN-like cells are also identified in human specimens from patients undergoing anti-androgen therapy [65], but whether they display bipotent differentiation capacity cannot be determined. The study of the mouse CARN cells demonstrates that the cells with a luminal phenotype also possess at least bipotent potential. A population of castration resistant Bmi-1-expressing (CARB) luminal cells was also identified in castrated mice [66]. CARB cells are different from CARN cells, but also possess a capacity for bipotent differentiation.

Subsequently, additional evidence supporting the existence of bipotent luminal progenitors emerged. A unique luminal cell population that express Sca-1 was identified in the proximal prostatic ducts of the mouse prostate [67]. These cells can generate prostate glands containing both basal and luminal cells in the prostate regeneration assay, but their regenerative capability is much less than that of the basal cells. It was shown that some prostate luminal cells can survive and expand to form colonies or organoids *in vitro* [68–70]. With the application of a WIT™ medium [12], human prostate luminal cells can be cultured *in vitro* as colonies at a low efficiency. Colonies containing cells expressing both the basal and luminal cells markers, and these cells can regenerate prostate glandular structures *in vivo* in the prostate regeneration assay [70]. The prostate luminal cells can also form organoids in prostate organoid assays oriented to culturing cells of the endodermal origin, in which Wnt agonists and inhibitors for TGFβ and BMP signaling are supplemented [68, 69]. The organoids contain both basal and luminal cells, demonstrating that these organoid-forming luminal cells are at least bipotent. Organoids generated from a CD38[+] luminal cell population can generate glands containing both basal and luminal cells [71].

However, as with previous studies, these studies are also not without their limitations. The CARN and CARB cells are identified in castrated adult mice. Their origin is unknown. It is possible that basal cells adapted a CARN/CARB cell phenotype upon castration. In addition, the Nkx3.1-CreER[T2] model was generated so that one allele of the endogenous Nkx3.1 is lost. Nkx3.1 plays a critical role in prostate epithelial cell lineage commitment [72]. It is unclear whether Nkx3.1 haploinsufficiency affects prostate epithelial cell differentiation. The regeneration, colony, and organoid assays only demonstrate the bipotent potential of the luminal cells under those experimental contexts. Whether this is an obligate or facultative property of luminal cells *in vivo* remains questionable.

Lineage tracing studies of the luminal cells in intact adult mice were also performed using transgenic mouse models harboring the expression of CreER[T2] driven by luminal cell promoters (Keratin 8 and PSA) [49, 53, 54, 73]. These studies unanimously demonstrated that the luminal cells in adult mice only generate luminal progeny. Collectively, these studies reach a conclusion like those studies on basal cell biology. Both mouse and human luminal cells possess bipotent differentiation capability but there is no compelling evidence that they can contribute to the maintenance of other cell lineages substantially under physiological conditions at adulthood.

Independent Lineage Theory and Intra-lineage Heterogeneity

The studies described above support an independent lineage theory, which argues that the basal and luminal cells are self-replicating [74, 75]. This theory appears to be accurate for the mouse prostate, but whether the theory holds true for the human prostate remains to be determined. The independent lineage theory implies the existence of progenitors within different cell lineages, which is supported by many studies demonstrating an intra-lineage heterogeneity with all the three cell lineages.

Basal Cell Heterogeneity

It has been well-documented that not all basal cells are the same phenotypically. Immunostaining of various basal cell markers and other antigens have indicated a heterogeneity. For example, not all basal cells express cytokeratin 14 [37]; basal cells express differential levels of the integrins $\alpha 2\beta 1$ [40]. In addition, by flow cytometric analyses of many stem cell-associated surface antigens, such as CD44, CD49f, CD133, c-Kit, Trop2 etc., have been demonstrated to be expressed only in a fraction of basal cells in both mouse and human prostate.

More interestingly, the phenotypic disparity reflects functional distinction in various *in vitro* and *ex vivo* prostate stem cell assays. Human basal cells expressing a higher level of $\alpha_2\beta_1$ are more potent in forming colonies *in vitro* and regeneration of prostate tissues *in vivo* [40]. Human basal cells enriched with EpCAM, CD44 and CD49f possess higher sphere-forming activity [76]. CD133+ human prostate basal cells also display higher colony-forming and regenerative capacities [52, 77]. Inside the human prostate sphere culture, the label-retaining sphere cells display more phenotypic stem cell features and higher functional activities [78]. *In vivo*, the Trop2+ human prostate basal cells possess a higher regenerative capacity in the prostate regeneration assay [20, 46]. Many mouse prostate studies also reached a similar conclusion. For instance, Trop2 also enriches for the mouse prostate basal stem cell activity in the prostate regeneration assay [20]. Single c-Kit-expressing mouse prostate basal cells can regenerate glandular structures *in vivo* [51]. These studies support the concept that a fraction of the basal cells in the prostate may represent progenitors that sustain the lineage.

In contrast, results from lineage tracing studies are less compelling. The CreER[T2] transgenic line driven by the stem cell markers such as Bmi-1, Lgr5, and CD133 can label a fraction of basal cells [66, 79]. However, these cells do not contribute significantly to the maintenance of the basal cell lineage even after extensive epithelial turnover induced by cycles of androgen deprivation and replacement. Some driver lines label only a very small percentage of the basal cells (<0.1%). So, it is possible they do not include most of the basal progenitors. But in another lineage tracing study, CD133+ basal cells (about 18% of total basal cells) were shown to contribute to basal cell homeostasis in a manner similar to CD133− basal cells [80], although the CD133+ basal cells are more efficient in forming colonies and spheres *in vitro*. These results question whether there is a distinct basal progenitor population *in vivo*.

The observations from the lineage tracing study and the *in vitro* and *ex vivo* studies highlight again the differences of the experimental conditions. In the *in vitro* and *ex vivo* assays, cells are removed from their niches and dissociated into single cells. Dissociation of cells from their native environment can induce apoptosis (anoikis). Therefore, these assays are probably measuring the potential of cells to survive anoikis. Anoikis can be inhibited by the ROCK kinase inhibitor Y-27632. We and others showed that when Y-27632 was supplemented in those *in vitro* and *ex vivo* stem cell assays (colony, sphere, and regeneration assays), the percentage of basal cells with stem cell activity can be increased by tenfold [81, 82]. Collectively, these studies implied that there may not necessarily be a specialized basal cell progenitor population that maintains the lineage via transit-amplifying cells *in vivo*, or alternatively there may be many different types of basal cell progenitors that can have the capacity to duplicate. This is further supported by an *in vivo* cell kinetic study showing that during normal prostate homeostasis most adult mouse prostate epithelial cells are formed by stochastic cell division, which implies an absence of the putative transit-amplifying cells [83].

Luminal Cell Heterogeneity

Several lines of evidence also demonstrate heterogeneity within the luminal cell lineage. First, not all luminal cells succumb to androgen ablation. Second, in mice some luminal cells are relatively quiescent and possess the capacity to retain labelling [84, 85]. In addition, luminal cells express varying levels of surface antigens. For example, some mouse luminal cells do not express Nkx3.1 but express Sca-1 [67]; only some mouse prostate luminal cells express Ly6d [86]; and CD38 is only expressed by a fraction of human luminal cells [71]. Like the studies on basal cells, these phenotypic differences predict functional disparity. Sca-1+ mouse prostate luminal cells and CD38+ human luminal cells both display more potent stem cell activity than their counterparts [67, 71].

There is also indirect evidence suggesting a heterogeneity of the luminal cell lineage. In a transgenic mouse model with prostate specific deletion of both Pten and p53, investigators identified two types of phenotypically luminal tumor cells that can give rise to adenocarcinoma alone, or a mixture of adenocarcinoma and squamous carcinoma [87]. However, because the origin of these two types of luminal tumor cells is unclear, this observation may only serve as indirect evidence for the existence of different types of luminal cells.

Phenotypic heterogeneity of luminal cells is also confirmed by *in vivo* lineage tracing studies. For example, Nkx3.1, CD133, Lgr5, and Bmi-1 are all capable of marking a fraction of the mouse luminal cells [64, 66, 79, 80]. However, *in vivo* evidence of a robust regeneration of the whole luminal cell lineage by a putative luminal progenitor has been missing. In all these studies, the labelled luminal cells are not functionally better than the non-labeled luminal cells in their contribution to the maintenance of luminal cells. Considering the study showing that most adult prostate epithelial cells are formed by stochastic cell division [83], it is also plausible to hypothesize that the luminal cells are not maintained by one specialized progenitor population, but are maintained by many types of progenitors or simply by cell duplication.

Neuroendocrine Cell Heterogeneity

Neuroendocrine cells are morphologically heterogeneous: the closed-type separated by other cells in lumen, and the open-type that can reach to the glandular lumen. In addition, they can also be subclassified based on their secretory products. For example, some express serotonin or chromogranin A and B, whereas others express calcitonin or bombesin. It is unknown whether these differences may reflect their origin, such as neural crest or embryonic urogenital sinus epithelial cells.

Heterogeneity Associated with Anatomy

Many studies in rodent prostate have highlighted an association between cellular heterogeneity and anatomy. The rodent prostate is composed of four different lobes: anterior, ventral, dorsal and lateral. Early studies have demonstrated that epithelial cells differentiate at different dynamics in these lobes [24]. In addition, luminal epithelial cells in the different lobes also produce different secretory proteins [88, 89]. The mouse prostate surrounds the urethra and buds away from the urethra. In an early study, investigators labelled prostate epithelial cells with BrdU, induced epithelial turnover by alternate androgen deprivation and replacement, and revealed that the cells that retained BrdU labeling resided at the proximal prostatic ducts [84]. These cells are considered to be the putative label-retaining quiescent stem cells. A similar study using H_2B-GFP labeling [85] corroborated that all the label-retaining cells were present in the proximal prostatic ducts and contained both basal and luminal cells. This is consistent with later studies showing that basal stem cells are enriched in the proximal region and that the basal cells in this region possess a higher stem cell activity than those in other regions [48, 50]. In addition, the Sca-1+ luminal cells that possess a bipotent differentiation capacity also localize in the proximal region [67]. Compared to the luminal cells at the distal prostatic ducts, the Sca-1+ luminal cells display less secretory function and may represent ductal cells rather than alveolar epithelial cells [67]. Finally, in mice neuroendocrine cells are much more frequent in the proximal region than in other areas [90]. Since the mouse prostate develops by budding outward from the urethra, these observations imply that cells with stem cell potential in the proximal region may represent the prostate stem cells that are left behind during tubule elongation. Or alternatively, the tissue microenvironment at the proximal ducts is unique and can keep the stem cell from further differentiation. We recently discovered that stromal cells in the mouse proximal prostate express higher levels of various Wnt ligands, including canonical and noncanonical Wnt ligands. These Wnt ligands regulate noncanonical Wnt signaling in the basal epithelial cells and canonical Wnt/β-Catenin signaling in the stromal cells, which together maintain the quiescent nature of the epithelial cells in this anatomic region [91].

Anatomy-associated cellular heterogeneity also exists in the human prostate. As described earlier, the human prostate is also divided into four zones, among which the transition zone and peripheral zone are the two major zones of endodermal origin. Interestingly, a recent single cell analysis revealed that in the collecting ducts around the urethra at the transition zone, there are two types of cells that express a high level of prostate stem cell antigen (PSCA) [92]. These cells resemble the club cells and hillock cells in the lung and are absent in the peripheral zone.

Lineage Plasticity Under Conditions of Stress

The discovery of induced pluripotent stem cells from the fibroblast cells [93] has made it clear that, given the right context, any type of cell can probably display lineage plasticity and be reprogrammed into a different type of cell. In fact, *in vitro* culture of prostate colonies, spheres, or organoids may represent artificial situations where the prostate basal or luminal cells are instructed to generate structures containing both the basal and luminal cells. Although basal and luminal cells are independently sustained in the adult mouse prostate under physiological conditions, studies have shown that under conditions

of stress, basal cells can generate luminal cells efficiently. For example, studies using a bacterial infection-induced mouse model for prostate inflammation in concert with the lineage tracing approach showed that the prostate basal cells can efficiently generate luminal cells during prostate inflammation [94]. Similarly, in a mouse model where luminal cells were induced to undergo anoikis via loss of E-Cadherin, basal cells were found to undergo basal-to-luminal differentiation [95]. Acute and chronic inflammation are frequently noted in the human prostate [96]. There are many potential sources for this affliction, including bacterial or virus infection, chemical and physical trauma caused by urine reflux, dietary factors, hormonal factors, or combinations thereof [97]. Therefore, it is probably not unexpected that in the human prostate, some of the luminal cells were derived from the basal cells during ageing. An interesting unresolved issue is whether the basal cell-derived luminal cells and preexisting luminal cells are phenotypically and functionally the same. Answering this question may help explain how cells of origin for prostate cancer may affect the clinical behavior of the resulting disease.

Cells of Origin for Prostate Cancer

Different subtypes of tumors with distinct histomorphological features or transcriptional profiles have been noted consistently in some tumors including leukemia, skin cancer, breast cancer, etc. [98]. Therefore, investigating whether the identity of cells-of-origin is a major determinant of histopathological and molecular subtypes is a natural question in these tumor models. In contrast, most human prostate cancer is acinar adenocarcinoma, and there is no distinct and consistent molecular subtype in prostate cancer based on transcriptional profiles [99–101]. Therefore, attributing the histomorphological and molecular features of tumors to the identity of cells of origin would not be a legitimate purpose for investigating the cells of origin for prostate cancer. Many prostate cancers are indolent and will not progress even if they are left untreated, whereas others are aggressive and need immediate therapeutic intervention. It is possible that the oncogenic signals driving indolent and aggressive prostate cancers are different. Alternatively, the aggressive feature of the disease could be determined by the nature of the cells of origin for the cancer. Therefore, the purpose of studying the cells of origin for prostate cancer is mainly to understand the cellular basis of aggressive prostate cancer. This should facilitate the development of novel prognostic markers for early detection of aggressive prostate and may provide insights into the therapeutic vulnerability of these tumors, leading to novel therapeutic strategies.

Histological Variants of Prostate Carcinoma

More than 90% of prostate cancers are acinar adenocarcinoma that exhibit glands and acini lined by a single layer of cuboidal cells. But there are also acinar adenocarcinoma variants that display a distinct histomorphological appearance, such as foamy, signet ring, atrophic acinar adenocarcinoma, etc. The presence of abnormal mitosis and absence of a basal cell layer has been considered to be standard criteria for diagnosis of prostate adenocarcinoma. Combinations of antibodies against the high molecular weight cytokeratins (34βE12) and P63, the markers for basal cells, have been employed in clinics to confirm the loss of the basal cell layer in prostate adenocarcinoma [102, 103].

Non-acinar carcinoma constitutes about 5–10% of human prostate carcinomas. There are also various histological variants among non-acinar carcinoma, such as ductal adenocarcinoma, basal cell carcinoma, neuroendocrine tumor etc. [104]. Ductal adenocarcinoma is characterized by pseudostratified columnar epithelium with papillary and cribriform patterns. Ductal adenocarcinoma is located mostly at the peripheral zone of the human prostate and is often intermingled with acinar adenocarcinoma. The outcome for patients with ductal differentiation is worse than that for men with acinar adeno-

carcinoma. Gene expression profiles of acinar and ductal adenocarcinoma are very similar [105], so ductal adenocarcinoma may represent a malignant trans-differentiation from acinar adenocarcinoma. Prostate basal cell carcinoma is extremely rare [106]. Cells within basal cell carcinomas express the basal cell markers P63 and high molecular weight cytokeratins. Most basal cell carcinomas occur in the transition zone but are also seen in the peripheral zone. Neuroendocrine tumors contain cells that are positive for the neuroendocrine cell markers to varying degrees. *De novo* NE tumors are extremely rare and are mostly small cell carcinomas [104]. These tumors are usually fatal, with most patients dying within 2 years of diagnosis. Besides *de novo* NE tumors, neuroendocrine differentiation is also seen in hormone naïve prostate adenocarcinoma [107] and recently have been observed more frequently in hormonally treated and castration resistant prostate cancer patients [108].

Interestingly, the histomorphological appearance of prostate cancer is also related to the anatomy. Peripheral zone tumors contain cubic cells with eosinophilic cytoplasm and vesicular nuclei, whereas transition zone tumors often have columnar cells with pale cytoplasm and dark nuclei [17].

Correlative Evidence for the Identity of Cells of Origin for Prostate Cancer

The identity of the cells of origin for cancer is often intuitively assumed as the cells that share the same phenotypic appearance with the tumor cells. Therefore, it is natural to consider the basal cells as the cell-of-origin for basal cell carcinoma and luminal cells as the cell-of-origin for acinar adenocarcinoma. The identity of the cells of origin for *de novo* NE tumors is unclear. However, because these tumor cells express stem cell-associated antigens such as c-Kit and P63, it was hypothesized that the multipotent stem cell is the cell of origin for NE prostate cancer [109]. In contrast, a consensus has been largely reached that NE tumors in castration resistant prostate cancer patients are the results of transdifferentiation of adenocarcinoma [108, 110, 111]. Since basal cell carcinoma and *de novo* NE tumors are rare, the debate of the identity of the cell-of-origin for prostate cancer is focused mainly on that for acinar adenocarcinoma.

Some believe that luminal cells are the cells of origin for prostate acinar adenocarcinoma because tumor cells display a luminal cell phenotype. Prostate luminal cells possess a low level of $H_2A.X$, hence are more vulnerable to oncogenic stress [112]. Luminal cells also express higher levels of 4EBP1, making them more resistant to inhibition of PI3K-AKT-mTOR signaling [113]. In addition, tumor initiating events such as TMPRSS2-ERG fusion, c-Myc upregulation, and telomere elongation are detected frequently in luminal cells in human prostatic intraepithelial neoplasia [114–116].

On the other hand, the phenotypic appearances and molecular profiles of cells of origin and their resulting tumors may be different [117]. Therefore, despite the observation that prostate acinar adenocarcinoma is devoid of basal cells, they may still serve as the cells of origin for the cancer. There is evidence to support this theory. For example, basal cells are less well-differentiated and proliferate more frequently than luminal cells in the human prostate, hence are more prone to accumulating genetic alterations [118]. During conditions of stress, such as prostate inflammation, basal cells are in closer contact with reactive stroma and are exposed directly to various cancer-promoting cytokines [119]. The TMPRSS2-ERG fusion is considered to be an early event in prostate cancer. The fusion event was also detected in the basal cells isolated from human prostate cancer biopsies [120], supporting the argument that basal cells are the cells of origin for prostate cancer. Over-expression of ERG and ETV1 in the prostate leads to a reduction of basal cells within the prostate [121, 122]. Since a hallmark of prostate acinar adenocarcinoma is loss of the basal cell layer, it is tempting to speculate that ETS fusion proteins drive basal-to-luminal differentiation and that the cellular origin for ETS fusion protein-positive prostate cancer is the basal cell.

Both the Basal and Luminal Cells Can Serve as Targets for Transformation

A prerequisite to understanding whether the identity of the cells of origin for prostate cancer can determine the clinical outcome of the resulting diseases is to determine which types of prostate cells can serve as targets for transformation. The first direct evidence for a cell type that could be targeted for transformation was obtained from early genetically engineered transgenic mouse models for prostate cancer. These models employ a truncated or composite promotor of the rat probasin gene to turn on oncogenic signaling in the prostate [123–125]. By deleting tumor suppressors such as Pten, P53, RB or activating oncogenes such as Myc, AKT, and FGFR1 these mouse models have developed cancer with various histological features, including acinar adenocarcinoma, neuroendocrine cancer and sarcomatoid carcinoma [123, 126–130]. Because Probasin is expressed in the luminal cells, it was believed that these studies demonstrated that the luminal cells can serve as a target for transformation to generate various types of prostate cancer. However, it was later realized that the Cre activity is also active in the basal cells, and even stromal cells in these transgenic models [125, 131]. Therefore, the conclusion of these studies became less definitive.

Early in vitro studies have shown that human prostate basal cells can be immortalized by virus DNA [132, 133] and transformed by a combination of oncogenes such as Myc and PI3K [134]. But the resulting tumor cells do not display a completely differentiated luminal cell phenotype. Early *in vivo* evidence demonstrating that basal cells can also serve as a target for transformation came from the prostate regeneration assay [48]. In this study, FACS-isolated mouse prostate basal cells were infected by lentivirus expressing constitutively active AKT1. Infected basal cells were able to generate prostatic intraepithelial neoplasia (PIN) lesions in the regeneration assay. Subsequent studies further demonstrated that not only mouse basal cells but also human basal cells are able to generate cancers in this system when various oncogenic signals were introduced in different combinations [135–140]. More recently, it was demonstrated that given the right combination of oncogenic signaling, basal cells can be transformed to generate neuroendocrine prostate cancer [141, 142]. Basal cells were also shown to form tumors under other conditions. Immortalized human prostate BPH-1 cells that display a basal cell phenotype were transformed when combined with cancer associated fibroblasts and stimulated with testosterone and estradiol. However, whether the resulting tumors displayed an adenocarcinoma phenotype was not reported [143]. It should be noted that the conditions of all these experiments are artificial and only provide a proof of principle. In the human, multiple oncogenic signaling may not occur in the same cells simultaneously. They may take place sequentially, in different orders, in the same cells, or even in different cells during basal-to-luminal differentiation.

More sophisticated mouse models have made it possible to test whether basal cells and luminal cells can serve as targets for transformation *in vivo* in their native environment. Using CreER driver lines driven by the basal cell-specific promoters (Keratin 14 and Keratin 5) or the luminal cell-specific promoter (Keratin 8), three independent groups disrupted Pten or Pten in combination with P53 specifically in the basal cells and luminal cells separately [49, 53, 54]. These studies showed that prostate cancer can arise from both basal and luminal cells. In addition, it was shown that the castration resistant Nkx3.1-expressing (CARN) and Bmi-1-expressing (CRAB) luminal cells can serve as the cells of origin for prostate [64, 66]. Together, these studies showed that both basal cells and luminal cells can serve as targets for transformation.

Preferred or More Efficient Target for Transformation, Basal or Luminal Cells?

After it was demonstrated that both the basal and luminal cells can serve as targets for transformation, an interesting debate emerged: which is the more efficient or favored target for transforma-

tion? There is evidence supporting the argument that basal cells are an efficient target for transformation. Most of these studies utilized the prostate regeneration technique. Isolated human and mouse basal cells, transduced with lentivirus mediating oncogenic signaling and incubated in the prostate regeneration assay, are more efficient in forming tumor tissues than luminal cells. There is no report showing that mouse luminal cells can be directly transformed in this experimental condition. As mentioned previously, a limitation of the studies utilizing the prostate regeneration model is that the luminal cells cannot survive well in this assay. Therefore, despite the conclusion from these studies that basal cells serve as an efficient target for transformation, a contribution of luminal cells cannot be ruled out.

There is also evidence supporting the argument that luminal cells are more susceptible to oncogenic transformation. When Pten was disrupted in luminal cells, they immediately underwent rapid proliferation and formed mild PIN lesions within one month [49, 53, 54]. In contrast, the basal cells did not respond to the loss of function of Pten and remained dormant for as long as 9 months after Pten deletion. However, as soon as these Pten-null basal cells underwent luminal differentiation, their luminal progeny reacted rapidly to the preexisting oncogenic signaling. Disease initiation from the basal cells can be accelerated by prostate inflammation since inflammation promotes basal-to-luminal differentiation [94]. In another study, investigators performed a lineage tracing study in various mouse models of prostate cancer driven by the probasin promoter. The study showed that most of the tumors in those models are derived from luminal cells, although a caveat for the study is whether or not the basal cells with activated oncogenic signaling were abundant and were effectively traced [144].

These literatures have caused much confusion especially to investigators outside the field who are not fully aware of the subtleties of the experimental conditions in those studies. In fact, these studies are not contradictory to each other: the basal cells being an efficient target does not exclude the luminal cells as a target for transformation; luminal cells being more responsive to oncogenic stimuli does not imply that they are the actual cells of origin for human prostate cancer, especially considering the fact that prostate cancer is an age-dependent disease, which occurs mostly in men over 50. In fact, which cellular compartment is the more efficient or preferred target for transformation is not an important question if they both give rise to a disease with the same histomorphological feature and clinical behavior. A question of more clinical relevance is whether the cells of origin for prostate cancer will determine its clinical behavior.

Cells of Origin for Aggressive Prostate Cancer

Although both the basal and luminal cells can serve as targets for transformation, it has been controversial whether the tumors originating from these two cell types are different with respect to aggressiveness. Some studies have suggested that either cell population can contain the cells of origin for aggressive prostate cancer. For example, human prostate basal cells preferentially express gene categories associated with stem cells and neurogenesis. The basal cell-specific gene signature is differentially enriched in various phenotypes of advanced prostate cancer such as anaplastic, metastatic and castration resistant prostate cancer [145, 146]. However, the gene signatures from the human prostate luminal progenitors are associated with higher mortality [70, 71]. These correlative observations probably reflect the fact that late stage tumor cells often turn on stemness-associated signaling to enhance their fitness during disease progression.

The best approach to investigate this question would be to introduce the same oncogenic signaling into the two epithelial cell types and evaluate the outcome. The prostate regeneration assay is not particularly useful to address this question because luminal cells cannot survive in this assay, hence cannot generate tumors. However, several recent studies managed to bypass this limitation by short-term expansion of the luminal cells in the prostate colony or organoid assay before

incubating them in the prostate regeneration assay. Investigators cultured human prostate luminal cells in the WIT™ media, transduced the cultured cells with activated AKT1, AR and ERG, and showed that the cells can regenerate prostate adenocarcinoma [70]. Other investigators transduced human basal and luminal cells with c-Myc and activated AKT1 and then expanded them separately in the organoid assay [140]. Organoid cells were then used for the regeneration assay. They showed that the organoid cells derived from the basal cells can generate aggressive tumors without acinar structures that do not express AR and PSA, whereas the organoid cells from the luminal cells that received the same oncogenic signaling generated well-differentiated PSA$^+$AR$^+$ acinar adenocarcinoma. Similarly, investigators showed that organoid cells derived from the CD38$^+$ human prostate luminal cells transduced with c-Myc, active AKT1, and AR also formed prostate adenocarcinoma [71]. Thus, *in vitro* expansion bypassed the limitation but also caused complexity as the condition of colonies and organoid culture may have reprogramed those cells in different ways. Otherwise, these studies may serve as strong evidence to support the concept that luminal cells may serve as a cell-of-origin for prostate adenocarcinoma and that basal cells and luminal cells can be transformed to generate tumors with distinct histologic features.

Genetically engineered mouse models may serve as a better approach to address this question. Three independent studies have employed this approach to disrupt the tumor suppressor Pten specifically in the mouse basal or luminal cells. One study showed that there is no distinct difference between the histomorphological features of the resulting tumors in the two groups, except that the PIN lesions started in the basal cell group with a long latency because basal-to-luminal differentiation had to occur first [53]. The limitation of this study is that the deletion of Pten in the basal cells took place very inefficiently. Another group reported that the tumors derived from the luminal cells are more aggressive [49]. However, this is probably because of the Nkx3.1 haploinsufficiency that was only introduced in the luminal cells as a result of the application of the Nkx3.1-CreER^{T2} model. A third group reported that the tumors derived from the basal cells showed more invasive features [54]. This is interesting as one would expect that loss of basal cells due to basal-to-luminal differentiation should facilitate growth of the tumor cells towards basement membrane. However, this observation was not noted in the other two studies and has not been confirmed by other groups since then. Besides these studies that investigated the lineage identity of the cells of origin for aggressive prostate cancer, an early study suggested that the anatomical location of cells-of-origin may also determine the aggressive feature of the resulting tumor. This study reported that the aggressive neuroendocrine tumor in a P53 and RB dual knockout model may arise from the cells in the mouse proximal prostatic ducts [147]. However, in contrast to this study, a recent study showed that Klf4 is expressed at a higher level in the mouse proximal prostate and that Klf4 suppresses transformation of prostate epithelial cells by oncogenic signaling such as activated AKT1 [148].

The mouse models also have limitations in addressing this question. First, oncogenic signaling introduced to the prostate basal and luminal cells may also be introduced to cells in other organs such as the basal cells in the skin and the epithelial cells in the small intestine. How to exclude the systemic influences of these undesired events on prostate initiation and progression should be considered. Second, the anatomical location and the number of the cells to be targeted are also critical. Ideally, one should compare tumor initiation and progression from the same number of basal and luminal cells at a similar anatomical location, which was not well-controlled in all the previous studies. This becomes even more complicated considering the intra-lineage heterogeneity described previously. Finally, the temporal regulation of introduced oncogenic signaling is also critical. It has been reported that tumors are initiated and progress faster when Pten is disrupted in younger mice than in older mice [149]. In the three studies mentioned above, the preexisting luminal cells were exposed to oncogenic signaling caused by loss of function of Pten earlier than those luminal cells derived from basal cells, which probably will have an impact on the initiation and progression of the resulting tumors. Therefore, it is still inconclusive whether the tumors derived from the two different cells of origin are different.

Summary and Future Directions

In summary, our knowledge of the prostate epithelial lineage hierarchy and the cells of origin remain incomplete. Although we have made much progress in understanding inter-lineage relationships, there are some missing connections. For example, are the basal cell-derived luminal cells in adults functionally equivalent to pre-existing luminal cells? What are the molecular signals that control basal-to-luminal differentiation? This signaling should play a role in the initiation of prostate cancer with a basal cell origin. Intra-lineage heterogeneity is an even more complicated issue. It remains a question whether there is a *bona fide* hierarchical structure within individual lineages, or whether the heterogeneities that we have discovered merely reflect phenotypic and functional plasticity of these cells in response to their different tissue microenvironments, such as anatomic location, interaction with the non-prostate cell lineages, inflammation, etc. Single cell RNA-seq analysis and high-resolution multiplex 3-D imaging may provide insights and lead to the identification of more specific markers that can be employed to test their functions *in vivo* using the lineage tracing approach. This may make it possible to identify definitively the cells of origin for prostate cancer and to determine their role in the clinical outcome of prostate cancer.

References

1. J.E. Visvader, Cells of origin in cancer. Nature **469**(7330), 314–322 (2011)
2. C. Blanpain, Tracing the cellular origin of cancer. Nat. Cell Biol. **15**(2), 126–134 (2013)

3. C. Liu, J.C. Sage, M.R. Miller, R.G. Verhaak, S. Hippenmeyer, H. Vogel, et al., Mosaic analysis with double markers reveals tumor cell of origin in glioma. Cell **146**(2), 209–221 (2011)

4. S.H. Lee, M.M. Shen, Cell types of origin for prostate cancer. Curr. Opin. Cell Biol. **37**, 35–41 (2015)

5. L. Xin, Cells of origin for cancer: an updated view from prostate cancer. Oncogene **32**(32), 3655–3663 (2013)

6. A.S. Goldstein, O.N. Witte, Does the microenvironment influence the cell types of origin for prostate cancer? Genes Dev. **27**(14), 1539–1544 (2013)

7. D. Zhang, S. Zhao, X. Li, J.S. Kirk, D.G. Tang, Prostate luminal progenitor cells in development and cancer. Trends Cancer. **4**(11), 769–783 (2018)

8. C.H. Jamieson, L.E. Ailles, S.J. Dylla, M. Muijtjens, C. Jones, J.L. Zehnder, et al., Granulocyte-macrophage progenitors as candidate leukemic stem cells in blast-crisis CML. N. Engl. J. Med. **351**(7), 657–667 (2004)

9. B.J. Huntly, H. Shigematsu, K. Deguchi, B.H. Lee, S. Mizuno, N. Duclos, et al., MOZ-TIF2, but not BCR-ABL, confers properties of leukemic stem cells to committed murine hematopoietic progenitors. Cancer Cell **6**(6), 587–596 (2004)

10. J.L. Kopp, G. von Figura, E. Mayes, F.F. Liu, C.L. Dubois, M. JPt, et al., Identification of Sox9-dependent acinar-to-ductal reprogramming as the principal mechanism for initiation of pancreatic ductal adenocarcinoma. Cancer Cell **22**(6), 737–750 (2012)

11. D. Yang, S.K. Denny, P.G. Greenside, A.C. Chaikovsky, J.J. Brady, Y. Ouadah, et al., Intertumoral heterogeneity in SCLC is influenced by the cell type of origin. Cancer Discov. **8**(10), 1316–1331 (2018)

12. T.A. Ince, A.L. Richardson, G.W. Bell, M. Saitoh, S. Godar, A.E. Karnoub, et al., Transformation of different human breast epithelial cell types leads to distinct tumor phenotypes. Cancer Cell **12**(2), 160–170 (2007)

13. M. El-Alfy, G. Pelletier, L.S. Hermo, F. Labrie, Unique features of the basal cells of human prostate epithelium. Microsc. Res. Tech. **51**(5), 436–446 (2000)

14. P.A. di Sant'Agnese, Neuroendocrine cells of the prostate and neuroendocrine differentiation in prostatic carcinoma: a review of morphologic aspects. Urology **51**(5A Suppl), 121–124 (1998)

15. M.A. Noordzij, G.J. van Steenbrugge, T.H. van der Kwast, F.H. Schroder, Neuroendocrine cells in the normal, hyperplastic and neoplastic prostate. Urol. Res. **22**(6), 333–341 (1995)

16. J.E. McNeal, E.A. Redwine, F.S. Freiha, T.A. Stamey, Zonal distribution of prostatic adenocarcinoma. Correlation with histologic pattern and direction of spread. Am. J. Surg. Pathol. **12**(12), 897–906 (1988)

17. A. Erbersdobler, H. Augustin, T. Schlomm, R.P. Henke, Prostate cancers in the transition zone: Part 1; pathological aspects. BJU Int. **94**(9), 1221–1225 (2004)

18. A. Staack, A.A. Donjacour, J. Brody, G.R. Cunha, P. Carroll, Mouse urogenital development: a practical approach. Differentiation **71**(7), 402–413 (2003)

19. J.C. Pignon, C. Grisanzio, Y. Geng, J. Song, R.A. Shivdasani, S. Signoretti, p63-expressing cells are the stem cells of developing prostate, bladder, and colorectal epithelia. Proc. Natl. Acad. Sci. U. S. A. **110**(20), 8105–8110 (2013)

20. A.S. Goldstein, D.A. Lawson, D. Cheng, W. Sun, I.P. Garraway, O.N. Witte, Trop2 identifies a subpopulation of murine and human prostate basal cells with stem cell characteristics. Proc. Natl. Acad. Sci. U. S. A. **105**(52), 20882–20887 (2008)

21. G. Aumuller, M. Leonhardt, M. Janssen, L. Konrad, A. Bjartell, P.A. Abrahamsson, Neurogenic origin of human prostate endocrine cells. Urology **53**(5), 1041–1048 (1999)

22. J. Szczyrba, A. Niesen, M. Wagner, P.M. Wandernoth, G. Aumuller, G. Wennemuth, Neuroendocrine cells of the prostate derive from the neural crest. J. Biol. Chem. **292**(5), 2021–2031 (2017)

23. Y. Wang, S. Hayward, M. Cao, K. Thayer, G. Cunha, Cell differentiation lineage in the prostate. Differentiation **68**(4–5), 270–279 (2001)

24. S.W. Hayward, L.S. Baskin, P.C. Haughney, A.R. Cunha, B.A. Foster, R. Dahiya, et al., Epithelial development in the rat ventral prostate, anterior prostate and seminal vesicle. Acta Anat. **155**(2), 81–93 (1996)

25. M. Ousset, A. Van Keymeulen, G. Bouvencourt, N. Sharma, Y. Achouri, B.D. Simons, et al., Multipotent and unipotent progenitors contribute to prostate postnatal development. Nat. Cell Biol. **14**(11), 1131–1138 (2012)

26. J. Wang, H.H. Zhu, M. Chu, Y. Liu, C. Zhang, G. Liu, et al., Symmetrical and asymmetrical division analysis provides evidence for a hierarchy of prostate epithelial cell lineages. Nat. Commun. **5**, 4758 (2014)

27. G.R. Cunha, The role of androgens in the epithelio-mesenchymal interactions involved in prostatic morphogenesis in embryonic mice. Anat. Rec. **175**(1), 87–96 (1973)

28. G.R. Cunha, Stromal induction and specification of morphogenesis and cytodifferentiation of the epithelia of the Mullerian ducts and urogenital sinus during development of the uterus and vagina in mice. J. Exp. Zool. **196**(3), 361–370 (1976)

29. B. Zhang, O.J. Kwon, G. Henry, A. Malewska, X. Wei, L. Zhang, et al., Non-cell-autonomous regulation of prostate epithelial homeostasis by androgen receptor. Mol. Cell **63**(6), 976–989 (2016)

30. J.T. Isaacs, Control of cell proliferation and death in normal and neoplastic prostate: a stem cell model, in *Benigh prostatic hyperplasia*, ed. by C. H. Rodgers, D. S. Coffey, G. R. Cunha, (National Institutes of Health, Bethesda, 1985), pp. 85–94

31. A.R. Uzgare, Y. Xu, J.T. Isaacs, In vitro culturing and characteristics of transit amplifying epithelial cells from human prostate tissue. J. Cell. Biochem. **91**(1), 196–205 (2004)
32. A.P. Verhagen, F.C. Ramaekers, T.W. Aalders, H.E. Schaafsma, F.M. Debruyne, J.A. Schalken, Colocalization of basal and luminal cell-type cytokeratins in human prostate cancer. Cancer Res. **52**(22), 6182–6187 (1992)
33. Y. Xue, F. Smedts, F.M. Debruyne, J.J. de la Rosette, J.A. Schalken, Identification of intermediate cell types by keratin expression in the developing human prostate. Prostate **34**(4), 292–301 (1998)
34. A.P. Verhagen, T.W. Aalders, F.C. Ramaekers, F.M. Debruyne, J.A. Schalken, Differential expression of keratins in the basal and luminal compartments of rat prostatic epithelium during degeneration and regeneration. Prostate **13**(1), 25–38 (1988)
35. E.J. Tokar, B.B. Ancrile, G.R. Cunha, M.M. Webber, Stem/progenitor and intermediate cell types and the origin of human prostate cancer. Differentiation **73**(9–10), 463–473 (2005)
36. J.K. Rane, A.P. Droop, D. Pellacani, E.S. Polson, M.S. Simms, A.T. Collins, et al., Conserved two-step regulatory mechanism of human epithelial differentiation. Stem Cell Rep. **2**(2), 180–188 (2014)
37. G. van Leenders, H. Dijkman, C. Hulsbergen-van de Kaa, D. Ruiter, J. Schalken, Demonstration of intermediate cells during human prostate epithelial differentiation in situ and in vitro using triple-staining confocal scanning microscopy. Lab. Invest. **80**(8), 1251–1258 (2000)
38. E.J. Robinson, D.E. Neal, A.T. Collins, Basal cells are progenitors of luminal cells in primary cultures of differentiating human prostatic epithelium. Prostate **37**(3), 149–160 (1998)
39. D.L. Hudson, M. O'Hare, F.M. Watt, J.R. Masters, Proliferative heterogeneity in the human prostate: evidence for epithelial stem cells. Lab. Invest. **80**(8), 1243–1250 (2000)
40. A.T. Collins, F.K. Habib, N.J. Maitland, D.E. Neal, Identification and isolation of human prostate epithelial stem cells based on alpha(2)beta(1)-integrin expression. J. Cell Sci. **114**(Pt 21), 3865–3872 (2001)
41. L.E. Lamb, B.S. Knudsen, C.K. Miranti, E-cadherin-mediated survival of androgen-receptor-expressing secretory prostate epithelial cells derived from a stratified in vitro differentiation model. J. Cell Sci. **123**(Pt 2), 266–276 (2010)
42. X. Shi, J. Gipp, W. Bushman, Anchorage-independent culture maintains prostate stem cells. Dev. Biol. **312**(1), 396–406 (2007)
43. L. Xin, R.U. Lukacs, D.A. Lawson, D. Cheng, O.N. Witte, Self-renewal and multilineage differentiation in vitro from murine prostate stem cells. Stem Cells **25**(11), 2760–2769 (2007)
44. L.W. Chung, G.R. Cunha, Stromal-epithelial interactions: II. Regulation of prostatic growth by embryonic urogenital sinus mesenchyme. Prostate **4**(5), 503–511 (1983)
45. L. Xin, H. Ide, Y. Kim, P. Dubey, O.N. Witte, In vivo regeneration of murine prostate from dissociated cell populations of postnatal epithelia and urogenital sinus mesenchyme. Proc. Natl. Acad. Sci. U. S. A. **100**(Suppl 1), 11896–11903 (2003)
46. A.S. Goldstein, J.M. Drake, D.L. Burnes, D.S. Finley, H. Zhang, R.E. Reiter, et al., Purification and direct transformation of epithelial progenitor cells from primary human prostate. Nat. Protoc. **6**(5), 656–667 (2011)
47. D.A. Lawson, L. Xin, R.U. Lukacs, D. Cheng, O.N. Witte, Isolation and functional characterization of murine prostate stem cells. Proc. Natl. Acad. Sci. U. S. A. **104**(1), 181–186 (2007)
48. L. Xin, D.A. Lawson, O.N. Witte, The Sca-1 cell surface marker enriches for a prostate-regenerating cell subpopulation that can initiate prostate tumorigenesis. Proc. Natl. Acad. Sci. U. S. A. **102**(19), 6942–6947 (2005)
49. Z.A. Wang, A. Mitrofanova, S.K. Bergren, C. Abate-Shen, R.D. Cardiff, A. Califano, et al., Lineage analysis of basal epithelial cells reveals their unexpected plasticity and supports a cell-of-origin model for prostate cancer heterogeneity. Nat. Cell Biol. **15**(3), 274–283 (2013)
50. P.E. Burger, X. Xiong, S. Coetzee, S.N. Salm, D. Moscatelli, K. Goto, et al., Sca-1 expression identifies stem cells in the proximal region of prostatic ducts with high capacity to reconstitute prostatic tissue. Proc. Natl. Acad. Sci. U. S. A. **102**(20), 7180–7185 (2005)
51. K.G. Leong, B.E. Wang, L. Johnson, W.Q. Gao, Generation of a prostate from a single adult stem cell. Nature **456**(7223), 804–808 (2008)
52. G.D. Richardson, C.N. Robson, S.H. Lang, D.E. Neal, N.J. Maitland, A.T. Collins, CD133, a novel marker for human prostatic epithelial stem cells. J. Cell Sci. **117**(Pt 16), 3539–3545 (2004)
53. N. Choi, B. Zhang, L. Zhang, M. Ittmann, L. Xin, Adult murine prostate basal and luminal cells are self-sustained lineages that can both serve as targets for prostate cancer initiation. Cancer Cell **21**(2), 253–265 (2012)
54. T.L. Lu, Y.F. Huang, L.R. You, N.C. Chao, F.Y. Su, J.L. Chang, et al., Conditionally ablated Pten in prostate basal cells promotes basal-to-luminal differentiation and causes invasive prostate cancer in mice. Am. J. Pathol. **182**(3), 975–991 (2013)
55. D.K. Lee, Y. Liu, L. Liao, F. Wang, J. Xu, The prostate basal cell (BC) heterogeneity and the p63-positive BC differentiation spectrum in mice. Int. J. Biol. Sci. **10**(9), 1007–1017 (2014)
56. T. Kurita, R.T. Medina, A.A. Mills, G.R. Cunha, Role of p63 and basal cells in the prostate. Development **131**(20), 4955–4964 (2004)
57. S. Signoretti, M.M. Pires, M. Lindauer, J.W. Horner, C. Grisanzio, S. Dhar, et al., p63 regulates commitment to the prostate cell lineage. Proc. Natl. Acad. Sci. U. S. A. **102**(32), 11355–11360 (2005)
58. S. Signoretti, D. Waltregny, J. Dilks, B. Isaac, D. Lin, L. Garraway, et al., p63 is a prostate basal

cell marker and is required for prostate development. Am. J. Pathol. **157**(6), 1769–1775 (2000)

59. J.K. Blackwood, S.C. Williamson, L.C. Greaves, L. Wilson, A.C. Rigas, R. Sandher, et al., In situ lineage tracking of human prostatic epithelial stem cell fate reveals a common clonal origin for basal and luminal cells. J. Pathol. **225**(2), 181–188 (2011)

60. M. Moad, E. Hannezo, S.J. Buczacki, L. Wilson, A. El-Sherif, D. Sims, et al., Multipotent basal stem cells, maintained in localized proximal niches, support directed long-ranging epithelial flows in human prostates. Cell Rep. **20**(7), 1609–1622 (2017)

61. C. Trapnell, D. Cacchiarelli, J. Grimsby, P. Pokharel, S. Li, M. Morse, et al., The dynamics and regulators of cell fate decisions are revealed by pseudotemporal ordering of single cells. Nat. Biotechnol. **32**(4), 381–386 (2014)

62. O.J. Kwon, J.M. Valdez, L. Zhang, B. Zhang, X. Wei, Q. Su, et al., Increased Notch signalling inhibits anoikis and stimulates proliferation of prostate luminal epithelial cells. Nat. Commun. **5**, 4416 (2014)

63. S.N. Salm, P.E. Burger, S. Coetzee, K. Goto, D. Moscatelli, E.L. Wilson, TGF-{beta} maintains dormancy of prostatic stem cells in the proximal region of ducts. J. Cell Biol. **170**(1), 81–90 (2005)

64. X. Wang, M. Kruithof-de Julio, K.D. Economides, D. Walker, H. Yu, M.V. Halili, et al., A luminal epithelial stem cell that is a cell of origin for prostate cancer. Nature **461**(7263), 495–500 (2009)

65. M. Germann, A. Wetterwald, N. Guzman-Ramirez, G. van der Pluijm, Z. Culig, M.G. Cecchini, et al., Stem-like cells with luminal progenitor phenotype survive castration in human prostate cancer. Stem Cells **30**(6), 1076–1086 (2012)

66. Y.A. Yoo, M. Roh, A.F. Naseem, B. Lysy, M.M. Desouki, K. Unno, et al., Bmi1 marks distinct castration-resistant luminal progenitor cells competent for prostate regeneration and tumour initiation. Nat. Commun. **7**, 12943 (2016)

67. O.J. Kwon, L. Zhang, L. Xin, Stem cell antigen-1 identifies a distinct androgen-independent murine prostatic luminal cell lineage with bipotent potential. Stem Cells **34**(1), 191–202 (2015)

68. W.R. Karthaus, P.J. Iaquinta, J. Drost, A. Gracanin, R. van Boxtel, J. Wongvipat, et al., Identification of multipotent luminal progenitor cells in human prostate organoid cultures. Cell **159**(1), 163–175 (2014)

69. C.W. Chua, M. Shibata, M. Lei, R. Toivanen, L.J. Barlow, S.K. Bergren, et al., Single luminal epithelial progenitors can generate prostate organoids in culture. Nat. Cell Biol. **16**(10), 951–961 (2014)., 1–4

70. D. Zhang, K. Lin, Y. Lu, K. Rycaj, Y. Zhong, H.P. Chao, et al., Developing a novel two-dimensional culture system to enrich human prostate luminal progenitors that can function as a cell of origin for prostate cancer. Stem Cells Transl. Med. **6**(3), 748–760 (2017)

71. X. Liu, T.R. Grogan, H. Hieronymus, T. Hashimoto, J. Mottahedeh, D. Cheng, et al., Low CD38 identifies progenitor-like inflammation-associated luminal cells that can initiate human prostate cancer and predict poor outcome. Cell Rep. **17**(10), 2596–2606 (2016)

72. A. Dutta, C. Le Magnen, A. Mitrofanova, X. Ouyang, A. Califano, C. Abate-Shen, Identification of an NKX3.1-G9a-UTY transcriptional regulatory network that controls prostate differentiation. Science **352**(6293), 1576–1580 (2016)

73. J. Liu, L.E. Pascal, S. Isharwal, D. Metzger, R. Ramos Garcia, J. Pilch, et al., Regenerated luminal epithelial cells are derived from preexisting luminal epithelial cells in adult mouse prostate. Mol. Endocrinol. **25**(11), 1849–1857 (2011)

74. G.S. Evans, J.A. Chandler, Cell proliferation studies in the rat prostate: II. The effects of castration and androgen-induced regeneration upon basal and secretory cell proliferation. Prostate **11**(4), 339–351 (1987)

75. G.S. Evans, J.A. Chandler, Cell proliferation studies in rat prostate. I. The proliferative role of basal and secretory epithelial cells during normal growth. Prostate **10**(2), 163–178 (1987)

76. C. Guo, H. Liu, B.H. Zhang, R.M. Cadaneanu, A.M. Mayle, I.P. Garraway, Epcam, CD44, and CD49f distinguish sphere-forming human prostate basal cells from a subpopulation with predominant tubule initiation capability. PLoS One **7**(4), e34219 (2012)

77. D.J. Vander Griend, W.L. Karthaus, S. Dalrymple, A. Meeker, A.M. DeMarzo, J.T. Isaacs, The role of CD133 in normal human prostate stem cells and malignant cancer-initiating cells. Cancer Res. **68**(23), 9703–9711 (2008)

78. W.Y. Hu, D.P. Hu, L. Xie, Y. Li, S. Majumdar, L. Nonn, et al., Isolation and functional interrogation of adult human prostate epithelial stem cells at single cell resolution. Stem Cell Res. **23**, 1–12 (2017)

79. B.E. Wang, X. Wang, J.E. Long, J. Eastham-Anderson, R. Firestein, M.R. Junttila, Castration-resistant Lgr5(+) cells are long-lived stem cells required for prostatic regeneration. Stem Cell Rep. **4**(5), 768–779 (2015)

80. X. Wei, A.V. Orjalo, L. Xin, CD133 does not enrich for the stem cell activity in vivo in adult mouse prostates. Stem Cell Res. **16**(3), 597–606 (2016)

81. L. Zhang, J.M. Valdez, B. Zhang, L. Wei, J. Chang, L. Xin, ROCK inhibitor Y-27632 suppresses dissociation-induced apoptosis of murine prostate stem/progenitor cells and increases their cloning efficiency. PLoS One **6**(3), e18271 (2011)

82. C. Zhang, H.J. Lee, A. Shrivastava, R. Wang, T.J. McQuiston, S.S. Challberg, et al., Long-term in vitro expansion of epithelial stem cells enabled by pharmacological inhibition of PAK1-ROCK-Myosin II and TGF-beta signaling. Cell Rep. **25**(3), 598–610.e5 (2018)

83. J.C. Pignon, C. Grisanzio, I. Carvo, L. Werner, M. Regan, E.L. Wilson, et al., Cell kinetic studies fail to identify sequentially proliferating progenitors as the major source of epithelial renewal in the adult murine prostate. PLoS One 10(5), e0128489 (2015)

84. A. Tsujimura, Y. Koikawa, S. Salm, T. Takao, S. Coetzee, D. Moscatelli, et al., Proximal location of mouse prostate epithelial stem cells: a model of prostatic homeostasis. J. Cell Biol. 157(7), 1257–1265 (2002)

85. D. Zhang, C. Jeter, S. Gong, A. Tracz, Y. Lu, J. Shen, et al., Histone 2B-GFP label-retaining prostate luminal cells possess progenitor cell properties and are intrinsically resistant to castration. Stem Cell Rep 10(1), 228–242 (2018)

86. J.D. Barros-Silva, D.E. Linn, I. Steiner, G. Guo, A. Ali, H. Pakula, et al., Single-cell analysis identifies LY6D as a marker linking castration-resistant prostate luminal cells to prostate progenitors and cancer. Cell Rep. 25(12), 3504–18.e6 (2018)

87. S. Agarwal, P.G. Hynes, H.S. Tillman, R. Lake, W.G. Abou-Kheir, L. Fang, et al., Identification of different classes of luminal progenitor cells within prostate tumors. Cell Rep. 13(10), 2147–2158 (2015)

88. D.E. Abbott, C. Pritchard, N.J. Clegg, C. Ferguson, R. Dumpit, R.A. Sikes, et al., Expressed sequence tag profiling identifies developmental and anatomic partitioning of gene expression in the mouse prostate. Genome Biol. 4(12), R79 (2003)

89. I.M. Berquin, Y. Min, R. Wu, H. Wu, Y.Q. Chen, Expression signature of the mouse prostate. J. Biol. Chem. 280(43), 36442–36451 (2005)

90. R.J. Cohen, G. Glezerson, L.F. Taylor, H.A. Grundle, J.H. Naude, The neuroendocrine cell population of the human prostate gland. J. Urol. 150(2 Pt 1), 365–368 (1993)

91. X. Wei, L. Zhang, Z. Zhou, O.J. Kwon, Y. Zhang, H. Nguyen, et al., Spatially restricted stromal Wnt signaling restrains prostate epithelial progenitor growth through direct and indirect mechanisms. Cell Stem Cell 24(5), 753–68.e6 (2019)

92. G.H. Henry, A. Malewska, D.B. Joseph, V.S. Malladi, J. Lee, J. Torrealba, et al., A cellular anatomy of the normal adult human prostate and prostatic urethra. Cell Rep. 25(12), 3530–42.e5 (2018)

93. K. Takahashi, S. Yamanaka, Induction of pluripotent stem cells from mouse embryonic and adult fibroblast cultures by defined factors. Cell 126(4), 663–676 (2006)

94. O.J. Kwon, L. Zhang, M.M. Ittmann, L. Xin, Prostatic inflammation enhances basal-to-luminal differentiation and accelerates initiation of prostate cancer with a basal cell origin. Proc. Natl. Acad. Sci. U. S. A. 111(5), E592–E600 (2014)

95. R. Toivanen, A. Mohan, M.M. Shen, Basal progenitors contribute to repair of the prostate epithelium following induced luminal anoikis. Stem Cell Rep. 6(5), 660–667 (2016)

96. A.M. De Marzo, Y. Nakai, W.G. Nelson, Inflammation, atrophy, and prostate carcinogenesis. Urol. Oncol. 25(5), 398–400 (2007)

97. A.M. De Marzo, E.A. Platz, S. Sutcliffe, J. Xu, H. Gronberg, C.G. Drake, et al., Inflammation in prostate carcinogenesis. Nat. Rev. Cancer 7(4), 256–269 (2007)

98. C.M. Perou, T. Sorlie, M.B. Eisen, M. van de Rijn, S.S. Jeffrey, C.A. Rees, et al., Molecular portraits of human breast tumours. Nature 406(6797), 747–752 (2000)

99. B.S. Taylor, N. Schultz, H. Hieronymus, A. Gopalan, Y. Xiao, B.S. Carver, et al., Integrative genomic profiling of human prostate cancer. Cancer Cell 18(1), 11–22 (2010)

100. G.V. Glinsky, A.B. Glinskii, A.J. Stephenson, R.M. Hoffman, W.L. Gerald, Gene expression profiling predicts clinical outcome of prostate cancer. J. Clin. Invest. 113(6), 913–923 (2004)

101. A.S. Goldstein, Y. Zong, O.N. Witte, A two-step toward personalized therapies for prostate cancer. Sci. Transl. Med. 3(72), 72ps7 (2011)

102. T.Z. Ali, J.I. Epstein, False positive labeling of prostate cancer with high molecular weight cytokeratin: p63 a more specific immunomarker for basal cells. Am. J. Surg. Pathol. 32(12), 1890–1895 (2008)

103. X.J. Yang, K. Lecksell, P. Gaudin, J.I. Epstein, Rare expression of high-molecular-weight cytokeratin in adenocarcinoma of the prostate gland: a study of 100 cases of metastatic and locally advanced prostate cancer. Am. J. Surg. Pathol. 23(2), 147–152 (1999)

104. P.A. Humphrey, Histological variants of prostatic carcinoma and their significance. Histopathology 60(1), 59–74 (2012)

105. S. Sanati, M.A. Watson, A.L. Salavaggione, P.A. Humphrey, Gene expression profiles of ductal versus acinar adenocarcinoma of the prostate. Mod. Pathol. 22(10), 1273–1279 (2009)

106. T.Z. Ali, J.I. Epstein, Basal cell carcinoma of the prostate: a clinicopathologic study of 29 cases. Am. J. Surg. Pathol. 31(5), 697–705 (2007)

107. S.Y. Nakada, P.A. di Sant' Agnese, R.A. Moynes, R.A. Hiipakka, S. Liao, A.T. Cockett, et al., The androgen receptor status of neuroendocrine cells in human benign and malignant prostatic tissue. Cancer Res. 53(9), 1967–1970 (1993)

108. A.H. Davies, H. Beltran, A. Zoubeidi, Cellular plasticity and the neuroendocrine phenotype in prostate cancer. Nat. Rev. Urol. 15(5), 271–286 (2018)

109. J.L. Yao, R. Madeb, P. Bourne, J. Lei, X. Yang, S. Tickoo, et al., Small cell carcinoma of the prostate: an immunohistochemical study. Am. J. Surg. Pathol. 30(6), 705–712 (2006)

110. M. Zou, R. Toivanen, A. Mitrofanova, N. Floch, S. Hayati, Y. Sun, et al., Transdifferentiation as a mechanism of treatment resistance in a mouse model of castration-resistant prostate cancer. Cancer Discov. 7(7), 736–749 (2017)

111. D. Lin, A.W. Wyatt, H. Xue, Y. Wang, X. Dong, A. Haegert, et al., High fidelity patient-derived xeno-

grafts for accelerating prostate cancer discovery and drug development. Cancer Res. **74**(4), 1272–1283 (2014)

112. S. Jaamaa, T.M. Af Hallstrom, A. Sankila, V. Rantanen, H. Koistinen, U.H. Stenman, et al., DNA damage recognition via activated ATM and p53 pathway in nonproliferating human prostate tissue. Cancer Res. **70**(21), 8630–8641 (2010)

113. A.C. Hsieh, H.G. Nguyen, L. Wen, M.P. Edlind, P.R. Carroll, W. Kim, et al., Cell type-specific abundance of 4EBP1 primes prostate cancer sensitivity or resistance to PI3K pathway inhibitors. Sci. Signal. **8**(403), ra116 (2015)

114. B. Gurel, T. Iwata, C.M. Koh, R.B. Jenkins, F. Lan, C. Van Dang, et al., Nuclear MYC protein overexpression is an early alteration in human prostate carcinogenesis. Mod. Pathol. **21**(9), 1156–1167 (2008)

115. A.K. Meeker, J.L. Hicks, E.A. Platz, G.E. March, C.J. Bennett, M.J. Delannoy, et al., Telomere shortening is an early somatic DNA alteration in human prostate tumorigenesis. Cancer Res. **62**(22), 6405–6409 (2002)

116. S.A. Tomlins, D.R. Rhodes, S. Perner, S.M. Dhanasekaran, R. Mehra, X.W. Sun, et al., Recurrent fusion of TMPRSS2 and ETS transcription factor genes in prostate cancer. Science **310**(5748), 644–648 (2005)

117. E. Lim, F. Vaillant, D. Wu, N.C. Forrest, B. Pal, A.H. Hart, et al., Aberrant luminal progenitors as the candidate target population for basal tumor development in BRCA1 mutation carriers. Nat. Med. **15**(8), 907–913 (2009)

118. H. Bonkhoff, U. Stein, K. Remberger, The proliferative function of basal cells in the normal and hyperplastic human prostate. Prostate **24**(3), 114–118 (1994)

119. J.A. Tuxhorn, G.E. Ayala, D.R. Rowley, Reactive stroma in prostate cancer progression. J. Urol. **166**(6), 2472–2483 (2001)

120. E.S. Polson, J.L. Lewis, H. Celik, V.M. Mann, M.J. Stower, M.S. Simms, et al., Monoallelic expression of TMPRSS2/ERG in prostate cancer stem cells. Nat. Commun. **4**, 1623 (2013)

121. Y. Zong, L. Xin, A.S. Goldstein, D.A. Lawson, M.A. Teitell, O.N. Witte, ETS family transcription factors collaborate with alternative signaling pathways to induce carcinoma from adult murine prostate cells. Proc. Natl. Acad. Sci. U. S. A. **106**(30), 12465–12470 (2009)

122. O. Klezovitch, M. Risk, I. Coleman, J.M. Lucas, M. Null, L.D. True, et al., A causal role for ERG in neoplastic transformation of prostate epithelium. Proc. Natl. Acad. Sci. U. S. A. **105**(6), 2105–2110 (2008)

123. N.M. Greenberg, F.J. DeMayo, P.C. Sheppard, R. Barrios, R. Lebovitz, M. Finegold, et al., The rat probasin gene promoter directs hormonally and developmentally regulated expression of a heterologous gene specifically to the prostate in transgenic mice. Mol. Endocrinol. **8**(2), 230–239 (1994)

124. J. Zhang, T.Z. Thomas, S. Kasper, R.J. Matusik, A small composite probasin promoter confers high levels of prostate-specific gene expression through regulation by androgens and glucocorticoids in vitro and in vivo. Endocrinology **141**(12), 4698–4710 (2000)

125. X. Wu, J. Wu, J. Huang, W.C. Powell, J. Zhang, R.J. Matusik, et al., Generation of a prostate epithelial cell-specific Cre transgenic mouse model for tissue-specific gene ablation. Mech. Dev. **101**(1–2), 61–69 (2001)

126. N. Masumori, T.Z. Thomas, P. Chaurand, T. Case, M. Paul, S. Kasper, et al., A probasin-large T antigen transgenic mouse line develops prostate adenocarcinoma and neuroendocrine carcinoma with metastatic potential. Cancer Res. **61**(5), 2239–2249 (2001)

127. S. Wang, J. Gao, Q. Lei, N. Rozengurt, C. Pritchard, J. Jiao, et al., Prostate-specific deletion of the murine Pten tumor suppressor gene leads to metastatic prostate cancer. Cancer Cell **4**(3), 209–221 (2003)

128. V.D. Acevedo, R.D. Gangula, K.W. Freeman, R. Li, Y. Zhang, F. Wang, et al., Inducible FGFR-1 activation leads to irreversible prostate adenocarcinoma and an epithelial-to-mesenchymal transition. Cancer Cell **12**(6), 559–571 (2007)

129. P.K. Majumder, J.J. Yeh, D.J. George, P.G. Febbo, J. Kum, Q. Xue, et al., Prostate intraepithelial neoplasia induced by prostate restricted Akt activation: the MPAKT model. Proc. Natl. Acad. Sci. U. S. A. **100**(13), 7841–7846 (2003)

130. K. Ellwood-Yen, T.G. Graeber, J. Wongvipat, M.L. Iruela-Arispe, J. Zhang, R. Matusik, et al., Myc-driven murine prostate cancer shares molecular features with human prostate tumors. Cancer Cell **4**(3), 223–238 (2003)

131. X. Wu, K. Xu, L. Zhang, Y. Deng, P. Lee, E. Shapiro, et al., Differentiation of the ductal epithelium and smooth muscle in the prostate gland are regulated by the Notch/PTEN-dependent mechanism. Dev. Biol. **356**(2), 337–349 (2011)

132. P.C. Weijerman, H.C. Romijn, D.M. Peehl, Human papilloma virus type 18 DNA immortalized cell lines from the human prostate epithelium. Prog. Clin. Biol. Res. **386**, 67–69 (1994)

133. P.C. Weijerman, J.J. Konig, S.T. Wong, H.G. Niesters, D.M. Peehl, Lipofection-mediated immortalization of human prostatic epithelial cells of normal and malignant origin using human papillomavirus type 18 DNA. Cancer Res. **54**(21), 5579–5583 (1994)

134. R. Berger, P.G. Febbo, P.K. Majumder, J.J. Zhao, S. Mukherjee, S. Signoretti, et al., Androgen-induced differentiation and tumorigenicity of human prostate epithelial cells. Cancer Res. **64**(24), 8867–8875 (2004)

135. A.S. Goldstein, J. Huang, C. Guo, I.P. Garraway, O.N. Witte, Identification of a cell of origin for human prostate cancer. Science **329**(5991), 568–571 (2010)

136. D.A. Lawson, Y. Zong, S. Memarzadeh, L. Xin, J. Huang, O.N. Witte, Basal epithelial stem cells are efficient targets for prostate cancer initiation. Proc. Natl. Acad. Sci. U. S. A. **107**(6), 2610–2615 (2010)

137. L. Xin, M.A. Teitell, D.A. Lawson, A. Kwon, I.K. Mellinghoff, O.N. Witte, Progression of prostate cancer by synergy of AKT with genotropic and nongenotropic actions of the androgen receptor. Proc. Natl. Acad. Sci. U. S. A. **103**(20), 7789–7794 (2006)

138. T. Stoyanova, M. Riedinger, S. Lin, C.M. Faltermeier, B.A. Smith, K.X. Zhang, et al., Activation of Notch1 synergizes with multiple pathways in promoting castration-resistant prostate cancer. Proc. Natl. Acad. Sci. U. S. A. **113**(42), E6457–E6E66 (2016)

139. T. Stoyanova, A.R. Cooper, J.M. Drake, X. Liu, A.J. Armstrong, K.J. Pienta, et al., Prostate cancer originating in basal cells progresses to adenocarcinoma propagated by luminal-like cells. Proc. Natl. Acad. Sci. U. S. A. **110**(50), 20111–20116 (2013)

140. J.W. Park, J.K. Lee, J.W. Phillips, P. Huang, D. Cheng, J. Huang, et al., Prostate epithelial cell of origin determines cancer differentiation state in an organoid transformation assay. Proc. Natl. Acad. Sci. U. S. A. **113**(16), 4482–4487 (2016)

141. J.K. Lee, J.W. Phillips, B.A. Smith, J.W. Park, T. Stoyanova, E.F. McCaffrey, et al., N-Myc drives neuroendocrine prostate cancer initiated from human prostate epithelial cells. Cancer Cell **29**(4), 536–547 (2016)

142. J.W. Park, J.K. Lee, K.M. Sheu, L. Wang, N.G. Balanis, K. Nguyen, et al., Reprogramming normal human epithelial tissues to a common, lethal neuroendocrine cancer lineage. Science **362**(6410), 91–95 (2018)

143. R.A. Taylor, R. Toivanen, M. Frydenberg, J. Pedersen, L. Harewood, Australian Prostate Cancer B, et al., Human epithelial basal cells are cells of origin of prostate cancer, independent of CD133 status. Stem Cells **30**(6), 1087–1096 (2012)

144. Z.A. Wang, R. Toivanen, S.K. Bergren, P. Chambon, M.M. Shen, Luminal cells are favored as the cell of origin for prostate cancer. Cell Rep. **8**(5), 1339–1346 (2014)

145. B.A. Smith, A. Sokolov, V. Uzunangelov, R. Baertsch, Y. Newton, K. Graim, et al., A basal stem cell signature identifies aggressive prostate cancer phenotypes. Proc. Natl. Acad. Sci. U. S. A. **112**(47), E6544–E6552 (2015)

146. D. Zhang, D. Park, Y. Zhong, Y. Lu, K. Rycaj, S. Gong, et al., Stem cell and neurogenic gene-expression profiles link prostate basal cells to aggressive prostate cancer. Nat. Commun. **7**, 10798 (2016)

147. Z. Zhou, A. Flesken-Nikitin, A.Y. Nikitin, Prostate cancer associated with p53 and Rb deficiency arises from the stem/progenitor cell-enriched proximal region of prostatic ducts. Cancer Res. **67**(12), 5683–5690 (2007)

148. X. Xiong, M. Schober, E. Tassone, A. Khodadadi-Jamayran, A. Sastre-Perona, H. Zhou, et al., KLF4, a gene regulating prostate stem cell homeostasis, is a barrier to malignant progression and predictor of good prognosis in prostate cancer. Cell Rep. **25**(11), 3006–20.e7 (2018)

149. C.K. Ratnacaram, M. Teletin, M. Jiang, X. Meng, P. Chambon, D. Metzger, Temporally controlled ablation of PTEN in adult mouse prostate epithelium generates a model of invasive prostatic adenocarcinoma. Proc. Natl. Acad. Sci. U. S. A. **105**(7), 2521–2526 (2008)

Prostate Cancer Genomic Subtypes

Michael Fraser and Alexandre Rouette

The Challenges of (Prostate) Cancer Genome Analysis

In 2000, the Human Genome Project published the first draft version of the human genome sequence [1]. Assembled through whole-genome shotgun sequencing using pairwise end Sanger sequencing over a period of 10 years, the Human Genome Project cost approximately $3 billion, and originally contained largely euchromatic regions of the genome with very low redundancy (*i.e.* read depth). The development of massively-parallel sequencing in the mid-2000s and the subsequent refinement of this technology had three primary effects: first, it became possible to sequence large target sequences, such as the human genome (~3000 Mb), in relatively short amounts of time; second, the cost of this sequencing was dramatically lower than Sanger-based approaches (a 10^5- to 10^6-fold reduction, as of 2019); third, the reduction in both throughput time and cost per base allowed for highly redundant sequencing, whereby each base can be sequenced 30 or more times, thereby increasing confidence in the called base and allowing for improved detection of low-frequency somatic variants.

However, sequencing of cancer genomes presents fundamental challenges relative to sequencing of germline genomes using DNA derived from purified leukocytes. This is due to the fact that cancer is characterized by the acquisition of mutations over time, as well as the divergence of clonal populations of cells during tumor evolution. While a mutation that arises early in tumor evolution will be present in the entire resultant cell clones, which arise from the progenitor, mutations that occur later during tumor evolution will be, necessarily, restricted to subclonal populations that arise following the acquisition of that mutation. This gives rise to a genomic heterogeneity that is absent from germline genomes. Moreover, cell populations from primary solid tumors are rarely free from contamination with non-malignant cell types (*e.g.* stroma, adjacent benign epithelium, etc.), thus reducing the relative abundance of tumor-associated mutant alleles—even those that are clonal—within mixed tumor-normal cell populations.

For these reasons, accurate analysis of cancer genomes typically requires two to three times higher sequencing coverage than analysis of germline genomes. Increasing the average number of reads for each base in the genome not only enhances confidence in each base call, but also increases the likelihood that low-frequency variants—either due to sub-clonality or to contamination with genomic DNA from non-malignant sources—will be detected. Naturally, there are

---

M. Fraser (✉) · A. Rouette
Computational Biology Program, Ontario Institute for Cancer Research, Toronto, ON, Canada
e-mail: Michael.Fraser@oicr.on.ca

© Springer Nature Switzerland AG 2019
S. M. Dehm, D. J. Tindall (eds.), *Prostate Cancer*, Advances in Experimental Medicine and Biology 1210, https://doi.org/10.1007/978-3-030-32656-2_5

trade-offs between depth of sequencing coverage, sample throughput, and total cost per cancer genome sequenced. Analysis of cancer genomes has also depended upon the presence of a germline genome sequence of the same individual. This is particularly important given that most tumor sequencing studies rely upon mapping of short (~100–150 bp) sequencing reads, with minimal overlap, and the fact that alignment to the reference genome remains computationally daunting [2].

Given these challenges, it is not surprising that the first cancer genomes to be sequenced were derived from liquid tumors such as acute myeloid leukemia [3], where normal contamination can be reduced or eliminated by purification of distinct cell populations using flow cytometry-based approaches, and where normal genomes can be readily sequenced using non-malignant lymphocytes from the same biospecimen. In these cancers, the depth of sequencing required to detect mutations of identical abundance is comparatively lower than in solid tumors, thus reducing the total cost per genome.

During the nascent phase of cancer genome sequencing, many groups—particularly those working on solid tumors—elected to sequence cancer whole exomes, rather than genomes; the Cancer Genome Atlas (TCGA) program typifies this approach. In whole exome sequencing, the coding regions of the genome are first purified from genomic DNA using an *in vitro* capture approach. Because they represent only ~2% of the whole genome, exomes can be sequenced to higher depths than genomes for the same cost (or, alternatively, equivalent depths for lower cost). Thus, whole exome sequencing allows for detection of low-frequency *coding* variants with a much greater sensitivity than whole genome sequencing. However, this comes at the cost of the inability to detect molecular aberrations outside of the coding region, or to assess genome-wide structural variation. As we will see, this is of great importance for the analysis of C-class cancers [4], such as prostate cancer, where noncoding variation, gene copy number aberrations, and structural variation are key drivers of tumorigenesis and disease progression.

The Prostate Cancer Genome

When discussing prostate cancer genomics, it is critical to distinguish between localized and metastatic disease. In the era of widespread prostate specific antigen (PSA) testing, the vast majority of new diagnoses are of localized, low grade disease with excellent prognosis [5], with comparatively few diagnoses of primary *de novo* metastatic disease. In some cases, metastatic, castration-resistant prostate cancer (mCRPC) develops following failure of both local therapy (radical prostatectomy and/or radiotherapy) and systemic androgen deprivation therapy (ADT), and may itself be treated with additional systemic therapies such as docetaxel [6]. These systemic therapies induce a strong selective pressure, which can alter the genomic make-up of metastatic lesions, relative to the primary tumor from which they were initially seeded. Indeed, recent evidence has demonstrated a clear selection—both positive and negative—on specific molecular alterations during ADT [7]. As such, it is appropriate to consider the molecular landscape of localized prostate cancer as fundamentally distinct from that of metastatic disease. Nevertheless, as we will see, important insights into prostate cancer development, progression, and clinical aggression can be inferred from a deep analysis of the localized, treatment-naïve disease state.

The Somatic Molecular Landscape of Localized Prostate Cancer

Molecular interrogation of localized prostate cancer has traditionally been hindered by the relatively slow growth of clinically-relevant cancers as well as focal heterogeneity and low tumor cellularity. These constraints have limited analysis to highly cellular tumors with sufficiently high DNA yields for deep next generation sequencing. Indeed, the first description of the whole genome landscape of localized prostate cancer did not emerge until 2011 [8], several years after the publication of the first tumor whole genomes [3] (Table 1).

Prostate Cancer Genomic Subtypes

Table 1 Summary of whole-genome sequencing studies of localized prostate cancer

References	Number and type of whole genomes sequenced	Major significance
Berger et al. [8]	7 primary high risk tumors	First whole-genome sequencing of localized prostate cancer. Identification of closed-loop chain rearrangements
Baca et al. [24]	55 primary tumors, 2 neuroendocrine metastases	Characterization of temporal changes in prostate cancer structural variation ("chromoplexy")
Weischenfeldt et al. [26]	11 early-onset primary tumors 7 elderly-onset primary tumors	Androgen-dependent structural variation enriched in prostate cancers arising in men <50 years of age
Cancer Genome Atlas Research Network [16]	19 primary tumors	Molecular subclasses of localized prostate cancers
Boutros et al. [57]	23 malignant foci from 5 primary tumors	Spatial heterogeneity of localized prostate cancer
Cooper et al. [89]	12 malignant foci from 3 primary tumors	Spatial-temporal heterogeneity of localized prostate cancer. Identification of aberrations in morphologically-normal prostate epithelium
Fraser et al. [25]	200 intermediate risk primary tumors	Largest study of prostate cancer whole genomes to date. Identification of recurrent driver aberrations linked to adverse clinical outcome
Taylor et al. [65]	19 disease foci from 14 germline *BRCA2* mutation carriers	Tumor genomes of *BRCA2* mutation carriers closely resemble those of castration-resistant metastatic disease. MED12/MED12L pathway as driver of clinical aggression
Camacho et al. [131]	103 primary tumors	Assessment of somatic genome-wide copy number aberrations and mechanism of copy number loss
Ren et al. [38]	65 primary tumors from Chinese men	Low frequency of *TMPRSS2:ERG* fusion in Chinese prostate cancers. Identification of novel tumor suppressor genes
Espiritu et al. [56]	93 intermediate risk primary tumors	Analysis of the temporal evolution of prostate cancer. Development of a clonality-aware multi-modal biomarker of adverse clinical outcome
Wedge et al. [37]	87 primary tumors, 20 metastatic lesions	Temporal evolution of prostate cancer. Identification of potential druggable targets in localized disease
Su et al. [132]	17 nuclei from 2 primary tumors	First report of single nucleus whole-genome sequencing in prostate cancer. Significant spatial heterogeneity within the same gland
Gerhauser et al. [96]	292 primary tumors from men <55 years of age	Characterization of mutational signatures impacting early evolution of prostate cancer in younger men. Identification of clinically-relevant subgroups based on multi-modal profiling

The origins of our understanding of the molecular determinants of prostate cancer tumorigenesis and aggression predates the current genomic era. The effectiveness of androgen ablation for the treatment of prostate cancer has been recognized for decades [9], which strongly (and accurately) suggested a link between AR activity and prostate cancer progression [10]. Clinical studies revealed that androgen receptor mutations and amplification events are common in advanced prostate cancer and may arise during the course of androgen-deprivation therapy [11, 12].

A seminal discovery was that the *ETS*-family oncogene, *ERG*, is over-expressed in 40–50% of primary prostate cancers [13] and that this is secondary to an androgen-dependent genomic rearrangement (GR) on chromosome 21q22.2-22.3, producing a fusion product involving the 5′ regulatory region of the *TMPRSS2* gene and the coding region of *ERG* [14] (Fig. 1). This fusion results either from deletion of the intervening ~2.8 Mbp between the *TMPRSS2* and *ERG* genes ('edel') or translocation of the intervening region to other locations ('esplit') [15]. All can be

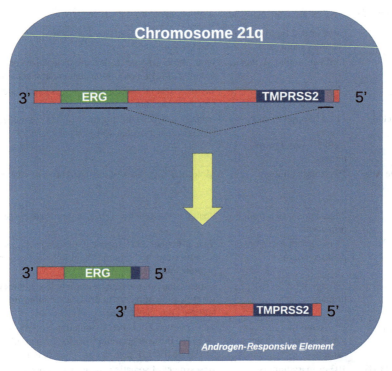

Fig. 1 Androgen-dependent fusion of the TMPRSS2 and ERG genes. The *TMPRSS2* and *ERG* genes are transcribed from the (−) strand of chromosome 21q22.3. *TMPRSS2* contains a strong Androgen-Responsive Element in its 5′ promoter/enhancer region (5′-ARE; purple). Fusion of the 5′-ARE (and often a small portion of the first exons of *TMPRSS2*) to the coding region of the *ERG* oncogene results in a fusion product (*TMPRSS2:ERG*; T2E) in which ERG expression is induced in an androgen-responsive manner. This can be directly detected by fluorescence in situ hybridization (FISH) [28], RNA- or DNA sequencing, qPCR, or indirectly through ERG immunohistochemistry. In approximately 50% of cases, the T2E fusion is accompanied by deletion of the intervening sequence ('edel' fusions), while in the other 50%, the intervening sequence is retained, often through integration into other chromosomal locations. These can be distinguished using break-apart FISH [15]

distinguished using three-color 'break-apart' fluorescence *in situ* hybridization (FISH) analysis, while edel fusions can also be inferred from copy number loss of the intervening genomic region (see below).

The *TMPRSS2:ERG* fusion (T2E) is present in ~45% of all localized PCs, while another ~5–10% harbor a fusion involving another *ETS*-family proto-oncogene (e.g. *ETV1*, *ETV4*) [16]. Fusion of the *TMPRSS2* promoter and enhancer region, which contain strong androgen-responsive elements (AREs), to the 5′ end of *ERG* leads to androgen-driven ERG overexpression, which can be detected in clinical specimens by immunohistochemistry against ERG [17, 18]. The *T2E* fusion is also readily detectable using fluorescence *in-situ* hybridization (FISH) and qPCR [15, 19]. Molecular heterogeneity and sub-clonality studies have consistently demonstrated that the *T2E* fusion is one of the earliest molecular events in prostate tumorigenesis [20]. Despite this, no clear picture has emerged regarding the precise function of *T2E* in this regard. Moreover, large studies have failed to establish a prognostic effect of *T2E* for differential clinical outcomes in localized disease [21, 22], although tumors harboring the fusion do show unique transcriptional programming resulting in a dependency on NOTCH signaling [23], which could make T2E a predictive biomarker for sensitivity to NOTCH pathway inhibitors.

Hallmarks of Prostate Cancer Genomes

Single Nucleotide Variation

Prostate cancer is a C-class tumor, typified by a paucity of driver single nucleotide variants (SNVs) [4]. Relative to other solid tumor types, localized prostate cancer harbors an intermediate global SNV burden of ~0.5–1 SNVs/megabase [8, 16, 24–26]. Meta-analyses of prostate cancer exome sequencing studies have revealed that the most frequently mutated genes in localized prostate cancer include *SPOP*, *FOXA1*, and *TP53* [25]. Strikingly, no single gene is mutated at >10–12% of localized prostate cancers [25], consistent with the hypothesis that prostate tumorigenesis is primarily driven by other classes of molecular aberration (see below).

In contrast, mCRPC harbors two- to threefold more non-synonymous SNVs per tumor than localized disease, consistent with acquisition of additional mutations and clonal selection during long-term androgen deprivation therapy [7, 27–29]. Indeed, most somatic driver SNVs are strongly enriched in mCRPC relative to localized prostate cancer. Interestingly, key exceptions to this are SNVs in *SPOP*, which are significantly enriched in localized prostate cancers, relative to mCRPC, suggesting that these mutations are key drivers of prostate tumorigenesis but are unlikely to drive progression [7]. Consistent with this hypothesis, *SPOP* mutation confers sensitivity to the CYP17A1 inhibitor abiraterone [30], suggesting that ADT may produce a strong negative selection against primary tumor clones harboring *SPOP* SNVs.

Conversely, SNVs in the androgen receptor (*AR*) gene are exceedingly rare in localized prostate cancer but ubiquitous in mCRPC [25, 27], consistent with the androgen-independent proliferation of mCRPC. The likely mechanism for this enrichment is positive selection of subclones within the primary tumor harboring AR mutations, although this has not been directly assessed in longitudinal studies of paired primary and metastatic tumor tissue, and the possibility of selection of *de novo* mutations arising during ADT cannot be ruled out. Such a study would require ultra-deep WGS or targeted re-sequencing to correct for tumor cellularity and low variant allele fractions. More generally, this represents a fundamental limitation of all current cancer sequencing studies: high rates of false negatives due to insufficient sequencing depth. As we shall see, however, one potential solution to this problem is the development of bespoke targeted sequencing panels informed both by existing whole-genome sequencing data and by basic prostate cancer cell biology.

Structural Variation

As a C-class tumor, the prostate cancer genome is characterized by extensive structural variation, gene duplications and deletions, translocations, inversions, and small insertions and deletions (*i.e.* Indels). In this section, we will summarize the recurrent driver GRs and copy number aberrations (CNAs) found in the prostate cancer genome and discuss the complex structural variation that typifies this disease.

Genomic Rearrangements

As noted above, the most frequent GR (and most commonly observed molecular aberration) in prostate cancer is a fusion between the 5′ untranslated region (UTR) of the *TMPRSS2* gene and the coding region and 3′-UTR of the *ERG* gene, a member of the *ETS* family of proto-oncogenes [14]. The fusion product, termed *TMPRSS2:ERG* (T2E), is generated through edel or esplit mechanisms described above [15]. The *TMPRSS2* 5′-UTR and upstream enhancer contain strong AREs which, when fused to the *ERG* coding region, drives overexpression of the ERG protein in an androgen-dependent manner [14]. The T2E fusion can be detected via breakpoint-spanning qPCR, RNA- or DNA sequencing, array comparative genomic hybridization (aCGH; in the case of edel fusions) or by fluorescence *in-situ* hybridization (FISH) [15], and can also be inferred from overexpression of ERG on immunohistochemistry [18].

T2E is a member of a larger family of gene fusions targeting *ETS* proto-oncogenes, which together, are found in ~50–60% of all prostate cancers. The family includes additional androgen-responsive 5′ partners driving *ERG* overexpression (e.g. *SLC45A3*, *NDRG1* and others), alternate ETS family 3′ partners for *TMPRSS2* (e.g. *ETV1*, *ETV4*, and *FLI1*), or unique androgen-driven fusions (e.g. *SLC45A3:ETV1*) [16, 24]. ETS fusions are among the very earliest aberrations to occur during prostate tumorigenesis, and overexpression of ERG can be used to definite unique clonal populations within tumor foci on immunohistochemistry (see below) [31]. While small molecule ERG inhibitors have been developed and proposed as a potential therapeutic strategy for prostate cancer harboring T2E fusions [32], it remains unclear whether T2E contributes to a differential clinical course for localized or metastatic prostate cancer. Evidence from cohorts of definitive, curative-intent radiotherapy- or surgery-treated patients shows that T2E does not predict a more aggressive clinical course for localized disease [21, 22]. Conversely, there is some evidence suggesting that *TMPRSS2:ERG* fusions may inhibit DNA double strand break repair by interfering with non-homologous end joining (NHEJ), thereby promoting PARP inhibitor-induced radiosensitivity [33]. Nevertheless, the preponderance of evidence suggests that ETS fusions are key drivers of tumorigenesis in more than half of all prostate cancers with only mild effects on the clinical progression of the disease. Interestingly, T2E fusions are strongly enriched in prostate cancer arising in men <50 years of age at diagnosis (i.e. 'early-onset') [26, 34], suggesting that these cancers represent a unique molecular subtype that is driven by androgen-dependent signaling. To that end, a key unresolved question is whether T2E and other fusions drive a unique evolutionary trajectory in nascent prostate cancers. This question may be resolved by emerging whole-genome sequencing of large, well-annotated patient cohorts coupled with novel computational algorithms allowing for inference of subclonal evolution from single tumor populations.

While *ETS*-family fusions represent the largest single class of prostate cancer-associated GRs, other aberrations have been described and may impact prostate cancer biology and clinical outcomes. For example, a recently-discovered, recurrent genomic inversion on chromosome 10 containing the *PTEN* tumor suppressor gene is associated with a significant reduction in *PTEN* mRNA abundance and *PTEN* function, similar to that observed in tumors harboring a *PTEN* deletion, which is among the most common DNA copy number aberrations (CNA) in prostate cancer (see below) [25]. A similar effect was also observed on a region of chromosome 3q, suggesting that this is a relatively common mechanism of mRNA abundance regulation. *PTEN* displays the behavior of a haploinsufficient tumor suppressor, whereby inactivation of a single *PTEN* allele by deletion (see below) or point mutation is sufficient to drive *PTEN*-mediated tumorigenesis [35]. However, this effect may, in fact, be explained by a copy-neutral loss of the second allele through methylation-induced silencing or genomic inversion. As the number of profiled tumors continues to increase (thus increasing statistical power), it will be important to evaluate if and how different classes of aberration interact to dysregulate *PTEN*.

Rearrangements have also been observed in the *RB1*, *GSK3B*, and *FOXO1* genes, leading to loss of function of these genes [24]. Other candidate driver events include *NRF1-BRAF* and *CRKL-MAPK1* fusions, and while these events are not common, they may be important driver events in individual tumors. Once again, a complete understanding of the frequency of these (and other) GRs, as well as their biological and clinical impact, will require a dramatic increase in the number of localized prostate cancer whole genomes available for analysis.

Additional non-*ETS* gene fusions have been identified in prostate cancers in populations of non-Caucasian ancestry, including fusions between the *USP9Y* and *TTTY15* genes on chromosome Y and a *CTAGE5:KHDRBS3* fusion resulting from a chr14:chr8 translocation [36]. In Chinese prostate cancer patients, these fusions are present at rates that exceed that of *TMPRSS2:ERG* [37, 38], which strongly suggests that genetic ancestry and environmental factors play a key role in the molecular progression of prostate tumorigenesis.

Chromoplexy

In general, cancer-associated gene fusions involve two genomic loci; often a regulatory region from a gene under the control of a strong promoter that drives constitutive expression of a proto-oncogene. Prototypical examples of this are the *BCR-ABL* fusion that defines chronic myeloid leukemia (CML) [39] and the *IGH-MYC* translocation found in various malignancies, including Burkitt Lymphoma and acute lymphoblastic leukemia [40, 41]. However, an improved understanding of the genomic architecture of cancer has revealed additional complex patterns of structural variation.

One such pattern is 'chromoplexy', which was initially identified based on deep whole genome sequencing of high grade localized prostate cancer [24]. Chromoplexy involves the formation of fusion products that include DNA from multiple chromosomes, in an 'end-to-end' fashion. Chromoplexy differs from chromothripsis in that it often involves disruption of specific loci across multiple chromosomes, which appears to occur in a single genomic event. Conversely, chromothripsis typically involves extensive 'shattering' and reassembly of a small number of chromosomes (typically one or two) (Fig. 2). Chromoplexy is important in some instances of T2E fusion in prostate cancer, and ETS fusion status is strongly related to the presence of chromoplexy events, the number of chromosomes involved, and the number of DNA fragments per fusion [24, 37]. This suggests that chromoplexy in prostate tumors may be related to androgen-driven transcription. While chromoplexy has been identified in other cancer types [42–46], its overall frequency in human cancer has not been firmly established, nor has its effects, if any, on the clinical evolution and aggression of these cancers.

Copy Number Aberrations

The prostate cancer genome harbors extensive recurrent copy number aberrations (CNAs) (Table 2). These events generally encompass large chromosomal regions spanning several megabases, although some focal CNAs (≤1 Mb) do occur. In localized disease, several well-established tumor suppressor genes are subject to mono-allelic or bi-allelic deletion. The most frequently observed deletion involves the *PTEN* locus on chromosome 10 [16, 25]. *PTEN* loss occurs in ~15–20% of localized prostate cancers, frequently in tumors that also harbor T2E fusions [16] and in hypoxic tumors [47]. Moreover, PTEN deletion is enriched in metastatic prostate cancers [7], suggest that PTEN loss contributes to disease aggression. Loss of *PTEN* leads to activation of the PI3K/AKT pathway, thereby promoting cell proliferation, survival, and migration [48]. Interestingly, $PTEN^{-/-}$ mouse embryonic fibroblasts are sensitive to poly-ADP ribose polymerase (PARP) inhibitors [49–52], possibly through decreased homologous recombination-mediated DNA repair via transcriptional down-regulation of RAD51 [53]. However, correlative studies have failed to establish any link between *PTEN* status and RAD51 expression in prostate cancer [54], and it is likely that this effect is tissue-specific, since follow-up clustered regularly interspaced short palindromic repeats (CRISPR)-based screens for PARP inhibitor sensitivity have failed to recapitulate the importance of *PTEN* in this regard [55]. Other recurrent somatic deletions in localized prostate cancer include *NKX3-1* (8p21.2; see below), *CHD1* (5q15-q21.1), *CDH1* (16q22.1), *RB1* (13q14.2), *CDKN1B* (12p13.1), *BRCA2* (13q13.1) and *TP53* (17p13.1) [16, 25, 56], and several of these may have significant prognostic value for adverse outcomes in localized disease [57].

CHD1 deletion and mutation of SPOP are mutually exclusive of T2E fusion [30, 58], and functional studies have demonstrated that *CHD1* is required for fusion formation [58], thereby explaining this mutual exclusivity. Conversely, SPOP is an E3 ubiquitin ligase for ERG and thus, in cells harboring *SPOP* mutations, ERG is stabilized, thereby reducing the selective pressure for T2E fusion as a means of driving ERG overexpression [59, 60]. The recent finding that *SPOP* mutations are predictive of abiraterone sensitivity in mCRPC [30] indirectly implies that fusion status may be a useful predictive biomarker in advanced prostate cancer.

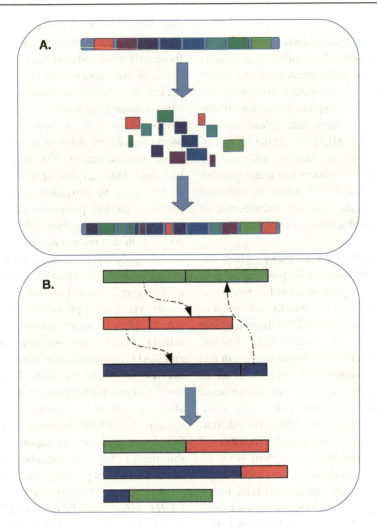

Fig. 2 Complex genomic rearrangements in prostate cancer. (**a**) Chromothripsis occurs as a result of chromosomal 'shattering' and massive rearrangement of single chromosomes or regions of single chromosomes, during a small number of cell divisions. The result is disruption of gene loci (shown as colored bars) within the chromothriptic region, with retention of gene copy number within the region [78]. (**b**) In contrast to chromothripsis, chromoplexy involves several distinct loci (shown as colored bars), often across different chromosomes. These loci exchange genomic segments with each other to form complex 'closed-loop' structures that may involve five or more independent loci [24]

Chromosome 8p is frequently subject to allelic losses, often affecting large portions of the entire p arm of chromosome 8 [25, 47, 56]. The major candidate gene in this region is *NKX3-1*, a homeobox-containing transcription factor, which acts as a transcriptional repressor of cell growth in prostate tissue [61], and an androgen-driven, haploinsufficient tumor suppressor gene that is deleted in up to 30% of localized prostate cancers. *NKX3-1* deletion has been associated with disease relapse following external-beam radiotherapy [62], and likely occurs early in prostate tumorigenesis, since most *NKX3-1* deletions are clonal events [56].

The most frequently amplified region of the somatic prostate cancer genome is chromosome 8q. Several significant amplification peaks exist on this chromosomal arm, the most prominent of which (8q24.21) harbors the *MYC* oncogene. The 8p24.21 region is amplified in 8–10% of localized cancers [16, 25, 56], and is prognostic of adverse clinical outcomes following definitive

Table 2 Recurrent copy number aberrations in localized prostate cancer

Gene	Aberration	Chromosomal locus	Frequency in localized prostate cancer [16, 25, 47, 56, 57, 65, 74, 89]
PTEN	Deletion	10q23.31	10–20%
NKX3-1	Deletion	8p21.2	5–30%
TP53	Deletion	17p13.1	4–20%
FOXO1	Deletion	13q14.11	5–15%
RB1	Deletion	13q14.2	5–15%
MYC	Amplification	8q24.21	6–10%
CHD1	Deletion	5q15-q21.1	8–10%
NBN	Amplification	8q21.3	5–7%
CDH1	Deletion	16q22.1	4–5%
BRCA2	Deletion	13q13.1	3–5%
CDKN1B	Deletion	12p13.1	2–5%
BRCA1	Deletion	17q21.31	1–2%

therapy [25, 63]. Interestingly, the rate of *MYC* amplification is substantially increased in localized prostate cancers in men who harbor a deleterious germline mutation in the *BRCA2* gene [64, 65]. Consistent with a role for *MYC* amplification as a determinant of poor clinical outcome, these familial cancers have an extremely aggressive clinical course, with 5-year overall survival rates approaching 50% [66].

Recurrent amplification affecting regions of chromosome 8q outside of the *MYC* gene are also commonly observed and may have important biological and clinical relevance. For example, amplification of the *NBN* gene (8q22.1), which encodes the DNA damage sensor protein NBS1, is predictive of poor response to external beam radiotherapy in low/intermediate risk prostate cancer [67]. Similarly, the *PCAT1* long noncoding RNA gene, which is implicated in aggressive localized and metastatic prostate cancer, is located on a frequently-amplified region of chromosome 8q immediately upstream of the *MYC* locus [68]. Intriguingly, *PCAT1* lies within a common fragile site (*FRA8C*) [69], which may help to explain the recurrent instability at this locus, and *PCAT1* promotes MYC stability and MYC-dependent transcription [70].

Amplification of the *MYCN* locus on chromosome 2p24 (encoding the c-Myc homolog n-Myc) is rare in localized prostate cancer, but is common in lethal neuroendocrine disease [71, 72]. Conversely, amplifications of the *MYCL* gene on chromosome 1p34 occurs in >20% of low/intermediate risk prostate cancers, although this amplification has only been observed in two independent studies [57, 73]. *MYCL* amplification is mutually exclusive from *MYC* gain and strongly associated with *TP53* deletion, and is highly focal (minimally amplified region = 8.5 kb), and rarely associated with amplification of >500 kb [57]. Moreover, FISH analysis demonstrated strong heterogeneity of *MYCL* amplification within malignant glands. This may explain why previous studies have failed to observe this amplification. While the relative contribution of the three MYC family homologs to prostate cancer progression has not been well-studied, it is likely that *MYCN* and *MYCL* play a subsidiary role, given their reduced incidence and their mutual exclusivity from MYC amplification (which strongly demarcates aggressive disease) [25, 63].

Recurrent amplifications have also been identified on chromosome 3q26, 11q13, and the entirety of chromosome 7 [16, 25]. Because these amplifications involve megabase stretches of the genome, identification of the putative driver genes in each region has proven difficult. However, it is clear that at least some of these large-scale amplifications have prognostic value in localized prostate cancer. Indeed, amplification of chromosome 7 defines a unique molecular cluster of prostate cancers based on CNA profiles, and is associated with decreased time to biochemical relapse in men treated with IGRT for intermediate risk prostate cancer [74].

The prognostic importance of somatic CNAs is now well-established. CNA burden, defined as the proportion of the genome affected by CNA events, is associated with reduced time to biochemical relapse in men with localized prostate cancer [74, 75]. Moreover, unbiased, machine learning-based approaches have identified CNA-based signatures that accurately classify patients for risk of biochemical and metastatic relapse following definitive, curative-intent therapy for localized prostate cancer [25, 74, 76]. These signatures significantly outperform RNA-based classifiers [74] as well as established clinical prognostic factors such as Gleason grade and pre-treatment PSA. Moreover, CNA-based classifiers can be derived from pre-treatment biopsy specimens, including formalin-fixed, paraffin-embedded (FFPE) specimens, and are readily adapted for clinical implementation via clinical laboratory improvement amendments-compatible platforms such as NanoString [76].

Chromothripsis

Chromothripsis is frequently observed in many human cancers, including prostate cancer [77, 78]. The phenomenon involves shattering of a single chromosome into thousands of small fragments during a very short time (one or, at most, a few cell divisions) which are randomly rejoined, leading to profound disruption of gene function across entire chromosomes or chromosomal arms (Fig. 2). Chromothripsis is associated with retention of heterozygosity and *TP53* deletion [78], although *TP53* allelic loss is not required for chromothripsis [79]. Chromothripsis has generally been identified in an *ad hoc* manner, although computational tools have been developed and validated to automate this process and provide an objective standard by which chromothripsis can be identified and quantified across tumor types [79].

Numerous studies have identified relatively high rates of chromothripsis in localized prostate cancer, consistent with the observed C-class nature of the disease [16, 24, 25, 80, 81]. Chromothripsis is associated with unique gene fusions and other molecular alterations, and thus may underlie a unique prostate cancer biology [25, 81], but is not associated with higher tumor grade or with differential outcomes in localized prostate cancer, suggesting that its function is related to tumor initiation and maintenance, rather than disease aggression. Chromothriptic tumors tend to be larger and have higher mutational burden across aberration classes, and chromothriptic regions are enriched for aberrant DNA methylation, raising the possibility that epigenetic mechanisms may help to initiate chromothripsis [25].

The DNA Damage Response

An effective DNA damage response (DDR) is critical for maintenance of genome integrity. As noted above, germline variants that increase prostate cancer risk are predominantly involved in DNA repair pathways, and variants in these genes are almost universally associated with genomic instability. DDR genes can be separated into distinct pathways mediating repair of particular types of DNA damage. For example, mismatch repair (MMR) is involved in correcting base substitutions, whereas homologous recombination (HR) uses the available sister chromatid as a template for error-free DNA double strand break repair during the S/G2 phases of the cell cycle. In addition to germline variants, multiple somatic alterations have been described in DDR genes that delineate distinct prostate cancer genomic subtypes in localized and metastatic tumors. A subset of prostate tumors show somatic alterations in genes of the MMR pathway (~5–10% in metastatic disease), which correlate with a higher mutational load [82].

Alterations in HR genes highlight a distinct subset of prostate tumors with high genomic instability and structural rearrangements. Approximately 10–20% localized tumors show alterations in HR genes, predominantly in *BRCA2* and *ATM*, but also in *BRCA1*, *ATR*, *FANCA*, *FANCD2*, *MRE11A*, *PALB2* and *RAD51C* [7, 16, 25]. Interestingly, the proportion of HR defects increases in metastatic tumors to 23–27% [28], which is associated with increased genomic instability in metastatic tumors [37]. In line with this, a higher DDR score is linked to decreased progression-free survival for localized prostate cancer [83]. Another subset of DDR-deficient prostate tumors, involves bi-allelic loss of *CDK12* in metastatic tumors, and is associated with high levels of focal tandem duplications and delineates an immunological-like subgroup [84]. Overall, alterations in genes of the DDR pathway form distinct subgroups of prostate tumors that generally have a poor prognosis.

Tumor Evolution and the Impact of Somatic Genome Heterogeneity

Multifocality

The vast majority of prostatectomy specimens show more than one disease focus [85, 86], although until recently it has not been clear whether these foci arise from a common precursor clone or represent independent cancer lineages. Macrodissection of 18 unique tumor foci from 5 patients with localized prostate cancer revealed dramatic intra-prostatic genomic heterogeneity across molecular aberration classes [57, 87]. For example, in some patients, global CNA counts varied by as much as 50-fold between tumor foci. This included heterogeneous CNAs in genes such as *PTEN, NKX3-1*, and *TP53*, which have established prognostic value [62, 63]. This implies that clinically-relevant molecular profiling is subject to significant sampling biases.

Importantly, in at least one patient, there were no shared SNVs or CNAs between two distinct tumor regions, at least at the available limit of detection. This suggests that a single prostate harbors multiple independent cancers.

In addition to genetic variability, multi-region sampling of five prostate tumors showed extensive epigenetic intra-tumor variability at several loci [88]. In particular, enhancer sites bound by androgen receptor exhibited substantial intra-tumoral DNA methylation heterogeneity. This implies that genetic heterogeneity is not solely responsible for phenotypic differences observed between foci.

Using T2E break apart FISH to define unique tumor regions in whole-mount specimens, a contemporaneous study [89] further demonstrated extensive genomic heterogeneity in multi-focal prostate cancer. While each tumor in this study was derived from a single ancestor clone, unique mutational signatures were identified across tumor foci from the same patient, suggesting that mutational processes vary during tumor evolution, consistent with subsequent reports showing that the acquisition of unique sub-clonal mutations is driven by temporal alterations in mutational signature composition [56].

More recently, whole exome sequencing of 153 specimens from 41 patients with localized prostate cancer, including normal prostate epithelium specimen from each patient, revealed extensive genomic heterogeneity, both between individuals and, strikingly, between tumor foci from the same individual [90]. Most foci from the same prostate contained no shared SNVs and very few shared CNAs. A limitation of this study is that the use of exome sequencing precluded analysis of non-coding mutations, although many, but not all, non-coding mutations are subject to less selection pressure than non-synonymous exonic mutations. Therefore, this study did not permit computation-based sub-clonal reconstruction or assessment of clonal origin (see below). Nevertheless, these data do suggest that, at minimum, independent tumor foci undergo substantial evolutionary divergence during prostate tumorigenesis. Moreover, a recent whole-genome interrogation of 21 metastases derived from 10 patients identified multiple distinct alterations in *RB1*, as late and sub-clonal events in related metastases, correlated with heterogeneous expression in a substantial 28% of mCRPC [91]. This finding is in line with another study showing that metastatic prostate tumors have three to four times more alterations in *RB1* [7], and implies that *RB1* inactivation is required for a selected set of patients to progress to mCRPC.

Thus, while it is now clear that multifocality is associated with significant molecular heterogeneity, the extent to which this contributes to differential biology, clinical course and metastatic progression remains unknown. For example, while the presence of high grade cancer (*i.e.* Gleason pattern ≥ 4) is associated with less favorable clinical outcomes, there is evidence that in tumors composed of heterogeneous pathological grades (*i.e.* Gleason pattern 3 adjacent to pattern 4), the lower grade component can, at least in some cases, seed distant metastases [92]. Unfortunately, these observations have been limited to a single "n of one" study. While all metastatic lesions in a given patient are seeded from a single progenitor clone arising within the primary tumor [93], there is an urgent need to clarify the source of that "lethal" clone in multi-focal primary tumors. This could be accomplished, in part, through longitudinal studies assessing the molecular profile of multiple primary foci and patient-matched distant metastatic disease. This would help to establish the proportion of metastases that are seeded from the largest/highest grade "index" lesion, and thereby suggest the extent to which multi-region profiling is required to inform clinical decision-making using prognostic and/or predictive molecular biomarkers (see below). Intriguingly, a very recent study showed that ~24% of patient-matched index lesions and lymph node metastases exhibited no common CNAs [94]. While this study was limited in sample size (n = 30 patients) and breadth of molecular interrogation, this finding does support the hypothesis that a small but significant percentage of metastases are seeded from tumor clones outside of the index lesion.

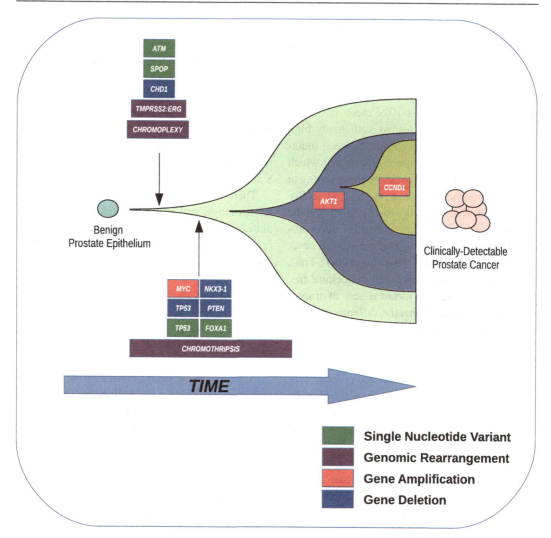

Fig. 3 Subclonal evolution of localized prostate cancer. Clonal evolution begins with initiating lesions such as *TMPRSS2:ERG* fusion (and other *ETS* family fusions), *SPOP* SNVs and *CHD1* deletions (in fusion-negative tumors), and broader genomic rearrangement events such as chromoplexy. Because these aberrations occur early in tumor development, they are enriched in tumor 'trunks', and are thus present in all tumor clones at clinical presentation. Other truncal events occurring prior to evolutionary divergence include *MYC* amplification, *NKX3-1*, *PTEN*, and *TP53* deletion, *TP53*, *FOXA1*, and *ATM* SNVs, and chromothripsis. In general, tumor trunks are enriched for large deletions and higher SNV burden. Over time, divergence from the trunk occurs, giving rise to subclonal populations containing both truncal aberrations and private 'branch' mutations such as *AKT1* and *CCND1* amplifications, which are present only in these subclones. Tumors with more extensive branching are more likely to recur following definitive therapy [56]. There is also significant switching of mutational signatures during tumor evolution (not shown) away from the early APOBEC-type and aging-related signatures and toward signatures of homologous recombination deficiency [37, 56, 96]

Sub-clonal Reconstruction of Localized Prostate Cancer

While molecular profiling of multiple cancer foci from the same prostate has dramatically enhanced our understanding of the spatial heterogeneity of localized disease, advancements in computational biology have allowed for reconstruction of the *temporal* evolution of individual cancer foci (Fig. 3), an approach that was pioneered through analysis of breast cancer whole genomes [95]. This approach relies upon three fundamental pieces of data: (1) a whole-genome sequence of patient-matched normal DNA to establish germline genotype, (2) an accurate measure of tumor cellularity and genome-wide copy number to establish variant allele frequency, and (3) sufficiently deep tumor whole genome sequencing ($\geq$80–100× mean coverage) for detection of rare alleles. At sufficiently high depths, this method distinguishes between mutations that arose early in tumorigenesis and which are thus present in clonal trunks vs. those arising in subclonal branches (*i.e.* high vs. low allele frequency). Subclonal branches can be inferred to be fully independent when the mutations that compose the branches are mutually exclusive (*i.e.* never observed on the same DNA fragment). Thus, tumors can be classified as either monoclonal (*i.e.* all cells contain largely identical sets of truncal mutant alleles) or polyclonal (*i.e.* different cell populations containing a mixture of truncal and branch mutations with more or less evolutionary "distance").

Approximately 40% of localized prostate cancers harbor a single clonal population while ~60% show strong evidence of significant evolutionary branching [56], though this is likely an underestimation, since clonal populations in this study were reconstructed from a single tumor region. Moreover, specific mutational classes showed strong biases toward either tumor trunks or branches. For example, while CNA burden was uniformly distributed between branches and trunks, truncal CNAs were strongly biased toward larger allelic losses. Conversely, CNAs arising in tumor branches tended to be smaller allelic gains. Similarly, SNVs are strongly enriched in tumor trunks. Indeed, virtually all established recurrent driver SNVs (e.g. *FOXA1*, *SPOP*, *ATM*, and *TP53*) are preferentially found in tumor trunks, suggesting that these aberrations play an important role in tumor initiation. Mutational signatures are also highly dynamic during tumor evolution, with ~2/3 of tumors showing strong evidence of signature switching. The earliest prostate cancer-associated mutations are driven by the APOBEC-related, 'clock-like' mutational signature, with subsequent accumulation of AR-related aberrations, DNA repair defects, and broader genomic instability [96].

Importantly, clonal complexity is a significant predictor of clinical outcome; polyclonal tumors have substantially shorter time to biochemical relapse than monoclonal tumors [56]. Moreover, tumor clonality strongly interacts with existing prognostic biomarkers based on single molecular analytes such as the percentage of genome altered (PGA) or multi-modal biomarkers (the utility of molecular biomarkers is discussed in detail in Section "Clinico-Genomics", below).

There are also clear differences in the temporal evolution and selective pressure at play in tumors harboring an *ETS* gene fusion, relative to fusion-negative cancers [37]. For example, *ETS* fusion-positive cancers are strongly enriched for specific deletions on chromosomes 5 and 10 (including the *PTEN* locus) but are mutually-exclusive from deletions of the chromosome 5 region containing *CHD1*, consistent with the requirement of CHD1 for formation of the fusion [58].

In the coming years, continued declines in cost per sequenced base will allow for deeper interrogation of tumor sub-clonality, thereby overcoming some of the inherent limitations in this process. For example, it is highly likely that putatively monoclonal prostate tumors do, in fact, contain small populations of cells harboring unique sets of private mutations, but are undetectable at the sequencing depths commonly employed today. In parallel, improvements in computational power will further increase sensitivity for detection of low frequency alleles. Furthermore, integration of sequencing data derived from multi-regional specimens followed by clonal reconstruction for each prostate tumor will allow a more comprehensive understanding of clonal dynamics [97, 98]. This has multiple clinical implications, including the identification and tracking of ultralow-frequency mutations that predict therapy resistance or relapse when sampled from tumor, blood or urine samples [99]. Finally, as the number of sequenced prostate tumors in the literature continues to rise—with a concomitant increase in the statistical power to detect rare-but-recurrent aberrations—the need for further whole genome sequencing analyses will fall, except in unique patient populations where the genomic landscape is not yet well-established (e.g. *BRCA2*-associated familial prostate cancer; see below) [64, 65].

Familial Prostate Cancer and Germline-Somatic Interactions

Germline Variation and Prostate Cancer Risk

Family history is a significant risk factor for the development of prostate cancer [100], suggesting that inherited genetic factors influence the propensity for prostate cancer initiation. While it is estimated that ~10% of prostate cancer is attributed to genetic factors [101], very few individual genes have been shown to contribute significantly. One prominent example is the *BRCA2* gene; carriers of deleterious germline *BRCA2* mutations are four- to fivefold more likely to develop prostate cancer than non-carriers [102], although penetrance is incomplete (*i.e.* not all carriers will develop prostate cancer). Moreover, these cancers are very aggressive, with 5-year overall survival rates of 50–60% for *BRCA2* mutation carriers diagnosed with prostate cancer [66]. Other mutations that may confer increased prostate cancer risk with moderate penetrance include *BRCA1*, *PALB2*, *HOXB13*, *CHEK2*, *ATM*, *KLK6* and *NBN* [103, 104]. Genes conferring comparatively large increases in risk are almost universally involved in the DNA damage response (DDR) and DNA repair, and patients harboring germline mutations in these genes may be candidates for treatment with poly(ADP)-ribose polymerase (PARP) inhibitors, which selectively target cells with deficient repair of DNA double strand breaks by homologous recombination [105]. Indeed, PARP inhibitors have shown promise for the treatment of patients with mCRPC who harbor germline and/or somatic mutations in DDR genes [106].

While deleterious mutations in these moderately penetrant genes confer a relatively large increase in prostate cancer risk, these mutations are rare in most populations. In contrast, well over 100 germline single nucleotide polymorphisms (SNPs) have been linked to increased prostate cancer risk. While each individual SNP increases risk by only a very small amount, collectively these SNPs account for >28% of the inherited risk of developing prostate cancer [107, 108]. Characterizing the effects of these SNPs requires genome-wide association studies (GWAS) on very large populations in order to attain sufficient statistical power to detect small changes in risk. To that end, the Prostate Cancer Association Group to Investigate Cancer Associated Alterations in the Genome (PRACTICAL) consortium has assembled over 120,000 prostate cancer cases and 100,000 controls, allowing for unprecedented insights into the effect of inherited variation on prostate cancer risk [107, 108].

Germline-Somatic Interaction: BRCA2

While germline genotype clearly influences the clinical course of prostate cancer, until recently, very little was known about the molecular underpinnings of disease aggression in men who carry a deleterious germline mutation. This gap in knowledge can be explained, in part, by the relatively low frequency of mutations that portend adverse clinical outcomes, which necessarily limits statistical power. Much of our understanding of the genomics of prostate cancer in *BRCA2* mutation carriers has come from correlations observed in incidentally collected cases in large whole exome studies of mCRPC [28], which, by definition, is a disease state wherein the somatic genomic profile has been subject to strong ADT-mediated selection pressures. Thus, if and how these *BRCA2*-associated cancers differ from sporadic (*i.e.* non-familial) prostate cancer while localized to the prostate—and prior to the administration of ADT—has not, until recently, been comprehensively addressed.

BRCA2-associated prostate cancers harbor significantly higher mutational burden across all molecular classes, relative to sporadic cancers of matched grade and stage [65]. In particular, amplifications of MYC and deletions of *GSK3B* and *MTOR* are enriched dramatically in *BRCA2*-associated prostate cancer, relative to sporadic localized disease [64, 65]. Indeed, hormone-naïve, localized, *BRCA2*-associated prostate cancer harbors a mutational profile that more closely resembles mCRPC than sporadic localized disease [16, 25, 28]. This includes CNAs and aberrant methylation of genes involved in *WNT* pathway activation (including *MED12*, *MED12L*, and *APC*), which are rare in sporadic, localized prostate cancer [65]. Intriguingly, *BRCA2*-associated prostate cancer is enriched for a unique histopathology known as intraductal carcinoma of the prostate (IDC-P), which is a major component of the clinical aggression of this disease state [66, 109]. Patients with IDC-P show increased genomic instability, are more likely to have hypoxic tumors, and show higher expression of SChLAP1 long noncoding RNA and increased proportions of somatic alterations such as point mutations in *TP53*, *SPOP* and *FOXA1* [110]. Despite this, it remains unclear how IDC-P differs from the surrounding invasive adenocarcinoma, even though the preponderance of evidence suggests that IDC-P-associated cancers are genomically-unstable entities. Additional statistically-powered studies, which directly compare the genomic alterations in microdissected regions of IDC-P and adjacent invasive adenocarcinoma, will be required to formally assess this question. Subclonal reconstruction of IDC-P in the presence or absence of a germline *BRCA2* mutation revealed a distinct evolutionary trajectory for these aggressive cancers. Moreover, IDC-P-associated sporadic cancers often harbor molecular aberrations that are enriched in BRCA2-associated disease, regardless of IDC-P status [65].

A number of questions remain unanswered with respect to *BRCA2*-associated prostate cancer. For example, it is unclear whether histologically normal prostate epithelium in *BRCA2* carriers harbors mutations that predispose to rapid malignant transformation, a phenomenon that has been observed in some sporadic cancers,

consistent with a prostate 'field cancerization' phenomenon [89].

Similarly, the molecular underpinnings of clinical aggression in prostate cancer associated with other germline mutations is largely unknown. Collection of specimens from these patients—particularly in early stage disease—will allow for creation of well-powered cohorts to dissect the unique biology of these familial cancers.

Clinico-Genomics

As the number of prostate cancer whole-genome and whole-exome datasets in the literature has increased, so too has our understanding of the key molecular events underlying the development and progression of the disease. However, the majority of these molecular studies have lacked long-term clinical follow up data, making it difficult to identify relationships between specific sets of genomic alterations and clinical outcome. Thus, while several RNA-based biomarkers of adverse outcome have entered into clinical practice (reviewed in the next chapter of this book) and while numerous genomic aberrations have been linked to adverse pathologic endpoints (*e.g.* Gleason upgrade at prostatectomy), comparatively little data exists to support the use of recurrent genomic alterations (or sets of alterations) as biomarkers of adverse clinical endpoints such as 3-year biochemical relapse, 10-year metastatic relapse, or prostate cancer-specific mortality. In this section, we review the literature with regards to DNA-based predictors of clinical outcome.

Prognostic Biomarkers: CNAs and Percentage of the Genome Altered

In their landmark paper 2010 paper, Taylor and colleagues performed molecular profiling of 181 localized prostate cancers and identified six patient clusters based on CNA profiling [111]. Patients with the highest number of CNAs (cluster 5) had significantly shorter time to biochemical relapse than other patients, although the clinical and pathological heterogeneity across clusters precluded a robust multivariable assessment of the relative contribution of tumor grade and CNA number on clinical outcomes. Subsequent studies revealed that PGA is a negative prognostic factor for disease relapse in localized prostate cancer, independent of tumor grade and other clinical prognostic factors [74, 75], a finding which has been validated across cancer types [112].

Nevertheless, the prognostic value of PGA does not appear to be randomly distributed across the genome, since some—but not all—CNAs are associated with poor outcome. For example, amplifications of chromosome 8q24—which contains the genes encoding both cMYC and the PCAT1 long non-coding RNA—are associated with poor clinical outcome [68], as are deletions of chromosome 8p21.2, which harbors the NKX3-1 tumor suppressor gene, chromosome 10q23.31, which contains the PTEN tumor suppressor gene [62, 63], and chromosome 17p13.1, which includes the TP53 gene [113, 114]. More recently, unbiased analysis of DNA copy number data from aCGH and SNP microarrays has identified multi-region signatures that accurately classified patients at high risk of biochemical and metastatic relapse. Using unbiased machine learning on aCGH data derived from a single index tumor biopsy, a 100-locus signature that accurately predicts relapse in patients treated with image-guided radiotherapy for low/intermediate risk was developed [74]. Importantly, this signature predicted 18-month biochemical relapse, which is a surrogate of lethal disease [115, 116], and synergized with other adverse clinical factors such as intratumoral hypoxia [74]. This signature was subsequently refined into a 30-locus version, which accurately predicts both biochemical and metastatic relapse in independent cohorts of localized prostate cancer, across the clinical risk and treatment spectrum [76].

Prognostic Biomarkers: Impact of Multi-modal Profiling

These DNA copy number-based biomarkers and signatures demonstrate that genomic aberrations can distinguish between indolent and aggressive localized disease. However, these signatures do not capture all of the observed heterogeneity in clinical outcomes, suggesting that information from other molecular aberration classes provides improved signature accuracy.

Indeed, a machine learning model trained on 40 genomic and epigenomic drivers in localized prostate cancer identified a six-feature, multi-modal signature that accurately predicts both early and overall biochemical relapse in intermediate risk disease [25]. Composed of two epigenetic aberrations (methylation of the ACTL6B and TCER1L gene loci), two structural variants (MYC amplification and a translocation involving chromosome 7), a non-synonymous mutation (ATM SNVs), and clinical T-stage, this signature outperformed other established single-class biomarkers (e.g. CNAs) for prediction of biochemical relapse, with a classification accuracy of ~83%. Indeed, addition of tumor subclonality data further improved the performance of this signature [56]. Thus, patients with monoclonal tumors and low signature scores were virtually free of biochemical relapse in the 8 years following definitive treatment, while polyclonal tumors with high signature scores were very aggressive, with 3-year biochemical relapse rates of ~40% and nearly 80% of patients experiencing biochemical relapse within 8 years.

Challenges for Implementation of Molecular Biomarkers

As the costs of DNA sequencing continue to fall, whole-genome sequencing of clinical and research tumor specimens will become viable for an increasing number of investigators. However, the decline in genomics cost has not been mirrored by a concomitant drop in the cost of computing and data storage hardware. Moreover, labor costs constitute a substantial portion of the total cost of analyzing and annotating a genome sequence. Finally, as we have seen, subclonal evolution and disease heterogeneity can profoundly influence the clinical course for an individual patient, thus necessitating comparatively deep sequencing to detect rare variant alleles. As such, routine whole-genome or whole-transcriptome sequencing may not be feasible for broad clinical implementation, even if direct sequencing costs are negligible.

As such, there is a need to develop strategies for clinical interrogation that effectively balance assay cost, test accuracy, and therapeutic impact. Moreover, multi-modal biomarkers are likely to outperform those developed based on a single molecular analyte, and thus clinically-relevant platforms must be able to effectively integrate different molecular aberration classes. One potential solution to this challenge is the development of bespoke targeted sequencing panels that include all coding and non-coding genomic regions that are of relevance to localized prostate cancer. As the number of sequenced prostate cancer genomes increases beyond a few thousand, the statistical power to discover molecular variants present at 1% or higher will quickly be reached (and has already been reached for some mutational classes such as SNVs), although further increases in discovery power will require exponentially larger patient numbers. The genomic loci encompassing these recurrent variants—in both the somatic and germline genomes—could be included in a custom capture assay and sequenced to extremely high depths (up to 500×). If such an assay contained 50 Mb of custom capture and were sequenced to 500× mean coverage, this would yield only ~13% of the total sequencing required for a typical 60× whole genome sequence. Such a platform would, therefore, dramatically reduce the storage and computing resources required for analysis, and, critically for clinically-useful biomarkers, would substantially reduce the time required to generate 'actionable' information. Moreover, the high sequencing depth would permit detection of rare variants, thus helping to ensure accuracy and clinical relevance.

There are several factors that currently limit the ability to design such a panel. A major factor is the current limit on our understanding of rare-but-recurrent variants in the prostate cancer genome. Maximizing clinical impact requires applicability to the largest possible patient population, and thus limiting genomic regions to those mutated at 1% or higher will, necessarily, fail to identify relevant mutations in up to 1% of cases. Nevertheless, capture panels can be designed in an iterative fashion, with new regions included (or old ones refined or removed) as additional validated research data is developed.

A second limitation is the compatibility with molecular aberration classes. For example, aberrant DNA methylation at unique loci is associated with differential clinical outcomes [25, 65, 117–126]. As such, an informative biomarker of aggressive disease is likely to include at least some information regarding DNA methylation, and therefore the compatibility of targeted sequencing panels with bisulfite-converted DNA should be considered.

Heterogeneity

As discussed, distinct prostate cancer foci can harbor extensive genomic heterogeneity. The clinical relevance of this is underscored by the discordance between CNAs in lymph node metastases and the matched index lesion [94]. In contrast, biomarkers based on a single biopsy or surgically-resected specimen can achieve very high prognostic accuracy, often exceeding 80% [25, 56, 74, 76]. It is unclear, however, whether this level of biomarker accuracy can be further extended by analysis of additional analytes and development of additional multi-modal signatures, or whether biomarker accuracy is fundamentally limited based on intra-prostate heterogeneity. This is a major unanswered question in translational prostate cancer genomics, and a robust understanding of the effects of heterogeneity will be required to optimize any tissue-based biomarker that ultimately achieves clinical implementation. One potential solution is the use of liquid-based biomarkers to complement tissue biomarkers, since these may better reflect the global mutational profile across tumor foci and appear to accurately reflect the mutational spectrum of the source tumor [127–130]. To date, however, these studies have been limited to large volume mCRPC, and the applicability to localized prostate cancers—particularly those of lower grade and stage—is not clear. Technological improvements will undoubtedly improve sensitivity of detection of circulating cell-free DNA in localized disease, and this may help to overcome the effects of heterogeneity of the primary tumor.

Conclusions

The fundamental molecular aberrations that characterize localized prostate cancer are now well-understood. It is clear that this disease state, in general, is typified by a low-to-moderate burden of driver SNVs, extensive genomic rearrangement, and epigenetic dysregulation. Moreover, as the number of well-annotated, sequenced cases with deep clinical history and follow-up data continues to increase, the development of prognostic and predictive biomarkers—informed by molecular profiling—will likewise increase. Analysis of disease heterogeneity—both between patients and within a single prostate gland—remains a major research challenge, and a full understanding of the impact of heterogeneity will be absolutely required to maximize the potential clinical impact of molecular biomarkers.

References

1. E.S. Lander, L.M. Linton, B. Birren, C. Nusbaum, M.C. Zody, J. Baldwin, et al., Initial sequencing and analysis of the human genome. Nature **409**, 860–921 (2001)
2. S. Baichoo, C.A. Ouzounis, Computational complexity of algorithms for sequence comparison, short-read assembly and genome alignment. Biosystems **156-157**, 72–85 (2017)
3. T.J. Ley, E.R. Mardis, L. Ding, B. Fulton, M.D. McLellan, K. Chen, et al., DNA sequencing of a cytogenetically normal acute myeloid leukaemia genome. Nature **456**, 66–72 (2008)
4. G. Ciriello, M.L. Miller, B.A. Aksoy, Y. Senbabaoglu, N. Schultz, C. Sander, Emerging landscape of onco-

genic signatures across human cancers. Nat. Genet. **45**, 1127–1133 (2013)
5. C.K. Zhou, D.P. Check, J. Lortet-Tieulent, M. Laversanne, A. Jemal, J. Ferlay, et al., Prostate cancer incidence in 43 populations worldwide: an analysis of time trends overall and by age group. Int. J. Cancer **138**, 1388–1400 (2016)
6. I.F. Tannock, R. de Wit, W.R. Berry, J. Horti, A. Pluzanska, K.N. Chi, et al., Docetaxel plus prednisone or mitoxantrone plus prednisone for advanced prostate cancer. N. Engl. J. Med. **351**, 1502–1512 (2004)
7. J. Armenia, S.A.M. Wankowicz, D. Liu, J. Gao, R. Kundra, E. Reznik, et al., The long tail of oncogenic drivers in prostate cancer. Nat. Genet. **50**, 645–651 (2018)
8. M.F. Berger, M.S. Lawrence, F. Demichelis, Y. Drier, K. Cibulskis, A.Y. Sivachenko, et al., The genomic complexity of primary human prostate cancer. Nature **470**, 214–220 (2011)
9. H. Muller, Androgen-control therapy in carcinoma of prostate. Arch. Chir. Neerl. **1**, 77–88 (1949)
10. T.H. van der Kwast, J. Schalken, J.A. Ruizeveld de Winter, C.C. van Vroonhoven, E. Mulder, W. Boersma, et al., Androgen receptors in endocrine-therapy-resistant human prostate cancer. Int. J. Cancer **48**, 189–193 (1991)
11. M.E. Taplin, G.J. Bubley, T.D. Shuster, M.E. Frantz, A.E. Spooner, G.K. Ogata, et al., Mutation of the androgen-receptor gene in metastatic androgen-independent prostate cancer. N. Engl. J. Med. **332**, 1393–1398 (1995)
12. J.P. Gaddipati, D.G. McLeod, H.B. Heidenberg, I.A. Sesterhenn, M.J. Finger, J.W. Moul, et al., Frequent detection of codon 877 mutation in the androgen receptor gene in advanced prostate cancers. Cancer Res. **54**, 2861–2864 (1994)
13. G. Petrovics, A. Liu, S. Shaheduzzaman, B. Furusato, C. Sun, Y. Chen, et al., Frequent overexpression of ETS-related gene-1 (ERG1) in prostate cancer transcriptome. Oncogene **24**, 3847–3852 (2005)
14. S.A. Tomlins, D.R. Rhodes, S. Perner, S.M. Dhanasekaran, R. Mehra, X.W. Sun, et al., Recurrent fusion of TMPRSS2 and ETS transcription factor genes in prostate cancer. Science **310**, 644–648 (2005)
15. M. Yoshimoto, A.M. Joshua, S. Chilton-Macneill, J. Bayani, S. Selvarajah, A.J. Evans, et al., Three-color FISH analysis of TMPRSS2/ERG fusions in prostate cancer indicates that genomic microdeletion of chromosome 21 is associated with rearrangement. Neoplasia **8**, 465–469 (2006)
16. Cancer Genome Atlas Research N, The molecular taxonomy of primary prostate cancer. Cell **163**, 1011–1025 (2015)
17. S.A. Tomlins, N. Palanisamy, J. Siddiqui, A.M. Chinnaiyan, L.P. Kunju, Antibody-based detection of ERG rearrangements in prostate core biopsies, including diagnostically challenging cases:

ERG staining in prostate core biopsies. Arch. Pathol. Lab. Med. **136**, 935–946 (2012)

18. K. Park, S.A. Tomlins, K.M. Mudaliar, Y.L. Chiu, R. Esgueva, R. Mehra, et al., Antibody-based detection of ERG rearrangement-positive prostate cancer. Neoplasia **12**, 590–598 (2010)

19. K.D. Mertz, S.R. Setlur, S.M. Dhanasekaran, F. Demichelis, S. Perner, S. Tomlins, et al., Molecular characterization of TMPRSS2-ERG gene fusion in the NCI-H660 prostate cancer cell line: a new perspective for an old model. Neoplasia **9**, 200–206 (2007)

20. S.J. Baker, E.P. Reddy, Understanding the temporal sequence of genetic events that lead to prostate cancer progression and metastasis. Proc. Natl. Acad. Sci. U. S. A. **110**, 14819–14820 (2013)

21. A. Dal Pra, E. Lalonde, J. Sykes, F. Warde, A. Ishkanian, A. Meng, et al., TMPRSS2-ERG status is not prognostic following prostate cancer radiotherapy: implications for fusion status and DSB repair. Clin. Cancer Res. **19**, 5202–5209 (2013)

22. S. Minner, M. Enodien, H. Sirma, A.M. Luebke, A. Krohn, P.S. Mayer, et al., ERG status is unrelated to PSA recurrence in radically operated prostate cancer in the absence of antihormonal therapy. Clin. Cancer Res. **17**, 5878–5888 (2011)

23. K.J. Kron, A. Murison, S. Zhou, V. Huang, T.N. Yamaguchi, Y.J. Shiah, et al., TMPRSS2-ERG fusion co-opts master transcription factors and activates NOTCH signaling in primary prostate cancer. Nat. Genet. **49**(9), 1336–1345 (2017)

24. S.C. Baca, D. Prandi, M.S. Lawrence, J.M. Mosquera, A. Romanel, Y. Drier, et al., Punctuated evolution of prostate cancer genomes. Cell **153**, 666–677 (2013)

25. M. Fraser, V.Y. Sabelnykova, T.N. Yamaguchi, L.E. Heisler, J. Livingstone, V. Huang, et al., Genomic hallmarks of localized, non-indolent prostate cancer. Nature **541**(7637), 359–364 (2017)

26. J. Weischenfeldt, R. Simon, L. Feuerbach, K. Schlangen, D. Weichenhan, S. Minner, et al., Integrative genomic analyses reveal an androgen-driven somatic alteration landscape in early-onset prostate cancer. Cancer Cell **23**, 159–170 (2013)

27. D.A. Quigley, H.X. Dang, S.G. Zhao, P. Lloyd, R. Aggarwal, J.J. Alumkal, et al., Genomic hallmarks and structural variation in metastatic prostate cancer. Cell **174**, 758–69.e9 (2018)

28. D. Robinson, E.M. Van Allen, Y.M. Wu, N. Schultz, R.J. Lonigro, J.M. Mosquera, et al., Integrative clinical genomics of advanced prostate cancer. Cell **161**, 1215–1228 (2015)

29. C.S. Grasso, Y.M. Wu, D.R. Robinson, X. Cao, S.M. Dhanasekaran, A.P. Khan, et al., The mutational landscape of lethal castration-resistant prostate cancer. Nature **487**, 239–243 (2012)

30. G. Boysen, D. Nava Rodrigues, P. Rescigno, G. Seed, D.I. Dolling, R. Riisnaes, et al., SPOP mutated/CHD1 deleted lethal prostate cancer and abiraterone sensitivity. Clin. Cancer Res. **24**(22), 5585–5593 (2018)

31. R. Mehra, B. Han, S.A. Tomlins, L. Wang, A. Menon, M.J. Wasco, et al., Heterogeneity of TMPRSS2 gene rearrangements in multifocal prostate adenocarcinoma: molecular evidence for an independent group of diseases. Cancer Res. **67**, 7991–7995 (2007)

32. X. Wang, Y. Qiao, I.A. Asangani, B. Ateeq, A. Poliakov, M. Cieslik, et al., Development of peptidomimetic inhibitors of the ERG gene fusion product in prostate cancer. Cancer Cell **31**, 532–48. e7 (2017)

33. J.C. Brenner, B. Ateeq, Y. Li, A.K. Yocum, Q. Cao, I.A. Asangani, et al., Mechanistic rationale for inhibition of poly(ADP-ribose) polymerase in ETS gene fusion-positive prostate cancer. Cancer Cell **19**, 664–678 (2011)

34. S. Steurer, P.S. Mayer, M. Adam, A. Krohn, C. Koop, D. Ospina-Klinck, et al., TMPRSS2-ERG fusions are strongly linked to young patient age in low-grade prostate cancer. Eur. Urol. **66**, 978–981 (2014)

35. L.C. Trotman, M. Niki, Z.A. Dotan, J.A. Koutcher, A. Di Cristofano, A. Xiao, et al., Pten dose dictates cancer progression in the prostate. PLoS Biol. **1**, E59 (2003)

36. S. Ren, Z. Peng, J.H. Mao, Y. Yu, C. Yin, X. Gao, et al., RNA-seq analysis of prostate cancer in the Chinese population identifies recurrent gene fusions, cancer-associated long noncoding RNAs and aberrant alternative splicings. Cell Res. **22**, 806–821 (2012)

37. D.C. Wedge, G. Gundem, T. Mitchell, D.J. Woodcock, I. Martincorena, M. Ghori, et al., Sequencing of prostate cancers identifies new cancer genes, routes of progression and drug targets. Nat. Genet. **50**, 682–692 (2018)

38. S. Ren, G.H. Wei, D. Liu, L. Wang, Y. Hou, S. Zhu, et al., Whole-genome and transcriptome sequencing of prostate cancer identify new genetic alterations driving disease progression. Eur. Urol. (2017)

39. N. Heisterkamp, K. Stam, J. Groffen, A. de Klein, G. Grosveld, Structural organization of the bcr gene and its role in the Ph' translocation. Nature **315**, 758–761 (1985)

40. J.L. Hecht, J.C. Aster, Molecular biology of Burkitt's lymphoma. J. Clin. Oncol. **18**, 3707–3721 (2000)

41. S.M. Kornblau, A. Goodacre, F. Cabanillas, Chromosomal abnormalities in adult non-endemic Burkitt's lymphoma and leukemia: 22 new reports and a review of 148 cases from the literature. Hematol. Oncol. **9**, 63–78 (1991)

42. G. Kaur, R. Gupta, N. Mathur, L. Rani, L. Kumar, A. Sharma, et al., Clinical impact of chromothriptic complex chromosomal rearrangements in newly diagnosed multiple myeloma. Leuk. Res. **76**, 58–64 (2019)

43. N.D. Anderson, R. de Borja, M.D. Young, F. Fuligni, A. Rosic, N.D. Roberts, et al., Rearrangement bursts generate canonical gene fusions in bone and soft tissue tumors. Science **361**, eaam8419 (2018)

44. A. Steininger, G. Ebert, B.V. Becker, C. Assaf, M. Mobs, C.A. Schmidt, et al., Genome-wide analy-

sis of interchromosomal interaction probabilities reveals chained translocations and overrepresentation of translocation breakpoints in genes in a cutaneous T-cell lymphoma cell line. Front. Oncol. **8**, 183 (2018)

45. Z. Wang, Y. Cheng, J.M. Abraham, R. Yan, X. Liu, W. Chen, et al., RNA sequencing of esophageal adenocarcinomas identifies novel fusion transcripts, including NPC1-MELK, arising from a complex chromosomal rearrangement. Cancer **123**, 3916–3924 (2017)

46. J.K. Lee, S. Louzada, Y. An, S.Y. Kim, S. Kim, J. Youk, et al., Complex chromosomal rearrangements by single catastrophic pathogenesis in NUT midline carcinoma. Ann. Oncol. **28**, 890–897 (2017)

47. V. Bhandari, C. Hoey, L.Y. Liu, E. Lalonde, J. Ray, J. Livingstone, et al., Molecular landmarks of tumor hypoxia across cancer types. Nat Genet **51**(2), 308–318 (2019)

48. V. Stambolic, A. Suzuki, J.L. de la Pompa, G.M. Brothers, C. Mirtsos, T. Sasaki, et al., Negative regulation of PKB/Akt-dependent cell survival by the tumor suppressor PTEN. Cell **95**, 29–39 (1998)

49. A.M. Mendes-Pereira, S.A. Martin, R. Brough, A. McCarthy, J.R. Taylor, J.S. Kim, et al., Synthetic lethal targeting of PTEN mutant cells with PARP inhibitors. EMBO Mol. Med. **1**, 315–322 (2009)

50. M.D. Forster, K.J. Dedes, S. Sandhu, S. Frentzas, R. Kristeleit, A. Ashworth, et al., Treatment with olaparib in a patient with PTEN-deficient endometrioid endometrial cancer. Nat. Rev. Clin. Oncol. **8**, 302–306 (2011)

51. K.J. Dedes, D. Wetterskog, A.M. Mendes-Pereira, R. Natrajan, M.B. Lambros, F.C. Geyer, et al., PTEN deficiency in endometrioid endometrial adenocarcinomas predicts sensitivity to PARP inhibitors. Sci. Transl. Med. **2**, 53ra75 (2010)

52. A. Gupta, Q. Yang, R.K. Pandita, C.R. Hunt, T. Xiang, S. Misri, et al., Cell cycle checkpoint defects contribute to genomic instability in PTEN deficient cells independent of DNA DSB repair. Cell Cycle **8**, 2198–2210 (2009)

53. B. McEllin, C.V. Camacho, B. Mukherjee, B. Hahm, N. Tomimatsu, R.M. Bachoo, et al., PTEN loss compromises homologous recombination repair in astrocytes: implications for glioblastoma therapy with temozolomide or poly(ADP-ribose) polymerase inhibitors. Cancer Res. **70**, 5457–5464 (2010)

54. M. Fraser, H. Zhao, K.R. Luoto, C. Lundin, C. Coackley, N. Chan, et al., PTEN deletion in prostate cancer cells does not associate with loss of RAD51 function: implications for radiotherapy and chemotherapy. Clin. Cancer Res. **18**, 1015–1027 (2012)

55. M. Zimmermann, O. Murina, M.A.M. Reijns, A. Agathanggelou, R. Challis, Z. Tarnauskaite, et al., CRISPR screens identify genomic ribonucleotides as a source of PARP-trapping lesions. Nature **559**, 285–289 (2018)

56. S.M.G. Espiritu, L.Y. Liu, Y. Rubanova, V. Bhandari, E.M. Holgersen, L.M. Szyca, et al., The evolutionary landscape of localized prostate cancers drives clinical aggression. Cell **173**, 1003–13.e15 (2018)

57. P.C. Boutros, M. Fraser, N.J. Harding, R. de Borja, D. Trudel, E. Lalonde, et al., Spatial genomic heterogeneity within localized, multifocal prostate cancer. Nat. Genet. **47**, 736–745 (2015)

58. L. Burkhardt, S. Fuchs, A. Krohn, S. Masser, M. Mader, M. Kluth, et al., CHD1 is a 5q21 tumor suppressor required for ERG rearrangement in prostate cancer. Cancer Res. **73**, 2795–2805 (2013)

59. W. Gan, X. Dai, A. Lunardi, Z. Li, H. Inuzuka, P. Liu, et al., SPOP promotes ubiquitination and degradation of the ERG oncoprotein to suppress prostate cancer progression. Mol. Cell **59**, 917–930 (2015)

60. J. An, S. Ren, S.J. Murphy, S. Dalangood, C. Chang, X. Pang, et al., Truncated ERG oncoproteins from TMPRSS2-ERG fusions are resistant to SPOP-mediated proteasome degradation. Mol. Cell **59**, 904–916 (2015)

61. E. Muhlbradt, E. Asatiani, E. Ortner, A. Wang, E.P. Gelmann, NKX3.1 activates expression of insulin-like growth factor binding protein-3 to mediate insulin-like growth factor-I signaling and cell proliferation. Cancer Res. **69**, 2615–2622 (2009)

62. J.A. Locke, G. Zafarana, A.S. Ishkanian, M. Milosevic, J. Thoms, C.L. Have, et al., NKX3.1 haploinsufficiency is prognostic for prostate cancer relapse following surgery or image-guided radiotherapy. Clin. Cancer Res. **18**, 308–316 (2012)

63. G. Zafarana, A.S. Ishkanian, C.A. Malloff, J.A. Locke, J. Sykes, J. Thoms, et al., Copy number alterations of c-MYC and PTEN are prognostic factors for relapse after prostate cancer radiotherapy. Cancer **118**, 4053–4062 (2012)

64. E. Castro, S. Jugurnauth-Little, Q. Karlsson, F. Al-Shahrour, E. Pineiro-Yanez, F. Van de Poll, et al., High burden of copy number alterations and c-MYC amplification in prostate cancer from BRCA2 germline mutation carriers. Ann. Oncol. **26**, 2293–2300 (2015)

65. R.A. Taylor, M. Fraser, J. Livingstone, S.M. Espiritu, H. Thorne, V. Huang, et al., Germline BRCA2 mutations drive prostate cancers with distinct evolutionary trajectories. Nat. Commun. **8**, 13671 (2017)

66. G.P. Risbridger, R.A. Taylor, D. Clouston, A. Sliwinski, H. Thorne, S. Hunter, et al., Patient-derived xenografts reveal that intraductal carcinoma of the prostate is a prominent pathology in BRCA2 mutation carriers with prostate cancer and correlates with poor prognosis. Eur. Urol. **67**, 496–503 (2015)

67. A. Berlin, E. Lalonde, J. Sykes, G. Zafarana, K.C. Chu, V.R. Ramnarine, et al., NBN gain is predictive for adverse outcome following image-guided radiotherapy for localized prostate cancer. Oncotarget **5**, 11081–11090 (2014)

68. H. Guo, M. Ahmed, F. Zhang, C.Q. Yao, S. Li, Y. Liang, et al., Modulation of long noncoding RNAs by risk SNPs underlying genetic predispositions to prostate cancer. Nat. Genet. **48**, 1142–1150 (2016)

69. B. Le Tallec, G.A. Millot, M.E. Blin, O. Brison, B. Dutrillaux, M. Debatisse, Common fragile site profiling in epithelial and erythroid cells reveals that most recurrent cancer deletions lie in fragile sites hosting large genes. Cell Rep. **4**, 420–428 (2013)

70. J.R. Prensner, W. Chen, S. Han, M.K. Iyer, Q. Cao, V. Kothari, et al., The long non-coding RNA PCAT-1 promotes prostate cancer cell proliferation through cMyc. Neoplasia **16**, 900–908 (2014)

71. W. Zhang, B. Liu, W. Wu, L. Li, B.M. Broom, S.P. Basourakos, et al., Targeting the MYCN-PARP-DNA damage response pathway in neuroendocrine prostate cancer. Clin. Cancer Res. **24**, 696–707 (2018)

72. J.M. Mosquera, H. Beltran, K. Park, T.Y. MacDonald, B.D. Robinson, S.T. Tagawa, et al., Concurrent AURKA and MYCN gene amplifications are harbingers of lethal treatment-related neuroendocrine prostate cancer. Neoplasia **15**, 1–10 (2013)

73. J. Edwards, N.S. Krishna, C.J. Witton, J.M. Bartlett, Gene amplifications associated with the development of hormone-resistant prostate cancer. Clin. Cancer Res. **9**, 5271–5281 (2003)

74. E. Lalonde, A.S. Ishkanian, J. Sykes, M. Fraser, H. Ross-Adams, N. Erho, et al., Tumor genomic and microenvironmental heterogeneity for integrated prediction of 5-year biochemical recurrence of prostate cancer: a retrospective cohort study. Lancet Oncol. **15**, 1521–1532 (2014)

75. H. Hieronymus, N. Schultz, A. Gopalan, B.S. Carver, M.T. Chang, Y. Xiao, et al., Copy number alteration burden predicts prostate cancer relapse. Proc. Natl. Acad. Sci. U. S. A. **111**, 11139–11144 (2014)

76. E. Lalonde, R. Alkallas, M.L. Chua, M. Fraser, S. Haider, A. Meng, et al., Translating a prognostic DNA genomic classifier into the clinic: retrospective validation in 563 localized prostate tumors. Eur. Urol. **72**(1), 22–31 (2017)

77. W.P. Kloosterman, V. Guryev, M. van Roosmalen, K.J. Duran, E. de Bruijn, S.C. Bakker, et al., Chromothripsis as a mechanism driving complex de novo structural rearrangements in the germline. Hum. Mol. Genet. **20**, 1916–1924 (2011)

78. P.J. Stephens, C.D. Greenman, B. Fu, F. Yang, G.R. Bignell, L.J. Mudie, et al., Massive genomic rearrangement acquired in a single catastrophic event during cancer development. Cell **144**, 27–40 (2011)

79. S.K. Govind, A. Zia, P.H. Hennings-Yeomans, J.D. Watson, M. Fraser, C. Anghel, et al., ShatterProof: operational detection and quantification of chromothripsis. BMC Bioinformatics **15**, 78 (2014)

80. J.R. Federer-Gsponer, C. Quintavalle, D.C. Muller, T. Dietsche, V. Perrina, T. Lorber, et al., Delineation of human prostate cancer evolution identifies chromothripsis as a polyclonal event and FKBP4 as a potential driver of castration resistance. J. Pathol. **245**, 74–84 (2018)

81. C. Wu, A.W. Wyatt, A. McPherson, D. Lin, B.J. McConeghy, F. Mo, et al., Poly-gene fusion transcripts and chromothripsis in prostate cancer. Genes Chromosomes Cancer **51**, 1144–1153 (2012)

82. D. Nava Rodrigues, P. Rescigno, D. Liu, W. Yuan, S. Carreira, M.B. Lambros, et al., Immunogenomic analyses associate immunological alterations with mismatch repair defects in prostate cancer. J. Clin. Invest. **128**, 4441–4453 (2018)

83. T.A. Knijnenburg, L. Wang, M.T. Zimmermann, N. Chambwe, G.F. Gao, A.D. Cherniack, et al., Genomic and molecular landscape of DNA damage repair deficiency across the cancer genome atlas. Cell Rep **23**, 239–54.e6 (2018)

84. Y.M. Wu, M. Cieslik, R.J. Lonigro, P. Vats, M.A. Reimers, X. Cao, et al., Inactivation of CDK12 delineates a distinct immunogenic class of advanced prostate cancer. Cell **173**, 1770–82.e14 (2018)

85. M. Noguchi, T.A. Stamey, J.E. McNeal, R. Nolley, Prognostic factors for multifocal prostate cancer in radical prostatectomy specimens: lack of significance of secondary cancers. J. Urol. **170**, 459–463 (2003)

86. A.M. Wise, T.A. Stamey, J.E. McNeal, J.L. Clayton, Morphologic and clinical significance of multifocal prostate cancers in radical prostatectomy specimens. Urology **60**, 264–269 (2002)

87. L. Wei, J. Wang, E. Lampert, S. Schlanger, A.D. DePriest, Q. Hu, et al., Intratumoral and intertumoral genomic heterogeneity of multifocal localized prostate cancer impacts molecular classifications and genomic prognosticators. Eur. Urol. **71**, 183–192 (2017)

88. D. Brocks, Y. Assenov, S. Minner, O. Bogatyrova, R. Simon, C. Koop, et al., Intratumor DNA methylation heterogeneity reflects clonal evolution in aggressive prostate cancer. Cell Rep. **8**, 798–806 (2014)

89. C.S. Cooper, R. Eeles, D.C. Wedge, P. Van Loo, G. Gundem, L.B. Alexandrov, et al., Analysis of the genetic phylogeny of multifocal prostate cancer identifies multiple independent clonal expansions in neoplastic and morphologically normal prostate tissue. Nat. Genet. **47**, 367–372 (2015)

90. M. Lovf, S. Zhao, U. Axcrona, B. Johannessen, A.C. Bakken, K.T. Carm, et al., Multifocal primary prostate cancer exhibits high degree of genomic heterogeneity. Eur. Urol. **75**(3), 498–505 (2019)

91. D. Nava Rodrigues, N. Casiraghi, A. Romanel, M. Crespo, S. Miranda, P. Rescigno, et al., RB1 heterogeneity in advanced metastatic castration-resistant prostate cancer. Clin. Cancer Res. **25**, 687–697 (2019)

92. M.C. Haffner, T. Mosbruger, D.M. Esopi, H. Fedor, C.M. Heaphy, D.A. Walker, et al., Tracking the clonal origin of lethal prostate cancer. J. Clin. Invest. **123**, 4918–4922 (2013)

93. G. Gundem, P. Van Loo, B. Kremeyer, L.B. Alexandrov, J.M. Tubio, E. Papaemmanuil,

et al., The evolutionary history of lethal metastatic prostate cancer. Nature **520**, 353–357 (2015)

94. J. Kneppers, O. Krijgsman, M. Melis, J. de Jong, D.S. Peeper, E. Bekers, et al., Frequent clonal relations between metastases and non-index prostate cancer lesions. JCI Insight. **4**, 124756 (2019)

95. S. Nik-Zainal, L.B. Alexandrov, D.C. Wedge, P. Van Loo, C.D. Greenman, K. Raine, et al., Mutational processes molding the genomes of 21 breast cancers. Cell **149**, 979–993 (2012)

96. C. Gerhauser, F. Favero, T. Risch, R. Simon, L. Feuerbach, Y. Assenov, et al., Molecular evolution of early-onset prostate cancer identifies molecular risk markers and clinical trajectories. Cancer Cell **34**, 996–1011.e8 (2018)

97. A. McPherson, A. Roth, E. Laks, T. Masud, A. Bashashati, A.W. Zhang, et al., Divergent modes of clonal spread and intraperitoneal mixing in high-grade serous ovarian cancer. Nat. Genet. **48**, 758–767 (2016)

98. A.W. Zhang, A. McPherson, K. Milne, D.R. Kroeger, P.T. Hamilton, A. Miranda, et al., Interfaces of malignant and immunologic clonal dynamics in ovarian cancer. Cell **173**, 1755–69.e22 (2018)

99. C. Abbosh, N.J. Birkbak, G.A. Wilson, M. Jamal-Hanjani, T. Constantin, R. Salari, et al., Phylogenetic ctDNA analysis depicts early-stage lung cancer evolution. Nature **545**, 446–451 (2017)

100. S. Roemeling, M.J. Roobol, S.H. de Vries, C. Gosselaar, T.H. van der Kwast, F.H. Schroder, Prevalence, treatment modalities and prognosis of familial prostate cancer in a screened population. J. Urol. **175**, 1332–1336 (2006)

101. S. Kommu, S. Edwards, R. Eeles, The clinical genetics of prostate cancer. Hered. Cancer Clin. Pract. **2**, 111–121 (2004)

102. C. Breast Cancer Linkage, Cancer risks in BRCA2 mutation carriers. J. Natl. Cancer Inst. **91**, 1310–1316 (1999)

103. C.C. Pritchard, J. Mateo, M.F. Walsh, N. De Sarkar, W. Abida, H. Beltran, et al., Inherited DNA-repair gene mutations in men with metastatic prostate cancer. N. Engl. J. Med. **375**, 443–453 (2016)

104. L. Briollais, H. Ozcelik, J. Xu, M. Kwiatkowski, E. Lalonde, D.H. Sendorek, et al., Germline mutations in the Kallikrein 6 region and predisposition for aggressive prostate cancer. J. Natl. Cancer Inst. **109** (2017)

105. N.C. Turner, C.J. Lord, E. Iorns, R. Brough, S. Swift, R. Elliott, et al., A synthetic lethal siRNA screen identifying genes mediating sensitivity to a PARP inhibitor. EMBO J. **27**, 1368–1377 (2008)

106. J. Mateo, S. Carreira, S. Sandhu, S. Miranda, H. Mossop, R. Perez-Lopez, et al., DNA-repair defects and olaparib in metastatic prostate cancer. N. Engl. J. Med. **373**, 1697–1708 (2015)

107. T. Dadaev, E.J. Saunders, P.J. Newcombe, E. Anokian, D.A. Leongamornlert, M.N. Brook, et al., Fine-mapping of prostate cancer susceptibility loci in a large meta-analysis identifies candidate causal variants. Nat. Commun. **9**, 2256 (2018)

108. F.R. Schumacher, A.A. Al Olama, S.I. Berndt, S. Benlloch, M. Ahmed, E.J. Saunders, et al., Association analyses of more than 140,000 men identify 63 new prostate cancer susceptibility loci. Nat. Genet. **50**, 928–936 (2018)

109. P. Isaacsson Velho, J.L. Silberstein, M.C. Markowski, J. Luo, T.L. Lotan, W.B. Isaacs, et al., Intraductal/ductal histology and lymphovascular invasion are associated with germline DNA-repair gene mutations in prostate cancer. Prostate **78**, 401–407 (2018)

110. R. Bottcher, C.F. Kweldam, J. Livingstone, E. Lalonde, T.N. Yamaguchi, V. Huang, et al., Cribriform and intraductal prostate cancer are associated with increased genomic instability and distinct genomic alterations. BMC Cancer **18**, 8 (2018)

111. B.S. Taylor, N. Schultz, H. Hieronymus, A. Gopalan, Y. Xiao, B.S. Carver, et al., Integrative genomic profiling of human prostate cancer. Cancer Cell **18**, 11–22 (2010)

112. H. Hieronymus, R. Murali, A. Tin, K. Yadav, W. Abida, H. Moller, et al., Tumor copy number alteration burden is a pan-cancer prognostic factor associated with recurrence and death. elife **7**, e37294 (2018)

113. M. Kluth, S. Harasimowicz, L. Burkhardt, K. Grupp, A. Krohn, K. Prien, et al., Clinical significance of different types of p53 gene alteration in surgically treated prostate cancer. Int. J. Cancer **135**, 1369–1380 (2014)

114. B. De Laere, S. Oeyen, M. Mayrhofer, T. Whitington, P.J. van Dam, P. Van Oyen, et al., TP53 outperforms other androgen receptor biomarkers to predict abiraterone or enzalutamide outcome in metastatic castration-resistant prostate cancer. Clin. Cancer Res. **25**(6), 1766–1773 (2019)

115. M.K. Buyyounouski, T. Pickles, L.L. Kestin, R. Allison, S.G. Williams, Validating the interval to biochemical failure for the identification of potentially lethal prostate cancer. J. Clin. Oncol. **30**, 1857–1863 (2012)

116. W.C. Jackson, K. Suresh, V. Tumati, S.G. Allen, R.T. Dess, S.S. Salami, et al., Intermediate endpoints after postprostatectomy radiotherapy: 5-year distant metastasis to predict overall survival. Eur. Urol. **74**(4), 413–419 (2018)

117. E. Olkhov-Mitsel, F. Siadat, K. Kron, L. Liu, A.J. Savio, J. Trachtenberg, et al., Distinct DNA methylation alterations are associated with cribriform architecture and intraductal carcinoma in Gleason pattern 4 prostate tumors. Oncol. Lett. **14**, 390–396 (2017)

118. S.N. Kamdar, L.T. Ho, K.J. Kron, R. Isserlin, T. van der Kwast, A.R. Zlotta, et al., Dynamic interplay between locus-specific DNA methylation and hydroxymethylation regulates distinct biological pathways in prostate carcinogenesis. Clin. Epigenetics **8**, 32 (2016)

119. N.M. White-Al Habeeb, L.T. Ho, E. Olkhov-Mitsel, K. Kron, V. Pethe, M. Lehman, et al., Integrated

analysis of epigenomic and genomic changes by DNA methylation dependent mechanisms provides potential novel biomarkers for prostate cancer. Oncotarget **5**, 7858–7869 (2014)

120. E. Olkhov-Mitsel, D. Zdravic, K. Kron, T. van der Kwast, N. Fleshner, B. Bapat, Novel multiplex MethyLight protocol for detection of DNA methylation in patient tissues and bodily fluids. Sci. Rep. **4**, 4432 (2014)

121. K. Kron, D. Trudel, V. Pethe, L. Briollais, N. Fleshner, T. van der Kwast, et al., Altered DNA methylation landscapes of polycomb-repressed loci are associated with prostate cancer progression and ERG oncogene expression in prostate cancer. Clin. Cancer Res. **19**, 3450–3461 (2013)

122. E. Olkhov-Mitsel, T. Van der Kwast, K.J. Kron, H. Ozcelik, L. Briollais, C. Massey, et al., Quantitative DNA methylation analysis of genes coding for kallikrein-related peptidases 6 and 10 as biomarkers for prostate cancer. Epigenetics **7**, 1037–1045 (2012)

123. K. Kron, L. Liu, D. Trudel, V. Pethe, J. Trachtenberg, N. Fleshner, et al., Correlation of ERG expression and DNA methylation biomarkers with adverse clinicopathologic features of prostate cancer. Clin. Cancer Res. **18**, 2896–2904 (2012)

124. L. Liu, K.J. Kron, V.V. Pethe, N. Demetrashvili, M.E. Nesbitt, J. Trachtenberg, et al., Association of tissue promoter methylation levels of APC, TGFbeta2, HOXD3 and RASSF1A with prostate cancer progression. Int. J. Cancer **129**, 2454–2462 (2011)

125. K.J. Kron, L. Liu, V.V. Pethe, N. Demetrashvili, M.E. Nesbitt, J. Trachtenberg, et al., DNA methylation of HOXD3 as a marker of prostate cancer progression. Lab. Investig. **90**, 1060–1067 (2010)

126. K. Kron, V. Pethe, L. Briollais, B. Sadikovic, H. Ozcelik, A. Sunderji, et al., Discovery of novel hypermethylated genes in prostate cancer using genomic CpG island microarrays. PLoS One **4**, e4830 (2009)

127. G. Vandekerkhove, W.J. Struss, M. Annala, K. HML, D. Khalaf, E.W. Warner, et al., Circulating tumor DNA abundance and potential utility in de novo metastatic prostate cancer. Eur. Urol. **75**(4), 667–675 (2019)

128. M. Annala, G. Vandekerkhove, D. Khalaf, S. Taavitsainen, K. Beja, E.W. Warner, et al., Circulating tumor DNA genomics correlate with resistance to abiraterone and enzalutamide in prostate cancer. Cancer Discov. **8**, 444–457 (2018)

129. E. Ritch, A.W. Wyatt, Predicting therapy response and resistance in metastatic prostate cancer with circulating tumor DNA. Urol. Oncol. **36**, 380–384 (2018)

130. G. Vandekerkhove, K.N. Chi, A.W. Wyatt, Clinical utility of emerging liquid biomarkers in advanced prostate cancer. Cancer Genet. **228-229**, 151–158 (2018)

131. N. Camacho, P. Van Loo, S. Edwards, J.D. Kay, L. Matthews, K. Haase, et al., Appraising the relevance of DNA copy number loss and gain in prostate cancer using whole genome DNA sequence data. PLoS Genet. **13**, e1007001 (2017)

132. F. Su, W. Zhang, D. Zhang, Y. Zhang, C. Pang, Y. Huang, et al., Spatial intratumor genomic heterogeneity within localized prostate cancer revealed by single-nucleus sequencing. Eur. Urol. **74**(5), 551–559 (2018)

Prostate Cancer Transcriptomic Subtypes

Daniel E. Spratt

Overview of Transcriptomics

Transcriptomics is the study of RNA molecules and is used to interrogate the activity of the genome in a cell or tumor by measuring its RNA makeup. Despite there being at least 11 types of known RNAs (e.g. mRNA, rRNA, tRNA, snRNA, snoRNA, siRNA, hnRNA, gRNA, tmRNA, telomerase RNA, catalytic RNA), the RNA of greatest interest in Oncology is currently messenger RNA (mRNA), which is actively transcribed from DNA and ultimately translated into protein. More recently, lncRNAs are also becoming of increasing interest [1, 2]. As discussed in the prior chapter, DNA is largely similar across cells of an organism, often with specific alterations that define specific genomic subtypes of cancer. mRNA in contrast is highly dynamic and is less binary and static compared to a DNA mutation. Gene expression typically reflects the functional activity of a cell more than DNA, as even if an upstream gene is mutated or has lost function, if alternative pathways become activated mRNA expression may remain constant or even increased.

Given that many genes have similar expression and are highly correlated with one another, transcriptomics often is synthesized into gene expression signatures to capture subtypes of a particular cancer. These signatures reflect a snapshot of the tumor in time, and despite their dynamic nature, can reproducibly capture more static genomic and biologic subtypes, and even serve as reliable prognostic and predictive biomarkers. In prostate cancer specifically, transcriptomics initially was often used to compliment genomics. However, especially in localized prostate cancer, the genomes of prostate cancers contain a relatively small number of somatic driver mutations and/or copy number alterations, and thus there is currently limited utility in routinely searching for DNA alterations. Thus, gene expression profiling alone is increasingly being studied in localized and recurrent prostate cancer.

Technology

The most common technologies used clinically for the assessment of gene expression include real-time PCR and microarray. Research studies have increasingly transitioned to RNA sequencing (RNA-seq), but commercial tests almost exclusively use PCR or microarray technology [3]. Each technology has its strengths and weaknesses that must be weighed, including costs, breadth of transcriptome covered, customizability, data-analysis, throughput, resolution, and

D. E. Spratt (✉)
Department of Radiation Oncology, University of Michigan Medical Center, Ann Arbor, MI, USA
e-mail: sprattda@med.umich.edu

© Springer Nature Switzerland AG 2019
S. M. Dehm, D. J. Tindall (eds.), *Prostate Cancer*, Advances in Experimental Medicine and Biology 1210, https://doi.org/10.1007/978-3-030-32656-2_6

dynamic range [3, 4]. Most whole transcriptome studies have used a discovery process with data generated using either microarray or RNA-seq technology, and the subsequent signature created of typically <50 genes is then recreated with either a targeted sequencing process or simply uses RT-PCR given the reduced costs. Table 1 summarizes the differences between microarray and RNA-seq technologies.

Methods of Subtyping

The goal of subtyping is to define a subgroup of prostate cancer that is unique using transcriptomics (Table 2). This could be to capture previously identified distinct genomic subtypes based on unique DNA profiles that can be captured using gene expression data. Alternatively, guided or semi-supervised subtyping methods can be performed, using gene expression to capture known biologic characteristics, such as basal- or luminal-ness, cell cycle activity, or neuroendocrine differentiation. More commonly, subgrouping by prognosis is performed with commercially available subtyping signatures, rather than looking at a biologic feature. In contrast, one can use gene expression data, which may or may not be rooted in known biologically driven mechanistic data, to identify patients who intrinsically are most- or least-likely to benefit from treatment. Finally, the least common is to perform unsupervised hierarchical clustering to determine what genes statistically form unbiased subgroups. To illustrate the relationship of many of the devel-

oped prognostic, predictive, and biological subtypes in prostate cancer, Fig. 1 shows a heatmap of many of the subtypes and transcriptomic signatures that will be discussed in this chapter.

Subtypes

Capture Genomic Subtypes

As described in the previous chapter, the most recognized subtypes of prostate cancer are classically defined based on DNA alterations. This is reflected in genomic data available from The Cancer Genome Atlas (TGCA) from localized prostate cancer specimens showing frequent ERG and ETS-family rearrangements and SPOP mutations. This is also reflected in genomic data from multiple large metastatic CRPC cohorts, which have shown common alterations in p53, RB loss, DNA repair alterations, PTEN loss, among a list of frequently occurring mutations. In many instances it is not practical to perform genomic sequencing to identify all of these alterations, especially given that it usually requires fresh-frozen tissue. Thus, investigators have developed methods to accurately and reliably capture these subtypes with gene expression data.

1. *ERG, ETS, SPINK1* [5]: A gene expression signature that accurately captures *ERG+* tumors was developed with a random forest supervised model to predict FISH-assessed *ERG* rearrangement status. The model was developed and trained (n = 252 samples) and

Table 1 Comparison of microarray and RNA-seq technology

	Microarray	RNA-seq
Principle	Hybridization	High-throughput sequencing
Thoughput	High	High
Background noise	Higher	Lower
Dynamic range	~100-fold	>8000-fold
Distinguish different isoforms	Limited	Easier
Cost (perform, store, and analyze data)	Lower	Higher
RNA content required	Higher	Lower
Heterogeneity of read coverage across expressed region	Yes	No
Analysis simplicity	Simple	Complex
Data portability (size of data)	Megabites	Gigabites

Prostate Cancer Transcriptomic Subtypes

Table 2 Common transcriptomic subtyping methods

Subtyping categories	Restriction of genes used?	Currently used clinically or in clinical research?	Examples
Unsupervised hierarchical clustering	No	No	TCGA RNA clusters
Capture genomic (DNA) subtypes/alterations	Yes	No	ERG, ETS, SPINK1
			SPOP mutant
			PTEN loss
Capture biologic characteristics	Yes	No	AR-Activity
			NEPC
Prognostic biomarkers	Either	Yes	Decipher
			Prolaris
			Oncotype Dx
Predictive biomarkers	Either	Yes	ADT-RS
			PAM50
			PORTOS

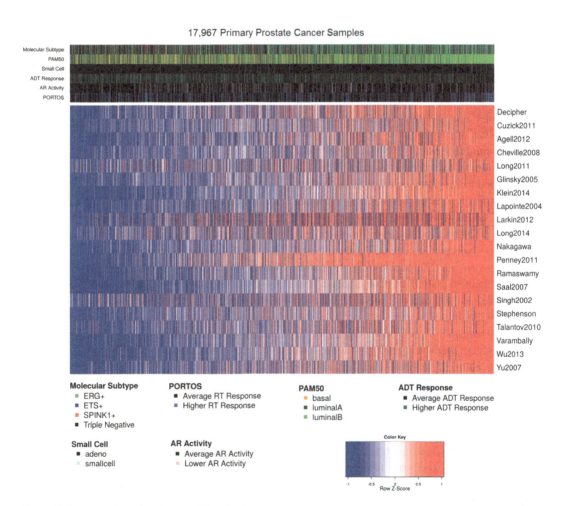

Fig. 1 Select transcriptomic subtypes of localized prostate cancer

validated with 155 tumors with known FISH-ERG status. Additionally, a classification method based on gene expression for *ETV1*, *ETV4*, *ETV5*, *FLI1*, and *SPINK1* was developed with an unsupervised outlier analysis using the *extremevalues* on the expression of core probe sets for each gene. Tumors were ultimately grouped into four subtypes (*ERG+* subtype, *ETS+*, *SPINK+*, or triple negative). Over 1500 patients were used to train and validate this signature. *ERG+* tumors typically had lower baseline serum PSA levels and lower Gleason scores compared to the triple negative subtype. *SPINK1* tumors typically had higher PSAs and were more common in African-Americans. Subsequently, these microarray expression-based signatures were analytically validated against established immunohistochemical and FISH assays. Despite these findings, there were no significant differences in time to biochemical recurrence or distant metastases, suggesting these subgroups are biologically based and not prognostic.

2. *SPOP mutant signature* [6]: A novel gene expression signature and decision tree was developed to accurately predict SPOP mutant cancers from gene expression data. Starting with TCGA data, including RNA-seq data and known SPOP mutant status, differential gene expression was performed and clustered based on SPOP status. 212 genes were ultimately used to define the SPOP mutant subclass. This signature, which was validated in a cohort from Weill Cornell Medicine (n = 68), found an 89% sensitivity and 95% specificity of SPOP mutant prediction compared with DNA mutation calling. Using the prior classifier for ERG+ and ETS+ status [5], which are mutually exclusive from the SPOP mutant subtype, This decision tree was able to identify tumors most likely to harbor SPOP mutations in samples without DNA data (n > 8000). It was found that SPOP predicted tumors were less likely to have higher grade tumors, positive surgical margins, or T3 disease. However, they were predicted to have higher PSAs. Therefore, despite the other clinicopathologic

factors being favorable, SPOP mutant tumors identified by this gene expression classifier had worse outcomes.

3. *PTEN loss signature* [7]: PI3K is frequently activated in prostate cancer, especially metastatic CRPC through PTEN loss. PTEN mRNA expression levels are the primary determinant of PTEN protein levels. A PTEN status signature was developed for breast cancer samples with microarray data to identify genes most significantly associated with PTEN IHC status. From this, a consensus ranked gene list was generated by sorting the average p-value from each cross-validation analysis. A total of 246 genes were ultimately included in the PTEN signature with a receiver-operator characteristic (ROC) of 0.758. This signature was also applied to other tumor types, including prostate cancer. In prostate cancer, this signature was shown to correlate with worse survival.

Unsupervised Hierarchical Clustering

One method of obtaining subtypes is to simply let the data determine what genes are differentially expressed across samples in a manner that clearly divides patients into a limited number of groups based on a list of genes. This usually requires a large panel (e.g. >1000) of genes to be assessed. Given that this method does not force or restrict the clustering to predict either an outcome (e.g. recurrence) or a feature (e.g. genomic subtype), the genes discovered may be of unclear importance in prostate cancer at first glance.

1. *TCGA- 3 clusters* [8]: The TCGA performed a multi-center study to interrogate primary prostate cancer comprehensively, at the molecular level Using 333 tumors, analyses were conducted on the genome, transcriptome, proteome, and epigenome, ultimately identifying seven molecularly defined subtypes (ERG, ETV1, ETV4, FLI1, SPOP, FOXA1, IDH1, and others). Integrative clustering based solely on mRNA data was also performed. This was done via unsupervised expression clustering of

prostate tumors using mRNA-seq data. The top 3000 most variable genes were used to develop mRNA subtypes. Three mRNA clusters were identified that largely grouped ERG and ETV positive tumors into one cluster, and SPOP, FOXA1, and IDH1 subtypes together into another cluster. The FLI1 genomic subgroup contained most of the third mRNA cluster. However, these mRNA subtypes did not optimally recapitulate the seven genomic subgroups. Thus, work from the Tomlins [5] and Barbieri [6] laboratories who derived the ERG+ and SPOP mutant signature appear to be more accurate than the TCGA subtypes, suggesting the original findings were likely over-fit, modeling error due to inference on a limited set of patient samples.

2. In another study tissue microarray profiles of 62 primary prostate cancer tumors, 41 normal prostate cancer specimens and 9 lymph node metastases captured >26,000 coding and non-coding genes [9]. Unsupervised hierarchical clustering was performed on all of the samples. Using 5153 cDNAs whose expression varied most across samples, tumor samples were distinguishable from normal samples. Additionally, three subtypes of prostate cancer were identified based on distinct gene expression patterns. However, the biological relevance or prognostic or predictive relevance of these molecular subtypes is unclear, which is one disadvantage to clustering performed in a completely unsupervised manner.

Supervised Clustering to Capture Specific Biologic Characteristics

1. *Prostate Cancer Subtypes 1-3* [10]: An integrated classification of prostate cancer was performed on a large training cohort of 1321 tumor samples and a validation set using 10 patient cohorts and 19 laboratory models of prostate cancer (cell lines and genetically engineered mouse models). Twenty two pathway-activation gene expression signatures relevant to prostate cancer were employed to perform the clustering. These were subsequently collapsed to 14 pathway signatures that were grouped into three categories: (1) AR, AR-V, EZH2, FOXA1, RAS, and PRC, (2) SPOP, TMPRSS2-ERG, PTEN, and (3) stemness, proliferation, epithelial-mesenchymal transition, pro-neural, and neuroendocrine differentiation. At this point unsupervised clustering was performed using the 14 pathways activation profiles, and three distinct clusters were identified and termed PC1, PC2, and PC3. These subtypes were validated in both localized and mCRPC. Interestingly, the TCGA subtypes, including ERG, ETV1/4, SPOP, and FOXA1 were found across all of the new subtypes identified, with differential enrichment by subtype. This study also looked at the association of basal and luminal expression and its correlation to the PC1-3 subtypes. They found a strong association between luminal genes with PC1 and PC2, and basal genes with PC3. The PC1-3 subtypes were also prognostic, in that the PC1 subtype had shorter metastasis-free survival than either PC2 or PC3. Ultimately, the subtypes were simplified into a 37-gene signature that could reasonably recapitulate the three subtypes. The clinical utility and clinical relevance of this signature is unclear, highlighting the immense biological heterogeneity of prostate cancer.

2. *AR-activity* [11, 12]: The androgen receptor (AR) gene, which is near ubiquitously expressed in prostate cancer, regulates thousands of genes. In localized prostate cancer AR expression has limited heterogeneity in expression, whereas in metastatic castration-resistant prostate cancer (mCRPC), there is more diversity in AR expression. However, the activity of the AR, or AR-signaling or AR-activity, which is measured by the expression of canonical AR-target genes, is significantly more heterogeneous in both localized and mCRPC. Recent work demonstrated that ~10% of localized prostate cancer has lower AR-activity measured by nine canonical AR-targets. This subset appears to closely resemble advanced mCRPC in that both have similar AR-activity. Furthermore, expression of

neuroendocrine markers and immunesignaling signatures are increased in this low AR-active subset. Not surprisingly, low AR-active localized prostate cancer has a poor prognosis with a more rapid progression to metastatic disease compared to high AR-activity tumors. Not only does low AR-active prostate cancer have a worse prognosis, it also appears to have unique treatment sensitivities. High AR-active prostate cancer is more sensitive to ADT and taxane chemotherapy, while low AR-active prostate cancer appears more sensitive to PARP inhibition and cisplatin chemotherapy. Further work is in development to assess if AR-activity can serve as a prognostic biomarker and a predictive biomarker to guide treatment selection.

3. *Neuroendocrine Prostate Cancer (NEPC) signature* [13]: A gene expression signature of neuroendocrine and primary small cell prostate cancer was developed using samples from eight cohorts to compare gene expression that is either up or down in NEPC compared to adenocarcinoma samples. A 69 gene signature was identified that captured at least 80% of NEPC patients. These genes generated three subgroups that were termed atypical small cell prostate cancer, prototypical adenocarcinoma, and prototypical small cell prostate cancer.

Subtypes Developed for Prognosis

Commercial Classifiers

1. *Decipher* [12, 14–19]: The Decipher assay is a clinical-grade transcriptome-wide gene expression profiling assay, based on the Human Exon 1.0 ST oligonucleotide microarray (GenomeDx, Inc). While the assay measures over 46,000 protein-coding and non-coding RNAs, the current Decipher clinical test result is a prognostic biomarker based on the expression of 22 genes. A cohort of radical prostatectomy samples was used to train the signature for the primary endpoint of clinical failure (e.g. metastases) between patients who did and did not develop failure post-treatment. Forty three RNA transcripts

were identified that were differentially expressed between groups. Through random forest machine learning 22-genes were ultimately identified that yielded the best performance for the prediction of metastatic disease. These 22 genes include both coding and noncoding genes that have roles in cell cycle progression, proliferation, immune response, cell adhesion and motility. The Decipher test has since been validated in over 3000 patients in >40 studies. Most notably the performance of the classifier was validated in a meta-analysis using 975 patients across five cohorts. Decipher was shown to independently predict for the development of metastatic disease, and had superior performance than currently used clinicopathologic variables (e.g. Gleason score, T-stage, margin status, PSA, etc). The C-index of the clinical model was 0.76, which increased to 0.81 from the addition of Decipher to the model. Furthermore, Decipher performed similarly across all subgroups by age, race, and treatment performed. Decipher has also been recently combined into an integrated clinical-genomic risk grouping system that mirrors NCCN risk groups. This study validated the superior performance of Decipher over clinical factors in both surgical samples as well as pre-treatment biopsy samples. The C-index for the combined clinical-genomic system was 0.84, and approximately 67% of patients were reclassified from NCCN risk groups to new clinical-genomic risk groups. The Decipher test has been used in prospective trials as well. The PRO-IMPACT trial assessed the clinical utility of changing management decisions based on the Decipher test. Furthermore, the G-MINOR trial has completed enrollment, and has randomized patients and providers to the receipt of the Decipher test as compared to the best available clinical nomogram (CAPRA-S model). This trial will be the first randomized trial to assess the clinical utility of any commercial genomic classifier in prostate cancer. Decipher is also being used in multiple ongoing national randomized trials, including NRG GU-002, which is stratifying patients by the use of

Decipher. Other trials, such as NRG GU-006 is leveraging the Decipher assay, rather than just the Decipher score, since the microarray used to assess genes in the Decipher test provides hundreds of additional signatures through the Decipher GRID, given that >46,000 genes are analyzed on every sample. Recently, the performance of Decipher in the first randomized trial of any commercial gene expression classifier has been reported. The SPARTAN trial, a randomized trial assessing the benefit of apalutamide in M0CRPC ran the Decipher test on a subset of the trial with banked tissue. They showed that Decipher was highly prognostic and predictive of first line ADT failure.

2. *Oncotype Dx* [20]: The Oncotype Dx Genomic Prostate Score (GPS) is a 17 gene signature designed for pre-treatment biopsy use. It is run on a RT-PCR platform. Its intended use is to help guide active surveillance decision making. To derive the signature, 198 genes were identified that correlated with recurrence, death from prostate cancer, and adverse pathology. This gene list was truncated to 81, which were associated with aggressive disease within the validation cohort. Ultimately, the signature was refined to 17 genes based on consistency of expression across cohorts. These genes are involved in four primary pathways, including stromal proliferation, androgen signaling, cellular organization, and proliferation. Although the GPS signature has been validated in multiple prostatectomy cohorts of patients eligible for active surveillance, until recently it had not been validated in actual active surveillance patients. Recently, the Canary PASS trial performed Oncotype Dx testing on 634 men entering active surveillance. Unfortunately, the Oncotype Dx test was not associated with subsequent biopsy upgrade on either uni- or multi-variable analysis. These results bring into question the clinical accuracy of the Oncotype Dx test. Future studies are needed to assess the role of GPS testing in prostate cancer.

3. *Prolaris* [21]: Prolaris, also known as the Cell Cycle Progression (CCP) score, measures 31 cell cycle progression genes and 15 housekeeping genes. It is run on a RT-PCR platform. CCP was initially developed in breast cancer patients and has since been tested and validated in prostate cancer patients. It remains unclear if the test was optimized fully for prostate cancer, but it has been validated in prostate cancer needle biopsies and also prostatectomy samples. It has been tested in patients undergoing active surveillance, prostatectomy, and radiotherapy. The CCP test has been tested for multiple outcomes, including biochemical recurrence, metastasis, and prostate cancer-specific mortality. The test has not been used in any randomized trials to date, and future prospective studies are needed to demonstrate its clinical utility and benefit in intact and post-treatment patients.

Non-commercial Classifiers

1. There have been dozens, if not hundreds, of prognostic gene expression signatures reported in the literature. A brief list is shown in Table 3. They have been developed for various indications with various degrees of validation. None have robust clinical data to support their use, and none are commercially available or covered by Medicare (in contrast to Decipher, Oncotype Dx, and Prolaris). When these signatures were optimized to predict for the development of metastatic disease, it was found that the Decipher 22-gene signature outperformed all of the other signatures when run on the same microarray platform [22].

Subtypes Developed for Predicting Treatment Response

1. *RSI* [23]: The Radiation Sensitivity Index (RSI) was developed to predict intrinsic sensitivity to ionizing radiotherapy. It claims to be a pan-cancer signature and was developed from the National Cancer Institute panel of 60 cell lines. Thirty five of these cell lines were ultimately used to determine which genes correlated with clonogenic survival after 2 Gy of radiation therapy. In cell line data the signature

Table 3 Select list of prognostic gene expression signatures in prostate cancer

Signature	Original technology[a]	Number of features
Erho 2013	Microarray	22
Penney 2011	Microarray	157
Wu 2013	RT-qPCR	30
Bibikova 2007	Microarray	16
Xie 2011	Microarray	71
Ramaswamy 2003	Microarray	17
Agell 2012	Microarray	12
LaPointe 2004	Microarray	22
Nakagawa 2008	Microarray	17
Bismar 2006	Microarray	13
Cheville 2008	Microarray	2
Cuzick 2011	RT-qPCR	31
Yu 2007	Microarray	87
Larkin 2012	RT-qPCR	3
Singh 2002	Microarray	12
Klein 2014	RT-qPCR	17
Larkin 2012	RT-qPCR	10
Saal 2007	Microarray	185
D. Antonio 2008	RT-qPCR	59
Glinsky 2005	Microarray	11
Varambally 2005	Microarray	50
Long 2011	Microarray	12
Stephenson 2005	Microarray	15
Talantov 2010	RT-qPCR	24
Yu 2007	Microarray	14
Roca 2012	Microarray	10
Glinsky 2004	Microarray	5
Stephenson 2005	Microarray	10
Ross 2012	RT-qPCR	6
Glinsky 2004	Microarray	5
Glinsky 2004	Microarray	4
Irshad 2013	Microarray	3
Olmos 2012	Microarray	9
Singh 2002	Microarray	5

[a]This refers to the technology used to discover/develop expression signature

only has a 62% accuracy of predicting cellular radiosensitivity. Ultimately, the RSI was developed, which is comprised of a linear algorithm of 11 genes (AR, cJun, STAT1, PKC, cABL, SUMO1, CDK1, HDAC1, and IRF1), each with its own weight, that are summed to yield a final score. There are limited data using this signature in prostate cancer, and it does not appear to be able to predict outcomes in patients treated with radiotherapy. Future work will be necessary to determine if RSI can be applied to patients with prostate cancer.

2. *PORTOS* [24]: Leveraging the Decipher GRID, a 24-gene Post-Operative Radiation Therapy Outcomes Score (PORTOS) was developed and validated to predict for benefit of post-operative radiotherapy. Using a training cohort of 198 patients, 1800 DNA damage repair and previously annotated radiation response genes were ranked based on outcomes after post-operative radiotherapy. Twenty-four genes were identified that predicted benefit from post-operative radiotherapy. Patients with high PORTOS scores derived a significant benefit, as measured as a reduction in distant metastasis, from receipt of post-operative radiotherapy. In contrast, patients with low PORTOS scores failed to derive benefit from the addition of post-operative radiotherapy. In the validation cohort (n = 330) it was confirmed that PORTOS was a predictive biomarker of post-operative radiotherapy benefit (p-interaction = 0.016). Importantly, it is probable that PORTOS may not be purely a measure of intrinsic radiation sensitivity, but rather a predictor of patients who harbor micrometastatic disease outside the radiation field.

3. *ADT-RS* [25]: The Decipher GRID was used to access 1212 patients that underwent a radical prostatectomy with adverse pathology. Patients who received early adjuvant ADT were matched to patients who did not receive early ADT. Rather than a purely unsupervised analysis, they limited genes to a curated gene list of 1632 genes identified from studies investigating neuroendocrine differentiation, castration resistance, and resistance to ADT. This gene list was then filtered based on feature ranking and model training. Ultimately 49 genes were identified and validated that

were predictive of early ADT benefit. This was demonstrated with a significant interaction test in those with a high ADT-RS score (p = 0.035), while low ADT-RS patients did not derive any benefit from early use of adjuvant ADT. Notably, ADT-RS was not prognostic, but was in fact highly predictive, and on multivariable analysis the interaction for ADT-RS was even stronger after adjusting for other clinicopathologic factors.

4. *PAM50* [26]: The PAM50 classifier was original developed in breast cancer. It is the basis for the commercially available Prosigna product run using NanoString. PAM50 successfully classifies breast cancers as luminal A, luminal B, HER2, and basal subtypes. These subtypes are not only prognostic, they are predictive of benefit of endocrine therapy and HER2 targeted therapy. Given that multiple cancers, including prostate cancer, also have luminal and basal subtypes, the PAM50 signature was applied to localized prostate cancer leveraging the Decipher GRID gene expression database. Notably, the HER2 subtype was removed, since ERBB2/HER2 amplification does not occur in prostate cancer as it does in breast cancer. The authors used the transcriptome-wide microarray Human Exon 1.0 ST microarray platform on 1567 retrospective samples with long-term follow up that was further divided into a training and validation cohort. Additionally, they used 2215 prospective samples to characterize the PAM50 subtypes in localized prostate cancer. All three subtypes, luminal A, luminal B, and basal, were identified in localized prostate cancer at similar distributions (~33% each). Known luminal markers, such as NKX3.1 and KRT18 were enriched in the luminal subtypes. Similarly, the basal marker CD49f was enriched in the basal subtype. Luminal B patients were the most likely to develop biochemical recurrence and distant metastasis, and display worse prostate cancer specific survival, and overall survival, as determined independently by multivariable analysis.

Luminal A patients had the most favorable outcomes. Given the ability of PAM50 to predict responses of breast cancer to endocrine therapy, the benefit of ADT was tested. It was demonstrated that luminal B patients derived a significant improvement in metastasis-free survival from the addition of post-operative ADT, while luminal A and basal patients did not. The interaction test was significant (p = 0.006), indicating that PAM50 appears to be a predictive biomarker of post-operative ADT benefit. These results have led to an open randomized phase 2 trial testing if the addition of apalutamide, a next generation anti-androgen, will improve outcomes over salvage radiotherapy alone (NRG GU006, NCT03371719).

Conclusions

The transcriptome of prostate cancer continues to be unraveled. This chapter primarily focused on gene expression signatures that are based on the expression of protein coding genes. It is clear that gene expression data can recapitulate many of the important genomic alterations. Perhaps more importantly, the transcriptome has been leveraged to provide unparalleled accuracy in assigning a personalized prognosis for a patient above and beyond routine clinicopathologic parameters. Many of these signatures are now in clinical practice, and randomized data will be reported over the next 1–2 years to validate some of these signatures. The most exciting area that is just beginning to unravel is the ability for gene expression classifiers to serve as true predictive biomarkers, which can identify patients most likely to benefit from standard of care treatments, such as radiotherapy or ADT. Some of these are currently in ongoing randomized trials and have the promise to change the clinical landscape of managing prostate cancer.

Disclosure Advisory board for Janssen and Blue Earth.

References

1. J.R. Prensner, M.K. Iyer, A. Sahu, et al., The long noncoding RNA SChLAP1 promotes aggressive prostate cancer and antagonizes the SWI/SNF complex. Nat. Genet. **45**(11), 1392 (2013)
2. J.R. Prensner, W. Chen, S. Han, et al., The long noncoding RNA PCAT-1 promotes prostate cancer cell proliferation through cMyc. Neoplasia **16**(11), 900–908 (2014)
3. Z. Wang, M. Gerstein, M. Snyder, RNA-Seq: a revolutionary tool for transcriptomics. Nat. Rev. Genet. **10**(1), 57 (2009)
4. A. Schulze, J. Downward, Navigating gene expression using microarrays—a technology review. Nat. Cell Biol. **3**(8), E190 (2001)
5. S.A. Tomlins, M. Alshalalfa, E. Davicioni, et al., Characterization of 1577 primary prostate cancers reveals novel biological and clinicopathologic insights into molecular subtypes. Eur. Urol. **68**(4), 555–567 (2015)
6. D. Liu, M. Takhar, M. Alshalalfa, et al., Impact of the SPOP mutant subtype on the interpretation of clinical parameters in prostate cancer. JCO Precis. Oncol. **2**, 1–13 (2018)
7. L.H. Saal, P. Johansson, K. Holm, et al., Poor prognosis in carcinoma is associated with a gene expression signature of aberrant PTEN tumor suppressor pathway activity. Proc. Natl. Acad. Sci. **104**(18), 7564–7569 (2007)
8. A. Abeshouse, J. Ahn, R. Akbani, et al., The molecular taxonomy of primary prostate cancer. Cell **163**(4), 1011–1025 (2015)
9. J. Lapointe, C. Li, J.P. Higgins, et al., Gene expression profiling identifies clinically relevant subtypes of prostate cancer. Proc. Natl. Acad. Sci. **101**(3), 811–816 (2004)
10. S. You, B.S. Knudsen, N. Erho, et al., Integrated classification of prostate cancer reveals a novel luminal subtype with poor outcome. Cancer Res. **76**(17), 4948–4958 (2016)
11. D. Spratt, R. Dess, H. Hartman, et al., Androgen receptor activity and radiotherapeutic sensitivity in African-American men with prostate cancer: a large scale gene expression analysis and meta-analysis of RTOG trials. Int. J. Radiat. Oncol. Biol. Phys. **102**(3), S3 (2018)
12. D.E. Spratt, M. Alshalalfa, A. Weiner, et al., Transcriptomic heterogeneity of androgen receptor activity in primary prostate cancer: identification and characterization of a low AR-active subclass. Proc. Am. Soc. Clin. Oncol. (2018)
13. H.K. Tsai, J. Lehrer, M. Alshalalfa, N. Erho, E. Davicioni, T.L. Lotan, Gene expression signatures of neuroendocrine prostate cancer and primary small cell prostatic carcinoma. BMC Cancer **17**(1), 759 (2017)
14. D.E. Spratt, K. Yousefi, S. Deheshi, et al., Individual patient-level meta-analysis of the performance of the decipher genomic classifier in high-risk men after prostatectomy to predict development of metastatic disease. J. Clin. Oncol. **35**(18), 1991–1998 (2017)
15. N. Erho, A. Crisan, I.A. Vergara, et al., Discovery and validation of a prostate cancer genomic classifier that predicts early metastasis following radical prostatectomy. PLoS One **8**(6), e66855 (2013)
16. D.E. Spratt, D.L. Dai, R.B. Den, et al., Performance of a prostate cancer genomic classifier in predicting metastasis in men with prostate-specific antigen persistence postprostatectomy. Eur. Urol. **74**(1), 107–114 (2018)
17. P.L. Nguyen, Z. Haddad, A.E. Ross, et al., Ability of a genomic classifier to predict metastasis and prostate cancer-specific mortality after radiation or surgery based on needle biopsy specimens. Eur. Urol. **72**(5), 845–852 (2017)
18. P.L. Nguyen, N.E. Martin, V. Choeurng, et al., Utilization of biopsy-based genomic classifier to predict distant metastasis after definitive radiation and short-course ADT for intermediate and high-risk prostate cancer. Prostate Cancer Prostatic Dis. **20**(2), 186–192 (Jun 2017)
19. R.B. Den, K. Yousefi, E.J. Trabulsi, et al., Genomic classifier identifies men with adverse pathology after radical prostatectomy who benefit from adjuvant radiation therapy. J. Clin. Oncol. **33**(8), 944–951 (2015)
20. M. Cooperberg, J. Simko, S. Falzarano, et al., Development and validation of the biopsy-based genomic prostate score (GPS) as a predictor of high grade or extracapsular prostate cancer to improve patient selection for active surveillance. J. Urol. **189**(4), e873 (2013)
21. S. Sommariva, R. Tarricone, M. Lazzeri, W. Ricciardi, F. Montorsi, Prognostic value of the cell cycle progression score in patients with prostate cancer: a systematic review and meta-analysis. Eur. Urol. **69**(1), 107–115 (2016)
22. A.E. Ross, M.H. Johnson, K. Yousefi, et al., Tissue-based genomics augments post-prostatectomy risk stratification in a natural history cohort of intermediate- and high-risk men. Eur. Urol. **69**(1), 157–165 (2016)
23. J.F. Torres-Roca, A molecular assay of tumor radiosensitivity: a roadmap towards biology-based personalized radiation therapy. Pers. Med. **9**(5), 547–557 (2012)
24. S.G. Zhao, S.L. Chang, D.E. Spratt, et al., Development and validation of a 24-gene predictor of response to postoperative radiotherapy in prostate cancer: a matched, retrospective analysis. Lancet Oncol. **17**(11), 1612–1620 (2016)
25. M. Alshalalfa, R.J. Karnes, V. Sharma, et al., Development and validation of a prostate cancer genomic signature that predicts early ADT treatment response following radical prostatectomy. Clin. Cancer Res. **24**(16), 3908–3916 (2018)
26. S.G. Zhao, S.L. Chang, N. Erho, et al., Associations of luminal and basal subtyping of prostate cancer with prognosis and response to androgen deprivation therapy. JAMA Oncol. **3**(12), 1663–1672 (2017)

Immunological Complexity of the Prostate Cancer Microenvironment Influences the Response to Immunotherapy

Nataliya Prokhnevska, Dana A. Emerson, Haydn T. Kissick, and William L. Redmond

Introduction

Current advances in cancer therapeutics have led to the development of immunotherapies that target the body's own immune system to combat cancer. Advances in immunotherapy have led to the development of checkpoint blockade, with anti-programmed cell death 1 (PD-1) and anti-cytotoxic T lymphocyte antigen-4 (CTLA-4) successfully treating many solid tumors. Checkpoint blockade can increase both the proliferative capacity and cytotoxicity of exhausted

Nataliya Prokhnevska and Dana A. Emerson contributed equally to this work.
Haydn T. Kissick and William L. Redmond are co-senior authors.

N. Prokhnevska · H. T. Kissick (✉)
Department of Urology, Emory University, Atlanta, GA, USA
e-mail: haydn.kissick@emory.edu

D. A. Emerson
Molecular Microbiology and Immunology, Oregon Health and Science University, Portland, OR, USA

Earle A. Chiles Research Institute, Providence Cancer Institute, Portland, OR, USA

W. L. Redmond (✉)
Earle A. Chiles Research Institute, Providence Cancer Institute, Portland, OR, USA
e-mail: william.redmond@providence.org

CD8[+] T cells, leading to disease regression [1, 2]. Even though these are very effective treatments, not all cancers have a high response rate to immunotherapy. These treatments depend on the presence of tumor-specific CD8[+] T cells, which can respond to checkpoint blockade. Therefore, understanding basic tumor immunology may help predict patient survival and response to checkpoint blockade.

CD8[+] T cells in Cancer

CD8[+] T cells are a crucial part of the immune response against tumors. Many studies have shown that CD8[+] T cells found within tumors acquire an exhausted phenotype. These exhausted cells lose their ability to proliferate, have increased expression of inhibitory receptors, and have lower effector function, including reduced interferon gamma (IFN-γ) production, granzyme B expression, and/or cytolytic activity [3]. Even though these cells have varying degrees of functionality, CD8[+] T cell infiltration into tumors predicts disease progression in melanoma, breast cancer, head and neck cancer, ovarian cancer, non-small cell lung cancer, esophageal cancer, small cell lung cancer, hepatocellular carcinoma, and renal cell carcinoma [4–6]. Recently, a group

© Springer Nature Switzerland AG 2019
S. M. Dehm, D. J. Tindall (eds.), *Prostate Cancer*, Advances in Experimental Medicine and Biology 1210, https://doi.org/10.1007/978-3-030-32656-2_7

developed an "immunoscore" for tumors based on T cell infiltration by looking at CD8[+] and CD45RO[+] T cells in both the tumor core and invasive margin. Higher immunoscores indicate more CD8[+] T cell infiltration, where patients with higher scores have better disease-free survival and overall survival compared to patients with low immunoscores [7, 8]. The number of CD8[+] T cells in the tumor can also predict response to PD-1 blockade, making it an important biomarker for both survival and response to current immunotherapies [9]. Understanding the mechanism behind the diversity of CD8[+] T cell infiltration in different cancers is crucial to improving current immunotherapies. Additionally, recent studies have changed our understanding of CD8[+] T cell differentiation and exhaustion in the context of chronic infections and cancer. Here, we will present evidence about the factors controlling the magnitude of an immune response in viral infections, discuss how this response differs in cancer, and why these responses are so variable.

Immune Response to Viruses

The primary role of the immune system is to protect against infections. To achieve this, the immune system possesses incredible intricacy and organization of many cell types that coordinate responses against perceived threats. In cancer, the immune system fails to function and organize a response in the same way. By understanding the successful immune response to a virus and applying that knowledge to tumor immunology we can better understand how the system fails to eradicate tumors.

Danger Sensing

A prototypical viral infection begins when a virus enters the body and infects permissive cells. Infected cells sense viral infections through pattern recognition receptors (PRRs) such as Toll-like receptors (TLR), Nod-like receptors, retinoic acid-inducible gene I (RIG-I)-like receptors, and the

stimulator of interferon genes (STING) pathway, which trigger the release of pro-inflammatory cytokines such as interleukin-1 (IL-1), IL-6, IL-8 and tumor necrosis factor (TNF)-α (Fig. 1a) [10, 11]. This pro-inflammatory reaction recruits immune cells including antigen presenting cells (APCs) that aid in clearing the infection, phagocytosis of antigen, and migration through the lymphatics to activate T and B cells (Fig. 1b). This is a crucial step that connects the innate immune system to the adaptive immune system, thus allowing immune-mediated clearance of the infection.

Antigen Presentation

Dendritic cells (DCs) are a critical part of the immune response. When DCs are activated by pathogen-associated molecular patterns (PAMPs) and take up antigen, they process it into peptides that are 8–10 amino acids in length and load them onto the major histocompatibility complex (MHC) for presentation to T cells [12]. DCs traffic to secondary and tertiary lymphoid organs through the lymphatics via expression of chemokine receptors (e.g., CC chemokine receptor-7; CCR7), wherein they interact with cognate antigen-specific T cells to facilitate T cell activation. In addition to T cell receptor (TCR) recognition of the MHC-peptide complex, T cells also need co-stimulation through the co-stimulatory receptor, CD28, which binds its ligands CD80/86 expressed by APCs [13]. Signaling through TCR and CD28 activates T cells and the presence of pro-inflammatory cytokines produced by DCs, such as IL-12 and type I IFN, further promote T cell activation and proliferation [14]. Once a T cell is activated, it upregulates effector molecules, chemokine receptors, and proliferates to produce a population of more antigen-specific CD8[+] T cells capable of combating the infection.

Lymphoid Organization in Viral Infections

Only antigen-specific CD8[+] T cells are activated by DCs presenting cognate MHC-peptides, and

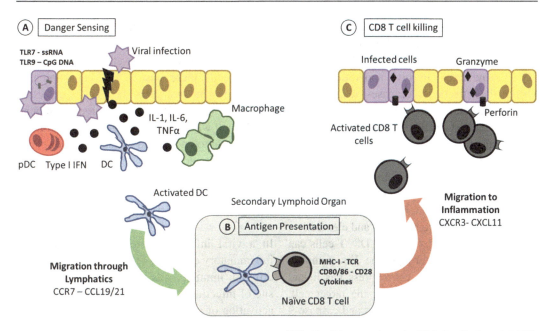

Fig. 1 Immune response to a viral infection. (**a**) Once a virus has infected cells the viral PAMPs induce the expression of pro-inflammatory cytokines, which can recruit macrophages and DCs. pDCs are activated through TLR7/9 and produce large quantities of type I IFN. The combination of viral PAMPs and pro-inflammatory cytokines leads to the activation of DCs and their migration to secondary lymphoid organs. (**b**) Once an activated APC finds an antigen-specific CD8$^+$ T cell it can activate the CD8$^+$ T cells through MHC-peptide interaction with TCR, co-stimulation with CD80/86 interacting with CD28, and through the production of Il-12. (**c**) Activated CD8$^+$ T cells upregulate chemokine receptors, effector molecules, and proliferate. Once the activated CD8$^+$ T cells leave the lymphoid organs and migrate to the site of infection, they can exert effector function by causing apoptosis of infected cells through perforin and granzyme B

for this to happen there needs to be an organization of CD8$^+$ T cells and DCs, which typically occurs in draining lymph nodes (dLN) or within tertiary lymphoid structures. The organization of immune cells is crucial to the anti-tumor response since conventional DCs (cDC1s, which are CD8α$^+$XCR1$^+$) and plasmacytoid DCs (pDCs) help orchestrate CD8$^+$ T cell activation and response to viral antigens. The expression of XCL1 by CD8$^+$ T cells recruits cDC1s, while the production of type I IFN by pDCs acts on DCs and CD8$^+$ T cells to further activation [15]. Thus, for optimal antigen-specific CD8$^+$ T cell activation, there must be collaboration of different DC subsets with each other and CD8$^+$ T cells. This level of organization of DCs and CD8$^+$ T cells is crucial for the immune response to viral infections.

CD8$^+$ T cell Effector Function

Once activated CD8$^+$ T cells upregulate chemokine receptors such as CXC chemokine receptor-3 (CXCR3), they can leave lymphoid organs and migrate to the site of infection and proliferate to produce many antigen-specific CD8$^+$ T cells (Fig. 1c). Activated CD8$^+$ T cells also upregulate effector molecules such as granzyme B, perforin, and death ligands such as Fas and TNF-related apoptosis inducing ligand (TRAIL). CD8$^+$ T cells selectively kill cells that express the MHC-peptide complex for which they are specific, leaving uninfected cells intact while targeting only infected cells. When activated CD8$^+$ T cells encounter an infected and antigen-specific cell, the engagement of its TCR leads to the release of cytotoxic granules. Perforin polymerizes to form a pore on the target cell, allowing different granzymes to enter the target cell. Granzyme B is a protease that can activate caspase-3 and cleave Bid. This induces an intrinsic apoptotic pathway wherein the activation of caspase-3 leads to DNA degradation and eventually apoptosis. The cleaved truncated Bid also interacts with Bax and Bad, leading to the release of cytochrome C from the mitochondria, which also leads to an intrinsic apoptosis. Overall, the selective killing of infected cells by cytotoxic CD8+ T cells prevents unnecessary inflammation and damage as the apoptotic cells are phagocytosed and cleared.

Another pathway by which CD8+ T cells can induce apoptosis is through the Fas pathway. Activated CD8+ T cells can express Fas ligand (FasL), which binds to Fas on the target cell, leading to the trimerization of Fas and caspase-9 activation. Similar to apoptosis through caspase 3, the activation of caspase-9 leads to DNA degradation and apoptosis [16]. The primary function of activated CD8+ T cells is to induce apoptosis of antigen-specific infected cells or transformed tumor cells that express tumor-specific antigens. In a viral infection when the antigen is cleared, some activated CD8+ T cells survive and become memory CD8+ T cells. This is an important function of CD8+ T cells, since having long-lived memory to a pathogen leads to fast recall responses upon reinfection. In the context of progressively growing tumors, CD8+ T cells do not become memory cells and instead gain an exhausted phenotype, upregulate expression of numerous inhibitory receptors, and lose their ability to proliferate.

The Immune Response to Cancer

During viral infections there are specific phases of the immune response: danger sensing, antigen presentation and then clearance of the infection. The immune response to cancer differs in how the immune system is activated and how it responds to the tumor antigens, due to the chronic nature of tumor development as well as differences in the types of antigens to which the immune system responds. Increased T cell infiltration in tumors is associated with improved survival as well as better response to checkpoint blockade. To generate activated tumor-specific CD8+ T cells, there must be APC activation and antigen presentation. How the CD8+ T cells are supported within the tumor microenvironment (TME) and which cells promote their infiltration is important to understand in the context of clinical outcomes and future immunotherapies.

Danger Sensing in Tumors

In a viral infection there are numerous pro-inflammatory cytokines and PAMPs that activate different immune cells and recruit APCs to the site of infection. In tumors, the recruitment and activation of immune cells often does not occur in the same pro-inflammatory environment. In tumors, the immune system relies on danger-associated molecular patterns (DAMPs) such as extracellular ATP, heat shock proteins, hydrophobic aggregates, reactive oxygen species and nucleic acids, which signal that there has been tissue damage and cell death, thereby eliciting an immune response [17]. The DAMPs released by necrotic cells can activate DCs and induce T cell proliferation, whereas apoptotic cell death does not induce the same type of DC activation and T cell activation [18]. This has been shown in both mouse and human DCs that are stimulated and activated by necrotic syngeneic cells or necrotic tumor cells, from melanoma, kidney adenocarcinoma, and thymoma cell lines. DCs activated by necrotic tumor cells are capable of activating antigen-specific CD8+ T cells [19]. Necrotic cells derived from prostate cancer cell lines, such as LNCaP and PC3, are also capable of activating DCs, which can then present antigen and activate CD8+ T cells. Studies with DCs from healthy donors and from stage IV prostate cancer patients, demonstrated that there is no intrinsic defect in DCs from prostate cancer patients [20]. This is a crucial component of generating a productive immune response against tumors, as is the composition of APCs within the TME capable of presenting tumor antigens to tumor-specific CD8+ T cells.

Antigen Presentation of Tumor Antigens

APCs are the interface between antigen and T cell activation. Understanding how this population of cells operates in cancer is key to understanding the initial generation of the anti-tumor T cell response.

DCs in Cancer

Several DC subsets have been classified based upon their phenotype and function in mice and humans. DCs can be broadly classified by the high expression of CD11c and MHC class II (MHC II). One crucial subset of DCs for CD8+ T cell activation is the cross-presenting DC, which refers to the processing and presentation of exogenous antigens on MHC class I molecules [21]. These cells are of interest in the context of cancer immunology since most tumor antigens are exogenous proteins and must be presented on MHC I in order to activate tumor-specific CD8+ T cells. The cross-presenting DC (cDC1) subset has been thoroughly characterized in mice. These cells are defined by CD8α and XCR1 expression and show increased antigen uptake, processing and presentation on MHC I [22]. This DC subset has also been characterized in human tissue, distinguished by the expression of CD141, and has been seen in the lung, liver, skin and blood compartments [23]. This DC subset is indispensable in the activation of CD8+ T cells in infection and tumor progression as cDC1-deficient mice do not control influenza infection or immunogenic tumors in a T cell-dependent manner [24, 25]. Overall, this is a key DC subset in the immune response that specializes in the activation of viral and tumor-specific CD8+ T cells. Another major DC subset (cDC2) in mice is characterized by the expression of CD11b on CD11c+ MHC II+ DCs. CD11b+ DC (mice) and CD1c (humans) are less efficient at cross-presentation of exogenous antigens and therefore thought to mainly activate CD4+ T cells through MHC II [26]. Even though there are many similarities between DC subsets in mice and humans the phenotype of the cells is not always translatable, which is an important consideration when analyzing and comparing DCs from mouse and human tissues.

The last major subset of DCs are pDCs, which are a crucial part of the antiviral immune response. In mouse and humans, this highly specialized subset of cells produces the largest quantities of type I IFN early after viral infection. These cells circulate in the periphery and express high levels of TLR7 and TLR9, which activate the pDCs and induce the expression of pro-inflammatory cytokines. pDCs are a crucial bridge between the innate and adaptive immune system during viral infections as they promote the activation of DCs and T cells. Although some studies have shown pDC infiltration in breast cancer predicts poorer overall survival, more studies are needed to determine how pDCs influence the CD8+ T cell response to tumors [27]. The activation of pDCs is a rapid response to viral infection that may not occur during tumor growth, and this lack of pDC activation may limit the generation of potent anti-tumor CD8+ T cells.

There have been numerous studies showing the prognostic power of DC infiltration in tumors. For example, one study used data from the cancer genome atlas (TCGA) and analyzed CD103/141-associated genes to determine cross-presenting DC infiltration (cDC1). The ratio of CD103/141+ signature genes to genes not associated with CD103/141 DCs acts as a prognostic marker that predicts overall survival in numerous cancer types including breast cancer, head-neck squamous cell carcinoma, and lung adenocarcinoma [26, 28]. This shows how the extent of DC infiltration alone can have prognostic power over a wide variety of cancers. Other studies using TCGA data have also determined that the CD141 gene signature correlated with CD8+ transcript levels [24]. This demonstrates the importance of DCs in the TME to promote and support CD8+ T cell infiltration and how DC and CD8+ T cell infiltration can be used to predict survival in patients with cancer (Fig. 2).

Studies have also shown in human melanoma tumors the crucial role of CD141+ DCs expressing CCR7, which allows for the migration into lymph nodes to present tumor antigens to tumor-specific CD8+ T cells. The tumors containing higher levels

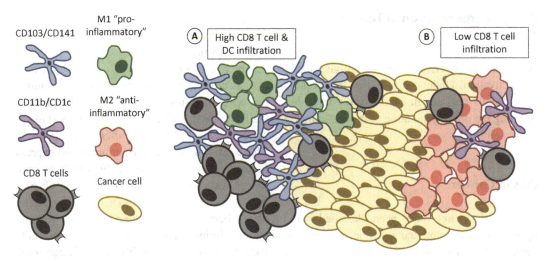

Fig. 2 Differences in immune cell composition in high and low infiltrated tumor. (**a**) Highly immune-infiltrated tumor with clusters of CD103+/CD141+ and CD11b+/CD1c+ DCs, M1-skewed macrophages and CD8+ T cells. These tumors have more pro-inflammatory APCs that can promote CD8+ T cell infiltration and effector function. (**b**) Poorly CD8+ T cell infiltrated tumors with more M2-skewed tumor-associated macrophages, which are immunosuppressive and prevent further CD8+ T cell activation

of CCR7 transcripts correlated with more CD3+ T cell infiltration and better survival [28]. DCs need to be able to bring tumor antigens into the lymphoid organs to activate tumor-specific CD8+ T cells more efficiently due to the higher concentration of CD8+ T cells and DCs in lymphoid tissues. In prostate cancer after androgen ablative therapy, there is an increase in DC and macrophage infiltration [29], but their phenotype before and after treatment remains poorly understood. Other studies, which also looked at DCs in the blood of prostate cancer patients before and after vaccination (GVAX) and checkpoint blockade (aCTLA-4; ipilimumab) treatment, revealed that increased CD1c+ DCs and CD11c+CD14lo DCs predicted better survival with the treatment [30]. Collectively, these studies show that the presence of DCs, especially CD141+ DCs in tumors, correlates with more T cell infiltration and better prognosis and survival in many tumor types. Understanding why DCs infiltrate certain tumors may help us understand the differences in CD8+ T cell infiltration.

Recent studies have also shown that murine DCs present within the TME are less efficient at presenting antigen to and activating CD8+ T cells. These DCs induce less CD8+ T cell proliferation, express lower levels of co-stimulatory molecules and produce less IL-12 [31]. Additional studies showed an inhibitory effect of IL-10-secreting macrophages in the TME, which suppresses DC activation and IL-12 production [32, 33]. Together, these studies show that it is important to not only have good DC infiltration in the tumor, but also to have functional DCs capable of activating T cells and promoting a pro-inflammatory immune response.

Lymphoid Organization in Tumors

Recent work has shown that in certain instances tertiary lymphoid structures (TLS), which contain CD8+ T cells, DCs, follicular DCs and high endothelial venules, can form near the tumor [34]. Tumors containing a TLS were associated with higher T cell infiltration and improved disease-free survival in both breast and colorectal cancer [35, 36]. In viral infections, the organization of APCs and CD8+ T cells is crucial for activating and promoting effector differentiation of antigen-specific CD8+ T cells. Interestingly, in non-small cell lung cancer, TLS-associated mature DCs correlate with CD8+ T cell infiltration and improved survival [37, 38]. This demonstrates the power of having an organized structure that supports T cells and DCs near the tumor that is comparable to the organization of lymphoid tissues during a viral infection. In prostate cancer, some immune structures contain T regulatory cells and other immunosuppressive cells [39]. In prostate cancer, TLS comprised of more pro-inflammatory Type 1 helper (Th1) and CD8[+] T cells are associated with improved tumor regression [40]. Understanding how these TLS form and how the composition and organization can affect patient outcomes is an important step towards developing novel therapeutics for patients that are refractory to immunotherapies and may have immunologically "cold" tumors, which are poorly infiltrated with CD8[+] T cells and/or DCs.

Macrophages in Cancer

Macrophages, which are APCs that are a critical component of the TME, are capable of presenting tumor antigens to T cells, phagocytosing apoptotic cells and secreting various cytokines [41]. Due to the plasticity of these cells, they can acquire different phenotypes based on the immune environment that influences them. A classically activated macrophage acquires an M1 pro-inflammatory phenotype, capable of secreting TNF, IL-1, IL-6, IL-8 to promote the activation of T cells. M1-skewed macrophages can also secrete reactive oxygen species (ROS) and chemokines to attract more pro-inflammatory immune cells to mediate the destruction of pathogens and tumor cells. The other phenotype of macrophages, that is more common in the TME, is a "wound repair" alternatively-activated M2 macrophage. These are a necessary part of the immune response for tissue homeostasis as well as wound repair but are thought to promote tumorigenesis within the TME. M2 macrophages can secrete anti-inflammatory cytokines such as IL-10 and TGF-β, and act as poor APCs. M2-skewed macrophages can also significantly hinder the CD8[+] T cell anti-tumor response through arginase-1 and ROS secretion, which limits CD8[+] T cell activation. They can also secrete growth factors to promote tumor growth and metastasis, such as epidermal growth factor (EGF), as well as suppress T cell anti-tumor activity [32, 42]. Macrophages can also express programmed cell death ligands 1/2 (PD-L1/2) and further hinder CD8[+] T cell anti-tumor activity. Tumor-infiltrating macrophages with an M2 phenotype can express PD-1 and respond to PD-1-blockade leading to reduced tumor burden in mouse models [43]. The location of macrophages in the TME can alter CD8[+] T cell responses, for example when macrophages are in the stroma of lung squamous-cell carcinomas they prevent the interaction of CD8[+] T cells with DCs that are present in the TME [5]. Overall, macrophages are a crucial component of the TME and can directly impact the tumor-specific CD8[+] T cell response.

When macrophages are incubated with conditioned media from prostate cancer cells, they are skewed towards an M2 phenotype and produce IL-10 [42]. Conditioned media from prostate cancer cells can also re-program M1 macrophages into an M2-like phenotype, demonstrating how the TME can alter the phenotype of immune cells. Especially in tumors that are not highly infiltrated by CD8[+] T cells and pro-inflammatory APCs, it is important to understand how the TME can skew the immune environment to promote, rather than inhibit, tumor growth. A higher macrophage density leads to poor prognosis in lung, hepatocellular carcinoma, and renal cell carcinomas [44, 45]. Macrophages are very plastic APCs that can promote immune responses as well as the growth and vascularization of tumors. Understanding what role macrophages play in the TME and how they can directly impact CD8[+] T cells as well as how they can be used to predict clinical outcomes is crucial for understanding the highly complex TME.

The TME also contains cells broadly classified as myeloid-derived suppressor cells (MDSCs), which can promote tumor growth and down-regulate CD8[+] T cell activity. Currently there are two subsets that have been classified, monocytic MDSCs (M-MDSCs, which originate from monocytes) and polymorphonuclear MDSCs (PMN-MDSCs, arising from granulocytic PMN precursors) [46]. In prostate cancer, mostly PMN-MDSCs have been identified and suggested to promote castrate-resistant prostate cancer (CRPC). M-MDSCs share many markers with monocytes and macrophages, making it difficult to distinguish between these cells. Similarly, PMN-MDSCs share many markers with other granulocytes and PMNs.

Understanding how the various APC populations influence CD8+ T cell responses is crucial to determine how to improve immunotherapy in cancers that have not responded to current therapies. In prostate tumors there is a large population of M2 macrophages, which could be the reason for less CD8+ T cell infiltration and reduced activation. Historically, prostate cancer has not responded to T cell-focused therapies, but when DCs pulsed with tumor antigen are given as a therapeutic intervention, such as with Sipuleucel-T treatment, CD8+ T cell infiltration increased and clinical outcomes were modestly improved [47, 48]. Therapies that focus on enhancing immune infiltration into prostate tumors are a current area of interest and could potentially be combined with T cell-focused therapies in order to achieve better clinical outcomes.

CD8+ T cells in Cancer

The extent of DC infiltration correlates with improved CD8+ T cell infiltration and better overall survival. Even though DCs can predict survival, CD8+ T cells are the effector cells capable of destroying tumors. The connection between DCs and CD8+ T cells is crucial in understanding the immune response to cancer. CD8+ T cells found within tumors typically have an exhausted phenotype, with increased inhibitory receptor expression, like PD-1 or T cell immunoglobulin and mucin domain-containing protein-3 (Tim-3), along with decreased proliferation and effector function. This does not explain the necessity of DCs within the TME, but the new stem-like model of CD8+ T cell exhaustion demonstrates

the need for a DC-rich nice to support stem-like CD8+ T cells.

LCMV Model of CD8+ T cell Exhaustion

The lymphocytic choriomeningitis virus (LCMV) model has been used to discover and understand many immunological phenomena, spanning from CD8+ T cell memory to exhaustion. Two strains of LCMV allow for the study of an acute versus chronic viral infection. LCMV Armstrong is an acute viral infection that is cleared via CD8+ T cells and elicits a strong memory CD8+ T cell response. The LCMV clone 13 strain models a chronic infection; by depleting CD4+ T cells and then infecting the mice with clone 13, the infection becomes truly chronic and is not cleared by the immune system [49, 50]. The early model of LCMV-induced T cell exhaustion described a gradual increase in inhibitory receptor expression and loss of effector function. CD8+ T cells first lose their ability to produce IL-2 and their cytotoxic function, followed by the loss of proliferation and ability to produce IFN-γ and TNF-α [3, 51]. PD-L1 blockade was shown to rescue antigen-specific exhausted CD8+ T cells in chronic LCMV clone 13 infected mice, thus restoring their ability to proliferate and produce IFN-γ [2]. This was the first example of PD-1 blockade restoring CD8+ T cell functionality in a chronic antigen setting.

CD8+ T cells found within tumors have the canonical exhausted phenotype. They have up-regulated numerous inhibitory receptors, such as PD-1, CTLA-4, lymphocyte activation gene 3 (Lag-3), and Tim-3. They also have a diminished ability to proliferate and produce effector cytokines [52, 53]. This shows that tumor-infiltrating CD8+ T cells have a comparable phenotype to the antigen-specific CD8+ T cells found in the chronic viral model of LCMV clone 13 infected mice. While this model of T cell exhaustion offers an understanding of why T cells in tumors have lost functionality, it does not explain why some tumors have very low T cell numbers, or more importantly why some patients do not respond to checkpoint therapy.

Stem-Like Model of CD8+ T cell Exhaustion

Recent work in the field of CD8+ T cell exhaustion has resulted in a revised model of how T cell exhaustion occurs [54]. The new working model of CD8+ T cell exhaustion has implications in how we think about CD8+ T cell exhaustion as well as modern immunotherapy approaches. This work described two populations of CD8+ T cells in a chronic viral infection that express PD-1. One is a stem-like CD8+ T cell that expresses CXCR5, T cell factor 1 (TCF1), and has higher expression of CD28, while the other is a terminally-differentiated Tim-3+ CD8+ T cell. The stem-like CD8+ T cells reside in a DC-rich location of the spleen (Fig. 3b), while the Tim-3+ subset localizes to sites of infection and is not restricted to lymphoid tissues and acts as effectors, expressing granzyme B and IFN-γ. The CXCR5+ TCF1+ stem-like CD8+ T cell subset can self-renew and give rise to Tim-3+ effector-like cells (Fig. 3c). During viral infection, PD-1 blockade promotes CXCR5+TCF1+ stem-like cell proliferation and differentiation into a large population of antigen-specific effector-like CD8+ T cells capable of killing infected cells. PD-1 blockade also affected Tim-3+ CD8+ T cells by blocking negative signaling and increasing their effector functions at the site of infection [54]. This new model of CD8+ T cell exhaustion can help us understand how the immune system responds to chronic antigen as well as how PD-1 blockade works within this model.

Since this model was described, several groups have observed a stem-like model of T cell exhaustion in numerous cancers and murine tumor models. Populations of TCF1+ and terminally-differentiated Tim-3+ CD8+ T cells have been described in MC38 sarcoma, B16 melanoma,

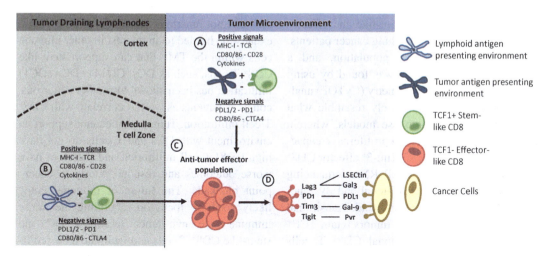

Fig. 3 Generation of an anti-tumor effector response against cancer. (**a**) Stem-like cells reside within tumors and their presence requires an antigen presenting niche for support: These cells receive their activating and inhibitory signals from a range of APCs including DCs, M1 and M2 macrophages, TAMs, and/or inflammatory monocytes. The rate of generation is an aggregate of many complex positive and negative signals. For example, PD-L1 blockade can increase the rate of effector generation from stem-like cells. Failure to generate an effector population due to lack of the correct signaling results in a tumor with low CD8+ T cell numbers, but no terminally exhausted cells. (**b**) It is unclear if a stem-like CD8+ T cell in lymphoid tissue gives rise to effector cells that migrate to areas of inflammation to kill target cells. One hypothesis is that these cells will be present in regions where DCs are most dense. (**c**) The total yield from the stem-like cell is an aggregate of what is produced by cells in the lymph nodes and tumor. The number of cells generated by this process may be critical for controlling tumor growth. High numbers of terminally exhausted CD8+ T cells implies the activation process of the stem-like cell is continuing effectively. (**d**) The anti-cancer effect caused by CD8+ T cells is proportional to the number of effector CD8+ T cells, the quality of these cells, and the negative signals sent back from tumor cells expressing inhibitory ligands. Inhibitory molecules including Tim-3, Lag-3, and others probably have minimal effect on the number of CD8+ T cells generated, but may have critical effects on the ability of the effector population to kill tumor cells

Transgenic adenocarcinoma of the mouse prostate (TRAMP)-C1 prostate cancer, Lewis Lung Carcinoma, and CT26 colon tumors [55, 56]. These cells closely resemble the originally identified CD8+ T cells from chronic viral infection. TCF1+ CD8+ T cells have higher expression of IL-7R, CCR7 and CD62L, like CXCR5+ cells. The effector TCF1-Tim-3+ CD8+ T cells were also found to express higher levels of effector molecules such as IFN-γ and granzyme B, which parallels their original description. An important part of the stem-like model of CD8+ T cell exhaustion is the ability of TCF1+ CD8+ T cells to proliferate and give rise to Tim-3+ effector cells. While Tim-3+ CD8+ T cells lack the ability to proliferate, they do express more effector molecules [54]. These studies demonstrate that the stem-like model of CD8+ T cell exhaustion can be applied to numerous mouse tumor models.

The stem-like model of CD8+ T cell exhaustion is also translatable to human cancer. Many recent studies have shown that in different human cancers there are similar stem-like and effector CD8+ T cell populations. In lung cancer patients, a CXCR5+ CD8+ T cell population, and a CXCR5-Tim-3+ population was found by using high dimensional mass cytometry (CyTOF) analysis. These two subsets closely resemble what has been described in mouse models, wherein CXCR5+ CD8+ T cells retain proliferative capacity and give rise to CXCR5-Tim-3+ effector CD8+ T cells [57]. In a single-cell RNA sequencing study of melanoma, a transitional and dysfunctional CD8+ T cell subset was found. Transitional CD8+ T cells present within tumors retain TCF1 expression, while dysfunctional CD8+ T cells have higher expression of inhibitory receptors such as PD-1 and Lag-3 [58]. Lack of functional CD8+ T cell response to tumors leads to poor T cell infiltration, there needs to be both a functional TCF1+ stem-like CD8+ T cell and it must produce Tim3+ effector-like CD8+ T cells in order to have a productive CD8+ T cell response to the tumor.

The environment in which these CD8+ T cell stem-cells are maintained is a crucial aspect of the stem-model of exhaustion, especially when considering that originally the stem-like TCF1+ CD8+ T cells were discovered in lymphoid tissues surrounded by DCs in a chronic viral infection. The organization of DCs and CD8+ T cells is crucial for the activation of CD8+ T cells in viral infections. These discoveries have led to an increasing interest in the immune environment that best supports stem-like CD8+ T cells and induces their proliferation and differentiation, especially within the TME. Since the amount of CD8+ T cells within tumors has clear positive prognostic power, it is important to understand the immune niche supporting CD8+ T cell infiltration into tumors.

Overall, the immune system is crucial in the response to cancer as CD8+ T cell infiltration into tumors can predict patient outcomes as well as response to immunotherapy. Understanding the stem-like model of CD8+ T cell exhaustion has led to a better understanding of the CD8+ T cell response to PD-1 blockade, which increases the differentiation of Tim-3+ CD8+ T cells as well as their effector function. This model of CD8+ T cell exhaustion has led to an interest in other immune cells within the TME that can support stem-like CD8+ T cells, such as DCs. CD141+ DC (cDC1) infiltration has been shown to predict better outcomes in patients as well as correlate with CD8+ T cell infiltration. Tumors that cannot support an environment with DCs and T cells do not have high CD8+ T cell infiltration and therefore have worse outcomes and respond poorly to checkpoint blockade. The future of immunotherapy likely needs to focus on rebuilding the tumor immune microenvironment to support DCs and stem-like CD8+ T cells and facilitate their proliferation and differentiation into effector-like CD8+ T cells capable of destroying tumor cells.

Prostate Cancer Tumor-Associated Antigens

Inducing an immune response against cancer is critical for producing an effective and long-lasting response. Integral to the anti-tumor immune response is the presentation of tumor-associated antigens (TAA) on APCs to CD8+ T cells for priming of tumor-specific cytotoxic T cells. Identifying tumor antigens present in prostate cancer and understanding their role in inducing an adaptive immune response is essential for developing effective vaccine strategies that enhance the generation of tumor-specific CD8+ T cells. Described below are the most common prostate cancer tumor antigens being utilized for immune therapy in preclinical models and in ongoing clinical trials (Table 1).

Prostate-Specific Antigen

Epithelial cells lining the acini and ducts of the prostate gland produce the serine protease prostate-specific antigen (PSA), and secrete it into the prostate lumen. PSA aids in liquefying semen coagulum and is excreted in seminal fluid. In healthy individuals, serum PSA levels are common at low concentrations (0–2.5 ng/ml), with serum concentrations above 2.5 ng/ml indicating a cause for biopsy [59]. Serum concentrations of PSA correlate with prostate cancer disease progression, which makes PSA a useful prognostic marker that aids in the grading and staging of prostate cancer, as well as an indicator of disease recurrence and progression [60, 61]. However, increased PSA levels are also associated with benign inflammatory conditions such as benign prostate hyperplasia (BPH) and prostatitis, which may result in a false positive screen that is not indicative of cancer progression [62].

In addition to its value as a prognostic marker, PSA is an immunogenic antigen that can drive immune responses [63]. PSA-specific CD8+ T cells are found in both healthy individuals and prostate cancer patients [64]. For this reason, PSA can serve as a target for prostate cancer vaccination to elicit an anti-tumor immune response from CD8+ T cells. For example, PROSTVAC is a poxvirus-based vaccine that expresses the PSA antigen as well as molecules to aid in T cell stimulation including CD80, intercellular adhesion molecule 1 (ICAM-1), and lymphocyte function-associated antigen 3 (LFA-3) [65, 66]. A phase II trial for the treatment of metastatic castration-resistant prostate cancer (mCRPC) showed that patients receiving PROSTVAC had an increase in their median overall survival of 8.5 months compared to control (25.1 months vs. 16.6 months) [67]. However, a recent phase III trial failed to demonstrate improved overall survival compared to controls (NCT01322490). Failures to respond to a PSA-expressing vaccine may be the result of a high frequency of PSA-specific CD8+ T cells expressing Tim-3, a marker of T cell exhaustion [68]. This suggests that while patients possess tumor-specific T cells needed to mount an anti-tumor response, those T cells may be unable to respond in a productive and effective manner. Thus, current ongoing studies are assessing the potential benefit of combination immunotherapy, specifically PD-1 checkpoint blockade, to improve vaccine efficacy.

Prostate Acid Phosphatase

The prostatic epithelium synthesizes prostate acid phosphatase (PAP), a prostate-specific, secreted glycoprotein enzyme that is involved in the liquefaction of semen [69]. Expression of PAP and its serum concentration correlates with testosterone levels, disease progression and the amount of bone metastases. Due to these prognostic correlations, PAP serum levels were originally used as a prostate tumor biomarker dating back to the 1940s, prior to the adoption of PSA screening in the 1980s. Despite the widespread use of PSA for prostate cancer screening, PAP still has prognostic value as a biomarker for determining and differentiating intermediate- and high-risk patients, and for predicting clinical recurrence and likelihood of developing distant metastases [70].

Like PSA, PAP also has properties as an immunogenic antigen recognized by T cells in humans and mice [71]. Evidence for this is provided by one study showing that intratumoral injection of a PAP-expressing vector resulted in decreased tumor growth in a xenograft model [72]. Additional studies have shown that PAP expressed by a DNA vaccine successfully induced PAP-specific CD8+ T cells, while also increasing PSA doubling time, suggesting that prostate cancer patients may benefit from targeting PAP using

Table 1 Clinical trials for prostate cancer immunotherapy

Therapy name	Description	NCT identifier	Ref number
PROSTVAC	PSA vaccine	NCT01322490	[65–68]
Sipuleucel-T	PAP vaccine	NCT00065442 NCT00005947	[48, 75, 76]
AGS-1C4D4	PSCA targeted mAb	NCT00519233	[95]
ETBX-061	MUC1 vaccine	NCT03481816	[99]
	NY-ESO-1 and MUC1 vaccine	NCT02692976	
	TARP vaccination	NCT02362451	[113]
CV9103	STEAP1 vaccination	NCT00831467	[121]
DSTP3086S	STEAP1 antibody-drug conjugate	NCT01283373	[123]
GVAX	Prostate cell line vaccination	NCT00089856 NCT00133224	[141]
Ipilimumab	CTLA-4 blocking mAb	NCT02113657 NCT01194271	[144, 145]
Ipilimumab	In combination with GVAX	NCT01510288	
Ipilimumab	In combination with PROSTVAC	NCT02506114	
Ipilimumab	In combination with GM-CSF	NCT00064129	[146]
Ipilimumab	In combination with ADT	NCT01377389 NCT02703623 NCT01498978 NCT00170157 NCT01688492 NCT02020070	
Atezolizumab	PD-L1 blocking mAb	NCT03016312	
Atezolizumab CPI-444	In combination with an A2A receptor agonist	NCT02655822	
Nivolumab	PD-1 blocking mAb, used in combination with Ipilimumab	NCT03061539 NCT02985957 NCT02601014	
Nivolumab	In combination with a PSA vaccine	NCT02933255	
Nivolumab	In combination with ADT	NCT03338790 NCT02484404	
Nivolumab AM0010	In combination with pegylated IL-10	NCT02009449	
Pembrolizumab ADXS31-142	PD-1 blocking mAb, used in combination with a PSA vaccine	NCT02325557	
Pembrolizumab pTVG-HP	PD-1 blocking mAb, used in combination with a PAP vaccine	NCT02499835	
Pembrolizumab	PD-1 blocking mAb, used in combination with a radium-223	NCT03093428	[151]
Durvalumab Tremelimumab	PD-1 blocking mAb, used in combination with a CTLA-4 blocking mAb	NCT03204812	
Durvalumab	In combination with a TLR3 agonist	NCT02643303	
	PSA and CD3 Bi-specific antibody	NCT01723475 NCT02262910	[153]
	CAR-T cell targeting PSMA	NCT01140373	

cancer vaccines [73, 74]. Sipuleucel-T, the first therapeutic cancer vaccine to receive Food and Drug Administration (FDA) approval works as an immunostimulant to trigger an anti-PAP immune response for the treatment of mCRPC [48, 75].

Delivery of Sipuleucel-T is achieved using leukapheresis of patient blood to remove and isolate primary DCs. The patient's DCs are loaded with PAP peptide in an incubation step and are additionally stimulated with granulocyte-macrophage colony stimulating factor (GM-CSF) to promote cell growth and survival in culture. These PAP peptide-loaded DCs are then re-infused back into the patient to promote a CD8[+] T cell-mediated anti-tumor response. The IMPACT clinical trial assessing the efficacy of Sipuleucel-T (NCT00065442) showed that patients receiving Sipuleucel-T experienced a statistically significant increase in median survival time of 4.1 months compared to the control arm. However, despite an increase in survival time, patients receiving Sipuleucel-T did not experience a significant decrease in tumor size [48]. The following D9901 trial (NCT00005947) confirmed these findings [76]. Sipuleucel-T was FDA-approved for mCRPC in 2010 and is in ongoing clinical trials to evaluate its efficacy when used in combination with other therapies.

Prostate-Specific Membrane Antigen

Prostate-specific membrane antigen (PSMA) is a membrane-bound zinc metalloenzyme expressed primarily on prostate epithelial cells, although low expression can be found in other tissues such as the kidney [77]. While PSMA is expressed in the healthy prostate epithelium compared to other tissues, it is also one of the most commonly and highly upregulated genes found in prostate cancers, including high staining on epithelial cells of prostatic intraepithelial neoplasia (PIN) and on malignant carcinomas [77]. Interestingly, PSMA expression appears to be linked to androgen deprivation therapy (ADT), with increased expression in response to therapy [78]. Like PSA and PAP, PSMA expression may be a useful prognostic biomarker. It has been shown that expression correlates with disease progression and time to recurrence [79], which could make PSMA useful as a prognostic marker accounting disease progression and tumor cell disease potential; yet, attempts to utilize PSMA serum levels alone as a prognostic marker have not been successful. However, other methods involving PSMA expression have had success, including the ProstaScint scan that combines CT and MRI scans. The ProstaScint technique utilizes PSMA-specific

antibodies bound to radioactive indium-111 to bind and identify prostate cancer metastases and has proven to be a valuable tool for identifying distant, remaining, and recurrent disease [80].

PSMA can also be targeted using various forms of monoclonal antibody (mAb) therapy. For example, unconjugated, radiolabeled, and drug-conjugated humanized mAbs against PSMA have all been utilized to induce antibody-dependent cellular cytotoxicity (ADCC) and cell death [81]. PSMA contains a cytoplasmic tail internalization signal that induces protein internalization to the endosome upon ligand binding, which leads to trafficking back to the cell surface or targeting to lysosomes for protein degradation [82]. This key property of PSMA internalization is exploited by treatments to introduce toxins specifically into prostate cancer cells. In treatments such as D7-PE40, a targeted immunotoxin consisting of an antibody fragment specific for PSMA is linked to exotoxin A, a dimer protein that blocks protein synthesis through elongation factor-2 inhibition. This conjugated therapy targeted against PSMA has shown efficacy in blocking tumor growth in pre-clinical models [83]. Other therapeutic strategies targeting PSMA have been developed, such as chimeric antigen receptor T cells (CAR-T) specific for PSMA. In a pre-clinical xenograft model of human prostate cancer, anti-PSMA CAR-T cells were highly effective at eradicating tumors and inducing tumor-specific lysis [84, 85].

Prostate Stem Cell Antigen

Prostate Stem Cell Antigen (PSCA) is a GPI-anchored surface protein involved in stem cell survival [86]. PSCA is highly expressed in prostate cancer, pancreatic cancer, and bladder cancer epithelial cells [87]. Expression of PSCA is found in 90% of primary prostate cancers and a high proportion of metastatic sites contain amplification of the PSCA gene [88]. Further, PSCA expression is associated with a higher Gleason score, high staging, and cancer progression to bone metastases [89]. In addition to these associations, PSCA detection also functions as a marker of response to therapy, as decreased

PSCA mRNA levels correlate with response to radiation therapy [90]. The PSCA gene is located about 3 kb downstream of an androgen-responsive enhancer, resulting in decreased PSCA expression in response to ADT [91]. However, despite these correlations, not much is known about the role and function of PSCA in prostate cancer or in normal prostate tissue.

Due to its high expression in primary prostate cancers and metastases, PSCA is being evaluated as a therapeutic target. Preclinical murine studies have examined the efficacy of a PSCA-based DNA vaccine and the use of engineered CAR-T cells specific for PSCA [92]. Clinical trials have included examinations of vaccine and mAb therapies targeting PSCA as well. A phase I/II clinical trial assessed the efficacy and safety of a PSCA peptide-loaded DCs vaccine, which induced an immune response against PSCA and showed promise for patients with mCRPC [93]. A six-patient phase I trial for the treatment of mCRPC utilized another DC-based vaccine that employed multiple binding epitopes of PSCA. This trial resulted in a beneficial increase in PSA doubling time and the generation of memory T cell responses against the peptides administered [94]. Finally, a phase I clinical trial for the treatment of mCRPC tested AGS-1C4D4, a mAb targeting PSCA [95]. In this trial, the therapy was well tolerated by the 13 patients treated, with six patients having stable disease for 24 months. Targeting PSCA through vaccine and antibody-based therapeutics has so far shown promise, which warrants further exploration in pre-clinical and clinical studies based upon its induction of memory T cells. It will be interesting to determine how PSCA-targeted therapeutics will combine and potentially synergize with established prostate cancer therapies and immunotherapies, such as PD-1 blockade.

Mucin-1

Mucin-1 (MUC1), a highly glycosylated protein, is expressed across a wide variety of epithelial cancers including prostate adenocarcinoma, but is not detected in healthy prostate tissue [96].

MUC1 is expressed on the apical borders of epithelial cells where it plays a role in cell adhesion and mucosal barrier protection. MUC1 exists in secreted and membrane-bound forms; however, in prostate adenocarcinomas and neoplasms, MUC1 is also expressed in the cytoplasm. Expression of MUC1 correlates with disease progression, tumor volume, and lymph node metastases [97]. However, MUC1 is suppressive of androgen receptor (AR) expression, leading to decreased androgen sensitivity and response to ADT [98].

Due to its high expression in adenocarcinomas, MUC1 is a target for antibody-based targeted therapy as well as vaccine therapy. Therapeutic targeting of MUC1 for prostate cancer is currently being tested in clinical trials. Researchers have developed an adenovirus-based vaccine that expresses MUC1 along with two other TAAs and contains specific gene modifications that decrease viral gene expression and prevent a host immune response against the viral protein components [99]. This vaccine, ETBX-061, is currently being tested for clinical trial use. This trial (NCT03481816) is testing the efficacy and safety of the MUC1 viral vaccine alongside other adenovirus-based vaccines expressing PSA and brachyury proteins for patients with mCRPC.

NY-ESO-1

NY-ESO-1 is a well-characterized cancer testis antigen of unknown function expressed in bladder, esophageal, liver, and breast cancer [100]. In transitional cell carcinoma, expression of NY-ESO-1 correlates with staging and seropositivity in high grade patients [101]. NY-ESO-1 is a MHC class I and class II antigen that induces cellular and humoral immune responses that are associated with anti-tumor immune activity [102] and the presence of these NY-ESO-1-specific T cells correlates with good prognosis. In one study, 10% of prostate cancer patients had autoantibodies against NY-ESO-1 in their serum, while healthy patients had none [103]. This supports evidence that NY-ESO-1 is expressed highly in cancerous tissue and lowly in healthy tissue, a property that makes NY-ESO-1 attractive as a potential cancer vaccine antigen. Over 30 clinical trials have been conducted with NY-ESO-1 vaccines or NY-ESO-1-targeted adoptive T cell therapy, several of which focus specifically on prostate cancer. Clinical trial NCT02692976 utilized a DC-based vaccine loaded with NY-ESO-1 and MUC1 peptides, with separate groups assessing the efficacy of plasmacytoid, myeloid, or the combination of both subtypes of DCs. The therapy was well-tolerated, with only 33% of patients experiencing low grade toxicity, and efficacious, with 67% of patients achieving 6 months of stable disease. To increase response to this vaccine, combinations of NY-ESO-1 vaccine therapy with immunotherapies are being considered. Preliminary work with CTLA-4 blockade/NY-ESO-1 vaccine combination therapy demonstrated an increase in the numbers of NY-ESO-1 specific T cells in metastatic melanoma, which correlated with clinical responses [104].

T-cell Receptor Alternate Reading Frame Protein

T-cell receptor alternate reading frame protein (TARP) is an MHC class II-restricted protein expressed on prostate and breast cancer epithelial cells [105]. TARP expression correlates with cancer progression and is found in primary and metastatic sites [106–108]. TARP is a 58-residue sequence product of an alternate reading frame of the TCR locus [109]. This alternate reading frame sequence contains two known CD8$^+$ T-cell binding epitopes that are presented on APCs and prostate cancer cells [110]. Testosterone upregulates TARP expression, which suggests a role for androgens in promoting its expression [111]. However, TARP expression has been observed in androgen-sensitive and androgen-insensitive prostate cancer [112]. Its high expression in prostate cancer makes it an ideal target antigen for cancer vaccine therapy.

Due to its immunogenicity, TARP-specific vaccination was explored in a phase I clinical trial. This trial consisted of 41 patients with asymptomatic hormone-sensitive prostate cancer receiving either emulsified TARP peptide with GM-CSF or DCs pulsed with TARP peptide [113]. The five-dose vaccination schedule was well tolerated and had an acceptable safety profile in both treatment groups. Clinical outcomes were assessed based upon PSA doubling time, which is a measure of the time it takes for serum concentrations of PSA to double. Both treatment groups showed a decrease in PSA doubling time, with 74.2% of total patients displaying a reduction in PSA doubling time 48 weeks post-treatment. Additionally, the TARP vaccine increased TARP peptide-specific IFN-γ production, with responses recorded for 80% of patients. These data suggest that TARP vaccine therapy may help in controlling micro-metastases and slowing disease progression. A trial with a second generation of this vaccine (NCT02362451) is underway with modifications that include the addition of MHC class II binding sites to the peptide sequence [113].

GRB2-Like Endophilin B2

The GRB2-Like Endophilin B2 (SH3GLB2) peptide was discovered in the TRAMP murine prostate model as a stimulator of prostatic adenocarcinoma specific T cells (SPAS1). TRAMP mice receiving vaccination with the TRAMP-C2 cell line, CTLA-4 blockade, and GM-CSF experienced a reduction in tumor growth and a reduction in spontaneously forming tumors. This treatment method was employed to identify unknown prostate cancer antigens, as had been done previously for melanoma [114, 115]. By testing T cell activity from the spleens of treated mice a 395 amino acid sequence was identified that induced T cell activation and shared 96% homology with human SH3GLB2, a protein of unknown function [116]. SH3GLB2 is highly expressed in prostate cancer metastases and is orthologous to SPAS1, the immunodominant model antigen in the TRAMP-C1 murine model. Analysis of SH3GLB2 expression in human prostate cancer has shown that it is highly expressed in lymph node metastases in patients with aggressive cancer [117]. Because T cell responses can be induced in a mouse vaccination model, it is thought that SH3GLB2 is a viable vaccination target antigen in humans. However, to our knowledge, there have been no attempts at this therapy as of this time.

Six Transmembrane Epithelial Antigen of the Prostate 1

Six transmembrane epithelial antigen of the prostate 1 (STEAP1) is a metalloreductase enzyme expressed at cell-cell junctions and is overexpressed in the cell membrane and cytoplasm of prostate cancer cells, with lower expression in healthy tissue [118, 119]. STEAP1 is highly expressed in PIN lesions, which suggests that it plays a role in early prostate cancer development and thus might be an attractive biomarker for early disease [120]. STEAP1 has been a target antigen for antibody-based therapy as well as vaccine therapy due to its high expression in PIN and more advanced forms of prostate cancer. The CV9103 vaccine utilized STEAP1 mRNA in a phase I/II trial of 44 patients (NCT00831467) [121]. This strategy effectively induces cytotoxic CD8$^+$ T lymphocytes with reactivity against STEAP1. Additionally, a vaccinia virus Ankara vector delivery system containing STEAP1 has shown promise as a vaccine therapy in murine models [122]. Developed mAbs against STEAP1 have also shown promise in murine models by inhibiting xenograft tumor growth. This success has led to the development of an antibody-drug conjugate (ADC). DSTP3086S is an ADC linked to monomethyl auristatin E, which inhibits cell growth by blocking tubulin polymerization. DSTP3086S has shown promise in a phase I clinical trial (NCT01283373) for mCRPC and has shown greater effectiveness at higher concentrations [123].

Models of Murine Prostate Cancer

TRAMP

The TRAMP murine model was originally developed in 1995, which was followed by further characterization and development of cell lines and modified models of TRAMP [124]. The TRAMP model of spontaneous prostate cancer was developed using the simian vacuolating virus 40 (SV40), a virus with oncogenic proteins, to induce cancer. Expression of the early region of the virus which is composed of the large and small T antigen was driven by a modified probasin-ARR2 promoter specific to the prostate. The large-T and small-t antigens of SV40 protein bind to and inhibit the activity of p53 and Rb tumor suppressors. Additionally, the small-t antigen binds to phosphatase PP2A along with several other oncogenic intracellular proteins. Inhibiting p53 and Rb tumor suppressors results in a predictable progression from PIN lesions to highly penetrant metastatic disease by week 28. All of these mice develop lymph node metastases and 67% go on to develop pulmonary metastases. However, these mice exhibit a variable response to ADT, although mice that are resistant to ADT are more likely to develop metastases [125]. Additionally, these mice develop phyllodes-like lesions that have a leaf-like structure, similar to that found in human breast cancer. It should also be noted that some studies have demonstrated that TRAMP may not actually be a form of adenocarcinoma and is instead an atypical epithelial hyperplasia that develops into neuroendocrine carcinoma [126].

Three prostate cancer cell lines were derived from tumor-bearing TRAMP mice [127]. Two of these cell lines (TRAMP-C1 and TRAMP-C2) readily form tumors when implanted subcutaneously into syngeneic wild-type (C57/BL6) mice. However, the third cell line (TRAMP-C3) only grows *in vitro*. These tumorigenic cell lines retained AR expression, but lost expression of the T antigen found in the transgenic mouse. TRAMP-C1 and TRAMP-C2 are amenable for studies investigating therapeutic efficacy of various immunotherapies, although it should be noted that TRAMP-C1 is considered poorly immunogenic with a low level of basal immune infiltration, while TRAMP-C2 is moderately immunogenic. TRAMP-C1 is responsive to CTLA-4 blockade in a T and natural killer (NK) cell-dependent manner. TRAMP-C2 cells are also responsive to checkpoint blockade, but can form metastases in the draining lymph nodes and lungs, thus making it a useful model for surgical resection and metastasis. For example, while 95% of mice with surgically resected primary TRAMP-C2 tumors develop metastases, this was reduced to 50% following CTLA-4 blockade [128].

LADY

The LADY model of prostate cancer is designed similarly to the TRAMP model; however, it has modifications in the probasin (PB) promoter region and a mutation to the small-t antigen, which renders it unable to bind phosphatase PP2A and other oncogenic intracellular proteins. Many different lines of the LADY model have been generated with variable tag transgene expression. These lines are named 12t − 1 through 11. The 12 in the name indicates the approximate 12 kb promoter length, t for the small-t antigen transgene, and 1–11 for the 11 lines generated. The mouse line that develops cancer at the fastest rate is 12t − 7, which progresses to locally invasive adenocarcinoma at 15–22 weeks, but does not become metastatic [129]. In contrast, Line 12t − 10 grows slowly, progressing to invasive neuroendocrine carcinoma at 33 weeks, and develops lymph node and lung metastases at 50 weeks-of-age. Considering the range in variation in metastatic potential seen across the model, these various lines may be helpful for comparing the genetics involved in metastatic progression.

The TRAMP and LADY transgenic mice each carry advantages and disadvantages as models of human prostate cancer. For example, both were generated by driving expression of SV40 T antigens off of a prostate-specific promoter to produce highly penetrant and well characterized disease, which is a mechanism of oncogenic transformation that is not mirrored in humans. Not all models of TRAMP and LADY are capable of progressing to metastatic disease, but both the TRAMP and the 12t − 10 LADY models have metastatic potential. Unfortunately, these models do not reciprocate human prostate cancer progression completely, due to being driven by an exogenous oncogene that does not exist in humans. This leads to accumulation of different mutations and subsequently altered disease progression than is found in patients with prostate cancer. This SV40 oncogene driver produces mostly neuroendocrine cancers and similar carcinoma subtypes in mice, which likely represents only 25–30% of cases of men with advanced prostate cancer. However, the development of neuroendocrine carcinomas depends upon the genetic background of the mouse. Mice that do develop neuroendocrine prostate cancer may be useful for studying human neuroendocrine prostate cancer.

c-Myc Models

c-Myc is a proto-oncogene commonly overexpressed or mutated in human prostate cancer. Furthermore, c-Myc overexpression is also found in PIN, which suggests a role in early cancer progression. Thus, to recapitulate the onset and progression of human prostate cancer, several mouse models utilize mutations and overexpression of c-Myc along with other commonly associated mutations of tumor suppressors, such as p53 and phosphatase and tensin homolog (PTEN). Two models were developed in which c-Myc expression is driven from either the Pb promoter or the ARR2PB promoter, similar to the TRAMP and LADY models. These two distinct models of c-Myc-driven prostate cancer are categorized as Lo-Myc and Hi-Myc. They are distinct in their responsiveness to androgens and ADT making them good models for studying human prostate cancer disease progression and responsiveness. The Lo-Myc model results in prostate cancer progression that is unresponsive to ADT, which makes this model castration resistant. In contrast, the Hi-Myc model exhibits androgen sensitivity, which reflects the regulation of c-Myc transgene expression by the androgen-regulated Pb promoter. However, in human prostate cancer, castration does not inhibit c-Myc expression. Hi-Myc mice progress to adenocarcinoma around 26 weeks-of-age, while the

Lo-Myc model progresses more slowly with adenocarcinoma developing at 56 weeks [130]. However, neither Hi-Myc nor Lo-Myc models develop spontaneous metastases [130]. Similar to what is seen in human prostate cancer, expression of the prostate-specific tumor suppressor NKX3.1 is reduced in Hi-Myc PIN and adenocarcinoma [131]. Further studies of this model have coupled it to genetic knockouts that are common in human prostate cancer.

PTEN Knockout

PTEN is a major tumor suppressor that functions to dephosphorylate activated AKT and 3-phosphoinositide-dependent protein kinase-1 (PDK1) [132, 133]. PTEN is commonly mutated or lost in many cancers, including a high percentage of human prostate cancers [134, 135]. Original models of PTEN heterozygous mice were not ideal for studying prostate cancer because many of the mice that survived embryonic development developed other types of cancers due to the non-specific knockout of PTEN. To overcome these limitations, prostate tissue-specific PTEN knockout mice were generated using the Cre recombinase system, driving Cre expression off of a prostate-specific promoter to flox out a section of PTEN flanked by loxP sites. These mice develop progressive prostate cancer from PIN to metastatic adenocarcinoma in a manner similar to human prostate cancer progression.

Two groups have developed PTEN knockout mice using this system. One group deleted exon 5 of PTEN and these mice develop PIN at 6 weeks and adenocarcinoma by 9–29 weeks-of-age with lung and lymph node metastases forming at 12–29 weeks in 45% of the mice [136]. Additionally, this model may be useful for studying androgen dependency as tumors initially regress in response to androgen ablation, but then become resistant mirroring what is observed in humans. Another group developed a PTEN model by deleting both exon 4 and 5. These mice developed prostate cancer more slowly compared to the exon 5 deleted mice, with observed lesions by

42 weeks [137]. An additional model utilizes a tamoxifen-inducible Cre transgene knocked in to the NKX3.1 locus, which simultaneously knocks out one allele of NKX3.1 and brings Cre expression under control of the NKX3.1 promoter [138]. This adds temporal control over PTEN deletion. Mice that receive tamoxifen at 2 months-of-age develop high grade PIN and micro-invasive adenocarcinoma [139]. PTEN knockout models have also been combined with targeted deletion of p53 as well. This combination of tumor suppressor knockouts results in an aggressive prostate cancer by week 11 and eventual death by week 29 [140].

Prostate Cancer Immunotherapy: Vaccines, Checkpoint Blockade, and Combination Therapy

Vaccines

As described previously, Sipuleucel-T was the first therapeutic cancer vaccine to garner FDA approval and uses PAP-loaded DCs to direct anti-tumor (PAP) responses in patients [48, 75]. In contrast, PROSTVAC is a poxvirus-based vaccine expressing PSA along with the co-stimulatory molecules CD80, ICAM-1, and LFA-3. In addition to these agents, GVAX is another prostate cancer vaccine platform, and instead of focusing on a single target, such as PSA or PAP, to induce an anti-tumor immune response, GVAX is a cellular vaccine that contains irradiated prostate cells from two different human prostate cancer cells lines, LNCaP and PC3 [141]. These cell lines have been additionally modified to produce GM-CSF in order to stimulate DCs for antigen presentation. This vaccine could potentially induce immune responses to multiple prostate cancer antigens at once, which is beneficial to eliminate a heterogeneous population of prostate cancer cells from the body. This vaccination method would bypass the need to human leukocyte antigen (HLA) match patients because it relies on the patient's own DCs to present antigen *in vivo* [142]. However, despite this potential for an effective and easier cancer vaccine design, GVAX has not seen success in clinical trials when compared to docetaxel in two separate phase III trials (NCT00089856, NCT00133224).

CTLA-4 Blockade

T cell costimulation by CD28 is critical for their activation and induction of cytotoxic activity. This pathway is negatively regulated by CTLA-4, which competes for the CD28 ligand. Ipilimumab is a CTLA-4 blocking mAb that prevents its activity and to increase CD28 signaling on T cells [143]. Initial studies of Ipilimumab in prostate cancer patients led to unacceptable adverse effects, including death among several patients being treated for CRPC, and unfavorable outcomes, with no increase in overall survival [144]. However, several patients did have a complete remission in response to this therapy [145]. Improvement in patient and biomarker selection, as well as new and improved CTLA-4 blocking antibodies may limit severe adverse effects in the future. Ipilimumab is being evaluated for its ability to elicit T cell responses against tumor-specific neoantigens as part of a phase II trial for CRPC (NCT02113657) and as neoadjuvant therapy prior to radical prostatectomy (NCT01194271).

While initial trials testing the efficacy of ipilimumab have not resulted in increased overall survival, ipilimumab may still be beneficial when applied in combination with other therapies. A phase I clinical trial of GVAX in combination with ipilimumab for patients with CRPC showed that the treatment had an acceptable safety profile. Additionally, the therapy led to favorable tumor responses and prolonged survival, especially among patients with higher peripheral blood expression of CTLA-4 and PD-1 on their CD4$^+$ T cells and lower frequencies of regulatory T cells prior to therapy. Overall, the best predictor of favorable outcome to this therapy was CTLA-4 expression by CD4$^+$ T cells, which might suggest CTLA-4 as a biomarker for selection of patients that would benefit from this therapy. A phase II trial is ongoing to determine the efficacy of the PROSTVAC vaccine in com-

bination with ipilimumab (NCT02506114) as a neoadjuvant therapy for patients with localized prostatic neoplasia.

Ipilimumab is also being tested in an ongoing phase I trial in combination with GM-CSF for the treatment of recurrent prostate carcinomas and stage IV prostate cancer (NCT00064129). It was found that patients with immune-related adverse events (IRAEs) had an increase in T cell clonality 2 weeks after initial ipilimumab treatment, and an increase in PSA-specific T cells was associated with increased T cell diversity [146]. Additional trials of ipilimumab include its use in combination with ADT for CRPC (NCT01377389, NCT02703623), stage IV and recurrent prostate cancer (NCT01498978, NCT00170157), for chemotherapy-naïve CRPC (NCT01688492), and for use prior to radical prostatectomy (NCT02020070).

PD-1/PD-L1 Blockade

PD-1 is a glycoprotein expressed on T cells that interacts with PD-L1 expressed in cancer cells and a range of other immune cells, to inhibit T cell activation. Targeting the PD-1/PD-L1 axis with blocking mAbs is an area of great focus in cancer therapy, having received FDA approval for several cancer types [147, 148]. However, PD-1-targeted therapy in prostate cancer may have a minimal impact due to lower PD-L1 expression in prostate cancer compared to other cancers [149]. Low PD-L1 expression may explain the minimal response seen in early clinical trials of nivolumab, which is a PD-1 blocking antibody, for mCRPC. Other studies looking at pembrolizumab (anti-PD-1 mAb) therapy in combination with enzalutamide for mCRPC have demonstrated meaningful clinical benefit for a subset of patients [150]. This may be a possible therapeutic avenue to follow after patients' progress towards androgen insensitivity because during the progression to androgen insensitivity, DCs increase expression of PD-L1 and CD8$^+$ T cells increase expression of PD-1. A phase III trial is underway to determine the efficacy of atezolizumab (a PD-L1 blocking antibody) in combination with enzalutamide compared to enzalutamide alone (NCT03016312). Additionally, the use of ipilimumab is being investigated in combination with nivolumab in mCRPC (NCT03061539, NCT02985957, NCT02601014).

A phase I/II trial is ongoing to investigate the efficacy of ADXS31-142, a PSA based vaccine, as a monotherapy or in combination with pembrolizumab for mCRPC (NCT02325557). The pTVG-HP PAP antigen-based DNA vaccine is being investigated in combination with pembrolizumab (NCT02499835). An ongoing phase II trial is looking at the efficacy of pembrolizumab in combination with Radium-223, which targets bone metastases (NCT03093428) and has shown efficacy in pre-clinical models [151]. Another phase II trial is investigating the efficacy of combination immunotherapy using durvalumab, an anti-PD-L1 antibody, in combination with tremelimumab, and anti-CTLA-4 antibody for the treatment of mCRPC (NCT03204812), as well as in combination with a TLR3 agonist (NCT02643303). Nivolumab is being looked at in combination with a poxvirus-based PSA expressing cancer vaccine (NCT02933255). An ongoing phase II trial is investigating the safety and efficacy of nivolumab in combination with multiple ADTs and poly-ADP ribose polymerase (PARP) inhibitors (NCT03338790, NCT02484404). Nivolumab is being investigated as part of a phase I trial looking at the safety of AM0010, a pegylated IL-10 immune stimulating agent (NCT02009449). Atezolizumab is being investigated as part of a phase I trial looking at the safety of CPI-444, an adenosine A2A receptor agonist for the treatment of mCRPC (NCT02655822).

Other Therapies

An alternative immunotherapy strategy is using bi-specific antibodies to link T cells to their cancer cell target [152]. A therapeutic method utilizing this strategy targeted PSMA and CD3 simultaneously to bring T cells in contact with their target cell to induce cytotoxicity. This therapy has shown promise in xenograft models with a reduction in PSA levels, and reduced tumor growth [153]. These promising findings have led to two phase I clinical trials (NCT01723475, NCT02262910). CAR-T cells have also been developed to target prostate cancer antigens. These CAR-T cells are modified by lentiviral- or retroviral-induced expression of an engineered TCR specific for a tumor surface antigen [154]. CAR-T targeting PSMA has shown promise in pre-clinical murine studies, as well as in a phase I clinical trial (NCT01140373). CAR-T therapies targeting other prostate tumor antigens such as MUC-1 and TARP have also shown promise in pre-clinical models [155].

Conclusions

Inducing immune responses against TAAs is a critical step in producing an effective and long-lasting anti-cancer response. Multiple prostate cancer antigens have been identified with varying degrees of expression in prostate cancer as well as other cancers and healthy tissues. Many of these associated antigens are capable of being identified by CD8[+] T cells and inducing an effective immune response. However, translating these findings into effective vaccine therapies has been challenging. The FDA approval of Sipuleucel-T has shown that vaccine therapy has efficacy in prostate cancer patients. The development of immune checkpoint blocking antibodies such as ipilimumab, nivolumab, and pembrolizumab have also shown promise as effective therapeutic strategies in treating prostate cancer. Ongoing and future approaches are looking at combinations of tumor antigen-specific vaccines along with checkpoint blockade to activate tumor-specific CD8[+] T cells and thereby enhance ant-tumor immunity.

References

1. A.O. Kamphorst, A. Wieland, T. Nasti, S. Yang, R. Zhang, D.L. Barber, et al., Rescue of exhausted CD8 T cells by PD-1–targeted therapies is CD28-dependent. Science **355**, 1423–1427 (2017)
2. D.L. Barber, E.J. Wherry, D. Masopust, B. Zhu, J.P. Allison, A.H. Sharpe, et al., Restoring function in exhausted CD8 T cells during chronic viral infection. Nature **439**(7077), 682–687 (2006)
3. E.J. Wherry, T cell exhaustion. Nat. Immunol. **12**, 492 (2011)
4. P. Savas, B. Virassamy, C. Ye, A. Salim, C.P. Mintoff, F. Caramia, et al., Single-cell profiling of breast cancer T cells reveals a tissue-resident memory subset associated with improved prognosis. Nat. Med. **24**(7), 986–993 (2018)
5. E. Peranzoni, J. Lemoine, L. Vimeux, V. Feuillet, S. Barrin, C. Kantari-Mimoun, et al., Macrophages impede CD8 T cells from reaching tumor cells and limit the efficacy of anti-PD-1 treatment. Proc. Natl. Acad. Sci. U. S. A. **115**(17), E4041–E4E50 (2018)
6. F. Azimi, R.A. Scolyer, P. Rumcheva, M. Moncrieff, R. Murali, S.W. McCarthy, et al., Tumor-infiltrating lymphocyte grade is an independent predictor of sentinel lymph node status and survival in patients with cutaneous melanoma. J. Clin. Oncol. **30**(21), 2678–2683 (2012)
7. B. Mlecnik, M. Tosolini, A. Kirilovsky, A. Berger, G. Bindea, T. Meatchi, et al., Histopathologic-based prognostic factors of colorectal cancers are associated with the state of the local immune reaction. J. Clin. Oncol. **29**(6), 610–618 (2011)
8. F. Pages, A. Berger, M. Camus, F. Sanchez-Cabo, A. Costes, R. Molidor, et al., Effector memory T cells, early metastasis, and survival in colorectal cancer. N. Engl. J. Med. **353**(25), 2654–2666 (2005)
9. P.C. Tumeh, C.L. Harview, J.H. Yearley, I.P. Shintaku, E.J. Taylor, L. Robert, et al., PD-1 blockade induces responses by inhibiting adaptive immune resistance. Nature **515**(7528), 568–571 (2014)
10. B. Beutler, E.T. Rietschel, Innate immune sensing and its roots: the story of endotoxin. Nat. Rev. Immunol. **3**, 169 (2003)
11. T. Kawai, S. Akira, The role of pattern-recognition receptors in innate immunity: update on Toll-like receptors. Nat. Immunol. **11**, 373 (2010)
12. A. Williams, C.A. Peh, T. Elliott, The cell biology of MHC class I antigen presentation. Tissue Antigens **59**(1), 3–17 (2002)
13. J.E. Smith-Garvin, G.A. Koretzky, M.S. Jordan, T cell activation. Annu. Rev. Immunol. **27**(1), 591–619 (2009)
14. P. Guermonprez, J. Valladeau, L. Zitvogel, C. Théry, S. Amigorena, Antigen presentation and T cell stimulation by dendritic cells. Annu. Rev. Immunol. **20**(1), 621–667 (2002)
15. A. Brewitz, S. Eickhoff, S. Dahling, T. Quast, S. Bedoui, R.A. Kroczek, et al., CD8(+) T cells orchestrate pDC-XCR1(+) dendritic cell spatial and functional cooperativity to optimize priming. Immunity **46**(2), 205–219 (2017)
16. J. Lieberman, The ABCs of granule-mediated cytotoxicity: new weapons in the arsenal. Nat. Rev. Immunol. **3**, 361 (2003)
17. P. Matzinger, The danger model: a renewed sense of self. Science **296**(5566), 301 (2002)
18. S. Gallucci, M. Lolkema, P. Matzinger, Natural adjuvants: endogenous activators of dendritic cells. Nat. Med. **5**, 1249 (1999)
19. B. Sauter, M.L. Albert, L. Francisco, M. Larsson, S. Somersan, N. Bhardwaj, Consequences of cell death. J. Exp. Med. **191**(3), 423 (2000)
20. D.E. Orange, M. Jegathesan, N.E. Blachère, M.O. Frank, H.I. Scher, M.L. Albert, et al., Effective antigen cross-presentation by prostate cancer patients'; dendritic cells: implications for prostate cancer immunotherapy. Prostate Cancer Prostatic Dis. **7**, 63 (2004)
21. A.L. Ackerman, P. Cresswell, Cellular mechanisms governing cross-presentation of exogenous antigens. Nat. Immunol. **5**, 678 (2004)
22. S. Jung, D. Unutmaz, P. Wong, G. Sano, K. De los Santos, T. Sparwasser, et al., In vivo depletion of CD11c+ dendritic cells abrogates priming of CD8+ T cells by exogenous cell-associated antigens. Immunity **17**(2), 211–220 (2002)
23. M. Haniffa, A. Shin, V. Bigley, N. McGovern, P. Teo, P. See, et al., Human tissues contain CD141hi cross-presenting dendritic cells with functional homology to mouse CD103+ nonlymphoid dendritic cells. Immunity **37**(1), 60–73 (2012)
24. S. Spranger, D. Dai, B. Horton, T.F. Gajewski, Tumor-residing Batf3 dendritic cells are required for effector T cell trafficking and adoptive T cell therapy. Cancer Cell **31**(5), 711 (2017)
25. C.H. GeurtsvanKessel, M.A.M. Willart, L.S. van Rijt, F. Muskens, M. Kool, C. Baas, et al., Clearance of influenza virus from the lung depends on migratory langerin+CD11b- but not plasmacytoid dendritic cells. J. Exp. Med. **205**(7), 1621–1634 (2008)
26. M.L. Broz, M. Binnewies, B. Boldajipour, A.E. Nelson, J.L. Pollack, D.J. Erle, et al., Dissecting the tumor myeloid compartment reveals rare activating antigen-presenting cells critical for T cell immunity. Cancer Cell **26**(5), 638–652 (2014)
27. I. Treilleux, J.-Y. Blay, N. Bendriss-Vermare, I. Ray-Coquard, T. Bachelot, J.-P. Guastalla, et al., Dendritic cell infiltration and prognosis of early stage breast cancer. Clin. Cancer Res. **10**(22), 7466 (2004)
28. E.W. Roberts, M.L. Broz, M. Binnewies, M.B. Headley, A.E. Nelson, D.M. Wolf, et al., Critical role for CD103+/CD141+dendritic cells bearing CCR7 for tumor antigen trafficking and priming of T cell immunity in melanoma. Cancer Cell **30**(2), 324–336 (2016)

29. M. Mercader, B.K. Bodner, M.T. Moser, P.S. Kwon, E.S.Y. Park, R.G. Manecke, et al., T cell infiltration of the prostate induced by androgen withdrawal in patients with prostate cancer. Proc. Natl. Acad. Sci. **98**(25), 14565 (2001)
30. S.J.A.M. Santegoets, A.G.M. Stam, S.M. Lougheed, H. Gall, K. Jooss, N. Sacks, et al., Myeloid derived suppressor and dendritic cell subsets are related to clinical outcome in prostate cancer patients treated with prostate GVAX and ipilimumab. J. Immunother. Cancer **2**(1), 31 (2014)
31. P. Stoitzner, L.K. Green, J.Y. Jung, K.M. Price, H. Atarea, B. Kivell, et al., Inefficient presentation of tumor-derived antigen by tumor-infiltrating dendritic cells. Cancer Immunology. Immunotherapy **57**(11), 1665–1673 (2008)
32. B.-Z. Qian, J.W. Pollard, Macrophage diversity enhances tumor progression and metastasis. Cell **141**(1), 39–51 (2010)
33. B. Ruffell, D. Chang-Strachan, V. Chan, A. Rosenbusch, C.M. Ho, N. Pryer, et al., Macrophage IL-10 blocks CD8+ T cell-dependent responses to chemotherapy by suppressing IL-12 expression in intratumoral dendritic cells. Cancer Cell **26**(5), 623–637 (2014)
34. L. de Chaisemartin, J. Goc, D. Damotte, P. Validire, P. Magdeleinat, M. Alifano, et al., Characterization of chemokines and adhesion molecules associated with T cell presence in tertiary lymphoid structures in human lung cancer. Cancer Res. **71**(20), 6391–6399 (2011)
35. E.N. McNamee, J. Rivera-Nieves, Ectopic tertiary lymphoid tissue in inflammatory bowel disease: protective or provocateur? Front. Immunol. **7**, 308 (2016)
36. M.C. Dieu-Nosjean, J. Goc, N.A. Giraldo, C. Sautes-Fridman, W.H. Fridman, Tertiary lymphoid structures in cancer and beyond. Trends Immunol. **35**(11), 571–580 (2014)
37. M.-C. Dieu-Nosjean, M. Antoine, C. Danel, D. Heudes, M. Wislez, V. Poulot, et al., Long-term survival for patients with non–small-cell lung cancer with intratumoral lymphoid structures. J. Clin. Oncol. **26**(27), 4410–4417 (2008)
38. J. Goc, W.H. Fridman, C. Sautes-Fridman, M.C. Dieu-Nosjean, Characteristics of tertiary lymphoid structures in primary cancers. Oncoimmunology. **2**(12), e26836 (2013)
39. M.D.L.L. Garcia-Hernandez, N.O. Uribe-Uribe, R. Espinosa-Gonzalez, W.M. Kast, S.A. Khader, J. Rangel-Moreno, A unique cellular and molecular microenvironment is present in tertiary lymphoid organs of patients with spontaneous prostate cancer regression. Front. Immunol. **8**, 563 (2017)
40. J. Goc, C. Germain, T.K. Vo-Bourgais, A. Lupo, C. Klein, S. Knockaert, et al., Dendritic cells in tumor-associated tertiary lymphoid structures signal a Th1 cytotoxic immune contexture and license the positive prognostic value of infiltrating CD8+ T cells. Cancer Res. **74**(3), 705–715 (2014)
41. L.A.M. Pozzi, J.W. Maciaszek, K.L. Rock, Both dendritic cells and macrophages can stimulate naive CD8 T cells in vivo to proliferate, develop effector function, and differentiate into memory cells. J. Immunol. **175**(4), 2071–2081 (2005)
42. G. Comito, E. Giannoni, C.P. Segura, P. Barcellos-de-Souza, M.R. Raspollini, G. Baroni, et al., Cancer-associated fibroblasts and M2-polarized macrophages synergize during prostate carcinoma progression. Oncogene **33**, 2423 (2013)
43. S.R. Gordon, R.L. Maute, B.W. Dulken, G. Hutter, B.M. George, M.N. McCracken, et al., PD-1 expression by tumour-associated macrophages inhibits phagocytosis and tumour immunity. Nature **545**(7655), 495–499 (2017)
44. Y. Komohara, H. Hasita, K. Ohnishi, Y. Fujiwara, S. Suzu, M. Eto, et al., Macrophage infiltration and its prognostic relevance in clear cell renal cell carcinoma. Cancer Sci. **102**(7), 1424–1431 (2011)
45. M.J. Campbell, N.Y. Tonlaar, E.R. Garwood, D.Z. Huo, D.H. Moore, A.I. Khramtsov, et al., Proliferating macrophages associated with high grade, hormone receptor negative breast cancer and poor clinical outcome. Br. Cancer Res. Treat. **128**(3), 703–711 (2011)
46. V. Bronte, S. Brandau, S.H. Chen, M.P. Colombo, A.B. Frey, T.F. Greten, et al., Recommendations for myeloid-derived suppressor cell nomenclature and characterization standards. Nat. Commun. **7**, 12150 (2016)
47. L. Fong, P. Carroll, V. Weinberg, S. Chan, J. Lewis, J. Corman, et al., Activated lymphocyte recruitment into the tumor microenvironment following preoperative sipuleucel-T for localized prostate cancer. J. Natl. Cancer Inst. **106**(11), dju372 (2014)
48. P.W. Kantoff, C.S. Higano, N.D. Shore, E.R. Berger, E.J. Small, D.F. Penson, et al., Sipuleucel-T immunotherapy for castration-resistant prostate cancer. N. Engl. J. Med. **363**(5), 411–422 (2010)
49. E.J. Wherry, S.J. Ha, S.M. Kaech, W.N. Haining, S. Sarkar, V. Kalia, et al., Molecular signature of CD8+ T cell exhaustion during chronic viral infection. Immunity **27**(4), 670–684 (2007)
50. A.J. Zajac, J.N. Blattman, K. Murali-Krishna, D.J.D. Sourdive, M. Suresh, J.D. Altman, et al., Viral immune evasion due to persistence of activated T cells without effector function. J. Exp. Med. **188**(12), 2205–2213 (1998)
51. G.J. Freeman, E.J. Wherry, R. Ahmed, A.H. Sharpe, Reinvigorating exhausted HIV-specific T cells via PD-1–PD-1 ligand blockade. J. Exp. Med. **203**(10), 2223–2227 (2006)
52. L. Baitsch, P. Baumgaertner, E. Devêvre, S.K. Raghav, A. Legat, L. Barba, et al., Exhaustion of tumor-specific CD8+ T cells in metastases from melanoma patients. J. Clin. Invest. **121**(6), 2350–2360 (2011)
53. A. Schietinger, M. Philip, V.E. Krisnawan, E.Y. Chiu, J.J. Delrow, R.S. Basom, et al., Tumor-specific T cell dysfunction is a dynamic antigen-driven differentia-

tion program initiated early during tumorigenesis. Immunity **45**(2), 389–401 (2016)

54. S.J. Im, M. Hashimoto, M.Y. Gerner, J. Lee, H.T. Kissick, M.C. Burger, et al., Defining CD8+ T cells that provide the proliferative burst after PD-1 therapy. Nature **537**(7620), 417–421 (2016)

55. I. Siddiqui, K. Schaeuble, V. Chennupati, S.A. Fuertes Marraco, S. Calderon-Copete, D. Pais Ferreira, et al., Intratumoral Tcf1(+)PD-1(+)CD8(+) T cells with stem-like properties promote tumor control in response to vaccination and checkpoint blockade immunotherapy. Immunity **50**(1), 195–211.e10 (2019)

56. S. Kurtulus, A. Madi, G. Escobar, M. Klapholz, J. Nyman, E. Christian, et al., Checkpoint blockade immunotherapy induces dynamic changes in PD-1(-)CD8(+) tumor-infiltrating T cells. Immunity **50**(1), 181–94.e6 (2019)

57. J. Brummelman, E.M.C. Mazza, G. Alvisi, F.S. Colombo, A. Grilli, J. Mikulak, et al., High-dimensional single cell analysis identifies stem-like cytotoxic CD8(+) T cells infiltrating human tumors. J. Exp. Med. **215**(10), 2520–2535 (2018)

58. H. Li, A.M. van der Leun, I. Yofe, Y. Lubling, D. Gelbard-Solodkin, A.C.J. van Akkooi, et al., Dysfunctional CD8 T cells form a proliferative, dynamically regulated compartment within human melanoma. Cell **176**(4), 775–89.e18 (2019)

59. W.J. Catalona, S. Loeb, Prostate cancer screening and determining the appropriate prostate-specific antigen cutoff values. J. Natl. Compr. Canc. Netw. **8**(2), 265–270 (2010)

60. J.A. Antenor, K.A. Roehl, S.E. Eggener, S.D. Kundu, M. Han, W.J. Catalona, Preoperative PSA and progression-free survival after radical prostatectomy for Stage T1c disease. Urology **66**(1), 156–160 (2005)

61. S. Loeb, C.M. Gonzalez, K.A. Roehl, M. Han, J.A. Antenor, R.L. Yap, et al., Pathological characteristics of prostate cancer detected through prostate specific antigen based screening. J. Urol. **175**(3 Pt 1), 902–906 (2006)

62. A. Amayo, W. Obara, Serum prostate specific antigen levels in men with benign prostatic hyperplasia and cancer of prostate. East Afr. Med. J. **81**(1), 22–26 (2004)

63. E. Elkord, A.W. Rowbottom, H. Kynaston, P.E. Williams, Correlation between CD8+ T cells specific for prostate-specific antigen and level of disease in patients with prostate cancer. Clin. Immunol. **120**(1), 91–98 (2006)

64. E. Elkord, P.E. Williams, H. Kynaston, A.W. Rowbottom, Differential CTLs specific for prostate-specific antigen in healthy donors and patients with prostate cancer. Int. Immunol. **17**(10), 1315–1325 (2005)

65. R.A. Madan, P.M. Arlen, M. Mohebtash, J.W. Hodge, J.L. Gulley, Prostvac-VF: a vector-based vaccine targeting PSA in prostate cancer. Expert Opin. Investig. Drugs **18**(7), 1001–1011 (2009)

66. J.W. Hodge, H. Sabzevari, A.G. Yafal, L. Gritz, M.G. Lorenz, J. Schlom, A triad of costimulatory molecules synergize to amplify T-cell activation. Cancer Res. **59**(22), 5800–5807 (1999)

67. P.W. Kantoff, T.J. Schuetz, B.A. Blumenstein, L.M. Glode, D.L. Bilhartz, M. Wyand, et al., Overall survival analysis of a phase II randomized controlled trial of a Poxviral-based PSA-targeted immunotherapy in metastatic castration-resistant prostate cancer. J. Clin. Oncol. **28**(7), 1099–1105 (2010)

68. A.S. Japp, M.A. Kursunel, S. Meier, J.N. Malzer, X. Li, N.A. Rahman, et al., Dysfunction of PSA-specific CD8+ T cells in prostate cancer patients correlates with CD38 and Tim-3 expression. Cancer Immunol. Immunother. **64**(11), 1487–1494 (2015)

69. S. Muniyan, N.K. Chaturvedi, J.G. Dwyer, C.A. Lagrange, W.G. Chaney, M.F. Lin, Human prostatic acid phosphatase: structure, function and regulation. Int. J. Mol. Sci. **14**(5), 10438–10464 (2013)

70. A. Taira, G. Merrick, K. Wallner, M. Dattoli, Reviving the acid phosphatase test for prostate cancer. Oncology (Williston Park) **21**(8), 1003–1010 (2007)

71. B.M. Olson, T.P. Frye, L.E. Johnson, L. Fong, K.L. Knutson, M.L. Disis, et al., HLA-A2-restricted T-cell epitopes specific for prostatic acid phosphatase. Cancer Immunol. Immunother. **59**(6), 943–953 (2010)

72. T. Igawa, F.F. Lin, P. Rao, M.F. Lin, Suppression of LNCaP prostate cancer xenograft tumors by a prostate-specific protein tyrosine phosphatase, prostatic acid phosphatase. Prostate **55**(4), 247–258 (2003)

73. L. Fong, C.L. Ruegg, D. Brockstedt, E.G. Engleman, R. Laus, Induction of tissue-specific autoimmune prostatitis with prostatic acid phosphatase immunization: implications for immunotherapy of prostate cancer. J. Immunol. **159**(7), 3113–3117 (1997)

74. D.G. McNeel, E.J. Dunphy, J.G. Davies, T.P. Frye, L.E. Johnson, M.J. Staab, et al., Safety and immunological efficacy of a DNA vaccine encoding prostatic acid phosphatase in patients with stage D0 prostate cancer. J. Clin. Oncol. **27**(25), 4047–4054 (2009)

75. M.A. Cheever, C.S. Higano, PROVENGE (Sipuleucel-T) in prostate cancer: the first FDA-approved therapeutic cancer vaccine. Clin. Cancer Res. **17**(11), 3520–3526 (2011)

76. E.J. Small, P.F. Schellhammer, C.S. Higano, C.H. Redfern, J.J. Nemunaitis, F.H. Valone, et al., Placebo-controlled phase III trial of immunologic therapy with sipuleucel-T (APC8015) in patients with metastatic, asymptomatic hormone refractory prostate cancer. J. Clin. Oncol. **24**(19), 3089–3094 (2006)

77. D.A. Silver, I. Pellicer, W.R. Fair, W.D. Heston, C. Cordon-Cardo, Prostate-specific membrane antigen expression in normal and malignant human tissues. Clin. Cancer Res. **3**(1), 81–85 (1997)

78. G.L. Wright Jr., B.M. Grob, C. Haley, K. Grossman, K. Newhall, D. Petrylak, et al., Upregulation of prostate-specific membrane antigen after androgen-deprivation therapy. Urology **48**(2), 326–334 (1996)
79. J.S. Ross, C.E. Sheehan, H.A. Fisher, R.P. Kaufman Jr., P. Kaur, K. Gray, et al., Correlation of primary tumor prostate-specific membrane antigen expression with disease recurrence in prostate cancer. Clin. Cancer Res. **9**(17), 6357–6362 (2003)
80. S.S. Taneja, ProstaScint(R) Scan: contemporary use in clinical practice. Rev. Urol. **6**(Suppl 10), S19–S28 (2004)
81. D. Ma, C.E. Hopf, A.D. Malewicz, G.P. Donovan, P.D. Senter, W.F. Goeckeler, et al., Potent antitumor activity of an auristatin-conjugated, fully human monoclonal antibody to prostate-specific membrane antigen. Clin. Cancer Res. **12**(8), 2591–2596 (2006)
82. S.A. Rajasekaran, G. Anilkumar, E. Oshima, J.U. Bowie, H. Liu, W. Heston, et al., A novel cytoplasmic tail MXXXL motif mediates the internalization of prostate-specific membrane antigen. Mol. Biol. Cell **14**(12), 4835–4845 (2003)
83. P. Wolf, K. Alt, D. Wetterauer, P. Buhler, D. Gierschner, A. Katzenwadel, et al., Preclinical evaluation of a recombinant anti-prostate specific membrane antigen single-chain immunotoxin against prostate cancer. J. Immunother. **33**(3), 262–271 (2010)
84. G. Zuccolotto, G. Fracasso, A. Merlo, I.M. Montagner, M. Rondina, S. Bobisse, et al., PSMA-specific CAR-engineered T cells eradicate disseminated prostate cancer in preclinical models. PLoS One **9**(10), e109427 (2014)
85. S.P. Santoro, S. Kim, G.T. Motz, D. Alatzoglou, C. Li, M. Irving, et al., T cells bearing a chimeric antigen receptor against prostate-specific membrane antigen mediate vascular disruption and result in tumor regression. Cancer Immunol. Res. **3**(1), 68–84 (2015)
86. R.E. Reiter, Z. Gu, T. Watabe, G. Thomas, K. Szigeti, E. Davis, et al., Prostate stem cell antigen: a cell surface marker overexpressed in prostate cancer. Proc. Natl. Acad. Sci. U. S. A. **95**(4), 1735–1740 (1998)
87. G. Bahrenberg, A. Brauers, H.G. Joost, G. Jakse, Reduced expression of PSCA, a member of the LY-6 family of cell surface antigens, in bladder, esophagus, and stomach tumors. Biochem. Biophys. Res. Commun. **275**(3), 783–788 (2000)
88. R.E. Reiter, I. Sato, G. Thomas, J. Qian, Z. Gu, T. Watabe, et al., Coamplification of prostate stem cell antigen (PSCA) and MYC in locally advanced prostate cancer. Genes Chromosomes Cancer **27**(1), 95–103 (2000)
89. Z. Gu, G. Thomas, J. Yamashiro, I.P. Shintaku, F. Dorey, A. Raitano, et al., Prostate stem cell antigen (PSCA) expression increases with high gleason score, advanced stage and bone metastasis in prostate cancer. Oncogene **19**(10), 1288–1296 (2000)
90. Z. Zhigang, S. Wenlu, Complete androgen ablation suppresses prostate stem cell antigen (PSCA) mRNA expression in human prostate carcinoma. Prostate **65**(4), 299–305 (2005)
91. A. Jain, A. Lam, I. Vivanco, M.F. Carey, R.E. Reiter, Identification of an androgen-dependent enhancer within the prostate stem cell antigen gene. Mol. Endocrinol. **16**(10), 2323–2337 (2002)
92. A. Morgenroth, M. Cartellieri, M. Schmitz, S. Gunes, B. Weigle, M. Bachmann, et al., Targeting of tumor cells expressing the prostate stem cell antigen (PSCA) using genetically engineered T-cells. Prostate **67**(10), 1121–1131 (2007)
93. A.K. Thomas-Kaskel, R. Zeiser, R. Jochim, C. Robbel, W. Schultze-Seemann, C.F. Waller, et al., Vaccination of advanced prostate cancer patients with PSCA and PSA peptide-loaded dendritic cells induces DTH responses that correlate with superior overall survival. Int. J. Cancer **119**(10), 2428–2434 (2006)
94. Y. Waeckerle-Men, E. Uetz-von Allmen, M. Fopp, R. von Moos, C. Bohme, H.P. Schmid, et al., Dendritic cell-based multi-epitope immunotherapy of hormone-refractory prostate carcinoma. Cancer Immunol. Immunother. **55**(12), 1524–1533 (2006)
95. E.S. Antonarakis, M.A. Carducci, M.A. Eisenberger, S.R. Denmeade, S.F. Slovin, K. Jelaca-Maxwell, et al., Phase I rapid dose-escalation study of AGS-1C4D4, a human anti-PSCA (prostate stem cell antigen) monoclonal antibody, in patients with castration-resistant prostate cancer: a PCCTC trial. Cancer Chemother. Pharmacol. **69**(3), 763–771 (2012)
96. P.J. Cozzi, J. Wang, W. Delprado, A.C. Perkins, B.J. Allen, P.J. Russell, et al., MUC1, MUC2, MUC4, MUC5AC and MUC6 expression in the progression of prostate cancer. Clin. Exp. Metastasis **22**(7), 565–573 (2005)
97. V. Genitsch, I. Zlobec, G.N. Thalmann, A. Fleischmann, MUC1 is upregulated in advanced prostate cancer and is an independent prognostic factor. Prostate Cancer Prostatic Dis. **19**(3), 242–247 (2016)
98. H. Rajabi, R. Ahmad, C. Jin, M.D. Joshi, M. Guha, M. Alam, et al., MUC1-C oncoprotein confers androgen-independent growth of human prostate cancer cells. Prostate **72**(15), 1659–1668 (2012)
99. E.S. Gabitzsch, K.Y. Tsang, C. Palena, J.M. David, M. Fantini, A. Kwilas, et al., The generation and analyses of a novel combination of recombinant adenovirus vaccines targeting three tumor antigens as an immunotherapeutic. Oncotarget **6**(31), 31344–31359 (2015)
100. A.A. Jungbluth, Y.T. Chen, E. Stockert, K.J. Busam, D. Kolb, K. Iversen, et al., Immunohistochemical analysis of NY-ESO-1 antigen expression in normal and malignant human tissues. Int. J. Cancer **92**(6), 856–860 (2001)
101. T. Kurashige, Y. Noguchi, T. Saika, T. Ono, Y. Nagata, A. Jungbluth, et al., Ny-ESO-1 expres-

sion and immunogenicity associated with transitional cell carcinoma: correlation with tumor grade. Cancer Res. **61**(12), 4671–4674 (2001)

102. M.T. Bethune, X.H. Li, J. Yu, J. McLaughlin, D. Cheng, C. Mathis, et al., Isolation and characterization of NY-ESO-1-specific T cell receptors restricted on various MHC molecules. Proc. Natl. Acad. Sci. U. S. A. **115**(45), E10702–E10E11 (2018)

103. Y. Oshima, H. Shimada, S. Yajima, T. Nanami, K. Matsushita, F. Nomura, et al., NY-ESO-1 autoantibody as a tumor-specific biomarker for esophageal cancer: screening in 1969 patients with various cancers. J. Gastroenterol. **51**(1), 30–34 (2016)

104. J. Yuan, S. Gnjatic, H. Li, S. Powel, H.F. Gallardo, E. Ritter, et al., CTLA-4 blockade enhances polyfunctional NY-ESO-1 specific T cell responses in metastatic melanoma patients with clinical benefit. Proc. Natl. Acad. Sci. U. S. A. **105**(51), 20410–20415 (2008)

105. M. Essand, G. Vasmatzis, U. Brinkmann, P. Duray, B. Lee, I. Pastan, High expression of a specific T-cell receptor gamma transcript in epithelial cells of the prostate. Proc. Natl. Acad. Sci. U. S. A. **96**(16), 9287–9292 (1999)

106. F.R. Fritzsche, C. Stephan, J. Gerhardt, M. Lein, I. Hofmann, K. Jung, et al., Diagnostic and prognostic value of T-cell receptor gamma alternative reading frame protein (TARP) expression in prostate cancer. Histol. Histopathol. **25**(6), 733–739 (2010)

107. S. Varambally, J. Yu, B. Laxman, D.R. Rhodes, R. Mehra, S.A. Tomlins, et al., Integrative genomic and proteomic analysis of prostate cancer reveals signatures of metastatic progression. Cancer Cell **8**(5), 393–406 (2005)

108. L. True, I. Coleman, S. Hawley, C.Y. Huang, D. Gifford, R. Coleman, et al., A molecular correlate to the Gleason grading system for prostate adenocarcinoma. Proc. Natl. Acad. Sci. U. S. A. **103**(29), 10991–10996 (2006)

109. S. Oh, M. Terabe, C.D. Pendleton, A. Bhattacharyya, T.K. Bera, M. Epel, et al., Human CTLs to wild-type and enhanced epitopes of a novel prostate and breast tumor-associated protein, TARP, lyse human breast cancer cells. Cancer Res. **64**(7), 2610–2618 (2004)

110. M. Epel, I. Carmi, S. Soueid-Baumgarten, S. Oh, T. Bera, I. Pastan, et al., Targeting TARP, a novel breast and prostate tumor-associated antigen, with T cell receptor-like human recombinant antibodies. Eur. J. Immunol. **38**(6), 1706–1720 (2008)

111. C.D. Wolfgang, M. Essand, B. Lee, I. Pastan, T-cell receptor gamma chain alternate reading frame protein (TARP) expression in prostate cancer cells leads to an increased growth rate and induction of caveolins and amphiregulin. Cancer Res. **61**(22), 8122–8126 (2001)

112. C.J. Best, J.W. Gillespie, Y. Yi, G.V. Chandramouli, M.A. Perlmutter, Y. Gathright, et al., Molecular alterations in primary prostate cancer after androgen ablation therapy. Clin. Cancer Res. **11**(19 Pt 1), 6823–6834 (2005)

113. L.V. Wood, A. Fojo, B.D. Roberson, M.S. Hughes, W. Dahut, J.L. Gulley, et al., TARP vaccination is associated with slowing in PSA velocity and decreasing tumor growth rates in patients with Stage D0 prostate cancer. Oncoimmunology. **5**(8), e1197459 (2016)

114. A. van Elsas, A.A. Hurwitz, J.P. Allison, Combination immunotherapy of B16 melanoma using anti-cytotoxic T lymphocyte-associated antigen 4 (CTLA-4) and granulocyte/macrophage colony-stimulating factor (GM-CSF)-producing vaccines induces rejection of subcutaneous and metastatic tumors accompanied by autoimmune depigmentation. J. Exp. Med. **190**(3), 355–366 (1999)

115. A. van Elsas, R.P. Sutmuller, A.A. Hurwitz, J. Ziskin, J. Villasenor, J.P. Medema, et al., Elucidating the autoimmune and antitumor effector mechanisms of a treatment based on cytotoxic T lymphocyte antigen-4 blockade in combination with a B16 melanoma vaccine: comparison of prophylaxis and therapy. J. Exp. Med. **194**(4), 481–489 (2001)

116. M. Fasso, R. Waitz, Y. Hou, T. Rim, N.M. Greenberg, N. Shastri, et al., SPAS-1 (stimulator of prostatic adenocarcinoma-specific T cells)/SH3GLB2: a prostate tumor antigen identified by CTLA-4 blockade. Proc. Natl. Acad. Sci. U. S. A. **105**(9), 3509–3514 (2008)

117. J. Lapointe, C. Li, J.P. Higgins, M. van de Rijn, E. Bair, K. Montgomery, et al., Gene expression profiling identifies clinically relevant subtypes of prostate cancer. Proc. Natl. Acad. Sci. U. S. A. **101**(3), 811–816 (2004)

118. R.S. Hubert, I. Vivanco, E. Chen, S. Rastegar, K. Leong, S.C. Mitchell, et al., STEAP: a prostate-specific cell-surface antigen highly expressed in human prostate tumors. Proc. Natl. Acad. Sci. U. S. A. **96**(25), 14523–14528 (1999)

119. I.M. Gomes, C.J. Maia, C.R. Santos, STEAP proteins: from structure to applications in cancer therapy. Mol. Cancer Res. **10**(5), 573–587 (2012)

120. I.M. Gomes, P. Arinto, C. Lopes, C.R. Santos, C.J. Maia, STEAP1 is overexpressed in prostate cancer and prostatic intraepithelial neoplasia lesions, and it is positively associated with Gleason score. Urol. Oncol. **32**(1), 53.e23–53.e29 (2014)

121. S. Rausch, C. Schwentner, A. Stenzl, J. Bedke, mRNA vaccine CV9103 and CV9104 for the treatment of prostate cancer. Hum. Vaccin. Immunother. **10**(11), 3146–3152 (2014)

122. M. Krupa, M. Canamero, C.E. Gomez, J.L. Najera, J. Gil, M. Esteban, Immunization with recombinant DNA and modified vaccinia virus Ankara (MVA) vectors delivering PSCA and STEAP1 antigens inhibits prostate cancer progression. Vaccine **29**(7), 1504–1513 (2011)

123. D.C. Danila, R.Z. Szmulewitz, A.D. Baron, C.S. Higano, H.I. Scher, M.J. Morris, et al., A phase I study of DSTP3086S, an antibody-drug conjugate (ADC) targeting STEAP-1, in patients (pts) with

124. metastatic castration-resistant prostate cancer (CRPC). J. Clin. Oncol. **32**(15 Suppl), 5024 (2014)
124. J.R. Gingrich, N.M. Greenberg, A transgenic mouse prostate cancer model. Toxicol. Pathol. **24**(4), 502–504 (1996)
125. J.R. Gingrich, R.J. Barrios, M.W. Kattan, H.S. Nahm, M.J. Finegold, N.M. Greenberg, Androgen-independent prostate cancer progression in the TRAMP model. Cancer Res. **57**(21), 4687–4691 (1997)
126. T. Chiaverotti, S.S. Couto, A. Donjacour, J.H. Mao, H. Nagase, R.D. Cardiff, et al., Dissociation of epithelial and neuroendocrine carcinoma lineages in the transgenic adenocarcinoma of mouse prostate model of prostate cancer. Am. J. Pathol. **172**(1), 236–246 (2008)
127. A.A. Hurwitz, B.A. Foster, J.P. Allison, N.M. Greenberg, E.D. Kwon, The TRAMP mouse as a model for prostate cancer. Curr. Protoc. Immunol. **Chapter 20**(Unit 20), 5 (2001)
128. E.D. Kwon, B.A. Foster, A.A. Hurwitz, C. Madias, J.P. Allison, N.M. Greenberg, et al., Elimination of residual metastatic prostate cancer after surgery and adjunctive cytotoxic T lymphocyte-associated antigen 4 (CTLA-4) blockade immunotherapy. Proc. Natl. Acad. Sci. U. S. A. **96**(26), 15074–15079 (1999)
129. S. Kasper, P.C. Sheppard, Y. Yan, N. Pettigrew, A.D. Borowsky, G.S. Prins, et al., Development, progression, and androgen-dependence of prostate tumors in probasin-large T antigen transgenic mice: a model for prostate cancer. Lab. Invest. **78**(6), i–xv (1998)
130. K. Ellwood-Yen, T.G. Graeber, J. Wongvipat, M.L. Iruela-Arispe, J. Zhang, R. Matusik, et al., Myc-driven murine prostate cancer shares molecular features with human prostate tumors. Cancer Cell **4**(3), 223–238 (2003)
131. T. Iwata, D. Schultz, J. Hicks, G.K. Hubbard, L.N. Mutton, T.L. Lotan, et al., MYC overexpression induces prostatic intraepithelial neoplasia and loss of Nkx3.1 in mouse luminal epithelial cells. PLoS One **5**(2), e9427 (2010)
132. M.S. Song, L. Salmena, P.P. Pandolfi, The functions and regulation of the PTEN tumour suppressor. Nat. Rev. Mol. Cell Biol. **13**(5), 283–296 (2012)
133. L.C. Cantley, B.G. Neel, New insights into tumor suppression: PTEN suppresses tumor formation by restraining the phosphoinositide 3-kinase/AKT pathway. Proc. Natl. Acad. Sci. U. S. A. **96**(8), 4240–4245 (1999)
134. M. Yoshimoto, J.C. Cutz, P.A. Nuin, A.M. Joshua, J. Bayani, A.J. Evans, et al., Interphase FISH analysis of PTEN in histologic sections shows genomic deletions in 68% of primary prostate cancer and 23% of high-grade prostatic intra-epithelial neoplasias. Cancer Genet. Cytogenet. **169**(2), 128–137 (2006)
135. I.N. Holcomb, J.M. Young, I.M. Coleman, K. Salari, D.I. Grove, L. Hsu, et al., Comparative analyses of chromosome alterations in soft-tissue metastases

within and across patients with castration-resistant prostate cancer. Cancer Res. **69**(19), 7793–7802 (2009)
136. S. Wang, J. Gao, Q. Lei, N. Rozengurt, C. Pritchard, J. Jiao, et al., Prostate-specific deletion of the murine Pten tumor suppressor gene leads to metastatic prostate cancer. Cancer Cell **4**(3), 209–221 (2003)
137. L.C. Trotman, M. Niki, Z.A. Dotan, J.A. Koutcher, A. Di Cristofano, A. Xiao, et al., Pten dose dictates cancer progression in the prostate. PLoS Biol. **1**(3), E59 (2003)
138. X. Wang, M. Kruithof-de Julio, K.D. Economides, D. Walker, H. Yu, M.V. Halili, et al., A luminal epithelial stem cell that is a cell of origin for prostate cancer. Nature **461**(7263), 495–500 (2009)
139. N. Floc'h, C.W. Kinkade, T. Kobayashi, A. Aytes, C. Lefebvre, A. Mitrofanova, et al., Dual targeting of the Akt/mTOR signaling pathway inhibits castration-resistant prostate cancer in a genetically engineered mouse model. Cancer Res. **72**(17), 4483–4493 (2012)
140. Z. Chen, L.C. Trotman, D. Shaffer, H.K. Lin, Z.A. Dotan, M. Niki, et al., Crucial role of p53-dependent cellular senescence in suppression of Pten-deficient tumorigenesis. Nature **436**(7051), 725–730 (2005)
141. J. Nemunaitis, Vaccines in cancer: GVAX, a GM-CSF gene vaccine. Expert Rev. Vaccines **4**(3), 259–274 (2005)
142. A.D. Simmons, B. Li, M. Gonzalez-Edick, C. Lin, M. Moskalenko, T. Du, et al., GM-CSF-secreting cancer immunotherapies: preclinical analysis of the mechanism of action. Cancer Immunol. Immunother. **56**(10), 1653–1665 (2007)
143. T.L. Walunas, D.J. Lenschow, C.Y. Bakker, P.S. Linsley, G.J. Freeman, J.M. Green, et al., CTLA-4 can function as a negative regulator of T cell activation. Immunity **1**(5), 405–413 (1994)
144. E.D. Kwon, C.G. Drake, H.I. Scher, K. Fizazi, A. Bossi, A.J. van den Eertwegh, et al., Ipilimumab versus placebo after radiotherapy in patients with metastatic castration-resistant prostate cancer that had progressed after docetaxel chemotherapy (CA184-043): a multicentre, randomised, double-blind, phase 3 trial. Lancet Oncol. **15**(7), 700–712 (2014)
145. L. Cabel, E. Loir, G. Gravis, P. Lavaud, C. Massard, L. Albiges, et al., Long-term complete remission with Ipilimumab in metastatic castrate-resistant prostate cancer: case report of two patients. J. Immunother. Cancer **5**, 31 (2017)
146. D.Y. Oh, J. Cham, L. Zhang, G. Fong, S.S. Kwek, M. Klinger, et al., Immune toxicities elicted by CTLA-4 blockade in cancer patients are associated with early diversification of the T-cell repertoire. Cancer Res. **77**(6), 1322–1330 (2017)
147. T. Powles, I. Duran, M.S. van der Heijden, Y. Loriot, N.J. Vogelzang, U. De Giorgi, et al., Atezolizumab versus chemotherapy in patients with platinum-treated locally advanced or metastatic

urothelial carcinoma (IMvigor211): a multicentre, open-label, phase 3 randomised controlled trial. Lancet **391**(10122), 748–757 (2018)

148. R.J. Motzer, B. Escudier, D.F. McDermott, S. George, H.J. Hammers, S. Srinivas, et al., Nivolumab versus everolimus in advanced renal-cell carcinoma. N. Engl. J. Med. **373**(19), 1803–1813 (2015)

149. A.M. Martin, T.R. Nirschl, C.J. Nirschl, B.J. Francica, C.M. Kochel, A. van Bokhoven, et al., Paucity of PD-L1 expression in prostate cancer: innate and adaptive immune resistance. Prostate Cancer Prostatic Dis. **18**(4), 325–332 (2015)

150. J.N. Graff, J.J. Alumkal, C.G. Drake, G.V. Thomas, W.L. Redmond, M. Farhad, et al., Early evidence of anti-PD-1 activity in enzalutamide-resistant prostate cancer. Oncotarget **7**(33), 52810–52817 (2016)

151. A.S. Malamas, S.R. Gameiro, K.M. Knudson, J.W. Hodge, Sublethal exposure to alpha radiation (223Ra dichloride) enhances various carcinomas' sensitivity to lysis by antigen-specific cytotoxic T lymphocytes through calreticulin-mediated immunogenic modulation. Oncotarget **7**(52), 86937–86947 (2016)

152. R.E. Kontermann, U. Brinkmann, Bispecific antibodies. Drug Discov. Today **20**(7), 838–847 (2015)

153. P. Buhler, P. Wolf, D. Gierschner, I. Schaber, A. Katzenwadel, W. Schultze-Seemann, et al., A bispecific diabody directed against prostate-specific membrane antigen and CD3 induces T-cell mediated lysis of prostate cancer cells. Cancer Immunol. Immunother. **57**(1), 43–52 (2008)

154. S.A. Rosenberg, N.P. Restifo, J.C. Yang, R.A. Morgan, M.E. Dudley, Adoptive cell transfer: a clinical path to effective cancer immunotherapy. Nat. Rev. Cancer **8**(4), 299–308 (2008)

155. V. Hillerdal, B. Nilsson, B. Carlsson, F. Eriksson, M. Essand, T cells engineered with a T cell receptor against the prostate antigen TARP specifically kill HLA-A2+ prostate and breast cancer cells. Proc. Natl. Acad. Sci. U. S. A. **109**(39), 15877–15881 (2012)

The Tumor Microenvironments of Lethal Prostate Cancer

William L. Harryman, Noel A. Warfel,
Raymond B. Nagle, and Anne E. Cress

Introduction

Prostate cancer (PCa) is the most common cancer, and the second leading cause of cancer deaths (after lung cancer), among men in the Western world [1]. For 2017, there were an estimated 161,360 new cases of PCa diagnosed, mostly in patients between 60 and 80 years old, as well as 26,730 deaths from the disease [2]. Tumor metastasis from the organ to a distant site is responsible for 90% of all cancer deaths [3, 4]. The incidence of metastatic disease in the United States increased 72% between 2004 and 2013 in a sample of 767,550 men diagnosed with PCa (from 1685 cases in 2004 to 2890 in 2013) [5]. Localized PCa (confined to the gland, T1 to T2 stage) generally is considered curable, with nearly a 100% 5-year-survival rate [2], but 5-year-survival drops to 29% with a tumor escaping the gland (T3 stage) [6]. The most common score, Gleason 6, is indeterminant, and controversy surrounds interpretation of the score to predict aggressive (high-risk) vs. indolent (low-risk) disease [7]. Recently the Gleason score (GS) grading pattern has been revised and patients are stratified into four groups: GS2-6, GS7, those with GS8 without GS5, and those with primary, secondary or tertiary GP5. Of these, the most common score in 563 patients (52.4%) was GS7 [8]. The prognostic benefit of the new prostate cancer grade-grouping was independently validated using surgical specimens. The greater predictive accuracy of the new system will improve risk stratification in the clinical setting and aid in patient counselling [9].

Currently there is an unmet clinical need to determine if or when localized disease has extended past the prostatic capsule, which requires invasion and migration into and through the smooth muscle stroma coordinated by tumor specific and stromal-specific signals. Research to discover molecular and biophysical signatures of aggressive PCa includes both genomic-based classification [reviewed in [10]] and proteomic-based biomarkers [reviewed in [11]] correlating with outcomes. The potential for blood-based and urinary biomarkers for early detection and treatment decisions [12] are attracting increased attention since sensitive and quantitative measurements are now achievable with an intent to avoid unnecessary biopsy and over-diagnosis of indolent disease [13]. The integration of histopathology [14], molecular genomic, proteomic, and transcriptomic ("omics") information [15], coupled with a clearer understanding of the changing microenvironments during PCa progression (illustrated in Fig. 1), may lead to the early detection of high

W. L. Harryman · N. A. Warfel · A. E. Cress (✉)
University of Arizona Cancer Center,
Tucson, AZ, USA
e-mail: cress@email.arizona.edu

R. B. Nagle
Department of Pathology, University of Arizona
Cancer Center, Tucson, AZ, USA

© Springer Nature Switzerland AG 2019
S. M. Dehm, D. J. Tindall (eds.), *Prostate Cancer*, Advances in Experimental Medicine
and Biology 1210, https://doi.org/10.1007/978-3-030-32656-2_8

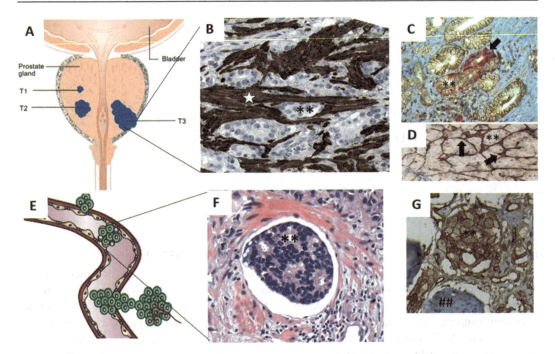

Fig. 1 Lethal PCa is found in different microenvironments. Panel (**a**) The prostate gland is located at the base of the bladder and is encapsulated by a band of smooth muscle (grey fill). Tumor stage is depicted as T1 or T2 within the gland or T3 when it has breached the smooth muscle capsule. Panel (**b**) Microscopically, the tumor (∗∗) moves though the smooth muscle (white star) which can be detected by staining desmin (white star). Panel (**c**) PCa (∗∗) stained for α6 integrin (brown) has disrupted and invaded a nerve (arrow) stained for PGP 9.5 (pink). Panel (**d**) PCa (∗∗) within a lymph node extracellular matrix (arrows) stained for laminin 511 (brown). Panel (**e**) Schematic of hematogenous spread of PCa clusters. Panel (**f**) Histological section of PCa cluster within a vessel. Panel (**g**) PCa clusters (∗∗) stained for α6 integrin (brown) within bone (##). (Panel **a** is from the Wikimedia Commons, Free Media Repository, File: CRUK278.svg. Panel **e** was adapted from Guo W, Giancotti FG. Integrin signalling during tumour progression. Nature reviews Molecular cell biology. 2004;5(10):816–26)

risk disease to inform treatment decisions and additional strategies for metastasis prevention.

Alteration of the microenvironment within the normal prostate gland occurs in males in the third and fourth decade of life and increases progressively with advancing age [16]. The microenvironment contextual changes are observed histologically and called prostatic intraepithelial neoplasia (PIN). PIN is defined by focal loss of normal morphological relationships of the epithelium to the stromal environment [17–19]. PIN has sporadic basal cells and primarily luminal cell phenotypes, expressing products of differentiation including the androgen receptor and prostate specific antigen (PSA). The luminal cells also have a loss of DNA damage sensors for DNA repair (i.e., using instead an error prone process of non-homologous end rejoining) [20], similar to differentiated cell populations that lose DNA repair capacity [21–24]. The loss or transformation of the normal basal cell component is a key step in PIN, resulting in epithelial cell exposure to a new microenvironment containing conserved developmental morphogens such as laminin 511 [25] and growth factors in the surrounding stroma [26–30]. Most evidence suggests PIN is the pre-malignant lesion [31]. A high-grade PIN (HG-PIN) lesion develops during the early period of PCa development, with increasing attenuation of the basal cell layer and extracellular matrix (ECM) [32, 33]. Since the goal of this chapter is to focus on the changing microenvironments of lethal PCa, it should be noted that while HG-PIN lesions are early forms of cancer, they are not synonymous with lethal PCa and do not elevate serum PSA concentration [31].

Different microenvironments (as shown in Fig. 1) are encountered during PCa invasion and escape from the gland. The environments include the smooth muscle, nerves, lymph nodes, vasculature, bone and hypoxic regions in the peripheral zone of the gland. Each of these environments are distinct, both in their biophysical and biochemical characteristics [34, 35]. Cohesive invasion of PCa through a band of laminin-lined smooth muscle, known as the prostatic capsule (Fig. 1, panel a), defines disease progression [36]. Aggressive PCa is called extracapsular extension and, clinically, the staging increases from a T2 to a T3 score. Muscle invasive tumors are clusters or groups of tumor cells (Fig. 1, panel b). Within the peripheral zone of the gland, tumors encounter and invade nerves (Fig. 1, panel c) and escape the gland as detected in the pelvic lymph nodes (Fig. 1, panel d) and as tumor clusters within the circulation (Fig. 1, panels e, f) and within the bone, as the primary metastatic site (Fig. 1, panel g).

In this chapter, we review evidence that prostate tumors change their microenvironment during the process of smooth muscle and neural invasion [37, 38]. Extracellular vesicles can condition the microenvironment and are extruded from PCa cells [39, 40]. Prostate tumors also adapt to their changing environments during treatment [41]. At least 20–30% of patients with low-risk tumors (Gleason sum score <6) treated with radical prostatectomy (RP), with or without pelvic lymph node dissection (PLND), will experience biochemical recurrence [42].

The adaptation of prostate tumors to hypoxic conditions during invasion is another physiologically relevant event that impacts therapeutic options and opportunities. The standard of care for tumors with intermediate to high risk of recurrence is androgen deprivation therapy (ADT) [42], often combined with radiation and/or chemotherapies [42]. When tumors progress and develop resistance to ADT, the disease is classified as castration-resistant PCa (CRPC), a lethal manifestation of the disease. Recently, more potent second-generation forms of ADT have been developed, which have become standard-of-care for patients with CRPC. For instance, enzalutamide binds to ligand-binding domain of the androgen receptor, displacing testosterone and dihydrotestosterone, thereby inhibiting androgen-receptor

signaling [43]. Abiraterone inhibits cytochrome P450 17A1 (CYP17A1), thereby disrupting androgen-receptor signaling by depriving the tumor of androgens [43]. Approximately 20–40% of CRPC-stage patients have no response to these agents with respect to prostate-specific antigen (PSA) levels (i.e., they have primary resistance) [44, 45], and among patients who initially have a response to enzalutamide or abiraterone, virtually all eventually acquire secondary resistance [43]. Clearly, prostate tumors have the capacity to evolve and adapt to extreme conditions. As these therapies were developed, an increased understanding of the disease has shifted to better use of existing therapies in early-stage, non-castrate disease and the need for new molecular phenotype markers of disease prognosis and treatment selection [reviewed in [46]].

The dynamic reciprocity between tumors and their microenvironments [reviewed in [47]] underscores the need for an increased understanding of the environmental cues encountered during PCa invasion into and through physiologically relevant environments. Understanding the microenvironment has the potential to provide alternative strategies for halting aggressiveness of high-risk disease by providing biochemical or mechanical contextual cues. This approach would have the advantage of avoiding the selection of aggressive disease that is inherent in survival-based therapeutics.

The Muscle Stroma and Pseudo-Capsule

PCa arises in the peripheral zone of the gland (Fig. 1) which contains smooth muscle stroma. The human gland is surrounded by a smooth muscle capsule [48] which is not observed in mice [49]. The capsule is a major partition, and breeching the barrier marks the transition from organ-confined disease (T1, T2) to more aggressive disease (T3). A major molecule that is present in the muscle ECM is laminin 511, a molecularly conserved protein trimer that self-assembles into a mesh-like network or partition [50].

The integrins are a family of tissue-specific ECM receptors that are mechano-transduction regulators for cell movement into and through the ECM while also inducing intracellular signaling. Integrins are non-covalently bound α,β heterodimeric cell surface receptors that mediate adhesion and migration via connections between extracellular adhesion molecules and the intracellular cytoskeleton [38, 51, 52]. There are 24 members found in vertebrates, comprised of 18α and 8β subunits, all programmed by distinct genes [53–55]. The laminin-binding integrin (LBIs) alpha subunits, which includes $\alpha6$ (CD49f), $\alpha3$ (CD49c), and $\alpha7$, are some of the most highly conserved integrin receptors [56, 57] and play important roles in both normal and pathological states. Of these, only $\alpha3\beta1$ and $\alpha6\beta1$ are found in prostate carcinoma [58, 59]. LBIs in PCa cells are important adhesion receptors for cell migration, invasion, and metastasis [60–62], while the stromal components (muscle cells) are likely to have an active role, as well. In this regard, another LBI, called $\alpha7$ integrin is muscle-specific, and contains identical and evolutionarily conserved amino acid residues [63] as described for the $\alpha6$ integrin [64] that are used for uPAR mediated cleavage. It remains to be determined if the muscle specific $\alpha7$ integrin cleavage would account for the known disruption of the smooth muscle stroma by prostate cancer [63].

Our group and others have documented the role of LBIs and laminin extracellular matrix proteins in regulating cell adhesion and migration during PCa progression [33, 58]. Epithelial cells attach to the substratum in normal prostate via integrin $\alpha6\beta4$ to laminin 332 in strongly adhesive structures [38]. The $\alpha6\beta4$ integrin is expressed mainly in stratified epithelial tissues of the prostate and is the predominant integrin heterodimer found in normal prostate epithelium [65]. However, $\alpha6\beta4$ is downregulated in PCa [58, 66, 67], enabling involvement of $\alpha6\beta1$ and $\alpha3\beta1$ laminin receptors and the pro-migratory phenotype of luminal epithelial cells during PCa progression [38].

Glandular escape occurs by tumor invasion of the smooth muscle capsule [18, 38], shown in Fig. 1, which is especially rich in laminin 511 [18]. Increased LBI expression ($\alpha6$ integrin) is significantly associated with risk of biochemical progression, clinically detectable metastasis, bone metastasis progression, and death from PCa [68]. The integrin marks a basal stem cell progenitor population of prostate cancer [69]. LBI-variant forms are tumor specific [64, 70], found on invasive tumors [71], elevated during invasion [72], and are required for successful PCa bone metastasis progression [64, 73]. Systemic blocking of LBI function significantly inhibits preconditioning of the metastatic niche [12], as well as metastatic lesion progression in mice [74], and blocks and sensitizes tumors to drug and radiation therapy [75, 76].

The prostate tumor invasive phenotype occurs primarily as a cohesive group of cancer cells, expressing the $\alpha6$ integrin at cellular interfaces, which contrasts significantly from its usual basal location [77]. Interestingly, $\alpha6$ integrin is also present at cell-cell interfaces in stem cell cultures and 3-D spheroid cultures [77]. In lower organisms, LBIs are required for morphogenesis and the migration of stem cell clusters on laminin [78]. Laminin is a potent morphogen, and a prominent component of the stiff smooth muscle of the prostatic capsule [18].

Invasion of human PCa into the muscle layer is depicted in Fig. 1, panel b. Figure 2 illustrates two *in vitro* models and one *in vivo* model of smooth muscle invasion. In the first model, 3D invasion imaging can be directly detected using tumor cells that have been modified to contain GFP- or RFP-tagged membrane or cytoskeletal proteins. Using time-lapse microscopy, the movement of tumor cells into and within the smooth muscle layer can be observed. In the second model, quantification and blocking of invasion can be determined using the modified Boyden chamber type assay. The invasion of tumor cells through the smooth muscle layer can be quantified with a plate reader, as this is adaptable to a 24-well format, and chemo-attractants or antagonists can be tested. In the third model, human tumor cells are injected directly into the peritoneal cavity of the NOD-SCID (immunodeficient) mouse, where colonies will grow on the undersurface of the diaphragm within 6–8 weeks. The diaphragm is a structure that contains a smooth

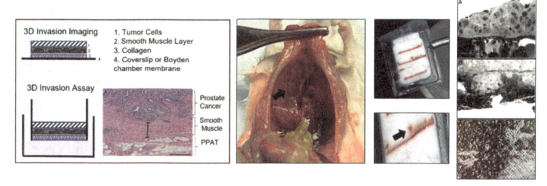

Fig. 2 *In vitro* and *In vivo* smooth muscle invasion models. Left Panel: Tumor cells (hatched bar) are placed on top of the 2–3 cell layer of smooth muscle (brown layer) grown on a collagen (blue layer) coated coverslip. For invasion assays, the same layering is done on Boyden chamber inserts. Histological section shows human PCa within the gland bounded by a smooth muscle layer. Middle Panel: In the SCID mouse, IP injection of human tumor cells will result in tumor colonies (black arrow) that grow on the underside of the diaphragm. Right Panels: The collected diaphragm is embedded so that the tumor colony (black arrow) is oriented on top of the muscle, and transverse sections detected the tumor displacing the myoepithelium (**a**) and tumor (brown cytokeratin stain) moved through the smooth muscle layer. (Right panels, adapted from McCandless [79]

muscle layer with skeletal muscle attachments and will support tumor colonization and invasion into and through the layer [79]. Using this model, the impact of genetic modifications of tumors or stroma on smooth muscle invasion or the effects of agonists and antagonists on the invasion process can be tested. In addition, it is possible to recover tumor cells that have breached the smooth muscle layer and are present on the superior surface of the diaphragm.

The mouse xenograft model allows detection and tracking of invasive cancer cells into smooth muscle (Fig. 3, panel a), and human needle biopsy specimens allow the observance of tumor invasion into muscle (Fig. 3, panel b). Together, these models can be used in an iterative process to validate candidate molecular events of smooth muscle invasion in PCa. In this way, promising new actionable targets may be discovered.

Our work supports a two-step process in which [1] early cohesive cluster invasion is mediated by LBI-dependent adhesion events, and [2] cohesive invasion through smooth muscle is mediated by membrane expression of LBIs that is regulated by recycling events. Specifically, the α6 integrin is contained in EEA1 vesicles in PCa cells, which supports both short-loop and long-loop recycling (perinuclear recycling compartment, PNRC, also called apical recycling endosomes, ARE) locations [80]. Other work has shown that FIP5 is part of the widely distributed tubulin network and vesicles [81]. Rab11, which is involved in the vesicular transport of many cellular membrane proteins (reviewed in [80]), plays an important role in the membrane flux of integrins [82], especially α6β4 integrin [83]. The selectivity for transport of "cargo" to the membrane is dictated by the Rab11 family interacting proteins (Rab11-FIPs) [84]. The selectivity of these adapters is an actionable step to block muscle invasion and can be tested using the models shown in Figs. 2 and 3.

Hypoxia in PCa Invasion and Metastasis

While laminin-binding integrins assist the tumor in escaping the prostate gland, the microenvironment within the gland can initiate growth and trigger pro-metastatic events. Hypoxia, a reduction in the normal level of tissue oxygen, is a hallmark of solid tumors that develops due to inflammation, cell remodeling, proliferating tumor cells, and abnormal tumor vasculature. Hypoxia is a key molecular feature of the tumor

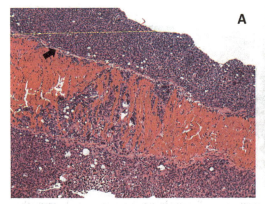

Fig. 3 Human prostate tumors are invasive into smooth muscle. Panel (**a**) DU145 prostate tumor cells were injected into the mouse model shown in Fig. 1. Eight weeks later the diaphragms were harvested and showed tumor (purple stain) invading the diaphragm (black arrow) and into the pink muscle. Panel (**b**) Human PCa obtained from a needle biopsy specimen was stained for E-cadherin (green), nuclei (DNA, blue) and the muscle was autofluorescent (red). The angulation of the prostate glands within the muscle is characteristic of invading cancer

microenvironment that controls the metastatic potential of solid tumors. Hypoxia is especially relevant to PCa because the prostate gland itself is hypoxic compared to most other soft tissues, and hypoxia increases with stage in PCa [85]. In this regard, hypoxia can be considered an early event during prostate carcinogenesis. The presence of hypoxia in PCa has been well documented through the detection of intrinsic hypoxia markers (HIF-1α, VEGF, LOX, and GLUT1 [86, 87]), medical imaging (PET with [^{18}F] fluoro-misonidazole [88–90]), and physical measurement (Eppendorf pO$_2$ microelectrodes [91, 92]). Low oxygen tension is associated with a worse clinical outcome in PCa patients. Hypoxia can be used to predict disease recurrence after radical prostatectomy [85], and hypoxia is associated with early biochemical relapse and local recurrence in the prostate gland [91]. Similarly, the expression of hypoxia-induced factors (HIF-1α and VEGF) predicts treatment failure, independent of tumor stage, Gleason score, or serum PSA levels [86].

Due to the strong correlation between hypoxia and PCa invasion, ongoing efforts are underway to develop new methods to assess hypoxia at the time of diagnosis to identify PCa patients at high risk of developing metastatic disease and avoid the wide-spread overtreatment of patients with indolent, low-risk disease. Advances in clinical imaging technologies allow for more resolved imaging of hypoxia and are currently being validated in preclinical and clinical models of PCa. A method was recently developed that integrates images related to oxygen consumption and supply into a single image (consumption and supply-based hypoxia (CSH) imaging). This CSH imaging method was then used to image hypoxia in patients administered pimonidazole prior to prostatectomy [93]. Another technology using ^{18}F-misonidazole (FMISO) uptake on PET/CT to determine the spatial distribution of hypoxia in the prostate prior to radiotherapy treatment found significant hypoxia in over one-third of patients [88]. While the study was too small to correlate pre-treatment hypoxia with outcome, recurrence and metastatic disease were more frequently observed in patients with hypoxic tumors.

Another study reported on the development and validation of a hypoxia-related prognostic signature for localized PCa [94]. A signature of 28 hypoxia-inducible genes was refined based on its ability to predict patient outcome after radical prostatectomy. This gene signature predicted the risk of biochemical recurrence and metastasis in over 1000 patients with primary tumors, regardless of treatment type, and the prognostic value was independent from existing clinic-pathological factors. These studies indicate that hypoxia is an important aspect of the prostate tumor microenvironment, and new methods to detect and therapeutically target tumor hypoxia are needed to improve how we detect and treat PCa patients at high risk of developing metastatic disease. Below, we will highlight HIF-dependent and HIF-independent hypoxia-induced signaling pathways, with a focus on how they alter the physiological properties of prostate tumors to promote invasion and metastasis.

Angiogenesis

Increased angiogenesis is one of the main outcomes of HIF signaling, and the connection between excessive vascularization and cancer progression is well established [95]. Hypoxia-induced angiogenesis plays a major role in PCa progression, as VEGF and HIF-1α are increased in PCa compared to benign prostatic hyperplasia (BPH) [96]. Despite this fact, anti-angiogenic agents that target VEGF or its receptors have provided no significant improvement in overall survival, even in combination chemo- and targeted therapies [97, 98]. The lack of efficacy of anti-angiogenic therapies can likely be attributed to the resultant increase in tumor hypoxia and hypoxia-mediated activation of alternative pro-angiogenic signaling pathways. Recent studies have found several VEGF-independent pathways that drive angiogenesis in the prostate and are activated in hypoxia, including PIM kinases [99], IL-6 [100], and p-REX [101], among others.

Intriguingly, PIM kinases are upregulated in an HIF-1-independent manner in hypoxia [99], and combined PIM inhibition and anti-angiogenic agents produced synergistic anti-vascular and anti-metastatic effects in orthotopic models of PCa [102]. In addition to increasing the expression and release of pro-angiogenic molecules, hypoxia increases vascular permeability by altering the integrity of blood vessels. Endothelial cells are the major structural component of blood vessels and serve as a barrier to the extravasation and intravasation processes [103]. Interestingly, depletion of HIF-1α in endothelial cells suppresses the migration of tumor cells through endothelial cells, whereas HIF-2α deple-

tion stimulates metastatic spread. These opposite effects of HIF-1α and HIF-2α on vessel formation and metastasis could explain the failure of HIF-1-targeting agents in clinical trials.

Epithelial-Mesenchymal Transition

A prevailing theory to explain tumor invasion and metastasis is that detachment of cancer cells from the primary tumor is initiated by the activation of epithelial to mesenchymal transition (EMT). Hypoxia can induce epithelial plasticity and a migratory phenotype through the direct and indirect regulation of the EMT transcription factors Snail, Slug, Twist, and ZEB1. One of the main events that enhances EMT during hypoxia is the repression of E-cadherin via HIF-1-dependent and -independent mechanisms [104]. Increased expression of LOX and LOXL2 under hypoxia is necessary and sufficient for hypoxic repression of E-cadherin, which is the result of increased expression of the SNAIL transcription factor [87]. Further, EMT in cancer is associated with a loss of epithelial-specific E-cadherin from adherens junctions and a switch from the expression of keratins as major intermediate filaments to the mesenchymal intermediate filament system composed of vimentin [105]. While this is a widely accepted transition in several cancers, in PCa, EMT occurs sporadically in patterns that are not understood. For example, in a majority of organ confined (low-risk) PCa, EMT is rarely observed as E-cadherin remains expressed [106].

E-cadherin is an important cell-to-cell adhesion molecule that is down-regulated in EMT and promotes invasive and metastatic tumors [107, 108]. ZEB1 decreases E-cadherin expression and, likewise, miR-200 targets and decreases expression of ZEB1 and ZEB2 [109], which, through the increased expression of E-cadherin, promotes tumor clusters rather than individual circulating tumor cells (CTCs) [77]. The miR-200 family of micro RNAs are central markers for epithelial cells, inhibiting expression of genes that trigger EMT and facilitate metastatic tumor growth [109, 110]. However, in human PCa tissue, E-cadherin is mostly up-regulated [18], providing a survival advantage for tumor cells [111, 112]. Accordingly, miR-200 is downregulated in PC3 cells [109]. A potential resolution to this apparent contradiction is that, while some cells in the tumor are negative for E-cadherin (appearing mesenchymal), others remain positive (appearing epithelial). There is a transient down-regulation of E-cadherin in gland-confined PCa, while metastatic PCa exhibits robust E-cadherin expression [106]. A mixed phenotype may be the model to resolve the EMT controversy and explain how the migration of tumor cell clusters provides a survival advantage over CTCs [77]. Accordingly, E-cadherin serves as an excellent marker of tumor cell clusters [77], and lymph node metastatic lesions contain prominent E-cadherin expression [77, 113, 114]. This suggests that prostate carcinoma invades and metastasizes by a collective type cell migration [115, 116] with prominent E-cadherin and cytokeratin expression and tumor cell-cell adhesion [106].

EMT (detected by E-cadherin loss) can be observed in occasional tumor cells within the periphery of tumor clusters in aggressive (high risk) tumors with a Gleason Grade 5 (Gleason sum scores >8) (reviewed in [18]). Collective migration can involve cell-cell adhesion via E-cadherin and the presence of "leader cells" without E-cadherin at the periphery of the clusters [117, 118]. EMT could account for a subpopulation of tumor cells within the estimated 30% of patients that have high-risk, aggressive disease prior to treatment [42]. Indeed, a switch from collective to amoeboid type migration in model systems is achieved with a hypoxic microenvironment [119].

Extracellular Matrix Degradation

The tumor ECM is shaped by cancer cells to provide structural and functional features that facilitate metastasis [120]. HIF-1 signaling mediates changes in ECM integrity through the upregulation and secretion of proteolytic enzymes, such as matrix metalloproteinases (MMPs), cathepsins, lysyl oxidases, and prolyl-4-hydroxylases (P4Hs). Cumulatively, these changes support the initial stages of metastasis through matrix degradation

and remodeling. Matrix metalloproteinases (MMPs) are a family of enzymes that cleave a broad range of components of the extracellular matrix (ECM), basement membrane, growth factors, and cell surface receptors. Stromal and inflammatory cells, rather than tumor cells, typically synthesize MMPs, which can then act on the stroma and regulate the tumor microenvironment, as well as acting on tumor cells themselves.

Several MMPs, including MMP-9 [121] and the urokinase receptor uPAR, which are known to increase in response to hypoxia, are overexpressed during PCa progression [122–124]. Hypoxia-driven expression of these MMPs can promote invasion and correlates with poor patient prognosis. Increased expression of MMP-2 in PCa cells is an independent predictor of shorter disease-free survival [125]. Moreover, expression of MMP-9 in prostatic carcinoma cells results in reduced lung metastases but does not affect the tumor growth rate [121, 126]. The most common site of metastasis for PCa is the bone where MMP expression promotes metastatic seeding. MMP-7 is highly expressed in PCa cells as well as osteoclasts at the tumor-bone interface, and overexpression of MMP-7 triggers bone metastasis in a rodent model of PCa [127]. Carbonic anhydrase (CA) IX is another hypoxia-induced gene that is critical for maintaining homeostasis and impacting the ECM. CAIX is an established marker of hypoxia, and expression of CAIX positively correlates with PCa staging and outcome [128]. CAIX is a membrane bound, zinc-dependent enzyme that catalyzes CO_2 hydration into bicarbonate with release of a proton [129]. Since CO_2 is elevated in hypoxic tissues, CAIX plays a major role in maintaining the acid/base homeostasis under low oxygen tension. The CAIX active site is located on the outer side of the plasma membrane, so the extrusion of hydrogen ions into the extracellular space by CAIX maintains intracellular pH, and at the same time promotes acidification of the prostate tumor and surrounding stroma. Thus, hypoxia-induced CAIX leads to degradation of the surrounding ECM and creates an environment conducive to invasion and metastasis.

Extracellular Matrix Remodeling

Alignment and stiffness of the ECM are critical regulators of tumor cell migration. Hypoxia is an important microenvironmental factor that stimulates tissue fibrosis. Collagen type I is the major structural ECM component in the prostate [130], and cancer cell invasion often takes place along its fibers. Hypoxia was recently shown to mediate collagen 1 fiber remodeling in the ECM of tumors, which may impact delivery of therapeutics as well as tumor cell dissemination [131]. HIF-1 activation in fibroblasts regulates ECM biogenesis to produce a stiff microenvironment that enhances cell adhesion, elongation, and motility. HIF-1 controls these events by increasing procollagen prolyl (P4HA1 and P4HA2) and lysyl (PLOD2) hydroxylase expression in hypoxia, resulting in increased fibrillar collagen deposition. HIF-1-dependent collagen modification is critical for dynamic matrix organization and stiffness, which enhances tumor cell invasion. Cells receive mechanistic signals from the alterations in the ECM via focal adhesions. Formation of focal adhesions requires the binding of ECM proteins to integrins and cell-surface ECM receptors. Recent studies in breast cancer have shown that hypoxia induces the expression of integrins that bind to collagen (ITGA1, ITGA11), fibronectin (ITGA5), and laminin (ITGA6). Notably, HIF-1 upregulates ITGA5 [132] and ITGA6 [133] under hypoxic conditions, both of which enhance migration and invasion, and silencing of ITGA5 or ITGA6 reduces cell motility and metastasis *in vivo*. Thus, the ability of hypoxia to promote metastasis relies heavily on interactions between tumor cells and factors within the tumor microenvironment.

Hypoxia alters a variety of physiological processes that directly enhance the invasive capacity of tumor cells, as well as changing aspects of the tumor microenvironment to facilitate metastasis. Thus, hypoxia-induced metastasis involves a complex network controlling the expression of cell-cell and cell-extracellular matrix (ECM) adhesion molecules, the secretion of extracellular proteases that break down the ECM, and the release of a plethora of growth factors and chemokines that promote tumor cell invasion. The broad effects of hypoxia have largely been attributed to activation of HIF-1. While HIF-1 remains a central player in the cellular response to hypoxia, it has become clear that HIF-1-independent signaling pathways are also responsive to hypoxia and contribute to PCa metastasis.

Perineural Invasion

While hypoxia creates the conditions for the tumor to become invasive of surrounding tissues, perineural invasion is a prominent mechanism for tumors to escape the capsule. Thus, a wider perspective can help contextualize how tumor uses the nerves as a route to escape.

A general step-model can be applied to tumor metastasis whereby tumor cells interact with their environment, including loss of cellular adhesion and detachment, local invasion of the ECM, intravasation into the lymph system or vasculature, and extravasation into the parenchyma of distant tissues [38]. Neurotropic tumors, including prostate, pancreatic [71], bladder [134], head and neck, stomach, and colorectal cancers [135–137], exhibit an affinity for utilizing the neuroanatomy of highly innervated organs as a pathway for the primary tumor to escape the local site [38, 135]. PCa cell migration along and invasion into prostatic nerves and neurovascular bundles, a process termed perineural invasion (PNI), is a common pathway of extracapsular escape during PCa metastasis [138, 139]. Across a variety of cancers, PNI is associated with increased tumor aggression, poorer outcome, and reduced survival [135, 137, 140, 141].

PNI is found in 22.4–65.4% of PCa biopsy tissue samples in patients with organ-confined disease [142, 143]. The presence of PNI is linked to a reduced occurrence of apoptosis in cancer cells, allowing for enhanced propagation of tumor cells [144]. While PNI in organ-confined disease is not an independent marker for the risk of biochemical recurrence in PCa [142] following treatment, previous research suggests that a greater diameter of the largest focus of PNI is correlated with an increased probability of progression after radical prostatectomy. Specifically, patients with a PNI < 0.25 mm demonstrated little adverse effect 5 years after surgery, patients with PNI diameter of 0.25–0.5 mm were considerably less likely to remain disease-free at 5 years, and only 36% of those with PNI of 0.5–0.75 mm and 14% of those with PNI ≥ 0.75 mm remained free of progression after 5 years [143].

PNI has been described in the literature, originally in head and neck cancer, since the mid-1800s [135]. Until recently, PNI was considered to be an extension of lymphatic metastasis because of the presence of lymphatic channels within the epineurial layer [136], the dense sheath of connective tissue surrounding the shaft of a nerve. However, conclusive studies revealed that there are no lymphatics within the nerve sheath [135]. Instead, there are three connective tissue layers comprising the nerve sheath—from the outside in, they are the epineurium, the perineurium, and the endoneurium [135]. The area just outside the nerve sheath is the perineurium, which surrounds the entire nerve. The epineurium connects one or multiple fascicles to create one nerve and contains two unique layers. The outer layer of areolar connective tissue and lightly joined collagen bundles and an inner layer of tightly joined collagen fibrils and elastin fibers [135, 145]. The innermost layer of the nerve sheath, the endoneurium, forms a barrier around discrete nerve fibers and encases the Schwann cells and individual axons of the nerve [135, 146].

The peripheral zone of the prostate gland contains primarily motor neurons as evidenced by the histopathology and the pathophysiology associated with post-prostatectomy incontinence and damage to the neurovascular bundles [147]. Peripheral nerves make up one of many complex tumor microenvironments and are comprised of multiple cell types, including Schwann cells, or peripheral nerve glial cells. Schwann cells produce a basal lamina that forms a protective layer surrounding the nerve; the basal lamina and extracellular collagen fibrils also form part of the endoneurial connective tissue environment [148].

The basal lamina of Schwann cells is composed mostly of glycoproteins, including collagens, laminins, entactin, fibronectin, and proteoglycans [149]. Cells interact with basal lamina elements via extracellular matrix surface receptors. These interactions generate signaling cascades necessary for cell orientation and function between laminin and integrins [149].

Integrins $\alpha3\beta1$, $\alpha6\beta1$, and $\alpha6\beta4$ are binding partners with laminins 10 and 11 in peripheral nerves [148]. Our group identified $\alpha6\beta1$ integrin as the primary laminin-binding integrin implicated in prostatic PNI [38, 150]. Macrophages can contribute to a tumor-specific form of $\alpha6\beta1$ integrin, $\alpha6p\beta1$, mediated by the urokinase plasminogen activator (uPA) [72], which cleaves the ligand-binding domain of full-length $\alpha6$ integrin while on the cell surface [72].

$\alpha6$ integrin is a 140-kDa laminin receptor, whereas $\alpha6p$ is a novel 70-kDa form of the $\alpha6$ integrin (the p is for the Latin word *parvus*, meaning small), and the $\alpha6p$ form is missing the extracellular β-propeller domain associated with ligand-binding [150, 151]. Macrophages significantly increased uPAR expression at the cell surface in the PC-3 cell-line (at 24, 48, and 72 h of co-culture) [72]. uPAR (the uPA receptor) is expressed in several tumor-associated cell types, including tumor cells, tumor associated stromal cells, neutrophils, and macrophages [124, 152], but by itself is insufficient to produce the $\alpha6p$ integrin variant. Rather, uPA is required for cleavage [153]. These results identified a new mechanism by which macrophages increase the invasive phenotype of prostate tumor cells through modulation of $\alpha6\beta1$ integrin cleavage, producing $\alpha6p\beta1$ [72]. Thus, cleavage of integrin $\alpha6\beta1$ to form $\alpha6p\beta1$ (Fig. 4) increases tumor cell motility, invasion, and osseous PCa metastasis [64, 71–73]. While the exact mechanism involving $\alpha6p\beta1$ in promoting perineural invasion is, thus far, undetermined, it is intriguing that uPAR is required for nerve regeneration [154] and uPA binding to uPAR promotes axonal regeneration via integrin activation [155].

Extracellular Vesicles: A Focus on Microvesicles in PCa

The term "extracellular vesicle" (EV) was introduced as a generic term to describe any type of membrane-enclosed particles released by any type of cell (including microorganisms) in the extracellular space. PCa EVs can include exosomes, microvesicles, microparticles, apoptotic particles, apoptotic bodies, oncosomes, and various derivative names, depending on the source tissue [reviewed in [156, 157]]. For this section, we will focus on microvesicles (MVs) since these are derived primarily from tumor cells using a process of budding of the plasma membrane [156]. In addition, their large size (0.2–1 mm diameter), ease of collection from both blood and urine samples [158–161], and quantification by high-resolution flow cytometry [162, 163] suggest a potential utility in meeting the need for clinical discriminators of low-risk (indolent) vs. high-risk (aggressive) disease at the time of diagnosis [157].

Several reports have documented the abundance of MVs or their composition in cell-based experiments of intravasation, extravasation, and metastasis [reviewed in [156]]. While their systemic presence in cancer patients can be observed, quantified, and are reported to be novel biomarkers correlating with increased risk of biochemical recurrence [39], their function appears to be quite varied.

Both the local tumor and systemic microenvironments in PCa can be influenced by extruded MVs that populate and change the microenvironment. EVs, including MVs, appear to have varying effects on the tumor cell population: promoting invasion into and through varied microenvironments [164, 165], metabolic reprogramming [166], conditioning niches for metastatic success [167], inducing a form of immune suppression by preventing the

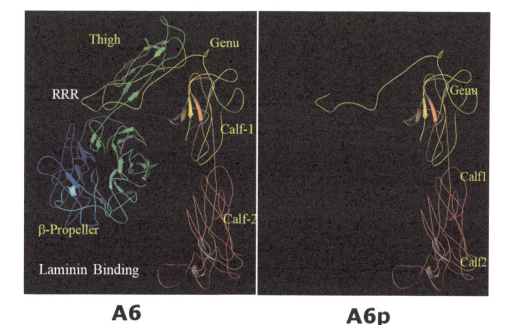

Fig. 4 Model of the α6p integrin variant. The α6 integrin extracellular domain (left panel) is shown, based on the structure of the Av integrin and contains a β-propeller region that is responsible for laminin binding. Mass spectrometry analysis of the α6p form revealed that the NH$_2$ terminal end of α6p integrin contained at least the 'genu' region and part of an exposed loop in the thigh domain, while amino acid residues corresponding to the β-propeller and most of the "thigh" region are not detectable. The location of the conserved RRR amino acids required for cleavage and production of α6p are shown

infiltration of immune cells [168], or providing surrogates of tumor progression [169]. EVs appear to be paracrine transcriptional activators since the nuclear transport of epidermal growth factor receptor in PCa cells occurs via EVs [170]. The role of EVs in cell-to-cell communication [171] has been observed in tumors subjected to therapies that are known to result in "bystander effects" such as cisplatin, ionizing radiation, and localized hyperthermia [172–174]. For example, cisplatin induces the release of EVs from epithelial cancer cells, which can prompt invasiveness and drug resistance in bystander cells [172]. Taken together these studies suggest that the EVs can reflect biological phenotypes of the tumor and represent a biologically active cell-cell communication tool.

The ability of extracellular vesicles and their contents to be acquired by the tumor cells may be a useful way to modify tumor behavior by selective cargo transfer [175]. The delivery of bioactive cargo opens the possibility for tumor cells to be modified for therapeutic purposes. The routes and mechanisms of EV uptake include both clathrin-dependent endocytosis pathways and clathrin-independent pathways such as micropinocytosis, phagocytosis, and lipid raft-mediated internalization [reviewed in [176]]. It is likely that the routes will be dependent upon the cell type and the type of vesicle. For example, EV uptake depends upon clathrin-based endocytosis in endothelial cells [177], whereas EVs from epithelial cancer cells depend upon clathrin-independent endocytosis and on Na^+/H^+ exchange and phosphoinositide 3-kinase activity, which are important for micropinocytosis [175].

The natural role of EVs in encapsulating and delivering cargo to modify cellular functions highlights the potential of these particles as therapeutic delivery vehicles [178]. In PCa, microvesicles have been used to enhance the cytotoxicity of paclitaxel in the LNCaP and PC3 cell lines. Importantly, EVs that were not loaded with paclitaxel increased cancer cell viability [179]. Advances in specific targeting of MVs likely will be required for effective delivery of the selective cargo and is an active area of emerging strategies for EV engineering [178].

While we know that the EVs are biologically active entities, and that their release can be stimulated by cytotoxic agents such as cisplatin [172], it is also important to know the physiologically relevant trigger(s) for EV production. This is especially true if the goal is to determine if EVs can be used to distinguish indolent disease that has not escaped the gland from aggressive disease that is escaping the gland at the time of diagnosis.

In vitro, hypoxia induces EVs in PCa cell lines [180]. In humans, hypoxia is a physiologically relevant microenvironment in the prostate gland that can be detected by immunohistochemical labeling using pimonidazole of radical prostatectomy samples [181] as well as non-invasive ^{18}F-fluoromisonidazole PET/CT of patients prior to radiotherapy treatment [88]. In these studies, tumor hypoxia was associated with radioresistance and poor prognosis after radiation therapy. More recently, magnetic resonance (MR) imaging, with an integration algorithm of oxygen consumption and supply, detected tumor hypoxia that correlated with increased stage and lymph node status [93]. Taken together, these studies suggest that hypoxia can be detected in the prostate gland using non-invasive imaging techniques, correlates with aggressive disease, and produces EVs released into the circulation. Future studies are needed to determine if a tractable EV signature exists that could serve as a biomarker of hypoxia and/or aggressive disease to distinguish indolent from aggressive disease at diagnosis.

Summary

Prostate tumors have the capacity to evolve and adapt to extreme conditions by utilizing complex adaptive systems [182]. We have examined some of the dynamic reciprocity between tumors and their microenvironments [reviewed in [47]], and this highlights the need for an increased and more comprehensive understanding of the environmental cues encountered during PCa invasion into and through physiologically relevant environments. The more we understand the microenvironmental context of tumor development, growth, and metastasis, the greater the possibility for developing new and innovative treatments to prevent aggressive disease. We note that this approach would have the advantage of avoiding the selection of more aggressive disease inherent in survival-based therapeutics.

References

1. Society AC, *Cancer Facts & Figures 2018* (Program SaHSR, Atlanta, GA, 2018). Report No. 500818
2. R.L. Siegel, K.D. Miller, A. Jemal, Cancer statistics, 2017. Cancer J. Clinicians **67**(1), 7–30 (2017). https://doi.org/10.3322/caac.21387. PubMed PMID: 28055103
3. S.H. Au, B.D. Storey, J.C. Moore, Q. Tang, Y.L. Chen, S. Javaid, A.F. Sarioglu, R. Sullivan, M.W. Madden, R. O'Keefe, D.A. Haber, S. Maheswaran, D.M. Langenau, S.L. Stott, M. Toner, Clusters of circulating tumor cells traverse capillary-sized vessels. Proc. Natl. Acad. Sci. U. S. A. **113**, 4947 (2016). https://doi.org/10.1073/pnas.1524448113. PubMed PMID: 27091969
4. G. Gundem, P. Van Loo, B. Kremeyer, L.B. Alexandrov, J.M. Tubio, E. Papaemmanuil, D.S. Brewer, H.M. Kallio, G. Hognas, M. Annala, K. Kivinummi, V. Goody, C. Latimer, S. O'Meara, K.J. Dawson, W. Isaacs, M.R. Emmert-Buck, M. Nykter, C. Foster, Z. Kote-Jarai, D. Easton, H.C. Whitaker, D.E. Neal, C.S. Cooper, R.A. Eeles, T. Visakorpi, P.J. Campbell, U. McDermott, D.C. Wedge, G.S. Bova, The evolutionary history of lethal metastatic prostate cancer. Nature **520**(7547), 353–357 (2015). https://doi.org/10.1038/nature14347. PubMed PMID: 25830880; PubMed Central PMCID: PMC4413032
5. A.B. Weiner, R.S. Matulewicz, S.E. Eggener, E.M. Schaeffer, Increasing incidence of metastatic prostate cancer in the United States (2004-2013). Prostate Cancer Prostatic Dis. **19**(4), 395–397 (2016). https://doi.org/10.1038/pcan.2016.30. PubMed PMID: 27431496
6. A.W. Partin, M.W. Kattan, E.N. Subong, P.C. Walsh, K.J. Wojno, J.E. Oesterling, P.T. Scardino, J.D. Pearson, Combination of prostate-specific antigen, clinical stage, and Gleason score to predict pathological stage of localized prostate cancer. A multi-institutional update. JAMA **277**(18), 1445–1451 (1997). PubMed PMID: 9145716
7. R. Knuchel, Gleason score 6 - prostate cancer or benign variant? Oncol. Res. Treat. **38**(12), 629–632 (2015). https://doi.org/10.1159/000441735. PubMed PMID: 26633167
8. W. Jackson, D.A. Hamstra, S. Johnson, J. Zhou, B. Foster, C. Foster, D. Li, Y. Song, G.S. Palapattu, L.P. Kunju, R. Mehra, F.Y. Feng, Gleason pattern 5 is the strongest pathologic predictor of recurrence, metastasis, and prostate cancer-specific death in patients receiving salvage radiation therapy following radical prostatectomy. Cancer **119**(18), 3287–3294 (2013). https://doi.org/10.1002/cncr.28215. PubMed PMID: 23821578
9. D.E. Spratt, A.I. Cole, G.S. Palapattu, A.Z. Weizer, W.C. Jackson, J.S. Montgomery, R.T. Dess, S.G. Zhao, J.Y. Lee, A. Wu, L.P. Kunju, E. Talmich, D.C. Miller, B.K. Hollenbeck, S.A. Tomlins, F.Y. Feng, R. Mehra, T.M. Morgan, Independent surgical validation of the new prostate cancer grade-grouping system. BJU Int. **118**(5), 763–769 (2016). https://doi.org/10.1111/bju.13488. PubMed PMID: 27009882
10. C. Quinello, L. Souza Ferreira, I. Picolli, M.L. Loesch, D.L. Portuondo, A. Batista-Duharte, I. Zeppone Carlos, Sporothrix schenckii cell wall proteins-stimulated BMDCs are able to induce a Th1-prone cytokine profile in vitro. J. Fungi **4**(3), E106 (2018). https://doi.org/10.3390/jof4030106. PubMed PMID: 30200530; PubMed Central PMCID: PMC6162427
11. A. Latosinska, M. Frantzi, A.S. Merseburger, H. Mischak, Promise and implementation of proteomic prostate cancer biomarkers. Diagnostics **8**(3), E57 (2018). https://doi.org/10.3390/diagnostics8030057. PubMed PMID: 30158500; PubMed Central PMCID: PMC6174350
12. R.J. Hendriks, I.M. van Oort, J.A. Schalken, Blood-based and urinary prostate cancer biomarkers: a review and comparison of novel biomarkers for detection and treatment decisions. Prostate Cancer Prostatic Dis. **20**(1), 12–19 (2017). https://doi.org/10.1038/pcan.2016.59. PubMed PMID: 27922627
13. M.G. Sanda, Z. Feng, D.H. Howard, S.A. Tomlins, L.J. Sokoll, D.W. Chan, M.M. Regan, J. Groskopf, J. Chipman, D.H. Patil, S.S. Salami, D.S. Scherr, J. Kagan, S. Srivastava, I.M. Thompson Jr., J. Siddiqui, J. Fan, A.Y. Joon, L.E. Bantis, M.A. Rubin, A.M. Chinnaiyan, J.T. Wei, M. Bidair, A. Kibel, D.W. Lin, Y. Lotan, A. Partin, S. Taneja, Association between combined TMPRSS2:ERG and PCA3 RNA urinary testing and detection of aggressive prostate cancer. JAMA Oncol. **3**(8), 1085–1093 (2017). https://doi.org/10.1001/jamaoncol.2017.0177. PubMed PMID: 28520829; PubMed Central PMCID: PMC5710334
14. J. Gordetsky, J. Epstein, Grading of prostatic adenocarcinoma: current state and prognostic implications. Diagn. Pathol. **11**, 25 (2016). https://doi.org/10.1186/s13000-016-0478-2. PubMed PMID: 26956509; PubMed Central PMCID: PMC4784293

15. M. Kim, I. Tagkopoulos, Data integration and predictive modeling methods for multi-omics datasets. Mol. Omics **14**(1), 8–25 (2018). https://doi.org/10.1039/c7mo00051k. PubMed PMID: 29725673

16. W.A. Sakr, D.J. Grignon, G.P. Haas, L.K. Heilbrun, J.E. Pontes, J.D. Crissman, Age and racial distribution of prostatic intraepithelial neoplasia. Eur. Urol. **30**(2), 138–144 (1996). PubMed PMID: 8875194

17. H. Bonkhoff, K. Remberger, Morphogenetic aspects of normal and abnormal prostatic growth. Pathol. Res. Pract. **191**(9), 833–835 (1995). PubMed PMID: 8606860

18. R.B. Nagle, A.E. Cress, Metastasis update: human prostate carcinoma invasion via tubulogenesis. Prostate Cancer. **2011**, 249290 (2011). https://doi.org/10.1155/2011/249290. PubMed PMID: 21949592; PubMed Central PMCID: 3177701

19. R. Montironi, R. Mazzucchelli, M. Scarpelli, Precancerous lesions and conditions of the prostate: from morphological and biological characterization to chemoprevention. Ann. N. Y. Acad. Sci. **963**, 169–184 (2002). PubMed PMID: 12095942

20. Z. Zhang, Z. Yang, S. Jaamaa, H. Liu, L.G. Pellakuru, T. Iwata, T.M. af Hallstrom, A.M. De Marzo, M. Laiho, Differential epithelium DNA damage response to ATM and DNA-PK pathway inhibition in human prostate tissue culture. Cell Cycle **10**(20), 3545–3553 (2011). https://doi.org/10.4161/cc.10.20.17841. PubMed PMID: 22030624

21. B.R. Adams, A.J. Hawkins, L.F. Povirk, K. Valerie, ATM-independent, high-fidelity nonhomologous end joining predominates in human embryonic stem cells. Aging **2**(9), 582–596 (2010). PubMed PMID: 20844317; PubMed Central PMCID: 2984607

22. H. Fung, D.M. Weinstock, Repair at single targeted DNA double-strand breaks in pluripotent and differentiated human cells. PLoS One **6**(5), e20514 (2011). https://doi.org/10.1371/journal.pone.0020514. PubMed PMID: 21633706; PubMed Central PMCID: 3102116

23. W.T. Lu, K. Lemonidis, R.M. Drayton, T. Nouspikel, The Fanconi anemia pathway is downregulated upon macrophage differentiation through two distinct mechanisms. Cell Cycle **10**(19), 3300–3310 (2011). https://doi.org/10.4161/cc.10.19.17178. PubMed PMID: 21926477

24. K. Naka, A. Hirao, Maintenance of genomic integrity in hematopoietic stem cells. Int. J. Hematol. **93**(4), 434–439 (2011). https://doi.org/10.1007/s12185-011-0793-z. PubMed PMID: 21384097

25. T. Rozario, D.W. DeSimone, The extracellular matrix in development and morphogenesis: a dynamic view. Dev. Biol. **341**(1), 126–140 (2010). https://doi.org/10.1016/j.ydbio.2009.10.026. S0012-1606(09)01285-8 [pii]. PubMed PMID: 19854168; PubMed Central PMCID: 2854274

26. A.A. Thomson, P.C. Marker, Branching morphogenesis in the prostate gland and seminal vesicles. Differentiation **74**(7), 382–392 (2006). https://doi.org/10.1111/j.1432-0436.2006.00101.x. S0301-4681(09)60225-5 [pii]. PubMed PMID: 16916376

27. S.J. Assinder, Q. Dong, Z. Kovacevic, D.R. Richardson, The TGF-beta, PI3K/Akt and PTEN pathways: established and proposed biochemical integration in prostate cancer. Biochem. J. **417**(2), 411–421 (2009). https://doi.org/10.1042/BJ20081610. PubMed PMID: 19099539

28. D. Bello-DeOcampo, D.J. Tindall, TGF-beta1/Smad signaling in prostate cancer. Curr. Drug Targets **4**(3), 197–207 (2003). PubMed PMID: 12643470

29. S. Kambhampati, G. Ray, K. Sengupta, V.P. Reddy, S.K. Banerjee, P.J. Van Veldhuizen, Growth factors involved in prostate carcinogenesis. Front. Biosci. **10**, 1355–1367 (2005). PubMed PMID: 15769631

30. M. Mimeault, S.K. Batra, Recent advances on multiple tumorigenic cascades involved in prostatic cancer progression and targeting therapies. Carcinogenesis **27**(1), 1–22 (2006). https://doi.org/10.1093/carcin/bgi229. PubMed PMID: 16195239

31. D.G. Bostwick, Progression of prostatic intraepithelial neoplasia to early invasive adenocarcinoma. Eur. Urol. **30**(2), 145–152 (1996). PubMed PMID: 8875195

32. M. Wang, R.B. Nagle, B.S. Knudsen, G.C. Rogers, A.E. Cress, A basal cell defect promotes budding of prostatic intraepithelial neoplasia. J. Cell Sci. **130**(1), 104–110 (2017). https://doi.org/10.1242/jcs.188177. PubMed PMID: 27609833; PubMed Central PMCID: PMC5394777

33. R.B. Nagle, J.D. Knox, C. Wolf, G.T. Bowden, A.E. Cress, Adhesion molecules, extracellular matrix, and proteases in prostate carcinoma. J. Cell. Biochem. Suppl. **19**, 232–237 (1994). PubMed PMID: 7823596

34. D.E. Discher, P. Janmey, Y.L. Wang, Tissue cells feel and respond to the stiffness of their substrate. Science **310**(5751), 1139–1143 (2005). https://doi.org/10.1126/science.1116995. PubMed PMID: 16293750

35. E. You, Y.H. Huh, A. Kwon, S.H. Kim, I.H. Chae, O.J. Lee, J.H. Ryu, M.H. Park, G.E. Kim, J.S. Lee, K.H. Lee, Y.S. Lee, J.W. Kim, S. Rhee, W.K. Song, SPIN90 depletion and microtubule acetylation mediate stromal fibroblast activation in breast cancer progression. Cancer Res. **77**(17), 4710–4722 (2017). https://doi.org/10.1158/0008-5472.Can-17-0657. PubMed PMID: 28652253

36. S. Eggener, The role of magnetic resonance image guided prostate biopsy in stratifying men for risk of extracapsular extension at radical prostatectomy. Raskolnikov D, George AK, Rais-Bahrami S, Turkbey B, Siddiqui MM, Shakir NA, Okoro C, Rothwax JT, Walton-Diaz A, Sankineni S, Su D, Stamatakis L, Merino MJ, Choyke PL, Wood BJ, Pinto PA. J Urol. 2015;194(1):105-11. Urol. Oncol. **35**(3), 121 (2017). https://doi.org/10.1016/j.urolonc.2016.12.013. PubMed PMID: 28159491

37. A.D. De Vivar, M. Sayeeduddin, D. Rowley, A. Cubilla, B. Miles, D. Kadmon, G. Ayala,

Histologic features of stromogenic carcinoma of the prostate (carcinomas with reactive stroma grade 3). Hum. Pathol. **63**, 202–211 (2017). https://doi.org/10.1016/j.humpath.2017.02.019. PubMed PMID: 28315427

38. I.C. Sroka, T.A. Anderson, K.M. McDaniel, R.B. Nagle, M.B. Gretzer, A.E. Cress, The laminin binding integrin α6β1 in prostate cancer perineural invasion. J. Cell. Physiol. **224**(2), 283–288 (2010). https://doi.org/10.1002/jcp.22149. PubMed PMID: 20432448; PubMed Central PMCID: PMC4816210

39. Y.H. Park, H.W. Shin, A.R. Jung, O.S. Kwon, Y.J. Choi, J. Park, J.Y. Lee, Prostate-specific extracellular vesicles as a novel biomarker in human prostate cancer. Sci. Rep. **6**, 30386 (2016). https://doi.org/10.1038/srep30386. PubMed PMID: 27503267; PubMed Central PMCID: PMC4977541

40. C.N. Biggs, K.M. Siddiqui, A.A. Al-Zahrani, S. Pardhan, S.I. Brett, Q.Q. Guo, J. Yang, P. Wolf, N.E. Power, P.N. Durfee, C.D. MacMillan, J.L. Townson, J.C. Brinker, N.E. Fleshner, J.I. Izawa, A.F. Chambers, J.L. Chin, H.S. Leong, Prostate extracellular vesicles in patient plasma as a liquid biopsy platform for prostate cancer using nanoscale flow cytometry. Oncotarget **7**(8), 8839–8849 (2016). https://doi.org/10.18632/oncotarget.6983. PubMed PMID: 26814433; PubMed Central PMCID: PMC4891008

41. I. Eke, A.Y. Makinde, M.J. Aryankalayil, J.L. Reedy, D.E. Citrin, S. Chopra, M.M. Ahmed, C.N. Coleman, Long-term tumor adaptation after radiotherapy: therapeutic implications for targeting integrins in prostate cancer. Mol. Cancer Res. **16**(12), 1855–1864 (2018). https://doi.org/10.1158/1541-7786.Mcr-18-0232. PubMed PMID: 30042176; PubMed Central PMCID: PMC6279542

42. W.W. Tan, Prostate Cancer Treatment Protocols [web page]. Medscape; 2018 [updated May 25, 2018; cited 2018 November 4]. Available from: https://emedicine.medscape.com/article/2007095-overview

43. E.S. Antonarakis, C. Lu, H. Wang, B. Luber, M. Nakazawa, J.C. Roeser, Y. Chen, T.A. Mohammad, Y. Chen, H.L. Fedor, T.L. Lotan, Q. Zheng, A.M. De Marzo, J.T. Isaacs, W.B. Isaacs, R. Nadal, C.J. Paller, S.R. Denmeade, M.A. Carducci, M.A. Eisenberger, J. Luo, AR-V7 and resistance to enzalutamide and abiraterone in prostate cancer. N. Engl. J. Med. **371**(11), 1028–1038 (2014). https://doi.org/10.1056/NEJMoa1315815. PubMed PMID: 25184630; PubMed Central PMCID: PMC4201502

44. H.I. Scher, K. Fizazi, F. Saad, M.E. Taplin, C.N. Sternberg, K. Miller, R. de Wit, P. Mulders, K.N. Chi, N.D. Shore, A.J. Armstrong, T.W. Flaig, A. Flechon, P. Mainwaring, M. Fleming, J.D. Hainsworth, M. Hirmand, B. Selby, L. Seely, J.S. de Bono, Increased survival with enzalutamide in prostate cancer after chemotherapy. N. Engl. J. Med. **367**(13), 1187–1197 (2012). https://doi.org/10.1056/NEJMoa1207506. PubMed PMID: 22894553

45. J.S. de Bono, C.J. Logothetis, A. Molina, K. Fizazi, S. North, L. Chu, K.N. Chi, R.J. Jones, O.B. Goodman Jr., F. Saad, J.N. Staffurth, P. Mainwaring, S. Harland, T.W. Flaig, T.E. Hutson, T. Cheng, H. Patterson, J.D. Hainsworth, C.J. Ryan, C.N. Sternberg, S.L. Ellard, A. Fléchon, M. Saleh, M. Scholz, E. Efstathiou, A. Zivi, D. Bianchini, Y. Loriot, N. Chieffo, T. Kheoh, C.M. Haqq, H.I. Scher, COU-AA-301 Investigators, Abiraterone and increased survival in metastatic prostate cancer. N. Engl. J. Med. **364**(21), 1995–2005 (2011). https://doi.org/10.1056/NEJMoa1014618. PubMed PMID: 21612468; PubMed Central PMCID: PMC3471149

46. M.Y. Teo, D.E. Rathkopf, P. Kantoff, Treatment of advanced prostate cancer. Annu. Rev. Med. **70**, 479–499 (2019). https://doi.org/10.1146/annurev-med-051517-011947. PubMed PMID: 30691365; PubMed Central PMCID: PMC6441973

47. C.M. Nelson, M.J. Bissell, Modeling dynamic reciprocity: engineering three-dimensional culture models of breast architecture, function, and neoplastic transformation. Semin. Cancer Biol. **15**(5), 342–352 (2005). https://doi.org/10.1016/j.semcancer.2005.05.001. PubMed PMID: 15963732; PubMed Central PMCID: PMC2933210

48. M. Wang, N. Janaki, C. Buzzy, L. Bukavina, A. Mahran, K. Mishra, G. MacLennan, L. Ponsky, Whole mount histopathological correlation with prostate MRI in Grade I and II prostatectomy patients. Int. Urol. Nephrol. **51**, 425 (2019). https://doi.org/10.1007/s11255-019-02083-8. PubMed PMID: 30671889

49. S.B. Shappell, G.V. Thomas, R.L. Roberts, R. Herbert, M.M. Ittmann, M.A. Rubin, P.A. Humphrey, J.P. Sundberg, N. Rozengurt, R. Barrios, J.M. Ward, R.D. Cardiff, Prostate pathology of genetically engineered mice: definitions and classification. The consensus report from the Bar Harbor meeting of the Mouse Models of Human Cancer Consortium Prostate Pathology Committee. Cancer Res. **64**(6), 2270–2305 (2004). PubMed PMID: 15026373

50. A. Pozzi, P.D. Yurchenco, R.V. Iozzo, The nature and biology of basement membranes. Matrix Biol. **57-58**, 1–11 (2017). https://doi.org/10.1016/j.matbio.2016.12.009. PubMed PMID: 28040522; PubMed Central PMCID: PMC5387862

51. M.J. Humphries, Integrin structure. Biochem. Soc. Trans. **28**(4), 311–339 (2000). PubMed PMID: 10961914

52. A. Sonnenberg, Integrins and their ligands. Curr. Top. Microbiol. Immunol. **184**, 7–35 (1993). PubMed PMID: 8313723

53. A. van der Flier, A. Sonnenberg, Function and interactions of integrins. Cell Tissue Res. **305**(3), 285–298 (2001). PubMed PMID: 11572082

54. R.O. Hynes, Integrins: bidirectional, allosteric signaling machines. Cell **110**(6), 673–687 (2002). PubMed PMID: 12297042

55. Y. Takada, X. Ye, S. Simon, The integrins. Genome Biol. **8**(5), 215 (2007). https://doi.org/10.1186/gb-2007-8-5-215. PubMed PMID: 17543136; PubMed Central PMCID: PMC1929136

56. A.L. Hughes, Evolution of the integrin alpha and beta protein families. J. Mol. Evol. **52**(1), 63–72 (2001). PubMed PMID: 11139295

57. M. Huhtala, J. Heino, D. Casciari, A. de Luise, M.S. Johnson, Integrin evolution: insights from ascidian and teleost fish genomes. Matrix Biol. **24**(2), 83–95 (2005). https://doi.org/10.1016/j.matbio.2005.01.003. PubMed PMID: 15890260

58. A.E. Cress, I. Rabinovitz, W. Zhu, R.B. Nagle, The alpha 6 beta 1 and alpha 6 beta 4 integrins in human prostate cancer progression. Cancer Metastasis Rev. **14**(3), 219–228 (1995). PubMed PMID: 8548870

59. M. Schmelz, A.E. Cress, K.M. Scott, F. Burger, H. Cui, K. Sallam, K.M. McDaniel, B.L. Dalkin, R.B. Nagle, Different phenotypes in human prostate cancer: alpha6 or alpha3 integrin in cell-extracellular adhesion sites. Neoplasia **4**(3), 243–254 (2002). https://doi.org/10.1038/sj/neo/7900223. PubMed PMID: 11988844; PubMed Central PMCID: 1531698

60. A.M. Mercurio, R.E. Bachelder, I. Rabinovitz, K.L. O'Connor, T. Tani, L.M. Shaw, The metastatic odyssey: the integrin connection. Surg. Oncol. Clin. N. Am. **10**(2), 313–328 (2001). viii-ix. PubMed PMID: 11382589

61. M. Fornaro, T. Manes, L.R. Languino, Integrins and prostate cancer metastases. Cancer Metastasis Rev. **20**(3-4), 321–331 (2001). PubMed PMID: 12085969

62. H.L. Goel, J. Li, S. Kogan, L.R. Languino, Integrins in prostate cancer progression. Endocr. Relat. Cancer **15**(3), 657–664 (2008). https://doi.org/10.1677/ERC-08-0019. ERC-08-0019 [pii]. PubMed PMID: 18524948; PubMed Central PMCID: 2668544

63. J. Liu, P.B. Gurpur, S.J. Kaufman, Genetically determined proteolytic cleavage modulates alpha-7beta1 integrin function. J. Biol. Chem. **283**(51), 35668–35678 (2008). https://doi.org/10.1074/jbc.M804661200. PubMed PMID: 18940796; PubMed Central PMCID: PMC2602887

64. M.O. Ports, R.B. Nagle, G.D. Pond, A.E. Cress, Extracellular engagement of alpha6 integrin inhibited urokinase-type plasminogen activator-mediated cleavage and delayed human prostate bone metastasis. Cancer Res. **69**(12), 5007–5014 (2009). https://doi.org/10.1158/0008-5472.can-09-0354. PubMed PMID: 19491258; PubMed Central PMCID: PMC2697270

65. T.L. Davis, A.E. Cress, B.L. Dalkin, R.B. Nagle, Unique expression pattern of the alpha6beta4 integrin and laminin-5 in human prostate carcinoma. Prostate **46**(3), 240–248 (2001). PubMed PMID: 11170153

66. M.V. Allen, G.J. Smith, R. Juliano, S.J. Maygarden, J.L. Mohler, Downregulation of the beta4 integrin subunit in prostatic carcinoma and prostatic intraepi-thelial neoplasia. Hum. Pathol. **29**(4), 311–318 (1998). PubMed PMID: 9563778

67. R.B. Nagle, J. Hao, J.D. Knox, B.L. Dalkin, V. Clark, A.E. Cress, Expression of hemidesmosomal and extracellular matrix proteins by normal and malignant human prostate tissue. Am. J. Pathol. **146**(6), 1498–1507 (1995). PubMed PMID: 7778688; PubMed Central PMCID: 1870922

68. E. Ricci, E. Mattei, C. Dumontet, C.L. Eaton, F. Hamdy, G. van der Pluije, M. Cecchini, G. Thalmann, P. Clezardin, M. Colombel, Increased expression of putative cancer stem cell markers in the bone marrow of prostate cancer patients is associated with bone metastasis progression. Prostate **73**(16), 1738–1746 (2013). https://doi.org/10.1002/pros.22689. PubMed PMID: 24115186

69. D.A. Lawson, Y. Zong, S. Memarzadeh, L. Xin, J. Huang, O.N. Witte, Basal epithelial stem cells are efficient targets for prostate cancer initiation. Proc. Natl. Acad. Sci. U. S. A. **107**(6), 2610–2615 (2010). https://doi.org/10.1073/pnas.0913873107. PubMed PMID: 20133806; PubMed Central PMCID: PMC2823887

70. M.C. Demetriou, K.A. Kwei, M.B. Powell, R.B. Nagle, G.T. Bowden, A.E. Cress, Integrin A6 cleavage in mouse skin tumors. Open Cancer J. **2**, 1–4 (2008). https://doi.org/10.2174/1874079000802010001. PubMed PMID: 20664806; PubMed Central PMCID: PMC2906811

71. I.C. Sroka, H. Chopra, L. Das, J.M. Gard, R.B. Nagle, A.E. Cress, Schwann cells increase prostate and pancreatic tumor cell invasion using laminin binding A6 integrin. J. Cell. Biochem. **117**(2), 491–499 (2016). https://doi.org/10.1002/jcb.25300. PubMed PMID: 26239765; PubMed Central PMCID: PMC4809241

72. I.C. Sroka, C.P. Sandoval, H. Chopra, J.M. Gard, S.C. Pawar, A.E. Cress, Macrophage-dependent cleavage of the laminin receptor alpha6beta1 in prostate cancer. Mol. Cancer Res. **9**(10), 1319–1328 (2011). https://doi.org/10.1158/1541-7786.mcr-11-0080. PubMed PMID: 21824975; PubMed Central PMCID: PMC3196809

73. T.E. King, S.C. Pawar, L. Majuta, I.C. Sroka, D. Wynn, M.C. Demetriou, R.B. Nagle, F. Porreca, A.E. Cress, The role of alpha 6 integrin in prostate cancer migration and bone pain in a novel xenograft model. PLoS One **3**(10), e3535 (2008). https://doi.org/10.1371/journal.pone.0003535. PubMed PMID: 18958175; PubMed Central PMCID: PMC2570216

74. T.H. Landowski, J. Gard, E. Pond, G.D. Pond, R.B. Nagle, C.P. Geffre, A.E. Cress, Targeting integrin alpha6 stimulates curative-type bone metastasis lesions in a xenograft model. Mol. Cancer Ther. **13**(6), 1558–1566 (2014). https://doi.org/10.1158/1535-7163.mct-13-0962. PubMed PMID: 24739392; PubMed Central PMCID: PMC4069206

75. S.C. Pawar, S. Dougherty, M.E. Pennington, M.C. Demetriou, B.D. Stea, R.T. Dorr, A.E. Cress, alpha6 integrin cleavage: sensitizing human prostate cancer to ionizing radiation. Int. J. Radiat.

Biol. **83**(11-12), 761–767 (2007). https://doi.org/10.1080/09553000701633135. 787782677 [pii]. PubMed PMID: 18058365; PubMed Central PMCID: 2732343

76. M.F. Emmons, A.W. Gebhard, R.R. Nair, R. Baz, M.L. McLaughlin, A.E. Cress, L.A. Hazlehurst, Acquisition of resistance toward HYD1 correlates with a reduction in cleaved alpha4 integrin expression and a compromised CAM-DR phenotype. Mol. Cancer Ther. **10**(12), 2257–2266 (2011). https://doi.org/10.1158/1535-7163.mct-11-0149. PubMed PMID: 21980133; PubMed Central PMCID: PMC3237739

77. W.L. Harryman, J.P. Hinton, C.P. Rubenstein, P. Singh, R.B. Nagle, S.J. Parker, B.S. Knudsen, A.E. Cress, The cohesive metastasis phenotype in human prostate cancer. Biochim. Biophys. Acta **1866**(2), 221–231 (2016). https://doi.org/10.1016/j.bbcan.2016.09.005. PubMed PMID: 27678419

78. M.R. Clay, D.R. Sherwood, Basement membranes in the worm: a dynamic scaffolding that instructs cellular behaviors and shapes tissues. Curr. Top. Membr. **76**, 337–371 (2015). https://doi.org/10.1016/bs.ctm.2015.08.001. PubMed PMID: 26610919; PubMed Central PMCID: PMC4697865

79. J. McCandless, A. Cress, I. Rabinovitz, C. Payne, G. Bowden, J. Knox, R. Nagle, A human xenograft model for testing early events of epithelial neoplastic invasion. Int. J. Oncol. **10**(2), 279–285 (1997). PubMed PMID: 21533373; PubMed Central PMCID: PMC5390482

80. T. Welz, J. Wellbourne-Wood, E. Kerkhoff, Orchestration of cell surface proteins by Rab11. Trends Cell Biol. **24**(7), 407–415 (2014). https://doi.org/10.1016/j.tcb.2014.02.004. PubMed PMID: 24675420

81. N.W. Baetz, J.R. Goldenring, Rab11-family interacting proteins define spatially and temporally distinct regions within the dynamic Rab11a-dependent recycling system. Mol. Biol. Cell **24**(5), 643–658 (2013). https://doi.org/10.1091/mbc.E12-09-0659. PubMed PMID: 23283983; PubMed Central PMCID: PMC3583667

82. P.T. Caswell, S. Vadrevu, J.C. Norman, Integrins: masters and slaves of endocytic transport. Nat. Rev. Mol. Cell Biol. **10**(12), 843–853 (2009). https://doi.org/10.1038/nrm2799. PubMed PMID: 19904298

83. S.O. Yoon, S. Shin, A.M. Mercurio, Hypoxia stimulates carcinoma invasion by stabilizing microtubules and promoting the Rab11 trafficking of the alpha-6beta4 integrin. Cancer Res. **65**(7), 2761–2769 (2005). https://doi.org/10.1158/0008-5472.can-04-4122. PubMed PMID: 15805276

84. C.P. Horgan, M.W. McCaffrey, The dynamic Rab11-FIPs. Biochem. Soc. Trans. **37**(Pt 5), 1032–1036 (2009). https://doi.org/10.1042/bst0371032. PubMed PMID: 19754446

85. B. Movsas, J.D. Chapman, R.E. Greenberg, A.L. Hanlon, E.M. Horwitz, W.H. Pinover, C. Stobbe, G.E. Hanks, Increasing levels of hypoxia in prostate carcinoma correlate significantly with increasing clinical stage and patient age: an Eppendorf pO(2) study. Cancer **89**(9), 2018–2024 (2000). PubMed PMID: 11064360

86. R. Vergis, C.M. Corbishley, A.R. Norman, J. Bartlett, S. Jhavar, M. Borre, S. Heebøll, A. Horwich, R. Huddart, V. Khoo, R. Eeles, C. Cooper, M. Sydes, D. Dearnaley, C. Parker, Intrinsic markers of tumour hypoxia and angiogenesis in localised prostate cancer and outcome of radical treatment: a retrospective analysis of two randomised radiotherapy trials and one surgical cohort study. Lancet Oncol. **9**(4), 342–351 (2008). https://doi.org/10.1016/s1470-2045(08)70076-7. PubMed PMID: 18343725

87. G.D. Stewart, K. Gray, C.J. Pennington, D.R. Edwards, A.C. Riddick, J.A. Ross, F.K. Habib, Analysis of hypoxia-associated gene expression in prostate cancer: lysyl oxidase and glucose transporter-1 expression correlate with Gleason score. Oncol. Rep. **20**(6), 1561–1567 (2008). PubMed PMID: 19020742

88. S. Supiot, C. Rousseau, M. Dore, C. Cheze-Le-Rest, C. Kandel-Aznar, V. Potiron, S. Guerif, F. Paris, L. Ferrer, L. Campion, P. Meingan, G. Delpon, M. Hatt, D. Visvikis, Evaluation of tumor hypoxia prior to radiotherapy in intermediate-risk prostate cancer using (18)F-fluoromisonidazole PET/CT: a pilot study. Oncotarget **9**(11), 10005–10015 (2018). https://doi.org/10.18632/oncotarget.24234. PubMed PMID: 29515786; PubMed Central PMCID: PMC5839367

89. M.R. Horsman, L.S. Mortensen, J.B. Petersen, M. Busk, J. Overgaard, Imaging hypoxia to improve radiotherapy outcome. Nat. Rev. Clin. Oncol. **9**(12), 674–687 (2012). https://doi.org/10.1038/nrclinonc.2012.171. PubMed PMID: 23149893

90. E.E. Parent, D.M. Schuster, Update on (18)F-fluciclovine PET for prostate cancer imaging. J. Nucl. Med. **59**(5), 733–739 (2018). https://doi.org/10.2967/jnumed.117.204032. PubMed PMID: 29523631

91. M. Milosevic, P. Warde, C. Menard, P. Chung, A. Toi, A. Ishkanian, M. McLean, M. Pintilie, J. Sykes, M. Gospodarowicz, C. Catton, R.P. Hill, R. Bristow, Tumor hypoxia predicts biochemical failure following radiotherapy for clinically localized prostate cancer. Clin. Cancer Res. **18**(7), 2108–2114 (2012). https://doi.org/10.1158/1078-0432.Ccr-11-2711. PubMed PMID: 22465832

92. M. Milosevic, P. Chung, C. Parker, R. Bristow, A. Toi, T. Panzarella, P. Warde, C. Catton, C. Menard, A. Bayley, M. Gospodarowicz, R. Hill, Androgen withdrawal in patients reduces prostate cancer hypoxia: implications for disease progression and radiation response. Cancer Res. **67**(13), 6022–6025 (2007). https://doi.org/10.1158/0008-5472.Can-07-0561. PubMed PMID: 17616657

93. T. Hompland, K.H. Hole, H.B. Ragnum, E.K. Aarnes, L. Vlatkovic, A.K. Lie, S. Patzke, B. Brennhovd, T. Seierstad, H. Lyng, Combined MR

imaging of oxygen consumption and supply reveals tumor hypoxia and aggressiveness in prostate cancer patients. Cancer Res. **78**(16), 4774–4785 (2018). https://doi.org/10.1158/0008-5472.Can-17-3806. PubMed PMID: 29945958

94. L. Yang, D. Roberts, M. Takhar, N. Erho, B.A.S. Bibby, N. Thiruthaneeswaran, V. Bhandari, W.C. Cheng, S. Haider, A.M.B. McCorry, D. McArt, S. Jain, M. Alshalalfa, A. Ross, E. Schaffer, R.B. Den, R. Jeffrey Karnes, E. Klein, P.J. Hoskin, S.J. Freedland, A.D. Lamb, D.E. Neal, F.M. Buffa, R.G. Bristow, P.C. Boutros, E. Davicioni, A. Choudhury, C.M.L. West, Development and validation of a 28-gene hypoxia-related prognostic signature for localized prostate cancer. EBioMedicine **31**, 182–189 (2018). https://doi.org/10.1016/j.ebiom.2018.04.019. PubMed PMID: 29729848; PubMed Central PMCID: PMC6014579

95. N. Grivas, A. Goussia, D. Stefanou, D. Giannakis, Microvascular density and immunohistochemical expression of VEGF, VEGFR-1 and VEGFR-2 in benign prostatic hyperplasia, high-grade prostate intraepithelial neoplasia and prostate cancer. C. Eur. J. Urol. **69**(1), 63–71 (2016). https://doi.org/10.5173/ceju.2016.726. PubMed PMID: 27123329; PubMed Central PMCID: PMC4846728

96. D.C. Weber, J.C. Tille, C. Combescure, J.F. Egger, M. Laouiti, K. Hammad, P. Granger, L. Rubbia-Brandt, R. Miralbell, The prognostic value of expression of HIF1alpha, EGFR and VEGF-A, in localized prostate cancer for intermediate- and high-risk patients treated with radiation therapy with or without androgen deprivation therapy. Rad. Oncol. **7**, 66 (2012). https://doi.org/10.1186/1748-717x-7-66. PubMed PMID: 22546016; PubMed Central PMCID: PMC3432017

97. A.C. Small, W.K. Oh, Bevacizumab treatment of prostate cancer. Expert Opin. Biol. Ther. **12**(9), 1241–1249 (2012). https://doi.org/10.1517/1471259 8.2012.704015. PubMed PMID: 22775507

98. I.F. Tannock, K. Fizazi, S. Ivanov, C.T. Karlsson, A. Flechon, I. Skoneczna, F. Orlandi, G. Gravis, V. Matveev, S. Bavbek, T. Gil, L. Viana, O. Aren, O. Karyakin, T. Elliott, A. Birtle, E. Magherini, L. Hatteville, D. Petrylak, B. Tombal, M. Rosenthal, Aflibercept versus placebo in combination with docetaxel and prednisone for treatment of men with metastatic castration-resistant prostate cancer (VENICE): a phase 3, double-blind randomised trial. Lancet Oncol. **14**(8), 760–768 (2013). https://doi.org/10.1016/s1470-2045(13)70184-0. PubMed PMID: 23742877

99. N.A. Warfel, A.G. Sainz, J.H. Song, A.S. Kraft, PIM kinase inhibitors kill hypoxic tumor cells by reducing Nrf2 signaling and increasing reactive oxygen species. Mol. Cancer Ther. **15**(7), 1637–1647 (2016). https://doi.org/10.1158/1535-7163.Mct-15-1018. PubMed PMID: 27196781; PubMed Central PMCID: PMC4936950

100. M. Hara, T. Nagasaki, K. Shiga, H. Takahashi, H. Takeyama, High serum levels of interleukin-6 in patients with advanced or metastatic colorectal cancer: the effect on the outcome and the response to chemotherapy plus bevacizumab. Surg. Today **47**(4), 483–489 (2017). https://doi.org/10.1007/s00595-016-1404-7. PubMed PMID: 27549777

101. H.L. Goel, B. Pursell, L.D. Shultz, D.L. Greiner, R.A. Brekken, C.W. Vander Kooi, A.M. Mercurio, P-Rex1 promotes resistance to VEGF/VEGFR-targeted therapy in prostate cancer. Cell Rep. **14**(9), 2193–2208 (2016). https://doi.org/10.1016/j.celrep.2016.02.016. PubMed PMID: 26923603; PubMed Central PMCID: PMC4791963

102. A.L. Casillas, R.K. Toth, A.G. Sainz, N. Singh, A.A. Desai, A.S. Kraft, N.A. Warfel, Hypoxia-inducible PIM kinase expression promotes resistance to antiangiogenic agents. Clin. Cancer Res. **24**(1), 169–180 (2018). https://doi.org/10.1158/1078-0432. Ccr-17-1318. PubMed PMID: 29084916; PubMed Central PMCID: PMC6214353

103. J.W. Franses, A.B. Baker, V.C. Chitalia, E.R. Edelman, Stromal endothelial cells directly influence cancer progression. Sci. Transl. Med. **3**(66), 66ra5 (2011). https://doi.org/10.1126/scitranslmed.3001542. PubMed PMID: 21248315; PubMed Central PMCID: PMC3076139

104. R. Schietke, C. Warnecke, I. Wacker, J. Schodel, D.R. Mole, V. Campean, K. Amann, M. Goppelt-Struebe, J. Behrens, K.U. Eckardt, M.S. Wiesener, The lysyl oxidases LOX and LOXL2 are necessary and sufficient to repress E-cadherin in hypoxia: insights into cellular transformation processes mediated by HIF-1. J. Biol. Chem. **285**(9), 6658–6669 (2010). https://doi.org/10.1074/jbc.M109.042424. PubMed PMID: 20026874; PubMed Central PMCID: PMC2825461

105. D.R. Hurst, D.R. Welch, Metastasis suppressor genes at the interface between the environment and tumor cell growth. Int. Rev. Cell Mol. Biol. **286**, 107–180 (2011). https://doi.org/10.1016/b978-0-12-385859-7.00003-3. PubMed PMID: 21199781; PubMed Central PMCID: PMC3575029

106. M.A. Rubin, N.R. Mucci, J. Figurski, A. Fecko, K.J. Pienta, M.L. Day, E-cadherin expression in prostate cancer: a broad survey using high-density tissue microarray technology. Hum. Pathol. **32**(7), 690–697 (2001). https://doi.org/10.1053/hupa.2001.25902. PubMed PMID: 11486167

107. L. Fan, H. Wang, X. Xia, Y. Rao, X. Ma, D. Ma, P. Wu, G. Chen, Loss of E-cadherin promotes prostate cancer metastasis via upregulation of metastasis-associated gene 1 expression. Oncol. Lett. **4**(6), 1225–1233 (2012). https://doi.org/10.3892/ol.2012.934. PubMed PMID: 23205121; PubMed Central PMCID: PMC3506747

108. T.T. Onder, P.B. Gupta, S.A. Mani, J. Yang, E.S. Lander, R.A. Weinberg, Loss of E-cadherin promotes metastasis via multiple downstream transcriptional pathways. Cancer Res. **68**(10), 3645–3654

(2008). https://doi.org/10.1158/0008-5472.Can-07-2938. PubMed PMID: 18483246

109. D. Kong, Y. Li, Z. Wang, S. Banerjee, A. Ahmad, H.R. Kim, F.H. Sarkar, miR-200 regulates PDGF-D-mediated epithelial-mesenchymal transition, adhesion, and invasion of prostate cancer cells. Stem Cells 27(8), 1712–1721 (2009). https://doi.org/10.1002/stem.101. PubMed PMID: 19544444; PubMed Central PMCID: PMC3400149

110. K.A. Pillman, C.A. Phillips, S. Roslan, J. Toubia, B.K. Dredge, A.G. Bert, R. Lumb, D.P. Neumann, X. Li, S.J. Conn, D. Liu, C.P. Bracken, D.M. Lawrence, N. Stylianou, A.W. Schreiber, W.D. Tilley, B.G. Hollier, Y. Khew-Goodall, L.A. Selth, G.J. Goodall, P.A. Gregory, miR-200/375 control epithelial plasticity-associated alternative splicing by repressing the RNA-binding protein Quaking. EMBO J. 37(13), e99016 (2018). https://doi.org/10.15252/embj.201899016. PubMed PMID: 29871889; PubMed Central PMCID: PMC6028027

111. L.E. Lamb, B.S. Knudsen, C.K. Miranti, E-cadherin-mediated survival of androgen-receptor-expressing secretory prostate epithelial cells derived from a stratified in vitro differentiation model. J. Cell Sci. 123(Pt 2), 266–276 (2010). https://doi.org/10.1242/jcs.054502. PubMed PMID: 20048343

112. B.S.M.C. Knudsen, The impact of cell adhesion changes on proliferation and survival during prostate cancer development and progression. J. Cell. Biochem. 99(2), 345–361 (2006). https://doi.org/10.1002/jcb.20934. PubMed PMID: 16676354

113. A.M. De Marzo, B. Knudsen, K. Chan-Tack, J.I. Epstein, E-cadherin expression as a marker of tumor aggressiveness in routinely processed radical prostatectomy specimens. Urology 53(4), 707–713 (1999). PubMed PMID: 10197845

114. J. Pontes-Junior, S.T. Reis, M. Dall'Oglio, L.C. Neves de Oliveira, J. Cury, P.A. Carvalho, L.A. Ribeiro-Filho, K.R. Moreira Leite, M. Srougi, Evaluation of the expression of integrins and cell adhesion molecules through tissue microarray in lymph node metastases of prostate cancer. J. Carcinogen. 8, 3 (2009). PubMed PMID: 19240373; PubMed Central PMCID: PMC2678866

115. P. Friedl, D. Gilmour, Collective cell migration in morphogenesis, regeneration and cancer. Nat. Rev. Mol. Cell Biol. 10(7), 445–457 (2009). https://doi.org/10.1038/nrm2720. PubMed PMID: 19546857

116. E.R. Shamir, A.J. Ewald, Adhesion in mammary development: novel roles for E-cadherin in individual and collective cell migration. Curr. Top. Dev. Biol. 112, 353–382 (2015). https://doi.org/10.1016/bs.ctdb.2014.12.001. PubMed PMID: 25733146; PubMed Central PMCID: PMC4696070

117. R. Mayor, S. Etienne-Manneville, The front and rear of collective cell migration. Nat. Rev. Mol. Cell Biol. 17(2), 97–109 (2016). https://doi.org/10.1038/nrm.2015.14. PubMed PMID: 26726037

118. A.A. Khalil, J. de Rooij, Cadherin mechanotransduction in leader-follower cell specification during collective migration. Exp. Cell Res. 376, 86 (2019). https://doi.org/10.1016/j.yexcr.2019.01.006. PubMed PMID: 30633881

119. S. Lehmann, V. Te Boekhorst, J. Odenthal, R. Bianchi, S. van Helvert, K. Ikenberg, O. Ilina, S. Stoma, J. Xandry, L. Jiang, R. Grenman, M. Rudin, P. Friedl, Hypoxia induces a HIF-1-dependent transition from collective-to-amoeboid dissemination in epithelial cancer cells. Curr. Biol. 27(3), 392–400 (2017). https://doi.org/10.1016/j.cub.2016.11.057. PubMed PMID: 28089517

120. M.F. Penet, S. Kakkad, A.P. Pathak, B. Krishnamachary, Y. Mironchik, V. Raman, M. Solaiyappan, Z.M. Bhujwalla, Structure and function of a prostate cancer dissemination-permissive extracellular matrix. Clin. Cancer Res. 23(9), 2245–2254 (2017). https://doi.org/10.1158/1078-0432.Ccr-16-1516. PubMed PMID: 27799248; PubMed Central PMCID: PMC5411337

121. D. Schveigert, K.P. Valuckas, V. Kovalcis, A. Ulys, G. Chvatovic, J. Didziapetriene, Significance of MMP-9 expression and MMP-9 polymorphism in prostate cancer. Tumori 99(4), 523–529 (2013). https://doi.org/10.1700/1361.15105. PubMed PMID: 24326842

122. M. Wood, K. Fudge, J.L. Mohler, A.R. Frost, F. Garcia, M. Wang, M.E. Stearns, In situ hybridization studies of metalloproteinases 2 and 9 and TIMP-1 and TIMP-2 expression in human prostate cancer. Clin. Exp. Metastasis 15(3), 246–258 (1997). PubMed PMID: 9174126

123. B.L. Lokeshwar, MMP inhibition in prostate cancer. Ann. N. Y. Acad. Sci. 878, 271–289 (1999). PubMed PMID: 10415736

124. P.J. Cozzi, J. Wang, W. Delprado, M.C. Madigan, S. Fairy, P.J. Russell, Y. Li, Evaluation of urokinase plasminogen activator and its receptor in different grades of human prostate cancer. Hum. Pathol. 37(11), 1442–1451 (2006). https://doi.org/10.1016/j.humpath.2006.05.002. PubMed PMID: 16949925

125. D. Trudel, Y. Fradet, F. Meyer, F. Harel, B. Tetu, Significance of MMP-2 expression in prostate cancer: an immunohistochemical study. Cancer Res. 63(23), 8511–8515 (2003). PubMed PMID: 14679018

126. G. Sehgal, J. Hua, E.J. Bernhard, I. Sehgal, T.C. Thompson, R.J. Muschel, Requirement for matrix metalloproteinase-9 (gelatinase B) expression in metastasis by murine prostate carcinoma. Am. J. Pathol. 152(2), 591–596 (1998). PubMed PMID: 9466586; PubMed Central PMCID: PMC1857976

127. C.C. Lynch, A. Hikosaka, H.B. Acuff, M.D. Martin, N. Kawai, R.K. Singh, T.C. Vargo-Gogola, J.L. Begtrup, T.E. Peterson, B. Fingleton, T. Shirai, L.M. Matrisian, M. Futakuchi, MMP-7 promotes prostate cancer-induced osteolysis via the solubilization of RANKL. Cancer Cell 7(5), 485–496 (2005). https://doi.org/10.1016/j.ccr.2005.04.013. PubMed PMID: 15894268

128. M.R. Ambrosio, C. Di Serio, G. Danza, B.J. Rocca, A. Ginori, I. Prudovsky, N. Marchionni, M.T. Del Vecchio, F. Tarantini, Carbonic anhydrase IX is a marker of hypoxia and correlates with higher Gleason scores and ISUP grading in prostate cancer. Diagn. Pathol. **11**(1), 45 (2016). https://doi.org/10.1186/s13000-016-0495-1. PubMed PMID: 27225200; PubMed Central PMCID: PMC4880832

129. V.A. Kobliakov, Role of proton pumps in tumorigenesis. Biochemistry **82**(4), 401–412 (2017). https://doi.org/10.1134/s0006297917040010. PubMed PMID: 28371597

130. Y. Zhang, S. Nojima, H. Nakayama, Y. Jin, H. Enza, Characteristics of normal stromal components and their correlation with cancer occurrence in human prostate. Oncol. Rep. **10**(1), 207–211 (2003). PubMed PMID: 12469170

131. S.M. Kakkad, M. Solaiyappan, B. O'Rourke, I. Stasinopoulos, E. Ackerstaff, V. Raman, Z.M. Bhujwalla, K. Glunde, Hypoxic tumor microenvironments reduce collagen I fiber density. Neoplasia **12**(8), 608–617 (2010). PubMed PMID: 20689755; PubMed Central PMCID: PMC2915405

132. J.A. Ju, I. Godet, I.C. Ye, J. Byun, H. Jayatilaka, S.J. Lee, L. Xiang, D. Samanta, M.H. Lee, P.H. Wu, D. Wirtz, G.L. Semenza, D.M. Gilkes, Hypoxia selectively enhances integrin alpha5beta1 receptor expression in breast cancer to promote metastasis. Mol. Cancer Res. **15**(6), 723–734 (2017). https://doi.org/10.1158/1541-7786.Mcr-16-0338. PubMed PMID: 28213554; PubMed Central PMCID: PMC5510543

133. D.L. Brooks, L.P. Schwab, R. Krutilina, D.N. Parke, A. Sethuraman, D. Hoogewijs, A. Schorg, L. Gotwald, M. Fan, R.H. Wenger, T.N. Seagroves, ITGA6 is directly regulated by hypoxia-inducible factors and enriches for cancer stem cell activity and invasion in metastatic breast cancer models. Mol. Cancer **15**, 26 (2016). https://doi.org/10.1186/s12943-016-0510-x. PubMed PMID: 27001172; PubMed Central PMCID: PMC4802728

134. P. Muppa, S. Gupta, I. Frank, S.A. Boorjian, R.J. Karnes, R.H. Thompson, P. Thapa, R.F. Tarrell, L.P. Herrera Hernandez, R.E. Jimenez, J.C. Cheville, Prognostic significance of lymphatic, vascular and perineural invasion for bladder cancer patients treated by radical cystectomy. Pathology **49**(3), 259–266 (2017). https://doi.org/10.1016/j.pathol.2016.12.347. PubMed PMID: 28259358

135. C. Liebig, G. Ayala, J.A. Wilks, D.H. Berger, D. Albo, Perineural invasion in cancer: a review of the literature. Cancer **115**(15), 3379–3391 (2009). https://doi.org/10.1002/cncr.24396. PubMed PMID: 19484787

136. F. Marchesi, L. Piemonti, A. Mantovani, P. Allavena, Molecular mechanisms of perineural invasion, a forgotten pathway of dissemination and metastasis. Cytokine Growth Factor Rev. **21**(1), 77–82 (2010). https://doi.org/10.1016/j.cytogfr.2009.11.001. PubMed PMID: 20060768

137. M. Amit, S. Na'ara, Z. Gil, Mechanisms of cancer dissemination along nerves. Nat. Rev. Cancer **16**(6), 399–408 (2016). https://doi.org/10.1038/nrc.2016.38. PubMed PMID: 27150016

138. A. Villers, J.E. McNeal, E.A. Redwine, F.S. Freiha, T.A. Stamey, The role of perineural space invasion in the local spread of prostatic adenocarcinoma. J. Urol. **142**(3), 763–768 (1989). PubMed PMID: 2769857

139. A. Olar, D. He, D. Florentin, Y. Ding, T. Wheeler, G. Ayala, Biological correlates of prostate cancer perineural invasion diameter. Hum. Pathol. **45**(7), 1365–1369 (2014). https://doi.org/10.1016/j.humpath.2014.02.011. PubMed PMID: 24768607; PubMed Central PMCID: PMC4492300

140. S.H. Azam, C.V. Pecot, Cancer's got nerve: Schwann cells drive perineural invasion. J. Clin. Invest. **126**(4), 1242–1244 (2016). https://doi.org/10.1172/JCI86801. PubMed PMID: 26999601; PubMed Central PMCID: PMC4811122

141. P. Zareba, R. Flavin, M. Isikbay, J.R. Rider, T.A. Gerke, S. Finn, A. Pettersson, F. Giunchi, R.H. Unger, A.M. Tinianow, S.-O. Andersson, O. Andrén, K. Fall, M. Fiorentino, L.A. Mucci, Perineural invasion and risk of lethal prostate cancer. Cancer Epidemiol. Biomarkers Prev. **26**(5), 719–726 (2017). https://doi.org/10.1158/1055-9965.Epi-16-0237. PubMed PMID: 28062398; PubMed Central PMCID: PMC5413395

142. R.D. Kraus, A. Barsky, L. Ji, P.M. Garcia Santos, N. Cheng, S. Groshen, N. Vapiwala, L.K. Ballas, The perineural invasion paradox: is perineural invasion an independent prognostic indicator of biochemical recurrence risk in patients with pT2N0R0 prostate cancer? A multi-institutional study. Adv. Rad. Oncol. **4**(1), 96–102 (2019). https://doi.org/10.1016/j.adro.2018.09.006. PubMed PMID: 30706016; PubMed Central PMCID: PMC6349660

143. N. Maru, M. Ohori, M.W. Kattan, P.T. Scardino, T.M. Wheeler, Prognostic significance of the diameter of perineural invasion in radical prostatectomy specimens. Hum. Pathol. **32**(8), 828–833 (2001). https://doi.org/10.1053/hupa.2001.26456. PubMed PMID: 11521227

144. G.E. Ayala, H. Dai, M. Ittmann, R. Li, M. Powell, A. Frolov, T.M. Wheeler, T.C. Thompson, D. Rowley, Growth and survival mechanisms associated with perineural invasion in prostate cancer. Cancer Res. **64**(17), 6082–6090 (2004). https://doi.org/10.1158/0008-5472.Can-04-0838. PubMed PMID: 15342391

145. C. Stolinski, Structure and composition of the outer connective tissue sheaths of peripheral nerve. J. Anat. **186**(Pt 1), 123–130 (1995). PubMed PMID: 7649808; PubMed Central PMCID: PMC1167278

146. Y. Olsson, Microenvironment of the peripheral nervous system under normal and pathological conditions. Crit. Rev. Neurobiol. **5**(3), 265–311 (1990). PubMed PMID: 2168810

147. J. Heesakkers, F. Farag, R.M. Bauer, J. Sandhu, D. De Ridder, A. Stenzl, Pathophysiology and contributing factors in postprostatectomy incontinence: a review. Eur. Urol. 71(6), 936–944 (2017). https://doi.org/10.1016/j.eururo.2016.09.031. PubMed PMID: 27720536

148. M.L. Feltri, L. Wrabetz, Laminins and their receptors in Schwann cells and hereditary neuropathies. J. Peripher. Nerv. Syst. 10(2), 128–143 (2005). https://doi.org/10.1111/j.1085-9489.2005.0010204.x. PubMed PMID: 15958125

149. G.J. Kidd, N. Ohno, B.D. Trapp, Biology of Schwann cells, in Handbook of Clinical Neurology, ed. by C. K. Gérard Said, (Elsevier B.V., Amsterdam, 2013), pp. 55–79

150. T.L. Davis, I. Rabinovitz, B.W. Futscher, M. Schnolzer, F. Burger, Y. Liu, M. Kulesz-Martin, A.E. Cress, Identification of a novel structural variant of the alpha 6 integrin. J. Biol. Chem. 276(28), 26099–26106 (2001). https://doi.org/10.1074/jbc.M102811200. PubMed PMID: 11359780; PubMed Central PMCID: PMC2824502

151. M.C. Demetriou, A.E. Cress, Integrin clipping: a novel adhesion switch? J. Cell. Biochem. 91(1), 26–35 (2004). https://doi.org/10.1002/jcb.10675. PubMed PMID: 14689578; PubMed Central PMCID: PMC2702438

152. P.A. Usher, O.F. Thomsen, P. Iversen, M. Johnsen, N. Brunner, G. Hoyer-Hansen, P. Andreasen, K. Dano, B.S. Nielsen, Expression of urokinase plasminogen activator, its receptor and type-1 inhibitor in malignant and benign prostate tissue. Int. J. Cancer 113(6), 870–880 (2005). https://doi.org/10.1002/ijc.20665. PubMed PMID: 15515049

153. M.C. Demetriou, M.E. Pennington, R.B. Nagle, A.E. Cress, Extracellular alpha 6 integrin cleavage by urokinase-type plasminogen activator in human prostate cancer. Exp. Cell Res. 294(2), 550–558 (2004). https://doi.org/10.1016/j.yexcr.2003.11.023. PubMed PMID: 15023541; PubMed Central PMCID: PMC2715336

154. C. Rivellini, G. Dina, E. Porrello, F. Cerri, M. Scarlato, T. Domi, D. Ungaro, U. Del Carro, A. Bolino, A. Quattrini, G. Comi, S.C. Previtali, Urokinase plasminogen receptor and the fibrinolytic complex play a role in nerve repair after nerve crush in mice, and in human neuropathies. PLoS One 7(2), e32059 (2012). https://doi.org/10.1371/journal.pone.0032059. PubMed PMID: 22363796; PubMed Central PMCID: PMC3283718

155. P. Merino, A. Diaz, V. Jeanneret, F. Wu, E. Torre, L. Cheng, M. Yepes, Urokinase-type plasminogen activator (uPA) binding to the uPA receptor (uPAR) promotes axonal regeneration in the central nervous system. J. Biol. Chem. 292(7), 2741–2753 (2017). https://doi.org/10.1074/jbc.M116.761650. PubMed PMID: 27986809; PubMed Central PMCID: PMC5314171

156. V. Vlaeminck-Guillem, Extracellular vesicles in prostate cancer carcinogenesis, diagnosis, and management. Front. Oncol. 8, 222 (2018). https://doi.org/10.3389/fonc.2018.00222. PubMed PMID: 29951375; PubMed Central PMCID: PMC6008571

157. V.R. Minciacchi, A. Zijlstra, M.A. Rubin, D. Di Vizio, Extracellular vesicles for liquid biopsy in prostate cancer: where are we and where are we headed? Prostate Cancer Prostatic Dis. 20(3), 251–258 (2017). https://doi.org/10.1038/pcan.2017.7. PubMed PMID: 28374743; PubMed Central PMCID: PMC5569339

158. K. Fujita, H. Kume, K. Matsuzaki, A. Kawashima, T. Ujike, A. Nagahara, M. Uemura, Y. Miyagawa, T. Tomonaga, N. Nonomura, Proteomic analysis of urinary extracellular vesicles from high Gleason score prostate cancer. Sci. Rep. 7, 42961 (2017). https://doi.org/10.1038/srep42961. PubMed PMID: 28211531; PubMed Central PMCID: PMC5314323

159. M. Rodriguez, C. Bajo-Santos, N.P. Hessvik, S. Lorenz, B. Fromm, V. Berge, K. Sandvig, A. Line, A. Llorente, Identification of non-invasive miRNAs biomarkers for prostate cancer by deep sequencing analysis of urinary exosomes. Mol. Cancer 16(1), 156 (2017). https://doi.org/10.1186/s12943-017-0726-4. PubMed PMID: 28982366; PubMed Central PMCID: PMC5629793

160. O.E. Bryzgunova, M.M. Zaripov, T.E. Skvortsova, E.A. Lekchnov, A.E. Grigor'eva, I.A. Zaporozhchenko, E.S. Morozkin, E.I. Ryabchikova, Y.B. Yurchenko, V.E. Voitsitskiy, P.P. Laktionov, Comparative study of extracellular vesicles from the urine of healthy individuals and prostate cancer patients. PLoS One 11(6), e0157566 (2016). https://doi.org/10.1371/journal.pone.0157566. PubMed PMID: 27305142; PubMed Central PMCID: PMC4909321

161. H. Shin, Y.H. Park, Y.G. Kim, J.Y. Lee, J. Park, Aqueous two-phase system to isolate extracellular vesicles from urine for prostate cancer diagnosis. PLoS One 13(3), e0194818 (2018). https://doi.org/10.1371/journal.pone.0194818. PubMed PMID: 29584777; PubMed Central PMCID: PMC5870972

162. A. Morales-Kastresana, B. Telford, T.A. Musich, K. McKinnon, C. Clayborne, Z. Braig, A. Rosner, T. Demberg, D.C. Watson, T.S. Karpova, G.J. Freeman, R.H. DeKruyff, G.N. Pavlakis, M. Terabe, M. Robert-Guroff, J.A. Berzofsky, J.C. Jones, Labeling extracellular vesicles for nanoscale flow cytometry. Sci. Rep. 7(1), 1878 (2017). https://doi.org/10.1038/s41598-017-01731-2. PubMed PMID: 28500324; PubMed Central PMCID: PMC5431945

163. A. Morales-Kastresana, J.C. Jones, Flow cytometric analysis of extracellular vesicles. Methods Mol. Biol. 1545, 215–225 (2017). https://doi.org/10.1007/978-1-4939-6728-5_16. PubMed PMID: 27943218

164. S.R. Krishn, A. Singh, N. Bowler, A.N. Duffy, A. Friedman, C. Fedele, S. Kurtoglu, S.K. Tripathi, K. Wang, A. Hawkins, A. Sayeed, C.P. Goswami, M.L. Thakur, R.V. Iozzo, S.C. Peiper, W.K. Kelly, L.R. Languino, Prostate cancer sheds the alphav-

beta3 integrin in vivo through exosomes. Matrix Biol. **77**, 41 (2019). https://doi.org/10.1016/j.matbio.2018.08.004. PubMed PMID: 30098419

165. H. Lu, N. Bowler, L.A. Harshyne, D. Craig Hooper, S.R. Krishn, S. Kurtoglu, C. Fedele, Q. Liu, H.Y. Tang, A.V. Kossenkov, W.K. Kelly, K. Wang, R.B. Kean, P.H. Weinreb, L. Yu, A. Dutta, P. Fortina, A. Ertel, M. Stanczak, F. Forsberg, D.I. Gabrilovich, D.W. Speicher, D.C. Altieri, L.R. Languino, Exosomal alphavbeta6 integrin is required for monocyte M2 polarization in prostate cancer. Matrix Biol. **70**, 20–35 (2018). https://doi.org/10.1016/j.matbio.2018.03.009. PubMed PMID: 29530483; PubMed Central PMCID: PMC6081240

166. I. Lazar, E. Clement, C. Attane, C. Muller, L. Nieto, A new role for extracellular vesicles: how small vesicles can feed tumors' big appetite. J. Lipid Res. **59**(10), 1793–1804 (2018). https://doi.org/10.1194/jlr.R083725. PubMed PMID: 29678957; PubMed Central PMCID: PMC6168303

167. F. Thuma, M. Zoller, Outsmart tumor exosomes to steal the cancer initiating cell its niche. Semin. Cancer Biol. **28**, 39–50 (2014). https://doi.org/10.1016/j.semcancer.2014.02.011. PubMed PMID: 24631836

168. J. Kim, S. Morley, M. Le, D. Bedoret, D.T. Umetsu, D. Di Vizio, M.R. Freeman, Enhanced shedding of extracellular vesicles from amoeboid prostate cancer cells: potential effects on the tumor microenvironment. Cancer Biol. Ther. **15**(4), 409–418 (2014). https://doi.org/10.4161/cbt.27627. PubMed PMID: 24423651; PubMed Central PMCID: PMC3979818

169. M.N. Theodoraki, T.K. Hoffmann, E.K. Jackson, T.L. Whiteside, Exosomes in HNSCC plasma as surrogate markers of tumour progression and immune competence. Clin. Exp. Immunol. **194**(1), 67–78 (2018). https://doi.org/10.1111/cei.13157. PubMed PMID: 30229863; PubMed Central PMCID: PMC6156813

170. J. Read, A. Ingram, H.A. Al Saleh, K. Platko, K. Gabriel, A. Kapoor, J. Pinthus, F. Majeed, T. Qureshi, K. Al-Nedawi, Nuclear transportation of exogenous epidermal growth factor receptor and androgen receptor via extracellular vesicles. Eur. J. Cancer **70**, 62–74 (2017). https://doi.org/10.1016/j.ejca.2016.10.017. PubMed PMID: 27886573

171. W. Stoorvogel, M.J. Kleijmeer, H.J. Geuze, G. Raposo, The biogenesis and functions of exosomes. Traffic **3**(5), 321–330 (2002). PubMed PMID: 11967126

172. P. Samuel, L.A. Mulcahy, F. Furlong, H.O. McCarthy, S.A. Brooks, M. Fabbri, R.C. Pink, D.R.F. Carter, Cisplatin induces the release of extracellular vesicles from ovarian cancer cells that can induce invasiveness and drug resistance in bystander cells. Philos. Trans. R. Soc. Lond. B Biol. Sci. **373**(1737), 20170065 (2018). https://doi.org/10.1098/rstb.2017.0065. PubMed PMID: 29158318; PubMed Central PMCID: PMC5717443

173. R. Yahyapour, E. Motevaseli, A. Rezaeyan, H. Abdollahi, B. Farhood, M. Cheki, M. Najafi, V. Villa, Mechanisms of radiation bystander and non-targeted effects: implications to radiation carcinogenesis and radiotherapy. Curr. Radiopharm. **11**(1), 34–45 (2018). https://doi.org/10.2174/1874471011666171229123130. PubMed PMID: 29284398

174. J.A. Majda, E.W. Gerner, B. Vanlandingham, K.R. Gehlsen, A.E. Cress, Heat shock-induced shedding of cell surface integrins in A549 human lung tumor cells in culture. Exp. Cell Res. **210**(1), 46–51 (1994). https://doi.org/10.1006/excr.1994.1007. PubMed PMID: 7505747

175. H. Costa Verdera, J.J. Gitz-Francois, R.M. Schiffelers, P. Vader, Cellular uptake of extracellular vesicles is mediated by clathrin-independent endocytosis and macropinocytosis. J. Control. Release **266**, 100–108 (2017). https://doi.org/10.1016/j.jconrel.2017.09.019. PubMed PMID: 28919558

176. L.A. Mulcahy, R.C. Pink, D.R. Carter, Routes and mechanisms of extracellular vesicle uptake. J. Extracell. Vesicles **3**, 24641 (2014). https://doi.org/10.3402/jev.v3.24641. PubMed PMID: 25143819; PubMed Central PMCID: PMC4122821

177. A.B. Banizs, T. Huang, R.K. Nakamoto, W. Shi, J. He, Endocytosis pathways of endothelial cell derived exosomes. Mol. Pharm. **15**, 5585 (2018). https://doi.org/10.1021/acs.molpharmaceut.8b00765. PubMed PMID: 30351959

178. D.M. Stranford, J.N. Leonard, Delivery of biomolecules via extracellular vesicles: a budding therapeutic strategy. Adv. Genet. **98**, 155–175 (2017). https://doi.org/10.1016/bs.adgen.2017.08.002. PubMed PMID: 28942793

179. H. Saari, E. Lazaro-Ibanez, T. Viitala, E. Vuorimaa-Laukkanen, P. Siljander, M. Yliperttula, Microvesicle- and exosome-mediated drug delivery enhances the cytotoxicity of Paclitaxel in autologous prostate cancer cells. J. Control. Release **220**(Pt B), 727–737 (2015). https://doi.org/10.1016/j.jconrel.2015.09.031. PubMed PMID: 26390807

180. G.K. Panigrahi, P.P. Praharaj, T.C. Peak, J. Long, R. Singh, J.S. Rhim, Z.Y. Abd Elmageed, G. Deep, Hypoxia-induced exosome secretion promotes survival of African-American and Caucasian prostate cancer cells. Sci. Rep. **8**(1), 3853 (2018). https://doi.org/10.1038/s41598-018-22068-4. PubMed PMID: 29497081; PubMed Central PMCID: PMC5832762

181. D.M. Carnell, R.E. Smith, F.M. Daley, M.I. Saunders, S.M. Bentzen, P.J. Hoskin, An immunohistochemical assessment of hypoxia in prostate carcinoma using pimonidazole: implications for radioresistance. Int. J. Radiat. Oncol. Biol. Phys. **65**(1), 91–99 (2006). https://doi.org/10.1016/j.ijrobp.2005.11.044. PubMed PMID: 16563659

182. E.D. Schwab, K.J. Pienta, Cancer as a complex adaptive system. Med. Hypotheses **47**(3), 235–241 (1996). PubMed PMID: 8898325

The Bone Microenvironment in Prostate Cancer Metastasis

Anthony DiNatale and Alessandro Fatatis

The Bone Marrow Microenvironment

Observations that the skeleton has an affinity for prostate cancer cells prompted Stephen Paget [1] to postulate that a receptive tissue is at least as important as the disseminating tumor cells in promoting metastatic colonization and progression [2]. It might be intuitive that the supporting features of organs frequently targeted by tumor cells should mimic those of the tissues in which the primary neoplasia originated. However, this is likely not the case, as malignant phenotypes acquire metastatic potential *via* somatic mutations or inherently retain the ability of adapting to foreign microenvironments [3]. It is widely recognized through experimentally testing that only a minority of cancer cells succeed in colonizing a specific tissue upon the conversion from Circulating Tumor Cells (CTCs) to Disseminated Tumor Cells (DTCs), which

occurs following extravasation [4, 5]. Thus, departure from the primary tumor by invasion and intravasation, survival during systemic circulation and recognition of adhesive and chemoattractant cues are events of the *metastatic cascade* that are required to seed distant organs, which would not lead to secondary lesions unless the 'seeds thrive in the soil' [6].

Cell-autonomous features may equip cancer cells to immediately benefit from the supporting conditions already present in the bone at the time of seeding; however, it is plausible—and convincing evidence exists—that cancer-induced conditioning of the local microenvironment also occurs *via* functional cross-talk [7, 8]. In this scenario, residing stromal cells are exposed to signaling molecules originating from the DTCs and would reciprocate by secreting trophic factors that render the bone marrow a hospitable ecosystem for tumor growth [9–11].

The implication of different resident cells of the bone microenvironment and role that local tissue factors play in the skeletal colonization of prostate cancer cells are discussed extensively in excellent reviews available in the literature.

This chapter will rather focus on (a) emerging concepts that deserve careful consideration and (b) established paradigms that should be critically reevaluated.

A. DiNatale · A. Fatatis (✉)
Department of Pharmacology and Physiology, Drexel University College of Medicine, Philadelphia, PA, USA

Program in Prostate Cancer, Sidney Kimmel Cancer Center, Thomas Jefferson University, Philadelphia, PA, USA
e-mail: Af39@drexel.edu

© Springer Nature Switzerland AG 2019
S. M. Dehm, D. J. Tindall (eds.), *Prostate Cancer*, Advances in Experimental Medicine and Biology 1210, https://doi.org/10.1007/978-3-030-32656-2_9

Hematopoietic Niche, Colonization and Dormancy

With the knowledge that hematopoietic stem cells (HSCs) are capable of homing to a specific niche in the bone marrow, both during development and HSC transplantation, it was correctly hypothesized that prostate cancer cells use a similar mechanism, thus explaining the common occurrence of skeletal metastases in advanced tumors [12, 13]. Additionally, DTCs can be detected in bone marrow aspirates following radical prostatectomy, in line with the fact that the disease can re-emerge at the skeletal level several years later following local therapy [14, 15]. This latency can be supported by the notion that the HSC niche is not only dormancy-supporting for HSCs, but also for prostate DTCs [16–18].

CXCL12 (stromal derived factor-1, SDF-1) and its receptors CXCR4 and CXCR7 are known to play a major role in HSC homing to the bone marrow [12, 19, 20]. These same chemokine receptor-ligand interactions are implicated functionally in prostate cancer metastasis to the bone, particularly CXCR4 [21–23]. Utilizing a mouse model of micrometastasis, it was shown that DTCs resulting from the spreading of a subcutaneous tumor competed for inhabiting the HSC niche with HSCs and prevented their engraftment [12]. These same studies indicated that HSCs and prostate cancer cells both home to the endosteal niche [24, 25]. Using parathyroid hormone to increase the number of cells of osteoblastic lineage and the number of osteoblastic niches, the number of prostate DTCs was increased. Additionally, a Col2.3Δ-TK transgene was used to deplete osteoblasts in a selective and inducible manner, which resulted in decreased arrival of prostate cancer cells to the skeleton, along with decreased growth in bone. Collectively, these findings provide compelling evidence for the role of the endosteal osteoblastic niche in prostate cancer metastasis.

HSCs can be mobilized into circulation by interfering with the CXCR4/CXCL12 signaling axis using AMD3100, a selective antagonist of CXCR4 that prevents CXCL12 binding [20]. In support of the concept that prostate cancer cells home to the bone marrow through hijacking the mechanism of HSC homing, AMD3100 was able to mobilize prostate cancer DTCs into circulation, similarly to what is observed for HSCs [20].

Annexin2 (ANXA2) plays a major role in HSC homing by aiding in the presentation of CXCL12 to HSCs by engaging this chemokine, as determined by *in vitro* binding assays and *in vivo* co-localization using immunofluorescent staining [26]. The role of ANXA2 was determined to be both cell intrinsic and extrinsic in HSC homing, as HSCs from Anxa2$^{-/-}$ mice express less CXCR4 mRNA, and wild-type bone marrow transplantation into Anxa2$^{-/-}$ mice results in decreased HSC engraftment. Initially, experimental evidence revealed that blocking ANXA2 can counteract prostate cancer localization to the osteoblastic niche [27]. Additionally, growth of bone-localized prostate cancer may be supported by ANXA2 through activation of the MAPK pathway. In line with the proposed role of ANXA2 in HSC homing, prostate cancer cells showed significantly greater binding to bone marrow stromal cells expressing ANXA2 when compared to those lacking this protein [26].

While the HSC niche has been found capable of orchestrating colonization of both HSCs and tumor cells in the skeleton, its impact might be even more complex by regulating cellular activities opposite to colonization and proliferation, such as dormancy. Cellular dormancy occurs when cells are retained in the G0 phase of the cell cycle and yet are capable of re-entering the cell cycle to resume proliferation [28]. HSCs undergo quiescence upon arriving to their bone niche, yet have the highest self-renewal activity and the ability to reversibly switch between dormancy and self-renewal [29, 30]. Overall, the function of the HSC niche is to maintain a reservoir of cells with self-renewal capacity, which are also protected from physiological stresses [31, 32]. DTCs can also enter a dormant state, an event that allows them to resist most cytotoxic therapeutics with mechanisms of action that interfere with either DNA replication and/or mitosis [33].

Notably, in addition to being implicated in homing to the HSC niche, ANXA2 also aids in dormancy induction, by binding to its receptor on prostate cancer cells and inducing Axl, a receptor tyrosine kinase highly expressed in multiple cancer types and associated with metastatic disease, poor prognosis, and drug resistance [34]. Axl binds to its ligand, growth-arrest specific factor (GAS6), which is expressed on osteoblasts, to induce dormancy and entry into the G0 phase of the cell cycle [18]. Additionally, GAS6 protects prostate cancer cells from docetaxel chemotherapy as a result of the induced dormancy [18]. Interestingly, if the expression of Tyro3, an additional receptor for GAS6, exceeds Axl expression, prostate cancer cells may revert to a highly proliferative state [16].

An additional mediator of cell dormancy is the p38 mitogen-activated protein kinase; a low extracellular signal-regulated kinase (ERK) to p38 ratio results in cell cycle arrest [35–37]. In prostate cancer cells, bone morphogenetic protein 7 (BMP7), a transforming growth factor-β (TGF-β) family member, is secreted by bone stromal cells in the metastatic niche and induces dormancy through activation of p38 [38]. The induction of p38 expression is responsible for upregulation of the metastasis suppressor NDRG1 (N-myc downstream regulated gene 1) and p21, a cell cycle inhibitor. BMP7 treatment potently suppresses cell growth and removal of this factor permits cell proliferation. This further supports the concept that the stromal factor BMP7 is capable of inducing dormancy in prostate cancer.

Subsequent to the evidence that prostate DTCs hijack HSC homing mechanisms to localize to the HSC niche, osteoblasts have been shown to contribute to dormancy in prostate cancer cells through induction of TANK binding kinase 1 (TBK1) [39]. Through TBK1 induction, osteoblasts inhibit the mTOR signaling pathway in prostate cancer cells, which can be rescued by shRNA-mediated TBK1 silencing. mTOR inhibition in prostate cancer cells by rapamycin increases dormancy, as determined by an increase in the Ki67 negative population, and cells are less susceptible to chemotherapy-induced apoptosis. To further support the involvement of TBK1, shRNA-mediated silencing of TBK1 increases the susceptibility of prostate cancer cells to docetaxel chemotherapy *in vitro*. Further, TBK1 knockdown *in vivo* prevents tumor recurrence following a 3-week regimen of docetaxel, while the scrambled shRNA fails to prevent recurrence.

Taken together, these studies reveal a skeletal metastatic niche far more complex than was previously recognized and identify additional, crucial events and mediators implicated in metastatic prostate cancer that cannot be neglected on the path of curative strategies for advanced patients.

Thus, the need for a paradigm shift in thinking with regard to cell-cooperative mechanisms in metastatic colonization and progression of prostate cancer should be seriously considered, as discussed in the next two paragraphs.

Osteoclasts and the Vicious-Cycle

Evidence that cancer cells disseminate to the skeleton and recruit osteoclasts to degrade mineralized bone matrix, thereby releasing trophic factors, and making these cells unwilling partners in crime, was first provided by Mundy [40]. This initial observation has been fueling collaborative efforts by experimental oncologists and bone physiologists, with the intent of understanding the events and mechanisms of the so called vicious cycle, by which cancer cells could sustain their survival and growth in the skeleton by benefiting from a forcefully-induced bone remodeling [41]. As expected, a further aim driving the research on the role of osteoclasts in skeletal metastasis was to identify means to interfere with the heterotypic interactions between these cells and disseminated tumor cells for therapeutic purposes [42]. The general paradigm of tumor-induced bone turnover entails the production of cytokines capable of inducing osteoclasts to resorb mineralized matrix, a physiological function that in this context is usurped by cancer cells to mobilize embedded growth factors normally intended for bone trophism and maintenance, such as transforming growth factor beta-1 (TGF-β1) [41, 43]. Cancer cells benefit from these factors by increasing proliferation, and the resulting surge in secreted cytokines would sustain a continuous cycle of osteoclast activation and tumor expansion. Normal bone turnover is paramount to maintaining the structural integrity and physiological functions of the skeleton; it is estimated that humans renew their skeleton several times over their lives [44]. This fine-tuned process requires functional cross-talk between osteoblasts—resident cells in charge of bone-matrix deposition—and osteoclasts, multinucleated syncytium cells derived from the fusion of macrophages that inhabit the bone marrow following their conversion from monocytes arriving *via* systemic blood [45]. Interestingly, osteoclastogenesis is promoted by IL-6 produced by osteoblasts [46], which then effectively orchestrate both the deposition and resorption phases of bone remodeling. The activation of osteoclasts requires a plethora of soluble factors including macrophage colony-stimulating factor (M-CSF) and receptor activator of nuclear factor κ-B ligand (RANKL), which are secreted by osteoblasts in response to parathyroid hormone (PTH) and active vitamin D3 [47], and by other bone-resident cells [48]. M-CSF enhances proliferation and survival of pre-osteoclast, deriving either from the circulating blood or upon differentiation of hematopoietic stem cells (HSCs) while also inducing the expression of RANK. This membrane receptor then responds to RANKL to promote both differentiation of pre-osteoclasts into osteoclasts and their functional activation [49]. Following the fusion of several activated precursors into a single cellular unit, osteoclasts will polarize and interact with the mineralized matrix at unique membrane domains, using mainly αvβ3 and α2β1 integrins forming a sealing zone and ruffled border, which create a cleft (resorption lacuna) into which acidic fluids and lysosomal enzymes, such as the cysteine protease cathepsin K, are discharged to dissolve the bone minerals and degrade type-I collagen [44, 50]. The number of osteoclasts is controlled by their limited lifespan as well as the demise occurring by apoptosis [51]. Conversely, their functional activity is constantly kept in check by osteoblasts either by direct cell-cell contacts involving the ephrin B receptors [52] or *via* secretion of osteoprotegerin (OPG), a soluble decoy receptor for RANKL

[53]. Skeletal metastases observed in prostate cancer patients have been historically defined as osteoblastic or osteosclerotic based on radiological imaging [54, 55]. However, histopathological evaluation consistently shows that the vast majority if not all these patients harbor skeletal tumors with a mixed pattern of osteolysis and osteosclerosis [55–57] and also present with markers of bone resorption in blood and urine [58, 59], indicating co-participation of osteoblasts and osteoclasts in the expansion of the metastatic tumors. Matrix deposition by osteoblasts follows osteolytic events in physiological conditions; skeletal tumors in metastatic prostate cancer display an increased formation of bone tissue, which exhibits altered composition and architecture as compared to normal bone generated during routine skeletal remodeling [55]. Major players in this process are endothelin-1 (ET-1) [60] and the prostate specific antigen (PSA), acting as an osteoblast mitogen and a protease activator of parathyroid-hormone related peptide (PTHrP) [61], respectively, and both actively secreted by prostate cancer cells. PTHrP would activate osteoclasts, thus leading to bone resorption and osteolysis [49]. An additional, and puzzling, observation, is that prostate cancer cells express transcription factors such as RUNX2 [62]. These transcription factors are usually expressed by osteoblasts during differentiation from mesenchymal stem cells and are responsible for the transcription of bone-specific matrix proteins such as osteocalcin [63] and matrix metalloprotease 9 (MMP9) [64]. A theorized explanation for this phenomenon is that the predilection of specific prostate malignant phenotypes for the skeleton depends on their osteomimetic properties, which would endow them with features characteristic of endogenous bone cells, thus facilitating their survival and growth in the skeleton. Furthermore, this osteomimicry [65] would accentuate the physiological bone-repairing mechanisms performed by osteoblasts, eventually leading to the osteosclerotic lesions that would nevertheless be characterized by poorly organized layers of collagen type-I fibrils and a woven bone with reduced mechanical resistance [66]. Another pathway counterbalancing the

osteolytic consequences of the *vicious cycle* is the secretion of ET-1 by cancer cells, as mentioned above, which binds the ET_A receptor on osteoblasts, promoting proliferation and bone-depositing activities [60, 67]. Understandably, the discovery of the close functional interactions between osteoclasts and prostate cancer cells, and the identification of several soluble mediators of osteosclerosis, generated high expectations for more effective strategies aimed to contain metastatic growth at the skeletal level. However, while counteracting the unbalanced activities of osteoclasts and osteoblasts pharmacologically has led to significant mitigation of the co-morbidities associated with skeletal metastases, to date these new approaches have not demonstrated relevant effects on the overall survival of prostate cancer patients. Bisphosphonates—analogues of naturally occurring inorganic pyrophosphate—and the RANKL inhibitor denosumab potently inhibit both bone-resorption and osteoclasts' viability [68], and are currently standards of care for patients with bone-metastatic prostate cancer [69]. These drugs reduce pain and improve quality of life, also effectively delaying the occurrence of skeletal-related events (SREs) such as pathological fractures and spinal compression [70–72], while failing to increase life expectancy [73]. Similarly, the ET_A antagonist Antrasentan [74] did not delay disease progression in CRPC patients [75]. Although the acquisition of additional palliative approaches for patients with metastatic prostate cancer should be always hailed, it is undeniable that the lack of effects on overall survival that was anticipated for these drugs has been disappointing.

A possible explanation for the ineffectiveness of osteoclasts-targeting strategies in controlling tumor progression in the skeleton may rest on the spatial distribution of areas of bone-resorption within the context of a metastatic lesion. As shown in Fig. 1a, osteoclasts are mostly confined to the edge of the tumor mass, at locations where they interact directly with the bone. As a result, the trophic factors released from the mineralized matrix during osteolytic events would most likely affect the cancer cells immediately adjacent to the areas of resorption. This scenario is effectively reproduced in an experimental animal model of metastasis, in which skeletal tumors, generated by human prostate cancer cells grafted in the systemic blood circulation of mice, display osteoclasts at the tumor-bone interface (Fig. 1b). However, in the innermost areas of the lesion, tumor cells grow with variable architecture and intermix with stromal cells but are devoid of osteoclasts or bone tissue. Therefore, it appears unrealistic to expect that soluble factors mobilized by osteoclasts may diffuse far enough within the tumor to effectively promote survival and proliferation of cancer cells except those situated immediately next to the bone-resorption areas.

On the other hand, it is easy to recognize how bisphosphonates and RANKL inhibitors limit pain from bone erosion, preventing fractures by avoiding the weakening of compact bone, and also indirectly counteracting osteosclerosis and spinal cord compression by depriving osteoblasts of the necessary osteolytic steps preceding bone-matrix deposition. Indeed, impairing osteolysis plausibly lessens the expansion of metastatic tumors within the bone tissue by limiting the space amenable for growth. However, the eradication of tumor cells growing as large foci from their skeletal lodging may require different strategies.

In light of the considerations above, one would anticipate that smaller bone tumors are more vulnerable to drugs targeting osteoclasts. For instance, at the initial stages of skeletal colonization, microscopic tumor foci generated by proliferating DTCs could be more reliant on osteolysis-derived trophic factors and their limited cross-section would expose the whole tumor mass to their supportive effect. Validating this hypothesis in human specimens may still be a daunting task, due to the tumor-size detection limits of current imaging modalities. Even if these limitations could be surmounted, validating this hypothesis would still be limited by the feasibility of performing core-needle biopsies on very small lesions, although improved protocols for CT-guided procedures are being successfully developed [76, 77]. A valid alternative is to

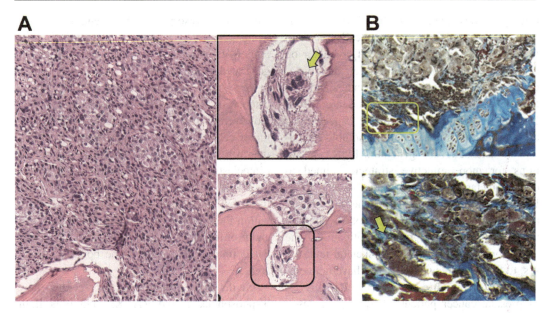

Fig. 1 Osteoclasts in skeletal metastases. (**a**) Tissue specimens obtained from skeletal metastases in prostate cancer patients. The Hematoxylin-Eosin stained tissue shows the typical adenocarcinoma architecture of the tumor cells growing within the bone marrow (light purple). The two insets show an osteoclast—visible as a syncytial, multinucleated cell—at the interface between the tumor mass at the compact bone (pink); (**b**) Experimental metastases generated in the skeleton of mice grafted with human prostate cancer cells display a similar scenario in a tissue section processed with a Masson's trichrome staining. The osteoclast (yellow area and arrow) is localized between the tumor cells in gray and the bone and cartilage stained in blue. The human tissue specimens were obtained through the bone biorepository established at the Sidney Kimmel Cancer Center of Thomas Jefferson University, Philadelphia [124]

employ animal models of metastasis, in which cancer cells are grafted in the arterial circulation to allow for their unbiased spreading to skeleton and soft-tissues [78, 79]. Since tumor seeding will occur *via* single cancer cells arriving at target organs and lodging as DTCs, the initial phases of skeletal colonization can be investigated by the appropriate combination of fluorescent and bioluminescent imaging approaches [11, 80, 81]. In an initial study, mice were grafted with human prostate cancer cells stably expressing GFP, which allowed the imaging of skeletal tumors while they were increasing in size following initial seeding by fluorescence stereomicroscopy. Osteoclasts were then visualized by staining serial tissue sections for Tartrate-resistant acid phosphatase (TRAP), an enzyme released by these cells when involved in active bone resorption [82]. The results showed that tumors with a cross-section area of ~3×10^3 μm^2 or larger were surrounded by layers of osteoclasts (Fig. 2a); however, smaller foci—routinely observed during the first 2 weeks following cancer cells seeding—were found to be spatially unrelated to osteoclasts [83]. There was a comparable paucity of osteoclasts in areas surrounding breast cancer cells grafted in mice by direct intratibial delivery [49]. An absence of bone resorption activities was observed when small colonies of prostate cancer cells in bone biopsies obtained from two different patients were studied [84]. These preclinical and *ex-vivo* observations are supported by clinical trials in patients with high-risk, non-metastatic prostate cancer having time to first metastasis as primary endpoint. A meta-analysis [85] of six randomized controlled trails, which accrued almost 6000 patients that were evaluated for clinical response to either bisphosphonates (five trials) or denosumab (1 trial), found that none of the trials evaluating bisphosphonates showed any difference in the incidence of skeletal metastases between placebo/standard-of-care

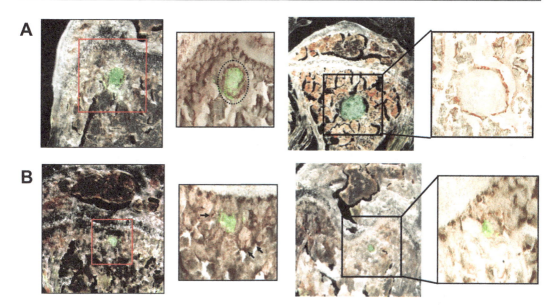

Fig. 2 Spatial relationship of osteoclasts with tumor cells in experimental metastases. Mice were grafted via intracardiac route with human prostate cancer cells stably expressing Green Fluorescent Protein (GFP) and sacrificed at either 1 week (**a**) or 2 weeks (**b**) post-grafting. Bone tissue samples harboring metastases were fixed, decalcified and frozen. Serial cryo-sections were examined using a fluorescence stereomicroscope. Calibrated digital analysis of fluorescent microscopy images indicated that the cross-section area of 1-week tumors measured $1.9 \pm 0.5 \times 10^3$ μm^2 and that of 2-week tumors measured $35 \pm 6 \times 10^3$ μm^2. The presence of active osteoclasts in the bone marrow regions colonized by cancer cells was established histologically by TRAP staining. Metastases with cross-section area larger than 28×10^3 μm^2—indicated by the green fluorescent signal and usually observed at 2 weeks post-grafting—were surrounded by an evident layer of active osteoclasts, as shown in the magnified panel (**b**), whereas smaller metastases (**a**) were instead spatially unrelated to osteoclasts, which appear sparsely distributed (black arrows). (*Reproduced with permission from Ref. [83].*)

and treated groups. The phase-III trial evaluating denosumab reported a significant delay in the appearance of bone metastases in treated patients, although it amounted to approximately 4 months [86]. A second meta-analysis of seven randomized controlled clinical trials for men with metastatic prostate cancer treated with bisphosphonates mirrored these results, also showing lack of improvement in overall survival [87].

In summary, the functional interplay between normal and malignant cell populations occurring in bone metastatic sites can lead to a noncanonical cycle of bone-resorption and matrix-deposition, and the histopathological effects of this process increase the occurrence of skeletal-related events (SREs), such as pathological fractures and spinal cord compression in prostate cancer patients. While the soluble factors released by osteolytic events should promote proliferation and survival of tumor cells, reducing their availability by pharmacologic inhibition of the bone-resorption events was found to produce minimal anti-tumor activity, despite the significant effects exerted by bisphosphonates and RANKL inhibitors on pain and the occurrence of SREs, which fully justify their FDA-approval as SOC for palliative treatment of patients with advanced bone-metastatic disease.

An additional—and seemingly more promising—strategy exploiting the tumor-induced remodeling of the bone tissue in the metastatic setting is represented by the use of the radioisotope Radium-223 dichloride (^{223}Ra). Due to its structure, ^{223}Ra behaves as a calcium analog and—when administered *via* a venous route—accumulates in sites of active mineralization, such as those controlled by osteoblasts countering the osteolytic activity of osteoclasts. At these sites and upon binding to hydroxyapatite, a major component of the mineralized matrix [88], ^{223}Ra is in close proximity to the cancer cells located at the interface between tumor and bone [89]. Due to its alpha-particle emitter properties, ^{223}Ra is then capable of inducing non-repairable DNA double-strand breaks in cancer cells [90], affecting their viability. Explorative clinical studies [91] were followed by a randomized phase-III clinical trial that compared ^{223}Ra to placebo in CRPC patients with symptomatic bone metastatic disease and demonstrated an increase in overall survival of 3.6 months [92]. Additional clinical studies are being conducted to explore potential therapeutic combinations of ^{223}Ra with chemotherapeutics and targeted therapies [89].

Cooperativity and AR Heterogeneity Among Metastatic Cells

The concept that malignant phenotypes inhabiting a tumor cooperate to foster their survival and growth was formulated in a comprehensive and convincing fashion more than a decade ago [93]. Initially based on game theory, the idea has been applied to events occurring at the primary tumor [94] and also extended to metastatic progression [95]. Indeed, clonal heterogeneity is widely observed in tumors [96], while pre-clinical and mathematical models have elegantly revealed how a minority of tumor-initiating or tumor-sustaining sub-populations can be instrumental in establishing inter-clonal interactions that can eventually benefit the entire spectrum of malignant phenotypes [97]. When comparing a neoplasia to an ecosystem, the gamut of cellular interactions includes negative events such as competition as well as positive events such as commensalism, synergism and mutualism, each with its own consequences for the more-fit or less-fit sub-clones [98]. A strong impetus to the adoption of this line of thinking for advanced prostate cancer derived from two landmark studies providing persuasive evidence that metastatic tumors harbor heterogenous populations of cancer cells as a result of metastasis-to-metastasis spreading [99, 100]. These observations added additional layers of complexity to the metastatic

process for a tumor that is often multifocal when residing in the prostate gland, with different clones displaying variable degrees of aggressiveness [101]. Indeed, a better delineation of the genetic composition of heterogenous prostate cancer populations will allow for a better understanding of the origin and evolution of primary tumors and their secondary lesions [102].

A parallel and compatible approach should be the recognition of the heterogeneous phenotypic composition of metastatic tumors in prostate cancer and the different functional interactions that can be established among tumor cells in the metastatic niche.

In this regard, a very compelling case can be made for the AR-mixed features of skeletal metastases in prostate cancer patients. While a major role for AR is undisputable, the foremost emphasis placed on this receptor—expressed either as a full-length protein or as its truncated versions—seems to have distracted from recognizing the fact that fractions of prostate cancer cells express very little, or lack, AR when growing at metastatic sites [103–105]. Metastatic patients indeed exhibit phenotypic heterogeneity when tumor cells are tested for PSA immunoreactivity. This scenario has been postulated to result from either the co-existence of tumor cells that differed in PSA expression *ab-initio* or from PSA-expressing cells that subsequently acquired somatic mutations leading to PSA loss [106]. The lack of AR and PSA expression in prostate epithelial cells has been long recognized as related to neuroendocrine (NE) cells, which are distributed throughout acini and ducts of the normal prostate gland, can be identified by morphological and histochemical means [107] and have a different embryonal origin as compared to the other epithelial prostate cells [108, 109]. The percentage of NE cells in the normal prostate is very low. For instance, focal areas of NE differentiation (NED) are detected in primary tumors ranging from 31% to 100% of patients, whereas uniformly diffuse NED is observed in less than 2% of cases [110, 111]. The latter tumors contain small and round cells that lack AR, disseminate predominantly to visceral organs and are rapidly fatal [112]. However, a more frequent scenario results from androgen-deprivation therapies (ADT) inducing an enrichment in NE cells observed in primary tumors of patients receiving ADT instead of, or prior to, surgery or radiotherapy [113, 114], as well as in metastatic lesions in advanced patients [115]. This NE trans-differentiation can be mimicked in human cell lines cultured in the absence of androgens [116], and is also observed in both castrated genetic-engineered mouse (GEM) models [117] and human xenografts [118, 119]. The study conducted in GEM models is of particular interest, as it uses lineage tracking to show that focal and overt regions of NE cancer in the prostate gland arise from trans-differentiation of luminal adenocarcinoma cells. Despite this body of evidence, it is becoming increasingly clear that NE phenotypes can only account for the fraction of prostate cancer cells lacking AR—or AR transcriptional activity [120]—in metastatic patients. Initial studies conducted by immunohistochemistry (IHC) analysis [121] were conclusively confirmed at the molecular level, using qRT-PCR interrogation of AR negative and AR positive tumor areas of bone metastatic specimens from different prostate cancer patients, harvested by Laser Capture Microdissection (LCM) guided by IHC staining [11] (Fig. 3). We found that AR negative prostate cancer cells failed to express NE markers, indicating the existence of malignant phenotypes in which the lack of AR was not attributable to NE trans-differentiation. This finding was concurrently confirmed in a separate study [122], which reported the relative increase of these AR-null/NE-null phenotypes, defined as double-negative prostate cancer (DNPC), in necropsy tissues obtained from 30 patients treated with the CYP17 inhibitor abiraterone and AR-antagonist enzalutamide, as compared to 56 patients that had been treated in the era prior the FDA approval of these two drugs. Furthermore, using a number of animal models the investiga-

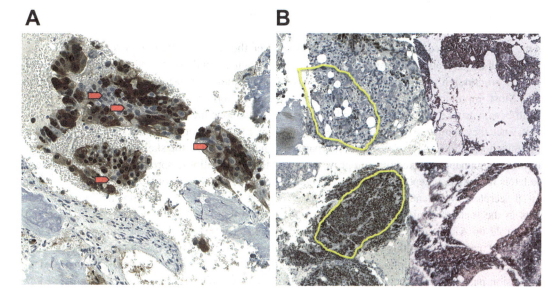

Fig. 3 Heterogeneous AR and IL-1β expression in human skeletal metastases. (**a**) Prostate cancer cells lacking nuclear AR staining (AR negative—red arrows) were commonly detected intermixed with AR positive cells. The fraction of AR_{NEG} cancer detected in bone metastases from ten different mCRPC patients was found to be 33 ± 14% [11]. (**b**) Bone metastasis specimens were selected for LCM-mediated harvesting of tumor areas negative or positive for AR. qRT-PCR detected the prostate-specific marker expressed at comparable levels in prostate cancer cells independently of their AR status, thus validating the lack of AR expression by immunohistochemistry at the transcriptional level. Finally, IL-1β mRNA was detected exclusively in AR negative prostate cancer cells, indicating an inverse association of this cytokine with the AR. (*Reproduced with permission from Ref. [11].*)

tors recapitulated the transition from tumors initially dependent on AR activity to tumors that could grow in an AR-independent fashion, without necessarily having to acquire NE characteristic to do so. The conclusion from these studies is that a continuum of differentiation stages most likely exists in the path from AR-dependent adenocarcinoma to DNPC or NE prostate cancer, and patients may very well present with all these phenotypes simultaneously.

While these latest studies provide support to the recognition of the existence of AR negative tumor cells in advanced prostate cancer, their functional role in metastatic lesions, in particular at the skeletal level, is still undefined.

Interestingly, AR expression in prostate cancer cells is inversely related to the expression of interleukin-1beta (IL-1β), both in bone metastases from patients and human cell lines [11]. In animal models of metastasis, AR negative/IL-1β positive PC3-ML cells progress in a very aggressive fashion generating large skeletal tumors, unless grafted in transgenic IL-1β receptor knockout mice. This phenomenon was due to IL-1β-induced over-expression of a set of genes in the tumor-associated bone stroma that could be implicated in generating a habitat favorable to skeletal progression to AR negative tumor cells [11]. Most importantly, AR positive tumor cells could benefit from these favorable conditions as well, since they colonized the bone marrow when intermixed with their AR negative counterparts, despite being unable to take hold and grow in the skeleton if grafted independently [11]. Indeed, IL-1β secreted by AR negative prostate cancer cells in the metastatic niche could also exert a direct role on AR positive cells, which express IL-1R [11] and respond to this cytokine by modulating a set of genes replicating basal gene expression patterns in AR negative cancer cells. This would suggest that tumor cells with a selective advantage for growing in the bone marrow and surviving AR-directed treatments could reprogram less robust cells to withstand unfavorable conditions [123].

Interestingly, FDA-approved therapeutics targeting IL-1β signaling are currently employed for non-oncological indications. Therefore, their successful testing for bone-metastatic prostate cancer could offer the opportunity for repurposing these drugs to target AR negative tumor cells, in combination with current standard-of-care aimed at impairing AR transcriptional activity in tumor cells that express AR.

Evidence indicates that AR negative phenotypes are likely to result from the selective pressure exerted by ADT and AR antagonists, an event that explains why mCRPC is—and could remain—incurable if the AR is the only therapeutic target pursued in these patients. In addition to IL-1β-induced recruitment of bone stroma discussed above, DNPC cells utilize a signaling program that relies on fibroblast growth factor (FGF). The autocrine stimulation of FGF receptor (FGFR) in these cells is responsible for the sustained activation of the MAPK signaling pathway, which bypasses the requirement for AR signaling and allows growth in androgen-deprived conditions [122].

Concluding Remarks

Over the last decade, our understanding of the mechanisms and events underpinning the skeletal colonization of prostate cancer cells has significantly improved. We can now fully appreciate that the bone metastatic niche is a far more complex microenvironment than previously thought. Tumor cells seed the skeleton as highly heterogenous populations, spreading from the primary tumor first and subsequently from already established metastatic tumors. Some malignant phenotypes hold the ability to initiate colonization and might also support survival and growth of less aggressive phenotypes. Notably, autochthonous cells of the bone are also crucially involved in promoting either tumor colonization or dormancy, activating molecular switches that are just beginning to be defined. Furthermore, normal and malignant cells do not behave like insulated contingents but rather interact and cooperate, following population dynamics that are as fascinating as they are menacing. Understanding the multiple interplays between tissue factors, molecular mediators and cellular populations involved in metastasis is a daunting task, but it is also the best way to identify targets worth pursuing for curative therapies, while dismissing those that could only lead, at best, to palliative measures.

References

1. S. Paget, The distribution of secondary growths in cancer of the breast. Cancer Metastasis Rev. **8**, 98–101 (1989)
2. I.J. Fidler, The pathogenesis of cancer metastasis: the 'seed and soil'hypothesis revisited. Nat. Rev. Cancer **3**, 453–458 (2003)
3. T. Celià-Terrassa, Y. Kang, Distinctive properties of metastasis-initiating cells. Genes Dev. **30**, 892–908 (2016)
4. T. Shibue, R.A. Weinberg, Metastatic colonization: settlement, adaptation and propagation of tumor cells in a foreign tissue environment. Semin. Cancer Biol. **21**, 99–106 (2011)
5. J. Massagué, A.C. Obenauf, Metastatic colonization by circulating tumour cells. Nature **529**, 298–306 (2016)
6. A.C. Obenauf, J. Massagué, Surviving at a distance: organ specific metastasis. Trends Cancer **1**, 76–91 (2015)
7. C.L. Chaffer, R.A. Weinberg, A perspective on cancer cell metastasis. Science **331**, 1559–1564 (2011)
8. M. Labelle, R.O. Hynes, The initial hours of metastasis: the importance of cooperative host-tumor cell interactions during hematogenous dissemination. Cancer Discov. **2**, 1091–1099 (2012)
9. D.F. Quail, J.A. Joyce, Microenvironmental regulation of tumor progression and metastasis. Nat. Med. **19**, 1423–1437 (2013)
10. M. Esposito, Y. Kang, Targeting tumor–stromal interactions in bone metastasis. Pharmacol. Ther. **141**, 222–233 (2014)
11. K. Shahriari et al., Cooperation among heterogeneous prostate cancer cells in the bone metastatic niche. Oncogene **36**, 2846–2856 (2017)
12. Y. Shiozawa et al., Human prostate cancer metastases target the hematopoietic stem cell niche to establish footholds in mouse bone marrow. J. Clin. Invest. **121**, 1298–1312 (2011)
13. A.D. Whetton, G.J. Graham, Homing and mobilization in the stem cell niche. Trends Cell Biol. **9**, 233–238 (1999)
14. C.L. Amling et al., Long-term hazard of progression after radical prostatectomy for clinically localized prostate cancer: continued risk of biochemical failure after 5 years. J. Urol. **164**, 101–105 (2000)
15. T.M. Morgan et al., Disseminated tumor cells in prostate cancer patients after radical prostatectomy and without evidence of disease predicts biochemi-

cal recurrence. Clin. Cancer Res. **15**, 677–683 (2009)
16. R.S. Taichman et al., GAS6 receptor status is associated with dormancy and bone metastatic tumor formation. PLoS One **8**, e61873 (2013)
17. P.E. Boulais, P.S. Frenette, Making sense of hematopoietic stem cell niches. Blood **125**, 2621–2629 (2015)
18. Y. Shiozawa et al., GAS6/AXL axis regulates prostate cancer invasion, proliferation, and survival in the bone marrow niche. Neoplasia **12**, 116–127 (2010)
19. T. Papayannopoulou, Current mechanistic scenarios in hematopoietic stem/progenitor cell mobilization. Blood **103**, 1580–1585 (2004)
20. H.E. Broxmeyer et al., Rapid mobilization of murine and human hematopoietic stem and progenitor cells with AMD3100, a CXCR4 antagonist. J. Exp. Med. **201**, 1307–1318 (2005)
21. J. Wang et al., The role of CXCR7/RDC1 as a chemokine receptor for CXCL12/SDF-1 in prostate cancer. J. Biol. Chem. **283**, 4283–4294 (2008)
22. Y.-X. Sun et al., Skeletal localization and neutralization of the SDF-1(CXCL12)/CXCR4 axis blocks prostate cancer metastasis and growth in osseous sites in vivo. J. Bone Miner. Res. **20**, 318–329 (2005)
23. R.S. Taichman et al., Use of the stromal cell-derived factor-1/CXCR4 pathway in prostate cancer metastasis to bone. Cancer Res. **62**, 1832–1837 (2002)
24. J. Zhang et al., Identification of the haematopoietic stem cell niche and control of the niche size. Nature **425**, 836–841 (2003)
25. L.M. Calvi et al., Osteoblastic cells regulate the haematopoietic stem cell niche. Nature **425**, 841–846 (2003)
26. Y. Jung et al., Annexin 2-CXCL12 interactions regulate metastatic cell targeting and growth in the bone marrow. Mol. Cancer Res. **13**, 197–207 (2015)
27. Y. Shiozawa et al., Annexin II/annexin II receptor axis regulates adhesion, migration, homing, and growth of prostate cancer. J. Cell. Biochem. **105**, 370–380 (2008)
28. J.A. Aguirre-Ghiso, Models, mechanisms and clinical evidence for cancer dormancy. Nat. Rev. Cancer **7**, 834–846 (2007)
29. A. Wilson et al., Hematopoietic stem cells reversibly switch from dormancy to self-renewal during homeostasis and repair. Cell **135**, 1118–1129 (2008)
30. T.H. Cheung, T.A. Rando, Molecular regulation of stem cell quiescence. Nat. Rev. Mol. Cell Biol. **14**, 329–340 (2013)
31. F. Arai et al., Tie2/angiopoietin-1 signaling regulates hematopoietic stem cell quiescence in the bone marrow niche. Cell **118**, 149–161 (2004)
32. P. Eliasson, J.-I. Jönsson, The hematopoietic stem cell niche: low in oxygen but a nice place to be. J. Cell. Physiol. **222**, 17–22 (2010)
33. V.T. DeVita, E. Chu, A history of cancer chemotherapy. Cancer Res. **68**, 8643–8653 (2008)

34. E.B. Rankin, A.J. Giaccia, The receptor tyrosine kinase AXL in cancer progression. Cancer **8**, E103 (2016)
35. J.A. Aguirre-Ghiso, D. Liu, A. Mignatti, K. Kovalski, L. Ossowski, Urokinase receptor and fibronectin regulate the ERK(MAPK) to p38(MAPK) activity ratios that determine carcinoma cell proliferation or dormancy in vivo. Mol. Biol. Cell **12**, 863–879 (2001)
36. J.A. Aguirre-Ghiso, Y. Estrada, D. Liu, L. Ossowski, ERK(MAPK) activity as a determinant of tumor growth and dormancy; regulation by p38(SAPK). Cancer Res. **63**, 1684–1695 (2003)
37. J.A. Aguirre-Ghiso, L. Ossowski, S.K. Rosenbaum, Green fluorescent protein tagging of extracellular signal-regulated kinase and p38 pathways reveals novel dynamics of pathway activation during primary and metastatic growth. Cancer Res. **64**, 7336–7345 (2004)
38. A. Kobayashi et al., Bone morphogenetic protein 7 in dormancy and metastasis of prostate cancer stem-like cells in bone. J. Exp. Med. **208**, 2641–2655 (2011)
39. J.K. Kim et al., TBK1 regulates prostate cancer dormancy through mTOR inhibition. Neoplasia **15**, 1064–1074 (2013)
40. G.R. Mundy, Mechanisms of bone metastasis. Cancer **80**, 1546–1556 (1997)
41. T.A. Guise, G.R. Mundy, Cancer and bone. Endocr. Rev. **19**, 18–54 (1998)
42. G.R. Mundy, Metastasis to bone: causes, consequences and therapeutic opportunities. Nat. Rev. Cancer **2**, 584–593 (2002)
43. L.A. Kingsley, P.G.J. Fournier, J.M. Chirgwin, T.A. Guise, Molecular biology of bone metastasis. Mol. Cancer Ther. **6**, 2609–2617 (2007)
44. J.C. Crockett, M.J. Rogers, F.P. Coxon, L.J. Hocking, M.H. Helfrich, Bone remodelling at a glance. J. Cell Sci. **124**, 991–998 (2011)
45. N. Takahashi, N. Udagawa, M. Takami, T. Suda, in *Principles of Bone Biology*, ed. by J. P. Bilezikian, L. G. Raisz, G. A. Rodan, vol. 1, (Elsevier, Amsterdam, 2002), pp. 109–126
46. K. Tawara, J.T. Oxford, C.L. Jorcyk, Clinical significance of interleukin (IL)-6 in cancer metastasis to bone: potential of anti-IL-6 therapies. Cancer Manag. Res. **3**, 177–189 (2011)
47. A. Leibbrandt, J.M. Penninger, RANK/RANKL: regulators of immune responses and bone physiology. Ann. N. Y. Acad. Sci. **1143**, 123–150 (2008)
48. A.E. Kearns, S. Khosla, P.J. Kostenuik, Receptor activator of nuclear factor kappaB ligand and osteoprotegerin regulation of bone remodeling in health and disease. Endocr. Rev. **29**, 155–192 (2008)
49. A. Maurizi, N. Rucci, The osteoclast in bone metastasis: player and target. Cancer **10**, E218 (2018)
50. M. Mulari, J. Vääräniemi, H.K. Väänänen, Intracellular membrane trafficking in bone resorbing osteoclasts. Microsc. Res. Tech. **61**, 496–503 (2003)
51. D.J. Mellis, C. Itzstein, M.H. Helfrich, J.C. Crockett, The skeleton: a multi-functional complex organ: the role of key signalling pathways in osteoclast differentiation and in bone resorption. J. Endocrinol. **211**, 131–143 (2011)
52. K. Matsuo, N. Otaki, Bone cell interactions through Eph/ephrin: bone modeling, remodeling and associated diseases. Cell Adh. Migr. **6**, 148–156 (2012)
53. P. Honore et al., Osteoprotegerin blocks bone cancer-induced skeletal destruction, skeletal pain and pain-related neurochemical reorganization of the spinal cord. Nat. Med. **6**, 521–528 (2000)
54. D.I. Rosenthal, Radiologic diagnosis of bone metastases. Cancer **80**, 1595–1607 (1997)
55. M.P. Roudier et al., Histopathological assessment of prostate cancer bone osteoblastic metastases. J. Urol. **180**, 1154–1160 (2008)
56. R.C. Percival et al., Biochemical and histological evidence that carcinoma of the prostate is associated with increased bone resorption. Eur. J. Surg. Oncol. **13**, 41–49 (1987)
57. N.W. Clarke, J. McClure, N.J. George, Morphometric evidence for bone resorption and replacement in prostate cancer. Br. J. Urol. **68**, 74–80 (1991)
58. R.E. Coleman, Skeletal complications of malignancy. Cancer **80**, 1588–1594 (1997)
59. P. Garnero et al., Markers of bone turnover for the management of patients with bone metastases from prostate cancer. Br. J. Cancer **82**, 858–864 (2000)
60. J.J. Yin et al., A causal role for endothelin-1 in the pathogenesis of osteoblastic bone metastases. Proc. Natl. Acad. Sci. U. S. A. **100**, 10954–10959 (2003)
61. M. Iwamura, J. Hellman, A.T. Cockett, H. Lilja, S. Gershagen, Alteration of the hormonal bioactivity of parathyroid hormone-related protein (PTHrP) as a result of limited proteolysis by prostate-specific antigen. Urology **48**, 317–325 (1996)
62. J. Akech et al., Runx2 association with progression of prostate cancer in patients: mechanisms mediating bone osteolysis and osteoblastic metastatic lesions. Oncogene **29**, 811–821 (2010)
63. P. Ducy, R. Zhang, V. Geoffroy, A.L. Ridall, G. Karsenty, Osf2/Cbfa1: a transcriptional activator of osteoblast differentiation. Cell **89**, 747–754 (1997)
64. S.K. Baniwal et al., Runx2 transcriptome of prostate cancer cells: insights into invasiveness and bone metastasis. Mol. Cancer **9**, 258 (2010)
65. K. Koeneman, F. Yeung, L.W. Chung, Osteomimetic properties of prostate cancer cells: a hypothesis supporting the predilection of prostate cancer metastasis and growth in the bone environment. Prostate **39**, 246–261 (1999)
66. G.A. Clines, T.A. Guise, Molecular mechanisms and treatment of bone metastasis. Expert Rev. Mol. Med. **10**, e7 (2008)
67. J.B. Nelson et al., Identification of endothelin-1 in the pathophysiology of metastatic adenocarcinoma of the prostate. Nat. Med. **1**, 944–949 (1995)

68. R.G.G. Russell et al., Bisphosphonates: an update on mechanisms of action and how these relate to clinical efficacy. Ann. N. Y. Acad. Sci. **1117**, 209–257 (2007)
69. L. Costa, P.P. Major, Effect of bisphosphonates on pain and quality of life in patients with bone metastases. Nat. Clin. Pract. Oncol. **6**, 163–174 (2009)
70. C.M. Bagi, Targeting of therapeutic agents to bone to treat metastatic cancer. Adv. Drug Deliv. Rev. **57**, 995–1010 (2005)
71. F. Saad et al., A randomized, placebo-controlled trial of zoledronic acid in patients with hormone-refractory metastatic prostate carcinoma. J. Natl. Cancer Inst. **94**, 1458–1468 (2002)
72. B.A. Gartrell et al., Metastatic prostate cancer and the bone: significance and therapeutic options. Eur. Urol. **68**, 850–858 (2015)
73. M. Wirth et al., Prevention of bone metastases in patients with high-risk nonmetastatic prostate cancer treated with zoledronic acid: efficacy and safety results of the Zometa European Study (ZEUS). Eur. Urol. **67**, 482–491 (2015)
74. M.A. Carducci, A. Jimeno, Targeting bone metastasis in prostate cancer with endothelin receptor antagonists. Clin. Cancer Res. **12**, 6296s–6300s (2006)
75. M.A. Carducci et al., A phase 3 randomized controlled trial of the efficacy and safety of atrasentan in men with metastatic hormone-refractory prostate cancer. Cancer **110**, 1959–1966 (2007)
76. M.G. Holmes et al., CT-guided bone biopsies in metastatic castration-resistant prostate cancer: factors predictive of maximum tumor yield. J. Vasc. Interv. Radiol. **28**, 1073–1081.e1 (2017)
77. V. Sailer et al., Bone biopsy protocol for advanced prostate cancer in the era of precision medicine. Cancer **124**, 1008–1015 (2018)
78. S.I. Park, S.-J. Kim, L.K. Mccauley, G.E. Gallick, Pre-clinical mouse models of human prostate cancer and their utility in drug discovery. Curr. Protoc. Pharmacol. **Chapter 14**(Unit 14), 15 (2010)
79. B.L. Eckhardt, Strategies for the discovery and development of therapies for metastatic breast cancer. Nat. Rev. Drug Discov. **11**, 479–497 (2012)
80. W.L. Jamieson-Gladney, Y. Zhang, A.M. Fong, O. Meucci, A. Fatatis, The chemokine receptor CX3CR1 is directly involved in the arrest of breast cancer cells to the skeleton. Breast Cancer Res. **13**, 1584 (2011)
81. F. Shen et al., Novel small-molecule CX3CR1 antagonist impairs metastatic seeding and colonization of breast cancer cells. Mol. Cancer Res. **14**, 518–527 (2016)
82. B. Kirstein, T.J. Chambers, K. Fuller, Secretion of tartrate-resistant acid phosphatase by osteoclasts correlates with resorptive behavior. J. Cell. Biochem. **98**, 1085–1094 (2006)
83. M.R. Russell, W.L. Jamieson, N.G. Dolloff, A. Fatatis, The alpha-receptor for platelet-derived growth factor as a target for antibody-mediated inhibition of skeletal metastases from prostate cancer cells. Oncogene **28**, 412–421 (2009)
84. N.W. Clarke, C.A. Hart, M.D. Brown, Molecular mechanisms of metastasis in prostate cancer. Asian J. Androl. **11**, 57–67 (2009)
85. A.R. Hayes, D. Brungs, N. Pavlakis, Osteoclast inhibitors to prevent bone metastases in men with high-risk, non-metastatic prostate cancer: a systematic review and meta-analysis. PLoS One **13**, e0191455 (2018)
86. M.R. Smith et al., Denosumab and bone-metastasis-free survival in men with castration-resistant prostate cancer: results of a phase 3, randomised, placebo-controlled trial. Lancet **379**, 39–46 (2012)
87. C.L. Vale et al., Addition of docetaxel or bisphosphonates to standard of care in men with localised or metastatic, hormone-sensitive prostate cancer: a systematic review and meta-analyses of aggregate data. Lancet Oncol. **17**, 243–256 (2016)
88. B. Clarke, Normal bone anatomy and physiology. Clin. J. Am. Soc. Nephrol. **3**(Suppl 3), S131–S139 (2008)
89. E. Deshayes et al., Radium 223 dichloride for prostate cancer treatment. Drug Des. Devel. Ther. **11**, 2643–2651 (2017)
90. M.A. Ritter, J.E. Cleaver, C.A. Tobias, High-LET radiations induce a large proportion of non-rejoining DNA breaks. Nature **266**, 653–655 (1977)
91. S. Nilsson et al., First clinical experience with alpha-emitting radium-223 in the treatment of skeletal metastases. Clin. Cancer Res. **11**, 4451–4459 (2005)
92. C. Parker et al., Alpha emitter radium-223 and survival in metastatic prostate cancer. N. Engl. J. Med. **369**, 213–223 (2013)
93. R. Axelrod, D.E. Axelrod, K.J. Pienta, Evolution of cooperation among tumor cells. Proc. Natl. Acad. Sci. U. S. A. **103**, 13474–13479 (2006)
94. T. Tsuji, S. Ibaragi, G.F. Hu, Epithelial-mesenchymal transition and cell cooperativity in metastasis. Cancer Res. **69**, 7135–7139 (2009)
95. F.-C. Bidard, J.-Y. Pierga, A. Vincent-Salomon, M.-F. Poupon, A 'class action' against the microenvironment: do cancer cells cooperate in metastasis? Cancer Metastasis Rev. **27**, 5–10 (2008)
96. H. Zhou, D. Neelakantan, H.L. Ford, Clonal cooperativity in heterogenous cancers. Semin. Cell Dev. Biol. **64**, 79–89 (2017)
97. A. Marusyk et al., Non-cell-autonomous driving of tumour growth supports sub-clonal heterogeneity. Nature **514**, 54–58 (2014)
98. D.P. Tabassum, K. Polyak, Tumorigenesis: it takes a village. Nat. Rev. Cancer **15**, 473–483 (2015)
99. G. Gundem et al., The evolutionary history of lethal metastatic prostate cancer. Nature **520**, 353–357 (2015)
100. M.K.H. Hong et al., Tracking the origins and drivers of subclonal metastatic expansion in prostate cancer. Nat. Commun. **6**, 6605 (2015)

101. M. Løvf et al., Multifocal primary prostate cancer exhibits high degree of genomic heterogeneity. Eur. Urol. **75**, 498–505 (2019)
102. J.L. Van Etten, S.M. Dehm, Clonal origin and spread of metastatic prostate cancer. Endocr. Relat. Cancer **23**, R207–R217 (2016)
103. A. Hobisch et al., Distant metastases from prostatic carcinoma express androgen receptor protein. Cancer Res. **55**, 3068–3072 (1995)
104. Z. Culig, H. Klocker, G. Bartsch, A. Hobisch, Androgen receptors in prostate cancer. Endocr. Relat. Cancer **9**, 155–170 (2002)
105. A. Queisser et al., Comparison of different prostatic markers in lymph node and distant metastases of prostate cancer. Mod. Pathol. **28**, 138–145 (2014)
106. M.P. Roudier et al., Phenotypic heterogeneity of end-stage prostate carcinoma metastatic to bone. Hum. Pathol. **34**, 646–653 (2003)
107. W. Wang, J. Epstein, I. Small cell carcinoma of the prostate. A morphologic and immunohistochemical study of 95 cases. Am. J. Surg. Pathol. **32**, 65–71 (2008)
108. G. Aumüller et al., Neurogenic origin of human prostate endocrine cells. Urology **53**, 1041–1048 (1999)
109. N. Vashchenko, P.-A. Abrahamsson, Neuroendocrine differentiation in prostate cancer: implications for new treatment modalities. Eur. Urol. **47**, 147–155 (2005)
110. S. Terry, H. Beltran, The many faces of neuroendocrine differentiation in prostate cancer progression. Front. Oncol. **4**, 60 (2014)
111. A.D. Grigore, E. Ben-Jacob, M.C. Farach-Carson, Prostate cancer and neuroendocrine differentiation: more neuronal, less endocrine? Front. Oncol. **5**, 37 (2015)
112. J.S. Palmgren, S.S. Karavadia, M.R. Wakefield, Unusual and underappreciated: small cell carcinoma of the prostate. Semin. Oncol. **34**, 22–29 (2007)
113. T. Jiborn, A. Bjartell, P.A. Abrahamsson, Neuroendocrine differentiation in prostatic carcinoma during hormonal treatment. Urology **51**, 585–589 (1998)
114. H. Beltran et al., Molecular characterization of neuroendocrine prostate cancer and identification of new drug targets. Cancer Discov. **1**, 487–495 (2011)
115. H. Beltran et al., Divergent clonal evolution of castration-resistant neuroendocrine prostate cancer. Nat. Med. **22**, 298–305 (2016)
116. S. Marchiani et al., Androgen-responsive and -unresponsive prostate cancer cell lines respond differently to stimuli inducing neuroendocrine differentiation. Int. J. Androl. **33**, 784–793 (2010)
117. M. Zou et al., Transdifferentiation as a mechanism of treatment resistance in a mouse model of castration-resistant prostate cancer. Cancer Discov. **7**, 736–749 (2017)
118. Z.G. Li et al., Androgen receptor–negative human prostate cancer cells induce osteogenesis in mice through FGF9-mediated mechanisms. J. Clin. Investig. **118**, 2697 (2008). https://doi.org/10.1172/JCI33093DS1
119. V. Tzelepi et al., Modeling a lethal prostate cancer variant with small-cell carcinoma features. Clin. Cancer Res. **18**, 666–677 (2012)
120. D.T. Miyamoto et al., Androgen receptor signaling in circulating tumor cells as a marker of hormonally responsive prostate cancer. Cancer Discov. **2**, 995–1003 (2012)
121. S. Crnalic et al., Nuclear androgen receptor staining in bone metastases is related to a poor outcome in prostate cancer patients. Endocr. Relat. Cancer **17**, 885–895 (2010)
122. E.G. Bluemn et al., Androgen receptor pathway-independent prostate cancer is sustained through FGF signaling. Cancer Cell **32**, 474–489.e6 (2017)
123. S.E. Thomas-Jardin et al., Identification of an IL-1-induced gene expression pattern in AR+ PCa cells that mimics the molecular phenotype of AR- PCa cells. Prostate **78**, 595–606 (2018)
124. K.R. Wang, J.A. Abraham, P. McCue, M.J. Schiewer, Characterization of a bone biorepository: comparison of bone metastases from breast, prostate, renal, lung cancers, and myeloma. J. Clin. Oncol. **36**, e24019–e24019 (2018)

Prostate Cancer Energetics and Biosynthesis

Chenchu Lin, Travis C. Salzillo, David A. Bader,
Sandi R. Wilkenfeld, Dominik Awad,
Thomas L. Pulliam, Prasanta Dutta,
Shivanand Pudakalakatti, Mark Titus,
Sean E. McGuire, Pratip K. Bhattacharya,
and Daniel E. Frigo

Introduction

Otto Warburg first described the ability of cancer cells to exhibit glycolysis even in the presence of oxygen (aerobic glycolysis), a phenomenon commonly referred to as the Warburg effect [1, 2]. Since this pioneering work, the altered metabolism of cancers relative to benign tissue has been more broadly recognized and is now considered one of the hallmarks of cancer [3]. Interestingly, while prostate cancers have distinct metabolic phenotypes from normal prostate, they also exhibit atypical metabolism compared to many other cancer types. Hence, many of the generalities established for cancer metabolism are not pertinent for prostate cancer. But like many diseases, the metabolism of prostatic tumors is context- and stage-dependent.

The prostate is a reproductive gland, generating and releasing fluid that nourishes sperm. Sperm are nourished in large part by citrate that is produced and secreted from the luminal epithelial cells of the prostate. Luminal epithelial cells are able to produce and secrete large amounts of citrate as the result of a truncated tricarboxylic acid (TCA) cycle (also referred to as the citric acid or Krebs cycle) that is caused by the extraordinarily high levels of zinc that accumulate in these cells. To that end, the luminal epithelial cells of the prostate contain amongst the highest levels of zinc (~.8–1.5 mM) of any cell in the human body [4–9]. The high intracellular levels of zinc disrupt the TCA cycle by inhibiting the enzyme m-aconitase, which converts citrate to isocitrate [7]. This truncated TCA cycle converts prostatic luminal epithelial cells from citrate-consuming to citrate-producing cells [7, 10]. One of the first events to occur during the malignant transformation of prostate epithelial cells is a decrease in the expression of zinc transporters

C. Lin · T. C. Salzillo · S. R. Wilkenfeld · D. Awad
Department of Cancer Systems Imaging, The
University of Texas MD Anderson Cancer Center,
Houston, TX, USA

The University of Texas MD Anderson Cancer Center
UTHealth Graduate School of Biomedical Sciences,
Houston, TX, USA

D. A. Bader
Department of Molecular and Cellular Biology,
Baylor College of Medicine, Houston, TX, USA

T. L. Pulliam
Department of Cancer Systems Imaging,
The University of Texas MD
Anderson Cancer Center,
Houston, TX, USA

Center for Nuclear Receptors and Cell Signaling,
University of Houston, Houston, TX, USA

Department of Biology and Biochemistry,
University of Houston,
Houston, TX, USA

© Springer Nature Switzerland AG 2019
S. M. Dehm, D. J. Tindall (eds.), *Prostate Cancer*, Advances in Experimental Medicine
and Biology 1210, https://doi.org/10.1007/978-3-030-32656-2_10

(ZIPs) [11, 12]. This results in decreased intracellular zinc accumulation, de-repression of aconitase, and concomitant de-repression of the TCA cycle. As such, in contrast to what has been described for most cancers, the majority of prostate cancers exhibit high levels of glucose oxidation through increased TCA cycle flux (Fig. 1).

While the shift towards increased glucose oxidation during transformation has been known for over 20 years, what has become clear is that prostate cancers co-opt a number of other important metabolic processes, described below, to help satisfy the increased energetic and biosynthetic demands of a rapidly growing tumor (Fig. 1). Further, these metabolic changes continue to change throughout disease progression. For example, many advanced, lethal prostate cancers will eventually demonstrate increased glycolytic flux, similar to the classic Warburg effect (Fig. 1). Importantly, cancer cells must also adapt to survive the harsh tumor microenvironment that evolves in part due to the increased metabolic waste produced from the cancer itself. Moreover, the increased uptake of many nutrients contributes directly to the synthesis of new signaling molecules that can function as oncogenic signals to reprogram cells and promote disease progression.

Our understanding of which nutrients are used by tumors, how and when they are metabolized, and the regulation of these metabolic processes, is required to translate these observations towards clinical utility. Importantly, the chemical nature of metabolism makes it possible to develop biomarkers (ex. imaging) that can assess which pathways have been altered in patients and therefore identify men who could benefit from emerging, metabolically targeted therapies.

Here, we describe the metabolic alterations that occur during the initiation and progression of prostate cancer. Further, we will highlight how key signaling pathways (ex. AR, PI3K, MYC) as well as other factors such as changes in the tumor microenvironment regulate these processes. Finally, we will discuss the clinical significance of this field. Accordingly, we will summarize the new metabolic-targeted therapies that are being tested for the treatment of prostate cancer. Importantly, we will also outline the emerging approaches being used to monitor metabolism in patients and how these could guide future clinical trials.

Metabolic Reprogramming in Prostate Cancer

Glucose Metabolism

The specific metabolic phenotype of normal prostate epithelial cells includes the accumulation of high zinc concentrations (~3–10-fold higher than in other tissues) that subsequently lead to a truncated TCA cycle and increased

---

P. Dutta · S. Pudakalakatti
Department of Cancer Systems Imaging,
The University of Texas MD Anderson Cancer
Center, Houston, TX, USA

M. Titus
Department of Genitourinary Medical Oncology,
The University of Texas MD Anderson Cancer
Center, Houston, TX, USA

S. E. McGuire
Department of Molecular and Cellular Biology,
Baylor College of Medicine, Houston, TX, USA

Department of Radiation Oncology, The University
of Texas MD Anderson Cancer Center,
Houston, TX, USA

P. K. Bhattacharya
Department of Cancer Systems Imaging,
The University of Texas MD Anderson Cancer Center,
Houston, TX, USA

The University of Texas Health Science Center at
Houston, Houston, TX, USA

D. E. Frigo (✉)
Department of Cancer Systems Imaging,
The University of Texas MD Anderson Cancer
Center, Houston, TX, USA

Center for Nuclear Receptors and Cell Signaling,
University of Houston, Houston, TX, USA

Department of Biology and Biochemistry, University
of Houston, Houston, TX, USA

Department of Genitourinary Medical Oncology,
The University of Texas MD
Anderson Cancer Center,
Houston, TX, USA

Molecular Medicine Program, The Houston
Methodist Research Institute, Houston, TX, USA
e-mail: frigo@mdanderson.org

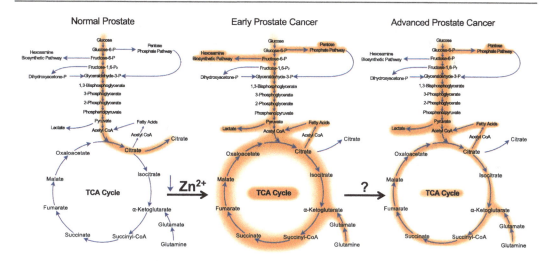

Fig. 1 Evolution of prostate cancer metabolism. Normal prostate epithelial cells exhibit a truncated TCA cycle that results in the increased production and secretion of citrate. During the initial transformation towards malignancy, intracellular concentrations of zinc drop causing a derepression of aconitase (the enzyme that converts citrate to isocitrate) and subsequent increased flux through the TCA cycle. Concurrently, cancer cells start to exhibit aerobic glycolysis, elevated glutaminolysis and increased flux through the hexosamine biosynthetic and pentose phosphate pathways. Interestingly, another hallmark of prostate cancers is the concurrent increases in both *de novo* lipogenesis and fatty acid oxidation. When prostate cancers progress into the late stages of the disease, the classic Warburg effect becomes more pronounced while some pathways, such as the hexosamine biosynthetic pathway, may reverse. While the initial metabolic transformation of prostatic cells has been well described to result from alterations such as the decreases in intracellular zinc concentrations, many of the drivers of the metabolic changes that occur in advanced prostate cancer remain poorly understood. Shown here is only a brief snapshot of central carbon metabolism

citrate production (~30–50-fold higher than other tissues), decreased oxidative phosphorylation and low energy metabolism [13]. Such inefficient metabolism cannot meet the energy requirements for rapidly growing prostate cancer cells. To adjust, prostate cancer cells are reprogrammed to have an efficient, energy-generating metabolism during their initial transformation. A notable metabolic shift during this transformation is an increased level of citrate oxidation as the malignant glands contain significantly lower concentrations of zinc compared to normal cells [14]. This shift allows cells to oxidize citrate and produce energy via a functional TCA cycle. This metabolic alteration can also protect prostate cancer cells from cell death [15]. In normal prostate epithelial cells, zinc accumulation facilitates Bax-associated mitochondrial pore formation, which promotes cytochrome c release from mitochondria and subsequent caspase cascades as well as an inhibition of the anti-apoptotic protein NFkB [16, 17]. Conversely, prostate cancer cells are less susceptible to mitochondrial-induced apoptosis in the presence of low zinc concentrations. As noted above, ZIPs are an important contributor to intracellular zinc regulation. The expression of ZIPs is decreased significantly or often absent altogether in prostate cancer [12, 18]. Interestingly, differences in ZIP expression may also explain in part some of the racial disparities observed for prostate cancer. A study comparing African American and Caucasians suggested that ZIPs are expressed at a lower level in prostate tumors from African Americans, preventing them from maintaining normal intracellular zinc concentrations [19]. Although it is not entirely clear how ZIPs are downregulated during prostate cancer progression, one potential explanation is epigenetic repression. In prostate cancer cells, higher methylation levels have been observed in the promoter region for the gene encoding activating enhancer binding protein 2 alpha (AP-2 alpha). AP-2 alpha is an important transcriptional regulator of the ZIPs. Therefore, methylation-mediated decreases in AP-2 alpha levels may contribute to ZIP deregulation [20].

Given the known role of zinc in prostate metabolism, there is interest in the use of zinc dietary supplements and the development of ZIP inhibitors to treat prostate cancer [18].

Due to the metabolic shift towards citrate oxidation, some metabolic intermediates and genes in the TCA cycle are also increased/hyperactivated during early transformation. A recent integrative proteomics study revealed that the TCA pathway proteins, citrate synthase, aconitate hydratase, 2-ketoglutarate dehydrogenase complex, succinate-CoA ligase, fumarate hydratase and malate dehydrogenase, are upregulated in primary prostate cancer compared to benign prostatic hyperplasia [21]. Furthermore, the corresponding intermediate metabolites in the TCA cycle like malate, fumarate, succinate and 2-hydroxyglutaric acid are significantly elevated in tumors, suggesting a dependence of primary prostate cancers on oxidative phosphorylation (OXPHOS). Interestingly, because of the high OXPHOS in primary prostate cancers, only modest levels of glucose uptake and the Warburg effect are observed [22, 23]. Consequently, [18F] fluorodeoxyglucose-positron emission tomography (FDG-PET) is a poor detector of primary prostate cancer.

Unlike most primary prostate cancers, late-stage prostate cancers can often be detected in FDG-PET scans. Accordingly, increased glycolytic metabolism has been correlated with disease progression and poor prognosis [24]. Though the exact mechanism of glucose metabolism regulation in prostate cancer has not been fully elucidated, emerging evidence suggests the regulation of multiple glycolytic enzymes in the advanced cancer stages.

Facilitative glucose transporters (GLUTs) control the first rate-limiting step of glucose metabolism by mediating glucose diffusion. To date, 14 members have been identified in the human GLUT family. Each is associated with different substrate affinities and tissue distributions. GLUT1 is overexpressed in several tumors including prostate cancer. High GLUT1 expression is elevated in prostate tissues compared to tumor-adjacent normal tissues, and its expression is correlated with a shorted time to recurrence after radical prostatectomy [25, 26]. Accordingly, higher expression of GLUT1 has been reported in androgen-independent prostate cancers [27]. Moreover, GLUT1 can be induced in a tissue-specific manner by androgens and glucose deprivation, which can help cancer cells survive in a low glucose environment [28]. Besides GLUT1, GLUT12 has also recently been shown to play a functional role in prostate cancer [29]. GLUT12 is required for androgen-induced glucose uptake and cell growth in LNCaP and VCaP cells. Further, it plays a similar role in AR-negative PC-3 cells. Interestingly, GLUT12 subcellular trafficking is increased in multiple prostate cancer cell models, suggesting a functional regulation beyond mRNA and protein expression. Of note, other GLUT members such as GLUT3, 7 and 11 are also overexpressed in prostate cancer [30]. However, the functional roles of these transporters are less clear.

Hexokinases (HKs) catalyze the first irreversible step in glycolysis by converting glucose to glucose-6-phosphate (G6P). Out of the four isoforms identified, HK2 is a major contributor to the Warburg effect and is required for tumor growth. HK2 is significantly elevated in prostate tumors relative to normal tissue and is significantly correlated with Gleason score [24, 31–33]. However, dramatic variability in expression levels occurs in individual castration-resistant prostate cancer (CRPC) patient samples, further indicating the heterogeneity of prostate cancer [33]. Though our understanding of the driving forces behind HK2 expression during disease evolution is incomplete, new findings have begun to reveal potential mechanisms. First, *PTEN* and *TP53* co-deletions/mutations have been correlated with high levels of HK2 in prostate cancer cell lines, xenograft and genetic mouse models, and prostate cancer patient samples. On one hand, *PTEN* loss leads to mTOR signaling pathway activation and 4EBP1 phosphorylation, triggering cap-dependent translation of HK2 through facilitating the dissociation of 4EBP1 and 4IF4E [33]. On the other hand, HK2 expression can be further increased by decreased miR-143 expression, a miRNA whose biogenesis is promoted by p53 [34]. In support of these findings, increased

HK2 expression and activity, caused by androgen deprivation, is associated with increased p-AKT in *Pten/Tp53* deficient mice [35]. AKT not only increases HK2 expression by mTORC1, but it also promotes the localization of HK2 to the mitochondria, thus increasing HK2 activity. Besides *PTEN/TP53*, HK2 expression is regulated by EZH2 and PKA. EZH2 can upregulate HK2 and glycolysis in PC-3 cells through the inhibition of miR-181b [36]. Additionally, PKA-CREB signaling can promote HK2 expression and glucose utilization following androgen treatment in LNCaP cells [37]. Importantly, systemic deletion of *Hk2* in a genetic mouse model results in a marked repression of tumor growth without any glucose homeostasis dysfunction in normal cells [38], suggesting a selective HK2 inhibitor could be exploited clinically with tolerant toxicities.

Pyruvate dehydrogenase complex (PDC) catalyzes the conversion of pyruvate to acetyl coenzyme A (acetyl-CoA), thereby regulating the carbon flux from glycolysis into the TCA cycle. PDC is a multi-enzyme complex composed of three enzymes: pyruvate dehydrogenase (E1), dihydrolipoamide acetyltransferase (E2), and lipoamide dehydrogenase (E3). PDHA1 (E1 alpha subunit) is a major component of PDC. As such, PDHA1, which is regulated by phosphorylation, plays a critical role in controlling the appropriate mitochondrial activity for the requirement of cell metabolism and growth [39, 40]. PDHA1 can be phosphorylated at serine 293 by pyruvate dehydrogenase kinases (PDKs). PDKs inhibit PDC activity and therefore the TCA cycle. This shifts the metabolism of a cell away from OXPHOS and towards a Warburg effect. Conversely, dehydrogenase phosphatases (PDPs) can reverse such inhibition. Typically, PDKs are overexpressed in glycolytic cancers [41], where the upregulated expression of PDKs partly inactivates PDC, and in turn re-routes pyruvate towards lactate and away from the mitochondrion for respiration. This metabolic alteration accelerates the Warburg effect and is frequently associated with increased tumorigenesis. In agreement with this, reduced expression of PDHA1 in prostate cancer is correlated with poor prognosis [42].

Accordingly, knockout of PDHA1 significantly decreases mitochondrial OXPHOS but increases glycolysis, which contributes to rapid tumor growth. Moreover, PDHA1 KO cells are able to expand a stem-like cell population, suggesting a potential role in chemotherapy resistance and migration. However, a different possibility is evident by the finding of amplification and overexpression of *PDHA1* and its phosphatase PDP1 in prostate cancer [43], where inhibition of PDC activity precludes the development of prostate cancer in mouse and human xenograft tumor models. Moreover, compartmentalized PDCs appear to have different functions, since distinct pools contribute to lipid biosynthesis for prostate cancer progression. For example, nuclear PDC promotes lipogenesis by regulating histone acetylation-mediated lipogenic gene expression [44]. Alternatively, mitochondrial PDC converts pyruvate to citrate for lipid anabolism. Taken together, it is still unclear whether PDHA1 and its regulators are tumor promoters or suppressors. Given the changing citrate-related metabolism during prostate cancer progression, it is very possible that PDC has context-dependent functions in prostate cancer.

Pyruvate kinase (PK) catalyzes the commitment step that transfers the phosphate group from phosphoenolpyruvate (PEP) to ADP and generates pyruvate and ATP. PK muscle isozymes M1 and M2 are different splice variants encoded by the same gene, *PKM*. PKM2 is correlated closely with tumorigenesis. While PKM1 is expressed in differentiated tissues, PKM2 is highly expressed in proliferating cells including embryonic and cancer cells [45]. Consistently, PKM2 is correlated with Gleason score and aggressive tumor types in prostate cancer [46, 47]. There are some potential explanations for such isoform differences. First, PKM2 is an allosterically regulated isoform. Tetrameric PKM2 actively promotes the conversion of PEP into ATP and pyruvate. Conversely, dimeric PKM2 has low catalytic activity and instead promotes the entry of glycolytic intermediates into glycolytic branch pathways such as the pentose phosphate pathway, through which cells can generate several key building blocks for their growth and proliferation

[48]. This activity change is most evident in rapidly proliferating cells that appear to adhere to the Warburg effect. Second, PKM2 has non-metabolic functions that can modulate signaling and transcriptional activity to promote prostate cancer progression. For example, in CRPC, PKM2 may partner with KDM8 and co-translocate to the nucleus to function as coactivators of HIF-1α, thereby upregulating glycolytic genes (*GLUT1, HK2, PKM2, LDHA*, etc.) and downregulating TCA cycle-related *PDHA1* and *B1* genes [49]. Moreover, the reciprocal regulation of PKM2 and HIF-1α could facilitate CRPC cell survival under hypoxic conditions and promote drug resistance [46, 50, 51]. Interestingly, PKM2 may also respond to extracellular signaling during prostate cancer metastasis. To that end, cancer-associated fibroblasts can induce PKM2 post-translational modifications and nuclear translocation. In the nucleus, PKM2 functions with HIF-1α and DEC1 to deregulate miR-205 expression and in turn promote an epithelial mesenchymal transition [52]. Several PKM2 inhibitors and PKM2 tetramerization activators have been developed and demonstrated to exhibit efficacy, including overcoming drug resistance, in preclinical models [45, 48]. Among the few studies that have been performed in prostate cancer, DASA-58, a PKM2 activator, did inhibit the lung metastasis of PC-3 cells in SCID mice [52], suggesting further studies are warranted in prostate cancer.

Lactate dehydrogenase (LDH) catalyzes the reversible conversion of pyruvate and lactate. LDH is a tetramer composed of two subunits, LDHA and LDHB. LDHA has a high affinity to pyruvate and thereby favors the reaction from pyruvate to lactate. In most tumors, increased LDHA is a hallmark for overactivated glycolysis and advanced progression and is a potential target for prostate cancer therapy [53, 54]. LDHA knockdown in prostate cancer cells both inhibits cell growth and sensitizes cells to radiotherapy [53, 55]. Conversely, LDHB exhibits context-dependent roles [56]. LDHA and LDHB have opposite roles in prostate cancer [57]. Abnormal LDH, hyperphosphorylation and high expression of LDHA and low expression of LDHB correlates with short overall survival and time to biochemical recurrence of patients with prostate cancer. Also, aberrant LDH status is regulated by fibroblast growth factor receptor 1 (FGFR1) signaling. Mechanistically, FGFR1 can phosphorylate four tyrosine residues on LDHA to stabilize the protein, thereby enhancing glycolysis and reducing oxygen consumption. However, FGFR can also repress transcription of the *LDHB* gene by inhibiting expression of the TET1 (ten-eleven translocation methylcytosine dioxygenase 1) demethylase and subsequently increasing DNA methylation at the *LDHB* promoter. Such coordinated regulation allows cancer cells to develop a robust glycolytic metabolism.

As noted above, late-stage prostate cancer cells often increase glycolysis. Thus, lactate builds up as a byproduct of excessive anaerobic metabolism. To avoid lactate-mediated toxicities and apoptosis, cancer cells express high levels of monocarboxylate transporters (MCTs) to ensure the rapid efflux of intracellular lactate. MCTs are membrane transport proteins that are responsible for the transmembrane shuttling of small carboxylates like lactate, pyruvate and short-chain fatty acids [58]. While the expression of MCTs vary in prostate cancer patients, more MCT has generally been correlated with more aggressive disease and poor prognosis [14].

Of the 14 isoforms identified, MCT1, 2 and 4 have been implicated in prostate cancer progression. MCT4 was first identified as a lactate exporter especially in highly glycolytic cells [59]. Elevated expression of MCT4 is strongly associated with high glycolytic rates in prostate cancer including CRPC and neuroendocrine prostate cancer [31, 55, 60]. However, a clinical pathology study argued against the relationship between increasing MCT4 with lactate efflux (also MCT2) in prostate cancer [61]. A significant increase of MCT4 (and MCT2) expression was observed in the cytoplasm of prostate cancer cells rather than at the plasma membrane, indicating that MCTs may be involved in organelle function instead of lactate shuttling in prostate cancer. Although further investigation is required to demonstrate why prostate cancer cells are reprogramed to express more MCT4, MCT4 does appear to be a useful prognostic factor and potential target for prostate cancer. Regarding the latter, knockdown of MCT4 significantly reduced PC-3 cell proliferation [55]. Moreover, targeting MCT4 by antisense oligonucleotides inhibits glycolysis, lactate production and cell proliferation in advanced prostate cancer [60]. Furthermore, high MCT4 expression is involved in cancer–stroma interactions thought to facilitate prostate cancer progression (discussed further below in the Section "Influence of the Tumor Microenvironment") [62].

In addition to MCT4, MCT2 promotes prostate cancer progression [63, 64]. Elevated MCT2 levels in prostate cancer have been linked to two differentially methylated regions at the *SLC16A7* locus (gene encoding MCT2). One locus, upstream of the promoter, is hypermethylated in patient samples and responsible for full length MCT2 expression. The other locus, within an internal promoter, is recurrently demethylated in patient samples and subsequently induces the expression of an alternative isoform of MCT2. This isoform contains a different set of 5′-UTR translation signals that are most likely related to the high MCT2 expression. In addition to expression, MCT2 has been demonstrated to translocate to peroxisomes during disease progression, where it binds to Pex19, thereby enhancing beta-oxidation for malignant transformation. Together, these findings suggest that prostate cancer cells are able to adaptively increase MCT2 by epigenetic and compartmentalized regulation to meet their metabolic demands.

MCT1 is a controversial target for prostate cancer therapy. In some glycolytic cancers, an MCT1 inhibitor is being investigated in clinical trials based on its potent ability to reduce tumor growth in preclinical studies [58]. In prostate cancer, unfortunately, the benefit is still undefined. Pharmacological inhibition of MCT1 by α-cyano-4-hydroxycinnamate (CHC), which has a tenfold selectivity for MCT1 compared to other MCTs, is associated with increased necrosis but does not affect xenograft tumor size [65]. However, another MCT1 inhibitor, AR-C155858, decreases cell proliferation and increases cell apoptosis in *Pten*-deficient mouse tumor tissues without substantial side effects on benign tissue [31]. Regardless, the relationship between MCT1 and prostate cancer is complicated. Though a continued decrease of MCT1 has been found from benign prostatic tissue to metastatic prostate cancer, the expression of MCT1, MCT4 and CD147 have been suggested to be markers of poor prognosis [61]. Interestingly, MCT1 expression is upregulated by hypoxia and glucose starvation [31, 62], two common features of the tumor microenvironment. However, it is unclear whether prostate cancer benefits from such MCT1 regulation.

The pentose phosphate pathway (PPP) is a glucose catabolic pathway that runs parallel to glycolysis. PPP appears to be important in prostate tumor growth, which is largely due to the contribution of the PPP in the production of nicotinamide adenine dinucleotide phosphate (NADPH) and ribose 5-phosphate for scavenging of reactive oxygen species, reductive biosynthesis and supplying of nucleotide precursors [66, 67]. Transketolase-like protein 1 (TKTL1), an enzyme involved in the non-oxidative phase of the PPP, is altered throughout prostate cancer progression [68]. Of note, metastatic prostate cancer tissue has elevated TKTL1 expression relative to benign or localized cancer, suggesting a potential role of TKTL1 in prostate cancer metastasis. Moreover, glucose-6-phosphate dehydrogenase (G6PD), the rate-limiting enzyme of the PPP, is also a critical regulator for prostate cancer. *G6PD* is upregulated in prostate cancer and is required for AR-mediated ROS modulation and cell growth [69]. At this time, the functional role of the PPP in metastasis is unknown.

Lipid Metabolism

Lipids are essential for many cell functions as they are used to form cell membranes and create membrane anchors, post-translationally modify proteins, promote various signaling pathways and are used for energy storage. Lipid metabolic pathways produce compounds such as fatty acids, steroids (including hormones and sterols like cholesterol), phospholipids and others. An important characteristic of prostate cancer cells is the dysregulation of lipid metabolic pathways. In normal cells, lipids are primarily acquired from extracellular sources, but can be synthesized in specific tissues such as liver and adipose. Fatty acids produced in these tissues can be stored or transported to other areas. In cancer cells, there is an increase in *de novo* lipogenesis, which is necessary to accommodate the excessive cell growth and proliferation that occurs in cancer [70]. Prostate cancer cells undergo this characteristic shift from a sole reliance on extracellular fatty acid uptake to augmenting *de novo* lipogenesis. This shift is accompanied by the upregulation of multiple enzymes in the fatty acid synthesis pathway and is regulated by AR signaling. Upregulation of AR signaling during prostate cancer progression is accompanied by an increase in lipogenic enzymes [71]. This is largely dependent on the major lipogenesis transcription factor sterol regulatory element-binding protein (SREBP) [72]. Interestingly, a feedback mechanism also exists where AR transcription is regulated by SREBP [73]. Beyond AR, additional genomic alterations in prostate cancer such as *PTEN* and *PML* co-deletion can also hyperactivate SREBP-mediated lipogenesis, which contributes to prostate cancer progression [74]. Thus, SREBP may represent a downstream conduit of several oncogenic signaling networks in prostate cancer.

SREBPs are important transcription factors regulating fatty acid metabolism. In mammals, SREBP exists as SREBP-1a and SREPB-1c (two variants of a single gene), and SREBP-2 [75, 76]. SREBP associates with SREBP cleavage-activating protein (SCAP) in the endoplasmic reticulum (ER) membrane. In response to a decrease in intracellular sterols, the SREBP-SCAP complex translocates to the Golgi apparatus, where SREBP is cleaved and activated, allowing it to move to the nucleus. SREBP controls the transcription of several genes involved in fatty acid and cholesterol synthesis, including *ACLY, ACACA, SCD1*, and *FASN* [77, 78]. SREBP is upregulated in various cancers and can be activated by androgens in prostate cancer, due

to the effect of AR on SCAP transcription [79]. SREBP can, in turn, further activate AR expression by binding to a sterol regulatory element (SRE) in the AR gene [80].

Lipids play an important role in forming the structure of outer and inner cell membranes. They are used to form the phospholipid bilayer as well as cholesterol-rich membrane rafts within the plasma membrane, which are important for intracellular signaling and trafficking. Since cancer cells experience rapid proliferation as well as increased synthesis and uptake of materials, membrane expansion is necessary. In addition, the metabolic shift to *de novo* lipogenesis in cancer increases lipid saturation, which may prevent cell death from oxidative damage [81]. Lipid rafts are microdomains within the plasma membrane that contain cholesterol, sphingolipids and transmembrane proteins [82]. Receptor tyrosine kinases in these lipid rafts can be activated by ligands to promote phosphorylation cascades, while the microdomain of the raft allows for the accumulation of intracellular scaffolding and adaptors and can protect against phosphatases [83, 84]. Cholesterol in lipid rafts regulates AKT signaling, a major driver of prostate cancer progression [85]. Finally, AR can interact with AKT at lipid rafts to promote oncogenic signaling [86]. However, the extragenomic roles of AR in prostate cancer are still debated.

Intracellular fat is stored in lipid droplets, which are composed of a phospholipid monolayer containing polar sterols and various transmembrane proteins, and internally, non-polar sterol esters and triacylglycerols [87]. In cancer, lipid droplets are increased due to increased lipid accumulation. Lipid droplet biogenesis also occurs in response to cell stress, such as when cancer cells undergo metabolic or oxidative imbalance due to conditions like hypoxia or nutrient starvation [88–91]. In addition to their roles in resistance to stress and as a source for excess lipid storage, lipid droplets can interact with mitochondria to regulate β-oxidation, and play a role in maintenance of ER homeostasis. Androgens increase lipid droplets, which correlate with prostate cancer aggressiveness [92]. Additionally, an increase in lipid droplets is related to the upregulation of the PI3K/AKT pathway in prostate cancer, often following *PTEN* loss, which causes the accumulation of intracellular cholesterol esterases [93].

PI3K/AKT signaling leads to the upregulation of fatty acid synthase (FASN), which is overexpressed in prostate cancer [94, 95]. FASN is also regulated by AR signaling and correlates with Gleason score and PSA levels [96, 97]. Inhibition of FASN with a small molecule inhibitor has been shown to suppress *de novo* fatty acid synthesis and tumor growth, in part by targeting AR signaling [98]. FASN expression can also be increased in response to oxidative stress such as hypoxia. This effect is via activation of the AKT pathway and SREBP-1, which regulates transcription of the *FASN* gene [99]. FASN expression is also regulated by the p300 acetyltransferase, which increases transcription of the *FASN* gene and contributes to lipid accumulation and prostate cancer growth [100]. It remains to be seen whether the effects of p300 are mediated through any of the above-mentioned transcription factors.

Lipids play important roles in the post-translational modification of proteins. Acylation by the covalent addition of palmitate or myristate to proteins can facilitate oncogenic signaling [101]. In prostate cancer, myristoylation of Src kinase increases oncogenic signaling and promotes tumor progression [101]. In addition, palmitoylation of Src contributes to prostate cancer initiation [102]. Prenylation, the addition of isoprenoids such as farnesyl or geranylgeranyl to proteins, often causes membrane association [103]. Inhibition of geranylgeranylation can induce autophagy in prostate cancer cells and reduce prostate cancer metastasis in murine models [104, 105].

While products from the TCA cycle are used for OXPHOS in normal cells, cancer cells upregulate alternative metabolic pathways to increase NADPH, in part for fatty acid production. In this context, the citrate that is produced in the TCA cycle can be shuttled to the cytoplasm to be used for fatty acid synthesis. Cytosolic citrate production can also result from increased glutamine uptake, another characteristic of prostate cancer

cells [29, 106]. Glutamine can be converted to glutamate and then to α-ketoglutarate, which can be used as a carbon source for fatty acid synthesis by conversion to cytosolic citrate via reductive metabolism by the enzymes IDH1 and ACO1 [107]. Alternately, α-ketoglutarate can replenish the carbon intermediates of the TCA cycle. This utilization of cytosolic glutamine for fatty acid synthesis and the TCA cycle is observed in cancer cells [108]. In addition, the oxidation of isocitrate to α-ketoglutarate can produce excess NADPH for fatty acid synthesis [109].

The process of fatty acid synthesis is mediated by a variety of enzymes that are involved in prostate cancer progression. Citrate is converted to acetyl-CoA in the cytoplasm by ATP citrate lyase (ACLY) [110]. ACLY activation can result from PI3K/AKT upregulation, which is common in prostate cancer [111]. Acetyl-CoA carboxylase (ACC1) converts acetyl-CoA to malonyl-CoA [112], which then is processed by fatty acid synthase (FASN), creating 16-carbon fatty acid palmitate [113]. Enzymes such as stearoyl-CoA desaturase 1 (SCD1) or elongation enzymes can function to either insert a double bond in the fatty acid chain to generate a mono-unsaturated fatty acid, or to lengthen the carbon chain, respectively. Desaturation of fatty acids is required for prostate cancer cell growth, an effect that can be blocked by inhibition of SCD1 [114]. These fatty acids are then used for energy production as well as various additional cellular functions including membrane biogenesis, protein post-translational modifications and regulation of oncogenic pathways. Cytosolic acetyl-CoA can alternately be used for the mevalonate pathway, in which it is converted to mevalonate by HMG-CoA reductase (HMGCR).

Mevalonate is a precursor for the synthesis of isoprenoids and cholesterol, which can be used to produce steroids such as androgen [115]. Statins act to reduce cholesterol levels by inhibiting HMGCR, thereby preventing the formation of mevalonate from acetyl-CoA. Statins have been shown to prevent prostate cancer progression, biochemical recurrence, and mortality [116, 117]. As such, they are under clinical evaluation for the treatment of prostate cancer. Statins cause an increase in the expression of low-density lipoprotein receptors (LDL-R) [118]. LDL-R is a plasma membrane protein that facilitates the intake of LDL into the cell via endocytosis, providing another source of cholesterol to the cell, thus regulating cholesterol homeostasis in prostate cancer cells [119]. The LDL-R pathway is regulated by various transcriptional and post-transcriptional factors. For example, the expression of LDL-R is activated by SREBP proteins, mainly SREBP-2, through interaction with the SRE, while SREBP-2 activation is inhibited by high levels of intracellular cholesterol [120]. SREBP-2 also regulates the transcription of HMGCR; together, this suggests that SREBP-2 contributes to cholesterol accumulation. SREBP-2 also has a role in regulating androgen production, which can be synthesized from cholesterol through this pathway. Interestingly, SREBP-2 can also be regulated by androgens, suggesting a feedback loop for androgen synthesis.

SREBP-2 contributes to cholesterol accumulation, which opposes the effect of the nuclear receptor liver X receptors (LXR), another important factor in cholesterol homeostasis [121]. LXRs respond to excess cholesterol by modulating the transcription of various intermediates in the cholesterol and fatty acid synthesis pathways. Two isoforms exist (LXRα and LXRβ), which differ in their tissue localization. Oxysterols, an oxidized derivative of cholesterol, can activate LXRs, which eliminate cholesterol by reducing LDL uptake [122], regulating ATP-binding cassette (ABC) transporter expression [123], and converting cholesterol to bile acid [123, 124]. ABC transporters can mediate the release of cholesterol in cells with increased levels of cholesterol by the process of reverse cholesterol transport (RCT) that is mediated by LXRs [125]. LXR has a tumor suppressive role in prostate cancer [125, 126]. In addition, the synthesis of steroids from cholesterol is regulated by the cytochrome P450 family of enzymes. *CYP27A1* encodes a cytochrome P450 oxidase that converts cholesterol into 27HC and other oxysterols. This creates a system in which 27HC inhibits prostate cancer cell growth by depleting cellular stores of cholesterol [127], possibly through the activation of LXR.

Interestingly, while prostate cancer has classically been characterized by increased fatty acid synthesis and lipid uptake, it is now clear that these tumors also exhibit and require high levels of beta oxidation [128]. While fatty acid synthesis provides lipids for various purposes and pathways in the cell, beta oxidation can provide energy from stored lipids, which can be used for cancer cell growth by providing ATP, as well as acetyl-CoA, which can be recycled into the TCA cycle and used as a second messenger or possibly rerouted for use in epigenetic regulation. The process of beta oxidation occurs mainly in the inner mitochondrial membrane, where fatty acids can be transported via carnitine transport. Carnitine palmitoyltransferase 1 (CPT1) is required for prostate cancer cell growth [129–131]. Blockade of beta oxidation at additional steps impairs prostate cancer growth and metastasis *in vivo* by inhibiting CaMKII activation [132]. Beta oxidation also occurs in peroxisomes, after which oxidized lipids can be transported to the mitochondria or can enter different cell pathways [133]. An increase in beta oxidation in prostate cancer is supported by the upregulation of α-methylacyl-CoA racemase (AMACR), an enzyme involved in catalyzing the peroxisomal and mitochondrial beta oxidation of branched-chain fatty acids [134, 135]. D-bifunctional protein (DBP), which is involved in peroxisomal beta oxidation, is also upregulated in prostate cancer [136]. Taken together, prostate cancers exhibit dynamic lipid metabolism, indicating that both the synthesis and breakdown of fats may represent therapeutic vulnerabilities for the treatment of the disease.

Amino Acid Metabolism

While glutamine is not an essential amino acid, it plays an important role during cancer cell starvation as a source of energy, carbon and nitrogen [137–139]. SLC1A5 (ASCT2) is the primary glutamine transporter in cancer [140]. Other transporters such as SLC1A4 (ASCT1) may also play a role in glutamine transport under conditions of stress in the tumor microenvironment

[141]. However, it is unclear if this is due to direct or indirect transport. Both SLC1A4 and SLC1A5 are upregulated in prostate cancer by androgens [142, 141]. However, *SLC1A4* and *SLC1A5* are not direct targets of AR [141]. While MYC is a major regulator of glutamine metabolism in many cancer types [139, 143, 144], MYC appears to regulate glutamine metabolism in prostate cancer in a context-dependent manner that is potentially influenced by the PTEN status of the cell [141]. Interestingly, glutamine uptake in prostate cancer cells is driven by several oncogenic networks such as AR, MYC and mTOR signaling.

Due to the dependence of cancer cells on glutamine, several attempts have been made to exploit this vulnerability. The drug CB-839 (Calithera Biosciences), which targets mitochondrial glutaminase, is currently being tested in early phase clinical trials for the treatment of solid tumors in combination with chemotherapeutic drugs [145]. The aggressive prostate cancer cell line PC-3 and the metastatic derivative PC-3M are sensitive to CB-839 due to their dependence on glutamine metabolism [106]. Although it is appealing to target the conversion of glutamine to glutamate, this approach has limitations as it does not consider the contributions of glutamine metabolism independent of glutaminolysis. First, glutamine can activate MAPK signaling independently of RAS [146]. Second, it was reported that amino acids, including glutamine, can activate CaMKK2 which in turn activates AMPK [147]. While AMPK was initially described as a tumor suppressor due to its upstream kinase LKB1 [148–150], a context-dependent, oncogenic role for AMPK has emerged in recent years [131, 151–153]. To take advantage of these additional aspects of glutamine metabolism, the first selective inhibitor of SLC1A5/ASCT2, V-9302, was developed [154]. This inhibitor was tested across a panel of 29 cancer cell models and *in vivo*, resulting in decreased growth, increased cell death and oxidative stress [154]. However, it should be mentioned that glutamine is not the only amino acid transported by ASCT2 [154]. Hence, the effect of V-9302 might not be due to glutamine uptake alone. Further, even though ASCT2 is the major glutamine transporter [155], additional transporters such as LAT1 and 2, as well as SNAT1-5 can also transport glutamine [154]. LAT1 and 3 are upregulated in prostate cancer, and while they are the main transporters for leucine, the transport of glutamine is also possible [142, 154]. Thus, inhibition of other amino acid transporters besides ASCT2, such as LAT1, may also decrease tumor growth [142]. An advantage of the transporters is their cell surface localization, making them theoretically accessible for other types of potentially more selective targeting (ex. antibody-mediated delivery).

Another non-essential amino acid that has been targeted for cancer treatment is arginine. Many advanced cancers, including prostate cancer, demonstrate a loss of components of the urea cycle such as argininosuccinate synthetase (ASS) [156–159]. ASS is needed for the eventual conversion of arginine from citrulline [160], which means that loss of ASS leads to a cellular dependence on the uptake of extracellular arginine. This vulnerability has been targeted by introducing the enzyme arginine deiminase (ADI) conjugated to polyethylene glycol 20 (PEG20). The PEG20 moiety decreases immune responses and increases serum half-life of ADI-PEG20, leading to depletion of circulating arginine from the serum [161]. Treatment of ASS-deficient prostate cancer cells such as CWR22Rv1 in mouse xenograft models, with ADI-PEG20 alone or in combination with docetaxel leads to reduced tumor growth [157]. As expected, cells expressing low levels of ASS (PC-3) are responsive to ADI-PEG20, while those expressing high levels of ASS (LNCaP) are resistant to ADI-PEG20 [157]. While this approach shows promising results [157], it should be considered that cells can develop resistance to arginine deprivation via a compensatory induction of ASS expression [156]. Another approach to starve cancer of arginine is the use of recombinant arginase, which converts arginine into ornithine [162]. While this approach is dependent on the levels of ornithine carbamoyl transferase (OCT) expression, several prostate cancer cell lines with low levels of OCT are responsive to arginase treatment [162].

However, this approach should be carefully considered due to the importance of the polyamine synthesis pathway that uses ornithine as a substrate. Several cancers display increased proliferation when the polyamine synthesis pathway is upregulated [163]. In prostate cancer, the enzyme ornithine decarboxylase (ODC), which converts ornithine to putrescine, is a direct target of AR [164]. Further, when ODC is overexpressed, non-malignant prostate epithelial cells can be transformed into prostate cancer cells [165]. Inhibition of the uptake of polyamines by N_1-spermine-L-lysine amide, or ODC itself by α-difluoromethylornithine, is sufficient to inhibit the growth of prostate cancer in culture and *in vivo* [166]. However, as noted below in the Section "Biofluids and Tissue Metabolism Biomarkers", the role of the polyamines in prostate cancer may not mirror that observed for other cancer types. For example, the levels of polyamines, and especially spermine, are often decreased in primary prostate cancer relative to benign prostate [167–169]. Hence, whether the polyamines, or perhaps specific polyamines, have unique roles in prostate cancer and whether these roles vary further in different disease stages, remains poorly understood.

One-carbon metabolism connects two important pathways: the folate and methionine cycles, which can then feed into the transsulfuration and polyamine synthesis pathways. As a result, one carbon metabolism regulates several cellular processes such as epigenetics via methylation (methionine cycle), DNA synthesis and repair (folate cycle), and protection against reactive oxygen species via glutathione (transsulfuration pathway) [170]. This intricate network is fueled by the amino acids serine and glycine [171]. Several key enzymes in these pathways are regulated by AR [171]. For example, the conversion of glycine to N-methylglycine (sarcosine) is facilitated by the glycine-N-methyl transferase (GNMT) and has been shown to be important for cell invasion [172]. Knockdown of GNMT in DU-145 cells results in lower amounts of sarcosine and a decrease in invasion [172]. In contrast, knockdown of the enzyme sarcosine dehydrogenase (SARDH), which is responsible for the conversion of sarcosine to glycine, leads to high levels of sarcosine and elevated invasion [173]. GNMT and SARDH are both regulated by AR and the TMPRSS2-ERG fusion [172, 174].

Another important aspect of the one-carbon cycle is its role as a source for methyl groups to perform epigenetic modifications. Methylation of histones depends on S-adenosyl-methionine (SAM), a product of the methionine cycle [170]. DNA is often hypermethylated in prostate cancer, regulating genes involved in cell cycle, DNA repair and apoptosis [175–184]. Thus, the main DNA methyltransferase 1 (DNMT1) is upregulated and activated in prostate cancer [185, 186]. Consequently, tumor formation can be altered with 5-azacitidine, an inhibitor of DNMT1 [187]. One-carbon metabolism is influenced by AR and fuels prostate epigenetic reprogramming by providing the substrate (SAM) to methyltransferases that are also upregulated by AR [170]. This is highly significant since epigenetic reprogramming leads to drug resistance in prostate cancer. Patients in phase I and II clinical trials exhibited increased chemosensitivity to docetaxel in combination with 5-azacitidine [188]. In addition to the important role of SAM in epigenetics as a substrate for methyltransferases, it can also be diverted into the polyamine synthesis pathway [170]. Given the importance of polyamines in the prostate and their association with cancer progression [163, 189], inhibition of methyltransferases may lead to an increase in polyamine synthesis. To that end, elevated levels of polyamines are associated with prostate cancer progression [190, 191] as the reduction of polyamines in CRPC patients leads to prolonged survival [192]. Interestingly, when SAM is utilized by methyltransferases it creates S-adenosyl-homocysteine (SAH), which moves along the methionine cycle to intersect with the folate cycle to create methionine and regenerate SAM [170]. Along the methionine cycle SAM is converted to SAH and further to homocysteine by the enzyme S-adenosylhomocysteine hydrolase [171]. Homocysteine can then feed into the transsulfuration pathway, resulting in the production of glutathione, thereby altering the cellular response to ROS [170]. Overall, the one-carbon cycle represents a complex network of several pathways, and we are in the very early stages of understanding the appropriate contexts for therapeutic targeting.

Hexosamine Biosynthetic Pathway

The Hexosamine Biosynthetic Pathway (HBP) branches from the traditional glycolytic pathway to synthesize UDP-GlcNAc, an essential substrate for N- and O-linked protein glycosylation and glycosaminoglycan, proteoglycan, and glycolipid production. While the HBP shunts only a small percentage of fructose 6-phosphate away from glycolysis, it is a surrogate of total cellular energy levels because it incorporates metabolites from several key metabolic pathways, including nucleotide (uracil), amino acid (glutamine), fatty acid (acetyl-CoA), and glucose (F6P) metabolism [193]. Additionally, the UDP in UDP-GlcNAc is an energetic compound that can serve as a non-ATP readout of cellular energy availability [193].

In prostate cancer, the HBP is upregulated in hormone-sensitive, localized tumor samples, as the mRNA and protein levels of the first and rate-limiting enzyme in the HBP, glutamine fructose-6-phosphate amidotransferase (GFAT/GFPT1), and the final enzyme in the HBP, UDP-N-Acetylglucosamine Pyrophosphorylase 1 (UAP1) are elevated in cancerous prostate tissues compared to matched benign samples. GFAT and UAP1 levels are also increased in response to androgens in LNCaP and VCaP cells [194, 195]. High-performance liquid chromatography (HPLC)-based evaluation of the levels of sugar nucleotides in cells revealed UDP-GlcNAc levels are high in early-stage, AR-positive cell lines (LNCaP and VCaP), but low in non-transformed human prostate cell lines (RWPE-1 and PNT2) as well as AR-negative PC-3 cells. In addition, AR-positive cell lines express ~50% more enzymes involved in the HBP compared to AR-negative cell lines. This indicates that AR mediates the upregulation of HBP enzymes, and, therefore, the flux of metabolites through the HBP in early, localized prostate cancer.

N-linked glycosylation utilizes UDP-GlcNAc to add complex sugar conjugates to proteins in the ER and Golgi bound for the outer membrane, which influence the localization and stability of those proteins [196]. The decrease in the expression of AR target genes *KLK3* and *CAMKK2* (cytosolic proteins that are not N-glycosylated) in the presence of the N-linked glycosylation inhibitor tunicamycin suggests that AR activity is dependent on positive cross talk with a membrane bound protein that relies on N-linked glycosylation for activity and/or stability. The main candidates for this crosstalk are RTKs, specifically insulin-like growth factor 1 receptor (IGF-1R). N-linked glycosylation increases RTK membrane retention time resulting in longer RTK signaling. In LNCaPs and VCaPs, glycosylation of IGF-1R increases AR-mediated transcription of IGF-1R indicating there is a positive feedback between the two that requires glycosylation-mediated IGF-1R membrane retention and subsequent IGF-1R-mediated AR activation.

O-linked β-N-acetylglucosamine transferase (OGT), the enzyme that utilizes UDP-GlcNAc to mediate O-GlycNAcylation (O-GlcNAc) of protein substrates, is also upregulated in clinical samples of localized prostate cancer and is associated with poor prognosis [195, 197, 198]. O-GlcNAc modification of Ser/Thr residues on proteins can act similarly to phosphorylation by changing the activity of a given target protein. Inhibition of OGT, either pharmacologically or molecularly, leads to a decrease in the viability of LNCaP, VCaP and PC-3 cells. It also results in a decrease in tumor size in mouse xenografts of prostate cancer and other cancer types. This could be due to the destabilization of c-MYC caused when OGT is lost. When glycosylated by OGT, c-MYC remains stable, but when c-MYC remains un-glycosylated, it becomes vulnerable to ubiquitin-mediated proteasomal degradation.

Paradoxically, in CRPC, expression of the HBP enzyme glucosamine-phosphate N-acetyltransferase 1 (GNPNAT1) is significantly decreased compared to localized prostate cancer [199]. Loss of GNPNAT1 expression increases the aggressiveness and proliferation of CRPC. This is mediated by the expression of several oncogenic cell cycle genes through activation of the PI3K-AKT pathway in cells expressing full-length AR or by specific protein 1 (SP1)-regulated expression of carbohydrate response element-binding protein (ChREBP) in cells containing the AR-V7 variant. Accordingly, addition of UDP-GlcNAc to CRPC cells decreases proliferation in cell culture and tumor growth *in vivo*. UDP-GlcNAc treatments also sensitize CRPC cells to enzalutamide. Taken together, activation of the HBP or addition of its end product UDP-GlcNAc, particularly in conjunction with a standard of care drug like enzalutamide, could have clinical efficacy in the treatment of CRPC.

Metabolic Scavenging

Autophagy is a cellular recycling process that can provide metabolites under conditions of cellular stress such as starvation. It can also help mitigate ROS by clearing out dysfunctional organelles (e.g., mitochondria), and clear the cell of protein aggregates and damaged proteins that interfere with normal cell operations. Initially, autophagy was categorized as a tumor suppressive process due to its ability to shut down proliferation and induce cell death when hyper-activated. However, recent studies indicate autophagy is contextual, exhibiting both pro-cancer and anti-cancer functions [200, 201]. In prostate cancer, autophagy has been shown to protect advanced prostate cancers against starvation and hypoxia while promoting resistance to cancer therapies [202]. Moving forward, the intricate relationship between autophagy core components, its upstream regulators, altered AR signaling, and conditions in the tumor microenvironment must be defined to determine the appropriate contexts for manipulating autophagy to treat prostate cancer.

A study of melanoma showed that knockout of the core autophagy gene *Atg7* led to a decrease in tumor growth due to DNA damage and activation of senescence [203]. Another study of melanoma showed monoallelic loss of *Atg5* led to tumor metastasis, whereas biallelic loss of *Atg5* led to increased sensitivity to *BRAF* inhibitors and decreased tumor burden [204]. These find-

ings suggest that there are dose-dependent effects of autophagy in cancer. Using genetic mouse models, deletion of *Atg7* specifically in prostate epithelia was sufficient to impair cancer progression in intact and castrated Pb-Cre, Pten$^{f/f}$ mice [205]. Interestingly, co-targeting of HK2 and ULK1-dependent autophagy suppressed the growth of *PTEN-* and *TP53-*deficient CRPC [185], indicating that autophagy inhibitors may have improved efficacy under induced conditions of metabolic stress.

It is becoming increasingly clear that autophagy plays a critical role in resistance to treatments of advanced prostate cancers [206, 207]. To that end, enzalutamide resistance in CRPC cells can be overcome by inhibition of autophagy [208–210]. Docetaxel induces autophagy through promoting Beclin1-Vps34-Atg14 complex formation in CRPC cells without affecting mTOR or p-mTOR expression [211]. Inhibition of autophagy boosts sensitivity to chemotherapy in CRPC cells, and interestingly, activation of STAT3 by IL-6 inhibits autophagy and improves chemotherapeutic efficacy [211–213]. Additionally, autophagy provides protection against the anti-cancer drug diindolylmethane (DIM) in LNCaP and C4-2B cell lines and PC-3 xenograft mouse models [214, 215]. Treatment of DIM in combination with the ULK1 inhibitor MRT 67307, chloroquine (CQ), or siRNAs against the oncogene AEG-1 or AMPK significantly reduces cell proliferation in culture and tumor growth *in vivo*. Further, when PC-3 cells are treated with the cyclooxygenase-2 (COX-2) inhibitor celecoxib, c-Jun N-terminal kinase (JNK) mediates the activation of autophagy to protect cells from celecoxib-induced apoptosis [216]. Finally, curcumin can activate autophagy and apoptosis in CRPC. Curcumin treatment in conjunction with inhibitors of autophagy increases apoptotic cell death and mitigates the protective effects of autophagy [217]. To date, however, significant anti-cancer efficacy of autophagy-targeted therapies has not been observed in prostate cancer patients. However, this could be due to the lack of potent and selective inhibitors of autophagy, a major limitation of the field.

Recently, loss of the transcription factor repressor element-1 silencing transcription factor (REST) combined with induction of autophagy was found to promote the neuroendocrine differentiation (NED) of prostate cancer. Monoamine oxidase A (MAOA), a mitochondrial enzyme, is downregulated by REST [218]. Downregulation of MAOA results in decreased autophagic flux (specifically mitophagy). Cells with downregulated MAOA have fewer neuroendocrine characteristics. Clinically, MAOA expression is correlated with prostate cancer relapse in patient samples. Thus, MAOA expression could induce NED in part through induction of autophagy/mitophagy. Like MAOA, tumor necrosis factor α-inducible protein 8 (TNFAIP8) is highly expressed in AR-negative PC-3 cells and associated with prostate cancer cell survival [219]. Autophagic flux and biomarkers of neuroendocrine differentiation are increased following the overexpression of TNFAIP8 in PC-3 cells. Another new potential regulator of autophagy is dCTP pyrophosphatase 1 (DCTPP1) [220]. DCTPP1 is typically involved in hydrolyzing dCTP to dCMP. High DCTPP1 levels track with prostate cancer progression and Gleason score as well as the progression of other cancer types. A bioinformatics study linked DCTPP1 tumor-promoting actions to autophagy, but further research is needed to determine how DCTPP1 is mechanistically linked to autophagy.

While still controversial, most data support a tumor suppressive role of autophagy in the early stages of prostate cancer and a tumor-supportive role of autophagy in later stages [202]. Current evidence suggests that inhibiting autophagy, or its upstream regulators, in combination with anti-androgens, taxanes, or other targeted therapies has clinical efficacy in advanced stages of prostate cancer. Recent findings point toward a link between autophagy and NED. How autophagy can induce this differentiation or if autophagy is just a bystander during NED is still unclear. As new ways to accurately measure autophagic flux *in vivo* become available, understanding what type of autophagy, when autophagy is activated, and where autophagy is activated will provide the needed insights into how autophagy is contributing to disease progression and therapy resistance. Further, discovery of new regulators of autophagy and inhibitors with *in vivo* efficacy are needed to enable therapeutic approaches in the clinic.

During a process known as macropinocytosis, cells engulf nearby extracellular substances and transport them to lysosomes for degradation to yield metabolites that provide cells with additional nutrients. In the context of cancer, macropinocytosis can provide starved and stressed cells with additional nutrients much like autophagy, even sharing some of the same components and regulatory pathways [221]. However, unlike autophagy, macropinocytosis can scavenge extracellular nutrients, easing the burden on cells to recycle their own components, which will eventually become depleted. Given that cancer cells can utilize macropinocytosis to survive in nutrient-starved environments, targeting its upstream regulators and core machinery could have therapeutic value.

Macropinocytosis requires PI3K-mediated production of PIP3 for membrane enclosure and RAC1 activation to prompt cytoskeletal remodeling and membrane ruffling necessary for engulfment of extracellular materials [222]. The PI3K inhibitor PTEN is the most commonly deleted tumor suppressor in prostate cancer. Concurrently, the RAC1 activator AMPK is often highly activated in prostate cancer due to the increased expression of the AR target gene and AMPK activator CaMKK2 [223]. Taken together, macropinocytosis can be highly activated in prostate cancer. Recently, it was found that macropinocytosis in PTEN-deficient prostate cancer differs from RAS-driven cancers as it did not account for all of the albumin uptake into the cell [222]. However, it was shown that necrotic debris was taken into the cell exclusively via macropinocytotic engulfment. This discovery was accomplished using fluorescently labeled murine hematopoietic cell "corpses" introduced into the media of prostate cancer cells. Necrotic debris is often available in the tumor microenvironment due to starvation-induced death initiated by the tumor cell itself or by cancer therapies that kill some but not all tumor cells. Engulfed necrotic tissue, containing organelles, proteins and other nutrients, is broken down into metabolites to sustain growth, proliferation, and survival. Proteins from necrotic debris provide amino acids for biomass. Besides increased biomass, the replenished amino acid pool activates mTOR signaling. Further, lipids are also derived from the necrotic tissue and supplement lipid biosynthesis and subsequent lipid metabolism that is critical to many prostate cancers. Hence, macropinocytosis is an emerging biological process in cancer that may warrant drug development efforts. To that end, the requirement for lysosomal activity in both autophagy and macropinocytosis may account for some of the efficacy of the reported autophagy inhibitors that function via disrupting lysosomal functions.

Regulation of Metabolic Reprogramming

Signal Transduction

The regulation of cancer metabolism by signal transduction pathways has garnered considerable attention over the past two decades. It is clear that aberrant signaling, from both intracellular and extracellular stimuli, converge to alter the central metabolism of a cancer cell to support the high demands for energy production and building blocks. In the context of prostate cancer, AR activation is tightly coupled with global metabolic alterations. Also, *MYC* amplification, *PTEN* loss and aberrant activation of PI3K/AKT/mTOR signaling, all common events in advanced prostate cancer, have profound effects on metabolic adaptation. Below, we summarize the association of several key signaling regulators with metabolic reprogramming in prostate cancer.

As the major driver of prostate cancer, the influence of AR on diverse metabolic pathways has significant implications for prostate cancer progression. Transcriptional upregulation of enzymatic genes is one of the important ways that AR works in metabolic rewiring. A common mechanism is AR directly binding to the promoters of these genes and increasing their transcrip-

tion. The expression of these critical enzymes promotes a metabolic shift that facilitates cell growth, survival and migration [153, 224–226]. A detailed description of known, direct AR metabolic target genes has been described previously [225]. Additionally, some important metabolic regulators are downstream targets of AR. For example, AR activates a CaMKK2-AMPK-mediated cascade. *CAMKK2* is a direct transcriptional target of AR and is overexpressed and highly active in prostate cancer [151–153, 227]. CaMKK2, the predominant upstream kinase of AMPK in the prostate, adapts cells to various energetic stresses. AMPK-mediated metabolic changes have been correlated with increasing intracellular ATP levels, glycolysis, glucose uptake and PGC-1α-mediated mitochondrial biogenesis [29, 131, 153, 227]. HIF-1α also coordinates with AR to mediate metabolic adaptation to hypoxia by maintaining redox balance and cell survival. In a low androgen environment, HIF-1α directly upregulates AR expression in the presence of hypoxia [228]. Meanwhile, AR can stabilize and activate HIF-1α through an autocrine loop of PI3K/AKT in a hypoxia-independent manner [229]. This crosstalk provides a rationale for the joint inhibition of AR and HIF-1α to treat prostate cancer by blocking metabolic adaption to varied androgen or oxygen levels. Of note, AR splice variants can also regulate prostate cancer cell metabolism. For example, AR-V7 can promote cell growth, migration, and glycolysis [173]. Like AR, in CRPC cells, AR-V7 can drive *de novo* lipogenesis [71]. However, AR-V7 exhibits some unique metabolic regulatory behavior. A metabolic profile showed that in AR-V7-stimulated cells, there were differences in the levels of TCA cycle intermediates [173]. Notably, AR-V7 promotes higher levels of citrate oxidation, similar to what was observed in CRPC patient samples [173]. Further, AR-V7 increases glutaminolysis and reductive carboxylation.

MYC is another common oncogene that drives prostate cancer tumorigenesis. Amplification and mutations of *MYC* are seen frequently in advanced prostate cancer and associated with poor prognosis in a subset of cases [230]. Similar to AR, MYC contributes to metabolic reprogramming partially through the activation and expression of metabolic enzymes. Mitochondrial glutaminase, GLS1, is a MYC downstream effector for glutaminolysis in PC-3 cells via miR-23a/b [143]. Additionally, glutamine uptake is regulated by MYC in a PTEN-dependent manner [141]. Many MYC-mediated effects are exerted through complex interactions. MYC-E2F1 has a greater regulation of nucleotide metabolism while MYC-HIF-1α is more involved in glucose metabolism [231]. Moreover, MYC may also play a role in lipid metabolism. Oncogene-mediated metabolic signatures in prostate cancer revealed that dysregulated lipid metabolism is induced by *MYC* overexpression [232].

PTEN loss and subsequent hyperactivation of PI3K/AKT/mTOR signaling are also common events in advanced prostate cancer. As a master regulator of metabolism, the PI3K/AKT/mTOR pathway controls nutrient uptake and utilization as well as metabolic scavenging. PI3K/AKT activation has been strongly linked to aerobic glycolytic metabolism [232]. Further, mTORC1 promotes glycolysis by increasing HK2 translation and upregulating the expression of HIF-1α [233, 234]. The mTORC2 complex further augments glycolysis through AKT-dependent HK2 activation [233]. In addition, activation of AKT via *PTEN*-deficiency increases glucose metabolism by increasing HK2 phosphorylation and expression which in turn increases intracellular ROS-mediated cell growth [235]. Moreover, AKT/mTORC1 has been suggested to influence fatty acid synthesis through the activation of SREBP and upregulation of FASN [95, 236]. Inhibition of AKT in *PTEN*-deficient cells modulates the activation of ACLY. This repression limits the conversion of citrate to acetyl-CoA which ultimately reduces histone acetylation and epigenetic regulation [234, 237]. *PTEN* loss also leads to cholesterol ester accumulation which has been linked with more aggressive diseases [93, 238]. Therapeutically, inhibition of mitochondrial complex I appears to be an effective strategy to decrease *PTEN* loss-induced cell growth [239]. This can be attributed to the fact that *PTEN*-null cells are more dependent on consuming ATP

through mitochondrial complex V. Additionally, mTOR signaling modulates amino acid metabolism in prostate cancer through its regulation of glutamine uptake, glutamine utilization, and polyamine biosynthesis [141, 240, 241]. Importantly, PI3K/AKT/mTOR signaling can function in part as a nutrient sensor by responding to changes in cellular energy status. For instance, leucine deprivation inhibits proliferation and induces apoptosis in CRPC via blocking mTORC1 signaling [242].

Of note, signaling pathways rarely work in isolation. Instead, they are greatly influenced by one another. AR regulates MYC expression in a context-dependent manner [141]. Blocking either AR or the PI3K/AKT/mTOR signaling pathway can mutually stimulate the other pathway to support cancer cell proliferation, particularly in the context of CRPC [243–245]. Moreover, AR activation induces mTOR nuclear localization and reprograms its genomic binding. In this scenario, mTOR acts as a transcriptional integrator to facilitate androgen-dependent metabolic rewiring [246]. In addition, AMPK activation is essential for *PTEN* loss-increased macropinocytosis [222]. Also, mTOR signaling promotes prostate cancer stem cell survival, an effect that is modulated by HIF-1α [247]. Considering the abundance of multiple feedback mechanisms, future therapeutic regimens may benefit from combinatorial treatment strategies, especially for overcoming drug resistance [244, 245, 248].

Non-coding RNAs

MicroRNAs (miRNAs) are small, endogenous non-coding RNAs of 18–25 nucleotides in length, which act as gene regulators. Different from transcription factors, miRNAs regulate gene expression by binding directly to the 3′-untranslated region (3′UTR) of mRNAs and inducing mRNA degradation and/or inhibiting translation. Therefore, miRNAs have been associated with a number of biological processes including proliferation, apoptosis and metabolism. Emerging evidence has revealed that altered metabolism in cancers, including prostate cancer, is regulated by miRNAs. They can either directly target the transporters, kinases and enzymes in established metabolic pathways or indirectly manipulate important signaling pathways that regulate cancer metabolic shifts. Here, we focus on the direct regulation and will briefly summarize miRNAs and their related metabolic targets in prostate cancer (Table 1).

In contrast to the largely global increase of miRNAs in prostate cancer [249], miRNAs directly regulating prostate cancer glucose metabolism are mostly downregulated. Such inhibition facilitates the stability and expression of metabolism-related mRNAs. Decreased expression of miR-132, observed in prostate cancer, can promote a metabolic shift towards glycolysis by increasing the expression of GLUT1, HK2 and PKM2 [250]. Inhibiting miR-132 is sufficient to stimulate glucose uptake, increase lactate secretion and boost cell proliferation. Similarly, miR-181b, 142, 421, 205 and 143 are also associated with the regulation of glycolysis (Table 1).

MiRNAs also regulate the TCA cycle and OXPHOS. Malate dehydrogenase 2 (MDH2), a TCA cycle enzyme, has been associated with miR-22 and miR-205 [21]. RNA-seq and proteomics data from benign prostate tissue, hormone-naïve primary prostate cancer, and CRPC samples showed that the protein expression of MDH2, which persistently increased during prostate cancer progression, was not correlated with its mRNA level. Strikingly, miR-22 and miR-205 bound directly to the MDH2 mRNA and suppressed its translation. MiR-205 also contributed to docetaxel resistance in prostate cancer by promoting a metabolic shift from glycolysis to OXPHOS. OXPHOS engagement may be a hallmark of docetaxel resistance in PC-3 cells. OXPHOS-related genes are upregulated in docetaxel-resistant PC-3 cells compared to parental PC-3 cells, while glycolytic genes are downregulated. Accordingly, restoration of miR-205 increased expression of HK2 and GLUT1 mRNA and promoted sensitivity to docetaxel. However, it remains unknown how miR-205 upregulates HK2 and GLUT1.

MiRNAs can target the PPP to provide building blocks for nucleotide biosynthesis as well as NADPH for anabolic metabolism and ROS homeostasis. miR-1 and its identical paralog miR-206 are associated with prostate cancer metabolic alterations through targeting three key PPP genes (*G6PD*, *PGD*, and *TKT*) and one carbohydrate/lipid metabolism regulation gene (glycerol-3-phosphate dehydrogenase, *GPD2*) [251]. Nuclear factor erythroid-2-related factor 2 (NRF2) promotes tumor growth by attenuating miR-1 and miR-206 expression, activating the PPP pathway, which in turn accelerates cell proliferation.

SREBP-1, -2, ACLY and *PPARA* are targets of miRNAs [252]. For instance, *ACLY* is a direct target of miR-22 [253]. MiR-22 binds to the seed sequence GGCAGCU in the 3′UTR of *ACLY* mRNA, thereby decreasing ACLY protein expression and inhibiting *de novo* lipid synthesis. As a result, miR-22 treatment is able to inhibit PC-3 cell growth and metastasis in cell and xenograft models. Moreover, SREBP-1 and -2, master transcription factors for lipogenesis and cholesterogenesis, are controlled by miR-185 and miR-342 in prostate cancer [254]. Through repressing *SREBP-1* and *-2*, these two miRNAs block the expression of *FASN* and *HMGCR*, two genes important for fatty acid and cholesterol synthesis. Compared to the non-cancerous prostate epithelial cell line RWPE-1, LNCaP and C4-2B cells have lower expression of both miR-185 and 342. Restoration of miR-185 and 342 decreases the amounts of fatty acid and cholesterol in these prostate cancer cells, and inhibits tumorigenesis, cell growth, migration and invasion.

MiR-23a and miR-23b are two miRNAs identified to regulate glutamine catabolism in prostate cancer. MiR-23a and miR-23b directly target mitochondrial glutaminase to influence cell survival [143]. Mechanistically, MYC transcriptionally impedes the expression of miR-23a and miR-23b, which increases the expression of glutaminase. This promotes the conversion of glutamine to glutamate. Glutamate serves as a substrate for ATP production or glutathione synthesis, both of which could impact cell prolifera-

Table 1 miRNAs regulating prostate cancer metabolism

miRNAs	Regulation	Target genes	Direct	miRNA function in metabolism	Tissues/cell lines	Reference (PMID)
miR-132	Down	GLUT1	Yes	Inhibit glucose uptake, lactate secretion and glycolysis	PC-3, DU-145, LNCaP, prostate cancer tissue	27398313
		PKM2 HK2	Unknown			
miR-181b	Down	HK2	Yes	Inhibit glycolysis	PC-3	28184935
miR-143	Down	HK2	Yes	Glucose metabolism	GSE21032 dataset, PC-3	26269764
miR-421	Down	PFKFB2	Yes	Inhibit glycolysis	LNCaP, 22Rv1, PC-3, LNCaP, prostate cancer tissue	26269764
miR-205	Down	HK2 GLUT1	Unknown	Promote metabolic shift from glycolysis to OXPHOS, Docetaxel resistance	Docetaxel-resistant PC-3 and DU-145	27542265
		MDH2	Yes	Inhibit MDH2 expression	Different stages of prostate cancer tissue	29563510
miR185 miR-342	Down	SREBP1 SREBP2	Yes	Inhibit the expression of FASN and HMGCT, inhibit lipogenesis and cholesterogenesis	LNCaP, C4-2B	23951060
miR23a/b	Down	GLS1	Yes	Increase glutamine catabolism	PC-3	19219026
miR22b-3p	Up	PRODH	Yes	Inhibit proline catabolism	PC-3	22615405
miR-22	Down	ACLY	Yes	Inhibit *de novo* lipid synthesis	PC-3	27317765
	Down	MDH2	Yes	Inhibit MDH2 expression	Primary prostate cancer, CRPC, PC-3	29563510
miR-17/92 cluster	Up	PPARA	Unknown	Increase lipogenesis	LNCaP	23059473
miR-1 miR-206	Down	G6PD TKT PGD GPD2	Unknown	Inhibit glycolysis	DU-145	23921124
miR-29c	Down	SLC2A3	Yes	Inhibit glucose metabolism	Prostate cancer tissue	29715514

tion. However, MYC upregulates miR-23b-3p and consequently proline dehydrogenase expression [255]. Since proline dehydrogenase can induce apoptosis, the decreased level of proline dehydrogenase protects cells against oxidative stress and increases cell survival. Clearly, additional work needs to be done to fully understand the regulation of this set of miRNAs by MYC.

Influence of the Tumor Microenvironment

While intrinsic mutations and signaling aberrations undoubtedly drive metabolic reprogramming in prostate cancer cells, it is now appreciated that the cancer microenvironment, including fibroblasts, adipocytes, immune cells as well as endothelial cells, can greatly influence metabolism and disease progression [256]. Cancer initiation, progression and metastasis all require adaptation to the harsh host microenvironments that can include a lack of nutrients, high oxidative pressure and hypoxia. Meanwhile, by interaction or signal secretion, cancer cells are able to remodel the extracellular matrix (ECM), repurpose the surrounding non-malignant cells and eventually leverage their neighbors to support their rapid proliferation. Therefore, the crosstalk between cancer cells and surrounding cells helps determine the fate of cancer and thus may provide an attractive target for cancer therapy. Here, we will focus on the metabolic interaction between prostate cancer and its microenvironment.

Hypoxia is an important factor that influences cancer progression, metastasis and drug resistance. Hypoxic areas in prostate cancer are linked to higher clinical disease stages [257]. Hypoxia promotes epigenetic and genetic adaptation and therefore induces corresponding biological changes, including metabolic reprograming, that support rapid cancer cell growth. The key regulator of this process is the HIF-1 complex, an oxygen-regulated transcription factor. Overexpression of HIF-1 has been detected in both primary and metastatic prostate cancer. HIF-1 can trigger a number of metabolic altera-

tions including the induction of glycolysis to maintain ATP levels and provide biosynthetic building blocks. We have discussed the regulatory role of HIF-1 on glycolytic genes including *HK2*, *PDK1*, *PKM2*, *LDHA* and *MCT4* above. These events ensure a glucose supply that is sufficient for anaerobic respiration, promotes lactate excretion to prevent the inhibition of glycolysis, and maintains the redox balance for proliferation and invasion. HIF-1-mediated glycolysis has been associated with resistance to androgen deprivation therapy (ADT) [258]. ADT decreases PPP in prostate cancer, while hypoxia and HIF-1 maintain glucose uptake and lactate production. Further, ADT increases the expression of glucose-6-phosphate isomerase (GPI) specifically under hypoxic conditions. These data suggest that targeting GPI or glycolysis is a viable strategy to overcome ADT resistance in hypoxic prostate tumors. In hypoxic conditions, glutamine also becomes a more significant carbon source for lipid synthesis. To that end, HIF-1-mediated PDK1 activation represses the production of citrate, which in turn strongly increases the α-ketoglutarate to citrate ratio and switches glutamine conversion from oxidative to reductive [107]. It is still not fully understood how hypoxia affects fatty acid metabolism in prostate cancer, but hypoxia induced-CPT1C expression and fatty acid oxidation may be one mechanism that helps explain how prostate cancer cells combat metabolic stress [259]. However, there are many other regulators of the hypoxic response that need to be considered including, but not limited to, p53, MYC and mTOR. To date, knowledge of their relative contributions to the hypoxic response in prostate cancer is incomplete.

Oxidative stress can also induce metabolic reprogramming and contribute to the progression of prostate cancer. Reactive oxygen species (ROS) commonly cause damage to normal cells, and facilitate tumor growth and malignant progression by inducing DNA damage and genetic alteration. Additionally, ROS promote reprogramming of cancer cell metabolism to adapt to the stressful tumor environment. Compared to normal prostate cells, prostate cancer cells harbor high levels of ROS. Several signaling pathways contribute to the observed increased ROS. First, androgens can promote ROS production by enhancing NADPH oxidase (NOX) expression, a major generator for extramitochondrial ROS, as well as the six transmembrane protein of prostate 2 (STAMP2) [260, 261]. Such ROS generation is essential for prostate cancer cell growth and invasion. Second, the loss of antioxidant proteins resulting from Nrf2 deregulation and inactivation of the glutathione S-transferase family members promotes increased ROS [262, 263]. The upregulation of thioredoxin-1 (TRX1) is one mechanism that protects CRPCs from oxidative stress following ADT [264]. Consequently, elevated ROS may lead to the accumulation of mtDNA mutations, which in turn changes cellular metabolism [263]. However, more direct evidence is needed to confirm this intriguing hypothesis. In contrast, high ROS makes prostate cancer cells more vulnerable to cell death. As such, further increasing ROS in prostate cancer has been tested as a treatment strategy [264, 265]. Interestingly, the inactivation of T cells in prostate cancer is partially due to the increased ROS that accumulates in T cells. In the tumor microenvironment, T cell activation is inhibited by nutrient deprivation and microenvironment acidification. Moreover, 1-pyrroline-5-carboxylate released by prostate cancer cells represses T cell proliferation and function by increasing ROS production and decreasing ATP production in T cells [256, 266].

Cancer associated fibroblasts (CAFs), which are derived from resident fibroblasts or other precursor cells, are the most abundant non-cancer cells in tumors and maintain a perpetually activated phenotype [267]. CAFs promote proliferation, invasion, and metastasis of prostate cancer, along with development of chemotherapy resistance. The reciprocal activation between cancer cells and CAFs occurs in prostate cancer, resulting in cancer "stemness" and EMT [268]. PKM2 is one of the key regulators of this crosstalk. The close contact of prostate cancer cells to CAFs triggers PKM2 translocation to the nucleus where it forms a trimeric complex with HIF-1α and DEC1 to stimulate EMT and an OXPHOS phenotype. In addition, cancer cells exert reciprocal effects on CAFs to influence their metabolism. The exposure of CAFs to cancer cells induces a metabolic shift in the CAFs mimicking the "Warburg effect" with higher glucose consumption and lactate production and export [269–271]. The expression of GLUT1 and MCT4 contributes to this metabolic reprogramming [62, 270]. Interestingly, lactate from the CAFs is an energy source to fuel cancer cell OXPHOS. Prostate cancer cells decrease glucose uptake via changes in GLUT1 expression but increase lactate influx via MCT1 when co-cultured with CAFs [62, 270]. Accordingly, MCT expression differences between prostate cancer cells and CAFs has been associated with poor clinical prognosis [272]. However, an IHC staining study in 96 node-negative prostate cancer specimens, while supporting the role of CAFs in promoting prostate cancers, suggested an opposite metabolic symbiotic relationship between prostate cancer and CAFs [273]. The cancer cells preferred to produce lactate and undergo anaerobic metabolism with high LDH5 (LDHA) expression while CAFs favored aerobic metabolism by LDH1 (LDHB) expression. Whether there is reversible metabolic switching that occurs between prostate cancer cells and CAFs is not known. The expression of carbonic anhydrase IX (CAIX) in CAFs is another important mediator for cancer progression. Cancer cells can stimulate the expression of CAIX in CAFs [272, 274, 275]. Since CAIX catalyzes the reversible reaction from carbon dioxide to bicarbonate and protons, it is involved in regulating the pH of the tumor microenvironment. An acidic microenvironment has been correlated with cancer invasion, dissemination and drug resistance [271]. Bicarbonate can serve as a one-carbon intermediate for growth needs [275]. Therefore, cancer induced–CAIX expression in CAFs could facilitate the establishment of an environment supportive of proliferation and metastasis. Finally, CAFs can protect prostate cancer cells against chemotherapy-induced cell death by increasing intracellular glutathione (GSH) levels [276]. CAF co-culture or conditioned medium can increase GSH levels ~30% in LNCaP cells. This prevented cells from doxorubicin-induced oxidative stress and ultimately apoptosis.

Similar to CAFs, adipocytes can influence cancer cell metabolism. While adipocytes are sparse in the prostate gland, the peri-prostatic adipose tissue (PPAT) and adipocytes in bone marrow are thought to provide a reciprocal interaction between adipocytes and prostate cancer cells that promotes disease progression [18, 277, 278]. In this regard, the prognostic value of PPAT quantity has been evaluated in several clinical studies and associated with prostate cancer aggressiveness [277]. Adipocytes may help fuel prostate cancer cells with glycerol and fatty acids via lipolysis. Using Fourier-transform infrared spectroscopy, marrow adipocytes were shown to provide lipids to CRPC metastases [279]. This work is supported by the finding that extracellular lipids provide a much greater fraction of carbons to intracellular lipid pools than previously appreciated [280]. This cell interaction may contribute directly to the growth, morphology and cytokine expression of prostate cancer metastases [281]. Interestingly, HIF-1α may play an important role in metabolic shifts after such lipid translocation. Paracrine lipids induce HIF-1α-mediated glycolysis in an oxygen-independent manner [282]. Co-culture or conditioned media induces a significant upregulation of glycolytic-associated genes including *PDK1*, *ENO2*, *HK2*, *GLUT1*, and *LDHA* in prostate cancer, which, in turn, promotes lactate and ATP production. The p62 scaffolding protein is also an important regulator of adipocyte/prostate cancer cell crosstalk [278]. Loss of p62 in adipocytes not only inhibits its own energy expenditure, but increases osteopontin secretion, which results in a lipid-rich environment that can be utilized by tumors. This metabolic reprogramming promotes tumorigenesis and invasiveness in the TRAMP mouse model. Interestingly, prostate cancer cells are capable of elevating the adipose triglyceride lipase in adipocytes, resulting in the activation of lipolysis and the subsequent supply of substrates to support cancer cell metabolism.

Exploiting Metabolic Alterations for Prostate Cancer Therapy

Diagnostic Imaging of Prostate Cancer

Prostate-specific antigen (PSA) level, Gleason score, and clinical stage are routinely used metrics to optimize prostate cancer patient treatments. Also, expression profiling applied to prostate cancer classification (ProlarisScore, OncotypeDx), provide valuable static molecular signatures, but do not fully account for the complexity of cancer progression. Since the biological behavior of prostate cancer varies widely, and markers for early disease progression are not clearly established, no consensus currently exists for the active surveillance of early onset disease. Selection criteria that control eligibility for early monitoring and the biometric changes that constitute cancer progression are lacking. Nonetheless, active surveillance has emerged as a viable management option for selected patients. However, given the physical and psychological burdens associated with active surveillance including the need for repeat biopsies [283], the development of non-invasive, *in vivo* molecular imaging for risk stratification and surveillance remains an unmet clinical need.

Molecular imaging is broadly defined as those techniques that visualize, characterize and measure the biological processes at the molecular and cellular levels in living systems. In recent years, there has been a growing interest among clinicians and basic researchers in imaging cancer metabolism as a viable diagnostic tool in oncology. This resurgent interest includes the need to understand the molecular and cellular changes that occur in metabolism, how the tumor microenvironment directly affects metabolism and how metabolic vulnerabilities can affect treatment response. Metabolic changes are emerging as hallmarks of cancer that can be exploited as imaging biomarkers and employed to determine cancer grade and severity [284]. Oncogenic mutations play a pivotal role in altering the metabolism in cancer. Real-time metabolic imaging provides an opportunity to visualize the perturbation of metabolism during cancer initiation and progression, as well as during treatment.

Hyperpolarized Imaging in Prostate Cancer

Hyperpolarized ^{13}C Magnetic Resonance (HP-MR) is emerging as a non-toxic, non-radioactive method for interrogating tissue metabolism [285]. Hyperpolarization allows for over 10,000-fold signal enhancement relative to conventional magnetic resonance imaging (MRI) or spectroscopy (MRS). After hyperpolarization, the signal enhancement of metabolites of the hyperpolarized molecules can be retained for several minutes. Techniques are being developed to extend this relaxation time so that more detailed metabolic studies can be considered. Dynamic HP-MR has been utilized for non-invasive assessment of the downstream metabolic product of glycolysis.

The Warburg effect is a hallmark of tumor growth, and detecting it provides useful information for the detection and characterization of cancer. Clinically, FDG-PET uptake is often used as a surrogate marker of the Warburg effect. In this regard, FDG-PET is useful for the diagnosis of a number of cancers in the clinic. However, slow-growing prostate tumors do not show an appreciable difference in FDG uptake compared to normal or abnormal prostate tissue; therefore, FDG-PET has poor diagnostic value in local staging of prostate cancer. Furthermore, location of the malignancy in a high background region (bladder) can hinder tumor detection. As such, an alternative metabolic imaging technique is needed that can be utilized where FDG-PET fails [286]. In addition, FDG-PET reveals only glucose uptake and phosphorylation; downstream metabolic processing of glucose cannot be detected. However, differences between indolent versus aggressive tumors are found in downstream glucose metabolism. Using hyperpolarization, small molecules important in key metabolic pathways can be imaged, as can their metabolites. One example is pyruvate and its breakdown product lactate, which are central to ATP production. Methods

of interrogating *in vivo* metabolic changes in real time would enable a more detailed understanding of tumor metabolism. Thus, hyperpolarized metabolic imaging has the potential of being a complementary diagnostic tool to FDG-PET.

The mostly widely used method for hyperpolarized metabolic imaging is dynamic nuclear polarization (DNP). DNP is a solid-state polarization method where the imaging compound is mixed with a matrix and irradiated with microwaves at low temperature (near absolute zero Kelvin) and in a high magnetic field. While performing imaging or spectroscopy at increasingly higher field strength increases the relative fraction of polarized nuclei to several parts per million (ppm), the hyperpolarization process amplifies the signal/noise 10,000–100,000-fold, thus increasing the polarized fraction to 80–90%. After the compound is hyperpolarized, it is dissolved in media, and the imaging compound is released and prepared for injection. The SPINLab polarizer (GE Healthcare) has been approved for clinical use and has been used in a Phase I trial for the diagnosis of prostate cancer.

To date, the most studied hyperpolarized ^{13}C compound is pyruvate. The utility of DNP polarized pyruvate in metabolic imaging has been explored extensively [106, 287–295]. Hyperpolarized pyruvate can be used to follow the metabolism of pyruvate to alanine, lactate and bicarbonate. The rate of hyperpolarized lactate production has been used as a marker for cancer in multiple animal studies. In addition to pyruvate, the utility of several metabolic imaging compounds in cancer diagnosis and other modalities of hyperpolarization are under investigation.

Although it is anticipated that hyperpolarized metabolic imaging could improve early diagnosis and determine the efficacy of therapies in many areas of oncology, to date most clinical efforts in this area have concentrated on prostate cancer, in part due to the above-described unique metabolism of the disease. There are two widely used methods for diagnosing prostate cancer: serum PSA levels and biopsy to determine the Gleason score. Both methods have limited sensitivity in determining the level of aggressiveness of the cancer. The PSA score can range significantly due to age or changes in prostate volume due to benign prostatic hyperplasia and inflammation. Further, PSA can be undetectable or low even in the malignant form of prostate cancer, and it cannot localize the disease if present. Determining the Gleason score from a biopsy is the preferred method of defining the aggressiveness of prostate cancer. Gleason score can be used to predict how that particular cancer will behave; however, due to the heterogeneous nature of prostate tumors, the diagnosis of the biopsied tissue might not reflect the overall aggressiveness of the disease. Sampling error can lead to high-grade cancer lesions being missed. Hyperpolarized pyruvate in mice has been shown to be able to accurately detect prostate cancer in mouse models. Moreover, the ratio of pyruvate to its metabolite lactate potentially distinguishes high-grade from low-grade disease and provides spatial information on the aggressive lesions.

Preclinical Hyperpolarized Metabolic Imaging in Prostate Cancer

The conversion of hyperpolarized [1-^{13}C] pyruvate to lactate has been used as a valuable tool to detect prostate cancer from surrounding healthy tissue as well as to differentiate between prostate cancer types and progression of the disease. Interestingly, a comparison of hyperpolarized pyruvate imaging between mice implanted with the CRPC cell line PC-3 and mice implanted with its more highly metastatic counterpart PC-3M revealed significantly lower lactate production in the PC-3M mice suggesting a greater role for OXPHOS in highly aggressive cancer cells [106]. Nuclear Magnetic Resonance (NMR) analysis of *ex vivo* tumor samples revealed significantly higher steady-state concentrations of lactate and taurine in the PC-3 tumors while aspartate, glutamate, glutamine, and succinate were significantly lower in PC-3 compared to PC-3M. Further analysis of cell culture media revealed significantly higher lactate production, but lower glutamine consumption in PC-3 cells compared to PC-3M. Thus, PC-3 cells appear more glycolytic,

while PC-3M cells heavily utilize glutamine as a source for the TCA cycle. Further, AR-positive PDX tumors produced significantly more lactate following hyperpolarized [1-^{13}C] pyruvate injections compared to AR-negative PDXs [290]. Follow-up NMR analysis of *ex vivo* tumor samples confirmed the significantly higher concentrations of lactate and succinate in AR-positive tumors compared to AR-negative tumors. In a separate study, the uptake of [1-^{13}C]pyruvate and its conversion into alanine and lactate were correlated with tumors of varying histological grade in the TRAMP mouse model of spontaneous prostate cancer [293]. Importantly, lactate production was able to stratify regions of interest into normal prostate, low-grade tumor, and high-grade tumor with high specificity and was verified via histopathology of excised tissue. While not as extensive, changes in alanine production followed the same trend as lactate across tumor grades. A follow-up study demonstrated similar findings using hyperpolarized lactate production to separate tumors into early and advanced disease groups [296]. In addition, a hyperpolarized dual agent of [1-^{13}C]pyruvate and ^{13}C urea was used to measure metabolic flux and blood perfusion in low- and high-grade prostate tumors of TRAMP mice [297]. Pyruvate and lactate signals were modeled to generate the rate constant between pyruvate and lactate (k_{PL}). The urea signal was modeled to generate the area under dynamic curve (AUC), which evaluates the distribution of the tracer in tissue, and the volume transfer constant (K^{trans}), which represents tissue permeability and perfusion. High-grade tumors were characterized by an AUC-k_{PL} mismatch possessing significantly lower urea AUC but significantly higher k_{PL} compared to low-grade tumors. High-grade tumors also possessed significantly higher K^{trans} than low-grade tumors. The contrast between high K^{trans} and low AUC suggests that high-grade tumors are very permeable with high intake and clearance rates. These metabolic and perfusion findings were supplemented with histopathologic analysis which revealed significantly higher expression of HIF-1α, LDHA/LDHB ratio, VEGF, MCT1, and MCT4 in the high-grade tumors.

The ability to test the efficacy of a multitude of different therapies is a major strength of hyperpolarized pyruvate imaging. Tumor properties of TRAMP mice were imaged at several time-points following high-intensity focused ultrasound (HIFU) ablation [298]. Metabolism was measured with hyperpolarized [1-^{13}C] pyruvate, perfusion with hyperpolarized ^{13}C urea and gadolinium dynamic contrast-enhanced (DCE) MRI, and cellularity with diffusion-weighted magnetic resonance imaging (DWI). In the fully ablated zone, the ratio of lactate/pyruvate, the mean ^{13}C urea signal, and gadolinium DCE parameters K^{trans} and AUC were all significantly reduced from baseline values by 3–4 h and remained so when measured at 1 day and 5 days post-treatment. In the partially ablated zones in the margin of the focused ultrasound beam, these values were initially reduced from baseline by 3–4 h post-treatment but had recovered by the 1-day and 5-day time-points. Hyperpolarized lactate production was significantly decreased by threefold in LNCaP prostate cancer cells treated with the AKT inhibitor MK2206 [299]. Platelet derived growth factor receptor (PDGFR) was inhibited with imatinib in mice implanted with the human PCa cell line PC-3MM2 [300]. Hyperpolarized lactate production in the tumor dropped after 2 days following treatment along with a drop of LDH activity and c-Myc protein levels. Notably, tumor volume did not change in this time although there was a reduction observed after a longer period of time. NAMPT, an enzyme needed to produce NAD, was targeted with the drug GNE-617 in PC-3 prostate cancer cells and interrogated with hyperpolarized [1-^{13}C]pyruvate and [^{18}F]FDG-PET [301]. Lactate production decreased by 6 h while FDG standardized uptake value (SUV) was reduced later by 24 h. Two human prostate cancer cell lines, DU145 and PC-3, were implanted in mice, and metabolic assays were performed before and after the administration of the LDH inhibitor FX-11 [302]. Following the injection of hyperpolarized [1-^{13}C] pyruvate, lactate production was higher in the mice harboring DU145 xenografts compared to mice harboring PC-3 xenografts. Interestingly, both *in vivo* and *ex vivo* analysis of steady-state metabolism revealed no differences in pyruvate and lactate concentrations between the cell line xenografts. FX-11 was administered to xenograft-bearing mice, and following hyperpolarized [1-^{13}C]pyruvate injections, lactate production was reduced in the mice harboring DU145 xenografts, but not the mice harboring PC-3 xenografts. This was accompanied by reduced tumor growth rate in the mice harboring DU145 xenografts but not in the mice harboring PC-3 xenografts. Thus, imaging with hyperpolarized [1-^{13}C] pyruvate is able to predict and evaluate treatment response to an LDH inhibitor. In another study, the MEK inhibitor U0126 was administered to prostate cancer PC-3 and breast cancer MCF-7 cell lines and followed with hyperpolarized [1-^{13}C]pyruvate perfusion [303]. While pyruvate-to-lactate flux decreased in the MCF-7 cells, it increased in PC-3 cells following treatment. The contrast in these findings was attributed to the combination of LDH, the enzyme that catalyzes the conversion between pyruvate and lactate, and MCT1, the membrane protein which transports monocarboxylates such as pyruvate across the cell membrane. Intracellular lactate concentration and LDH activity were increased following treatment in both cell lines, but MCT1 levels decreased in MCF-7 cells while remaining unchanged in PC-3. Thus, it is important to understand the mechanism behind these molecular therapies to accurately interpret readouts of hyperpolarized MRI assays.

In addition to pyruvate, prostate cancer models have been studied with many alternative hyperpolarized ^{13}C agents. The role of glutaminolysis in prostate cancer was interrogated by monitoring the production of glutamate following the perfusion of hyperpolarized [5-^{13}C]glutamine in PC-3 and DU-145 cell lines [304]. DU-145 cells appeared to upregulate glutaminolysis, as four times as much glutamate production was observed in these cells compared to the PC-3 cell line. The natural anticancer drugs resveratrol and sulforaphane, which can act on glutamine-dependent cellular regulators such as PI3K, were administered. Both treatments reduced the cell count of each cell line by approximately 50% which correlated with a reduction of hyperpolarized glutamate production. Alternatively, the redox status of prostate tumors in TRAMP mice was probed with the injection of hyperpolarized [1-^{13}C]dehydroascorbate (DHA) and monitoring its reduction to vitamin C [305]. This reaction was increased by 2.5-fold in the tumor compared to surrounding healthy prostate tissue, and correlated with an increase in glutathione concentration. Additionally, there was a threefold increase of [^{18}F]-FDG SUV in the tumor compared to surrounding healthy prostate tissue imaged with PET. In a proof-of-concept study, [2-^{13}C]fructose was polarized and injected into TRAMP mice [306]. Its metabolic product β-fructofuranose-6-phosphate was observed in the tumor with sufficient signal-to-noise ratio to be accurately quantified. D-[1,2,3,4,5,6,6-^{13}C$_6$] glucose-d$_7$ solution was polarized and perfused into MCF7 breast cancer and PC-3 prostate cancer cells to measure downstream metabolic products in the glycolytic and pentose phosphate pathways [307]. The pentose phosphate products 6-phophogluconate and 6-phosphogluconolactone were observed as well as the glycolytic products dihydroxyacetone phosphate, pyruvate, lactate, and bicarbonate. These resonances were fit to a kinetic model, and the free cytosolic [NAD$^+$]/[NADH] ratio was calculated, which was found to be approximately threefold higher in PC-3 cells compared to MCF-7. Further, simultaneous hyperpolarization and imaging of multiple ^{13}C-enriched compounds was demonstrated in TRAMP mice [308]. A solution containing co-polarized [1-^{13}C]pyruvate and ^{13}C-sodium bicarbonate was injected into mice, after which pyruvate, lactate, bicarbonate, and carbon dioxide peaks could all be detected. Through incorporation into the Henderson–Hasselbalch equation, the bicarbonate and carbon dioxide signals could be used to form voxel-based pH maps in the tumor. Moreover, it was shown that it is possible to simultaneously polarize [1-^{13}C]pyruvate, ^{13}C sodium bicarbonate, [1,1-^{13}C]fumaric acid and [1-^{13}C]urea. Injection of a solution containing these hyperpolarized compounds into TRAMP mice resulted in spectra where pyruvate, lactate, bicarbonate, carbon dioxide, fumarate, and urea could all be resolved. Thus, through advanced data processing and modeling, simultaneous assays of pyruvate-to-lactate flux, pH, necrosis, and perfusion could be performed, all of which are relevant to prostate cancer detection, staging, and therapeutic response.

Clinical Hyperpolarized Metabolic Imaging in Prostate Cancer

The first clinical study using HP-MR evaluated the safety and feasibility of hyperpolarized [1-^{13}C]pyruvate as an agent for noninvasively characterizing alterations in tumor metabolism for patients with prostate cancer [294]. The study population consisted of patients with biopsy-proven prostate cancer, with 31 subjects injected with hyperpolarized [1-^{13}C]pyruvate. No dose-limiting toxicities were observed, and the highest dose (0.43 ml/kg of 230 mM agent) gave the best signal-to-noise ratio for hyperpolarized [1-^{13}C]pyruvate. The results were promising because they confirmed the safety of the agent and also showed elevated [1-^{13}C]lactate/[1-^{13}C]pyruvate in regions of biopsy-proven cancer. A follow-up study with prostate cancer patients employed hyperpolarized [1-^{13}C]-pyruvate MRI to detect an early metabolic response to androgen deprivation therapy (ADT) [295]. After 6 weeks of ADT, the patient tumors showed reduced lactate production following the injection of hyperpolarized pyruvate. Although there was negligible change in tumor size on T$_2$-weighted MRI and only a modest change on apparent diffusion coefficient (ADC) imaging, this study demonstrated the ability of hyperpolarized ^{13}C MRI to detect early metabolic responses. Translation of this technology into humans encouraged additional clinical trials in prostate cancer. To date, seven clinical hyperpolarization trials with the SPINlab polarizer have been performed, and more than 20 such polarizers have been installed around the world.

Challenges of Clinical Translation of Hyperpolarized Metabolic Imaging

One challenge of clinical translation of HP-MR is dynamic HP-MRI data acquisition. The high signal-to-noise ratio provided by this imaging technique makes high-resolution acquisitions feasible. However, rapid metabolism and short longitudinal relaxation times (T_1) of the HP-MR imaging agents can limit the matrix size, and thus the spatial resolution and coverage possible, with conventional MRI. The MR acquisition techniques used for initial animal studies and the first human trial were limited in spatial coverage (typically <8 cm) and dynamic temporal information. New acquisition and analysis techniques are needed for volumetric, dynamic HP MR data with improved spatial coverage and temporal resolution.

A second challenge is post-image data processing and quantification. Dynamic MRS data can be resolved in three spatial dimensions, the spectral dimension, and time with adequate coverage and speed to enable reliable quantification of metabolic parameters. Spatial data reduction strategies, including parallel and constrained imaging methods, are crucial for reducing the number of excitations that are necessary to reconstruct data that are changing dynamically. It will be important moving forward to develop acquisition strategies that minimize uncertainty while maximizing the efficiency of spatial, temporal, and spectral encoding.

A third challenge is the reproducibility and kinetic modeling that comes from analyzing the real-time kinetic data of metabolic conversions, interpreting the underlying biochemical meaning of these parameters, and understanding the linkage of measured hyperpolarized metabolite fluxes to underlying fluxes. Clinical studies for HP imaging, like FDG-PET studies, should include tests of repeatability and precision to measure the qualitative and quantitative variability of HP imaging results using test/retest methods. There are a few kinetic modeling techniques (both unidirectional and bidirectional) that provide insights about underlying biology. But more fine-tuning is needed with these kinetic modeling approaches before the data can be presented to radiologists in a meaningful, reliable manner.

Despite these challenges, clinical oncology practice relies increasingly on anatomic imaging at different stages of patient care. HP-MR has the potential to provide a new dimension and understanding of the underlying tumor biology, thus allowing a more personalized, patient-centric approach. Despite its proven feasibility in humans and its significant potential in clinical oncology, HP metabolic imaging will still have to prove itself against established and emerging clinical techniques such as PET and demonstrate its added value in clinical practice.

PET Imaging in Prostate Cancer

PET imaging with the glucose analogue 2-fluorodeoxyglucose (FDG) verified Warburg's hypothesis of altered glucose metabolism in cancer cells. As described above, [[18F]FDG-PET is not particularly effective in imaging patients with prostate cancer [309]. Alternatively, other PET-based molecular agents including [11C-choline, [11C-acetate, [18F-fluciclovine, and [[18F]-PSMA have been successfully employed for prostate cancer imaging (Table 2). Prostate cancer patients exhibit elevated total choline, primarily due to an increase in phosphocholine and glycerophosphocholine. As a PET tracer, [11C-choline is readily incorporated into cells through phosphorylcholine synthesis and is integrated largely into membrane phospholipids [310, 311]. [11C-acetate is primarily viewed as an indirect biomarker of fatty acid synthesis, which is also upregulated in prostate cancer. Imaging of prostate cancer with

Table 2 PET-imaging agents currently employed in the diagnostic imaging of prostate cancer

PET-imaging agents for prostate cancer	Reference (PMID)
[18F-FDG	12209157
[11C-choline	9627331
	11007527
[11C-acetate	15235071
[18F-fluciclovine	28267449
[18F-PSMA	27789722

^{11}C-acetate thus provides information about biosynthesis [312]. [^{18}F]-fluciclovine and [^{18}F]-prostate specific membrane antigen (PSMA) are also currently being used in PET imaging of prostate cancer [313, 314]. Fluciclovine, which is a synthetic analog of the amino acid L-leucine that is preferentially taken up by prostate cancer cells, can predict disease relapse following ADT. PSMA is a transmembrane protein that is overexpressed in many prostate cancers. PSMA-PET appears to be a promising new agent, especially for the detection of metastatic prostate cancer. Table 2 highlights some of the PET-based molecular imaging agents employed for diagnostic imaging of prostate cancer.

Biofluid and Tissue Metabolite Biomarkers

Although LDH is an intracellular enzyme involved in metabolism, it can also be released into the bloodstream when tissues are injured or during progression of diseases like cancer. Therefore, serum LDH, often detected with an enzyme-linked immunosorbent assay (ELISA), is an established prognostic indicator for progression and overall survival probability in many malignancies including prostate cancer [315]. While LDH levels cannot monitor the early stages of prostate cancer [22], elevated LDH levels strongly associate with poor outcome in patients with metastatic CRPC. A clinical study collected data from 1101 CRPC patients from 1991 to 2001 and established a prognostic model that consisted of LDH, PSA, alkaline phosphatase, Gleason score, Eastern Cooperative Oncology Group performance status, hemoglobin, and the presence of visceral disease to predict patient survival. These parameters were found to predict patient survival and stratified mCRPC patients [316]. Another study investigated 165 patients from 1998 to 2003 and noticed abnormally high serum LDH levels in patients with bone metastasis. Notably, a combination of biomarkers that includes LDH is now being tested as a surrogate for survival. One example is a phase III trial of abiraterone acetate plus pred-

nisone versus prednisone alone in patients with metastatic CRPC previously treated with docetaxel. Circulating tumor cell counts combined with LDH levels were found to satisfy the statistical surrogacy requirements [317]. However, in a clinical study of radium-223 safety and efficacy with CRPC patients, total alkaline phosphatase and LDH were indicated to be correlated with overall survival but did not meet the surrogacy requirements [318]. Hence, while undoubtedly useful as a clinical biomarker, the value of LDH as a bona fide surrogate marker is still unresolved.

NMR spectroscopy is a routinely employed analytical technique in metabolomics. NMR has been used extensively in biomarker discovery to detect, grade, and intervene in the therapy of prostate cancer. Researchers generally use blood serum and tissue biopsies for NMR-based metabolomics to better understand prostate cancer metabolism. In NMR-based metabolomics studies, detection of proton (^{1}H) nuclei is used with the next detected nuclei in descending order being ^{13}C, ^{31}P, and ^{23}Na. Studies involve extracting metabolites from liquid biopsies such as blood serum, fluids and urine or intact tissues in rotors for high resolution-magic angle spinning (HR-MAS) NMR spectroscopy. The metabolite extraction, data acquisition and analysis are explained in detail in many NMR-based metabolomics reviews, which we recommend for further reading [319–322].

In vivo magnetic resonance spectroscopy (MRS) and *ex vivo* HR-MAS have shown altered total choline, creatine, polyamines, myo-inositol, and citrate as biomarkers to detect prostate cancer [167, 323, 324]. In addition, lactate and alanine are elevated in prostate tumors compared to normal tissues [325]. Conversely, citrate and polyamines decrease in concentration within primary prostate cancers compared to healthy prostate. To that end, citrate and spermine can be used to distinguish indolent from malignant prostate cancer [167–169]. This is consistent with the proposal that spermine is an endogenous inhibitor of primary prostate cancer. In another NMR-based metabolomics study, analyses of 102 serum samples, consisting of 40 low-grade prostate cancers, 30 high-grade prostate cancers and 32 healthy control samples, demonstrated that the combination of biomarkers sarcosine, pyruvate, alanine, and glycine detects and differentiates low-grade prostate cancer, high-grade prostate cancer, and healthy controls [326]. Using HR-MAS spectroscopy, ethanolamine was also reported to be decreased in prostate cancer compared to benign tissues [327]. An *in vitro* ^{31}P NMR study has shown that phosphocholine, glycerophosphocholine, phosphoethanolamine and glycerophosphoethanolamine were significantly altered in prostate cancer tissues compared to benign prostatic hyperplasia [328].

Mass spectrometry is an effective tool to analyze the molecular composition of clinical prostate cancer biofluids and biospecimens [329–331]. Liquid chromatography/mass spectrometry (LC/MS) is frequently used for targeted metabolomics studies because of high sample throughput, assay sensitivity and ability to measure multiple molecules in a complex biologic matrix. A common approach for LC/MS analysis is multiple reaction monitoring (MRM) using a triple quadrupole mass spectrometer [332, 333]. An LC/MS MRM platform can identify and quantify prostate cancer metabolites unambiguously in metabolic pathways such as glycolysis, respiration, the TCA cycle, steroid and lipid metabolism, amino acid and nucleotide metabolism [159, 334]. To determine individual metabolites, there are various open-source and commercial software packages, such as Mass Profiler (Agilent), available for raw LC/MS data analysis. Alternatively, an internal reference library including m/z ratios, retention time and parent/product spectra is used from authentic chemical standards and isotopically labeled standards to characterize specific metabolites. Sample normalization is achieved by adding internal standards, thereby limiting sample variation due to batch effects [335]. As described above, there are a number of unique metabolic alterations in prostate cancer that can be readily detected using MS-based approaches. Thus, small molecule biomarkers determined by LC/MS show promise in differentiating aggressive prostate cancer phenotypes and may help guide future treatment strategies.

Drug Development

Prostate tumors are metabolically distinct among solid tumors owing to their enhanced reliance on mitochondrial oxidative phosphorylation [9] and a marked lipogenic character [336]. While current approaches for the management of metastatic disease are focused primarily on the inhibition of AR, new insights into prostate tumor metabolism are emerging by identifying metabolic dependencies that may be leveraged for therapeutic benefit. Moreover, AR signaling itself has dramatic effects on cellular metabolism [153], suggesting that AR-driven metabolic processes may be viable targets for treating AR-positive CRPC. While clinical translation of metabolic inhibitors is an active area of interest, to date, no drug specifically targeting metabolic endpoints has received FDA approval for the treatment of prostate cancer. This section will review briefly the emerging strategies to perturb prostate tumor metabolism pharmacologically with a focus on oxidative phosphorylation, lipogenesis, and glutaminolysis (Table 3).

Metformin has been a frontline agent for the management of type 2 diabetes for decades. A variety of epidemiological studies indicated that metformin use is associated with decreased cancer incidence in diabetic patients, sparking interest in metformin as an anticancer agent [337, 338]. However, subsequent retrospective studies found no link between metformin use and decreased cancer risk [339–350]. In fact, some studies found that increased metformin use correlated with more aggressive prostate cancer [339, 344]. As of January 2019, on ClinicalTrials.gov, there are nine actively recruiting clinical trials for the use of metformin in prostate cancer. Importantly, the first prospective clinical trials directly testing the effect of metformin on prostate cancer have recently been completed and shown to have limited efficacy [351, 352]. Although disappointing, these results may be due to an incomplete understanding of how best to select a patient population that would benefit from such an approach. Metformin acts pleiotropically, though the major molecular target is thought to be complex I of the electron transport chain [353]. At this time, it is unclear

Table 3 Emerging treatments for prostate cancer that target metabolism

Compound	Target	Preclinical model	Effect	Reference (PMID)
Oxidative phosphorylation				
Metformin	ETC Complex I, Pleiotropic	Multiple preclinical studies, retrospective analyses, and multiple ongoing clinical trials	Inhibits tumor proliferation in multiple tumor types, potential survival benefit in large retrospective prostate tumor cohort, ongoing clinical trials.	29940252 30150001 28444639 29075616 27746051
IACS-010759	ETC Complex I	Brain cancer and acute myeloid leukemia	Inhibits proliferation, depletes macromolecule pools, and induces apoptosis. *In vitro* and *in vivo* data, ongoing clinical trial.	29892070
MSDC-0160	Mitochondrial Pyruvate Carrier	Hormone-responsive and castrate-resistant AR-positive prostate cancer	Decreases mitochondrial oxygen consumption, depletes TCA intermediates, inhibits lipogenesis, activates integrated stress response, suppresses cell proliferation and tumor growth. *In vitro* and *in vivo* data.	31198906
Lipogenesis				
IPI-9119	FASN	Hormone responsive and castrate resistant AR positive prostate cancer	Antagonizes growth through metabolic reprogramming and results in reduced protein expression and transcriptional activity of full-length AR and splice variant AR-V7. *In vitro* and *in vivo* data.	30578319
ND-646	Acetyl-CoA Carboxylase	Non-small-cell lung cancer	Suppresses fatty acid synthesis, inhibits tumor growth in KRAS p53 and KRAS Lkb1 autochthonous mouse models.	27643638
MT 63–78	AMPK	Hormone responsive and castrate resistant prostate cancer. AR positive and AR negative.	Inhibits cell proliferation, induces mitotic arrest and apoptosis. Constitutively activates AMPK and suppresses lipogenesis. *In vitro* and *in vivo* data.	24497570
Fatostatin	SREBP-SCAP	Metastatic and non-metastatic autochthonous models of mouse prostate cancer	Inhibits lipogenesis, blocks tumor growth and metastatic spread.	29335545
Glutaminolysis				
CB839	Glutaminase	triple-negative breast cancer, lung cancer	Antiproliferative activity and decreased glutamine consumption in triple negative breast cancer models. *In vitro* and *in vivo* data. Radiosensitization in lung tumor models. *In vitro* and *in vivo* data	24523301 30557074
V-9302	SLC1A5/ ASCT2	Multiple cell line models, colon cancer xenografts	Attenuates proliferation, increases apoptosis and oxidative stress. *In vitro* and *in vivo* data.	29334372

whether the potential anti-cancer effects of metformin are due to direct tumor effects or tumor-indirect/host effects. Recent data suggests metformin can exert direct anti-cancer effects in ovarian cancer through inhibition of cell-intrinsic mitochondrial metabolism [354]. Whether this will also be the case in prostate cancer is currently unknown.

But, given the dependency of prostate cancer on mitochondrial OXPHOS, a similar mechanism of anti-cancer effects is possible [355]. Certainly, prospective, randomized controlled clinical trials will be necessary to evaluate these associations and define the clinical utility of metformin in prostate cancer.

OXPHOS is an emerging target in cancer therapy [356] and the metabolic properties of prostate cancer suggest agents inhibiting oxidative phosphorylation may be a particularly effective strategy for the treatment of this disease. Though OXPHOS is required for cellular proliferation, the molecular mechanisms underpinning this requirement remained unclear until recently. One major role of respiration in proliferating cells is to provide electron acceptors for aspartate biosynthesis, which in turn enables nucleotide biosynthesis [357, 358]. Likewise, respiration is intimately coupled with TCA function, and TCA cycling supplies a variety of anabolic pathways via cataplerosis [359]. OXPHOS can be inhibited directly by targeting components of the electron transport chain, or indirectly by preventing generation of the reducing equivalents required to power the electron transport chain. For example, IACS-010759 is a complex I inhibitor that dramatically inhibits OXPHOS and, by extension, tumor growth in a variety of pre-clinical models of brain cancer and acute myeloid leukemia [360]. Ongoing efforts to examine the efficacy of IACS-010759 in prostate cancer are underway. Indirect inhibition of OXPHOS with MSDC-0160, which inhibits the mitochondrial pyruvate carrier, restricts oxidative phosphorylation by depleting metabolic intermediates in the TCA cycle and dramatically decreasing reducing equivalents. The net effect of MSDC-0160 administration is growth restriction in pre-clinical models of hormone-responsive and castrate-resistant prostate adenocarcinoma [361]. Moreover, the mitochondrial pyruvate carrier is directly regulated by AR, suggesting that mitochondrial pyruvate import is a required metabolic process for AR-mediated proliferation in the hormone-responsive and castrate-resistant settings. Regardless of the way in which it is achieved, inhibition of mitochondrial oxidative phosphorylation in prostate cancer is expected to hold therapeutic promise, and therefore, clinical trials to gauge efficacy are warranted.

Another prominent metabolic feature of prostate cancer is the *de novo* synthesis and accumulation of lipids. Therapeutic inhibition of lipogenesis in prostate cancer is feasible because multiple points along the lipogenic pathway are targets for drug design. For example, direct inhibition of fatty acid synthase in prostate cancer with a novel small molecule, IPI-9119, repressed xenograft growth and elicited endoplasmic reticulum stress, which resulted in the downregulation of AR protein expression. Similarly, direct inhibition of acetyl-CoA carboxylase, a rate-limiting lipogenic enzyme, with the allosteric inhibitor ND-646 resulted in the repression of tumor growth in preclinical models of non-small-cell lung cancer. The lipogenic pathway can also be suppressed by preventing the master transcriptional regulator of lipogenesis, SREBP, from interacting with its activating protein, SCAP, using Fatostatin. Fatostatin blocks prostate tumor growth and metastasis in *Pten*-deficient autochthonous mouse models of prostate cancer [74]. Lipogenesis may also be repressed by constitutively activating AMPK (which in turn applies an inhibitory phosphorylation to acetyl-CoA carboxylase) or by depleting the cellular citrate pool (which serves as the primary lipogenic substrate). AMPK activation has been achieved with the direct AMPK activator MT 63–78, while citrate depletion has been demonstrated using MSDC-0160. In both cases, cellular lipid pools were depleted and prostate tumor growth suppressed. Inhibition of lipogenesis may be achieved at multiple points along the pathway, suggesting anti-tumor efficacy in the preclinical setting with the expectation of clinical trials on the horizon.

Studies have demonstrated therapeutic efficacy of inhibiting glutamine uptake in prostate cancer models [141, 362]. Additional studies using current generation glutaminase inhibitors (e.g. CB-839) or glutamine uptake inhibitors (e.g. V-9302) are warranted to gauge the translational potential of glutamine restriction in prostate cancer. Curiously, some AR-positive cell line models do not require glutamine for growth in culture, while AR-negative cells such as PC-3 and DU-145 are heavily reliant on this amino acid. These data suggest fundamental metabolic differences between AR-positive and

AR-negative prostate cancer. However, glutamine can be used to maintain the TCA cycle during impaired mitochondrial pyruvate transport [363]. In accordance, inhibition of the mitochondrial pyruvate carrier sensitizes AR-positive prostate tumor models to glutamine restriction [361]. Additional work in this area is expected to clarify the requirement for glutamine in prostate tumors *in vivo*.

Metformin is currently being used in several clinical trials as an adjuvant to current therapies. In addition, multiple drugs that inhibit OXPHOS have been demonstrated to be efficacious in the preclinical setting. Early phase clinical trials are testing IACS-010759 for the treatment of leukemia and solid tumors including prostate cancer. Lipogenesis remains a promising therapeutic target, and multiple new agents targeting various points along this pathway demonstrate encouraging preclinical results. Although glutamine reliance in prostate cancer is heterogeneous, current observations indicate it may be most prominent in aggressive models of the disease, suggesting that agents which restrict glutamine metabolism may be useful for the management of treatment emergent, aggressive variant prostate cancer. Prostate tumor metabolism is complex, dynamic, and not yet fully understood. Nevertheless, multiple new drugs targeting critical pathways including OXPHOS, lipogenesis, and glutaminolysis have been described and are nearing or entering clinical trials.

Conclusions and Future Directions

The field of cancer metabolism has seen a resurgence in recent years. This is in part due to a realization that many of the alterations observed in oncogenic and tumor suppressive pathways impact tumorigenesis through shifting the metabolism of tumor cells. These changes allow a cancer to use a greater array of nutrients to feed anabolic tumor processes and withstand the harsh tumor microenvironment. Importantly, the unique metabolism of a cancer relative to benign tissue offers new opportunities for therapeutic intervention. The most direct benefit to patient care will likely come from the establishment of new clinical biomarkers and development of novel treatment modalities. Tumors that become overly dependent on select metabolic pathways should be more susceptible to disruption of these pathways. Further, the differential use of nutrients and production of new metabolites should also yield new biomarkers that can distinguish patient populations based on their underlying biology. At a minimum, these biomarkers should provide correlative data points that are prognostic. However, biomarkers that provide an accurate readout of known pathways could also be used as pharmacodynamic markers of new treatments that target these same pathways. For example, monitoring hypoxia using ^{18}F-FAZA-PET imaging may hold promise as a non-invasive approach to assess the response to novel OXPHOS inhibitors that are currently under clinical investigation [360]. Importantly, if biomarkers can be developed and validated to detect the causal metabolic changes that drive the disease, this would create a powerful new precision medicine approach that could guide the selection of patients that would benefit the most from new metabolic-targeted therapies.

Several major challenges lie ahead for the field of prostate cancer metabolism. First, linking correlative changes in metabolism to causal driver events remains a major obstacle. Metabolite levels assessed in large patient cohorts have detected changes that correlate with prostate cancer [364–367]. However, finding causal events within these studies has been elusive. It should be noted that many large clinical studies rely heavily on the measurement of metabolites from serum, which will be heavily influenced by other systemic differences between patients. Alternatively, the direct analysis of metabolites from patient tumor samples is more technically challenging.

A second major challenge is our lack of ability to separate cell type-specific metabolism from a heterogeneous tumor cell population. This inability to account for the metabolic crosstalk between different cell types within a tumor (ex. a metabolite secreted from a tumor-associated macrophage is taken up and used by the cancer cell) limits our understanding of the metabolic relationship a cancer cell can have with the tumor microenvironment. To overcome this, researchers could focus on transcriptomics to assess alterations in the expression of metabolic genes. This would allow the application of laser capture dissection and single cell sequencing to overcome issues regarding heterogeneity. A major drawback of this approach is that it would miss metabolic changes that occur as the result of posttranscriptional alterations. Clearly, the development of new approaches that can define the different metabolic phenotypes in a heterogeneous sample would be a major breakthrough.

A third challenge in the field is the difficultly in delineating the etiology of an observed change in metabolism. One of the first questions that often arise following the determination of a significant change in metabolite levels is whether the change resulted from alterations in the synthesis or breakdown of the metabolite. To date, much of the metabolomics data reported for prostate cancer has been limited to steady state metabolomics. This is because such studies can be performed on flash frozen tissues without the prior need for additional labeling. While amenable to the study of archived samples, a limitation of this approach is that it provides only a snapshot in time of the metabolite levels. To overcome this limitation, the use of labeled isotopes, either radioactive or stable (the latter becoming increasingly common in metabolomics), can provide additional information on the flux of a particular metabolite, and more specifically the labeled atoms on the metabolite, through a specific metabolic pathway. As a result, techniques such as stable isotope tracing measured using NMR or MS can provide additional context to any changes observed using steady state metabolomics. Therefore, these two approaches can complement one another to generate a deeper understanding of cancer metabolism.

A fourth challenge, which is related to the previous point, is that isotope tracing studies require, by definition, a prior labeling step. Although this is relatively straightforward for cell culture experiments, it is significantly more difficult to do *in vivo*. This is due to the constant exchange of nutrients and waste products throughout the body. Despite the technical challenges that *in vivo* isotope tracing pose, the potential differences in tumor metabolism *in vivo* compared to cell culture [368, 369] indicate that these types of analyses will still be critical to complete our understanding of prostate cancer metabolism in the important context of the tumor microenvironment. To overcome these issues, new methods have been developed to perform stable isotope tracing in preclinical models of cancer [370]. These methods use constant infusions of a labeled nutrient of interest (ex. ^{13}C-labeled glucose) over time until steady state concentrations are achieved in both the blood and tumor. Then tumors/tissues are rapidly harvested/frozen for subsequent NMR or MS analysis. Importantly, because these tracers are not radioactive, they are amenable for use in patients where they have begun to yield important insights and importantly, challenge previous dogma [369, 371, 372].

Finally, while the general metabolic changes that occur during the initiation of prostate cancer have been well described, our understanding of the shifts that occur in the later, lethal stages of the disease is limited. Importantly, new subtypes of prostate cancer have emerged (ex. neuroendocrine-like prostate cancer) as the unwanted by-product/mechanism of resistance of new and improved AR-targeted drugs such as enzalutamide [373]. Our understanding of the underlying biology of these emerging subtypes, including their metabolic dependencies, is still in its infancy (Fig. 2). Given that these cancers are among the deadliest forms of the disease and that they are becoming more common, efforts are ongoing to define their molecular drivers. The significant genomic, transcriptomic and even morphological differences between many of these aggressive cancers and their potential adenocarcinoma origins suggest that they have unique metabolic requirements. Thus, these new forms of prostate cancer may be excellent targets for the development of novel metabolic-based biomarkers and/or therapies to identify and treat this advanced stage of the disease.

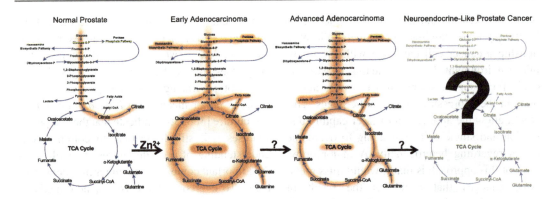

Fig. 2 Continued evolution of prostate cancer metabolism. The emergence of a new subtype of advanced prostate cancer termed neuroendocrine-like prostate cancer (NEPC; also commonly referred to as small cell-like prostate cancer (SCPC) and/or aggressive variant prostate cancer (AVPC) depending on its features) raises new questions regarding the metabolic phenotype of this form of the disease. Is the metabolism of NEPCs similar to that of advanced adenocarcinomas? Are there unique features that could be exploited for detection or treatment purposes?

Acknowledgements We thank Kelly Kage (UT MD Anderson Cancer Center) for assistance with the figures. This work was supported by grants from the National Institutes of Health (R01CA184208 to D.E.F.; P50CA094056, U54CA151668 and R21CA185536 to P.B.), American Cancer Society (RSG-16-084-01-TBE to D.E.F.), an Institutional Research Grant (to P.B.), startup grants from the University of Texas MD Anderson Cancer Center (to P.B. and D.E.F.), a grant from the Gulf Coast Consortium (to P.B.) and generous philanthropic contributions to The University of Texas MD Anderson Moon Shots Program (to D.E.F.) and Koch Foundation Genitourinary Medical Oncology Funds (to P.B.). This work was also supported by an Antje Wuelfrath Gee and Harry Gee, Jr. Family Legacy Scholarship (to C.L.), an American Legion Auxiliary Fellowship (to D.A.), a CPRIT Research Training Grant Award (RP170067 to T.C.S.) and a GCC/Keck Center CCBTP postdoctoral fellowship (CPRIT RP170593 to S.P.). M.T. also acknowledges support from the Neubauer Family Foundation.

References

1. O. Warburg, On the origin of cancer cells. Science **123**(3191), 309–314 (1956). https://doi.org/10.1126/science.123.3191.309
2. O. Warburg, F. Wind, E. Negelein, The metabolism of tumors in the body. J. Gen. Physiol. **8**(6), 519–530 (1927). https://doi.org/10.1085/jgp.8.6.519
3. D. Hanahan, R.A. Weinberg, Hallmarks of cancer: the next generation. Cell **144**(5), 646–674 (2011). https://doi.org/10.1016/j.cell.2011.02.013
4. L.C. Costello, R.B. Franklin, A comprehensive review of the role of zinc in normal prostate function and metabolism; and its implications in prostate cancer. Arch. Biochem. Biophys. **611**, 100–112 (2016). https://doi.org/10.1016/j.abb.2016.04.014
5. L.C. Costello, R.B. Franklin, P. Feng, Mitochondrial function, zinc, and intermediary metabolism relationships in normal prostate and prostate cancer. Mitochondrion **5**(3), 143–153 (2005). https://doi.org/10.1016/j.mito.2005.02.001
6. L.C. Costello, R.B. Franklin, Zinc is decreased in prostate cancer: an established relationship of prostate cancer. J. Biol. Inorg. Chem. **16**(1), 3–8 (2011). https://doi.org/10.1007/s00775-010-0736-9
7. L.C. Costello, R.B. Franklin, Aconitase activity, citrate oxidation, and zinc inhibition in rat ventral prostate. Enzyme **26**(6), 281–287 (1981). https://doi.org/10.1159/000459195
8. L.C. Costello, R.B. Franklin, Novel role of zinc in the regulation of prostate citrate metabolism and its implications in prostate cancer. Prostate **35**(4), 285–296 (1998). https://doi.org/10.1002/(SICI)1097-0045(19980601)35:4<285::AID-PROS8>3.0.CO;2-F
9. L.C. Costello, R.B. Franklin, The clinical relevance of the metabolism of prostate cancer; zinc and tumor suppression: connecting the dots. Mol. Cancer **5**, 17 (2006). https://doi.org/10.1186/1476-4598-5-17
10. L.C. Costello, Y. Liu, R.B. Franklin, M.C. Kennedy, Zinc inhibition of mitochondrial aconitase and its importance in citrate metabolism of prostate epithelial cells. J. Biol. Chem. **272**(46), 28875–28881 (1997). https://doi.org/10.1074/jbc.272.46.28875
11. M.M. Desouki, J. Geradts, B. Milon, R.B. Franklin, L.C. Costello, hZip2 and hZip3 zinc transporters are down regulated in human prostate adenocarcinomatous glands. Mol. Cancer **6**, 37 (2007). https://doi.org/10.1186/1476-4598-6-37
12. R.B. Franklin, P. Feng, B. Milon, M.M. Desouki, K.K. Singh, A. Kajdacsy-Balla, O. Bagasra, L.C. Costello, hZIP1 zinc uptake transporter down regulation and zinc depletion in prostate cancer. Mol. Cancer **4**, 32 (2005). https://doi.org/10.1186/1476-4598-4-32

13. M.C. Franz, P. Anderle, M. Burzle, Y. Suzuki, M.R. Freeman, M.A. Hediger, G. Kovacs, Zinc transporters in prostate cancer. Mol. Aspects Med. **34**(2-3), 735–741 (2013). https://doi.org/10.1016/j.mam.2012.11.007

14. E. Eidelman, J. Twum-Ampofo, J. Ansari, M.M. Siddiqui, The metabolic phenotype of prostate cancer. Front. Oncol. **7**, 131 (2017). https://doi.org/10.3389/fonc.2017.00131

15. P. Feng, J.Y. Liang, T.L. Li, Z.X. Guan, J. Zou, R. Franklin, L.C. Costello, Zinc induces mitochondria apoptogenesis in prostate cells. Mol. Urol. **4**(1), 31–36 (2000)

16. P. Feng, T.L. Li, Z.X. Guan, R.B. Franklin, L.C. Costello, Direct effect of zinc on mitochondrial apoptogenesis in prostate cells. Prostate **52**(4), 311–318 (2002). https://doi.org/10.1002/pros.10128

17. R.B. Franklin, L.C. Costello, Zinc as an anti-tumor agent in prostate cancer and in other cancers. Arch. Biochem. Biophys. **463**(2), 211–217 (2007). https://doi.org/10.1016/j.abb.2007.02.033

18. F. Cutruzzola, G. Giardina, M. Marani, A. Macone, A. Paiardini, S. Rinaldo, A. Paone, Glucose metabolism in the progression of prostate cancer. Front. Physiol. **8**, 97 (2017). https://doi.org/10.3389/fphys.2017.00097

19. I. Rishi, H. Baidouri, J.A. Abbasi, R. Bullard-Dillard, A. Kajdacsy-Balla, J.P. Pestaner, M. Skacel, R. Tubbs, O. Bagasra, Prostate cancer in African American men is associated with downregulation of zinc transporters. Appl. Immunohistochem. Mol. Morphol. **11**(3), 253–260 (2003)

20. P.B. Makhov, K.V. Golovine, A. Kutikov, D.J. Canter, V.A. Rybko, D.A. Roshchin, V.B. Matveev, R.G. Uzzo, V.M. Kolenko, Reversal of epigenetic silencing of AP-2alpha results in increased zinc uptake in DU-145 and LNCaP prostate cancer cells. Carcinogenesis **32**(12), 1773–1781 (2011). https://doi.org/10.1093/carcin/bgr212

21. L. Latonen, E. Afyounian, A. Jylha, J. Nattinen, U. Aapola, M. Annala, K.K. Kivinummi, T.T.L. Tammela, R.W. Beuerman, H. Uusitalo, M. Nykter, T. Visakorpi, Integrative proteomics in prostate cancer uncovers robustness against genomic and transcriptomic aberrations during disease progression. Nat. Commun. **9**(1), 1176 (2018). https://doi.org/10.1038/s41467-018-03573-6

22. K. Naruse, Y. Yamada, S. Aoki, T. Taki, K. Nakamura, M. Tobiume, K. Zennami, R. Katsuda, S. Sai, Y. Nishio, Y. Inoue, H. Noguchi, N. Hondai, Lactate dehydrogenase is a prognostic indicator for prostate cancer patients with bone metastasis. Hinyokika Kiyo **53**(5), 287–292 (2007)

23. I. Elia, R. Schmieder, S. Christen, S.M. Fendt, Organ-specific cancer metabolism and its potential for therapy. Handb. Exp. Pharmacol. **233**, 321–353 (2016). https://doi.org/10.1007/164_2015_10

24. J. Wang, J. Li, X. Li, S. Peng, J. Li, W. Yan, Y. Cui, H. Xiao, X. Wen, Increased expression of glycolytic enzymes in prostate cancer tissues and association with Gleason score. Int. J. Clin. Exp. Pathol. **10**(11), 11080–11089 (2017)

25. H. Xiao, J. Wang, W. Yan, Y. Cui, Z. Chen, X. Gao, X. Wen, J. Chen, GLUT1 regulates cell glycolysis and proliferation in prostate cancer. Prostate **78**(2), 86–94 (2018). https://doi.org/10.1002/pros.23448

26. J. Jans, J.H. van Dijk, S. van Schelven, P. van der Groep, S.H. Willems, T.N. Jonges, P.J. van Diest, J.L. Bosch, Expression and localization of hypoxia proteins in prostate cancer: prognostic implications after radical prostatectomy. Urology **75**(4), 786–792 (2010). https://doi.org/10.1016/j.urology.2009.08.024

27. C.V. Vaz, M.G. Alves, R. Marques, P.I. Moreira, P.F. Oliveira, C.J. Maia, S. Socorro, Androgen-responsive and nonresponsive prostate cancer cells present a distinct glycolytic metabolism profile. Int. J. Biochem. Cell Biol. **44**(11), 2077–2084 (2012). https://doi.org/10.1016/j.biocel.2012.08.013

28. P. Gonzalez-Menendez, D. Hevia, R. Alonso-Arias, A. Alvarez-Artime, A. Rodriguez-Garcia, S. Kinet, I. Gonzalez-Pola, N. Taylor, J.C. Mayo, R.M. Sainz, GLUT1 protects prostate cancer cells from glucose deprivation-induced oxidative stress. Redox Biol. **17**, 112–127 (2018). https://doi.org/10.1016/j.redox.2018.03.017

29. M.A. White, E. Tsouko, C. Lin, K. Rajapakshe, J.M. Spencer, S.R. Wilkenfeld, S.S. Vakili, T.L. Pulliam, D. Awad, F. Nikolos, R.R. Katreddy, B.A. Kaipparettu, A. Sreekumar, X. Zhang, E. Cheung, C. Coarfa, D.E. Frigo, GLUT12 promotes prostate cancer cell growth and is regulated by androgens and CaMKK2 signaling. Endocr. Relat. Cancer **25**(4), 453–469 (2018). https://doi.org/10.1530/ERC-17-0051

30. P. Gonzalez-Menendez, D. Hevia, J.C. Mayo, R.M. Sainz, The dark side of glucose transporters in prostate cancer: are they a new feature to characterize carcinomas? Int. J. Cancer **142**(12), 2414–2424 (2018). https://doi.org/10.1002/ijc.31165

31. N. Pertega-Gomes, S. Felisbino, C.E. Massie, J.R. Vizcaino, R. Coelho, C. Sandi, S. Simoes-Sousa, S. Jurmeister, A. Ramos-Montoya, M. Asim, M. Tran, E. Oliveira, A. Lobo da Cunha, V. Maximo, F. Baltazar, D.E. Neal, L.G. Fryer, A glycolytic phenotype is associated with prostate cancer progression and aggressiveness: a role for monocarboxylate transporters as metabolic targets for therapy. J. Pathol. **236**(4), 517–530 (2015). https://doi.org/10.1002/path.4547

32. Y. Deng, J. Lu, Targeting hexokinase 2 in castration-resistant prostate cancer. Mol. Cell. Oncol. **2**(3), e974465 (2015). https://doi.org/10.4161/23723556.2014.974465

33. L. Wang, H. Xiong, F. Wu, Y. Zhang, J. Wang, L. Zhao, X. Guo, L.J. Chang, Y. Zhang, M.J. You, S. Koochekpour, M. Saleem, H. Huang, J. Lu, Y. Deng, Hexokinase 2-mediated Warburg effect is required for PTEN- and p53-deficiency-driven prostate cancer growth. Cell Rep. **8**(5), 1461–1474 (2014). https://doi.org/10.1016/j.celrep.2014.07.053

34. P. Zhou, W.G. Chen, X.W. Li, MicroRNA-143 acts as a tumor suppressor by targeting hexokinase 2 in human prostate cancer. Am. J. Cancer Res. **5**(6), 2056–2063 (2015)

35. P.L. Martin, J.J. Yin, V. Seng, O. Casey, E. Corey, C. Morrissey, R.M. Simpson, K. Kelly, Androgen deprivation leads to increased carbohydrate metabolism and hexokinase 2-mediated survival in Pten/Tp53-deficient prostate cancer. Oncogene **36**(4), 525–533 (2017). https://doi.org/10.1038/onc.2016.223

36. T. Tao, M. Chen, R. Jiang, H. Guan, Y. Huang, H. Su, Q. Hu, X. Han, J. Xiao, Involvement of EZH2 in aerobic glycolysis of prostate cancer through miR-181b/HK2 axis. Oncol. Rep. **37**(3), 1430–1436 (2017). https://doi.org/10.3892/or.2017.5430

37. J.S. Moon, W.J. Jin, J.H. Kwak, H.J. Kim, M.J. Yun, J.W. Kim, S.W. Park, K.S. Kim, Androgen stimulates glycolysis for de novo lipid synthesis by increasing the activities of hexokinase 2 and 6-phosphofructo-2-kinase/fructose-2,6-bisphosphatase 2 in prostate cancer cells. Biochem. J. **433**(1), 225–233 (2011). https://doi.org/10.1042/BJ20101104

38. K.C. Patra, Q. Wang, P.T. Bhaskar, L. Miller, Z. Wang, W. Wheaton, N. Chandel, M. Laakso, W.J. Muller, E.L. Allen, A.K. Jha, G.A. Smolen, M.F. Clasquin, B. Robey, N. Hay, Hexokinase 2 is required for tumor initiation and maintenance and its systemic deletion is therapeutic in mouse models of cancer. Cancer Cell **24**(2), 213–228 (2013). https://doi.org/10.1016/j.ccr.2013.06.014

39. P.W. Stacpoole, Therapeutic targeting of the pyruvate dehydrogenase complex/pyruvate dehydrogenase kinase (PDC/PDK) axis in cancer. J. Natl. Cancer Inst. **109**(11) (2017). https://doi.org/10.1093/jnci/djx071

40. E. Kolobova, A. Tuganova, I. Boulatnikov, K.M. Popov, Regulation of pyruvate dehydrogenase activity through phosphorylation at multiple sites. Biochem. J. **358**(1), 69–77 (2001). https://doi.org/10.1042/bj3580069

41. G. Sutendra, E.D. Michelakis, Pyruvate dehydrogenase kinase as a novel therapeutic target in oncology. Front. Oncol. **3**, 38 (2013). https://doi.org/10.3389/fonc.2013.00038

42. Y. Zhong, X. Li, Y. Ji, X. Li, Y. Li, D. Yu, Y. Yuan, J. Liu, H. Li, M. Zhang, Z. Ji, D. Fan, J. Wen, M.A. Goscinski, L. Yuan, B. Hao, J.M. Nesland, Z. Suo, Pyruvate dehydrogenase expression is negatively associated with cell stemness and worse clinical outcome in prostate cancers. Oncotarget **8**(8), 13344–13356 (2017). https://doi.org/10.18632/oncotarget.14527

43. J. Chen, I. Guccini, D. Di Mitri, D. Brina, A. Revandkar, M. Sarti, E. Pasquini, A. Alajati, S. Pinton, M. Losa, G. Civenni, C.V. Catapano, J. Sgrignani, A. Cavalli, R. D'Antuono, J.M. Asara, A. Morandi, P. Chiarugi, S. Crotti, M. Agostini, M. Montopoli, I. Masgras, A. Rasola, R. Garcia-Escudero, N. Delaleu, A. Rinaldi, F. Bertoni, J. Bono, A. Carracedo, A. Alimonti, Compartmentalized activities of the pyruvate dehydrogenase complex sustain lipogenesis in prostate cancer. Nat. Genet. **50**(2), 219–228 (2018). https://doi.org/10.1038/s41588-017-0026-3

44. G. Sutendra, A. Kinnaird, P. Dromparis, R. Paulin, T.H. Stenson, A. Haromy, K. Hashimoto, N. Zhang, E. Flaim, E.D. Michelakis, A nuclear pyruvate dehydrogenase complex is important for the generation of acetyl-CoA and histone acetylation. Cell **158**(1), 84–97 (2014). https://doi.org/10.1016/j.cell.2014.04.046

45. M.C. Hsu, W.C. Hung, Pyruvate kinase M2 fuels multiple aspects of cancer cells: from cellular metabolism, transcriptional regulation to extracellular signaling. Mol. Cancer **17**(1), 35 (2018). https://doi.org/10.1186/s12943-018-0791-3

46. F.A. Siddiqui, G. Prakasam, S. Chattopadhyay, A.U. Rehman, R.A. Padder, M.A. Ansari, R. Irshad, K. Mangalhara, R.N.K. Bamezai, M. Husain, S.M. Ali, M.A. Iqbal, Curcumin decreases Warburg effect in cancer cells by down-regulating pyruvate kinase M2 via mTOR-HIF1alpha inhibition. Sci. Rep. **8**(1), 8323 (2018). https://doi.org/10.1038/s41598-018-25524-3

47. N. Wong, J. Yan, D. Ojo, J. De Melo, J.C. Cutz, D. Tang, Changes in PKM2 associate with prostate cancer progression. Cancer Invest. **32**(7), 330–338 (2014). https://doi.org/10.3109/07357907.2014.919306

48. G. Dong, Q. Mao, W. Xia, Y. Xu, J. Wang, L. Xu, F. Jiang, PKM2 and cancer: the function of PKM2 beyond glycolysis. Oncol. Lett. **11**(3), 1980–1986 (2016). https://doi.org/10.3892/ol.2016.4168

49. H.J. Wang, M. Pochampalli, L.Y. Wang, J.X. Zou, P.S. Li, S.C. Hsu, B.J. Wang, S.H. Huang, P. Yang, J.C. Yang, C.Y. Chu, C.L. Hsieh, S.Y. Sung, C.F. Li, C.G. Tepper, D.K. Ann, A.C. Gao, C.P. Evans, Y. Izumiya, C.P. Chuu, W.C. Wang, H.W. Chen, H.J. Kung, KDM8/JMJD5 as a dual coactivator of AR and PKM2 integrates AR/EZH2 network and tumor metabolism in CRPC. Oncogene **38**, 17 (2019). https://doi.org/10.1038/s41388-018-0414-x

50. Y. Yasumizu, H. Hongo, T. Kosaka, S. Mikami, K. Nishimoto, E. Kikuchi, M. Oya, PKM2 under hypoxic environment causes resistance to mTOR inhibitor in human castration resistant prostate cancer. Oncotarget **9**(45), 27698–27707 (2018). https://doi.org/10.18632/oncotarget.25498

51. D. Hasan, E. Gamen, N. Abu Tarboush, Y. Ismail, O. Pak, B. Azab, PKM2 and HIF-1alpha regulation in prostate cancer cell lines. PLoS One **13**(9), e0203745 (2018). https://doi.org/10.1371/journal.pone.0203745

52. E. Giannoni, M.L. Taddei, A. Morandi, G. Comito, M. Calvani, F. Bianchini, B. Richichi, G. Raugei, N. Wong, D. Tang, P. Chiarugi, Targeting stromal-induced pyruvate kinase M2 nuclear translocation impairs oxphos and prostate cancer metastatic spread. Oncotarget **6**(27), 24061–24074 (2015). https://doi.org/10.18632/oncotarget.4448

53. Z.Y. Xian, J.M. Liu, Q.K. Chen, H.Z. Chen, C.J. Ye, J. Xue, H.Q. Yang, J.L. Li, X.F. Liu, S.J. Kuang, Inhibition of LDHA suppresses tumor progression in prostate cancer. Tumour Biol. **36**(10), 8093–8100 (2015). https://doi.org/10.1007/s13277-015-3540-x

54. M.I. Koukourakis, A. Giatromanolaki, M. Panteliadou, S.E. Pouliliou, P.S. Chondrou, S. Mavropoulou, E. Sivridis, Lactate dehydrogenase 5 isoenzyme overexpression defines resistance of prostate cancer to radiotherapy. Br. J. Cancer **110**(9), 2217–2223 (2014). https://doi.org/10.1038/bjc.2014.158

55. S.Y. Choi, H. Xue, R. Wu, L. Fazli, D. Lin, C.C. Collins, M.E. Gleave, P.W. Gout, Y. Wang, The MCT4 gene: a novel, potential target for therapy of advanced prostate cancer. Clin. Cancer Res. **22**(11), 2721–2733 (2016). https://doi.org/10.1158/1078-0432.CCR-15-1624

56. J.R. Doherty, J.L. Cleveland, Targeting lactate metabolism for cancer therapeutics. J. Clin. Invest. **123**(9), 3685–3692 (2013). https://doi.org/10.1172/JCI69741

57. J. Liu, G. Chen, Z. Liu, S. Liu, Z. Cai, P. You, Y. Ke, L. Lai, Y. Huang, H. Gao, L. Zhao, H. Pelicano, P. Huang, W.L. McKeehan, C.L. Wu, C. Wang, W. Zhong, F. Wang, Aberrant FGFR tyrosine kinase signaling enhances the warburg effect by reprogramming LDH isoform expression and activity in prostate cancer. Cancer Res. **78**(16), 4459–4470 (2018). https://doi.org/10.1158/0008-5472.CAN-17-3226

58. R.S. Jones, M.E. Morris, Monocarboxylate transporters: therapeutic targets and prognostic factors in disease. Clin. Pharmacol. Ther. **100**(5), 454–463 (2016). https://doi.org/10.1002/cpt.418

59. M.C. Wilson, V.N. Jackson, C. Heddle, N.T. Price, H. Pilegaard, C. Juel, A. Bonen, I. Montgomery, O.F. Hutter, A.P. Halestrap, Lactic acid efflux from white skeletal muscle is catalyzed by the monocarboxylate transporter isoform MCT3. J. Biol. Chem. **273**(26), 15920–15926 (1998). https://doi.org/10.1074/jbc.273.26.15920

60. S.Y.C. Choi, S.L. Ettinger, D. Lin, H. Xue, X. Ci, N. Nabavi, R.H. Bell, F. Mo, P.W. Gout, N.E. Fleshner, M.E. Gleave, C.C. Collins, Y. Wang, Targeting MCT4 to reduce lactic acid secretion and glycolysis for treatment of neuroendocrine prostate cancer. Cancer Med. **7**, 3385 (2018). https://doi.org/10.1002/cam4.1587

61. N. Pertega-Gomes, J.R. Vizcaino, V. Miranda-Goncalves, C. Pinheiro, J. Silva, H. Pereira, P. Monteiro, R.M. Henrique, R.M. Reis, C. Lopes, F. Baltazar, Monocarboxylate transporter 4 (MCT4) and CD147 overexpression is associated with poor prognosis in prostate cancer. BMC Cancer **11**, 312 (2011). https://doi.org/10.1186/1471-2407-11-312

62. P. Sanita, M. Capulli, A. Teti, G.P. Galatioto, C. Vicentini, P. Chiarugi, M. Bologna, A. Angelucci, Tumor-stroma metabolic relationship based on lactate shuttle can sustain prostate cancer progression. BMC Cancer **14**, 154 (2014). https://doi.org/10.1186/1471-2407-14-154

63. N. Pertega-Gomes, J.R. Vizcaino, S. Felisbino, A.Y. Warren, G. Shaw, J. Kay, H. Whitaker, A.G. Lynch, L. Fryer, D.E. Neal, C.E. Massie, Epigenetic and oncogenic regulation of SLC16A7 (MCT2) results in protein over-expression, impacting on signalling and cellular phenotypes in prostate cancer. Oncotarget **6**(25), 21675–21684 (2015). https://doi.org/10.18632/oncotarget.4328

64. I. Valenca, N. Pertega-Gomes, J.R. Vizcaino, R.M. Henrique, C. Lopes, F. Baltazar, D. Ribeiro, Localization of MCT2 at peroxisomes is associated with malignant transformation in prostate cancer. J. Cell. Mol. Med. **19**(4), 723–733 (2015). https://doi.org/10.1111/jcmm.12481

65. H.S. Kim, E.M. Masko, S.L. Poulton, K.M. Kennedy, S.V. Pizzo, M.W. Dewhirst, S.J. Freedland, Carbohydrate restriction and lactate transporter inhibition in a mouse xenograft model of human prostate cancer. BJU Int. **110**(7), 1062–1069 (2012). https://doi.org/10.1111/j.1464-410X.2012.10971.x

66. P. Jiang, W. Du, M. Wu, Regulation of the pentose phosphate pathway in cancer. Protein Cell **5**(8), 592–602 (2014). https://doi.org/10.1007/s13238-014-0082-8

67. E.S. Cho, Y.H. Cha, H.S. Kim, N.H. Kim, J.I. Yook, The pentose phosphate pathway as a potential target for cancer therapy. Biomol. Ther. **26**(1), 29–38 (2018). https://doi.org/10.4062/biomolther.2017.179

68. I.A. da Costa, J. Hennenlotter, V. Stuhler, U. Kuhs, M. Scharpf, T. Todenhofer, A. Stenzl, J. Bedke, Transketolase like 1 (TKTL1) expression alterations in prostate cancer tumorigenesis. Urol. Oncol. **36**(10), 472.e421–472.e427 (2018). https://doi.org/10.1016/j.urolonc.2018.06.010

69. E. Tsouko, A.S. Khan, M.A. White, J.J. Han, Y. Shi, F.A. Merchant, M.A. Sharpe, L. Xin, D.E. Frigo, Regulation of the pentose phosphate pathway by an AR-mTOR-mediated mechanism and its role in prostate cancer cell growth. Oncogenesis **3**, e103 (2014). https://doi.org/10.1038/oncsis.2014.18

70. M. Ookhtens, R. Kannan, I. Lyon, N. Baker, Liver and adipose tissue contributions to newly formed fatty acids in an ascites tumor. Am. J. Physiol. Regul. Integr. Compar. Physiol. **247**(1), R146–R153 (1984). https://doi.org/10.1152/ajpregu.1984.247.1.R146

71. W. Han, S. Gao, D. Barrett, M. Ahmed, D. Han, J.A. Macoska, H.H. He, C. Cai, Reactivation of AR-regulated lipid biosynthesis drives the progression of castration-resistant prostate cancer. Oncogene **37**(6), 710–721 (2018). https://doi.org/10.1038/onc.2017.385

72. J.V. Swinnen, H. Heemers, T.V. de Sande, E.D. Schrijver, K. Brusselmans, W. Heyns, G. Verhoeven, Androgens, lipogenesis and prostate cancer. J. Steroid Biochem. Mol. Biol. **92**(4), 273–279 (2004). https://doi.org/10.1016/j.jsbmb.2004.10.013

73. W.-C. Huang, X. Li, J. Liu, J. Lin, L.W.K. Chung, Activation of AR, lipogenesis, and oxidative stress converged by SREBP-1 is responsible for regulating growth and progression of prostate cancer cells. Mol. Cancer Res. **10**(1), 133–142 (2012). https://doi.org/10.1158/1541-7786.MCR-11-0206

74. M. Chen, J. Zhang, K. Sampieri, J.G. Clohessy, L. Mendez, E. Gonzalez-Billalabeitia, X.S. Liu, Y.R. Lee, J. Fung, J.M. Katon, A.V. Menon, K.A. Webster, C. Ng, M.D. Palumbieri, M.S. Diolombi, S.B. Breitkopf, J. Teruya-Feldstein, S. Signoretti, R.T. Bronson, J.M. Asara, M. Castillo-Martin, C. Cordon-Cardo, P.P. Pandolfi, An aberrant SREBP-dependent lipogenic program promotes metastatic prostate cancer. Nat. Genet. **50**(2), 206–218 (2018). https://doi.org/10.1038/s41588-017-0027-2

75. C. Yokoyama, X. Wang, M.R. Briggs, A. Admon, J. Wu, X. Hua, J.L. Goldstein, M.S. Brown, SREBP-1, a basic-helix-loop-helix-leucine zipper protein that controls transcription of the low density lipoprotein receptor gene. Cell **75**(1), 187–197 (1993). https://doi.org/10.1016/S0092-8674(05)80095-9

76. W. Shao, P.J. Espenshade, Expanding roles for SREBP in metabolism. Cell Metab. **16**(4), 414–419 (2012). https://doi.org/10.1016/j.cmet.2012.09.002

77. S.L. Ettinger, R. Sobel, T.G. Whitmore, M. Akbari, D.R. Bradley, M.E. Gleave, C.C. Nelson, Dysregulation of sterol response element-binding proteins and downstream effectors in prostate cancer during progression to androgen independence. Cancer Res. **64**(6), 2212–2221 (2004)

78. H. Heemers, G. Verrijdt, S. Organe, F. Claessens, W. Heyns, G. Verhoeven, J.V. Swinnen, Identification of an androgen response element in intron 8 of the sterol regulatory element-binding protein cleavage-activating protein gene allowing direct regulation by the AR. J. Biol. Chem. **279**(29), 30880–30887 (2004). https://doi.org/10.1074/jbc.M401615200

79. J.V. Swinnen, W. Ulrix, W. Heyns, G. Verhoeven, Coordinate regulation of lipogenic gene expression by androgens: evidence for a cascade mechanism involving sterol regulatory element binding proteins. Proc. Natl. Acad. Sci. **94**(24), 12975–12980 (1997). https://doi.org/10.1073/pnas.94.24.12975

80. W.-C. Huang, H.E. Zhau, L.W.K. Chung, AR survival signaling is blocked by anti-β2-microglobulin monoclonal antibody via a MAPK/lipogenic pathway in human prostate cancer cells. J. Biol. Chem. **285**(11), 7947–7956 (2010). https://doi.org/10.1074/jbc.M109.092759

81. E. Rysman, K. Brusselmans, K. Scheys, L. Timmermans, R. Derua, S. Munck, P.P. Van Veldhoven, D. Waltregny, V.W. Daniëls, J. Machiels, F. Vanderhoydonc, K. Smans, E. Waelkens, G. Verhoeven, J.V. Swinnen, De novo lipogenesis protects cancer cells from free radicals and chemotherapeutics by promoting membrane lipid saturation. Cancer Res. **70**(20), 8117 (2010). https://doi.org/10.1158/0008-5472.CAN-09-3871

82. K. Simons, D. Toomre, Lipid rafts and signal transduction. Nat. Rev. Mol. Cell Biol. **1**, 31 (2000). https://doi.org/10.1038/35036052

83. L. Li, C.H. Ren, S.A. Tahir, C. Ren, T.C. Thompson, Caveolin-1 maintains activated Akt in prostate cancer cells through scaffolding domain binding site interactions with and inhibition of serine/threonine protein phosphatases PP1 and PP2A. Mol. Cell. Biol. **23**(24), 9389–9404 (2003). https://doi.org/10.1128/MCB.23.24.9389-9404.2003

84. R.M. Young, D. Holowka, B. Baird, A lipid raft environment enhances lyn kinase activity by protecting the active site tyrosine from dephosphorylation. J. Biol. Chem. **278**(23), 20746–20752 (2003). https://doi.org/10.1074/jbc.M211402200

85. K.R. Rice, M.O. Koch, L. Cheng, T.A. Masterson, Dyslipidemia, statins and prostate cancer. Expert Rev. Anticancer Ther. **12**(7), 981–990 (2012). https://doi.org/10.1586/era.12.75

86. B. Cinar, N.K. Mukhopadhyay, G. Meng, M.R. Freeman, Phosphoinositide 3-kinase-independent non-genomic signals transit from the AR to Akt1 in membrane raft microdomains. J. Biol. Chem. **282**(40), 29584–29593 (2007). https://doi.org/10.1074/jbc.M703310200

87. R.V. Farese, T.C. Walther, Lipid droplets finally get a little R-E-S-P-E-C-T. Cell **139**(5), 855–860 (2009). https://doi.org/10.1016/j.cell.2009.11.005

88. T. Petan, E. Jarc, M. Jusović, Lipid droplets in cancer: guardians of fat in a stressful world. Molecules **23**(8), E1941 (2018). https://doi.org/10.3390/molecules23081941

89. K. Bensaad, E. Favaro, A. Lewis Caroline, B. Peck, S. Lord, M. Collins Jennifer, E. Pinnick Katherine, S. Wigfield, M. Buffa Francesca, J.-L. Li, Q. Zhang, J.O. Wakelam Michael, F. Karpe, A. Schulze, L. Harris Adrian, Fatty Acid uptake and lipid storage induced by HIF-1α contribute to cell growth and survival after hypoxia-reoxygenation. Cell Rep. **9**(1), 349–365 (2014). https://doi.org/10.1016/j.celrep.2014.08.056

90. A.G. Cabodevilla, L. Sánchez-Caballero, E. Nintou, V.G. Boiadjieva, F. Picatoste, A. Gubern, E. Claro, Cell survival during complete nutrient deprivation depends on lipid droplet-fueled β-oxidation of fatty acids. J. Biol. Chem. **288**(39), 27777–27788 (2013). https://doi.org/10.1074/jbc.M113.466656

91. S. Koizume, Y. Miyagi, Lipid droplets: a key cellular organelle associated with cancer cell survival under normoxia and hypoxia. Int. J. Mol. Sci. **17**(9), 1430 (2016). https://doi.org/10.3390/ijms17091430

92. J.V. Swinnen, G. Verhoeven, Androgens and the control of lipid metabolism in human prostate cancer cells. J. Steroid Biochem. Mol. Biol. **65**(1), 191–198 (1998). https://doi.org/10.1016/S0960-0760(97)00187-8

93. S. Yue, J. Li, S.Y. Lee, H.J. Lee, T. Shao, B. Song, L. Cheng, T.A. Masterson, X. Liu, T.L. Ratliff, J.X. Cheng, Cholesteryl ester accumulation induced by PTEN loss and PI3K/AKT activation underlies

human prostate cancer aggressiveness. Cell Metab. **19**(3), 393–406 (2014). https://doi.org/10.1016/j.cmet.2014.01.019

94. J.V. Swinnen, T. Roskams, S. Joniau, H. Van Poppel, R. Oyen, L. Baert, W. Heyns, G. Verhoeven, Overexpression of fatty acid synthase is an early and common event in the development of prostate cancer. Int. J. Cancer **98**(1), 19–22 (2002). https://doi.org/10.1002/ijc.10127

95. T. Van de Sande, T. Roskams, E. Lerut, S. Joniau, H. Van Poppel, G. Verhoeven, J.V. Swinnen, High-level expression of fatty acid synthase in human prostate cancer tissues is linked to activation and nuclear localization of Akt/PKB. J. Pathol. **206**(2), 214–219 (2005). https://doi.org/10.1002/path.1760

96. T. Migita, S. Ruiz, A. Fornari, M. Fiorentino, C. Priolo, G. Zadra, F. Inazuka, C. Grisanzio, E. Palescandolo, E. Shin, C. Fiore, W. Xie, A.L. Kung, P.G. Febbo, A. Subramanian, L. Mucci, J. Ma, S. Signoretti, M. Stampfer, W.C. Hahn, S. Finn, M. Loda, Fatty acid synthase: a metabolic enzyme and candidate oncogene in prostate cancer. J. Natl. Cancer Inst. **101**(7), 519–532 (2009). https://doi.org/10.1093/jnci/djp030

97. S. Hamada, A. Horiguchi, K. Kuroda, K. Ito, T. Asano, K. Miyai, K. Iwaya, Increased fatty acid synthase expression in prostate biopsy cores predicts higher Gleason score in radical prostatectomy specimen. BMC Clin. Pathol. **14**(1), 3 (2014). https://doi.org/10.1186/1472-6890-14-3

98. G. Zadra, C.F. Ribeiro, P. Chetta, Y. Ho, S. Cacciatore, X. Gao, S. Syamala, C. Bango, C. Photopoulos, Y. Huang, S. Tyekucheva, D.C. Bastos, J. Tchaicha, B. Lawney, T. Uo, L. D'Anello, A. Csibi, R. Kalekar, B. Larimer, L. Ellis, L.M. Butler, C. Morrissey, K. McGovern, V.J. Palombella, J.L. Kutok, U. Mahmood, S. Bosari, J. Adams, S. Peluso, S.M. Dehm, S.R. Plymate, M. Loda, Inhibition of de novo lipogenesis targets AR signaling in castration-resistant prostate cancer. Proc. Natl. Acad. Sci. **116**(2), 631–640 (2019). https://doi.org/10.1073/pnas.1808834116

99. E. Furuta, S.K. Pai, R. Zhan, S. Bandyopadhyay, M. Watabe, Y.-Y. Mo, S. Hirota, S. Hosobe, T. Tsukada, K. Miura, S. Kamada, K. Saito, M. Iiizumi, W. Liu, J. Ericsson, K. Watabe, Fatty acid synthase gene is up-regulated by hypoxia via activation of Akt and sterol regulatory element binding protein-1. Cancer Res. **68**(4), 1003 (2008). https://doi.org/10.1158/0008-5472.CAN-07-2489

100. X. Gang, Y. Yang, J. Zhong, K. Jiang, Y. Pan, R.J. Karnes, J. Zhang, W. Xu, G. Wang, H. Huang, P300 acetyltransferase regulates fatty acid synthase expression, lipid metabolism and prostate cancer growth. Oncotarget **7**(12), 15135–15149 (2016). https://doi.org/10.18632/oncotarget.7715

101. S. Kim, X. Yang, Q. Li, M. Wu, L. Costyn, Z. Beharry, M.G. Bartlett, H. Cai, Myristoylation of Src kinase mediates Src-induced and high-fat diet–accelerated prostate tumor progression in mice. J. Biol. Chem. **292**(45), 18422–18433 (2017). https://doi.org/10.1074/jbc.M117.798827

102. H. Cai, D.A. Smith, S. Memarzadeh, C.A. Lowell, J.A. Cooper, O.N. Witte, Differential transformation capacity of Src family kinases during the initiation of prostate cancer. Proc. Natl. Acad. Sci. **108**(16), 6579–6584 (2011). https://doi.org/10.1073/pnas.1103904108

103. F.L. Zhang, P.J. Casey, Protein prenylation: molecular mechanisms and functional consequences. Annu. Rev. Biochem. **65**(1), 241–269 (1996). https://doi.org/10.1146/annurev.bi.65.070196.001325

104. B.M. Wasko, A. Dudakovic, R.J. Hohl, Bisphosphonates induce autophagy by depleting geranylgeranyl diphosphate. J. Pharmacol. Exp. Ther. **337**(2), 540 (2011). https://doi.org/10.1124/jpet.110.175521

105. J.E. Reilly, J.D. Neighbors, R.J. Hohl, Targeting protein geranylgeranylation slows tumor development in a murine model of prostate cancer metastasis. Cancer Biol. Ther. **18**(11), 872–882 (2016). https://doi.org/10.1080/15384047.2016.1219817

106. N.M. Zacharias, C. McCullough, S. Shanmugavelandy, J. Lee, Y. Lee, P. Dutta, J. McHenry, L. Nguyen, W. Norton, L.W. Jones, P.K. Bhattacharya, Metabolic differences in glutamine utilization lead to metabolic vulnerabilities in prostate cancer. Sci. Rep. **7**(1), 16159 (2017). https://doi.org/10.1038/s41598-017-16327-z

107. S.M. Fendt, E.L. Bell, M.A. Keibler, B.A. Olenchock, J.R. Mayers, T.M. Wasylenko, N.I. Vokes, L. Guarente, M.G. Vander Heiden, G. Stephanopoulos, Reductive glutamine metabolism is a function of the alpha-ketoglutarate to citrate ratio in cells. Nat. Commun. **4**, 2236 (2013). https://doi.org/10.1038/ncomms3236

108. A.R. Mullen, W.W. Wheaton, E.S. Jin, P.H. Chen, L.B. Sullivan, T. Cheng, Y. Yang, W.M. Linehan, N.S. Chandel, R.J. DeBerardinis, Reductive carboxylation supports growth in tumour cells with defective mitochondria. Nature **481**(7381), 385–388 (2011). https://doi.org/10.1038/nature10642

109. C.M. Metallo, P.A. Gameiro, E.L. Bell, K.R. Mattaini, J. Yang, K. Hiller, C.M. Jewell, Z.R. Johnson, D.J. Irvine, L. Guarente, J.K. Kelleher, M.G. Vander Heiden, O. Iliopoulos, G. Stephanopoulos, Reductive glutamine metabolism by IDH1 mediates lipogenesis under hypoxia. Nature **481**(7381), 380–384 (2011). https://doi.org/10.1038/nature10602

110. N. Zaidi, J.V. Swinnen, K. Smans, ATP-citrate lyase: a key player in cancer metabolism. Cancer Res. **72**(15), 3709–3714 (2012). https://doi.org/10.1158/0008-5472.CAN-11-4112

111. S. Shah, W.J. Carriveau, J. Li, S.L. Campbell, P.K. Kopinski, H.W. Lim, N. Daurio, S. Trefely, K.J. Won, D.C. Wallace, C. Koumenis, A. Mancuso, K.E. Wellen, Targeting ACLY sensitizes castration-resistant prostate cancer cells to AR antagonism by impinging on an ACLY-AMPK-AR feedback mechanism. Oncotarget **7**(28), 43713–43730 (2016). https://doi.org/10.18632/oncotarget.9666

112. A. Beckers, S. Organe, L. Timmermans, K. Scheys, A. Peeters, K. Brusselmans, G. Verhoeven, J.V. Swinnen, Chemical inhibition of acetyl-CoA carboxylase induces growth arrest and cytotoxicity selectively in cancer cells. Cancer Res. **67**(17), 8180–8187 (2007). https://doi.org/10.1158/0008-5472.CAN-07-0389

113. J.V. Swinnen, P.P. Van Veldhoven, L. Timmermans, E. De Schrijver, K. Brusselmans, F. Vanderhoydonc, T. Van de Sande, H. Heemers, W. Heyns, G. Verhoeven, Fatty acid synthase drives the synthesis of phospholipids partitioning into detergent-resistant membrane microdomains. Biochem. Biophys. Res. Commun. **302**(4), 898–903 (2003). https://doi.org/10.1016/S0006-291X(03)00265-1

114. B. Peck, Z.T. Schug, Q. Zhang, B. Dankworth, D.T. Jones, E. Smethurst, R. Patel, S. Mason, M. Jiang, R. Saunders, M. Howell, R. Mitter, B. Spencer-Dene, G. Stamp, L. McGarry, D. James, E. Shanks, E.O. Aboagye, S.E. Critchlow, H.Y. Leung, A.L. Harris, M.J.O. Wakelam, E. Gottlieb, A. Schulze, Inhibition of fatty acid desaturation is detrimental to cancer cell survival in metabolically compromised environments. Cancer Metab. **4**(1), 6 (2016). https://doi.org/10.1186/s40170-016-0146-8

115. R.B. Montgomery, E.A. Mostaghel, R. Vessella, D.L. Hess, T.F. Kalhorn, C.S. Higano, L.D. True, P.S. Nelson, Maintenance of intratumoral androgens in metastatic prostate cancer: a mechanism for castration-resistant tumor growth. Cancer Res. **68**(11), 4447–4454 (2008). https://doi.org/10.1158/0008-5472.CAN-08-0249

116. D.J. Mener, Prostate specific antigen reduction following statin therapy: mechanism of action and review of the literature. IUBMB Life **62**(8), 584–590 (2010). https://doi.org/10.1002/iub.355

117. A.D. Raval, D. Thakker, H. Negi, A. Vyas, M.W. Salkini, Association between statins and clinical outcomes among men with prostate cancer: a systematic review and meta-analysis. Prostate Cancer Prostatic Dis. **19**, 151 (2016). https://doi.org/10.1038/pcan.2015.58

118. C.R. Sirtori, The pharmacology of statins. Pharmacol. Res. **88**, 3–11 (2014). https://doi.org/10.1016/j.phrs.2014.03.002

119. Y. Chen, M. Hughes-Fulford, Human prostate cancer cells lack feedback regulation of low-density lipoprotein receptor and its regulator, SREBP2. Int. J. Cancer **91**(1), 41–45 (2001). https://doi.org/10.1002/1097-0215(20010101)91:1<41::AID-IJC1009>3.0.CO;2-2

120. Y. Zhang, K.L. Ma, X.Z. Ruan, B.C. Liu, Dysregulation of the low-density lipoprotein receptor pathway is involved in lipid disorder-mediated organ injury. Int. J. Biol. Sci. **12**(5), 569–579 (2016). https://doi.org/10.7150/ijbs.14027

121. J.R. Krycer, A.J. Brown, Cross-talk between the AR and the liver X receptor. J. Biol. Chem. **286**(23), 20637–20647 (2011). https://doi.org/10.1074/jbc.M111.227082

122. N. Zelcer, C. Hong, R. Boyadjian, P. Tontonoz, LXR regulates cholesterol uptake through idol-dependent ubiquitination of the LDL Receptor. Science **325**(5936), 100 (2009). https://doi.org/10.1126/science.1168974

123. J.J. Repa, S.D. Turley, J.M.A. Lobaccaro, J. Medina, L. Li, K. Lustig, B. Shan, R.A. Heyman, J.M. Dietschy, D.J. Mangelsdorf, Regulation of absorption and ABC1-mediated efflux of cholesterol by RXR heterodimers. Science **289**(5484), 1524 (2000). https://doi.org/10.1126/science.289.5484.1524

124. D.J. Peet, S.D. Turley, W. Ma, B.A. Janowski, J.-M.A. Lobaccaro, R.E. Hammer, D.J. Mangelsdorf, Cholesterol and bile acid metabolism are impaired in mice lacking the nuclear oxysterol receptor LXRα. Cell **93**(5), 693–704 (1998). https://doi.org/10.1016/S0092-8674(00)81432-4

125. A.J.C. Pommier, G. Alves, E. Viennois, S. Bernard, Y. Communal, B. Sion, G. Marceau, C. Damon, K. Mouzat, F. Caira, S. Baron, J.M.A. Lobaccaro, Liver X receptor activation downregulates AKT survival signaling in lipid rafts and induces apoptosis of prostate cancer cells. Oncogene **29**, 2712 (2010). https://doi.org/10.1038/onc.2010.30

126. C.-Y. Lin, J.-Å. Gustafsson, Targeting liver X receptors in cancer therapeutics. Nat. Rev. Cancer **15**, 216 (2015). https://doi.org/10.1038/nrc3912

127. M.A. Alfaqih, E.R. Nelson, W. Liu, R. Safi, J.S. Jasper, E. Macias, J. Geradts, J.W. Thompson, L.G. Dubois, M.R. Freeman, C.Y. Chang, J.T. Chi, D.P. McDonnell, S.J. Freedland, CYP27A1 loss dysregulates cholesterol homeostasis in prostate cancer. Cancer Res. **77**(7), 1662–1673 (2017). https://doi.org/10.1158/0008-5472.CAN-16-2738

128. Y. Liu, Fatty acid oxidation is a dominant bioenergetic pathway in prostate cancer. Prostate Cancer Prostatic Dis. **9**(3), 230–234 (2006)

129. T.W. Flaig, M. Salzmann-Sullivan, L.J. Su, Z. Zhang, M. Joshi, M.A. Gijon, J. Kim, J.J. Arcaroli, A. Van Bokhoven, M.S. Lucia, F.G. La Rosa, I.R. Schlaepfer, Lipid catabolism inhibition sensitizes prostate cancer cells to antiandrogen blockade. Oncotarget **8**(34), 56051–56065 (2017). https://doi.org/10.18632/oncotarget.17359

130. I.R. Schlaepfer, L. Rider, L.U. Rodrigues, M.A. Gijon, C.T. Pac, L. Romero, A. Cimic, S.J. Sirintrapun, L.M. Glode, R.H. Eckel, S.D. Cramer, Lipid catabolism via CPT1 as a therapeutic target for prostate cancer. Mol. Cancer Ther. **13**(10), 2361–2371 (2014). https://doi.org/10.1158/1535-7163.MCT-14-0183

131. J.B. Tennakoon, Y. Shi, J.J. Han, E. Tsouko, M.A. White, A.R. Burns, A. Zhang, X. Xia, O.R. Ilkayeva, L. Xin, M.M. Ittmann, F.G. Rick, A.V. Schally, D.E. Frigo, Androgens regulate prostate cancer cell growth via an AMPK-PGC-1α-mediated metabolic switch. Oncogene **33**, 5251–5261 (2014). https://doi.org/10.1038/onc.2013.463

132. G. Yu, C.J. Cheng, S.C. Lin, Y.C. Lee, D.E. Frigo, L.Y. Yu-Lee, G.E. Gallick, M.A. Titus, L.K. Nutt,

S.H. Lin, Organelle-derived acetyl-CoA promotes prostate cancer cell survival, migration, and metastasis via activation of calmodulin kinase II. Cancer Res. **78**(10), 2490–2502 (2018). https://doi.org/10.1158/0008-5472.CAN-17-2392

133. R.J.A. Wanders, Metabolic functions of peroxisomes in health and disease. Biochimie **98**, 36–44 (2014). https://doi.org/10.1016/j.biochi.2013.08.022

134. J. Luo, S. Zha, W.R. Gage, T.A. Dunn, J.L. Hicks, C.J. Bennett, C.M. Ewing, E.A. Platz, S. Ferdinandusse, R.J. Wanders, J.M. Trent, W.B. Isaacs, A.M. De Marzo, α-Methylacyl-CoA racemase a new molecular marker for prostate cancer. Cancer Res. **62**(8), 2220–2226 (2002)

135. M.D. Lloyd, M. Yevglevskis, G.L. Lee, P.J. Wood, M.D. Threadgill, T.J. Woodman, α-Methylacyl-CoA racemase (AMACR): metabolic enzyme, drug metabolizer and cancer marker P504S. Prog. Lipid Res. **52**(2), 220–230 (2013). https://doi.org/10.1016/j.plipres.2013.01.001

136. S. Zha, S. Ferdinandusse, J.L. Hicks, S. Denis, T.A. Dunn, R.J. Wanders, J. Luo, A.M. De Marzo, W.B. Isaacs, Peroxisomal branched chain fatty acid β-oxidation pathway is upregulated in prostate cancer. Prostate **63**(4), 316–323 (2005). https://doi.org/10.1002/pros.20177

137. D.R. Wise, C.B. Thompson, Glutamine addiction: a new therapeutic target in cancer. Trends Biochem. Sci. **35**(8), 427–433 (2010). https://doi.org/10.1016/j.tibs.2010.05.003

138. D. Daye, K.E. Wellen, Metabolic reprogramming in cancer: unraveling the role of glutamine in tumorigenesis. Semin. Cell Dev. Biol. **23**(4), 362–369 (2012). https://doi.org/10.1016/j.semcdb.2012.02.002

139. D.R. Wise, R.J. DeBerardinis, A. Mancuso, N. Sayed, X.Y. Zhang, H.K. Pfeiffer, I. Nissim, E. Daikhin, M. Yudkoff, S.B. McMahon, C.B. Thompson, Myc regulates a transcriptional program that stimulates mitochondrial glutaminolysis and leads to glutamine addiction. Proc. Natl. Acad. Sci. **105**(48), 18782–18787 (2008). https://doi.org/10.1073/pnas.0810199105

140. B.J. Altman, Z.E. Stine, C.V. Dang, From Krebs to clinic: glutamine metabolism to cancer therapy. Nat. Rev. Cancer **16**(10), 619–634 (2016). https://doi.org/10.1038/nrc.2016.71

141. M.A. White, C. Lin, K. Rajapakshe, J. Dong, Y. Shi, E. Tsouko, R. Mukhopadhyay, D. Jasso, W. Dawood, C. Coarfa, D.E. Frigo, Glutamine transporters are targets of multiple oncogenic signaling pathways in prostate cancer. Mol. Cancer Res. **15**(8), 1017–1028 (2017). https://doi.org/10.1158/1541-7786.MCR-16-0480

142. Q. Wang, J. Tiffen, C.G. Bailey, M.L. Lehman, W. Ritchie, L. Fazli, C. Metierre, Y.J. Feng, E. Li, M. Gleave, G. Buchanan, C.C. Nelson, J.E. Rasko, J. Holst, Targeting amino acid transport in metastatic castration-resistant prostate cancer: effects on cell cycle, cell growth, and tumor development. J. Natl. Cancer Inst. **105**(19), 1463–1473 (2013). https://doi.org/10.1093/jnci/djt241

143. P. Gao, I. Tchernyshyov, T.C. Chang, Y.S. Lee, K. Kita, T. Ochi, K.I. Zeller, A.M. De Marzo, J.E. Van Eyk, J.T. Mendell, C.V. Dang, c-Myc suppression of miR-23a/b enhances mitochondrial glutaminase expression and glutamine metabolism. Nature **458**(7239), 762–765 (2009). https://doi.org/10.1038/nature07823

144. K. Ellwood-Yen, T.G. Graeber, J. Wongvipat, M.L. Iruela-Arispe, J. Zhang, R. Matusik, G.V. Thomas, C.L. Sawyers, Myc-driven murine prostate cancer shares molecular features with human prostate tumors. Cancer Cell **4**(3), 223–238 (2003). https://doi.org/10.1016/s1535-6108(03)00197-1

145. J.J. Harding, M.L. Telli, P.N. Munster, M.H. Le, C. Molineaux, M.K. Bennett, E. Mittra, H.A. Burris, A.S. Clark, M. Dunphy, F. Meric-Bernstam, M.R. Patel, A. DeMichele, J.R. Infante, Safety and tolerability of increasing doses of CB-839, a first-in-class, orally administered small molecule inhibitor of glutaminase, in solid tumors. J. Clin. Oncol. **33**(15 Suppl), 2512–2512 (2015). https://doi.org/10.1200/jco.2015.33.15_suppl.2512

146. J.M. Rhoads, R.A. Argenzio, W. Chen, L.M. Graves, L.L. Licato, A.T. Blikslager, J. Smith, J. Gatzy, D.A. Brenner, Glutamine metabolism stimulates intestinal cell MAPKs by a cAMP-inhibitable, RAF-independent mechanism. Gastroenterology **118**(1), 90–100 (2000). https://doi.org/10.1016/S0016-5085(00)70417-3

147. P. Dalle Pezze, S. Ruf, A.G. Sonntag, M. Langelaar-Makkinje, P. Hall, A.M. Heberle, P. Razquin Navas, K. van Eunen, R.C. Tölle, J.J. Schwarz, H. Wiese, B. Warscheid, J. Deitersen, B. Stork, E. Fäßler, S. Schäuble, U. Hahn, P. Horvatovich, D.P. Shanley, K. Thedieck, A systems study reveals concurrent activation of AMPK and mTOR by amino acids. Nat. Commun. **7**, 13254 (2016). https://doi.org/10.1038/ncomms13254

148. S.A. Hawley, J. Boudeau, J.L. Reid, K.J. Mustard, L. Udd, T.P. Mäkelä, D.R. Alessi, D.G. Hardie, Complexes between the LKB1 tumor suppressor, STRAD alpha/beta and MO25 alpha/beta are upstream kinases in the AMP-activated protein kinase cascade. J. Biol. **2**(4), 28 (2003). https://doi.org/10.1186/1475-4924-2-28

149. A. Woods, S.R. Johnstone, K. Dickerson, F.C. Leiper, L.G. Fryer, D. Neumann, U. Schlattner, T. Wallimann, M. Carlson, D. Carling, LKB1 is the upstream kinase in the AMP-activated protein kinase cascade. Curr. Biol. **13**(22), 2004–2008 (2003). https://doi.org/10.1016/j.cub.2003.10.031

150. R.J. Shaw, M. Kosmatka, N. Bardeesy, R.L. Hurley, L.A. Witters, R.A. DePinho, L.C. Cantley, The tumor suppressor LKB1 kinase directly activates AMP-activated kinase and regulates apoptosis in response to energy stress. Proc. Natl. Acad. Sci. **101**(10), 3329–3335 (2004). https://doi.org/10.1073/pnas.0308061100

151. D.E. Frigo, M.K. Howe, B.M. Wittmann, A.M. Brunner, I. Cushman, Q. Wang, M. Brown, A.R. Means, D.P. McDonnell, CaM kinase kinase

beta-mediated activation of the growth regulatory kinase AMPK is required for androgen-dependent migration of prostate cancer cells. Cancer Res. **71**(2), 528–537 (2011). https://doi.org/10.1158/0008-5472. CAN-10-2581

152. L.G. Karacosta, B.A. Foster, G. Azabdaftari, D.M. Feliciano, A.M. Edelman, A Regulatory feedback loop between Ca(2+)/calmodulin-dependent protein kinase kinase 2 (CaMKK2) and the AR in prostate cancer progression. J. Biol. Chem. **287**(29), 24832–24843 (2012). https://doi.org/10.1074/jbc. M112.370783

153. C.E. Massie, A. Lynch, A. Ramos-Montoya, J. Boren, R. Stark, L. Fazli, A. Warren, H. Scott, B. Madhu, N. Sharma, H. Bon, V. Zecchini, D.M. Smith, G.M. Denicola, N. Mathews, M. Osborne, J. Hadfield, S. Macarthur, B. Adryan, S.K. Lyons, K.M. Brindle, J. Griffiths, M.E. Gleave, P.S. Rennie, D.E. Neal, I.G. Mills, The AR fuels prostate cancer by regulating central metabolism and biosynthesis. EMBO J. **30**(13), 2719–2733 (2011). https://doi.org/10.1038/emboj.2011.158

154. M.L. Schulte, A. Fu, P. Zhao, J. Li, L. Geng, S.T. Smith, J. Kondo, R.J. Coffey, M.O. Johnson, J.C. Rathmell, J.T. Sharick, M.C. Skala, J.A. Smith, J. Berlin, M.K. Washington, M.L. Nickels, H.C. Manning, Pharmacological blockade of ASCT2-dependent glutamine transport leads to antitumor efficacy in preclinical models. Nat. Med. **24**(2), 194–202 (2018). https://doi.org/10.1038/ nm.4464

155. Y. Liu, T. Zhao, Z. Li, L. Wang, S. Yuan, L. Sun, The role of ASCT2 in cancer: a review. Eur. J. Pharmacol. **837**, 81–87 (2018). https://doi.org/10.1016/j. ejphar.2018.07.007

156. L. Feun, M. You, C.J. Wu, M.T. Kuo, M. Wangpaichitr, S. Spector, N. Savaraj, Arginine deprivation as a targeted therapy for cancer. Curr. Pharm. Des. **14**(11), 1049–1057 (2008). https://doi. org/10.2174/138161208784246199

157. R.H. Kim, J.M. Coates, T.L. Bowles, G.P. McNerney, J. Sutcliffe, J.U. Jung, R. Gandour-Edwards, F.Y.S. Chuang, R.J. Bold, H.-J. Kung, Arginine deiminase as a novel therapy for prostate cancer induces autophagy and caspase-independent apoptosis. Cancer Res. **69**(2), 700 (2009). https://doi. org/10.1158/0008-5472.can-08-3157

158. R. Keshet, P. Szlosarek, A. Carracedo, A. Erez, Rewiring urea cycle metabolism in cancer to support anabolism. Nat. Rev. Cancer **18**(10), 634–645 (2018). https://doi.org/10.1038/s41568-018-0054-z

159. J.S. Lee, L. Adler, H. Karathia, N. Carmel, S. Rabinovich, N. Auslander, R. Keshet, N. Stettner, A. Silberman, L. Agemy, D. Helbling, R. Eilam, Q. Sun, A. Brandis, S. Malitsky, M. Itkin, H. Weiss, S. Pinto, S. Kalaora, R. Levy, E. Barnea, A. Admon, D. Dimmock, N. Stern-Ginossar, A. Scherz, S.C.S. Nagamani, M. Unda, D.M. Wilson 3rd, R. Elhasid, A. Carracedo, Y. Samuels, S. Hannenhalli, E. Ruppin, A. Erez, Urea cycle

dysregulation generates clinically relevant genomic and biochemical signatures. Cell **174**(6), 1559–1570.e1522 (2018). https://doi.org/10.1016/j. cell.2018.07.019

160. A. Husson, C. Brasse-Lagnel, A. Fairand, S. Renouf, A. Lavoinne, Argininosuccinate synthetase from the urea cycle to the citrulline–NO cycle. Eur. J. Biochem. **270**(9), 1887–1899 (2003). https://doi. org/10.1046/j.1432-1033.2003.03559.x

161. C.M. Ensor, F.W. Holtsberg, J.S. Bomalaski, M.A. Clark, Pegylated arginine deiminase (ADI-SS PEG20,000 mw) inhibits human melanomas and hepatocellular carcinomas in vitro and in vivo. Cancer Res. **62**(19), 5443–5450 (2002)

162. E.C. Hsueh, S.M. Knebel, W.-H. Lo, Y.-C. Leung, P.N.-M. Cheng, C.-T. Hsueh, Deprivation of arginine by recombinant human arginase in prostate cancer cells. J. Hematol. Oncol. **5**, 17 (2012). https://doi. org/10.1186/1756-8722-5-17

163. S.L. Nowotarski, P.M. Woster, R.A. Casero, Polyamines and cancer: implications for chemoprevention and chemotherapy. Expert Rev. Mol. Med. **15**, e3 (2013). https://doi.org/10.1017/erm.2013.3

164. G.E. Bai, S. Kasper, R.J. Matusik, P.S. Rennie, J.A. Moshier, A. Krongrad, Androgen regulation of the human ornithine decarboxylase promoter in prostate cancer cells. J. Androl. **19**(2), 127–135 (1998). https://doi.org/10.1002/j.1939-4640.1998. tb01981.x

165. A. Shukla-Dave, M. Castillo-Martin, M. Chen, J. Lobo, N. Gladoun, A. Collazo-Lorduy, F.M. Khan, V. Ponomarev, Z. Yi, W. Zhang, P.P. Pandolfi, H. Hricak, C. Cordon-Cardo, Ornithine decarboxylase is sufficient for prostate tumorigenesis via AR signaling. Am. J. Pathol. **186**, 3131 (2016). https:// doi.org/10.1016/j.ajpath.2016.08.021

166. B.H. Devens, R.S. Weeks, M.R. Burns, C.L. Carlson, M.K. Brawer, Polyamine depletion therapy in prostate cancer. Prostate Cancer Prostatic Dis. **3**, 275 (2000). https://doi.org/10.1038/sj.pcan.4500420

167. G.F. Giskeødegård, H. Bertilsson, K.M. Selnæs, A.J. Wright, T.F. Bathen, T. Viset, J. Halgunset, A. Angelsen, I.S. Gribbestad, M.-B. Tessem, Spermine and citrate as metabolic biomarkers for assessing prostate cancer aggressiveness. PLoS One **8**(4), e62375 (2013). https://doi.org/10.1371/journal. pone.0062375

168. M.G. Swanson, A.S. Zektzer, Z.L. Tabatabai, J. Simko, S. Jarso, K.R. Keshari, L. Schmitt, P.R. Carroll, K. Shinohara, D.B. Vigneron, J. Kurhanewicz, Quantitative analysis of prostate metabolites using 1H HR-MAS spectroscopy. Magn. Reson. Med. **55**(6), 1257–1264 (2006). https://doi. org/10.1002/mrm.20909

169. L.L. Cheng, C.-l. Wu, M.R. Smith, R.G. Gonzalez, Non-destructive quantitation of spermine in human prostate tissue samples using HRMAS 1H NMR spectroscopy at 9.4 T. FEBS Lett. **494**(1-2), 112–116 (2001). https://doi.org/10.1016/ S0014-5793(01)02329-8

Prostate Cancer Energetics and Biosynthesis

170. M.J. Corbin, J.M. Ruiz-Echevarría, One-carbon metabolism in prostate cancer: the role of androgen signaling. Int. J. Mol. Sci. **17**(8), E1208 (2016). https://doi.org/10.3390/ijms17081208

171. O. Shuvalov, A. Petukhov, A. Daks, O. Fedorova, E. Vasileva, N.A. Barlev, One-carbon metabolism and nucleotide biosynthesis as attractive targets for anticancer therapy. Oncotarget **8**(14), 23955–23977 (2017). https://doi.org/10.18632/oncotarget.15053

172. A. Sreekumar, L.M. Poisson, T.M. Rajendiran, A.P. Khan, Q. Cao, J. Yu, B. Laxman, R. Mehra, R.J. Lonigro, Y. Li, M.K. Nyati, A. Ahsan, S. Kalyana-Sundaram, B. Han, X. Cao, J. Byun, G.S. Omenn, D. Ghosh, S. Pennathur, D.C. Alexander, A. Berger, J.R. Shuster, J.T. Wei, S. Varambally, C. Beecher, A.M. Chinnaiyan, Metabolomic profiles delineate potential role for sarcosine in prostate cancer progression. Nature **457**, 910 (2009). https://doi.org/10.1038/nature07762

173. A.A. Shafi, V. Putluri, J.M. Arnold, E. Tsouko, S. Maity, J.M. Roberts, C. Coarfa, D.E. Frigo, N. Putluri, A. Sreekumar, N.L. Weigel, Differential regulation of metabolic pathways by AR and its constitutively active splice variant, AR-V7, in prostate cancer cells. Oncotarget **6**(31), 31997–32012 (2015). https://doi.org/10.18632/oncotarget.5585

174. S. Ottaviani, G.N. Brooke, C. O'Hanlon-Brown, J. Waxman, S. Ali, L. Buluwela, Characterisation of the androgen regulation of glycine N-methyltransferase in prostate cancer cells. J. Mol. Endocrinol. **51**(3), 301–312 (2013). https://doi.org/10.1530/jme-13-0169

175. W.G. Nelson, A.M. De Marzo, S. Yegnasubramanian, Epigenetic alterations in human prostate cancers. Endocrinology **150**(9), 3991–4002 (2009). https://doi.org/10.1210/en.2009-0573

176. S. Yegnasubramanian, J. Kowalski, M.L. Gonzalgo, M. Zahurak, S. Piantadosi, P.C. Walsh, G.S. Bova, A.M. De Marzo, W.B. Isaacs, W.G. Nelson, Hypermethylation of CpG islands in primary and metastatic human prostate cancer. Cancer Res. **64**(6), 1975 (2004). https://doi.org/10.1158/0008-5472.can-03-3972

177. C. Jerónimo, R. Henrique, M.O. Hoque, E. Mambo, F.R. Ribeiro, G. Varzim, J. Oliveira, M.R. Teixeira, C. Lopes, D. Sidransky, A quantitative promoter methylation profile of prostate cancer. Clin. Cancer Res. **10**(24), 8472 (2004). https://doi.org/10.1158/1078-0432.ccr-04-0894

178. R. Maruyama, S. Toyooka, K.O. Toyooka, A.K. Virmani, S. Zöchbauer-Müller, A.J. Farinas, J.D. Minna, J. McConnell, E.P. Frenkel, A.F. Gazdar, Aberrant promoter methylation profile of prostate cancers and its relationship to clinicopathological features. Clin. Cancer Res. **8**(2), 514 (2002)

179. A.R. Florl, C. Steinhoff, M. Müller, H.H. Seifert, C. Hader, R. Engers, R. Ackermann, W.A. Schulz, Coordinate hypermethylation at specific genes in prostate carcinoma precedes LINE-1 hypomethyl-

ation. Br. J. Cancer **91**(5), 985–994 (2004). https://doi.org/10.1038/sj.bjc.6602030

180. A. Padar, U.G. Sathyanarayana, M. Suzuki, R. Maruyama, J.T. Hsieh, E.P. Frenkel, J.D. Minna, A.F. Gazdar, Inactivation of cyclin D2 gene in prostate cancers by aberrant promoter methylation. Clin. Cancer Res. **9**(13), 4730–4734 (2003)

181. M. Yamanaka, M. Watanabe, Y. Yamada, A. Takagi, T. Murata, H. Takahashi, H. Suzuki, H. Ito, H. Tsukino, T. Katoh, Y. Sugimura, T. Shiraishi, Altered methylation of multiple genes in carcinogenesis of the prostate. Int. J. Cancer **106**(3), 382–387 (2003). https://doi.org/10.1002/ijc.11227

182. J. Ellinger, P.J. Bastian, T. Jurgan, K. Biermann, P. Kahl, L.C. Heukamp, N. Wernert, S.C. Müller, A. von Ruecker, CpG island hypermethylation at multiple gene sites in diagnosis and prognosis of prostate cancer. Urology **71**(1), 161–167 (2008). https://doi.org/10.1016/j.urology.2007.09.056

183. D. Lodygin, A. Epanchintsev, A. Menssen, J. Diebold, H. Hermeking, Functional epigenomics identifies genes frequently silenced in prostate cancer. Cancer Res. **65**(10), 4218 (2005). https://doi.org/10.1158/0008-5472.can-04-4407

184. S.P. Antoinette, F. Ruth, W. Karen, L. Mark, The emerging roles of DNA methylation in the clinical management of prostate cancer. Endoc. Relat. Cancer **13**(2), 357–377 (2006). https://doi.org/10.1677/erc.1.01184

185. E. Lee, J. Wang, K. Yumoto, Y. Jung, F.C. Cackowski, A.M. Decker, Y. Li, R.T. Franceschi, K.J. Pienta, R.S. Taichman, DNMT1 Regulates epithelial-mesenchymal transition and cancer stem cells, which promotes prostate cancer metastasis. Neoplasia **18**(9), 553–566 (2016). https://doi.org/10.1016/j.neo.2016.07.007

186. S.R. Morey Kinney, M.T. Moser, M. Pascual, J.M. Greally, B.A. Foster, A.R. Karpf, Opposing roles of Dnmt1 in early- and late-stage murine prostate cancer. Mol. Cell. Biol. **30**(17), 4159 (2010). https://doi.org/10.1128/mcb.00235-10

187. M.T. McCabe, J.A. Low, S. Daignault, M.J. Imperiale, K.J. Wojno, M.L. Day, Inhibition of DNA methyltransferase activity prevents tumorigenesis in a mouse model of prostate cancer. Cancer Res. **66**(1), 385 (2006). https://doi.org/10.1158/0008-5472.can-05-2020

188. R. Singal, K. Ramachandran, E. Gordian, C. Quintero, W. Zhao, I.M. Reis, Phase I/II study of azacitidine, docetaxel, and prednisone in patients with metastatic castration-resistant prostate cancer previously treated with docetaxel-based therapy. Clin. Genitourin. Cancer **13**(1), 22–31 (2015). https://doi.org/10.1016/j.clgc.2014.07.008

189. R.A. Casero Jr., L.J. Marton, Targeting polyamine metabolism and function in cancer and other hyperproliferative diseases. Nat. Rev. Drug Discov. **6**, 373 (2007). https://doi.org/10.1038/nrd2243

190. E.W. Gerner, F.L. Meyskens Jr., Polyamines and cancer: old molecules, new understanding. Nat.

191. K. Soda, The mechanisms by which polyamines accelerate tumor spread. J. Exp. Clin. Cancer Res. **30**(1), 95–95 (2011). https://doi.org/10.1186/1756-9966-30-95

192. B.G. Cipolla, R. Havouis, J.-P. Moulinoux, Polyamine reduced diet (PRD) nutrition therapy in hormone refractory prostate cancer patients. Biomed. Pharmacother. **64**(5), 363–368 (2010). https://doi.org/10.1016/j.biopha.2009.09.022

193. I.G. Fantus, H.J. Goldberg, C.I. Whiteside, D. Topic, The hexosamine biosynthesis pathway, in *The Diabetic Kidney*, ed. by P. Cortes, C. E. Mogensen, (Humana Press, Totowa, NJ, 2006), pp. 117–133. https://doi.org/10.1007/978-1-59745-153-6_7

194. H.M. Itkonen, N. Engedal, E. Babaie, M. Luhr, I.J. Guldvik, S. Minner, J. Hohloch, M.C. Tsourlakis, T. Schlomm, I.G. Mills, UAP1 is overexpressed in prostate cancer and is protective against inhibitors of N-linked glycosylation. Oncogene **34**(28), 3744–3750 (2015). https://doi.org/10.1038/onc.2014.307

195. H.M. Itkonen, S. Minner, I.J. Guldvik, M.J. Sandmann, M.C. Tsourlakis, V. Berge, A. Svindland, T. Schlomm, I.G. Mills, O-GlcNAc transferase integrates metabolic pathways to regulate the stability of c-MYC in human prostate cancer cells. Cancer Res. **73**(16), 5277–5287 (2013). https://doi.org/10.1158/0008-5472.CAN-13-0549

196. H.M. Itkonen, I.G. Mills, N-linked glycosylation supports cross-talk between receptor tyrosine kinases and AR. PLoS One **8**(5), e65016 (2013). https://doi.org/10.1371/journal.pone.0065016

197. T. Kamigaito, T. Okaneya, M. Kawakubo, H. Shimojo, O. Nishizawa, J. Nakayama, Overexpression of O-GlcNAc by prostate cancer cells is significantly associated with poor prognosis of patients. Prostate Cancer Prostatic Dis. **17**(1), 18–22 (2014). https://doi.org/10.1038/pcan.2013.56

198. T.P. Lynch, C.M. Ferrer, S.R. Jackson, K.S. Shahriari, K. Vosseller, M.J. Reginato, Critical role of O-linked beta-N-acetylglucosamine transferase in prostate cancer invasion, angiogenesis, and metastasis. J. Biol. Chem. **287**(14), 11070–11081 (2012). https://doi.org/10.1074/jbc.M111.302547

199. A.K. Kaushik, A. Shojaie, K. Panzitt, R. Sonavane, H. Venghatakrishnan, M. Manikkam, A. Zaslavsky, V. Putluri, V.T. Vasu, Y.Q. Zhang, A.S. Khan, S. Lloyd, A.T. Szafran, S. Dasgupta, D.A. Bader, F. Stossi, H.W. Li, S. Samanta, X.H. Cao, E. Tsouko, S.X. Huang, D.E. Frigo, L. Chan, D.P. Edwards, B.A. Kaipparettu, N. Mitsiades, N.L. Weigel, M. Mancini, S.E. McGuire, R. Mehra, M.M. Ittmann, A.M. Chinnaiyan, N. Putluri, G.S. Palapattu, G. Michailidis, A. Sreekumar, Inhibition of the hexosamine biosynthetic pathway promotes castration-resistant prostate cancer. Nat. Commun. **7**, 11612 (2016). https://doi.org/10.1038/ncomms11612

200. L. Galluzzi, F. Pietrocola, J.M. Bravo-San Pedro, R.K. Amaravadi, E.H. Baehrecke, F. Cecconi, P. Codogno, J. Debnath, D.A. Gewirtz, V. Karantza, A. Kimmelman, S. Kumar, B. Levine, M.C. Maiuri, S.J. Martin, J. Penninger, M. Piacentini, D.C. Rubinsztein, H.U. Simon, A. Simonsen, A.M. Thorburn, G. Velasco, K.M. Ryan, G. Kroemer, Autophagy in malignant transformation and cancer progression. EMBO J. **34**(7), 856–880 (2015). https://doi.org/10.15252/embj.201490784

201. J. Goldsmith, B. Levine, J. Debnath, Autophagy and cancer metabolism. Methods Enzymol. **542**, 25–57 (2014). https://doi.org/10.1016/B978-0-12-416618-9.00002-9

202. E. Ziparo, S. Petrungaro, E.S. Marini, D. Starace, S. Conti, A. Facchiano, A. Filippini, C. Giampietri, Autophagy in prostate cancer and androgen suppression therapy. Int. J. Mol. Sci. **14**(6), 12090–12106 (2013). https://doi.org/10.3390/ijms140612090

203. X. Xie, J.Y. Koh, S. Price, E. White, J.M. Mehnert, Atg7 overcomes senescence and promotes growth of BrafV600E-driven melanoma. Cancer Discov. **5**(4), 410–423 (2015). https://doi.org/10.1158/2159-8290.CD-14-1473

204. M. Garcia-Fernandez, P. Karras, A. Checinska, E. Canon, G.T. Calvo, G. Gomez-Lopez, M. Cifdaloz, A. Colmenar, L. Espinosa-Hevia, D. Olmeda, M.S. Soengas, Metastatic risk and resistance to BRAF inhibitors in melanoma defined by selective allelic loss of ATG5. Autophagy **12**(10), 1776–1790 (2016). https://doi.org/10.1080/15548627.2016.1199301

205. U. Santanam, W. Banach-Petrosky, C. Abate-Shen, M.M. Shen, E. White, R.S. DiPaola, Atg7 cooperates with Pten loss to drive prostate cancer tumor growth. Genes Dev. **30**(4), 399–407 (2016). https://doi.org/10.1101/gad.274134.115

206. J.M. Farrow, J.C. Yang, C.P. Evans, Autophagy as a modulator and target in prostate cancer. Nat. Rev. Urol. **11**(9), 508–516 (2014). https://doi.org/10.1038/nrurol.2014.196

207. X. Chen, J. Lu, L. Xia, G. Li, Drug resistance of enzalutamide in CRPC. Curr. Drug Targets **19**(6), 613–620 (2018). https://doi.org/10.2174/1389450118666170417144250

208. H.G. Nguyen, J.C. Yang, H.J. Kung, X.B. Shi, D. Tilki, P.N. Lara Jr., R.W. DeVere White, A.C. Gao, C.P. Evans, Targeting autophagy overcomes Enzalutamide resistance in castration-resistant prostate cancer cells and improves therapeutic response in a xenograft model. Oncogene **33**(36), 4521–4530 (2014). https://doi.org/10.1038/onc.2014.25

209. M. Parikh, J.C. Hyunh, P. Lara, C.X. Pan, D. Robles, C.P. Evans, Enzalutamide and metformin combination therapy to overcome autophagy resistance in castration resistant prostate cancer (CRPC): current results from a phase I study. J. Clin. Oncol. **36**(6), 281 (2018). https://doi.org/10.1200/JCO.2018.36.6_suppl.281

210. B. Kranzbuhler, S. Salemi, A. Mortezavi, T. Sulser, D. Eberli, Combined N-terminal AR and autophagy inhibition increases the antitumor effect in enzalutamide sensitive and enzalutamide resistant prostate cancer cells. Prostate **79**(2), 206–214 (2019). https://doi.org/10.1002/pros.23725

211. F. Hu, Y. Zhao, Y. Yu, J.M. Fang, R. Cui, Z.Q. Liu, X.L. Guo, Q. Xu, Docetaxel-mediated autophagy promotes chemoresistance in castration-resistant prostate cancer cells by inhibiting STAT3. Cancer Lett. **416**, 24–30 (2018). https://doi.org/10.1016/j.canlet.2017.12.013

212. R. Cristofani, M. Montagnani Marelli, M.E. Cicardi, F. Fontana, M. Marzagalli, P. Limonta, A. Poletti, R.M. Moretti, Dual role of autophagy on docetaxel-sensitivity in prostate cancer cells. Cell Death Dis. **9**(9), 889 (2018). https://doi.org/10.1038/s41419-018-0866-5

213. Q. Wang, W.Y. He, Y.Z. Zeng, A. Hossain, X. Gou, Inhibiting autophagy overcomes docetaxel resistance in castration-resistant prostate cancer cells. Int. Urol. Nephrol. **50**(4), 675–686 (2018). https://doi.org/10.1007/s11255-018-1801-5

214. H. Draz, A.A. Goldberg, V.I. Titorenko, E.S. Tomlinson Guns, S.H. Safe, J.T. Sanderson, Diindolylmethane and its halogenated derivatives induce protective autophagy in human prostate cancer cells via induction of the oncogenic protein AEG-1 and activation of AMP-activated protein kinase (AMPK). Cell. Signal. **40**, 172–182 (2017). https://doi.org/10.1016/j.cellsig.2017.09.006

215. H. Draz, A.A. Goldberg, E.S. Tomlinson Guns, L. Fazli, S. Safe, J.T. Sanderson, Autophagy inhibition improves the chemotherapeutic efficacy of cruciferous vegetable-derived diindolymethane in a murine prostate cancer xenograft model. Invest. New Drugs **36**(4), 718–725 (2018). https://doi.org/10.1007/s10637-018-0595-8

216. X. Zhu, M. Zhou, G.Y. Liu, X.L. Huang, W.Y. He, X. Gou, T. Jiang, Autophagy activated by the c-Jun N-terminal kinase-mediated pathway protects human prostate cancer PC3 cells from celecoxib-induced apoptosis. Exp. Ther. Med. **13**(5), 2348–2354 (2017). https://doi.org/10.3892/etm.2017.4287

217. C. Yang, X. Ma, Z. Wang, X. Zeng, Z. Hu, Z. Ye, G. Shen, Curcumin induces apoptosis and protective autophagy in castration-resistant prostate cancer cells through iron chelation. Drug Des. Devel. Ther. **11**, 431–439 (2017). https://doi.org/10.2147/DDDT.S126964

218. Y.C. Lin, Y.T. Chang, M. Campbell, T.P. Lin, C.C. Pan, H.C. Lee, J.C. Shih, P.C. Chang, MAOA-a novel decision maker of apoptosis and autophagy in hormone refractory neuroendocrine prostate cancer cells. Sci. Rep. **7**, 46338 (2017). https://doi.org/10.1038/srep46338

219. S. Niture, M. Ramalinga, H. Kedir, D. Patacsil, S.S. Niture, J. Li, H. Mani, S. Suy, S. Collins, D. Kumar, TNFAIP8 promotes prostate cancer cell survival by inducing autophagy. Oncotarget **9**(42), 26884–26899 (2018). https://doi.org/10.18632/oncotarget.25529

220. J. Lu, W. Dong, H. He, Z. Han, Y. Zhuo, R. Mo, Y. Liang, J. Zhu, R. Li, H. Qu, L. Zhang, S. Wang, R. Ma, Z. Jia, W. Zhong, Autophagy induced by overexpression of DCTPP1 promotes tumor progression and predicts poor clinical outcome in prostate cancer. Int. J. Biol. Macromol. **118**(Pt A), 599–609 (2018). https://doi.org/10.1016/j.ijbiomac.2018.06.005

221. C. Commisso, J. Debnath, Macropinocytosis fuels prostate cancer. Cancer Discov. **8**(7), 800–802 (2018). https://doi.org/10.1158/2159-8290.CD-18-0513

222. S.M. Kim, T.T. Nguyen, A. Ravi, P. Kubiniok, B.T. Finicle, V. Jayashankar, L. Malacrida, J. Hou, J. Robertson, D. Gao, J. Chernoff, M.A. Digman, E.O. Potma, B.J. Tromberg, P. Thibault, A.L. Edinger, PTEN deficiency and AMPK activation promote nutrient scavenging and anabolism in prostate cancer cells. Cancer Discov. **8**(7), 866–883 (2018). https://doi.org/10.1158/2159-8290.CD-17-1215

223. A.S. Khan, D.E. Frigo, A spatiotemporal hypothesis for the regulation, role, and targeting of AMPK in prostate cancer. Nat. Rev. Urol. **14**(3), 164–180 (2017). https://doi.org/10.1038/nrurol.2016.272

224. C.V. Vaz, R. Marques, M.G. Alves, P.F. Oliveira, J.E. Cavaco, C.J. Maia, S. Socorro, Androgens enhance the glycolytic metabolism and lactate export in prostate cancer cells by modulating the expression of GLUT1, GLUT3, PFK, LDH and MCT4 genes. J. Cancer Res. Clin. Oncol. **142**(1), 5–16 (2016). https://doi.org/10.1007/s00432-015-1992-4

225. D. Awad, T.L. Pulliam, C. Lin, S.R. Wilkenfeld, D.E. Frigo, Delineation of the androgen-regulated signaling pathways in prostate cancer facilitates the development of novel therapeutic approaches. Curr. Opin. Pharmacol. **41**, 1–11 (2018). https://doi.org/10.1016/j.coph.2018.03.002

226. L.M. Butler, M.M. Centenera, J.V. Swinnen, Androgen control of lipid metabolism in prostate cancer: novel insights and future applications. Endocr. Relat. Cancer **23**(5), R219–R227 (2016). https://doi.org/10.1530/ERC-15-0556

227. U.C. Dadwal, E.S. Chang, U. Sankar, AR-CaMKK2 axis in prostate cancer and bone microenvironment. Front. Endocrinol. **9**, 335 (2018). https://doi.org/10.3389/fendo.2018.00335

228. T. Mitani, R. Yamaji, Y. Higashimura, N. Harada, Y. Nakano, H. Inui, Hypoxia enhances transcriptional activity of AR through hypoxia-inducible factor-1alpha in a low androgen environment. J. Steroid Biochem. Mol. Biol. **123**(1-2), 58–64 (2011). https://doi.org/10.1016/j.jsbmb.2010.10.009

229. N.J. Mabjeesh, M.T. Willard, C.E. Frederickson, H. Zhong, J.W. Simons, Androgens stimulate hypoxia-inducible factor 1 activation via autocrine loop of tyrosine kinase receptor/phosphatidylinositol 3′-kinase/protein kinase B in prostate cancer cells. Clin. Cancer Res. **9**(7), 2416–2425 (2003)

230. R.J. Rebello, R.B. Pearson, R.D. Hannan, L. Furic, Therapeutic approaches targeting MYC-driven prostate cancer. Genes (Basel) **8**(2), E71 (2017). https://doi.org/10.3390/genes8020071

231. C.V. Dang, A. Le, P. Gao, MYC-induced cancer cell energy metabolism and therapeutic opportunities. Clin. Cancer Res. **15**(21), 6479–6483 (2009). https://doi.org/10.1158/1078-0432.CCR-09-0889

232. C. Priolo, S. Pyne, J. Rose, E.R. Regan, G. Zadra, C. Photopoulos, S. Cacciatore, D. Schultz, N. Scaglia, J. McDunn, A.M. De Marzo, M. Loda, AKT1 and MYC induce distinctive metabolic fingerprints in human prostate cancer. Cancer Res. **74**(24), 7198–7204 (2014). https://doi.org/10.1158/0008-5472.CAN-14-1490

233. D. Mossmann, S. Park, M.N. Hall, mTOR signalling and cellular metabolism are mutual determinants in cancer. Nat. Rev. Cancer **18**(12), 744–757 (2018). https://doi.org/10.1038/s41568-018-0074-8

234. J.S. Yu, W. Cui, Proliferation, survival and metabolism: the role of PI3K/AKT/mTOR signalling in pluripotency and cell fate determination. Development **143**(17), 3050–3060 (2016). https://doi.org/10.1242/dev.137075

235. V. Nogueira, K.C. Patra, N. Hay, Selective eradication of cancer displaying hyperactive Akt by exploiting the metabolic consequences of Akt activation. Elife **7**, e32213 (2018). https://doi.org/10.7554/eLife.32213

236. F. Giunchi, M. Fiorentino, M. Loda, The metabolic landscape of prostate cancer. Eur. Urol. Oncol. **2**, 28 (2019). https://doi.org/10.1016/j.euo.2018.06.010

237. J.V. Lee, A. Carrer, S. Shah, N.W. Snyder, S. Wei, S. Venneti, A.J. Worth, Z.F. Yuan, H.W. Lim, S. Liu, E. Jackson, N.M. Aiello, N.B. Haas, T.R. Rebbeck, A. Judkins, K.J. Won, L.A. Chodosh, B.A. Garcia, B.Z. Stanger, M.D. Feldman, I.A. Blair, K.E. Wellen, Akt-dependent metabolic reprogramming regulates tumor cell histone acetylation. Cell Metab. **20**(2), 306–319 (2014). https://doi.org/10.1016/j.cmet.2014.06.004

238. P. Toren, A. Zoubeidi, Targeting the PI3K/Akt pathway in prostate cancer: challenges and opportunities (review). Int. J. Oncol. **45**(5), 1793–1801 (2014). https://doi.org/10.3892/ijo.2014.2601

239. A. Naguib, G. Mathew, C.R. Reczek, K. Watrud, A. Ambrico, T. Herzka, I.C. Salas, M.F. Lee, N. El-Amine, W. Zheng, M.E. Di Francesco, J.R. Marszalek, D.J. Pappin, N.S. Chandel, L.C. Trotman, Mitochondrial complex I inhibitors expose a vulnerability for selective killing of Pten-null cells. Cell Rep. **23**(1), 58–67 (2018). https://doi.org/10.1016/j.celrep.2018.03.032

240. A. Csibi, S.M. Fendt, C. Li, G. Poulogiannis, A.Y. Choo, D.J. Chapski, S.M. Jeong, J.M. Dempsey, A. Parkhitko, T. Morrison, E.P. Henske, M.C. Haigis, L.C. Cantley, G. Stephanopoulos, J. Yu, J. Blenis, The mTORC1 pathway stimulates glutamine metabolism and cell proliferation by repressing SIRT4. Cell **153**(4), 840–854 (2013). https://doi.org/10.1016/j.cell.2013.04.023

241. A. Zabala-Letona, A. Arruabarrena-Aristorena, N. Martin-Martin, S. Fernandez-Ruiz, J.D. Sutherland, M. Clasquin, J. Tomas-Cortazar, J. Jimenez, I. Torres, P. Quang, P. Ximenez-Embun, R. Bago, A. Ugalde-Olano, A. Loizaga-Iriarte, I. Lacasa-Viscasillas, M. Unda, V. Torrano, D. Cabrera, S.M. van Liempd, Y. Cendon, E. Castro, S. Murray, A. Revandkar, A. Alimonti, Y. Zhang, A. Barnett, G. Lein, D. Pirman, A.R. Cortazar, L. Arreal, L. Prudkin, I. Astobiza, L. Valcarcel-Jimenez, P. Zuniga-Garcia, I. Fernandez-Dominguez, M. Piva, A. Caro-Maldonado, P. Sanchez-Mosquera, M. Castillo-Martin, V. Serra, N. Beraza, A. Gentilella, G. Thomas, M. Azkargorta, F. Elortza, R. Farras, D. Olmos, A. Efeyan, J. Anguita, J. Munoz, J.M. Falcon-Perez, R. Barrio, T. Macarulla, J.M. Mato, M.L. Martinez-Chantar, C. Cordon-Cardo, A.M. Aransay, K. Marks, J. Baselga, J. Tabernero, P. Nuciforo, B.D. Manning, K. Marjon, A. Carracedo, mTORC1-dependent AMD1 regulation sustains polyamine metabolism in prostate cancer. Nature **547**(7661), 109–113 (2017). https://doi.org/10.1038/nature22964

242. Q. Wang, C.G. Bailey, C. Ng, J. Tiffen, A. Thoeng, V. Minhas, M.L. Lehman, S.C. Hendy, G. Buchanan, C.C. Nelson, J.E. Rasko, J. Holst, AR and nutrient signaling pathways coordinate the demand for increased amino acid transport during prostate cancer progression. Cancer Res. **71**(24), 7525–7536 (2011). https://doi.org/10.1158/0008-5472.CAN-11-1821

243. B.S. Carver, C. Chapinski, J. Wongvipat, H. Hieronymus, Y. Chen, S. Chandarlapaty, V.K. Arora, C. Le, J. Koutcher, H. Scher, P.T. Scardino, N. Rosen, C.L. Sawyers, Reciprocal feedback regulation of PI3K and AR signaling in PTEN-deficient prostate cancer. Cancer Cell **19**(5), 575–586 (2011). https://doi.org/10.1016/j.ccr.2011.04.008

244. M.P. Edlind, A.C. Hsieh, PI3K-AKT-mTOR signaling in prostate cancer progression and androgen deprivation therapy resistance. Asian J. Androl. **16**(3), 378–386 (2014). https://doi.org/10.4103/1008-682X.122876

245. M. Crumbaker, L. Khoja, A.M. Joshua, AR signaling and the PI3K pathway in prostate cancer. Cancers (Basel) **9**(4), E34 (2017). https://doi.org/10.3390/cancers9040034

246. E. Audet-Walsh, C.R. Dufour, T. Yee, F.Z. Zouanat, M. Yan, G. Kalloghlian, M. Vernier, M. Caron, G. Bourque, E. Scarlata, L. Hamel, F. Brimo, A.G. Aprikian, J. Lapointe, S. Chevalier, V. Giguere, Nuclear mTOR acts as a transcriptional integrator of the androgen signaling pathway in prostate cancer. Genes Dev. **31**(12), 1228–1242 (2017). https://doi.org/10.1101/gad.299958.117

247. M. Marhold, E. Tomasich, A. El-Gazzar, G. Heller, A. Spittler, R. Horvat, M. Krainer, P. Horak, HIF1alpha regulates mTOR signaling and viability

of prostate cancer stem cells. Mol. Cancer Res. **13**(3), 556–564 (2015). https://doi.org/10.1158/1541-7786. MCR-14-0153-T

248. S.J. Barfeld, H.M. Itkonen, A. Urbanucci, I.G. Mills, Androgen-regulated metabolism and biosynthesis in prostate cancer. Endocr. Relat. Cancer **21**(4), T57–T66 (2014). https://doi.org/10.1530/ERC-13-0515

249. L. Fabris, Y. Ceder, A.M. Chinnaiyan, G.W. Jenster, K.D. Sorensen, S. Tomlins, T. Visakorpi, G.A. Calin, The potential of microRNAs as prostate cancer biomarkers. Eur. Urol. **70**(2), 312–322 (2016). https://doi.org/10.1016/j.eururo.2015.12.054

250. W. Qu, S.M. Ding, G. Cao, S.J. Wang, X.H. Zheng, G.H. Li, miR-132 mediates a metabolic shift in prostate cancer cells by targeting Glut1. FEBS Open Bio. **6**(7), 735–741 (2016). https://doi.org/10.1002/2211-5463.12086

251. A. Singh, C. Happel, S.K. Manna, G. Acquaah-Mensah, J. Carrerero, S. Kumar, P. Nasipuri, K.W. Krausz, N. Wakabayashi, R. Dewi, L.G. Boros, F.J. Gonzalez, E. Gabrielson, K.K. Wong, G. Girnun, S. Biswal, Transcription factor NRF2 regulates miR-1 and miR-206 to drive tumorigenesis. J. Clin. Invest. **123**(7), 2921–2934 (2013). https://doi.org/10.1172/JCI66353

252. J.W. Shih, L.Y. Wang, C.L. Hung, H.J. Kung, C.L. Hsieh, Non-coding RNAs in castration-resistant prostate cancer: regulation of AR signaling and cancer metabolism. Int. J. Mol. Sci. **16**(12), 28943–28978 (2015). https://doi.org/10.3390/ijms161226138

253. M. Xin, Z. Qiao, J. Li, J. Liu, S. Song, X. Zhao, P. Miao, T. Tang, L. Wang, W. Liu, X. Yang, K. Dai, G. Huang, miR-22 inhibits tumor growth and metastasis by targeting ATP citrate lyase: evidence in osteosarcoma, prostate cancer, cervical cancer and lung cancer. Oncotarget **7**(28), 44252–44265 (2016). https://doi.org/10.18632/oncotarget.10020

254. X. Li, Y.T. Chen, S. Josson, N.K. Mukhopadhyay, J. Kim, M.R. Freeman, W.C. Huang, MicroRNA-185 and 342 inhibit tumorigenicity and induce apoptosis through blockade of the SREBP metabolic pathway in prostate cancer cells. PLoS One **8**(8), e70987 (2013). https://doi.org/10.1371/journal.pone.0070987

255. W. Liu, A. Le, C. Hancock, A.N. Lane, C.V. Dang, T.W. Fan, J.M. Phang, Reprogramming of proline and glutamine metabolism contributes to the proliferative and metabolic responses regulated by oncogenic transcription factor c-MYC. Proc. Natl. Acad. Sci. **109**(23), 8983–8988 (2012). https://doi.org/10.1073/pnas.1203244109

256. C.A. Lyssiotis, A.C. Kimmelman, Metabolic interactions in the tumor microenvironment. Trends Cell Biol. **27**(11), 863–875 (2017). https://doi.org/10.1016/j.tcb.2017.06.003

257. L. Marignol, K. Rivera-Figueroa, T. Lynch, D. Hollywood, Hypoxia, notch signalling, and prostate cancer. Nat. Rev. Urol. **10**(7), 405–413 (2013). https://doi.org/10.1038/nrurol.2013.110

258. H. Geng, C. Xue, J. Mendonca, X.X. Sun, Q. Liu, P.N. Reardon, Y. Chen, K. Qian, V. Hua, A. Chen, F. Pan, J. Yuan, S. Dang, T.M. Beer, M.S. Dai, S.K. Kachhap, D.Z. Qian, Interplay between hypoxia and androgen controls a metabolic switch conferring resistance to androgen/AR-targeted therapy. Nat. Commun. **9**(1), 4972 (2018). https://doi.org/10.1038/s41467-018-07411-7

259. K. Zaugg, Y. Yao, P.T. Reilly, K. Kannan, R. Kiarash, J. Mason, P. Huang, S.K. Sawyer, B. Fuerth, B. Faubert, T. Kalliomaki, A. Elia, X. Luo, V. Nadeem, D. Bungard, S. Yalavarthi, J.D. Growney, A. Wakeham, Y. Moolani, J. Silvester, A.Y. Ten, W. Bakker, K. Tsuchihara, S.L. Berger, R.P. Hill, R.G. Jones, M. Tsao, M.O. Robinson, C.B. Thompson, G. Pan, T.W. Mak, Carnitine palmitoyltransferase 1C promotes cell survival and tumor growth under conditions of metabolic stress. Genes Dev. **25**(10), 1041–1051 (2011). https://doi.org/10.1101/gad.1987211

260. Y. Jin, L. Wang, S. Qu, X. Sheng, A. Kristian, G.M. Maelandsmo, N. Pallmann, E. Yuca, I. Tekedereli, K. Gorgulu, N. Alpay, A. Sood, G. Lopez-Berestein, L. Fazli, P. Rennie, B. Risberg, H. Waehre, H.E. Danielsen, B. Ozpolat, F. Saatcioglu, STAMP2 increases oxidative stress and is critical for prostate cancer. EMBO Mol. Med. **7**(3), 315–331 (2015). https://doi.org/10.15252/emmm.201404181

261. N.N. Tam, Y. Gao, Y.K. Leung, S.M. Ho, Androgenic regulation of oxidative stress in the rat prostate: involvement of NAD(P)H oxidases and antioxidant defense machinery during prostatic involution and regrowth. Am. J. Pathol. **163**(6), 2513–2522 (2003). https://doi.org/10.1016/S0002-9440(10)63606-1

262. D.A. Frohlich, M.T. McCabe, R.S. Arnold, M.L. Day, The role of Nrf2 in increased reactive oxygen species and DNA damage in prostate tumorigenesis. Oncogene **27**(31), 4353–4362 (2008). https://doi.org/10.1038/onc.2008.79

263. L. Khandrika, B. Kumar, S. Koul, P. Maroni, H.K. Koul, Oxidative stress in prostate cancer. Cancer Lett. **282**(2), 125–136 (2009). https://doi.org/10.1016/j.canlet.2008.12.011

264. G.J. Samaranayake, C.I. Troccoli, M. Huynh, R.D.Z. Lyles, K. Kage, A. Win, V. Lakshmanan, D. Kwon, Y. Ban, S.X. Chen, E.R. Zarco, M. Jorda, K.L. Burnstein, P. Rai, Thioredoxin-1 protects against AR-induced redox vulnerability in castration-resistant prostate cancer. Nat. Commun. **8**(1), 1204 (2017). https://doi.org/10.1038/s41467-017-01269-x

265. K. Li, Q. Zheng, X. Chen, Y. Wang, D. Wang, J. Wang, Isobavachalcone induces ROS-mediated apoptosis via targeting thioredoxin reductase 1 in human prostate cancer PC-3 cells. Oxid. Med. Cell. Longev. **2018**, 1915828 (2018). https://doi.org/10.1155/2018/1915828

266. Y. Yan, L. Chang, H. Tian, L. Wang, Y. Zhang, T. Yang, G. Li, W. Hu, K. Shah, G. Chen, Y. Guo, 1-Pyrroline-5-carboxylate released by pros-

266. tate cancer cell inhibit T cell proliferation and function by targeting SHP1/cytochrome c oxidoreductase/ROS axis. J. Immunother. Cancer **6**(1), 148 (2018). https://doi.org/10.1186/s40425-018-0466-z

267. K. Shiga, M. Hara, T. Nagasaki, T. Sato, H. Takahashi, H. Takeyama, Cancer-associated fibroblasts: their characteristics and their roles in tumor growth. Cancers (Basel) **7**(4), 2443–2458 (2015). https://doi.org/10.3390/cancers7040902

268. E. Giannoni, F. Bianchini, L. Masieri, S. Serni, E. Torre, L. Calorini, P. Chiarugi, Reciprocal activation of prostate cancer cells and cancer-associated fibroblasts stimulates epithelial-mesenchymal transition and cancer stemness. Cancer Res. **70**(17), 6945–6956 (2010). https://doi.org/10.1158/0008-5472.CAN-10-0785

269. F. Lopes-Coelho, S. Gouveia-Fernandes, J. Serpa, Metabolic cooperation between cancer and noncancerous stromal cells is pivotal in cancer progression. Tumour Biol. **40**(2), 1010428318756203 (2018). https://doi.org/10.1177/1010428318756203

270. T. Fiaschi, A. Marini, E. Giannoni, M.L. Taddei, P. Gandellini, A. De Donatis, M. Lanciotti, S. Serni, P. Cirri, P. Chiarugi, Reciprocal metabolic reprogramming through lactate shuttle coordinately influences tumor-stroma interplay. Cancer Res. **72**(19), 5130–5140 (2012). https://doi.org/10.1158/0008-5472.CAN-12-1949

271. P. Chiarugi, P. Paoli, P. Cirri, Tumor microenvironment and metabolism in prostate cancer. Semin. Oncol. **41**(2), 267–280 (2014). https://doi.org/10.1053/j.seminoncol.2014.03.004

272. N. Pertega-Gomes, J.R. Vizcaino, J. Attig, S. Jurmeister, C. Lopes, F. Baltazar, A lactate shuttle system between tumour and stromal cells is associated with poor prognosis in prostate cancer. BMC Cancer **14**, 352 (2014). https://doi.org/10.1186/1471-2407-14-352

273. A. Giatromanolaki, M.I. Koukourakis, A. Koutsopoulos, S. Mendrinos, E. Sivridis, The metabolic interactions between tumor cells and tumor-associated stroma (TAS) in prostatic cancer. Cancer Biol. Ther. **13**(13), 1284–1289 (2012). https://doi.org/10.4161/cbt.21785

274. T. Fiaschi, E. Giannoni, M.L. Taddei, P. Cirri, A. Marini, G. Pintus, C. Nativi, B. Richichi, A. Scozzafava, F. Carta, E. Torre, C.T. Supuran, P. Chiarugi, Carbonic anhydrase IX from cancer-associated fibroblasts drives epithelial-mesenchymal transition in prostate carcinoma cells. Cell Cycle **12**(11), 1791–1801 (2013). https://doi.org/10.4161/cc.24902

275. A. Santi, A. Caselli, P. Paoli, D. Corti, G. Camici, G. Pieraccini, M.L. Taddei, S. Serni, P. Chiarugi, P. Cirri, The effects of CA IX catalysis products within tumor microenvironment. Cell Commun. Signal **11**, 81 (2013). https://doi.org/10.1186/1478-811X-11-81

276. E.H. Cheteh, M. Augsten, H. Rundqvist, J. Bianchi, V. Sarne, L. Egevad, V.J. Bykov, A. Ostman, K.G. Wiman, Human cancer-associated fibroblasts enhance glutathione levels and antagonize drug-induced prostate cancer cell death. Cell Death Dis. **8**(6), e2848 (2017). https://doi.org/10.1038/cddis.2017.225

277. Z.D. Nassar, A.T. Aref, D. Miladinovic, C.Y. Mah, G.V. Raj, A.J. Hoy, L.M. Butler, Peri-prostatic adipose tissue: the metabolic microenvironment of prostate cancer. BJU Int. **121**(Suppl 3), 9–21 (2018). https://doi.org/10.1111/bju.14173

278. J. Huang, A. Duran, M. Reina-Campos, T. Valencia, E.A. Castilla, T.D. Muller, M.H. Tschop, J. Moscat, M.T. Diaz-Meco, Adipocyte p62/SQSTM1 suppresses tumorigenesis through opposite regulations of metabolism in adipose tissue and tumor. Cancer Cell **33**(4), 770–784.e776 (2018). https://doi.org/10.1016/j.ccell.2018.03.001

279. E. Gazi, J. Dwyer, N.P. Lockyer, P. Gardner, J.H. Shanks, J. Roulson, C.A. Hart, N.W. Clarke, M.D. Brown, Biomolecular profiling of metastatic prostate cancer cells in bone marrow tissue using FTIR microspectroscopy: a pilot study. Anal. Bioanal. Chem. **387**(5), 1621–1631 (2007). https://doi.org/10.1007/s00216-006-1093-y

280. S. Balaban, Z.D. Nassar, A.Y. Zhang, E. Hosseini-Beheshti, M.M. Centenera, M. Schreuder, H.M. Lin, A. Aishah, B. Varney, F. Liu-Fu, L.S. Lee, S.R. Nagarajan, R.F. Shearer, R.A. Hardie, N.L. Raftopulos, M.S. Kakani, D.N. Saunders, J. Holst, L.G. Horvath, L.M. Butler, A.J. Hoy, Extracellular fatty acids are the major contributor to lipid synthesis in prostate cancer. Mol. Cancer Res. **17**, 949 (2019). https://doi.org/10.1158/1541-7786.MCR-18-0347

281. Y. Tokuda, Y. Satoh, C. Fujiyama, S. Toda, H. Sugihara, Z. Masaki, Prostate cancer cell growth is modulated by adipocyte-cancer cell interaction. BJU Int. **91**(7), 716–720 (2003). https://doi.org/10.1046/j.1464-410X.2003.04218.x

282. J.D. Diedrich, E. Rajagurubandara, M.K. Herroon, G. Mahapatra, M. Huttemann, I. Podgorski, Bone marrow adipocytes promote the Warburg phenotype in metastatic prostate tumors via HIF-1alpha activation. Oncotarget **7**(40), 64854–64877 (2016). https://doi.org/10.18632/oncotarget.11712

283. K. Fujita, P. Landis, B.K. McNeil, C.P. Pavlovich, Serial prostate biopsies are associated with an increased risk of erectile dysfunction in men with prostate cancer on active surveillance. J. Urol. **182**(6), 2664–2669 (2009). https://doi.org/10.1016/j.juro.2009.08.044

284. M.G. Vander Heiden, Targeting cancer metabolism: a therapeutic window opens. Nat. Rev. Drug Discov. **10**(9), 671–684 (2011). https://doi.org/10.1038/nrd3504

285. J. Kurhanewicz, D.B. Vigneron, K. Brindle, E.Y. Chekmenev, A. Comment, C.H. Cunningham,

R.J. DeBerardinis, G.G. Green, M.O. Leach, S.S. Rajan, Analysis of cancer metabolism by imaging hyperpolarized nuclei: prospects for translation to clinical research. Neoplasia **13**(2), 81–97 (2011). https://doi.org/10.1593/neo.101102

286. A.G. Wibmer, I.A. Burger, E. Sala, H. Hricak, W.A. Weber, H.A. Vargas, Molecular imaging of prostate cancer. Radiographics **36**(1), 142–159 (2015). https://doi.org/10.1148/rg.2016150059

287. J.H. Ardenkjaer-Larsen, Hyperpolarized 13C magnetic resonance imaging-principles and applications, in *Molecular Imaging: Principles and Practice*, (People's Medical Publishing House, Shelton, CT, 2009), pp. 377–388

288. P. Bhattacharya, B.D. Ross, R. Bünger, Cardiovascular applications of hyperpolarized contrast media and metabolic tracers. Exp. Biol. Med. **234**(12), 1395–1416 (2009). https://doi.org/10.3181/0904-MR-135

289. K. Brindle, New approaches for imaging tumour responses to treatment. Nat. Rev. Cancer **8**(2), 94 (2008)

290. N. Zacharias, J. Lee, S. Ramachandran, S. Shanmugavelandy, J. McHenry, P. Dutta, S. Millward, S. Gammon, E. Efstathiou, P. Troncoso, D.E. Frigo, D. Piwnica-Worms, C.J. Logothetis, S.N. Maity, M.A. Titus, P. Bhattacharya, AR signaling in castration-resistant prostate cancer alters hyperpolarized pyruvate to lactate conversion and lactate levels in vivo. Mol. Imaging Biol. **21**, 86 (2019). https://doi.org/10.1007/s11307-018-1199-6

291. P. Dutta, G.V. Martinez, R.J. Gillies, A new horizon of DNP technology: application to in-vivo 13 C magnetic resonance spectroscopy and imaging. Biophys. Re. **5**(3), 271–281 (2013)

292. P. Dutta, A. Le, D.L. Vander Jagt, T. Tsukamoto, G.V. Martinez, C.V. Dang, R.J. Gillies, Evaluation of LDH-A and glutaminase inhibition in vivo by hyperpolarized 13C-pyruvate magnetic resonance spectroscopy of tumors. Cancer Res. **73**, 4190 (2013). https://doi.org/10.1158/0008-5472.CAN-13-0465

293. M.J. Albers, R. Bok, A.P. Chen, C.H. Cunningham, M.L. Zierhut, V.Y. Zhang, S.J. Kohler, J. Tropp, R.E. Hurd, Y.-F. Yen, Hyperpolarized 13C lactate, pyruvate, and alanine: noninvasive biomarkers for prostate cancer detection and grading. Cancer Res. **68**(20), 8607–8615 (2008). https://doi.org/10.1158/0008-5472.CAN-08-0749

294. S.J. Nelson, J. Kurhanewicz, D.B. Vigneron, P.E. Larson, A.L. Harzstark, M. Ferrone, M. van Criekinge, J.W. Chang, R. Bok, I. Park, Metabolic imaging of patients with prostate cancer using hyperpolarized [1-13C] pyruvate. Sci. Transl. Med. **5**(198), 198ra108 (2013). https://doi.org/10.1126/scitranslmed.3006070

295. R. Aggarwal, D.B. Vigneron, J. Kurhanewicz, Hyperpolarized 1-[13C]-pyruvate magnetic resonance imaging detects an early metabolic response to androgen ablation therapy in prostate cancer.

Eur. Urol. **72**(6), 1028–1029 (2017). https://doi.org/10.1016/j.eururo.2017.07.022

296. J.M. Lupo, A.P. Chen, M.L. Zierhut, R.A. Bok, C.H. Cunningham, J. Kurhanewicz, D.B. Vigneron, S.J. Nelson, Analysis of hyperpolarized dynamic 13C lactate imaging in a transgenic mouse model of prostate cancer. Magn. Reson. Imaging **28**(2), 153–162 (2010). https://doi.org/10.1016/j.mri.2009.07.007

297. H.-Y. Chen, P.E. Larson, R.A. Bok, C. von Morze, R. Sriram, R.D. Santos, J.D. Santos, J.W. Gordon, N. Bahrami, M. Ferrone, Assessing prostate cancer aggressiveness with hyperpolarized dual-agent 3D dynamic imaging of metabolism and perfusion. Cancer Res. **77**, 3207 (2017). https://doi.org/10.1158/0008-5472.CAN-16-2083

298. J.E. Lee, C.J. Diederich, R. Bok, R. Sriram, R.D. Santos, S.M. Noworolski, V.A. Salgaonkar, M.S. Adams, D.B. Vigneron, J. Kurhanewicz, Assessing high-intensity focused ultrasound treatment of prostate cancer with hyperpolarized 13C dual-agent imaging of metabolism and perfusion. NMR Biomed. **32**, e3962 (2018). https://doi.org/10.1002/nbm.3962

299. S.S. Tee, I. Suster, S. Truong, S. Jeong, R. Eskandari, V. DiGialleonardo, J.A. Alvarez, H.N. Aldeborgh, K.R. Keshari, Targeted AKT inhibition in prostate cancer cells and spheroids reduces aerobic glycolysis and generation of hyperpolarized [1-(13)C] lactate. Mol. Cancer Res. **16**(3), 453–460 (2018). https://doi.org/10.1158/1541-7786.MCR-17-0458

300. H. Dafni, P.E. Larson, S. Hu, H.A. Yoshihara, C.S. Ward, H.S. Venkatesh, C. Wang, X. Zhang, D.B. Vigneron, S.M. Ronen, Hyperpolarized 13C spectroscopic imaging informs on hypoxia-inducible factor-1 and myc activity downstream of platelet-derived growth factor receptor. Cancer Res. **70**, 7400 (2010). https://doi.org/10.1158/0008-5472.CAN-10-0883

301. K.R. Keshari, D.M. Wilson, M. Van Criekinge, R. Sriram, B.L. Koelsch, Z.J. Wang, H.F. VanBrocklin, D.M. Peehl, T. O'brien, D. Sampath, Metabolic response of prostate cancer to nicotinamide phophoribosyltransferase inhibition in a hyperpolarized MR/PET compatible bioreactor. Prostate **75**(14), 1601–1609 (2015). https://doi.org/10.1002/pros.23036

302. B.T. Scroggins, M. Matsuo, A.O. White, K. Saito, J.P. Munasinghe, C. Sourbier, K. Yamamoto, V. Diaz, Y. Takakusagi, K. Ichikawa, Hyperpolarized [1-13C]-pyruvate magnetic resonance spectroscopic imaging of prostate cancer in vivo predicts efficacy of targeting the Warburg effect. Clin. Cancer Res. **24**, 3137 (2018). https://doi.org/10.1158/1078-0432.CCR-17-1957

303. A. Lodi, S.M. Woods, S.M. Ronen, Treatment with the MEK inhibitor U0126 induces decreased hyperpolarized pyruvate to lactate conversion in breast, but not prostate, cancer cells. NMR Biomed. **26**(3), 299–306 (2013). https://doi.org/10.1002/nbm.2848

304. C. Canapè, G. Catanzaro, E. Terreno, M. Karlsson, M.H. Lerche, P.R. Jensen, Probing treatment response of glutaminolytic prostate cancer cells to natural drugs with hyperpolarized [5-13C] glutamine. Magn. Reson. Med. **73**(6), 2296–2305 (2015). https://doi.org/10.1002/mrm.25360

305. K.R. Keshari, V. Sai, Z.J. Wang, H.F. VanBrocklin, J. Kurhanewicz, D.M. Wilson, Hyperpolarized [1-13C] dehydroascorbate MR spectroscopy in a murine model of prostate cancer: comparison with 18F-FDG PET. J. Nucl. Med. **54**(6), 922–928 (2013). https://doi.org/10.2967/jnumed.112.115402

306. K.R. Keshari, D.M. Wilson, A.P. Chen, R. Bok, P.E. Larson, S. Hu, M.V. Criekinge, J.M. Macdonald, D.B. Vigneron, J. Kurhanewicz, Hyperpolarized [2-13C]-fructose: a hemiketal DNP substrate for in vivo metabolic imaging. J. Am. Chem. Soc. **131**(48), 17591–17596 (2009). https://doi.org/10.1021/ja9049355

307. C.E. Christensen, M. Karlsson, J.R. Winther, P.R. Jensen, M.H. Lerche, Non-invasive in-cell determination of free cytosolic [NAD+]/[NADH] ratios using hyperpolarized glucose show large variations in metabolic phenotypes. J. Biol. Chem. **289**(4), 2344–2352 (2014). https://doi.org/10.1074/jbc.M113.498626

308. D.M. Wilson, K.R. Keshari, P.E. Larson, A.P. Chen, S. Hu, M. Van Criekinge, R. Bok, S.J. Nelson, J.M. Macdonald, D.B. Vigneron, Multi-compound polarization by DNP allows simultaneous assessment of multiple enzymatic activities in vivo. J. Magn. Reson. **205**(1), 141–147 (2010). https://doi.org/10.1016/j.jmr.2010.04.012

309. S.S. Gambhir, Molecular imaging of cancer with positron emission tomography. Nat. Rev. Cancer **2**(9), 683 (2002)

310. T. Hara, N. Kosaka, H. Kishi, PET imaging of prostate cancer using carbon-11-choline. J. Nucl. Med. **39**(6), 990–995 (1998)

311. J. Kotzerke, J. Prang, B. Neumaier, B. Volkmer, A. Guhlmann, K. Kleinschmidt, R. Hautmann, S.N. Reske, Experience with carbon-11 choline positron emission tomography in prostate carcinoma. Eur. J. Nucl. Med. **27**(9), 1415–1419 (2000). https://doi.org/10.1007/s002590000309

312. M.A. Seltzer, S.A. Jahan, R. Sparks, D.B. Stout, N. Satyamurthy, M. Dahlbom, M.E. Phelps, J.R. Barrio, Radiation dose estimates in humans for 11C-acetate whole-body PET. J. Nucl. Med. **45**(7), 1233–1236 (2004)

313. B. Savir-Baruch, L. Zanoni, D.M. Schuster, Imaging of prostate cancer using fluciclovine. Urol. Clin. North Am. **45**(3), 489–502 (2018). https://doi.org/10.1016/j.ucl.2018.03.015

314. J. Cardinale, M. Schäfer, M. Benešová, U. Bauder-Wüst, K. Leotta, M. Eder, O.C. Neels, U. Haberkorn, F.L. Giesel, K. Kopka, Preclinical evaluation of 18F-PSMA-1007, a new prostate-specific membrane antigen ligand for prostate cancer imaging. J. Nucl. Med. **58**(3), 425–431 (2017). https://doi.org/10.2967/jnumed.116.181768

315. R. Liu, J. Cao, X. Gao, J. Zhang, L. Wang, B. Wang, L. Guo, X. Hu, Z. Wang, Overall survival of cancer patients with serum lactate dehydrogenase greater than 1000 IU/L. Tumour Biol. **37**(10), 14083–14088 (2016). https://doi.org/10.1007/s13277-016-5228-2

316. S. Halabi, E.J. Small, P.W. Kantoff, M.W. Kattan, E.B. Kaplan, N.A. Dawson, E.G. Levine, B.A. Blumenstein, N.J. Vogelzang, Prognostic model for predicting survival in men with hormone-refractory metastatic prostate cancer. J. Clin. Oncol. **21**(7), 1232–1237 (2003). https://doi.org/10.1200/JCO.2003.06.100

317. H.I. Scher, G. Heller, A. Molina, G. Attard, D.C. Danila, X. Jia, W. Peng, S.K. Sandhu, D. Olmos, R. Riisnaes, R. McCormack, T. Burzykowski, T. Kheoh, M. Fleisher, M. Buyse, J.S. de Bono, Circulating tumor cell biomarker panel as an individual-level surrogate for survival in metastatic castration-resistant prostate cancer. J. Clin. Oncol. **33**(12), 1348–1355 (2015). https://doi.org/10.1200/JCO.2014.55.3487

318. O. Sartor, R.E. Coleman, S. Nilsson, D. Heinrich, S.I. Helle, J.M. O'Sullivan, N.J. Vogelzang, O. Bruland, S. Kobina, S. Wilhelm, L. Xu, M. Shan, M.W. Kattan, C. Parker, An exploratory analysis of alkaline phosphatase, lactate dehydrogenase, and prostate-specific antigen dynamics in the phase 3 ALSYMPCA trial with radium-223. Ann. Oncol. **28**(5), 1090–1097 (2017). https://doi.org/10.1093/annonc/mdx044

319. L.R. Euceda, M.K. Andersen, M.-B. Tessem, S.A. Moestue, M.T. Grinde, T.F. Bathen, NMR-based prostate cancer metabolomics, in *Prostate Cancer*, (Springer, New York, NY, 2018), pp. 237–257

320. N.V. Reo, NMR-based metabolomics. Drug Chem. Toxicol. **25**(4), 375–382 (2002). https://doi.org/10.1081/DCT-120014789

321. J.L. Markley, R. Brüschweiler, A.S. Edison, H.R. Eghbalnia, R. Powers, D. Raftery, D.S. Wishart, The future of NMR-based metabolomics. Curr. Opin. Biotechnol. **43**, 34–40 (2017). https://doi.org/10.1016/j.copbio.2016.08.001

322. H.K. Kim, Y.H. Choi, R. Verpoorte, NMR-based metabolomic analysis of plants. Nat. Protoc. **5**(3), 536 (2010)

323. M.L. García-Martín, M. Adrados, M. Ortega, I. Fernández González, P. López-Larrubia, J. Viano, J. García-Segura, Quantitative 1H MR spectroscopic imaging of the prostate gland using LCModel and a dedicated basis-set: correlation with histologic findings. Magn. Reson. Med. **65**(2), 329–339 (2011). https://doi.org/10.1002/mrm.22631

324. N.J. Serkova, E.J. Gamito, R.H. Jones, C. O'donnell, J.L. Brown, S. Green, H. Sullivan, T. Hedlund, E.D. Crawford, The metabolites citrate, myo-inositol, and spermine are potential age-independent markers of prostate cancer in human expressed pros-

tatic secretions. Prostate **68**(6), 620–628 (2008). https://doi.org/10.1002/pros.20727

325. M.B. Tessem, M.G. Swanson, K.R. Keshari, M.J. Albers, D. Joun, Z.L. Tabatabai, J.P. Simko, K. Shinohara, S.J. Nelson, D.B. Vigneron, Evaluation of lactate and alanine as metabolic biomarkers of prostate cancer using 1H HR-MAS spectroscopy of biopsy tissues. Magn. Reson. Med. **60**(3), 510–516 (2008). https://doi.org/10.1002/mrm.21694

326. D. Kumar, A. Gupta, A. Mandhani, S.N. Sankhwar, Metabolomics-derived prostate cancer biomarkers: fact or fiction? J. Proteome Res. **14**(3), 1455–1464 (2015). https://doi.org/10.1021/pr5011108

327. M.G. Swanson, K.R. Keshari, Z.L. Tabatabai, J.P. Simko, K. Shinohara, P.R. Carroll, A.S. Zektzer, J. Kurhanewicz, Quantification of choline- and ethanolamine-containing metabolites in human prostate tissues using 1H HR-MAS total correlation spectroscopy. Magn. Reson. Med. **60**(1), 33–40 (2008). https://doi.org/10.1002/mrm.21647

328. R.A. Komoroski, J.C. Holder, A.A. Pappas, A.E. Finkbeiner, 31P NMR of phospholipid metabolites in prostate cancer and benign prostatic hyperplasia. Magn. Reson. Med. **65**(4), 911–913 (2011). https://doi.org/10.1002/mrm.22677

329. G. Zadra, M. Loda, Metabolic vulnerabilities of prostate cancer: diagnostic and therapeutic opportunities. Cold Spring Harb. Perspect. Med. **8**(10), a030569 (2018). https://doi.org/10.1101/cshperspect.a030569

330. T. Cajka, O. Fiehn, Toward merging untargeted and targeted methods in mass spectrometry-based metabolomics and lipidomics. Anal. Chem. **88**(1), 524–545 (2015). https://doi.org/10.1021/acs.analchem.5b04491

331. D.S. Wishart, Emerging applications of metabolomics in drug discovery and precision medicine. Nat. Rev. Drug Discov. **15**(7), 473 (2016)

332. Y.-F. Xu, W. Lu, J.D. Rabinowitz, Avoiding misannotation of in-source fragmentation products as cellular metabolites in liquid chromatography–mass spectrometry-based metabolomics. Anal. Chem. **87**(4), 2273–2281 (2015). https://doi.org/10.1021/ac504118y

333. N.R. Kitteringham, R.E. Jenkins, C.S. Lane, V.L. Elliott, B.K. Park, Multiple reaction monitoring for quantitative biomarker analysis in proteomics and metabolomics. J. Chromatogr. B **877**(13), 1229–1239 (2009)

334. R. Thapar, M.A. Titus, Recent advances in metabolic profiling and imaging of prostate cancer. Curr. Metab. **2**(1), 53–69 (2014)

335. B. Guo, B. Chen, A. Liu, W. Zhu, S. Yao, Liquid chromatography-mass spectrometric multiple reaction monitoring-based strategies for expanding targeted profiling towards quantitative metabolomics. Curr. Drug Metab. **13**(9), 1226–1243 (2012)

336. G. Zadra, C. Photopoulos, M. Loda, The fat side of prostate cancer. Biochim. Biophys. Acta **1831**(10), 1518–1532 (2013). https://doi.org/10.1016/j.bbalip.2013.03.010

337. J. Whitburn, C.M. Edwards, P. Sooriakumaran, Metformin and prostate cancer: a new role for an old drug. Curr. Urol. Rep. **18**, 1–7 (2017). https://doi.org/10.1007/s11934-017-0693-8

338. K.A. Richards, L. Ji, V.L. Cryns, T.M. Downs, E.J. Abel, D.F. Jarrard, Metformin use is associated with improved survival for patients with advanced prostate cancer on androgen deprivation therapy. J. Urol. **200**, 1256–1263 (2018). https://doi.org/10.1016/j.juro.2018.06.031

339. E.H. Allott, M.R. Abern, L. Gerber, C.J. Keto, W.J. Aronson, M.K. Terris, C.J. Kane, C.L. Amling, M.R. Cooperberg, P.G. Moorman, S.J. Freedland, Metformin does not affect risk of biochemical recurrence following radical prostatectomy: results from the SEARCH database. Prostate Cancer Prostatic Dis. **16**(4), 391–397 (2013). https://doi.org/10.1038/pcan.2013.48

340. M. Rieken, E. Xylinas, L. Kluth, J.J. Crivelli, J. Chrystal, T. Faison, Y. Lotan, P.I. Karakiewicz, H. Fajkovic, M. Babjuk, A. Kautzky-Willer, A. Bachmann, D.S. Scherr, S.F. Shariat, Association of diabetes mellitus and metformin use with oncological outcomes of patients with non-muscle-invasive bladder cancer. BJU Int. **112**(8), 1105–1112 (2013). https://doi.org/10.1111/bju.12448

341. S.Y. Wang, C.S. Chuang, C.H. Muo, S.T. Tu, M.C. Lin, F.C. Sung, C.H. Kao, Metformin and the incidence of cancer in patients with diabetes: a nested case-control study. Diabetes Care **36**(9), e155–e156 (2013). https://doi.org/10.2337/dc13-0708

342. D. Soranna, L. Scotti, A. Zambon, C. Bosetti, G. Grassi, A. Catapano, C. La Vecchia, G. Mancia, G. Corrao, Cancer risk associated with use of metformin and sulfonylurea in type 2 diabetes: a meta-analysis. Oncologist **17**(6), 813–822 (2012). https://doi.org/10.1634/theoncologist.2011-0462

343. L. Azoulay, S. Dell'Aniello, B. Gagnon, M. Pollak, S. Suissa, Metformin and the incidence of prostate cancer in patients with type 2 diabetes. Cancer Epidemiol. Biomarkers Prev. **20**(2), 337–344 (2011). https://doi.org/10.1158/1055-9965.EPI-10-0940

344. L. Bensimon, H. Yin, S. Suissa, M.N. Pollak, L. Azoulay, The use of metformin in patients with prostate cancer and the risk of death. Cancer Epidemiol. Biomarkers Prev. **23**(10), 2111–2118 (2014). https://doi.org/10.1158/1055-9965.EPI-14-0056

345. C.J. Currie, C.D. Poole, E.A. Gale, The influence of glucose-lowering therapies on cancer risk in type 2 diabetes. Diabetologia **52**(9), 1766–1777 (2009). https://doi.org/10.1007/s00125-009-1440-6

346. D. Kaushik, R.J. Karnes, M.S. Eisenberg, L.J. Rangel, R.E. Carlson, E.J. Bergstralh, Effect of metformin on prostate cancer outcomes after radical prostatectomy. Urol. Oncol. **32**(1), 43.e41–43.e47 (2014). https://doi.org/10.1016/j.urolonc.2013.05.005

347. D. Margel, D. Urbach, L.L. Lipscombe, C.M. Bell, G. Kulkarni, P.C. Austin, N. Fleshner, Association between metformin use and risk of prostate cancer

and its grade. J. Natl. Cancer Inst. **105**(15), 1123–1131 (2013). https://doi.org/10.1093/jnci/djt170

348. T. Patel, G. Hruby, K. Badani, C. Abate-Shen, J.M. McKiernan, Clinical outcomes after radical prostatectomy in diabetic patients treated with metformin. Urology **76**(5), 1240–1244 (2010). https://doi.org/10.1016/j.urology.2010.03.059

349. K.K. Tsilidis, D. Capothanassi, N.E. Allen, E.C. Rizos, D.S. Lopez, K. van Veldhoven, C. Sacerdote, D. Ashby, P. Vineis, I. Tzoulaki, J.P. Ioannidis, Metformin does not affect cancer risk: a cohort study in the U.K. Clinical Practice Research Datalink analyzed like an intention-to-treat trial. Diabetes Care **37**(9), 2522–2532 (2014). https://doi.org/10.2337/dc14-0584

350. T. Feng, X. Sun, L.E. Howard, A.C. Vidal, A.R. Gaines, D.M. Moreira, R. Castro-Santamaria, G.L. Andriole, S.J. Freedland, Metformin use and risk of prostate cancer: results from the REDUCE study. Cancer Prev. Res. (Phila.) **8**(11), 1055–1060 (2015). https://doi.org/10.1158/1940-6207.CAPR-15-0141

351. A.M. Joshua, V.E. Zannella, M.R. Downes, B. Bowes, K. Hersey, M. Koritzinsky, M. Schwab, U. Hofmann, A. Evans, T. van der Kwast, J. Trachtenberg, A. Finelli, N. Fleshner, J. Sweet, M. Pollak, A pilot 'window of opportunity' neoadjuvant study of metformin in localised prostate cancer. Prostate Cancer Prostatic Dis. **17**(3), 252–258 (2014). https://doi.org/10.1038/pcan.2014.20

352. Medicine UNLo, ClinicalTrials.gov (2015). https://clinicaltrials.gov/ct2/show/NCT01433913

353. A. Vancura, P. Bu, M. Bhagwat, J. Zeng, I. Vancurova, Metformin as an anticancer agent. Trends Pharmacol. Sci. **39**, 867–878 (2018). https://doi.org/10.1016/j.tips.2018.07.006

354. X. Liu, I.L. Romero, L.M. Litchfield, E. Lengyel, J.W. Locasale, Metformin targets central carbon metabolism and reveals mitochondrial requirements in human cancers. Cell Metab. **24**, 728–739 (2016). https://doi.org/10.1016/j.cmet.2016.09.005

355. V. Zingales, A. Distefano, M. Raffaele, A. Zanghi, I. Barbagallo, L. Vanella, Metformin: a bridge between diabetes and prostate cancer. Front. Oncol. **7**, 1–7 (2017). https://doi.org/10.3389/fonc.2017.00243

356. T.M. Ashton, W.G. McKenna, L.A. Kunz-Schughart, G.S. Higgins, Oxidative phosphorylation as an emerging target in cancer therapy. Clin. Cancer Res. **24**, 2482–2490 (2018). https://doi.org/10.1158/1078-0432.CCR-17-3070

357. L.B. Sullivan, D.Y. Gui, A.M. Hosios, L.N. Bush, E. Freinkman, M.G. Vander Heiden, Supporting aspartate biosynthesis is an essential function of respiration in proliferating cells. Cell **162**(3), 552–563 (2015). https://doi.org/10.1016/j.cell.2015.07.017

358. K. Birsoy, T. Wang, W.W. Chen, E. Freinkman, M. Abu-Remaileh, D.M. Sabatini, An essential role of the mitochondrial electron transport chain in cell proliferation is to enable aspartate synthesis. Cell **162**(3), 540–551 (2015). https://doi.org/10.1016/j.cell.2015.07.016

359. R.J. DeBerardinis, N.S. Chandel, Fundamentals of cancer metabolism. Sci. Adv. **2**, e1600200 (2016). https://doi.org/10.1126/sciadv.1600200

360. J.R. Molina, Y. Sun, M. Protopopova, S. Gera, M. Bandi, C. Bristow, T. McAfoos, P. Morlacchi, J. Ackroyd, A.-nA. Agip, G. Al-Atrash, J. Asara, J. Bardenhagen, C.C. Carrillo, C. Carroll, E. Chang, S. Ciurea, J.B. Cross, B. Czako, A. Deem, N. Daver, J.F. de Groot, J.-w. Dong, N. Feng, G. Gao, J. Gay, M.G. Do, J. Greer, V. Giuliani, J. Han, L. Han, V.K. Henry, J. Hirst, S. Huang, Y. Jiang, Z. Kang, T. Khor, S. Konoplev, Y.-h. Lin, G. Liu, A. Lodi, T. Lofton, H. Ma, M. Mahendra, P. Matre, R. Mullinax, M. Peoples, A. Petrocchi, J. Rodriguez-Canale, R. Serreli, T. Shi, M. Smith, Y. Tabe, J. Theroff, S. Tiziani, Q. Xu, Q. Zhang, F. Muller, R.A. DePinho, C. Toniatti, G.F. Draetta, T.P. Heffernan, M. Konopleva, P. Jones, M.E. Di Francesco, J.R. Marszalek, An inhibitor of oxidative phosphorylation exploits cancer vulnerability. Nat. Med. **24**, 1036–1046 (2018). https://doi.org/10.1038/s41591-018-0052-4

361. D.A. Bader, S.M. Hartig, V. Putluri, C. Foley, M.P. Hamilton, E.A. Smith, P.K. Saha, A. Panigrahi, C. Walker, L. Zong, H. Martini-Stoica, R. Chen, K. Rajapakshe, C. Coarfa, A. Sreekumar, N. Mitsiades, J.A. Bankson, M.M. Ittmann, B.W. O'Malley, N. Putluri, S.E. McGuire, Mitochondrial pyruvate import is a metabolic vulnerability in AR-driven prostate cancer. Nat. Metab. **1**, 70–85 (2019). https://doi.org/10.1038/s42255-018-0002-y

362. Q. Wang, R.A. Hardie, A.J. Hoy, M. Van Geldermalsen, D. Gao, L. Fazli, M.C. Sadowski, S. Balaban, M. Schreuder, R. Nagarajah, J.J.L. Wong, C. Metierre, N. Pinello, N.J. Otte, M.L. Lehman, M. Gleave, C.C. Nelson, C.G. Bailey, W. Ritchie, J.E. Rasko, J. Holst, Targeting ASCT2-mediated glutamine uptake blocks prostate cancer growth and tumour development. J. Pathol. **236**, 278–289 (2015). https://doi.org/10.1002/path.4518

363. C. Yang, B. Ko, C.T. Hensley, L. Jiang, A.T. Wasti, J. Kim, J. Sudderth, M.A. Calvaruso, L. Lumata, M. Mitsche, J. Rutter, M.E. Merritt, R.J. DeBerardinis, Glutamine oxidation maintains the TCA cycle and cell survival during impaired mitochondrial pyruvate transport. Mol. Cell **56**, 414–424 (2014). https://doi.org/10.1016/j.molcel.2014.09.025

364. C. Adams, R.C. Richmond, D.L. Santos Ferreira, W. Spiller, V.Y. Tan, J. Zheng, P. Wurtz, J.L. Donovan, F.C. Hamdy, D.E. Neal, J.A. Lane, G. Davey Smith, C.L. Relton, R.A. Eeles, B.E. Henderson, C.A. Haiman, Z. Kote-Jarai, F.R. Schumacher, A. Amin Al Olama, S. Benlloch, K. Muir, S.I. Berndt, D.V. Conti, F. Wiklund, S.J. Chanock, S.M. Gapstur, V.L. Stevens, C.M. Tangen, J. Batra, J.A. Clements, H. Gronberg, N. Pashayan, J. Schleutker, D. Albanes,

A. Wolk, C.M.L. West, L.A. Mucci, G. Cancel-Tassin, S. Koutros, K.D. Sorensen, L. Maehle, R.C. Travis, R. Hamilton, S.A. Ingles, B.S. Rosenstein, Y.J. Lu, G.G. Giles, A.S. Kibel, A. Vega, M. Kogevinas, K.L. Penney, J.Y. Park, J.L. Stanford, C. Cybulski, B.G. Nordestgaard, H. Brenner, C. Maier, J. Kim, E.M. John, M.R. Teixeira, S.L. Neuhausen, K. DeRuyck, A. Razack, L.F. Newcomb, D. Lessel, R.P. Kaneva, N. Usmani, F. Claessens, P. Townsend, M. Gago Dominguez, M.J. Roobol, F. Menegaux, K.T. Khaw, L.A. Cannon-Albright, H. Pandha, S.N. Thibodeau, R.M. Martin, Circulating metabolic biomarkers of screen-detected prostate cancer in the ProtecT study. Cancer Epidemiol. Biomarkers Prev. **28**, 208 (2019). https://doi.org/10.1158/1055-9965. EPI-18-0079

365. J. Huang, A.M. Mondul, S.J. Weinstein, S. Koutros, A. Derkach, E. Karoly, J.N. Sampson, S.C. Moore, S.I. Berndt, D. Albanes, Serum metabolomic profiling of prostate cancer risk in the prostate, lung, colorectal, and ovarian cancer screening trial. Br. J. Cancer **115**(9), 1087–1095 (2016). https://doi.org/10.1038/bjc.2016.305

366. A.M. Mondul, S.C. Moore, S.J. Weinstein, E.D. Karoly, J.N. Sampson, D. Albanes, Metabolomic analysis of prostate cancer risk in a prospective cohort: the alpha-tocolpherol, beta-carotene cancer prevention (ATBC) study. Int. J. Cancer **137**(9), 2124–2132 (2015). https://doi.org/10.1002/ijc.29576

367. A.M. Mondul, S.C. Moore, S.J. Weinstein, S. Mannisto, J.N. Sampson, D. Albanes, 1-stea-roylglycerol is associated with risk of prostate cancer: results from serum metabolomic profiling. Metabolomics **10**(5), 1036–1041 (2014). https://doi.org/10.1007/s11306-014-0643-0

368. J.R. Mayers, M.G. Vander Heiden, Famine versus feast: understanding the metabolism of tumors in vivo. Trends Biochem. Sci. **40**(3), 130–140 (2015). https://doi.org/10.1016/j.tibs.2015.01.004

369. S.M. Davidson, T. Papagiannakopoulos, B.A. Olenchock, J.E. Heyman, M.A. Keibler, A. Luengo, M.R. Bauer, A.K. Jha, J.P. O'Brien, K.A. Pierce, D.Y. Gui, L.B. Sullivan, T.M. Wasylenko, L. Subbaraj, C.R. Chin, G. Stephanopolous, B.T. Mott, T. Jacks, C.B. Clish, M.G. Vander Heiden, Environment impacts the metabolic dependencies of Ras-driven non-small cell lung cancer. Cell Metab. **23**(3), 517–528 (2016). https://doi.org/10.1016/j.cmet.2016.01.007

370. J.A. Reisz, A. D'Alessandro, Measurement of metabolic fluxes using stable isotope tracers in whole animals and human patients. Curr. Opin. Clin. Nutr. Metab. Care **20**(5), 366–374 (2017). https://doi.org/10.1097/MCO.0000000000000393

371. B. Faubert, K.Y. Li, L. Cai, C.T. Hensley, J. Kim, L.G. Zacharias, C. Yang, Q.N. Do, S. Doucette, D. Burguete, H. Li, G. Huet, Q. Yuan, T. Wigal, Y. Butt, M. Ni, J. Torrealba, D. Oliver, R.E. Lenkinski, C.R. Malloy, J.W. Wachsmann, J.D. Young, K. Kernstine, R.J. DeBerardinis, Lactate metabolism in human lung tumors. Cell **171**(2), 358–371.e359 (2017). https://doi.org/10.1016/j.cell.2017.09.019

372. C.T. Hensley, B. Faubert, Q. Yuan, N. Lev-Cohain, E. Jin, J. Kim, L. Jiang, B. Ko, R. Skelton, L. Loudat, M. Wodzak, C. Klimko, E. McMillan, Y. Butt, M. Ni, D. Oliver, J. Torrealba, C.R. Malloy, K. Kernstine, R.E. Lenkinski, R.J. DeBerardinis, Metabolic heterogeneity in human lung tumors. Cell **164**(4), 681–694 (2016). https://doi.org/10.1016/j.cell.2015.12.034

373. A.H. Davies, H. Beltran, A. Zoubeidi, Cellular plasticity and the neuroendocrine phenotype in prostate cancer. Nat. Rev. Urol. **15**(5), 271–286 (2018). https://doi.org/10.1038/nrurol.2018.22

Canonical and Noncanonical Androgen Metabolism and Activity

Karl-Heinz Storbeck and Elahe A. Mostaghel

Introduction

Androgen deprivation therapy (ADT) remains the primary treatment for metastatic prostate cancer (PCa) since the seminal recognition of the disease as androgen-dependent by Huggins and Hodges in 1941. However, castration does not eliminate androgens from the prostate tumor microenvironment. Castration resistant prostate cancer (CRPC) is characterized by elevated tumor androgen levels that are well within the range capable of activating the androgen receptor (AR) and AR-mediated gene expression, as well as by alterations in the expression levels of steroid metabolizing enzymes that may potentiate *de novo* androgen biosynthesis and utilization of circulating adrenal androgen precursors. Indeed, residual intratumoral androgens are implicated in

nearly every mechanism by which AR-mediated signaling promotes castration-resistant disease, and the importance of residual ligands in disease progression is supported by the clinical efficacy of new drugs targeting the AR axis such as abiraterone and enzalutamide. These observations suggest that tissue-based alterations in steroid metabolism contribute to the development of CRPC and underscore these steroidogenic pathways as critical targets of therapy.

The best-studied pathways are those contributing to the uptake and intratumoral (intracrine) conversion of circulating canonical adrenal androgen precursors such as dehydroepiandrosterone (DHEA) and its sulfate (DHEA-S) to the potent androgen 5α-dihydrotestosterone (DHT) which is a recognized driver of CRPC through activation of the wild type AR. However, the less characterized adrenal steroid, 11β-hydroxyandrostenedione (11OHA4) may play a previously unrecognized role in promoting AR activation. In particular, 11OHA4 is efficiently converted within PCa cells to 11-ketotestosterone (11KT), which is a potent and efficacious activator of the wild type AR. Thus, in the low androgen environment of CRPC, alternative sources of androgens may supplement AR activation normally mediated by the canonical 5α-reduced agonist, DHT.

Herein, we review the accumulated body of evidence which supports intracrine androgen biosynthesis as an important mechanism underlying

K.-H. Storbeck
Department of Biochemistry, Stellenbosch University, Stellenbosch, South Africa

E. A. Mostaghel (✉)
Clinical Research Division, Fred Hutchinson Cancer Research Center, Seattle, WA, USA

Department of Medicine, University of Washington, Seattle, WA, USA

Geriatric Research, Education and Clinical Center S-182, VA Puget Sound Health Care System, Seattle, WA, USA
e-mail: emostagh@fredhutch.org

© Springer Nature Switzerland AG 2019
S. M. Dehm, D. J. Tindall (eds.), *Prostate Cancer*, Advances in Experimental Medicine and Biology 1210, https://doi.org/10.1007/978-3-030-32656-2_11

PCa progression, starting with the presence and significance of residual prostate tumor androgens in the progression of CRPC. We review the classical and non-classical pathways of androgen metabolism, and how dysregulated expression of steroidogenic enzymes is likely to potentiate tumor androgen production in the progression to CRPC, including a role for the enzymes mediating pre-receptor control of DHT metabolism in modulating intra-tumoral androgen levels, and recent findings on the role of oncogenic splicing of steroidogenic enzymes in tumor androgen metabolism. We review the *in vitro* and *in vivo* data in human tumors, xenografts, and cell line models, which demonstrates the capacity of prostate tumors to utilize cholesterol and adrenal androgens in the production of testosterone (T) and DHT, and briefly review the potential role of exogenous influences on this process.

We next discuss an emerging literature on non-canonical aspects of androgens and steroidogenic enzymes in PCa. We review the data regarding the role of 11-oxygenated androgens of adrenal origin in activating wild-type AR and serving as an under-recognized reservoir of active androgens. We discuss the enzymatic pathways and novel downstream metabolites of 11OHA4 that mediate these effects, and review evidence that these non-canonical androgens are better substrates for conversion to their active forms by aldo-keto reductase 1C3 (AKR1C3, also known as 17β-hydroxysteroid dehydrogenase type 5/HSD17B5) and less susceptible to glucuronidation and inactivation than the canonical androgens. We also discuss an emerging literature on the potential noncanonical activity of androgen metabolizing enzymes in mediating AR activity and driving PCa, thereby playing an unrecognized role in CRPC progression separate from any role in androgen metabolism.

Finally, we discuss data regarding mechanisms of response and resistance to potent ligand synthesis inhibitors entering clinical practice, and conclude by discussing the implications of these findings for understanding resistance and optimizing response to new agents targeting the AR-axis for PCa therapy.

Residual Prostate Tumor Androgens in the Progression of CRPC

The efficacy of ADT is routinely based on achieving castrate levels of serum T, defined as <20 ng/dl (0.69 nM). However, tissue androgen levels in the setting of benign prostatic hyperplasia (BPH), primary prostate tumors, locally recurrent PCa, or metastatic CRPC have consistently demonstrated that castration does not eliminate androgens from the prostate tumor microenvironment.

Geller et al. examined prostatic DHT levels by radioimmunoassay (RIA) and demonstrated that castration by orchiectomy (or Megace plus DES) reduced prostatic DHT levels by 75–80% to 1 ng/g in some but not all patients. Epithelial and stromal cell protein biosynthesis was strongly correlated with tissue DHT levels, and prostatic DHT levels were further reduced when castration was combined with adrenal androgen blockade by ketoconazole [1–6], suggesting the goal of therapy should be to decrease prostatic DHT to as low as possible, a concept similarly framed in early studies by Labrie [7].

Incomplete suppression of prostate tissue androgens by castration has been subsequently confirmed in numerous studies of short and long term castration therapy [8]. Treatment of BPH patients for 3 months with an LHRH agonist decreased intraprostatic T levels by 75%, to about 0.1 ng/g, and DHT levels by 90%, to 0.48 ng/g [9]. In men with PCa, 6 months of neoadjuvant ADT with castration and flutamide reduced prostatic DHT levels by 75% to about 1.35 ng/g [10]. Notably, tumor differentiation based on Gleason grading correlated with changes in tissue DHT, with an 85% decrease measured in Gleason 6 cancers, but only a 60% decrement in Gleason 7–10 tumors [11]. This indicates that tumor type-specific changes in androgen metabolism may impact responses to systemic T suppression.

Residual androgens have also been demonstrated in both locally recurrent and metastatic castration resistant tumors. T levels in locally recurrent tumors from castrate patients were equivalent to those of BPH patients, where DHT levels were reduced 80%, to about 0.4 ng/g [12].

Compared to primary prostate tumors from untreated patients (T 0.25 ng/g, DHT 2.75 ng/g) androgen levels in metastatic CRPC tumors obtained via rapid autopsy showed threefold higher T levels and an inverted ratio of T to DHT (T 0.74 ng/g; DHT 0.25 ng/g) [13]. Adrenal androgen precursors have also been detected at significant levels in prostate tissue of castrate men. Prostatic levels of DHEA, DHEA-S, and androstenedione (A4) were decreased by about 50% in castrate patients and far exceeded values of T and DHT in recurrent tumors [12]. No decrease in prostatic levels of 5-androstenediol (A5-diol) were found after castration [14], which is of particular significance as this androgen has been shown to bind wild type AR without being inhibited by flutamide or bicalutamide [15].

Two studies of men with localized PCa demonstrated that the addition of adrenal androgen biosynthesis inhibitors to castration therapy can lower prostate androgens below that achieved with standard androgen blockade, suggesting a role for circulating adrenal androgens in post-castration tissue androgen levels. The addition of dutasteride and ketoconazole to combined androgen blockade (CAB) for 3 months prior to prostatectomy lowered prostate DHT from 0.92 ng/g (in the CAB arm) to 0.03 ng/g [16]. In a second study, the potent cytochrome P450 17α-hydroxylase/17,20-lyase (CYP17A1) inhibitor, abiraterone, was added to LHRH agonist therapy for 3 or 6 months prior to prostatectomy. Abiraterone decreased prostate tissue DHT from 1.3 ng/g (in men treated with LHRH agonist therapy alone) to 0.18 ng/g and also decreased prostate levels of A4 and DHEA [17].

These findings clearly demonstrate that achieving castrate levels of circulating T does not eliminate androgens from the prostate tumor microenvironment. The ability of DHT in the range observed in castrate tumors (~1 nm, 0.5–1.0 ng/g) to activate the AR, stimulate expression of AR-regulated genes, and promote androgen mediated tumor growth has been convincingly demonstrated in both *in vitro* and *in vivo* studies [12, 18–21], and is evidenced by the nearly universal rise in serum prostate specific antigen (PSA) that accompanies CRPC progression.

Residual tissue androgens are implicated in the majority of mechanisms whereby persistent AR-mediated signaling drives castration resistant disease. These mechanisms include AR overexpression, AR mutations that broaden ligand specificity and/or confer sensitivity to adrenal androgens, alterations in AR coactivators and/or corepressors that modulate AR stability and ligand sensitivity, and activation of the AR or downstream regulatory molecules by "cross talk" with other signaling pathways. Restoration of AR expression and signaling in a xenograft model was both necessary and sufficient to drive progression from androgen-dependent to castration resistant growth, allowing tumor cell proliferation in 80% lower androgen concentrations [22]. Importantly, ligand binding was required for castration-resistant growth, and modest increases in AR expression were sufficient to support signaling in a low androgen environment.

The clinical relevance of intratumoral androgens in promoting CRPC tumor growth is confirmed by the clinical responses to agents targeting residual androgen pathway activity. These include historical responses described in response to adrenalectomy and/or hypophysectomy [23, 24]; the limited but consistent ~5% overall survival benefit seen in meta-analyses of CAB [25–27]; the observation that nearly 30% of recurrent prostate tumors demonstrate at least transient clinical responses to secondary or tertiary hormonal manipulation [28]; and most recently, the striking clinical response observed with novel inhibitors of androgen biosynthesis, such as abiraterone, and potent AR inhibitors such as enzalutamide [29, 30]. Perhaps most importantly, emerging studies suggest that response and resistance to abiraterone is associated with upregulated androgen biosynthesis in tumors, clearly demonstrating the importance of intratumoral androgen metabolism in CRPC tumor survival [31–33].

Pathways of Androgen Biosynthesis

The source of residual androgens within prostate tumors of castrate men has not been fully elucidated, but is generally attributed to the uptake and conversion of circulating adrenal androgen precursors [14, 34], and potentially *de novo* biosynthesis of androgens from progesterone or cholesterol [35]. Here we review the various pathways of *de novo* androgen biosynthesis in adrenal and peripheral tissues including testis and prostate (Fig. 1, reviewed in [36]), the enzymatic pathways mediating prostate androgen metabolism, and the 'backdoor' and '5α-dione' pathways of androgen biosynthesis. A general outline of the canonical and non-canonical steroidogenic pathways is provided in Fig. 2.

Androgen Biosynthesis in the Adrenal Gland and Peripheral Tissues

Steroid hormone biosynthesis begins with transfer of a 27-carbon (C-27) cholesterol molecule from the outer mitochondrial membrane to the inner membrane by steroidogenic acute regulatory protein (StAR), followed by its conversion to the C-21 steroid, pregnenolone catalyzed by cytochrome P450 side-change cleavage (CYP11A1). Subsequent metabolism to proges-

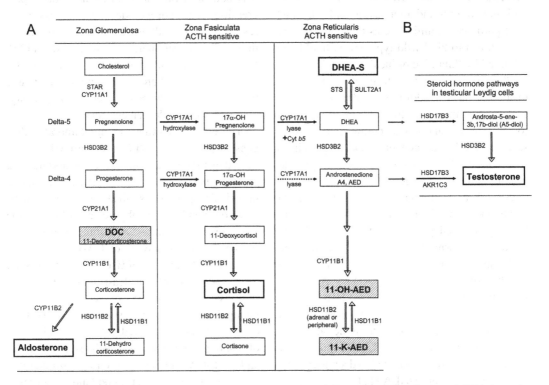

Fig. 1 Steroid hormone biosynthesis pathways in the adrenal gland and testis. (**a**) Steroid biosynthesis in the adrenal cortex occurs in three zones, each with a specific complement of enzymes. The zona glomerulosa contains the enzymes necessary to produce aldosterone. The zona fasciculata and reticularis additionally express CYP17A1. The 17α-hydroxylase activity of CYP17A1 is active in the zona fasiculata resulting in the production of cortisol. Due to tissue-specific expression of the cytochrome *b5* cofactor, the 17,20-lyase activity of CYP17A1 is only present in the zona reticularis and drives efficient production of DHEA which is then sulfated to DHEA-S. 17α-hydroxyprogesterone is a poor substrate for CYP17A1 17,20-lyase activity (dotted arrow) and thus androstenedione is formed at lower levels from this substrate than 17α-hydroxypregnenolone. (**b**) Testicular androgen biosynthesis follows a similar pathway to DHEA formation as that in the zona reticularis, but due to the absence of SULT2A1, and the presence of HSD3B2 and HSD17B3, DHEA is efficiently converted to testosterone. The primary product of each zone is denoted within the darker squares (aldosterone, cortisol, DHEA-S in the adrenal gland, and testosterone from the testis). The adrenal derived 11-oxygenated androgen precursors are depicted in the hatched boxes (11β-hydroxyandrostenedione, 11OHA4; 11-ketoandrostenedione, 11KA4). (Modified from International Journal of Biological Science Volume 22, Issue 10, Elahe A. Mostaghel. Beyond T and DHT – Novel Steroid Derivatives Capable of Wild Type Androgen Receptor Activation. Pages No. 602–613, Copyright (2014), with permission from Ivyspring International Publisher)

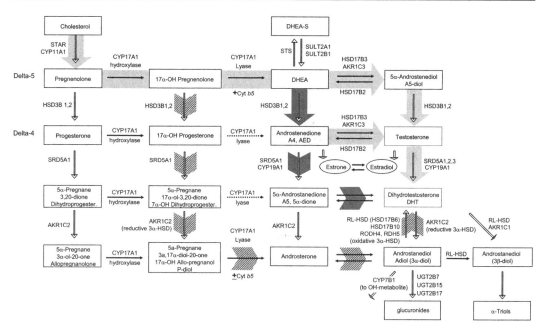

Fig. 2 Classical and non-classical pathways of androgen biosynthesis. Cholesterol is converted pregnenolone and by the action of StAR and CYP11A1. In the classical pathway (light gray arrows) pregnenolone and progesterone are converted to the adrenal androgens DHEA and androstenedione (A4) by the sequential 17α-hydroxylase and 17,20-lyase activity of CYP17A1. Due to the substrate preference the 17,20-lyase activity of CYP17A1 (which requires the cytochrome *b5* cofactor) favors production of DHEA. DHEA (from intrinsic or circulating sources depending on the tissue) is subsequently acted on by HSD3B and HSD17B3 or AKR1C3 to form testosterone, which is converted to DHT via SRD5A. In the backdoor pathway (hatched arrows) the progesterone intermediates are acted on first by the activity of SRD5A and the reductive activity of AKR1C2 prior to the 17,20-lyase activity of CYP17A1. Androsterone is then acted on by HSD17B3 or AKR1C3 and must undergo an oxidative step mediated by RL-HSD (or others) to generate DHT. In a third pathway, termed the 5α-Androstanedione pathway (dark gray arrows) DHEA and A4 are produced as in the classical pathway. However, instead of conversion of A4 to testosterone followed by the activity of SRD5A to produce DHT, the enzymatic sequence is reversed such that A4 is converted first by SRD5A to 5α-Androstanedione and then by HSD17B3 or AKR1C3 to DHT. (Modified from Best Practice & Research Clinical Endocrinology & Metabolism, Volume 22, Issue 2, Elahe A. Mostaghel and Peter S. Nelson. Intracrine androgen metabolism in prostate cancer progression: mechanisms of castration resistance and therapeutic implications. Pages No. 243-258, Copyright (2008), with permission from Elsevier)

terone, mineralocorticoids, glucocorticoids (all C-21 steroids), androgens (C-19) or estrogens (C-18) is dictated in a tissue-specific manner, driven by the expression of specific enzymes and catalytic cofactors.

CYP17A1, expressed in the adrenal gland, testis and ovary, is a single enzyme with one active site, which catalyzes sequential but independent 17α-hydroxylase and 17,20-lyase reactions, both of which are required for converting C-21 progestogens to C-19 androgen precursors along the delta-5 pathway from pregnenolone or the delta-4 pathway from progesterone. The 17α-hydroxylase activity of CYP17A1 for pregnenolone and progesterone is similar, but its 17,20-lyase activity for delta-5 and delta-4 substrates requires the activity of the cytochrome *b5* cofactor, and is approximately 50 times more efficient for converting the delta-5 substrate 17α-hydroxypregnenolone to DHEA than the delta-4 substrate 17α-hydroxyprogesterone to A4 [36]. Androgen biosynthesis in the human therefore favors the delta-5 pathway to DHEA. 3β-Hydroxysteroid dehydrogenase/delta5-4 isomerase (HSD3B) enzymes catalyze the conversion of delta-5 to delta-4 steroids. Whereas HSD3B2 is the primary isoform expressed in adrenal, testis and ovary (all sites of *de novo* steroidogenesis), HSD3B1 (tenfold more efficient) is the isoform expressed in peripheral tissues such as prostate, breast, skin, placenta and brain [36].

In adrenal steroidogenesis (Fig. 1a, reviewed in [37]), the zona glomerulosa lacks CYP17A1 expression and produces aldosterone via the sequential activity of HSD3B2, cytochrome 21-hydroxylase (CYP21A2), and cytochrome P450 11β-hydroxylase (CYP11B) on pregnenolone. Both the zona fasciculata and zona reticularis express CYP17A1, but the zona fasciculata does not express cytochrome $b5$ and therefore channels precursors to production of glucocorticoids. The differential expression of cytochrome $b5$ in the zona reticularis augments the 17,20-lyase activity of CYP17A1 tenfold, leading to robust production of DHEA, followed by conversion to DHEA-S via the sulfotransferase activity of SULT2A1. The zona reticularis is also characterized by low expression of HSD3B2, favoring conversion of pregnenolone to DHEA and DHEA-S, although small amounts are converted to A4 [38].

Less recognized is that the human zona reticularis also expresses AKR1C3, which mediates the final step in T biosynthesis from A4. Notably, in a small study of eight women, adrenal vein T levels increased sixfold (18.5–116 ng/dl) before and after ACTH stimulation [39]. In a much earlier study, selective adrenal vein catheterization in men also demonstrated adrenal to peripheral venous T gradients, although a compensatory increase in adrenal production of T was not observed in castrate vs. intact men [40].

The adrenal gland also produces less recognized non-canonical androgens. 11OHA4 is an abundant product of the human adrenal derived from the CYP11B mediated 11β-hydroxylation of A4. CYP11B1 and 2 have also been shown to 11β-hydroxylate T, yielding 11β-hydroxytestosterone (11OHT), though the levels produced by the adrenal are low due to the limited availability of adrenal derived T [41, 42]. Low levels of 11-ketoandrostenedione (11KA4) and 11KT are also produced from 11OHA4 and 11OHT, respectively due to the low levels of 11β-hydroxysteroid dehydrogenase type 2

(HSD11B2) expressed in the adrenal, with 11KA4 also serving as a substrate for AKR1C3 [41]. Adrenal vein sampling before and after ACTH stimulation showed a 5.2-fold increase in 11OHA4 (157 ± 96.2 nM to 811 ± 260 nM), a 3.2-fold increase in 11KA4 (0.99 ± 0.33 nM to 3.18 ± 0.63 nM), a 5.5-fold increase in 11OHT (0.48 ± 0.17 nM to 2.62 ± 0.74 nM) and a 1.3-fold increase in 11KT (0.39 ± 0.09 nM to 0.49 ± 0.11 nM). Adrenal output does not, however, account for the circulating levels of 11KA4 and 11KT, which are instead likely achieved by peripheral conversion of abundant adrenal derived 11OHA4 to 11KA4 by peripheral tissue expressing HSD11B2, with 11KA4 subsequently serving as a substrate for the peripheral AKR1C3 mediated conversion to 11KT [43].

Leydig cells of the testis (Fig. 1b) express metabolic machinery that is similar to the adrenal gland, including StAR and CYP11A1, and also display preference of CYP17A1 for delta-5 substrates. This allows Leydig cells to produce DHEA from cholesterol, but with several key differences compared to the adrenal gland, including absence of SULT2A1, preventing conversion of DHEA to DHEA-S, and abundant expression of HSD3B2, which mediates the delta-5 to delta-4 conversion required to generate T. The final steps in T biosynthesis are catalyzed by 17β-hydroxysteroid dehydrogenase type 3 (HSD17B3) and/or AKR1C3. HSD17B3 is expressed primarily in testicular Leydig cells, while AKR1C3 mediates production of T in peripheral tissues. The activity of HSD3B2 and HSD17B3 thus drives the stepwise conversion of DHEA to T, via either A4 or A5-diol.

Androgen Biosynthesis in the Prostate and Pre-receptor Control of DHT Metabolism

The uptake of circulating androgen precursors and the local biosynthesis of active steroids in peripheral target tissues such as prostate, breast and skin has been termed intracrinology [44], a process that is complemented by the paracrine diffusion and conversion of steroid substrates among neighboring cell types with different enzyme capacities. In the prostate, circulating T from the Leydig cells is converted to DHT by steroid 5α-reductase type 2 (SRD5A2) present in both basal and luminal epithelial cells. Circulating DHEA-S must be de-sulfated by the activity of steroid sulfatase (STS) which has been detected in normal and cancer prostate tissue [45–47], and can then be converted to A4, T and DHT via the activity of HSD3B1, AKR1C3 and SRD5A2 present in basal epithelial cells [48, 49]. Circulating T or T produced in the basal cells diffuses to the AR positive luminal cells where it is then converted to DHT by SRD5A2 [44].

Prostate tissue also demonstrates epithelial cell expression of phase I (reducing) and phase II (conjugating) DHT catabolizing enzymes that act in concert to regulate access of DHT to the AR. Aldo-keto reductase 1C1 (AKR1C1) is the primary enzyme responsible for the irreversible reduction of DHT to the weak metabolite, 5α-androstane-3α,17β-diol (3α-androstanediol or 3α-diol, a low affinity AR ligand), whereas aldo-keto reductase 1C2 (AKR1C2) catalyzes the reversible conversion of DHT to 5α-androstane-3β,17β-diol (3β-diol, a pro-apoptotic ligand of estrogen receptor beta, ERβ) [50]. The reductase activity of AKR1C2, coupled with the reverse oxidative activity of specific 3α-HSD enzymes is a critical molecular switch regulating access of DHT to the AR [50–53].

Candidate enzymes mediating the reversible conversion of 3α-diol to DHT include RL-HSD (HSD17B6), HSD17B10, RODH4, RDH5, and DHRS9. Transcripts of RL-HSD and HSD17B10 are highly expressed in the prostate, however several studies suggest RL-HSD is more active in converting 3α-diol to DHT in prostate cells [54, 55]. Basal epithelial cell expression of RL-HSD is present at the protein level, while transcript profiling of cultured epithelial and stromal cells detects stromal expression as well [54, 56]. RL-HSD also acts as an epimerase to convert 3β-diol to 3α-diol, although at much higher substrate concentrations [57]. RL-HSD also catalyzes conversion of physiologic levels of DHT to 3β-diol, suggesting RL-HSD is involved in maintaining the intraprostatic balance of DHT, 3α-diol and 3β-diol [56].

The glucuronidating enzymes (UDP-glucuronosyltransferases) UGT2B15 and UTG2B17 located in prostate luminal and basal epithelial cells, respectively, irreversibly terminate the androgen signal by glucuronidation of 3α-diol (as well as T, DHT and other metabolites), and are major determinants of the androgen signal in PCa cell lines [58–60]. UDP-glucose 6-dehydrogenase (UGDH) is required to generate the substrate for glucuronide conjugation (UDP-glucuronate), and over-expression of UGDH increases the generation of glucuronidated androgens [61]. Thus, the relative activity of AKR1C2 in converting DHT to 3β-diol, and of RL-HSD and UGT2B17 in competing for conversion of 3β-diol back to DHT or to 3βdiol-G, respectively, will collectively determine the amount of active steroid available for AR ligand occupancy.

Classical, Backdoor and 5α-Dione Pathways of Androgen Metabolism

In the classical pathway of androgen biosynthesis discussed above (Fig. 2, light gray arrows), pregnenolone is generated from cholesterol and is then converted to DHEA by sequential 17α-hydroxylase and 17,20-lyase reactions catalyzed by CYP17A1. DHEA is converted by HSD3B1/2 to A4, which is then acted on by HSD17B3 or peripherally expressed AKR1C3 to generate T. In target tissues the subsequent conversion of T to DHT is carried out by SRD5A2, thereby amplifying the androgen signal as DHT is the more potent androgen. However, in steroidogenic tissues in which both CYP17A1 and SRD5A are co-expressed, an alternate route to DHT, called the 'backdoor' pathway (Fig. 2, hatched arrows) is possible wherein adrenal derived C-21 steroids undergo 5α-reduction by SRD5A *prior* to being acted upon by the 17,20-lyase activity of CYP17A1 [62]. In fact, 17α-hydroxyprogesterone is a better substrate for SRD5A (especially SRD5A1) than either A4 or T [63]. Since 17OH-dihydroprogesterone (the 5α-reduced product of 17α-hydroxyprogesterone) is a poor substrate for the 17,20-lyase activity of

CYP17A1, biosynthesis proceeds via the 3α-reduction of 17OH-dihydroprogesterone by ARK1C2, which yields 17OH-allopregnanolone, an excellent substrate for CYP17A1 lyase activity that is minimally dependent on cytochrome $b5$ [64]. Androsterone generated by the 17,20-lyase activity of CYP17A1 is then acted upon by HSD17B3 or AKR1C3 to generate 3α-diol, followed by a reverse oxidative step (not required in the classical pathway) to generate DHT [36]. Candidate enzymes mediating the reversible conversion of 3α-diol to DHT are the enzymes RL-HSD (HSD17B6), HSD17B10, RODH4, RDH5, and DHRS9, involved in the pre-receptor metabolism of DHT discussed above [36, 54, 55]. This pathway, e.g. 5α-reduction of C-21 steroids prior to the action of CYP17A1 17,20-lyase, occurs in the testis of the immature mouse and the tammar wallaby. This pathway is also hypothesized to occur in ovarian hyperandrogenism and polycystic ovarian syndrome, as the human ovary expresses both CYP17A1 and SRD5A [62].

Interestingly, production of DHT in mouse testis via this mechanism is specifically mediated by type 1 and not the type 2 isoform of SRD5A [65]. This observation is of relevance to prostatic androgen metabolism in that a clear shift from SRD5A2 to SRD5A1 expression occurs in the transition from benign to neoplastic prostate tissue (discussed below). Moreover, human CYP17A1 displays markedly more robust 17,20-lyase activity for the 5α-reduced progesterone intermediate 17OH-allopregnanolone than for the classical substrates 17α-hydroxypregnenolone or 17α-hydroxyprogesterone, such that the combination of increased SRD5A1 activity in conjunction with expression of CYP17A1 in PCa tissue may favor *de novo* biosynthesis via the backdoor pathway over the classical pathway [66].

In a third, and currently most accepted route to DHT in CRPC, termed the 5α-androstanedione (5α-dione) pathway (Fig. 2, dark gray arrows) DHEA and A4 are produced as in the classical pathway. However, A4 is converted first by SRD5A to 5α-dione and then by AKR1C3 (or HSD17B3) to DHT, rather than conversion of A4

to T by AKR1C3 followed by the activity of SRD5A to produce DHT. Consistent with prior findings that SRD5A activity in PCa cells has a preference for A4 rather than T [67–69], the group of Sharifi demonstrated that (1) the 5α-reduction of A4 to 5α-dione is a required step for DHT biosynthesis in PCa cells (rather than direct 5α-reduction of T to DHT); (2) this conversion is specifically mediated by SRD5A1; and (3) that in PCa cells T and A4 are actually negligible substrates for SRD5A2 [66] (possibly related to the altered redox environment of tumor cells as SRD5A1 and 2 have different pH optima). Storbeck et al. have shown that A4 is also a poor substrate for conversion to T by AKR1C3 and that it is the combination of SRD5A1 favouring A4 over T; also, A4 is a very poor substrate for AKR1C3, which directs the flux via 5α-dione [43]. The subtotal loss of SRD5A2, upregulation of SRD5A1 and increased expression of AKR1C3 observed in the transition from benign prostate to CRPC may thus reflect selection of tumors cells capable of efficiently synthesizing DHT via this pathway [34, 45, 49, 70–72].

Importantly, while the non-classical pathways to DHT bypass conventional intermediates of A4 and T, it is worth emphasizing that *the backdoor and 5α-dione pathways still include the same enzymatic conversions that produce DHT via the conventional pathway*; all that differs is the order in which the enzymes mediate the reactions.

Altered Expression of Steroidogenic Enzymes in Progression to CRPC

Primary PCa and CRPC tumors are characterized by a number of changes in steroidogenic gene expression that are consistent with either promoting conversion of adrenal androgen precursors to DHT, inhibiting conversion of DHT to inactive metabolites, or in case of CRPC tumors, mediating *de novo* biosynthesis of androgens from cholesterol and/or progestogen precursors. Here we review the alterations observed in prostate tumors during the progression to CRPC and discuss implications of these changes for determining intra-tumor androgen levels.

Altered Expression of Steroidogenic Genes in Primary Prostate Cancer

Perhaps the most consistently observed alteration in prostate tumors is a subtotal loss of tumoral SRD5A2, the principle steroid 5α-reductase isoform expressed in benign prostate tissue [45], and a relative shift in primary and recurrent prostate tumors to expression of SRD5A1 [34, 71, 72] (although some studies have shown Gleason grade-related increases in both SRD5A1 and SRD5A2 [73]). As discussed above, in the 5α-dione pathway, the 5α-reduction of A4 to 5α-dione is a required step for DHT biosynthesis in PCa cells and is specifically mediated by SRD5A1, suggesting that the upregulation of SRD5A1 observed in the transition to CRPC reflects selection of tumor cells capable of efficiently synthesizing DHT via this pathway. Interestingly, a recent report demonstrated that progression to CRPC was correlated with a higher pre-treatment ratio of T to DHT in prostate biopsies taken before the start of ADT (T:DHT ratio 0.19 [0.98–4.92 pg/mg] vs. 0.05 [0.45–16.89 pg/mg] in patients who did not develop CRPC) [74]. It is tempting to speculate that this elevated ratio of T to DHT reflects tumor cells with pre-treatment loss of SRD5A2 activity, followed by induction of SRD5A1-mediated DHT production via 5α-dione under the selective pressure of ADT. Altered expression of a third SRD5A isozyme, SRD5A3, has also been reported, with increased expression observed in primary and castration recurrent prostate tumors [75]. The importance and/or activity of this enzyme in PCa progression awaits further evaluation [76].

Differential changes in the expression of reductive and oxidative enzyme pairs, which favors the conversion of inactive precursors to active androgens (e.g. A4 to T, or 5α-dione to DHT), has been observed in primary prostate tumors, including increased tumor expression of the reductive enzymes HSD17B3 [77] and AKR1C3 [34, 49, 70], and decreased expression of the oxidative enzyme catalyzing the reverse reaction, HSD17B2 [77, 78], suggesting a shift in tumoral androgen metabolism to the formation of T and DHT. Silencing of HSD17B2 occurs via DNA methylation, as well as generation of two alternatively spliced, catalytic-deficient isoforms that bind wild type HSD17B2 and promote its degradation [79]. Increased prostate tumor expression of HSD17B4 has also been observed, but a role in androgen metabolism was only recently revealed. This enzyme (also known as D-bifunctional protein or DBP) has a unique peroxisomal targeting sequence and acts primarily in peroxisomal β-chain oxidation of fatty acids [80]. However, HSD17B4 has five splice isoforms, of which isoform 2 is capable of metabolizing T and DHT. Despite an increase in expression of the other isoforms in PCa, expression of isoform 2 is specifically lost in CRPC, consistent with a shift in tumoral androgen metabolism to the formation of T and DHT. This finding highlights the role of tumor-based changes that prevent androgen inactivation in maintaining CRPC tumor androgen levels [81].

Primary PCa also demonstrates a selective loss of both AKR1C2 and AKR1C1 versus paired benign tissues, accompanied by a reduced capacity to metabolize DHT to 3α-diol, resulting in increased tumoral DHT levels [53]. Increased expression of HSD17B10, one of the oxidative enzymes capable of mediating the back conversion of 3α-diol to DHT, has also been observed in malignant vs. benign prostate epithelial cells, similarly consistent with an increased capacity to generate DHT in tumor tissue [82]. In contrast, epithelial expression of RL-HSD (which can mediate either conversion of 3β-diol to DHT or of DHT to 3α-diol) is lost in primary PCa, which is hypothesized to reflect loss of the 3β-diol/ERβ mediated growth inhibition pathway during malignant transformation [56].

Another mechanism that may modulate prostate tissue androgen levels is sulfation. While SULT2A1 is the primary phase II enzyme responsible for sulfonation in the adrenal gland, SULT2B1b is highly expressed in the prostate and may limit the pool of unconjugated DHEA available for conversion to A4 [83]. Notably, SULT2B1 shows selective loss of expression in tumor vs. benign prostate epithelial cells [84]. This is consistent with a report demonstrating increased DHEA-stimulated LNCaP proliferation in cells with knockdown of SULT2B1 [85].

Notably, the expression of enzymes involved in *de novo* steroidogenesis, including MLN64 (homolog of StAR), CYP11A1, CYP17A1, HSD3B1 and HSD3B2 has also been demonstrated in primary prostate tumor tissues [86–89]. While a role for *de novo* steroidogenesis (e.g. from cholesterol) *per se* in primary prostate tumors is less likely, these observations suggest that the selection pressure of androgen deprivation therapy leads to upregulated expression of these enzymes and reconstitution of tumor androgen levels in CRPC.

Altered Expression of Steroidogenic Genes in Castration Resistant Prostate Tumors

CRPC tumors demonstrate altered expression of numerous genes in the steroid biosynthetic pathway, including genes involved in cholesterol metabolism, *de novo* steroidogenesis, as well as utilization of adrenal androgen precursors, suggesting that castration resistant tumors have the ability to utilize cholesterol, progesterone and/or adrenal precursors for conversion to T and DHT [13, 34]. Changes related to cholesterol metabolism include increased expression of squalene epoxidase (SQLE), the rate-limiting enzyme in cholesterol synthesis, as well other genes in this pathway such as HMG-CoA synthase, squalene synthetase and lanosterol synthase [35]. In a study comparing CRPC with primary tumors, the relative expression of numerous transcripts involved in *de novo* androgen biosynthesis and adrenal androgen precursor utilization were altered, including increased expression of HSD3B2 (1.8), AKR1C3 (5.3), SRD5A1 (2.1), SRD5A2 (0.54), AKR1C2 (3.4), AKR1C1 (3.1) and UGT2B15 (3.5). Another study of CRPC metastases in which elevated levels of tumor T and DHT were also measured (T 0.74 ng/g, DHT 0.25 ng/g), showed elevated expression of StAR, CYP17A1, HSD3B1/2, HSD17B3, AKR1C3, SRD5A1/2, UGT2B15/17, CYP19A1 and decreased SRD5A2 [13, 86, 88, 90].

Other studies have not found increased expression of CYP17A1 specifically in CRPC tumors but have demonstrated findings suggestive of intracrine utilization of adrenal androgens, including increased expression of HSD17B3 and AKR1C3 [34, 91, 92]. Interestingly, CYP17A1 has squalene epoxidase activity in assays using recombinant CYP17A1 and in a mouse Leydig tumor cell line [93], suggesting that it has a dual role in CRPC steroid metabolism. Also of note, AKR1C3 has recently been identified as an AR coactivator and thus may play dual roles in promoting ligand biosynthesis and AR activation [94].

A naturally-occurring single nucleotide polymorphism (SNP) in HSD3B1 (1245C; N367T, population frequency 22%) has been identified as a gain of function somatic mutation in CRPC tumors [95]. Three of 25 CRPC tumors from patients that were homozygous for the major allele had acquired a somatic N367T gain of function mutation in the tumor. Moreover, 3 of 11 CRPC tumors with heterozygous germline DNA showed loss of heterozygosity of the major allele. Expression of the N367T form of HSD3B1 resulted in increased protein levels of HSD3B1, rendered the protein resistant to ubiquitination and degradation, and led to increased intratumoral DHT production. Compared to the poor conversion of DHEA to A4 in LAPC4 cells, which do not have this mutation, the N367T form of HSD3B1 was shown to account for the efficient flux of DHEA to A4 in LNCaP cells and was also detected in the VCaP cell line.

Alternative Splicing of Androgen Metabolizing Enzymes in Prostate Cancer

As discussed, both primary PCa and CRPC are characterized by differential changes in the expression of reductive and oxidative enzyme pairs favoring the conversion of inactive precursors (e.g. A4 and 5α-dione) to active androgens (e.g. T and DHT, respectively). Data also demonstrate a role for alternative splicing of androgen metabolizing enzymes in modulating tissue androgen levels, including functional silencing of HSD17B2 and HSD17B4, which are responsible for the reverse metabolism of the active androgens, T and DHT, to their inactive precursors, A4 and 5α-dione, respectively [79]. HSD17B2 protein levels can be reduced by generation of two alternatively spliced, catalytic-deficient isoforms that bind wild type HSD17B2 and promote its degradation. HSD17B4 has five alternatively spliced iso-forms, of which isoform 2 metabolizes T and DHT and is specifically lost in CRPC [81]. While the factors involved in alternative splicing of HSD17B4 were not investigated, one study found that overexpression of SRSF1 (also known as ASF1/SF2) or SRSF5 (also known as SRp40) resulted in generation of the truncated HSD17B2 isoforms.

Notably, SRSF1 and SRSF5 are involved in the oncogenic splicing of multiple genes, including BCL-X, CCND1, KLF6, VEGF, and perhaps of most interest, AR, which are implicated in driving PCa progression [96, 97]. Splice variants of the AR which lack the C terminal ligand binding domain (LBD) but retain the DNA binding and N terminal domains required for AR dimerization, DNA binding and transcriptional regulation have been described [96, 98]. Among these, the ligand independent, constitutively active AR variant 7 (AR-V7) is the most common variant in CRPC and has been shown to be an adverse prognostic and predictive marker [99]. The splicing factors U2AF55 and SRSF1 act as pioneer factors, specifically recruiting the spliceosome to the 3′ splice site of AR-V7, thus increasing the expression of AR-V7 mRNA [96]. Thus, the novel observation that SRSF1 and SRSF5 are involved in the alternative splicing of HSD17B2 adds steroidogenic enzymes to the known suite of drivers regulated by these splicing factors.

Functional Evidence of Intracrine Steroidogenesis in Prostate Cancer

The ability of prostate tissue and prostate tumors to mediate the intracrine conversion of adrenally derived androgen precursors or cholesterol to the downstream androgens T and DHT has been evaluated in normal rat and human prostate, in primary prostate tumors, in CRPC tumors, and *in vitro* and *in vivo* models of CRPC. Here we review the evidence in each of these settings that demonstrate the activity of steroidogenic pathways in the continuum from normal prostate to CRPC.

Evaluation of Steroidogenesis in Normal Prostate and Prostate Cancer Tissue

A number of early studies attempted to directly examine the steroidogenic ability of rat and human prostate tissue by evaluating the conversion of exogenous radiolabeled-adrenal androgen precursors to T or DHT. Bruchovsky administered radioactively labeled androgens including T, DHT, and the adrenal androgen precursors DHEA and A4 to castrated male rats and evaluated prostatic metabolites 60 min after injection [100]. Following administration of DHEA, approximately 1% and 8% of the recovered radioactivity was found in T and DHT respectively. With A4 it was 2% and 12%, respectively. In comparison, 37% of exogenous labeled T was converted into DHT. Labrie et al. demonstrated that administering DHEA or A4 to castrate adult rats at levels found in the serum of adult men led to increased prostatic DHT levels and increases in ventral prostate weight [101]. In the Dunning R3327 prostate carcinoma model, administration of adrenal androgen precursors to castrate male rats increased tumor DHT levels and stimulated tumor growth to the level of intact controls [102].

In studies of human prostate tissue, one study evaluated prostate androgen metabolism by infusing eugonadal men with ^{3}H-T, ^{3}H-A4 or ^{3}H-DHEA-sulfate (DHEA-S) 30 min prior to performing radical prostatectomy for BPH [103]. The major metabolite present in prostate tissue after ^{3}H-T infusion was DHT (about 65% conversion). Infusion of ^{3}H-A4 resulted in approximately 7–10% radioactivity associated with either T or DHT. ^{3}H-DHEA-S was primarily converted to DHEA (70–90%), with 1–3% conversion to T, DHT and A4. Consistent with these observations, a more recent study using mass spectrometry to identify metabolites formed from incubation of human prostate homogenate *ex vivo* with DHEA demonstrated production of A5-diol, T, DHT and androsterone [104]. Together, these studies in rat and human prostate tissues suggest that while the most efficient substrate for DHT production in non-tumor prostate tissue is T, a limited amount of DHT is also formed from exogenous DHEA or A4, consistent with intracrine steroid metabolism.

Metabolism of ^{14}C progesterone was investigated in primary PCa tissues, but no significant metabolic conversion beyond the formation of immediate progesterone derivatives was observed [105]. This finding is not necessarily unexpected, as studies have now clearly demonstrated that it is CRPC tumors in which steroidogenic genes capable of *de novo* biosynthesis are upregulated. One study evaluated the presence of adrenal androgen precursors and steroid metabolizing activity (including SRD5A, HSD3B, and HSD17B) *ex vivo* in hormone naive tumors and lymph node metastases. Although malignant tissue had a subtotal loss of SRD5A activity, primary tumors and metastases were found to have the capacity to metabolize adrenal androgen precursors to DHT [45]. Another study demonstrated the conversion of DHEA-S to DHEA within PCa tissue extracts from both eugonadal and castrate men [106]. A third study subsequently confirmed the presence of the steroid sulfatase required for conversion of DHEA-S to DHEA within prostate epithelial tissue [107], which was later confirmed by others [46, 47]. Consistent with the discovery that the primary route to DHT in PCa cells is from A4 to 5α-dione rather than from A4 to T, Sharifi's group demonstrated robust conversion of A4 to 5α-dione and low/no metabolism of A4 to T in biopsy tissue from two patients with CRPC [66]. However, in contrast to their findings in CRPC tissue where T is a poor substrate for conversion to DHT by SRD5A, this group subsequently reported that in primary PCa both A4 and T underwent 5-α reduction, leading them to suggest that the transition to CRPC coincides with a metabolic switch toward A4 as the favored substrate [108].

Experimental Models of De Novo Steroidogenesis in CRPC

Studies using *in vitro* and *in vivo* models of CRPC support the concept of intratumoral androgen biosynthesis, including both adrenal androgen precursor utilization and *de novo* androgen biosynthesis [109]. Notably, circulating levels of exogenously administered cholesterol were associated with tumor size (R = 0.3957, p = 0.0049) and intratumoral T levels (R = 0.41, p = 0.0023) in subcutaneous LNCaP tumors grown in hormonally intact mice, and were directly correlated with tumoral expression of CYP17A1 (R = 0.4073, p = 0.025). Since the hypercholesterolemia did not raise circulating androgen levels, these data suggest the administered cholesterol led to increased intratumoral androgens via *de novo* steroidogenesis. Consistent with these observations, increased serum cholesterol was associated with elevated intraprostatic levels of DHEA, T, and A4 in a PTEN-null transgenic mouse model of PCa, and inhibition of serum cholesterol levels slowed tumor growth and was associated with a decrease in intraprostatic androgens [110].

Numerous studies using CRPC xenografts in castrate mice have demonstrated measurement of substantial intratumor androgen levels [13, 31, 32, 111–115]. As it has been commonly believed that rodent adrenal glands do not biosynthesize significant amounts of adrenal androgen precursors, these findings were considered suggestive of *de novo* steroidogenesis from cholesterol or progesterone precursors. However, more recent studies have demonstrated the presence of circulating adrenal androgen precursors in mice, and that adrenalectomy reduced both serum and tumor androgen levels and slowed growth of CRPC xenograft tumors [116, 117]. Notably, a subset of tumors recurred after adrenalectomy with increased steroid levels and/or induction of AR, truncated ligand-independent AR variants, and glucocorticoid receptor (GR), suggesting *de novo* steroidogenesis remains among the resistance mechanisms employed by CRPC tumors [116].

A number of groups have addressed this question more directly by carrying out *in vitro* studies with radiolabeled cholesterol precursors to demonstrate intratumoral conversion to downstream metabolites. The androgen-independent LNCaP derivative (C81) showed higher expression of StAR, CYP11A1 and CYP17A1 compared to its androgen-dependent counterpart (C33) and was shown to directly convert radioactive cholesterol into T [118]. Increases in expression of genes responsible for accumulation of free cholesterol and cholesterol biosynthesis including LDLR, SRB1, ABCA1, STAR, ACAT, HMG-CoA and CYP11A1 were demonstrated in a xenograft LNCaP model [112, 113, 119], as well as increases in transcripts encoding CYP17A1, AKR1C1, AKR1C2, AKR1C3, HSD17B2, and SRD5A1. Conversion of ^{14}C-acetic acid to DHT was observed in these xenografts, and tumors were shown to metabolize ^{3}H-progesterone to six different intermediates upstream of DHT, suggesting occurrence of steroidogenesis via both classic and "backdoor" pathways [120]. In a study of six prostate cell lines (LnCaP, 22Rv1, DU145, RWPE1, PC3 and ALVA4), expression of CYP11A1, CYP17A1, HSD3B2, HSD17B3 was detected in all, with conversion of ^{14}C-labled cholesterol to T and DHT demonstrated in each cell line, albeit with different efficiencies [86]. It should be noted that other studies have not detected expression of CYP17A1 nor demonstrated clear evidence for *de novo* steroidogenesis in PCa cell lines [121–123]. Interestingly, one study in the LNCaP sub-line C4-2 as well as VCaP cells showed CYP17A1-independent metabolism of pregnenolone to its 5α-pregnane metabolite, allopregnanolone (mediated in three steps by HSD3B, SRD5A and AKR1C in the backdoor pathway), which has been associated with proliferation, mitogenesis and metastasis in other malignant cell types such as ovarian, testis, breast and leukemia [124].

Drivers of Intratumoral Androgen Biosynthesis

A number of exogenous and endogenous factors including cytokines, growth factors, nuclear transcription factors, and paracrine cellular interaction have been found to promote steroid production in PCa cell lines and tissues. IL-6 is implicated in cross-talk and regulation of AR activity and PCa growth but may also play a role in modulating androgen biosynthesis. Treatment of LNCaP cells with IL-6 induced the expression of steroidogenic enzymes including CYP11A1, HSD3B2, AKR1C3 and HSD17B3, and increased levels of T twofold in lysates of cells grown in serum free media [125].

In a study designed to evaluate the effects of insulin on steroidogenesis, exposure of LNCaP cells to insulin caused an increase in transcript levels of cholesterol and steroid synthesizing genes, including SREBP1, StAR, CYP11A1, CYP17A1, HSD3B2, HSD17B3, and SRD5A1, which were confirmed at the protein level for a number of genes including CYP11A1 and CYP17A1. In parallel, insulin increased intracellular levels of pregnenolone, 17α-hydroxyprogesterone, DHEA and T, and incubation of insulin-treated LNCaP and VCaP cells with ^{14}C-acetate resulted in detection of radiolabeled pregnan-3,20-dione, A4, T and androsterone [114]. In similar studies evaluating the effect of IGF2 on steroidogenesis, these authors demonstrated increased conversion of ^{14}C-acetate to pregnan-3,20-dione, pregnan-3,17-diol-20-one, androsterone, A4, and T [126].

Receptors for luteinizing hormone (LH), the target of LH releasing hormone (LHRH) agonist therapy in the brain, have also been demonstrated in PCa specimens and may play a role in steroidogenesis [127]. Exposure of both androgen-sensitive (LNCaP) and androgen-independent (22RV1 and C4-2B) PCa cell lines to LH increased the protein expression of steroidogenic enzymes including StAR, CYB5B, CYP11A1, and HSD3B, and a 2.5-fold increase in progesterone synthesis was observed in LH treated C4-2B cells compared to controls [128]. These data suggest that LH may have a role in the regulation of steroid biosynthesis in PCa cells and identify the LH receptor as a potential therapeutic target.

Nuclear receptor (NR) liver receptor homolog-1 (LRH-1, NR5A2), an orphan nuclear receptor without a previously known role in steroidogenesis, was recently shown to promote *de novo* androgen biosynthesis via its direct transactivation of several key steroidogenic enzyme genes, including CYP17A1, CYP11A1, StAR, HSD3B1, HSD3B2, and SRD5A2 [129]. LNCaP xenografts with overexpression of LRH-1 grew faster than controls in intact mice, were resistant to castration, and had significantly higher levels of intra-tumoral T and DHT, accompanied by upregulated expression of transcripts encoding multiple steroidogenic genes. Notably, in VCaP cells (which have high endogenous LRH-1) LRH-1 knockdown or treatment with an LRH-1 inverse agonist ML-180 suppressed the expression of CYP17A1 and other steroidogenic genes (StAR, HSD3B1, and AKR1C3) and sensitized cells to androgen deprivation. Analysis of clinical specimens demonstrated increased LRH-1 expression in CRPC tissues compared to hormone naïve PCa tissues or benign prostate tissue, identifying LRH-1 as a potential therapeutic target.

Steroidogenic factor 1 (SF1, NR5A1) is a transcription factor that potently regulates steroidogenesis within the adrenal glands and gonads by driving expression of genes involved in cholesterol metabolism and conversion to steroid hormones [130]. Abnormal SF1 expression has been implicated in promoting aberrant steroidogenesis in ovarian and adrenal cancers and endometriosis. While the expression of SF1 in CRPC has not been delineated, gain- and loss-of-function experiments showed that the presence of SF1 increased steroid biosynthesis in PCa cell lines and stimulated expression of steroidogenic enzymes, most notably, CYP17A1, HSD3B1, HSD17B3, and CYP19A1, each a known target of SF1 regulation [131]. Other factors reported to promote intratumoral androgen levels in PCa cell lines include inactivation of ID4 (via an unknown mechanism), and loss of the beta2-adrenergic receptor (ADRB2) which appeared to associate with a decrease in expression of the glucuronidating enzymes UGT2B15 and UGT2B17 [132, 133], although the clinical significance of these observation remains to be determined [134, 135].

Long-chain acyl-coenzyme A (CoA) synthetase 3 (ACSL3), an androgen-responsive gene involved in the generation of fatty acyl-CoA esters, is expressed in hormone naive and CRPC tumor cells and was recently shown to promote intratumoral steroidogenesis in PCa cells [136]. ACSL3 overexpression in LNCaP significantly upregulated steroidogenesis related genes, including SLCO1B3, which encodes an uptake transporter of DHEA-S, and AKR1C3 and HSD3B1, involved in T biosynthesis, while reducing expression of SRD5A1, UGT2B15 and UGT2B17, involved in metabolism of T and DHT. Treatment with DHEA-S, a substrate for conversion to downstream steroids by AKR1C3, significantly increased T levels and proliferation in ACSL3-overexpressing cells. Genes involved in de novo steroidogenesis such as StAR and CYP11A1 were either downregulated or remained unchanged, suggesting ACSL3 drives intratumoral steroidogenesis via utilization of adrenal androgens, and not through cholesterol anabolism.

In total, these findings demonstrate multiple mechanisms by which intratumoral androgen biosynthesis may be modulated in CRPC tumors, confirming the importance of this pathway and suggesting potential candidates for therapeutic targeting.

Impact of Stromal Cells and the Bone Microenvironment on Intratumoral Androgen Concentrations

The stromal and bone microenvironment may be particularly important in promoting intratumor androgen levels in PCa cells. In particular, steroidogenesis in PCa and bone marrow-derived stromal cells may play a paracrine role in determining intratumor androgen levels and AR activation in PCa cells, and factors in the bone microenvironment may directly stimulate intratumoral androgen biosynthesis within PCa cells. Whereas DHEA induced little or no PSA expression in monocultures of LAPC-4 PCa cells, coculture with PCa-associated stromal cells resulted in marked stimulation of PSA expression, likely mediated by stromal cell generation of T from DHEA (as T was detected in a time and dose-dependent manner in PCa-stromal cell monocultures treated with DHEA) [137]. Similarly, the impact of DHEA on PSA promoter activity in LNCaP cells was markedly enhanced in the presence of PCa-derived stromal cells [138]. Knockdown of AR in the LNCaP cells abrogated this effect, while coculture with PCa-stromal cells transfected with AR shRNA did not, suggesting paracrine factors secreted by the stromal cells act on the LNCaP AR. Furthermore, following DHEA treatment, T and DHT concentrations were ~5-fold higher in the PCa-stromal/LNCaP coculture vs. the LNCaP monoculture. Interestingly, normal prostate stroma, bone-marrow stroma, lung stroma and bone-derived stromal cells also induced an increase in PSA expression, although the strongest effects were noted with PCa-associated stromal cells. In a separate study of bone-marrow stromal cells, resting mesenchymal cells were found to express HSD3B and SRD5A protein, while incubation with DHEA resulted in the additional expression of AKR1C3 [139].

Consistent with these findings, and in accord with literature showing that the Hedgehog (Hh) signaling pathway modulates steroidogenesis in multiple endocrine tissues including testis, ovary, adrenal cortex and placenta [140], the group of Buttyan demonstrated that the steroidogenic activity of primary benign human prostate stromal cells (PrSCs) is significantly increased by activation of the Hh signaling pathway [141, 142]. Exposure to a Hh agonist or transduction of PrSCs with lentiviruses expressing active Gli2, a transcription factor that is triggered by Hh signaling, resulted in the upregulation of multiple steroidogenic genes and increased T output from DHEA supplemented PrSCs. Moreover, primary bone-marrow stromal cells became more steroidogenic and produced T under the influence of a Hh agonist, and treatment of mice bearing LNCaP xenografts with a Hh antagonist, TAK-441, delayed the onset of CRPC after castration and substantially reduced androgen levels in residual tumors [142]. These findings suggest that paracrine signaling maintains androgen levels in the primary or metastatic tumor microenvironment by promoting steroidogenesis within tumor-associated stromal cells, and that Hh antagonists may be useful for targeting prostate tumor stromal cell-derived steroid production.

In addition to a paracrine role played by steroidogenesis in stromal cells, factors in the bone tumor microenvironment may also directly enhance intratumoral steroidogenesis. The increased expression of genes encoding steroidogenic enzymes found in bone metastatic tissue from patients suggests that up-regulated steroidogenesis contributes to tumor growth at the metastatic site. One study described significantly higher levels of SRD5A1, AKR1C2, AKR1C3, and HSD17B10 mRNA in bone metastases than benign prostate or primary PCa [143] Consistent with this another study suggested that osteoblasts promote CRPC by altering intratumoral steroidogenesis [144]. LNCaP-19 cells treated with osteoblast conditioned media displayed an increased expression of genes encoding steroidogenic enzymes (CYP11A1, HSD3B1, and AKR1C3), estrogen pathway-related genes (CYP19A1, and ESR2), and genes for DHT-inactivating enzymes (UGT2B7, UGT2B15, and UGT2B17). The osteoblast-induced effect was exclusive to osteogenic CRPC cells (LNCaP-19) in contrast to osteolytic PC3 and androgen-dependent LNCaP cells, and the steroidogenic effect was reflected in increased levels of progesterone and T in serum from castrated mice harboring intratibial xenografts. Consistent with increased expression of the UGT enzymes mediating irreversible glucuronidation of DHT, levels of DHT were decreased in serum from castrated mice with intratibial tumors, which is in accord with the decreased intratumoral DHT/T ratio shown in metastatic CRPC tissue compared to primary PCa or benign prostate tissue [12, 13].

Several findings converge to suggest that osteoblast-secreted osteocalcin (OCN) is a driver of intratumoral steroidogenesis in PCa cells. OCN (also known as bone g-carboxyglutamic acid protein, encoded by BGLAP) is a bone-secreted hormone that binds to G-protein coupled receptor family C group 6 member A (GPRC6A) to regulate the endocrine function of multiple target tissues, including pancreas, adipocytes, intestinal cells, skeletal muscle, and Leydig cells of the testis [145–147]. Circulating OCN exists in two forms, carboxylated and uncarboxylated, of which only the uncarboxylated form binds GPRC6A to function as a hormone. OCN is post-translationally carboxylated on three glutamate residues, increasing its affinity for hydroxyapatite crystals and keeping most secreted OCN embedded in the bone matrix. However, the acidic environment generated during bone resorption promotes decarboxylation of OCN, which reduces its affinity for bone, promoting release of uncarboxylated OCN into the circulation where it then exerts activity in peripheral tissues via binding to GPRC6A.

Studies in knockout models have established that binding of GPRC6A by OCN regulates T biosynthesis in Leydig cells via expression of steroidogenic enzymes [145], and mutations in GPRC6A which prevent membrane localization have been identified as a susceptibility locus for primary testicular failure in humans [148, 149]. GPRC6A, which has also been identified as a T binding membrane receptor that mediates nongenomic androgen signaling [150], is expressed in prostate tissue and cell lines [151] and has been associated with PCa risk [152]. Studies have demonstrated the ability of OCN to mediate GPRC6A-dependent intracellular T biosynthesis via induction of CYP11A1 and CYP17A1 in PCa cell lines *in vitro* [153, 154]. Other studies have demonstrated a tumor promoting role of OCN-mediated GPRC6A activation in PCa cell line models and xenografts [151, 153]. Whether this was related to a change in intracellular androgens was not reported, although one study found reduced androgen-mediated induction of transcripts encoding steroidogenic enzymes, including HSD3B1 and AKR1C3 in a GPRC6A knockdown model [153]. Notably, OCN is a well-established target gene transcriptionally regulated by RUNX2, and it was recently demonstrated that knockdown of Runx2 in the Pten null mouse model of PCa decreased intratumoral Cyp11a1 and Cyp17a1 expression, T levels, and tumor growth in castrated mice [154]. These observations suggest that intratumoral androgen biosynthesis in PCa cells could be promoted by bone-derived OCN (either locally within the bone microenvironment or at other sites of metastasis via circulating OCN) or by OCN that is ectopically expressed within the prostate tumor itself, due to osteoblast or bone-marrow stromal cell-induction of osteomimicry [155, 156].

In total, these findings suggest that the maintenance of intratumoral androgen levels in the CRPC tumor microenvironment is facilitated by paracrine-stimulated steroidogenesis within tumor-associated stromal cells in the primary or metastatic tumor microenvironment, as well as by paracrine and intracrine stimulation of steroidogenesis within the tumor itself, supporting the role of steroidogenesis in reactivating AR signaling in CRPC, and highlighting the interplay between paracrine stromal and epithelial cell interactions in this mechanism.

Noncanonical Androgens as Unrecognized Drivers of Prostate Cancer Progression

While the contribution of canonical androgen precursors of adrenal origin (DHEA-S, DHEA and A4) towards the androgen pool in CRPC is established, the involvement of the 11-oxygenated androgens in activating wild-type AR and serving as an under-recognized reservoir of active androgens has only recently been elucidated. 11OHA4 is an abundant product of the human adrenal derived from the CYP11B1 mediated hydroxylation of A4 (Fig. 2) [41, 42]. Despite circulating concentrations of 11OHA4 similar to or exceeding that of A4, this steroid has historically been of little interest due to its inability to activate the AR [157–159]. In fact, until the critical observation that the 5α-reduced metabolites of 11OHA4 are potent activators of wild type AR, its production from A4 was primarily viewed as a mechanism to inactivate A4 and regulate adrenal androgen output [160].

Circulating Levels of Non-canonical 11-Oxygenated Androgens

Studies in adrenal cell-line models and primary adrenal cultures confirmed earlier studies that 11OHA4 is an abundant product of adrenal steroidogenesis [42, 161, 162]. Subsequently, these findings were confirmed using adrenal vein sampling where it was found that the concentration of 11OHA4 in the adrenal vein (159 nM) exceeded that of A4 (79 nM) and DHEA (125 nM) under basal conditions and increased fivefold upon stimulation with ACTH. However, more importantly it was shown that the adrenal also produces low levels of 11KA4 (0.99 nM), 11OHT (0.48 nM) and 11KT (0.39 nM) [41].

Recent studies have further confirmed that 11OHA4, 11KA4, 11OHT and 11KT are present in circulation in both men and women [157, 159, 163]. Although reference levels are yet to be determined, studies consistently show that 11OHA4 (nM levels) is the most abundant of these in circulation, followed by 11KA4 (nM levels) and 11KT (nM levels), with only low levels of 11OHT detected (sub nM levels) [157].

While 11OHA4 is produced in the adrenal by the CYP11B1 catalyzed 11β-hydroxylation of A4, the production of 11KA4 and 11KT is likely peripheral. Conversion of 11OHA4 to 11KA4 is efficiently catalyzed by HSD11B2, the enzyme responsible for the inactivation of cortisol to cortisone within mineralocorticoid target tissue such as the kidney [42, 164]. Both AKR1C3 and HSD17B3 can catalyze the subsequent conversion of 11KA4 to 11KT (Fig. 3). It should, however, be noted that 11OHA4 cannot be converted to 11OHT by HSD17B3 or AKR1C3 and, as a result the low levels of 11OHT observed in circulation, are likely the result of the 11β-hydroxylation of adrenal derived T [43]. The conversion of 11OHA4 to 11KA4 by HSD11B2 is therefore an absolute requirement for the production of 11KT. 11KT and 11OHT levels do not appear to be higher in men than in women despite significantly higher levels of circulating T [163]. Furthermore, a recent study showed that the addition of exogenous T did not lead to an increase in the circulating levels of 11-oxygenated androgens thereby

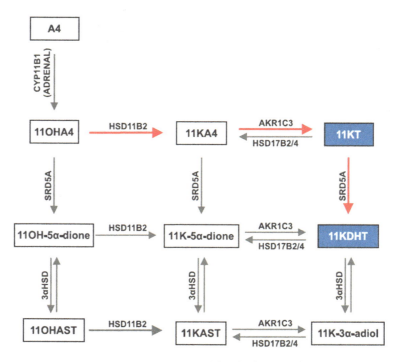

Fig. 3 Conversion of 11β-hydroxyandrostendione (11OHA4) to the potent 11-oxygenated androgens 11-ketotestosterone (11KT) and 11-keto-5α-dihydrotestosterone (11KDHT) in the 11-oxygenated androgen pathway. 11OHA4 is produced in the adrenal cortex by the CYP11B1 catalyzed 11β-hydroxylation of A4. 11OHA4 is in turn converted to 11-ketoandrostenedione (11KA4) by HSD11B2 expressed in peripheral tissue such as the kidney. 11KA4 in turn serves as a substrate for AKR1C3, yielding 11KT. 11KT can be acted on by SRD5A, yielding the 5α-reduced form 11KDHT

Activation of Wild-Type AR by 11-Oxygenated Derivatives of Adrenal Steroids

Importantly, activity assays revealed that both 11OHT and 11KT demonstrated activity toward the wild type AR at nM concentrations, thereby suggesting that these 11-oxygenated androgens may be physiologically relevant. These 11-oxygenated androgens also serve as substrates for SRD5A1 and SRD5A2, yielding novel 5α-reduced 11-oxygenated androgens [164]. Of these, 11keto-5α-dihydrotestosterone (11KDHT) demonstrated the highest androgenic activity. Studies from different laboratories have confirmed the androgenic activity of 11-oxygenated androgens, with the T derivatives consistently demonstrating the highest activity. Of these 11OHT and 11β-hydroxy-5α-testosterone (11OHDHT) are partial AR agonists, while 11KT and 11KDHT are full AR agonists [41, 166, 167]. Indeed, both 11KT and 11KDHT bind to the wild type AR with affinities comparable to that of T and DHT, and 11KT and 11KDHT are equipotent to T and DHT, respectively [167]. This finding therefore challenges the paradigm that T and DHT are the only potent androgens in human physiology.

Contribution of 11-Oxygenated Androgens to the Androgen Pool in CRPC

Given that the progression of CRPC is dependent on intracrine activation of adrenal androgen precursors, the identification of 11OHA4 as a novel androgen precursor raised the question as to whether 11-oxygenated androgens could contribute to the intratumoral androgen pool. The metabolic pathways downstream of 11OHA4 have therefore been the subject of recent investigation. As described above, the first step in the conversion of 11OHA4 to 11KT requires HSD11B2 to produce 11KA4 [164]. Expression of HSD11B2 has been observed in PCa cells [168, 169], in prostate tissue [170, 171]. Accordingly, conversion of 11OHA4 to 11KA4 in LNCaP cells confirmed the intracellular activity of HSD11B2 [42]. The same study showed that 11KT was also produced from 11OHA4 confirming the activity of a reductive HSD17B (presumably AKR1C3). One study demonstrated the 5α-reduction of 11OHA4, yielding 11β-hydroxy-5α-dione (11OH-5α-dione). These results were confirmed and extended by the demonstration that both 11KT and 11KDHT are produced from 11OHA4 in PCa cells. While the production of 11KT from 11OHA4 required HSD11B2 and AKR1C3 activity, 11KDHT production required the activity of SRD5A1, HSD11B2 and AKR1C3, although the sequence of enzymatic reactions was not determined [164].

Altered Sensitivity of 11-Oxygenated Steroids to Activation by AKR1C3 and Inactivation by UGT2B Enzymes

Irrespective of the reaction sequence, AKR1C3 remains a key enzyme in the activation of both 11-oxygenated and canonical androgens. When the activity of AKR1C3 towards all available substrates was characterized [43], it revealed that 11-oxygenated androgen precursors are preferred. Significantly, AKR1C3 catalyzes the conversion of 11KA4 and 11keto-5α-androstenedione (11K5α-dione) 8- and 24-fold more efficiently than their canonical equivalents, A4 and 5α-dione, respectively (Fig. 3). Moreover, after also characterizing the activity of the oxidative enzyme, =HSD17B2, which catalyzes the reverse reactions, a computational model was constructed which included that activities of both AKR1C3 and HSD17B2. The model was validated by its ability to predict metabolism in PCa cell lines which revealed that it is the ratio of AKR1C3:HSD17B2 that controls the activation/inactivation of 11-oxygenated and canonical androgens.

Strikingly, the activation of canonical androgens (A4 to T or 5α-dione to DHT) required significantly higher ratios of AKR1C3:HSD17B2 than did activation of the non-canonical androgens. Both A4 and 5α-dione were shown to be poor substrates for AKR1C3, while even low levels of HSD17B2 efficiently converted T and DHT back to A4 and 5α-dione thereby preventing the accumulation of T and DHT. In contrast, AKR1C3 efficiently catalyzed activation of the 11-oxygenated androgens, with even low levels of AKR1C3 overcoming the HSD17B2 catalyzed reverse reaction. Increased AKR1C3:HSD17B2 ratios (as observed in CRPC) therefore had a substantially larger effect on the activation of 11-oxygenated androgens over that of canonical androgens in their model system, thereby suggesting that the activation of 11-oxygenated androgens is preferred over that of canonical androgens. As discussed above, CRPC is characterized by functional suppression of oxidative HSD17Bs (HSD17B2 and HSD17B4) catalyzing the inactivation of androgens. While the absence of oxidative HSD17B enzymes would allow for the production of potent canonical AR ligands, the significant substrate preference of AKR1C3 for 11-oxygenated androgens still suggests that intracrine biosynthesis of these androgens would be favored.

Coupled to this observation, PCa cell lines inactivate both 11KT and 11KDHT at a significantly lower rate than T and DHT, respectively [167]. Further, 11KDHT and 11KT are less efficiently glucuronidated than DHT and T, respectively [172, 173]. Taken together, these results suggest that 11-oxygenated androgens accumulate in CRPC tissues due to a higher rate of production coupled to a lower rate of inactivation. While comprehensive analyses of intratumoral tissue levels are currently underway, higher levels of 11-oxygenated androgens vs. canonical androgens were observed in two PCa tissue samples (one treatment naïve and the second following treatment with an AR antagonist) [172]. The implications of these findings are that 11-oxygenated androgens may play a previously unrecognized role in CRPC by facilitating AR activation in the low androgen environment of CRPC.

Despite the accumulating evidence presented above, much still remains to be determined regarding the contribution of the 11-oxygenated androgens towards the development of CRPC. Nonetheless, available data are consistent with the proposed intracrine generation and activity of 11-oxygenated metabolites. Extrapolation from available data suggests that serum levels of 11OHA4 are likely to be unchanged after castration. Moreover, a recent study has shown that adrenal 11-oxygenated androgen production does not decrease with age in women despite the decrease in adrenal output of canonical androgen precursors due to the involution of the zona reticularis [159], with similar findings in men reported in abstract form. Intracrine production of potent 11-oxygenated androgens would therefore be dependent on the steroidogenic machinery expressed in the tumor, with all available data clearly demonstrating the expression of all key enzymes (HSD11B2, AKR1C3 and SRD5A1). Furthermore, some expression of the CYP11B1 and CYP11B2 enzymes (which mediate conversion of A4 and T to the 11-oxygenated derivatives) in CRPC tissue and cell lines has been reported [82, 163], suggesting uptake of adrenally-derived 11OHA4 is supplemented by intracrine generation of 11OHA4 and 11OHT (from A4 and T, respectively). However, the relatively high serum level of 11OHA4 suggests tissue levels will be primarily contributed by circulating adrenal sources. Taken together, these data strongly suggest that a larger pool of residual androgens are available to activate the AR in CRPC than previously anticipated, and that measurements of residual T and DHT may underestimate the contribution of ligand-mediated activation of wild type AR.

Inhibition of Steroidogenesis in CRPC

Collectively, these studies demonstrate the capacity of primary and castration resistant prostate tumors to carry out the intracrine conversion of adrenal androgen precursors to DHT and active

11-oxygenated androgens, while the *in vitro* and *in vivo* experimental models clearly show that PCa cells are capable of *de novo* steroidogenesis starting from cholesterol and/or progesterone precursors. These findings cannot address the efficiency with which these pathways are active in human CRPC tumors *in situ*, but strongly support the premise that the residual androgens measured in CRPC tumors reflect the increased expression and activity of enzymes mediating *de novo* steroidogenesis and adrenal androgen utilization. These data provide mechanistic support for the role of intracrine androgen production in maintaining the tumor androgen microenvironment in CRPC and underscore these metabolic pathways as critical therapeutic targets.

Mechanisms of Response and Resistance to Inhibition of CYP17A1

Given its central role in the production of either adrenal or tumor-derived androgens, CYP17A1 has emerged as a primary target of novel therapeutics. Abiraterone, a pregnenolone derivative that acts as a selective irreversible inhibitor of both the 17α-hydroxylase and 17,20-lyase activity CYP17A1, is the first of these agents to enter clinical practice. While clinical responses have been impressive, not all patients respond, the duration of response is variable, and a majority of men eventually progress with a rising PSA. Although the mechanisms determining response and mediating resistance to CYP17A1 inhibition have not been fully elucidated, emerging clinical and pre-clinical data suggest several possibilities.

Perhaps most importantly, pre-clinical studies provided the first *in vivo* confirmation that the clinical effect of abiraterone is associated with suppression of tumor androgen levels. Clinical studies have clearly demonstrated abiraterone-mediated suppression of serum androgens, including suppression of DHEA by approximately 75% and of DHEA-S, A4, and T to essentially undetectable [33, 174, 175], while neoadjuvant studies have demonstrated suppression of prostate tissue A4, T and DHT to unde-

tectable in response to abiraterone [17, 176]. However, the efficacy of abiraterone in suppressing tumor androgens in men with CRPC remains to be demonstrated.

In this regard, treatment of castration resistant LuCaP35 and LuCaP23 xenografts significantly inhibited tumor growth, serum PSA, and intratumoral androgen levels, supporting the hypothesis that abiraterone's primary mechanism of action is through effects on tissue androgens [31]. Seven days after starting treatment, levels of T and DHT decreased from 0.49 to 0.03 pg/mg and 2.65–0.23 pg/mg, respectively in LuCaP23, and from 0.69 to 0.02 pg/mg and 3.5–0.24 pg/mg in LuCaP35. Notably, while androgen levels remained suppressed in LuCaP23 tumors recurring after therapy, increasing levels of T and DHT were observed in LuCaP35 tumors recurring on abiraterone. A similar impact of abiraterone on T and DHT levels was observed in separate studies of castration resistant VCaP or LAPC4 tumors [32, 177]. Among four LuCaP models treated with abiraterone, the ultra-responder LuCaP136 was characterized by significant decreases in intratumoral levels of T, DHT and A4, while three lines with intermediate or no response showed markedly less significant changes in tumor androgens [178].

Further evaluation demonstrated that these CRPC models responded to CYP17A1 inhibition with multiple mechanisms directed at maintaining AR signaling. This included upregulated expression of full-length AR and ligand independent AR variants, as well as induction of steroidogenic genes (including the target gene, CYP17A1), several of which showed strong correlations with DHT levels in recurrent tumors [31, 32, 178, 179]. Other potential mechanisms of abiraterone resistance observed in pre-clinical models included induction of GR expression [178], and increased ErbB2 signaling [177]. Mechanistically, one study demonstrated that ErbB2 signaling and subsequent activation of the PI3K/AKT signaling stabilizes AR protein in abiraterone resistant LAPC4 cells, and that concomitant treatment of LAPC4 xenografts with abiraterone and an ErbB2 inhibitor, lapatinib, blocked AR reactivation and suppressed tumor progression.

Clinical studies have also supported the importance of androgen levels and persistent AR activity in mediating response and resistance to CYP17A1 inhibition. In particular, higher pre-treatment levels of circulating adrenal androgen precursors have been associated with better response to CYP17A1 inhibition in men with CRPC [180–182]. As well, higher levels of AR and CYP17A1 staining in pre-treatment tumor-infiltrated bone marrow biopsies from men with CRPC were associated with longer responses to abiraterone treatment, supporting CYP17A1 mediated androgen production as the target of abiraterone activity [33]. Development of resistance to abiraterone has not been associated with a rise in serum androgen levels or in bone marrow aspirate T levels (although 5α-dione may be more appropriate to assess if the route to DHT bypasses T). However, biopsies from patients treated with the CYP17A1 inhibitor ketoconazole demonstrated increased expression of transcripts encoding CYP17A1 compared to biopsies from CRPC patients not treated with ketoconazole [32], suggesting local induction of steroidogenesis, and numerous studies (reviewed above) show that circulating androgen levels do not necessarily reflect tumor cell androgen concentrations. Moreover, while markedly decreased in castrate men, DHEA-S levels in abiraterone-treated men remain substantial (from ~5 μM to 424 nM in serum, and 1 μM to 42.4 nM in prostate tissue), and are likely to serve as a continuing depot for uptake and intra-tumoral conversion to downstream androgens [17, 183].

AR signaling is also critical in mediating response and resistance to CYP17A1 inhibition in clinical studies. Notably, presence of the ligand-independent constitutively active AR variant ARV7 in circulating tumor cells (CTCs) has been associated with resistance to AR-directed therapies such as abiraterone and enzalutamide [184, 185], and instances of CTCs converting from ARV7 negative to ARV7 positive status during first-line treatment with abiraterone have been reported [186]. Abiraterone, which increases serum levels of progesterone by blocking CYP17A1, was also shown to select for on-going AR activity via generation of progesterone-responsive mutant ARs. Targeted sequencing of tumor biopsies from 18 CRPC patients progressing on abiraterone demonstrated the presence of the progesterone-activated T878A-mutant AR at high allele frequency in three cases [187]. As reviewed above, consistent with the finding that ErbB2 signaling and subsequent activation of the PI3K/AKT signaling stabilizes AR protein, prostatectomy samples from a neoadjuvant trial of abiraterone plus leuprolide showed staining for the ErbB2 target site on ErbB3 (Tyr1289) that mediates PI3K activation in 9 of 48 cases [177].

These findings are consistent with clinical observations that patients progressing on abiraterone have a rise in PSA, suggesting reactivation of AR signaling. Thus, in the setting of tumor progression on abiraterone, the rationale for focusing further therapeutic efforts on more potent AR antagonists (against AR mutations and truncated AR variants) and agents suppressing AR ligands remains strong.

Metabolism of Abiraterone by Steroidogenic Enzymes and Implications for Treatment

While the inhibition of steroidogenic machinery is a logical target for drug development, many compounds targeting these enzymes share a steroidal structure and therefore have the potential to be metabolized by the very machinery that they are designed to inhibit. The best documented example is that of abiraterone, which has been demonstrated to undergo conversion to distinct metabolites with capacity to inhibit steroidogenic enzymes besides CYP17A1, as well as demonstrating AR antagonist and AR agonist activity. The delta-5, 3β-hydroxyl-structure of abiraterone makes it susceptible to enzymatic conversion by the HSD3B isoenzymes yielding delta-4-abiraterone (D4A) [188]. D4A was detected in the serum of mice following the administration of abiraterone acetate, as well as in the serum of CRPC patients treated with abiraterone. While D4A maintained the ability to inhibit the activity of CYP17A1, it also inhibited both HSD3B isoenzymes, HSD3B1 and HSD3B2, which are essential in producing the delta-4, 3-keto moiety shared by all active androgens. Furthermore, D4A inhibited the activity of SRD5A enzymes, which are required for the production of the potent androgens DHT and 11KDHT. D4A therefore has the potential to inhibit androgen biosynthesis at multiple sites downstream of CYP17A1, which ensures a more comprehensive enzymatic blockade. Moreover, D4A was shown to bind to and antagonize both the wild type and T877A mutant AR with IC50 values comparable to that of enzalutamide (D4A, IC50 for mutant AR = 5.3 nM; D4A, IC50 for wild type AR = 7.9 nM; enzalutamide IC50 for mutant AR = 24 nM; enzalutamide IC50 for mutant AR = 23 nM). This suggests that D4A is a more suitable treatment option likely to result in a greater clinical benefit than its parent compound abiraterone.

However, a more complex picture has emerged from studies showing that D4A is metabolized to three 5α-reduced and three 5β-reduced metabolites. All of these metabolites were detectable in the serum of CRPC patients treated with abiraterone and one was demonstrated to have AR agonist activity [189]. The delta-4, 3-keto moiety of D4A is a target for irreversible 5α- or 5β-reduction by SRD5A and AKR1D1, respectively, yielding 5α-abiraterone (5α-Abi) and 5β-abiraterone (5β-Abi). Subsequent 3-keto-reduction of 5α-Abi or 5β-Abi reversibly yield their respective 3α-hydroxy and 3β-hydroxy congeners. The hepatic 5β-reduction of D4A leads to the cessation of inhibitory activity, thereby reducing the concentration of active D4A, leading to a conclusion that inhibition of 5β-reduction maintains higher levels of D4A. Similarly, 5α-reduction of D4A yielding 5α-Abi and 3α-OH-5α-Abi, the more abundant of the 3-hydroxy congeners, was accompanied by loss of inhibition of CYP17A1, HSD3B and SRD5A.

Notably, 5α-Abi was shown to bind to both the wild type and T877A mutant AR with affinities similar to that of D4A. However, instead or acting as an antagonist like D4A, 5α-Abi activated the AR leading to the expression of androgen responsive genes in PCa cell models. 5α-Abi also significantly shortened progression-free survival in CRPC xenograft models. Moreover, it was demonstrated that propagating PCa cells in the presence of abiraterone or D4A for 6 months resulted in increased SRD5A1 expression, suggesting that the conversion of abiraterone to 5α-Abi via D4A serves as a drug resistance mechanism. Indeed, SRD5A is one of the most upregulated steroidogenic enzymes observed during abiraterone resistance [31]. It was therefore proposed that treatment with abiraterone together with dutasteride, a dual SRD5A isoenzyme inhibitor, would reduce the production of the 5α-Abi. Analysis of serum from 16 CRPC patients first treated with abiraterone and later with abiraterone plus dutasteride revealed an 89% decrease in the mean concentration of 5α-Abi following the addition of dutasteride (25.8 nM vs. 2.9 nM). Unsurprisingly the reduction in 5α-Abi was accompanied by a nearly double mean concentration of D4A (9.9 nM vs. 18.2 nM) [189]. This suggests that the metabolism of abiraterone can be fine-tuned to yield optimal anti-androgen outcomes.

Taken together it is clear that metabolism of steroidal drugs needs to be taken into account when developing new treatments. Steroidogenic metabolism of galeterone, another steroidal CYP17A1 inhibitor, revealed a similar diversity of biochemical activities [190]. Understanding of the steroid machinery and metabolism can aid in the manipulation of these pathways to ensure optimal therapeutic outcomes.

Targeting Steroid Sulphatase

To serve as a depot for intra-tumoral conversion to downstream androgens, DHEA-S taken up from the serum by prostate tumors must first undergo desulfation by STS [45–47], in a manner analogous to the desulfation of estrogen that occurs in normal and malignant endocrine tissues such as breast and endometrium [191]. Accordingly, development of STS inhibitors has been pursued as a therapeutic approach for hormone-dependent diseases including prostate cancer [47, 192]. To date, the non-steroidal sulphatase inhibitor STX64 (Irosustat) has been evaluated clinically in prostate cancer, breast cancer and endometrial cancer. A phase I dose escalation study of Irosustat in 17 chemo-naïve CRPC patients demonstrated pharmacodynamic proof of concept, with notable suppression of non-sulphated androgens (DHEA, Adiol and testosterone), and the DHEA:DHEA-S ratio in serum and was well tolerated [193]. While a phase II study of this agent in men with CRPC has not been reported, further clinical evaluation of this and other novel STS inhibitors currently in development is warranted [192].

Targeting HSD11B2 and CYP11B1, Enzymes Unique to the Generation of 11-Oxygenated Androgens

While clinical responses to enzalutamide and abiraterone have been impressive, not all patients respond, the duration of response is variable, and a majority of men eventually progress with a rising PSA suggestive of AR axis reactivation [194, 195]. The mechanisms of resistance have not been fully elucidated, and involvement of 11-oxygenated androgens in activating the AR has yet to be determined. Recent data evaluating serum and tissue steroid levels in abiraterone-treated patient suggests generation of androgenic metabolites is not completely abrogated. In particular, although residual levels of DHEA-S in abiraterone-treated men are markedly decreased compared to eugonadal levels (from ~5 μM to 424 nM in serum, and 1 μM to 42.4 nM in tissue), a substantial reservoir of this precursor clearly remains available for the peripheral conversion to potent androgens [17]. While one would assume that abiraterone would significantly reduce the levels of circulating 11OHA4, it remains to be determined if abrogation is complete as studies measuring circulating steroid concentrations of CRPC patients treated with abiraterone fail to include the 11-oxygenated androgens in the panels of steroids that are measured.

While progression on abiraterone is not associated with an increase in serum levels of canonical adrenal androgens, studies in xenograft models demonstrate that increased tissue androgens and steroidogenesis partially underlie resistance to abiraterone [16–18]. Moreover, the ability of new AR antagonists such as enzalutamide to inhibit AR activation by these ligands cannot be assumed, as A5-diol has been shown to activate wild type AR with nM potency without being inhibited by flutamide or bicalutamide [15].

Collectively, these observations suggest that simultaneously targeting multiple nodes of androgen production will be required to truly abrogate generation of all downstream metabolites. Moreover, the significant potency of 11KT and 11KDHT in activating the AR suggests that inhibiting the enzymes unique to the production of these metabolites might significantly decrease the overall androgenicity of the CRPC tumor microenvironment. Enzymatic targets specific to these metabolites include the adrenal activity of CYP11B1 in converting A4 to 11OHA4, and the peripheral action of HSD11B2 in yielding the potent 11keto-derivatives.

While endogenous selective inhibitors of 11BHSD isoforms have been described [196, 197], inhibition of HSD11B2, which metabolizes cortisol to its inactive form cortisone, may not be a viable therapeutic target. Aldosterone and cortisol have similar affinity for the mineralocorticoid receptor, and the presence of this enzyme in mineralocorticoid target tissues prevents excess stimulation of the receptor by cortisol [196, 197]. Notably, mutations in HSD11B2 which affect enzyme activity result in a rare autosomal dominant disorder, the syndrome of apparent mineralocorticoid excess (AME), and inhibition of this enzyme by chronic licorice ingestion results in pseudohyperaldosteronism [198]. However, in CRPC tissues, enzalutamide treatment leads to increased activity of the ligase responsible for ubiquitin-mediated degradation of HSD11B2, resulting in increased cortisol levels and facilitating activation of GR as a mechanism of resistance [199]. The extent to which this effect of enzalutamide might be important in suppressing generation of 11-hydroxygenated androgens is unknown, but is would depend on whether it also decreases 11BHSD2 activity in peripheral tissues such as the kidney where conversion of 11OHA4 to 11KA4 is believed to occur [42, 164]. As enzalutamide is well tolerated clinically, this suggests it does not have the same effect on HSD11B2 activity in the adrenal gland.

Inhibition of CYP11B1 is likely to be a clinically feasible target. In fact, metyrapone is a CYP11B1 inhibitor already in use for treatment of hypercortisolism associated with Cushing's syndrome. Of more relevance to the treatment of PCa, however, is the recent development of dual CYP17A1 and CYP11B inhibitors [200, 201]. A dual CYP17A1/CYP11B2 inhibitor (Novartis, CFG920), intended to ameliorate the mineralocorticoid side effects of CYP17A1 inhibition, is currently under clinical evaluation (NCT01647789). Although this agent would not be anticipated to have efficacy in suppressing generation of 11OHA4, specific inhibitors of CYP17A1 and CYP11B1 have been reported [200, 201], and the androgenicity associated with the downstream metabolites of 11OHA4 would support clinical evaluation of these dual inhibitors as well.

Targeting AKR1C3

AKR1C3 activity is an absolute requirement for the production of potent androgens from adrenal precursors irrespective of the intracrine pathway followed or precursor pool and is therefore an attractive drug target to inhibit that local biosynthesis of potent androgens [202, 203]. AKR1C3 is significantly elevated in enzalutamide resistant PCa cells and enzalutamide-resistant prostate xenograft tumors [204]. While overexpression of AKR1C3 conferred resistance to enzalutamide, inhibition of AKR1C3 by indomethacin or knockdown of AKR1C3 expression both resensitized enzalutamide-resistant PCa cells to enzalutamide treatment *in vitro* and *in vivo*. Similarly, overexpression of AKR1C3 in PCa cells confers resistance to abiraterone, with AKR1C3 inhibition by indomethacin overcoming abiraterone-resistance and enhancing the effectiveness of abiraterone both *in vitro* and *in vivo* [205]. Both studies suggest that AKR1C3

inhibitors used in conjunction with abiraterone or enzalutamide will increase the effectiveness of the latter.

The development of AKR1C3 specific inhibitors is challenging given the >86% sequence identity shared by AKR1C3, AKR1C1 and AKR1C2, which are all expressed in the prostate [206]. Unlike AKR1C3, AKR1C1 and AKR1C2 catalyze reactions that inactivate potent androgens. Therefore, their inhibition would lead to the unwanted accumulation of potent androgens. Numerous studies have set out to develop selective AKR1C3 inhibitors [207]. Despite ongoing drug development efforts, some of the most striking preclinical proof-of-principle studies of AKR1C3 inhibition have employed the nonsteroidal anti-inflammatory drug (NSAID), indomethacin, a nonselective inhibitor of cyclooxygenase (COX) 1 and 2 which also inhibits AKR1C3 activity [208, 209].

A group recently reported on two indomethacin analogues, which target AKR1C3 in a selective manner and demonstrate potencies higher than that of indomethacin in PCa cells. Their hydroxyfurazan derivative in particular demonstrated 90-times higher selectivity over AKR1C2 and no activity towards COX enzymes [210]. This group used a scaffold hopping approach to develop a series of potent and selective AKR1C3 inhibitors which have demonstrated synergistic effects PCa cells when used in combination with abiraterone or enzalutamide [211]. Another group recently reported on a potent, isoform-selective and hydrolytically stable AKR1C3 inhibitor known as KV-37, which when used in combination with enzalutamide demonstrated a >200-fold potentiation of enzalutamide action in drug-resistant PCa cells [212].

Given the beneficial effects of combined AKR1C3 inhibition and AR antagonism, one emerging strategy is to develop specific AKR1C3 inhibitors with dual AR antagonist activity. An example of this strategy is provided by N-naphthylaminobenzoate, which inhibits AKR1C3 activity, while at the same time acting as a direct AR antagonist [213]. The stage is therefore set for further preclinical optimization of these AKR1C3 inhibitors followed by clinical trials in which their benefit to the treatment of CRPC will be determined.

Precision Predictors of Response to Abiraterone

Precision medicine has emerged as a critical approach for identifying prognostic and predictive biomarkers in men with CRPC [214–216]. Next generation sequencing of CRPC tumors has identified frequent aberrations in multiple genes including AR, TP53, PTEN, and DNA repair pathway genes such as BRCA1, 2 and ATM, and the potential treatment implications of these findings are being actively explored [217]. A number of genes involved in these pathways, which may portend AR pathway independence, as well as genes involved in the uptake and metabolism of steroids and abiraterone have been explored as predictive biomarkers of response to abiraterone, including SLCO2B1, HSD3B1, AR, DNA repair genes, TP53, PTEN, and SPOP [218–229].

Germline Variation in SLCO Transport Genes

Germline variation in solute carrier organic anion (SLCO) genes influences cellular uptake of various steroids including DHEA-S and T and has been associated with PCa outcomes including the duration of response to ADT [230–232]. Based on its steroidal structure, it has been hypothesized that abiraterone may undergo transport by SLCO-encoded transporters and that SLCO gene variation may influence intracellular abiraterone levels and outcomes. In support of this, LNCaP cells expressing SLCO2B1 showed two- to fourfold higher abiraterone levels compared with vector control [222]. In a cohort of men with intermediate- or high-risk localized PCa randomized to neoadjuvant ADT or neoadjuvant ADT plus abiraterone acetate, the AA/AG genotypes of the SLCO2B1 rs12422149 variant were associated with higher mean tissue abiraterone levels than the GG genotype (258 pg/mg vs. 99 pg/mg;

$P = 0.03$), and higher tissue abiraterone levels were associated with improved PSA and pathologic response after radical prostatectomy. These findings suggest that variation in SLCO genes can serve as predictors of response to abiraterone treatment [222].

In the first clinical validation of this hypothesis, 401 men with CRPC were treated with abiraterone acetate. Men heterozygous for rs12422149 (AG, 19%) had significantly improved median progression-free survival (PFS) on first-line abiraterone acetate compared with the homozygous wild-type group (GG, 81%) [8.9 months vs. 6.3 months; HR, 0.46; 95% confidence interval (CI), 0.23–0.94, $P = 0.03$] [221]. Importantly, rs12422149 is a predictive biomarker of response to ADT in men with castration sensitive PCa, raising the possibility that SLCO2B1 is simply a prognostic biomarker of outcomes in CRPC and not necessarily a predictive marker of response to abiraterone [231, 233]. Thus, while these findings require independent prospective validation in randomized data sets with matched controls, they suggest genetic variation in SLCO2B1 can serve as a biomarker of response to abiraterone acetate.

Germline Variation in HSD3B1

The common germline variant of HSD3B1 (1245A>C), described previously in this chapter, encodes a proteolysis-resistant enzyme resulting in increased metabolic flux of adrenal androgen precursors to DHT biosynthesis within CRPC tumors and is a predictive biomarker of resistance to castration [234–236] These patients may therefore benefit from treatments blocking the biosynthesis of adrenal androgen precursors. Indeed, patients with the HSD3B1 (1245C) variant achieved better clinical responses to ketoconazole, a nonsteroidal CYP17A1 inhibitor than those without the variant [220]. An increase in the number of inherited HSD3B1 (1245C) variant alleles from 0 to 2 increased the median duration of therapy from 5.0 months (95% CI, 3.4–10.4) with 0 variant alleles to 12.3 months (95% CI, 1.8-not reached) with 2 variant alleles. Patients with a single variant allele had a median duration of therapy of 7.5 months (95% CI, 4.9–19.2). Similarly, the median PFS times were 5.4 months (95% CI, 3.7–7.5), 9.7 months (95% CI, 5.6–32.9) and 15.2 months (95% CI, 7.8-not reached) for 0, 1 and 2 variant alleles respectively.

However, in the case of abiraterone treatment the stable HSD3B1 (1245C) variant may lead to increased conversion of abiraterone to D4A, the precursor to the AR agonist 5α-Abi, thereby partially negating the treatment benefits in these patients who are otherwise more likely to benefit from CYP17A1 inhibition [237]. Results from one study confirmed that patients who inherit 0, 1, and 2 copies of HSD3B1 (1245C) demonstrate a stepwise increase in 5α-Abi after correcting for pharmacokinetics (0.04 ng/ml, 2.60 ng/ml, and 2.70 ng/ml, respectively) [219]. Accordingly, in a small study of 76 men with metastatic CRPC the HSD3B1 (1245C) variant did not predict response to first-line abiraterone acetate, likely due to the opposing effects of the D4A and 5α-Abi metabolites on androgen signaling [218]. Whether inhibiting generation of the 5α-Abi metabolite by combining abiraterone with a 5α-reductase inhibitor reveals the positive predictive value of this biomarker that was observed in patients treated with the non-steroidal CYP17A1 inhibitor ketoconazole remains to be seen [220].

Noncanonical Role for Steroid Metabolizing Enzymes in Prostate Cancer Progression

A significant body of data supports the hypothesis that genes mediating pre-receptor control of DHT metabolism play an important role in determining intra-tumoral androgen levels in primary and castrate resistant prostate tumors. However, the increased expression of DHT catabolizing enzymes such as AKR1C2, UGT2B15 and UGT2B17 in CRPC (which would theoretically lower ligand levels available for AR activation) are not entirely congruent with this hypothesis. Several potential explanations for these observations exist, including the joint regulation of multiple steroidogenic genes by single transcription

factors [238], the potential engagement of metabolism pathways that bypass T biosynthesis [66], and the possibility that putatively 'steroidogenic' enzymes may have cancer-related functions beyond their steroidogenic potential.

The AKR1C family is an important reminder that many steroidogenic enzymes have alternative substrates, and have capacity to modify non-steroidal metabolites, which can influence disease progression or response to therapy independently of their steroid metabolizing function. For example, AKR1C1 is involved in detoxification of lipid peroxidation products [239], which may influence responses to oxidative stress, and AKR1C3 and AKR1C2 are critical regulators of prostaglandin (PG) synthesis [240]. In particular, AKR1C3 forms PGF2α and 11beta-PGF2α which stimulate the prostaglandin F (FP) receptor, and prevent the activation of PPARγ, resulting in a pro-proliferative signal that may stimulate PCa growth independently of an effect on steroidogenesis [241]. Increased expression of AKR1C2 *in vitro* and its associated increase in levels of prostaglandin F2α has also been associated with resistance to several chemotherapy drugs [242], illustrating another mechanism by which these genes may influence treatment response independent of androgen signaling.

Alternatively, these proteins may also have functions independent of any enzymatic activity. For example, AKR1C3 has recently been identified as an AR coactivator and thus may play dual roles in promoting ligand biosynthesis as well as AR activation [94]. AKR1C3 has also been shown to bind and stabilize the ubiquitin ligase Siah2, inhibiting its degradation and thereby enhancing Siah2-dependent regulation of AR activity in PCa cells [243]. Notably, AKR1C3 may play a role in modulating epigenetic susceptibility in PCa cells independently of an effect on AR. Knockdown of AKR1C3 was accompanied by a significantly reduced expression of a range of histone deacetylases, transcriptional co-regulators, and increased sensitivity towards SAHA, a clinically approved histone deacetylase inhibitor [244].

Overexpression of UGT2B17 has been associated with more aggressive PCa growth *in vivo*, potentially via activation of SRC kinase [245]. Looking beyond PCa, UGT2B17 has also been identified as a disease accelerator in chronic lymphocytic leukemia [246], and knockdown of UGT2B17 in an endometrial carcinoma cell line increased apoptosis in association with downregulation of the anti-apoptotic protein Mcl-1, and upregulation of the pro-apoptotic target of Mcl-1, Puma [247]. While the mechanism of UGT2B17 involvement in these tumors remains to be elucidated, these reports underscore the potential role of these enzymes in non-steroid metabolizing capacities.

Importantly, it remains to be established whether the increased expression of these genes is truly pathogenic, or merely a bystander of altered CRPC signaling. For example, while UGT genes are generally repressed by AR regulated signaling [248, 249], UGT2B17 has been identified as a positively regulated gene target of the constitutively active AR splice variants present in many CRPC tumors [248]. Thus, its presence in CRPC tumors may simply be a reflection of an altered, AR-variant associated transcriptional profile rather than an inherently pathogenic alteration.

Conclusions

Data regarding the molecular response of PCa to hormone therapy continues to emerge, providing critical insight into cellular growth and signaling pathways that may be exploited as therapeutic targets. The presence of residual androgens and persistent activation of the AR signaling axis in CRPC suggest that a multi-targeted treatment approach to ablate all contributions to AR signaling within the prostate tumor will be required for optimal anti-tumor efficacy. The introduction of potent steroidogenic inhibitors such as abiraterone and novel AR inhibitors such as enzalutamide holds promise for improving the treatment of men with CRPC, although to date these therapies are characterized by eventual disease progression. Importantly, while it is tempting to focus on steroid metabolic pathways as drivers of PCa biology, alternative hypotheses remain to be explored,

including the capacity of metabolic enzymes to modify non-steroidal substrates with pro or anti-carcinogenic activity, and their potential to act in roles independent of their catalytic functions.

The molecular alterations occurring in CRPC tumors following abiraterone treatment suggest tumor-specific methods of addressing resistance, either through optimizing steroidogenic blockade or by inhibiting AR signaling. Importantly, a two- to threefold increase in AR expression can render low androgen levels (in the range detected in the abiraterone-treated tumors) physiologically relevant in promoting AR driven growth [22]. Combining CYP17A1 blockade with inhibitors of other critical components of the pathway such as HSD3B1 or SRD5A2 or with AR inhibitors could offset adaptive upregulation of CYP17A1 [250], and, in the case of abiraterone, prevent generation of downstream metabolites with AR agonist activity [189]. Abiraterone at higher (but clinically achievable) concentrations can strongly inhibit HSD3B1 and 2 [251], and can antagonize the promiscuous T877A mutant AR [252], providing a rationale for dose-escalation of abiraterone at time of progression. To date, however, small studies of high (2000 mg) or low (250 mg) dose abiraterone in men with CRPC have not shown a significant impact on clinical outcomes, suggesting this approach does not improve clinical efficacy [253, 254]. Importantly, the induction of full length and ligand-independent AR splice variants (lacking the C terminal LBD) in abiraterone-treated tumors suggests strategies directed at targeting ligand synthesis combined with novel AR inhibitors capable of targeting the N terminal AR domain may have the greatest efficacy [255, 256].

While clinical efforts have focused on inhibition of ligand biosynthesis, the specific loss of androgen inactivating enzymes such as HSD17B2 and HSD17B4 by alternative splicing demonstrates that CRPC tumors harbor pre-existing mechanisms that conserve residual androgens in their active form and amplify the effect of low androgen levels in driving ligand-mediated AR signaling, contributing to development of resistant disease. It remains to be determined whether loss of HSD17B2 or HSD17B4 is associated with higher androgen levels in CRPC tumors, and whether these tumors are more likely to respond to continued AR and ligand synthesis directed therapy vs. tumors that are driven by ligand-independent mechanisms. Identification of the splicing factors driving loss of functional isoforms of these enzymes suggest approaches that can downregulate activity of these factors and reverse these changes [96]. For example, small molecular inhibitors of SRPK1, the splice factor kinase that phosphorylates and activates SRSF1, are already in pre-clinical development in PCa models of angiogenesis and would appear to warrant testing in more diverse PCa tumor models [257].

Emerging data demonstrate that adrenal derived 11-oxygenated androgens can activate the wild type AR and suggest that a larger pool of residual androgens may be available to activate the AR in CRPC than previously anticipated. Although these pathways remain to be more fully elucidated, available data are consistent with the proposed intracrine generation and activity of these potent androgens, especially considering the significant substrate preference that AKR1C3 exhibits towards these steroids. Moreover, these data suggest that measurements of residual T and DHT can underestimate the contribution of ligand-mediated activation of wild type AR. Importantly, the significant potency of 11KT and 11KDHT in activating the wild type AR suggests that inhibiting the enzymes unique to the production of these metabolites significantly decreases the overall androgenicity of the CRPC tumor microenvironment and thereby can improve clinical outcomes.

In conclusion, primary PCa and castration resistant tumors are characterized by a number of steroid enzyme alterations acting to enhance utilization of circulating adrenal androgens, inhibit metabolism of T and DHT to inactive metabolites, and in the case of CRPC tumors, promote *de novo* androgen biosynthesis. These observations strongly suggest that tissue-based alterations in steroid metabolism contribute to the development of CRPC and underscore these metabolic pathways as critical targets of therapy. The optimal timing, sequence, and potential combinatorial strategies using new AR pathway and ligand biosynthetic modulators are critical unanswered questions in optimizing treatment of men with PCa. Delineating mechanisms and biomarkers of resistance will be critical for rational trial design and for the stratification of men to treatment strategies with the highest likelihood of durable efficacy.

Acknowledgement NIH Pacific Northwest Prostate Cancer SPORE P50 CA97186
NIH P01 CA163227
Department of Defense CDMRP W81XWH-12-1-0208
Department of Veterans Affairs Puget Sound Health Care System
Damon Runyon Cancer Research Foundation (Damon Runyon-Genentech Clinical Investigator Award CI-40-08).
National Research Foundation of South Africa (CPRR 98886)
Medical Research Council of South Africa
Department of Defense CDMRP W81XWH-11-2-0154
Department of Defense CDMRP W81XWH-15-1-0150

References

1. J. Geller, J. Albert, D. Loza, S. Geller, W. Stoeltzing, D. de la Vega, DHT concentrations in human prostate cancer tissue. J. Clin. Endocrinol. Metab. **46**(3), 440–444 (1978)
2. J. Geller, J. Albert, S.S. Yen, S. Geller, D. Loza, Medical castration of males with megestrol acetate and small doses of diethylstilbestrol. J. Clin. Endocrinol. Metab. **52**(3), 576–580 (1981)
3. J. Liu, J. Geller, J. Albert, M. Kirshner, Acute effects of testicular and adrenal cortical blockade on protein synthesis and dihydrotestosterone content of human prostate tissue. J. Clin. Endocrinol. Metab. **61**(1), 129–133 (1985)
4. J. Liu, J. Albert, J. Geller, Effects of androgen blockade with ketoconazole and megestrol acetate on human prostatic protein patterns. Prostate **9**(2), 199–205 (1986)
5. J. Geller, J. Albert, Effects of castration compared with total androgen blockade on tissue dihydrotestosterone (DHT) concentration in benign prostatic hyperplasia (BPH). Urol. Res. **15**(3), 151–153 (1987)
6. J. Geller, J. Liu, J. Albert, W. Fay, C.C. Berry, P. Weis, Relationship between human prostatic epithelial cell protein synthesis and tissue dihydrotestosterone level. Clin. Endocrinol. (Oxf) **26**(2), 155–161 (1987)
7. F. Labrie, A. Dupont, A. Belanger, L. Cusan, Y. Lacourciere, G. Monfette, et al., New hormonal therapy in prostatic carcinoma: combined treatment

with an LHRH agonist and an antiandrogen. Clin. Invest. Med. **5**(4), 267–275 (1982)

8. S.T. Page, D.W. Lin, E.A. Mostaghel, D.L. Hess, L.D. True, J.K. Amory, et al., Persistent intraprostatic androgen concentrations after medical castration in healthy men. J. Clin. Endocrinol. Metab. **91**(10), 3850–3856 (2006)

9. G. Forti, R. Salerno, G. Moneti, S. Zoppi, G. Fiorelli, T. Marinoni, et al., Three-month treatment with a long-acting gonadotropin-releasing hormone agonist of patients with benign prostatic hyperplasia: effects on tissue androgen concentration, 5 alpha-reductase activity and androgen receptor content. J. Clin. Endocrinol. Metab. **68**(2), 461–468 (1989)

10. T. Nishiyama, Y. Hashimoto, K. Takahashi, The influence of androgen deprivation therapy on dihydrotestosterone levels in the prostatic tissue of patients with prostate cancer. Clin. Cancer Res. **10**(21), 7121–7126 (2004)

11. T. Nishiyama, T. Ikarashi, Y. Hashimoto, K. Wako, K. Takahashi, The change in the dihydrotestosterone level in the prostate before and after androgen deprivation therapy in connection with prostate cancer aggressiveness using the Gleason score. J. Urol. **178**(4 Pt 1), 1282–1288 (2007). discussion 8-9

12. J.L. Mohler, C.W. Gregory, O.H. Ford 3rd, D. Kim, C.M. Weaver, P. Petrusz, et al., The androgen axis in recurrent prostate cancer. Clin. Cancer Res. **10**(2), 440–448 (2004)

13. R.B. Montgomery, E.A. Mostaghel, R. Vessella, D.L. Hess, T.F. Kalhorn, C.S. Higano, et al., Maintenance of intratumoral androgens in metastatic prostate cancer: a mechanism for castration-resistant tumor growth. Cancer Res. **68**(11), 4447–4454 (2008)

14. A. Mizokami, E. Koh, H. Fujita, Y. Maeda, M. Egawa, K. Koshida, et al., The adrenal androgen androstenediol is present in prostate cancer tissue after androgen deprivation therapy and activates mutated androgen receptor. Cancer Res. **64**(2), 765–771 (2004)

15. H. Miyamoto, S. Yeh, H. Lardy, E. Messing, C. Chang, Delta5-androstenediol is a natural hormone with androgenic activity in human prostate cancer cells. Proc. Natl. Acad. Sci. U. S. A. **95**(19), 11083–11088 (1998)

16. E.A. Mostaghel, P. Nelson, P.H. Lange, D.W. Lin, M. Taplin, S.P. Balk, et al., Neoadjuvant androgen pathway suppression prior to prostatectomy. J. Clin. Oncol. **30**, 4520 (2012)

17. M.E. Taplin, B. Montgomery, C.J. Logothetis, G.J. Bubley, J.P. Richie, B.L. Dalkin, et al., Intense androgen-deprivation therapy with abiraterone acetate plus leuprolide acetate in patients with localized high-risk prostate cancer: results of a randomized phase II neoadjuvant study. J. Clin. Oncol. **32**(33), 3705–3715 (2014)

18. Z. Culig, J. Hoffmann, M. Erdel, I.E. Eder, A. Hobisch, A. Hittmair, et al., Switch from antagonist to agonist of the androgen receptor bicalutamide is associated with prostate tumour progression in a new model system. Br. J. Cancer **81**(2), 242–251 (1999)

19. C.W. Gregory, R.T. Johnson Jr., J.L. Mohler, F.S. French, E.M. Wilson, Androgen receptor stabilization in recurrent prostate cancer is associated with hypersensitivity to low androgen. Cancer Res. **61**(7), 2892–2898 (2001)

20. C.W. Gregory, K.G. Hamil, D. Kim, S.H. Hall, T.G. Pretlow, J.L. Mohler, et al., Androgen receptor expression in androgen-independent prostate cancer is associated with increased expression of androgen-regulated genes. Cancer Res. **58**(24), 5718–5724 (1998)

21. J.L. Mohler, T.L. Morris, O.H. Ford 3rd, R.F. Alvey, C. Sakamoto, C.W. Gregory, Identification of differentially expressed genes associated with androgen-independent growth of prostate cancer. Prostate **51**(4), 247–255 (2002)

22. C.D. Chen, D.S. Welsbie, C. Tran, S.H. Baek, R. Chen, R. Vessella, et al., Molecular determinants of resistance to antiandrogen therapy. Nat. Med. **10**(1), 33–39 (2004)

23. E. Greenberg, Endocrine therapy in the management of prostatic cancer. Clin. Endocrinol. Metab. **9**(2), 369–381 (1980)

24. M.R. Robinson, R.J. Shearer, J.D. Fergusson, Adrenal suppression in the treatment of carcinoma of the prostate. Br. J. Urol. **46**(5), 555–559 (1974)

25. D.J. Samson, J. Seidenfeld, B. Schmitt, V. Hasselblad, P.C. Albertsen, C.L. Bennett, et al., Systematic review and meta-analysis of monotherapy compared with combined androgen blockade for patients with advanced prostate carcinoma. Cancer **95**(2), 361–376 (2002)

26. B. Schmitt, C. Bennett, J. Seidenfeld, D. Samson, T. Wilt, Maximal androgen blockade for advanced prostate cancer. Cochrane Database Syst. Rev. **2**, CD001526 (2000)

27. J.F. Caubet, T.D. Tosteson, E.W. Dong, E.M. Naylon, G.W. Whiting, M.S. Ernstoff, et al., Maximum androgen blockade in advanced prostate cancer: a meta-analysis of published randomized controlled trials using nonsteroidal antiandrogens. Urology **49**(1), 71–78 (1997)

28. E.J. Small, C.J. Ryan, The case for secondary hormonal therapies in the chemotherapy age. J. Urol. **176**(6 Suppl 1), S66–S71 (2006). Innovations and Challenges in Prostate Cancer: Recommendations for Defining and Treating High Risk Disease. 2006;176(6, Supplement 1):S66-S71

29. J. S. de Bono (ed.), *Abiraterone acetate improves survival in metastatic castration-resistant prostate cancer: Phase III results* (European Society for Medical Oncology, Milan, 2010)

30. H.I. Scher, T.M. Beer, C.S. Higano, A. Anand, M.E. Taplin, E. Efstathiou, et al., Antitumour activity of MDV3100 in castration-resistant prostate cancer: a phase 1-2 study. Lancet **375**(9724), 1437–1446 (2010)

31. E.A. Mostaghel, B.T. Marck, S.R. Plymate, R.L. Vessella, S. Balk, A.M. Matsumoto, et al., Resistance to CYP17A1 inhibition with abiraterone in castration-resistant prostate cancer: induction of steroidogenesis and androgen receptor splice variants. Clin. Cancer Res. **17**(18), 5913–5925 (2011)

32. C. Cai, S. Chen, P. Ng, G.J. Bubley, P.S. Nelson, E.A. Mostaghel, et al., Intratumoral de novo steroid synthesis activates androgen receptor in castration-resistant prostate cancer and is upregulated by treatment with CYP17A1 inhibitors. Cancer Res. **71**(20), 6503–6513 (2011)

33. E. Efstathiou, M. Titus, D. Tsavachidou, V. Tzelepi, S. Wen, A. Hoang, et al., Effects of abiraterone acetate on androgen signaling in castrate-resistant prostate cancer in bone. J. Clin. Oncol. **30**(6), 637–643 (2012)

34. M. Stanbrough, G.J. Bubley, K. Ross, T.R. Golub, M.A. Rubin, T.M. Penning, et al., Increased expression of genes converting adrenal androgens to testosterone in androgen-independent prostate cancer. Cancer Res. **66**(5), 2815–2825 (2006)

35. J. Holzbeierlein, P. Lal, E. LaTulippe, A. Smith, J. Satagopan, L. Zhang, et al., Gene expression analysis of human prostate carcinoma during hormonal therapy identifies androgen-responsive genes and mechanisms of therapy resistance. Am. J. Pathol. **164**(1), 217–227 (2004)

36. W.L. Miller, R.J. Auchus, The molecular biology, biochemistry, and physiology of human steroidogenesis and its disorders. Endocr. Rev. **32**(1), 81–151 (2011)

37. W.E. Rainey, B.R. Carr, H. Sasano, T. Suzuki, J.I. Mason, Dissecting human adrenal androgen production. Trends Endocrinol. Metab. **13**(6), 234–239 (2002)

38. A. Endoh, S.B. Kristiansen, P.R. Casson, J.E. Buster, P.J. Hornsby, The zona reticularis is the site of biosynthesis of dehydroepiandrosterone and dehydroepiandrosterone sulfate in the adult human adrenal cortex resulting from its low expression of 3 beta-hydroxysteroid dehydrogenase. J. Clin. Endocrinol. Metab. **81**(10), 3558–3565 (1996)

39. Y. Nakamura, P.J. Hornsby, P. Casson, R. Morimoto, F. Satoh, Y. Xing, et al., Type 5 17beta-hydroxysteroid dehydrogenase (AKR1C3) contributes to testosterone production in the adrenal reticularis. J. Clin. Endocrinol. Metab. **94**(6), 2192–2198 (2009)

40. E.J. Sanford, D.F. Paulson, T.J. Rohner Jr., R.J. Santen, C.W. Bardin, The effects of castration on adrenal testosterone secretion in men with prostatic carcinoma. J. Urol. **118**(6), 1019–1021 (1977)

41. J. Rege, Y. Nakamura, F. Satoh, R. Morimoto, M.R. Kennedy, L.C. Layman, et al., Liquid chromatography-tandem mass spectrometry analysis of human adrenal vein 19-carbon steroids before and after ACTH stimulation. J. Clin. Endocrinol. Metab. **98**(3), 1182–1188 (2013)

42. A.C. Swart, L. Schloms, K.H. Storbeck, L.M. Bloem, T. Toit, J.L. Quanson, et al., 11beta-hydroxyandrostenedione, the product of androstenedione metabolism in the adrenal, is metabolized in LNCaP cells by 5alpha-reductase yielding 11beta-hydroxy-5alpha-androstanedione. J. Steroid Biochem. Mol. Biol. **138**, 132–142 (2013)

43. M. Barnard, J.L. Quanson, E. Mostaghel, E. Pretorius, J.L. Snoep, K.H. Storbeck, 11-Oxygenated androgen precursors are the preferred substrates for aldo-keto reductase 1C3 (AKR1C3): Implications for castration resistant prostate cancer. J. Steroid Biochem. Mol. Biol. **183**, 192–201 (2018)

44. F. Labrie, V. Luu-The, S. Lin, J. Simard, C. Labrie, M. El-Alfy, et al., Intracrinology: role of the family of 17 beta-hydroxysteroid dehydrogenases in human physiology and disease. J. Mol. Endocrinol. **25**, 1):1–1)16 (2000)

45. H. Klein, M. Bressel, H. Kastendieck, K.D. Voigt, Androgens, adrenal androgen precursors, and their metabolism in untreated primary tumors and lymph node metastases of human prostatic cancer. Am. J. Clin. Oncol. **11**(Suppl 2), S30–S36 (1988)

46. Y. Nakamura, T. Suzuki, T. Fukuda, A. Ito, M. Endo, T. Moriya, et al., Steroid sulfatase and estrogen sulfotransferase in human prostate cancer. Prostate **66**(9), 1005–1012 (2006)

47. J.M. Day, A. Purohit, H.J. Tutill, P.A. Foster, L.W. Woo, B.V. Potter, et al., The development of steroid sulfatase inhibitors for hormone-dependent cancer therapy. Ann. N. Y. Acad. Sci. **1155**, 80–87 (2009)

48. V. Luu-The, A. Belanger, F. Labrie, Androgen biosynthetic pathways in the human prostate. Best Pract. Res. Clin. Endocrinol. Metab. **22**(2), 207–221 (2008)

49. K.M. Fung, E.N. Samara, C. Wong, A. Metwalli, R. Krlin, B. Bane, et al., Increased expression of type 2 3alpha-hydroxysteroid dehydrogenase/type 5 17beta-hydroxysteroid dehydrogenase (AKR1C3) and its relationship with androgen receptor in prostate carcinoma. Endocr. Relat. Cancer **13**(1), 169–180 (2006)

50. T.M. Penning, D.R. Bauman, Y. Jin, T.L. Rizner, Identification of the molecular switch that regulates access of 5[alpha]-DHT to the androgen receptor. Mol. Cell. Endocrinol. **265-266**(Adrenal/Molecular Steroidogenesis Conference 2006), 77–82 (2007)

51. T.L. Rizner, H.K. Lin, D.M. Peehl, S. Steckelbroeck, D.R. Bauman, T.M. Penning, Human type 3 3{alpha}-hydroxysteroid dehydrogenase (aldo-keto reductase 1c2) and androgen metabolism in prostate cells. Endocrinology **144**(7), 2922–2932 (2003). https://doi.org/10.1210/en.2002-0032

52. Q. Ji, L. Chang, D. VanDenBerg, F.Z. Stanczyk, A. Stolz, Selective reduction of AKR1C2 in prostate cancer and its role in DHT metabolism. Prostate **54**(4), 275–289 (2003)

53. Q. Ji, L. Chang, F.Z. Stanczyk, M. Ookhtens, A. Sherrod, A. Stolz, Impaired dihydrotestosterone catabolism in human prostate cancer: critical role of AKR1C2 as a pre-receptor regulator of androgen

receptor signaling. Cancer Res. **67**(3), 1361–1369 (2007)

54. D.R. Bauman, S. Steckelbroeck, M.V. Williams, D.M. Peehl, T.M. Penning, Identification of the major oxidative 3{alpha}-hydroxysteroid dehydrogenase in human prostate that converts 5{alpha}-androstane-3{alpha},17{beta}-diol to 5{alpha}-dihydrotestosterone: a potential therapeutic target for androgen-dependent disease. Mol. Endocrinol. **20**(2), 444–458 (2006)

55. J.L. Mohler, M.A. Titus, S. Bai, B.J. Kennerley, F.B. Lih, K.B. Tomer, et al., Activation of the androgen receptor by intratumoral bioconversion of androstanediol to dihydrotestosterone in prostate cancer. Cancer Res. **71**(4), 1486–1496 (2011)

56. S. Muthusamy, S. Andersson, H.J. Kim, R. Butler, L. Waage, U. Bergerheim, et al., Estrogen receptor beta and 17beta-hydroxysteroid dehydrogenase type 6, a growth regulatory pathway that is lost in prostate cancer. Proc. Natl. Acad. Sci. U. S. A. **108**(50), 20090–20094 (2011)

57. X.F. Huang, V. Luu-The, Molecular characterization of a first human 3(alpha-->beta)-hydroxysteroid epimerase. J. Biol. Chem. **275**(38), 29452–29457 (2000)

58. C. Guillemette, E. Levesque, M. Beaulieu, D. Turgeon, D.W. Hum, A. Belanger, Differential regulation of two uridine diphospho-glucuronosyltransferases, UGT2B15 and UGT2B17, in human prostate LNCaP cells. Endocrinology **138**(7), 2998–3005 (1997)

59. S. Chouinard, O. Barbier, A. Belanger, UDP-glucuronosyltransferase 2B15 (UGT2B15) and UGT2B17 enzymes are major determinants of the androgen response in prostate cancer LNCaP cells. J. Biol. Chem. **282**(46), 33466–33474 (2007)

60. S. Chouinard, G. Pelletier, A. Belanger, O. Barbier, Cellular specific expression of the androgen-conjugating enzymes UGT2B15 and UGT2B17 in the human prostate epithelium. Endocr. Res. **30**(4), 717–725 (2004)

61. Q. Wei, R. Galbenus, A. Raza, R.L. Cerny, M.A. Simpson, Androgen-stimulated UDP-glucose dehydrogenase expression limits prostate androgen availability without impacting hyaluronan levels. Cancer Res. **69**(6), 2332–2339 (2009)

62. R.J. Auchus, The backdoor pathway to dihydrotestosterone. Trends Endocrinol. Metab. **15**(9), 432–438 (2004)

63. R.J. Auchus, Non-traditional metabolic pathways of adrenal steroids. Rev. Endocr. Metab. Disord. **10**(1), 27–32 (2009)

64. M.K. Gupta, O.L. Guryev, R.J. Auchus, 5alpha-reduced C21 steroids are substrates for human cytochrome P450c17. Arch. Biochem. Biophys. **418**(2), 151–160 (2003)

65. M. Mahendroo, J.D. Wilson, J.A. Richardson, R.J. Auchus, Steroid 5alpha-reductase 1 promotes 5alpha-androstane-3alpha,17beta-diol synthesis in

immature mouse testes by two pathways. Mol. Cell. Endocrinol. **222**(1-2), 113–120 (2004)

66. K.H. Chang, R. Li, M. Papari-Zareei, L. Watumull, Y.D. Zhao, R.J. Auchus, et al., Dihydrotestosterone synthesis bypasses testosterone to drive castration-resistant prostate cancer. Proc. Natl. Acad. Sci. U. S. A. **108**(33), 13728–13733 (2011)

67. P. Negri-Cesi, M. Motta, Androgen metabolism in the human prostatic cancer cell line LNCaP. J. Steroid Biochem. Mol. Biol. **51**(1-2), 89–96 (1994)

68. A.E. Thigpen, K.M. Cala, D.W. Russell, Characterization of Chinese hamster ovary cell lines expressing human steroid 5 alpha-reductase isozymes. J. Biol. Chem. **268**(23), 17404–17412 (1993)

69. M. Samson, F. Labrie, C.C. Zouboulis, Luu-The V. Biosynthesis of dihydrotestosterone by a pathway that does not require testosterone as an intermediate in the SZ95 sebaceous gland cell line. J. Invest. Dermatol. **130**(2), 602–604 (2010)

70. H.-K. Lin, S. Steckelbroeck, K.-M. Fung, A.N. Jones, T.M. Penning, Characterization of a monoclonal antibody for human aldo-keto reductase AKR1C3 (type 2 3[alpha]-hydroxysteroid dehydrogenase/type 5 17[beta]-hydroxysteroid dehydrogenase); immunohistochemical detection in breast and prostate. Steroids **69**(13-14), 795–801 (2004)

71. J. Luo, T.A. Dunn, C.M. Ewing, P.C. Walsh, W.B. Isaacs, Decreased gene expression of steroid 5 alpha-reductase 2 in human prostate cancer: implications for finasteride therapy of prostate carcinoma. Prostate **57**(2), 134–139 (2003)

72. M.A. Titus, C.W. Gregory, O.H. Ford III, M.J. Schell, S.J. Maygarden, J.L. Mohler, Steroid 5{alpha}-Reductase Isozymes I and II in Recurrent Prostate Cancer. Clin. Cancer Res. **11**(12), 4365–4371 (2005)

73. L.N. Thomas, R.C. Douglas, C.B. Lazier, R. Gupta, R.W. Norman, P.R. Murphy, et al., Levels of 5[alpha]-reductase type 1 and type 2 are increased in localized high grade compared to low grade prostate cancer. J. Urol. **179**(1), 147–151 (2007). In Press, Corrected Proof

74. Y. Shibata, K. Suzuki, S. Arai, Y. Miyoshi, S. Umemoto, N. Masumori, et al., Impact of pretreatment prostate tissue androgen content on the prediction of castration-resistant prostate cancer development in patients treated with primary androgen deprivation therapy. Andrology **1**(3), 505–511 (2013)

75. A. Godoy, E. Kawinski, Y. Li, D. Oka, B. Alexiev, F. Azzouni, et al., 5alpha-reductase type 3 expression in human benign and malignant tissues: a comparative analysis during prostate cancer progression. Prostate **71**(10), 1033–1046 (2011)

76. F. Azzouni, A. Godoy, Y. Li, J. Mohler, The 5 alpha-reductase isozyme family: a review of basic biology and their role in human diseases. Adv. Urol. **2012**, 530121 (2012)

77. E. Koh, T. Noda, J. Kanaya, M. Namiki, Differential expression of 17beta-hydroxysteroid dehydrogenase

isozyme genes in prostate cancer and noncancer tissues. Prostate **53**(2), 154–159 (2002)

78. J.P. Elo, L.A. Akinola, M. Poutanen, P. Vihko, A.P. Kyllonen, O. Lukkarinen, et al., Characterization of 17beta-hydroxysteroid dehydrogenase isoenzyme expression in benign and malignant human prostate. Int. J. Cancer **66**(1), 37–41 (1996)

79. X. Gao, C. Dai, S. Huang, J. Tang, G. Chen, J. Li, Z. Zhu, X. Zhu, S. Zhou, Y. Gao, Z. Hou, Z. Fang, C. Xu, J. Wang, D. Wu, N. Sharifi, Z. Li, Functional silencing of HSD17B2 in prostate cancer promotes disease progression. Clin. Cancer Res. **25**, 1291 (2018)

80. S. Zha, S. Ferdinandusse, J.L. Hicks, S. Denis, T.A. Dunn, R.J. Wanders, et al., Peroxisomal branched chain fatty acid beta-oxidation pathway is upregulated in prostate cancer. Prostate **63**(4), 316–323 (2005)

81. H.K. Ko, M. Berk, Y.M. Chung, B. Willard, R. Bareja, M. Rubin, et al., Loss of an androgen-inactivating and isoform-specific HSD17B4 splice form enables emergence of castration-resistant prostate cancer. Cell Rep. **22**(3), 809–819 (2018)

82. X.Y. He, Y.Z. Yang, D.M. Peehl, A. Lauderdale, H. Schulz, S.Y. Yang, Oxidative 3alpha-hydroxysteroid dehydrogenase activity of human type 10 17beta-hydroxysteroid dehydrogenase. J. Steroid Biochem. Mol. Biol. **87**(2-3), 191–198 (2003)

83. C.N. Falany, D. He, N. Dumas, A.R. Frost, J.L. Falany, Human cytosolic sulfotransferase 2B1: isoform expression, tissue specificity and subcellular localization. J. Steroid Biochem. Mol. Biol. **102**(1-5), 214–221 (2006)

84. Y.K. Seo, N. Mirkheshti, C.S. Song, S. Kim, S. Dodds, S.C. Ahn, et al., SULT2B1b sulfotransferase: induction by vitamin D receptor and reduced expression in prostate cancer. Mol. Endocrinol. **27**(6), 925–939 (2013)

85. D. He, C.N. Falany, Inhibition of SULT2B1b expression alters effects of 3beta-hydroxysteroids on cell proliferation and steroid hormone receptor expression in human LNCaP prostate cancer cells. Prostate **67**(12), 1318–1329 (2007)

86. N.C. Bennett, J.D. Hooper, D. Lambie, C.S. Lee, T. Yang, D.A. Vesey, et al., Evidence for steroidogenic potential in human prostate cell lines and tissues. Am. J. Pathol. **181**(3), 1078–1087 (2012)

87. A. Stigliano, O. Gandini, L. Cerquetti, P. Gazzaniga, S. Misiti, S. Monti, et al., Increased metastatic lymph node 64 and CYP17 expression are associated with high stage prostate cancer. J. Endocrinol. **194**(1), 55–61 (2007)

88. M. Sakai, D.B. Martinez-Arguelles, A.G. Aprikian, A.M. Magliocco, V. Papadopoulos, De novo steroid biosynthesis in human prostate cell lines and biopsies. Prostate **76**(6), 575–587 (2016)

89. E. Neubauer, M. Latif, J. Krause, A. Heumann, M. Armbrust, C. Luehr, et al., Up regulation of the steroid hormone synthesis regulator HSD3B2 is

linked to early PSA recurrence in prostate cancer. Exp. Mol. Pathol. **105**(1), 50–56 (2018)

90. S. Paquet, L. Fazli, L. Grosse, M. Verreault, B. Tetu, P.S. Rennie, et al., Differential expression of the androgen-conjugating UGT2B15 and UGT2B17 enzymes in prostate tumor cells during cancer progression. J. Clin. Endocrinol. Metab. **97**(3), E428–E432 (2012)

91. J. Hofland, W.M. van Weerden, N.F. Dits, J. Steenbergen, G.J. van Leenders, G. Jenster, et al., Evidence of limited contributions for intratumoral steroidogenesis in prostate cancer. Cancer Res. **70**(3), 1256–1264 (2010)

92. N. Mitsiades, C.C. Sung, N. Schultz, D.C. Danila, B. He, V.K. Eedunuri, et al., Distinct patterns of dysregulated expression of enzymes involved in androgen synthesis and metabolism in metastatic prostate cancer tumors. Cancer Res. **72**(23), 6142–6152 (2012)

93. Y. Liu, Z.X. Yao, V. Papadopoulos, Cytochrome P450 17alpha hydroxylase/17,20 lyase (CYP17) function in cholesterol biosynthesis: identification of squalene monooxygenase (epoxidase) activity associated with CYP17 in Leydig cells. Mol. Endocrinol. **19**(7), 1918–1931 (2005)

94. M. Yepuru, Z. Wu, A. Kulkarni, F. Yin, C.M. Barrett, J. Kim, et al., Steroidogenic enzyme AKR1C3 is a novel androgen receptor-selective coactivator that promotes prostate cancer growth. Clin. Cancer Res. **19**, 5613 (2013)

95. K. Chang, R. Li, K. B, Y. Lotan, C. Roehrborn, J. Liu, et al., A gain-of-function mutation in DHTSynthesis in castration resistant prostate cancer. Cell **154**, 1074–1084 (2013)

96. A. Paschalis, A. Sharp, J.C. Welti, A. Neeb, G.V. Raj, J. Luo, et al., Alternative splicing in prostate cancer. Nat. Rev. Clin. Oncol. **15**(11), 663–675 (2018)

97. C. Sette, Alternative splicing programs in prostate cancer. Int. J. Cell Biol. **2013**, 458727 (2013)

98. J. Luo, G. Attard, S.P. Balk, C. Bevan, K. Burnstein, L. Cato, et al., Role of androgen receptor variants in prostate cancer: report from the 2017 mission androgen receptor variants meeting. Eur. Urol. **73**, 715 (2017)

99. D.A. Bastos, E.S. Antonarakis, CTC-derived AR-V7 detection as a prognostic and predictive biomarker in advanced prostate cancer. Expert Rev. Mol. Diagn. **18**(2), 155–163 (2018)

100. N. Bruchovsky, Comparison of the metabolites formed in rat prostate following the in vivo administration of seven natural androgens. Endocrinology **89**(5), 1212–1222 (1971)

101. C. Labrie, J. Simard, H.F. Zhao, A. Belanger, G. Pelletier, F. Labrie, Stimulation of androgen-dependent gene expression by the adrenal precursors dehydroepiandrosterone and androstenedione in the rat ventral prostate. Endocrinology **124**(6), 2745–2754 (1989)

102. C.D. Schiller, M.R. Schneider, H. Hartmann, A.H. Graf, H. Klocker, G. Bartsch, Growth-

stimulating effect of adrenal androgens on the R3327 Dunning prostatic carcinoma. Urol. Res. **19**(1), 7–13 (1991)

103. M.E. Harper, A. Pike, W.B. Peeling, K. Griffiths, Steroids of adrenal origin metabolized by human prostatic tissue both in vivo and in vitro. J. Endocrinol. **60**(1), 117–125 (1974)

104. K. Mitamura, T. Nakagawa, K. Shimada, M. Namiki, E. Koh, A. Mizokami, et al., Identification of dehydroepiandrosterone metabolites formed from human prostate homogenate using liquid chromatography-mass spectrometry and gas chromatography-mass spectrometry. J. Chromatogr. A **961**(1), 97–105 (2002)

105. H.F. Acevedo, J.W. Goldzieher, The metabolism of [4-14C] progesterone by hypertrophic and carcinomatous human prostate tissue. Biochim. Biophys. Acta **111**(1), 294–298 (1965)

106. F. Di Silverio, V. Gagliardi, G. Sorcini, F. Sciarra, Biosynthesis and metabolism of androgenic hormones in cancer of the prostate. Invest. Urol. **13**(4), 286–288 (1976)

107. H. Klein, T. Molwitz, W. Bartsch, Steroid sulfate sulfatase in human benign prostatic hyperplasia: characterization and quantification of the enzyme in epithelium and stroma. J. Steroid Biochem. **33**(2), 195–200 (1989)

108. C. Dai, Y.M. Chung, E. Kovac, Z. Zhu, J. Li, C. Magi-Galluzzi, et al., Direct metabolic interrogation of dihydrotestosterone biosynthesis from adrenal precursors in primary prostatectomy tissues. Clin. Cancer Res. **23**(20), 6351–6362 (2017)

109. E. Koh, J. Kanaya, M. Namiki, Adrenal steroids in human prostatic cancer cell lines. Arch. Androl. **46**(2), 117–125 (2001)

110. E.H. Allott, E.M. Masko, A.R. Freedland, E. Macias, K. Pelton, K.R. Solomon, et al., Serum cholesterol levels and tumor growth in a PTEN-null transgenic mouse model of prostate cancer. Prostate Cancer Prostatic Dis. **21**(2), 196–203 (2018)

111. E.A. Mostaghel, K.R. Solomon, K. Pelton, M.R. Freeman, R.B. Montgomery, Impact of circulating cholesterol levels on growth and intratumoral androgen concentration of prostate tumors. PLoS One **7**(1), e30062 (2012)

112. C.G. Leon, J.A. Locke, H.H. Adomat, S.L. Etinger, A.L. Twiddy, R.D. Neumann, et al., Alterations in cholesterol regulation contribute to the production of intratumoral androgens during progression to castration-resistant prostate cancer in a mouse xenograft model. Prostate **70**(4), 390–400 (2009)

113. J.A. Locke, E.S. Guns, A.A. Lubik, H.H. Adomat, S.C. Hendy, C.A. Wood, et al., Androgen levels increase by intratumoral de novo steroidogenesis during progression of castration-resistant prostate cancer. Cancer Res. **68**(15), 6407–6415 (2008)

114. A.A. Lubik, J.H. Gunter, S.C. Hendy, J.A. Locke, H.H. Adomat, V. Thompson, et al., Insulin increases de novo steroidogenesis in prostate cancer cells. Cancer Res. **71**, 5754 (2011)

115. M. Knuuttila, E. Yatkin, J. Kallio, S. Savolainen, T.D. Laajala, T. Aittokallio, et al., Castration induces up-regulation of intratumoral androgen biosynthesis and androgen receptor expression in an orthotopic VCaP human prostate cancer xenograft model. Am. J. Pathol. **184**(8), 2163–2173 (2014)

116. E.A. Mostaghel, A. Zhang, S. Hernandez, B.T. Marck, X. Zhang, D. Tamae, et al., Contribution of adrenal glands to intratumor androgens and growth of castration-resistant prostate cancer. Clin. Cancer Res. **25**(1), 426–439 (2019)

117. R. Huhtaniemi, R. Oksala, M. Knuuttila, A. Mehmood, E. Aho, T.D. Laajala, et al., Adrenals contribute to growth of castration-resistant VCaP prostate cancer xenografts. Am. J. Pathol. **188**(12), 2890–2901 (2018)

118. P.R. Dillard, M.F. Lin, S.A. Khan, Androgen-independent prostate cancer cells acquire the complete steroidogenic potential of synthesizing testosterone from cholesterol. Mol. Cell. Endocrinol. **295**(1-2), 115–120 (2008)

119. J.A. Locke, K.M. Wasan, C.C. Nelson, E.S. Guns, C.G. Leon, Androgen-mediated cholesterol metabolism in LNCaP and PC-3 cell lines is regulated through two different isoforms of acyl-coenzyme A:Cholesterol Acyltransferase (ACAT). Prostate **68**(1), 20–33 (2008)

120. J.A. Locke, C.C. Nelson, H.H. Adomat, S.C. Hendy, M.E. Gleave, E.S. Guns, Steroidogenesis inhibitors alter but do not eliminate androgen synthesis mechanisms during progression to castration-resistance in LNCaP prostate xenografts. J. Steroid Biochem. Mol. Biol. **115**, 126 (2009)

121. C.W. Jeong, C.Y. Yoon, S.J. Jeong, S.K. Hong, S.S. Byun, S.E. Lee, Limited expression of cytochrome p450 17alpha-hydroxylase/17,20-lyase in prostate cancer cell lines. Korean J. Urol. **52**(7), 494–497 (2011)

122. J. Kumagai, J. Hofland, S. Erkens-Schulze, N.F. Dits, J. Steenbergen, G. Jenster, et al., Intratumoral conversion of adrenal androgen precursors drives androgen receptor-activated cell growth in prostate cancer more potently than de novo steroidogenesis. Prostate **73**, 1636 (2013)

123. S. Deb, S. Pham, D.S. Ming, M.Y. Chin, H. Adomat, A. Hurtado-Coll, et al., Characterization of precursor-dependent steroidogenesis in human prostate cancer models. Cancers **10**(10), E343 (2018)

124. A.G.G. de Mello Martins, G. Allegretta, G. Unteregger, J. Haupenthal, J. Eberhard, M. Hoffmann, et al., CYP17A1-independent production of the neurosteroid-derived 5alpha-pregnan-3beta,6alpha-diol-20-one in androgen-responsive prostate cancer cell lines under serum starvation and inhibition by Abiraterone. J. Steroid Biochem. Mol. Biol. **174**, 183–191 (2017)

125. J.Y. Chun, N. Nadiminty, S. Dutt, W. Lou, J.C. Yang, H.J. Kung, et al., Interleukin-6 regulates androgen synthesis in prostate cancer cells. Clin. Cancer Res. **15**(15), 4815–4822 (2009)

126. A.A. Lubik, J.H. Gunter, B.G. Hollier, S. Ettinger, L. Fazli, N. Stylianou, et al., IGF2 increases de novo steroidogenesis in prostate cancer cells. Endocr. Relat. Cancer **20**(2), 173–186 (2013)

127. S.V. Liu, A.V. Schally, D. Hawes, S. Xiong, L. Fazli, M. Gleave, et al., Expression of receptors for luteinizing hormone-releasing hormone (LH-RH) in prostate cancers following therapy with LH-RH agonists. Clin. Cancer Res. **16**(18), 4675–4680 (2010)

128. J.K. Pinski, S. Xiong, Q. Wang, F. Stanczyk, S. Liu, Effect of luteinizing hormone on the steroid biosynthesis pathway in prostate cancer. 2010 Genitourinary Cancers Symposium 2010

129. L. Xiao, Y. Wang, K. Xu, H. Hu, Z. Xu, D. Wu, et al., Nuclear receptor LRH-1 functions to promote castration-resistant growth of prostate cancer via its promotion of intratumoral androgen biosynthesis. Cancer Res. **78**(9), 2205–2218 (2018)

130. F.W. Buaas, J.R. Gardiner, S. Clayton, P. Val, A. Swain, In vivo evidence for the crucial role of SF1 in steroid-producing cells of the testis, ovary and adrenal gland. Development **139**(24), 4561–4570 (2012)

131. S.R. Lewis, C.J. Hedman, T. Ziegler, W.A. Ricke, J.S. Jorgensen, Steroidogenic factor 1 promotes aggressive growth of castration-resistant prostate cancer cells by stimulating steroid synthesis and cell proliferation. Endocrinology **155**(2), 358–369 (2014)

132. P.R. Braadland, H.H. Grytli, H. Ramberg, B. Katz, R. Kellman, L. Gauthier-Landry, et al., Low beta(2)-adrenergic receptor level may promote development of castration resistant prostate cancer and altered steroid metabolism. Oncotarget **7**(2), 1878–1894 (2016)

133. D. Patel, A.E. Knowell, M. Korang-Yeboah, P. Sharma, J. Joshi, S. Glymph, et al., Inhibitor of differentiation 4 (ID4) inactivation promotes de novo steroidogenesis and castration-resistant prostate cancer. Mol. Endocrinol. **28**(8), 1239–1253 (2014)

134. H. Ramberg, T. Eide, K.A. Krobert, F.O. Levy, N. Dizeyi, A.S. Bjartell, et al., Hormonal regulation of beta2-adrenergic receptor level in prostate cancer. Prostate **68**(10), 1133–1142 (2008)

135. J.B. Joshi, D. Patel, D.J. Morton, P. Sharma, J. Zou, D. Hewa Bostanthirige, et al., Inactivation of ID4 promotes a CRPC phenotype with constitutive AR activation through FKBP52. Mol. Oncol. **11**(4), 337–357 (2017)

136. T. Migita, K.I. Takayama, T. Urano, D. Obinata, K. Ikeda, T. Soga, et al., ACSL3 promotes intratumoral steroidogenesis in prostate cancer cells. Cancer Sci. **108**(10), 2011–2021 (2017)

137. J.T. Arnold, N.E. Gray, K. Jacobowitz, L. Viswanathan, P.W. Cheung, K.K. McFann, et al., Human prostate stromal cells stimulate increased PSA production in DHEA-treated prostate cancer epithelial cells. J. Steroid Biochem. Mol. Biol. **111**(3-5), 240–246 (2008)

138. A. Mizokami, E. Koh, K. Izumi, K. Narimoto, M. Takeda, S. Honma, et al., Prostate cancer stromal cells and LNCaP cells coordinately activate the androgen receptor through synthesis of testosterone and dihydrotestosterone from dehydroepiandrosterone. Endocr. Relat. Cancer **16**(4), 1139–1155 (2009)

139. T. Sillat, R. Pöllänen, J.R. Lopes, P. Porola, G. Ma, M. Korhonen, et al., Intracrine androgenic apparatus in human bone marrow stromal cells. J. Cell. Mol. Med. **13**(9B), 3296–3302 (2009)

140. I. Finco, C.R. LaPensee, K.T. Krill, G.D. Hammer, Hedgehog signaling and steroidogenesis. Annu. Rev. Physiol. **77**, 105–129 (2015)

141. E. Levina, M. Chen, R. Carkner, M. Shtutman, R. Buttyan, Paracrine Hedgehog increases the steroidogenic potential of prostate stromal cells in a Gli-dependent manner. Prostate **72**(8), 817–824 (2012)

142. A.A. Lubik, M. Nouri, S. Truong, M. Ghaffari, H.H. Adomat, E. Corey, et al., Paracrine sonic hedgehog signaling contributes significantly to acquired steroidogenesis in the prostate tumor microenvironment. Int. J. Cancer **140**(2), 358–369 (2017)

143. E. Jernberg, E. Thysell, E. Bovinder Ylitalo, S. Rudolfsson, S. Crnalic, A. Widmark, et al., Characterization of prostate cancer bone metastases according to expression levels of steroidogenic enzymes and androgen receptor splice variants. PLoS One **8**(11), e77407 (2013)

144. M. Hagberg Thulin, M.E. Nilsson, P. Thulin, J. Ceraline, C. Ohlsson, J.E. Damber, et al., Osteoblasts promote castration-resistant prostate cancer by altering intratumoral steroidogenesis. Mol. Cell. Endocrinol. **422**, 182–191 (2016)

145. M.C. Diaz-Franco, R. Franco-Diaz de Leon, J.R. Villafan-Bernal, OsteocalcinGPRC6A: An update of its clinical and biological multiorganic interactions (Review). Mol. Med. Rep. **19**(1), 15–22 (2019)

146. S.C. Moser, B.C.J. van der Eerden, Osteocalcin-A versatile bone-derived hormone. Front. Endocrinol. **9**, 794 (2018)

147. L. De Toni, A. Di Nisio, M.S. Rocca, M. De Rocco Ponce, A. Ferlin, C. Foresta, Osteocalcin, a bone-derived hormone with important andrological implications. Andrology **5**(4), 664–670 (2017)

148. F. Oury, M. Ferron, W. Huizhen, C. Confavreux, L. Xu, J. Lacombe, et al., Osteocalcin regulates murine and human fertility through a pancreas-bone-testis axis. J. Clin. Invest. **123**(6), 2421–2433 (2013)

149. G. Karsenty, F. Oury, Regulation of male fertility by the bone-derived hormone osteocalcin. Mol. Cell. Endocrinol. **382**(1), 521–526 (2014)

150. P. Thomas, Membrane androgen receptors unrelated to nuclear steroid receptors. Endocrinology **160**, 772 (2019)

151. M. Pi, L.D. Quarles, GPRC6A regulates prostate cancer progression. Prostate **72**(4), 399–409 (2012)

152. Q.Z. Long, Y.F. Du, X.Y. Ding, X. Li, W.B. Song, Y. Yang, et al., Replication and fine mapping for association of the C2orf43, FOXP4, GPRC6A and RFX6 genes with prostate cancer in the Chinese population. PLoS One **7**(5), e37866 (2012)

153. R. Ye, M. Pi, J.V. Cox, S.K. Nishimoto, L.D. Quarles, CRISPR/Cas9 targeting of GPRC6A suppresses prostate cancer tumorigenesis in a human xenograft model. J. Exp. Clin. Cancer Res. **36**(1), 90 (2017)

154. Y. Yang, Y. Bai, Y. He, Y. Zhao, J. Chen, L. Ma, et al., PTEN loss promotes intratumoral androgen synthesis and tumor microenvironment remodeling via aberrant activation of RUNX2 in castration-resistant prostate cancer. Clin. Cancer Res. **24**(4), 834–846 (2018)

155. W.C. Huang, Z. Xie, H. Konaka, J. Sodek, H.E. Zhau, L.W. Chung, Human osteocalcin and bone sialoprotein mediating osteomimicry of prostate cancer cells: role of cAMP-dependent protein kinase A signaling pathway. Cancer Res. **65**(6), 2303–2313 (2005)

156. M. Hagberg Thulin, K. Jennbacken, J.E. Damber, K. Welen, Osteoblasts stimulate the osteogenic and metastatic progression of castration-resistant prostate cancer in a novel model for in vitro and in vivo studies. Clin. Exp. Metastasis **31**(3), 269–283 (2014)

157. M.W. O'Reilly, P. Kempegowda, C. Jenkinson, A.E. Taylor, J.L. Quanson, K.H. Storbeck, et al., 11-Oxygenated C19 steroids are the predominant androgens in polycystic ovary syndrome. J. Clin. Endocrinol. Metab. **102**(3), 840–848 (2017)

158. E. Pretorius, W. Arlt, K.H. Storbeck, A new dawn for androgens: Novel lessons from 11-oxygenated C19 steroids. Mol. Cell. Endocrinol. **441**, 76–85 (2017)

159. A.T. Nanba, J. Rege, J. Ren, R.J. Auchus, W.E. Rainey, A.F. Turcu, 11-Oxygenated C19 steroids do not decline with age in women. J. Clin. Endocrinol. Metab. **104**, 2615 (2019)

160. J.W. Goldzieher, A. de la Pena, M.M. Aivaliotis, Radioimmunoassay of plasma androstenedione, testosterone and 11beta-hydroxyandrostenedione after chromatography on Lipidex-5000 (hydroxyalkoxypropyl Sephadex). J. Steroid Biochem. **9**(2), 169–173 (1978)

161. Y. Xing, M.A. Edwards, C. Ahlem, M. Kennedy, A. Cohen, C.E. Gomez-Sanchez, et al., The effects of ACTH on steroid metabolomic profiles in human adrenal cells. J. Endocrinol. **209**(3), 327–335 (2011)

162. L. Schloms, K.H. Storbeck, P. Swart, W.C. Gelderblom, A.C. Swart, The influence of Aspalathus linearis (Rooibos) and dihydrochalcones on adrenal steroidogenesis: quantification of steroid intermediates and end products in H295R cells. J. Steroid Biochem. Mol. Biol. **128**(3-5), 128–138 (2012)

163. A.F. Turcu, A.T. Nanba, R. Chomic, S.K. Upadhyay, T.J. Giordano, J.J. Shields, et al., Adrenal-derived 11-oxygenated 19-carbon steroids are the dominant androgens in classic 21-hydroxylase deficiency. Eur. J. Endocrinol. **174**(5), 601–609 (2016)

164. K.H. Storbeck, L.M. Bloem, D. Africander, L. Schloms, P. Swart, A.C. Swart, 11beta-Hydroxydihydrotestosterone and 11-ketodihydrotestosterone, novel C19 steroids with androgenic activity: a putative role in castration resistant prostate cancer? Mol. Cell. Endocrinol. **377**(1-2), 135–146 (2013)

165. P.J. Robinson, R.J. Bell, S.R. Davis, A.F. Turcu, Exogenous testosterone does not influence 11-oxygenated C19 steroid concentrations in healthy postmenopausal women. J. Endoc. Soc. **3**(3), 670–677 (2019)

166. C. Campana, J. Rege, A.F. Turcu, V. Pezzi, C.E. Gomez-Sanchez, D.M. Robins, et al., Development of a novel cell based androgen screening model. J. Steroid Biochem. Mol. Biol. **156**, 17–22 (2016)

167. E. Pretorius, D.J. Africander, M. Vlok, M.S. Perkins, J. Quanson, K.H. Storbeck, 11-Ketotestosterone and 11-ketodihydrotestosterone in castration resistant prostate cancer: potent androgens which can no longer be ignored. PLoS One **11**(7), e0159867 (2016)

168. A. Dovio, M.L. Sartori, S. De Francia, S. Mussino, P. Perotti, L. Saba, et al., Differential expression of determinants of glucocorticoid sensitivity in androgen-dependent and androgen-independent human prostate cancer cell lines. J. Steroid Biochem. Mol. Biol. **116**(1-2), 29–36 (2009)

169. N. Page, N. Warriar, M.V. Govindan, 11 beta-Hydroxysteroid dehydrogenase and tissue specificity of androgen action in human prostate cancer cell LNCaP. J. Steroid Biochem. Mol. Biol. **49**(2-3), 173–181 (1994)

170. G. Pelletier, V. Luu-The, S. Li, J. Ouellet, F. Labrie, Cellular localization of mRNA expression of enzymes involved in the formation and inactivation of hormonal steroids in the mouse prostate. J. Histochem. Cytochem. **52**(10), 1351–1356 (2004). https://doi.org/10.1369/jhc.4A6311.2004

171. A.L. Albiston, V.R. Obeyesekere, R.E. Smith, Z.S. Krozowski, Cloning and tissue distribution of the human 11 beta-hydroxysteroid dehydrogenase type 2 enzyme. Mol. Cell. Endocrinol. **105**(2), R11–R17 (1994)

172. T. du Toit, L.M. Bloem, J.L. Quanson, R. Ehlers, A.M. Serafin, A.C. Swart, Profiling adrenal 11beta-hydroxyandrostenedione metabolites in prostate cancer cells, tissue and plasma: UPC(2)-MS/MS quantification of 11beta-hydroxytestosterone, 11keto-testosterone and 11keto-dihydrotestosterone. J. Steroid Biochem. Mol. Biol. **166**, 54–67 (2017)

173. T. du Toit, A.C. Swart, Inefficient UGT-conjugation of adrenal 11beta-hydroxyandrostenedione metabolites highlights C11-oxy C19 steroids as the predominant androgens in prostate cancer. Mol. Cell. Endocrinol. **461**, 265 (2017)

174. C.J. Ryan, M.R. Smith, L. Fong, J.E. Rosenberg, P. Kantoff, F. Raynaud, et al., Phase I clinical trial of the CYP17 inhibitor abiraterone acetate demon-

strating clinical activity in patients with castration-resistant prostate cancer who received prior ketoconazole therapy. J. Clin. Oncol. **28**(9), 1481–1488 (2010)

175. G. Attard, A.H. Reid, T.A. Yap, F. Raynaud, M. Dowsett, S. Settatree, et al., Phase I clinical trial of a selective inhibitor of CYP17, abiraterone acetate, confirms that castration-resistant prostate cancer commonly remains hormone driven. J. Clin. Oncol. **26**(28), 4563–4571 (2008)

176. E. Cho, E.A. Mostaghel, K.J. Russell, J.J. Liao, M.A. Konodi, B.F. Kurland, et al., External beam radiation therapy and abiraterone in men with localized prostate cancer: safety and effect on tissue androgens. Int. J. Radiat. Oncol. Biol. Phys. **92**(2), 236–243 (2015)

177. S. Gao, H. Ye, S. Gerrin, H. Wang, A. Sharma, S. Chen, et al., ErbB2 signaling increases androgen receptor expression in abiraterone-resistant prostate cancer. Clin. Cancer Res. **22**(14), 3672–3682 (2016)

178. H.M. Lam, R. McMullin, H.M. Nguyen, I. Coleman, M. Gormley, R. Gulati, et al., Characterization of an abiraterone ultraresponsive phenotype in castration-resistant prostate cancer patient-derived xenografts. Clin. Cancer Res. **23**(9), 2301–2312 (2017)

179. Z. Yu, S. Chen, A.G. Sowalsky, O.S. Voznesensky, E.A. Mostaghel, P.S. Nelson, et al., Rapid induction of androgen receptor splice variants by androgen deprivation in prostate cancer. Clin. Cancer Res. **20**(6), 1590–1600 (2014)

180. W. Kim, L. Zhang, J.H. Wilton, G. Fetterly, J.L. Mohler, V. Weinberg, et al., Sequential use of the androgen synthesis inhibitors ketoconazole and abiraterone acetate in castration-resistant prostate cancer and the predictive value of circulating androgens. Clin. Cancer Res. **20**(24), 6269–6276 (2014)

181. C.J. Ryan, A. Molina, J. Li, T. Kheoh, E.J. Small, C.M. Haqq, et al., Serum androgens as prognostic biomarkers in castration-resistant prostate cancer: results from an analysis of a randomized phase III trial. J. Clin. Oncol. **31**(22), 2791–2798 (2013)

182. C.J. Ryan, W. Peng, T. Kheoh, E. Welkowsky, C.M. Haqq, D.W. Chandler, et al., Androgen dynamics and serum PSA in patients treated with abiraterone acetate. Prostate Cancer Prostatic Dis. **17**(2), 192–198 (2014)

183. D. Tamae, E. Mostaghel, B. Montgomery, P.S. Nelson, S.P. Balk, P.W. Kantoff, et al., The DHEA-sulfate depot following P450c17 inhibition supports the case for AKR1C3 inhibition in high risk localized and advanced castration resistant prostate cancer. Chem. Biol. Interact. **234**, 332–338 (2015)

184. E.S. Antonarakis, C. Lu, B. Luber, H. Wang, Y. Chen, Y. Zhu, et al., Clinical significance of androgen receptor splice variant-7 mRNA detection in circulating tumor cells of men with metastatic castration-resistant prostate cancer treated with first- and second-line abiraterone and enzalutamide. J. Clin. Oncol. **35**(19), 2149–2156 (2017)

185. M. Kohli, Y. Ho, D.W. Hillman, J.L. Van Etten, C. Henzler, R. Yang, et al., Androgen receptor variant AR-V9 is coexpressed with AR-V7 in prostate cancer metastases and predicts abiraterone resistance. Clin. Cancer Res. **23**(16), 4704–4715 (2017)

186. M. Nakazawa, C. Lu, Y. Chen, C.J. Paller, M.A. Carducci, M.A. Eisenberger, et al., Serial blood-based analysis of AR-V7 in men with advanced prostate cancer. Ann. Oncol. **26**(9), 1859–1865 (2015)

187. E.J. Chen, A.G. Sowalsky, S. Gao, C. Cai, O. Voznesensky, R. Schaefer, et al., Abiraterone treatment in castration-resistant prostate cancer selects for progesterone responsive mutant androgen receptors. Clin. Cancer Res. **21**(6), 1273–1280 (2015)

188. Z. Li, A.C. Bishop, M. Alyamani, J.A. Garcia, R. Dreicer, D. Bunch, et al., Conversion of abiraterone to D4A drives anti-tumour activity in prostate cancer. Nature **523**(7560), 347–351 (2015)

189. Z. Li, M. Alyamani, J. Li, K. Rogacki, M. Abazeed, S.K. Upadhyay, et al., Redirecting abiraterone metabolism to fine-tune prostate cancer anti-androgen therapy. Nature **533**(7604), 547–551 (2016)

190. M. Alyamani, Z. Li, M. Berk, J. Li, J. Tang, S. Upadhyay, et al., Steroidogenic metabolism of galeterone reveals a diversity of biochemical activities. Cell Chem. Biol. **24**(7), 825–32.e6 (2017)

191. J.W. Mueller, L.C. Gilligan, J. Idkowiak, W. Arlt, P.A. Foster, The regulation of steroid action by sulfation and desulfation. Endocr. Rev. **36**(5), 526–563 (2015)

192. B.V.L. Potter, SULFATION PATHWAYS: Steroid sulphatase inhibition via aryl sulphamates: clinical progress, mechanism and future prospects. J. Mol. Endocrinol. **61**(2), T233–Tt52 (2018)

193. S. Denmeade, D. George, G. Liu, C. Peraire, A. Geniaux, F. Baton, et al., A phase I pharmacodynamics dose escalation study of steroid sulphatase inhibitor Irosustat in patients with prostate cancer. Eur. J. Cancer **47**, S499 (2011)

194. E.A. Mostaghel, Abiraterone in the treatment of metastatic castration-resistant prostate cancer. Cancer Manag. Res. **6**, 39–51 (2014)

195. H.I. Scher, K. Fizazi, F. Saad, M.E. Taplin, C.N. Sternberg, K. Miller, et al., Increased survival with enzalutamide in prostate cancer after chemotherapy. N. Engl. J. Med. **367**(13), 1187–1197 (2012)

196. R.S. Ge, Q. Dong, E.M. Niu, C.M. Sottas, D.O. Hardy, J.F. Catterall, et al., 11{beta}-Hydroxysteroid dehydrogenase 2 in rat leydig cells: its role in blunting glucocorticoid action at physiological levels of substrate. Endocrinology **146**(6), 2657–2664 (2005)

197. S.A. Latif, H.A. Pardo, M.P. Hardy, D.J. Morris, Endogenous selective inhibitors of 11beta-hydroxysteroid dehydrogenase isoforms 1 and 2

of adrenal origin. Mol. Cell. Endocrinol. **243**(1-2), 43–50 (2005)

198. F.B. Coeli, L.F. Ferraz, S.H. Lemos-Marini, S.Z. Rigatto, V.M. Belangero, M.P. de-Mello, Apparent mineralocorticoid excess syndrome in a Brazilian boy caused by the homozygous missense mutation p.R186C in the HSD11B2 gene. Arq. Bras. Endocrinol. Metabol. **52**(8), 1277–1281 (2008)

199. J. Li, M. Alyamani, A. Zhang, K.H. Chang, M. Berk, Z. Li, et al., Aberrant corticosteroid metabolism in tumor cells enables GR takeover in enzalutamide resistant prostate cancer. Elife **6**, e20183 (2017)

200. Q. Hu, C. Jagusch, U.E. Hille, J. Haupenthal, R.W. Hartmann, Replacement of imidazolyl by pyridyl in biphenylmethylenes results in selective CYP17 and dual CYP17/CYP11B1 inhibitors for the treatment of prostate cancer. J. Med. Chem. **53**(15), 5749–5758 (2010)

201. M.A. Pinto-Bazurco Mendieta, Q. Hu, M. Engel, R.W. Hartmann, Highly potent and selective nonsteroidal dual inhibitors of CYP17/CYP11B2 for the treatment of prostate cancer to reduce risks of cardiovascular diseases. J. Med. Chem. **56**(15), 6101–6107 (2013)

202. A.O. Adeniji, M. Chen, T.M. Penning, AKR1C3 as a target in castrate resistant prostate cancer. J. Steroid Biochem. Mol. Biol. **137**, 136–149 (2013)

203. T.M. Penning, AKR1C3 (type 5 17beta-hydroxysteroid dehydrogenase/prostaglandin F synthase): Roles in malignancy and endocrine disorders. Mol. Cell. Endocrinol. **489**, 82 (2018)

204. C. Liu, W. Lou, Y. Zhu, J.C. Yang, N. Nadiminty, N.W. Gaikwad, et al., Intracrine androgens and AKR1C3 activation confer resistance to enzalutamide in prostate cancer. Cancer Res. **75**(7), 1413–1422 (2015)

205. C. Liu, C.M. Armstrong, W. Lou, A. Lombard, C.P. Evans, A.C. Gao, Inhibition of AKR1C3 activation overcomes resistance to abiraterone in advanced prostate cancer. Mol. Cancer Ther. **16**(1), 35–44 (2017)

206. T.M. Penning, M.E. Burczynski, J.M. Jez, C.F. Hung, H.K. Lin, H. Ma, et al., Human 3alpha-hydroxysteroid dehydrogenase isoforms (AKR1C1-AKR1C4) of the aldo-keto reductase superfamily: functional plasticity and tissue distribution reveals roles in the inactivation and formation of male and female sex hormones. Biochem. J. **351**(Pt 1), 67–77 (2000)

207. T.M. Penning, Aldo-Keto reductase (AKR) 1C3 inhibitors: a patent review. Expert Opin. Ther. Pat. **27**(12), 1329–1340 (2017)

208. A.J. Liedtke, A.O. Adeniji, M. Chen, M.C. Byrns, Y. Jin, D.W. Christianson, et al., Development of potent and selective indomethacin analogues for the inhibition of AKR1C3 (Type 5 17beta-hydroxysteroid dehydrogenase/prostaglandin F synthase) in castrate-resistant prostate cancer. J. Med. Chem. **56**(6), 2429–2446 (2013)

209. A.L. Lovering, J.P. Ride, C.M. Bunce, J.C. Desmond, S.M. Cummings, S.A. White, Crystal structures of prostaglandin D(2) 11-ketoreductase (AKR1C3) in complex with the nonsteroidal anti-inflammatory drugs flufenamic acid and indomethacin. Cancer Res. **64**(5), 1802–1810 (2004)

210. A.C. Pippione, A. Giraudo, D. Bonanni, I.M. Carnovale, E. Marini, C. Cena, et al., Hydroxytriazole derivatives as potent and selective aldo-keto reductase 1C3 (AKR1C3) inhibitors discovered by bioisosteric scaffold hopping approach. Eur. J. Med. Chem. **139**, 936–946 (2017)

211. A.C. Pippione, I.M. Carnovale, D. Bonanni, M. Sini, P. Goyal, E. Marini, et al., Potent and selective aldo-keto reductase 1C3 (AKR1C3) inhibitors based on the benzoisoxazole moiety: application of a bioisosteric scaffold hopping approach to flufenamic acid. Eur. J. Med. Chem. **150**, 930–945 (2018)

212. K. Verma, N. Gupta, T. Zang, P. Wangtrakluldee, S.K. Srivastava, T.M. Penning, et al., AKR1C3 inhibitor KV-37 exhibits antineoplastic effects and potentiates enzalutamide in combination therapy in prostate adenocarcinoma cells. Mol. Cancer Ther. **17**(9), 1833–1845 (2018)

213. P. Wangtrakuldee, A.O. Adeniji, T. Zang, L. Duan, B. Khatri, B.M. Twenter, et al., A 3-(4-nitronaphthen-1-yl) amino-benzoate analog as a bifunctional AKR1C3 inhibitor and AR antagonist: Head to head comparison with other advanced AKR1C3 targeted therapeutics. J. Steroid Biochem. Mol. Biol. **192**, 105283 (2019)

214. D. Robinson, E.M. Van Allen, Y.M. Wu, N. Schultz, R.J. Lonigro, J.M. Mosquera, et al., Integrative clinical genomics of advanced prostate cancer. Cell **161**(5), 1215–1228 (2015)

215. H.H. Cheng, N. Klemfuss, B. Montgomery, C.S. Higano, M.T. Schweizer, E.A. Mostaghel, et al., A pilot study of clinical targeted next generation sequencing for prostate cancer: consequences for treatment and genetic counseling. Prostate **76**(14), 1303–1311 (2016)

216. C.C. Pritchard, J. Mateo, M.F. Walsh, N. De Sarkar, W. Abida, H. Beltran, et al., Inherited DNA-repair gene mutations in men with metastatic prostate cancer. N. Engl. J. Med. **375**(5), 443–453 (2016)

217. P. Nuhn, J.S. De Bono, K. Fizazi, S.J. Freedland, M. Grilli, P.W. Kantoff, et al., Update on systemic prostate cancer therapies: management of metastatic castration-resistant prostate cancer in the era of precision oncology. Eur. Urol. **75**(1), 88–99 (2019)

218. A.W. Hahn, D.M. Gill, R.H. Nussenzveig, A. Poole, J. Farnham, L. Cannon-Albright, et al., Germline variant in HSD3B1 (1245 A > C) and response to abiraterone acetate plus prednisone in men with new-onset metastatic castration-resistant prostate cancer. Clin. Genitourin. Cancer **16**(4), 288–292 (2018)

219. M. Alyamani, H. Emamekhoo, S. Park, J. Taylor, N. Almassi, S. Upadhyay, et al., HSD3B1(1245A>C)

219. variant regulates dueling abiraterone metabolite effects in prostate cancer. J. Clin. Invest. **128**(8), 3333–3340 (2018)
220. N. Almassi, C. Reichard, J. Li, C. Russell, J. Perry, C.J. Ryan, et al., HSD3B1 and response to a non-steroidal CYP17A1 inhibitor in castration-resistant prostate cancer. JAMA Oncol. **4**(4), 554–557 (2018)
221. A.W. Hahn, D.M. Gill, A. Poole, R.H. Nussenzveig, S. Wilson, J.M. Farnham, et al., Germline variant in SLCO2B1 and response to abiraterone acetate plus prednisone (AA) in new-onset metastatic castration-resistant prostate cancer (mCRPC). Mol. Cancer Ther. **18**(3), 726–729 (2019)
222. E.A. Mostaghel, E. Cho, A. Zhang, M. Alyamani, A. Kaipainen, S. Green, et al., Association of tissue abiraterone levels and SLCO genotype with intraprostatic steroids and pathologic response in men with high-risk localized prostate cancer. Clin. Cancer Res. **23**(16), 4592–4601 (2017)
223. A. Romanel, D. Gasi Tandefelt, V. Conteduca, A. Jayaram, N. Casiraghi, D. Wetterskog, et al., Plasma AR and abiraterone-resistant prostate cancer. Sci. Transl. Med. **7**(312), 312re10 (2015)
224. V. Conteduca, D. Wetterskog, M.T.A. Sharabiani, E. Grande, M.P. Fernandez-Perez, A. Jayaram, et al., Androgen receptor gene status in plasma DNA associates with worse outcome on enzalutamide or abiraterone for castration-resistant prostate cancer: a multi-institution correlative biomarker study. Annals of oncology: official journal of the European Society for. Med. Oncol. **28**(7), 1508–1516 (2017)
225. M. Annala, G. Vandekerkhove, D. Khalaf, S. Taavitsainen, K. Beja, E.W. Warner, et al., Circulating tumor DNA genomics correlate with resistance to abiraterone and enzalutamide in prostate cancer. Cancer Discov. **8**(4), 444–457 (2018)
226. B. De Laere, S. Oeyen, M. Mayrhofer, T. Whitington, P.J. van Dam, P. Van Oyen, et al., TP53 outperforms other androgen receptor biomarkers to predict abiraterone or enzalutamide outcome in metastatic castration-resistant prostate cancer. Clin. Cancer Res. **25**(6), 1766–1773 (2019)
227. B.L. Maughan, L.B. Guedes, K. Boucher, G. Rajoria, Z. Liu, S. Klimek, et al., p53 status in the primary tumor predicts efficacy of subsequent abiraterone and enzalutamide in castration-resistant prostate cancer. Prostate Cancer Prostatic Dis. **21**(2), 260–268 (2018)
228. R. Ferraldeschi, D. Nava Rodrigues, R. Riisnaes, S. Miranda, I. Figueiredo, P. Rescigno, et al., PTEN protein loss and clinical outcome from castration-resistant prostate cancer treated with abiraterone acetate. Eur. Urol. **67**(4), 795–802 (2015)
229. G. Boysen, D.N. Rodrigues, P. Rescigno, G. Seed, D. Dolling, R. Riisnaes, et al., SPOP-Mutated/CHD1-deleted lethal prostate cancer and abiraterone sensitivity. Clin. Cancer Res. **24**(22), 5585–5593 (2018)
230. E. Cho, R.B. Montgomery, E.A. Mostaghel, Minireview: SLCO and ABC transporters: a role for steroid transport in prostate cancer progression. Endocrinology **155**(11), 4124–4132 (2014)
231. X. Wang, L.C. Harshman, W. Xie, M. Nakabayashi, F. Qu, M.M. Pomerantz, et al., Association of SLCO2B1 genotypes with time to progression and overall survival in patients receiving androgen-deprivation therapy for prostate cancer. J. Clin. Oncol. **34**(4), 352–359 (2016)
232. T. Terakawa, E. Katsuta, L. Yan, N. Turaga, K.-A. McDonald, M. Fujisawa, et al., High expression of SLCO2B1 is associated with prostate cancer recurrence after radical prostatectomy. Oncotarget **9**, 14207 (2018)
233. M. Yang, W. Xie, E. Mostaghel, M. Nakabayashi, L. Werner, T. Sun, et al., SLCO2B1 and SLCO1B3 may determine time to progression for patients receiving androgen deprivation therapy for prostate cancer. J. Clin. Oncol. **29**(18), 2565–2573 (2011)
234. J.W.D. Hearn, G. AbuAli, C.A. Reichard, C.A. Reddy, C. Magi-Galluzzi, K.H. Chang, et al., HSD3B1 and resistance to androgen-deprivation therapy in prostate cancer: a retrospective, multicohort study. Lancet Oncol. **17**(10), 1435–1444 (2016)
235. J.W.D. Hearn, W. Xie, M. Nakabayashi, N. Almassi, C.A. Reichard, M. Pomerantz, et al., Association of HSD3B1 genotype with response to androgen-deprivation therapy for biochemical recurrence after radiotherapy for localized prostate cancer. JAMA Oncol. **4**(4), 558–562 (2018)
236. G. Wu, S. Huang, K.L. Nastiuk, J. Li, J. Gu, M. Wu, et al., Variant allele of HSD3B1 increases progression to castration-resistant prostate cancer. Prostate **75**(7), 777–782 (2015)
237. D. Hettel, N. Sharifi, HSD3B1 status as a biomarker of androgen deprivation resistance and implications for prostate cancer. Nat. Rev. Urol. **15**(3), 191–196 (2018)
238. K.A. Jung, B.H. Choi, C.W. Nam, M. Song, S.T. Kim, J.Y. Lee, et al., Identification of aldo-keto reductases as NRF2-target marker genes in human cells. Toxicol. Lett. **218**(1), 39–49 (2013)
239. M.E. Burczynski, G.R. Sridhar, N.T. Palackal, T.M. Penning, The reactive oxygen species--and Michael acceptor-inducible human aldo-keto reductase AKR1C1 reduces the alpha,beta-unsaturated aldehyde 4-hydroxy-2-nonenal to 1,4-dihydroxy-2-nonene. J. Biol. Chem. **276**(4), 2890–2897 (2001)
240. K. Matsuura, H. Shiraishi, A. Hara, K. Sato, Y. Deyashiki, M. Ninomiya, et al., Identification of a principal mRNA species for human 3alpha-hydroxysteroid dehydrogenase isoform (AKR1C3) that exhibits high prostaglandin D2 11-ketoreductase activity. J. Biochem. **124**(5), 940–946 (1998)
241. T.M. Penning, M.C. Byrns, Steroid hormone transforming aldo-keto reductases and cancer. Ann. N. Y. Acad. Sci. **1155**, 33–42 (2009)
242. K.H. Huang, S.H. Chiou, K.C. Chow, T.Y. Lin, H.W. Chang, I.P. Chiang, et al., Overexpression of aldo-keto reductase 1C2 is associated with dis-

ease progression in patients with prostatic cancer. Histopathology 57(3), 384–394 (2010)

243. L. Fan, G. Peng, A. Hussain, L. Fazli, E. Guns, M. Gleave, et al., The steroidogenic enzyme AKR1C3 regulates stability of the ubiquitin ligase siah2 in prostate cancer cells. J. Biol. Chem. 290(34), 20865–20879 (2015)

244. C.L. Doig, S. Battaglia, F.L. Khanim, C.M. Bunce, M.J. Campbell, Knockdown of AKR1C3 exposes a potential epigenetic susceptibility in prostate cancer cells. J. Steroid Biochem. Mol. Biol. 155(Pt A), 47–55 (2016)

245. H. Li, N. Xie, R. Chen, M. Verreault, L. Fazli, M.E. Gleave, et al., UGT2B17 expedites progression of castration-resistant prostate cancers by promoting ligand-independent AR signaling. Cancer Res. 76(22), 6701–6711 (2016)

246. M. Gruber, J. Bellemare, G. Hoermann, A. Gleiss, E. Porpaczy, M. Bilban, et al., Overexpression of uridine diphospho glucuronosyltransferase 2B17 in high-risk chronic lymphocytic leukemia. Blood 121(7), 1175–1183 (2013)

247. H. Hirata, Y. Hinoda, M.S. Zaman, Y. Chen, K. Ueno, S. Majid, et al., Function of UDP-glucuronosyltransferase 2B17 (UGT2B17) is involved in endometrial cancer. Carcinogenesis 31(9), 1620–1626 (2010)

248. L. Gauthier-Landry, A. Belanger, O. Barbier, Multiple roles for UDP-glucuronosyltransferase (UGT)2B15 and UGT2B17 enzymes in androgen metabolism and prostate cancer evolution. J. Steroid Biochem. Mol. Biol. 145, 187–192 (2015)

249. B.Y. Bao, B.F. Chuang, Q. Wang, O. Sartor, S.P. Balk, M. Brown, et al., Androgen receptor mediates the expression of UDP-glucuronosyltransferase 2 B15 and B17 genes. Prostate 68(8), 839–848 (2008)

250. K. Evaul, R. Li, M. Papari-Zareei, R.J. Auchus, N. Sharifi, 3beta-hydroxysteroid dehydrogenase is a possible pharmacological target in the treatment of castration-resistant prostate cancer. Endocrinology 151(8), 3514–3520 (2010)

251. R. Li, K. Evaul, K.K. Sharma, K.H. Chang, J. Yoshimoto, J. Liu, et al., Abiraterone inhibits 3beta-hydroxysteroid dehydrogenase: a rationale for increasing drug exposure in castration-resistant prostate cancer. Clin. Cancer Res. 18(13), 3571–3579 (2012)

252. J. Richards, A.C. Lim, C.W. Hay, A.E. Taylor, A. Wingate, K. Nowakowska, et al., Interactions of abiraterone, eplerenone, and prednisolone with wild-type and mutant androgen receptor: a rationale for increasing abiraterone exposure or combining with MDV3100. Cancer Res. 72(9), 2176–2182 (2012)

253. T.W. Friedlander, J.N. Graff, K. Zejnullahu, A. Anantharaman, L. Zhang, R. Paz, et al., High-dose abiraterone acetate in men with castration resistant prostate cancer. Clin. Genitourin. Cancer 15(6), 733–41 e1 (2017)

254. R.Z. Szmulewitz, C.J. Peer, A. Ibraheem, E. Martinez, M.F. Kozloff, B. Carthon, et al., Prospective international randomized phase II study of low-dose abiraterone with food versus standard dose abiraterone in castration-resistant prostate cancer. J. Clin. Oncol. 36(14), 1389–1395 (2018)

255. Y.C. Yang, C.A. Banuelos, N.R. Mawji, J. Wang, M. Kato, S. Haile, et al., Targeting androgen receptor activation function-1 with EPI to overcome resistance mechanisms in castration-resistant prostate cancer. Clin. Cancer Res. 22(17), 4466–4477 (2016)

256. J.K. Myung, C.A. Banuelos, J.G. Fernandez, N.R. Mawji, J. Wang, A.H. Tien, et al., An androgen receptor N-terminal domain antagonist for treating prostate cancer. J. Clin. Invest. 123(7), 2948–2960 (2013)

257. E. Antonopoulou, M. Ladomery, Targeting splicing in prostate cancer. Int. J. Mol. Sci. 19(5), 1287 (2018)

Germline and Somatic Defects in DNA Repair Pathways in Prostate Cancer

Sara Arce, Alejandro Athie, Colin C. Pritchard, and Joaquin Mateo

Abbreviations

ADT	Androgen deprivation therapy
alt-NHEJ	Alternative non-homologous end joining pathway
AP	Apurinic-apyrimidinic sites
AR	Androgen receptor
BER	Base Excision Repair
CRPC	Castration-resistant prostate cancer
DDR	DNA damage repair
DSB	Double strand break
ETS	E26 transformation-specific
GG-NER	Global genome NER pathways
HR	Homologous recombination, or Hazard Ratio
ICI	Immune checkpoint inhibitors
IR	Ionizing radiation
mCRPC	Metastatic castration resistant prostate cancer
MMR	Mismatch DNA Repair
NEPC	Neuroendocrine prostate cancer
NER	Nucleotide Excision Repair
NGS	Next-generation sequencing
NHEJ	Non-homologous end joining
PARP	Poly(ADP) ribose polymerase
PCF	Prostate Cancer Foundation
PFS	Progression-free survival
ROS	Reactive oxygen species
SNP	Single nucleotide polymorphism
SSB	Single strand break
SU2C	Stand-up-to-Cancer
TC-NER	Transcription-coupled NER
UV	Ultraviolet
UV-DDB	Ultraviolet damaged DNA-binding protein

S. Arce · A. Athie · J. Mateo (✉)
Vall d'Hebron Institute of Oncology,
Barcelona, Spain
e-mail: jmateo@vhio.net

C. C. Pritchard
University of Washington, Seattle, WA, USA

The DNA Damage Repair Machinery: An Overview

Cells are constantly exposed to different sources of damage, both endogenous (such as those resulting from normal metabolism, DNA replication and cell division) and exogenous (such as UV exposure, ionizing radiation or chemical agents) [1]. Such processes cause direct damage to DNA in our cells. To counteract the deleterious effects of such insults, all cells can activate a number of pathways responsible for repairing the damage and restoring genomic integrity [2]. These pathways are referred to as DNA damage repair (DDR) pathways.

Deficient DDR leads to accumulation of DNA damage, resulting in genomic instability, which is one of the hallmarks of cancer [3]. Depending on the exact type of damage, different signaling pathways are activated that (1) recognize damage

© Springer Nature Switzerland AG 2019
S. M. Dehm, D. J. Tindall (eds.), *Prostate Cancer*, Advances in Experimental Medicine and Biology 1210, https://doi.org/10.1007/978-3-030-32656-2_12

and prevent further cell replication, with cell cycle check-point promoting cell cycle arrest until damage is resolved and (2) initiate a signaling cascade that activates effector proteins to repair damaged DNA and restore genome integrity. If repair is not successful, then cells trigger programmed cell death responses to sacrifice themselves and prevent perpetuation of aberrant cells.

Damage to DNA may result in alteration of a single-DNA strand, referred to as a single-strand break (SSB) or both DNA strands, referred to as a double-strand break (DSB). In the case of a SSB, the code in the non-mutated strand is preserved and can be used as a template for repair. A more severe type of lesion affects both DNA stands, resulting in a DSB. A DSB can occur directly, or as the result of an unrepaired SSB, which occurs in proliferating cells when a SSB leads to the collapse of the DNA replication forks. If a DSB is not properly repaired, there is risk for intra- or inter-chromosomal rearrangements between broken ends of chromosomes.

SSBs are repaired primarily by two systems: Nucleotide Excision Repair (NER) and Base Excision Repair (BER). Sources of SSBs include UV exposure, high temperature, carcinogens and IR. Intracellular processes such as DNA replication and recombination can generate mispaired DNA bases that will be identified by the mismatch repair (MMR) mechanism. When a DSB appears, repair is mainly dependent on Homologous Recombination (HR) and Non-Homologous End Joining (NHEJ) pathways.

SSBs are normally generated by endogenous damage, such as that from reactive oxygen species (ROS). When a SSB is generated during the S phase of the cell cycle and is not correctly repaired, the SSB will progress to a DSB, and cells will activate the HR pathway. SSBs can also be generated in non-proliferating cells, leading to polymerase stalling during the transcription process.

The primary steps to repair a SSB include: DNA broken end detection, end processing, DNA gap filling and ligation. The NER pathway is involved in the repair of complex lesions such as pyrimidine dimers and crosslinks. Damage rec-

ognition can be performed by the transcription-coupled NER (TC-NER) and Global Genome NER pathways (GG-NER). Different sensor molecules are involved in this step such as UV-DDB and XPC in the case of GG-NER and Cockayne syndrome factors B (CSB) during TC-NER. After recognition, both pathways merge and RPA, together with XPA is recruited to the chromatin. Several molecules participate downstream such as XPF/ERCC1 endonuclease complex that will perform incisions $5'$ and $3'$ from the damaged sites. After that, other molecules, such as PCNA, RFC and RPA, will be recruited and bound to the site. Finally, the generated DNA gap will be filled by a DNA polymerase and ligases I and III [4, 5].

Genomic damage can also result from generation of apurinic-apyrimidinic (AP) sites, either in the form of a base loss or by enzymatic excision of a damaged base. In this case, APE1 acts as the main sensor. If SSB are directly generated, they are first recognized by PARP1, a sensor which binds the chromatin and recruits other DDR components such as XRCC1 and LIG3. After damage sensing, the end processing step is carried out by different proteins such as XRCC1, LIG3, PNKP and APTX, with some overlap among the pathways. This step generates a gap in the chromatin that needs to be filled with activities of other proteins such as PCNA. Finally, and depending on the length of the generated patch (short, 1 nt or long >2 nt), DNA ligation will be driven by either LIG3 or LIG1 respectively [6].

Notably, PARP1 also plays a role in recognition of DSBs. Upon damage, PARP1 promotes the alternative Non-Homologous End Joining (alt-NHEJ) pathway. When acting as DSBs sensor, PARP1 competes with Ku in binding the broken chromosomal ends [7]. PARP1 binding will trigger either HR or alt-NHEJ, depending on the presence of DSB resection. Moreover, PARP1 is relevant for the accumulation of the MRN complex to chromatin and has been proposed to facilitate activation of ATM substrates [2].

Cellular processes such as DNA replication and recombination can also induce generation of mis-paired bases, which are mainly repaired by the MMR pathway [8]. Mispaired bases can result from exposure to UV radiation, environmental carcinogens or DNA alkylators, leading to MMR activation. The majority of proteins that participate in this repair mechanism belong to the MSH and MLH family (i.e. MSH3, MSH2 and MSH6). These proteins are sensors that recognize the mismatched bases. Following recognition, PMS2 is recruited and, together with EXO1 and MLH1, the excision step is completed. Next, LIG1 and a DNA-pol are responsible for DNA re-synthesis and ligation.

The Homologous Recombination pathway, which is activated primarily by the generation of double-strand breaks during S/G2 phase, is an error-free mechanism of DNA repair, and hence the preferred system for DSB repair. HR is based on the use of the sister chromatid (complementary to the damaged strand) as a template for *de novo* synthesis of the broken ends, thus preserving the original sequence.

HR sensor proteins first recognize the damage. In humans, MRE11, RAD50 and NBS1, the building blocks of the MRN complex, are the main sensors of double strand breaks. Following damage recognition, the ATM and ATR kinases are recruited. ATM binds to a DSB and becomes activated by auto-phosphorylation, in parallel it activates other proteins downstream by phosphorylation. Additional ATM targets include histone H2AX, an indicator of DNA damage. The activation of these targets triggers a signaling cascade that involves effector and mediator proteins, such as CHEK2 [9]. In the case in which Replication Protein A (RPA) recognizes the damage, the ATR kinase gets recruited. CHEK1 is among the different ATR substrates [10]. Even though the first step on the HR pathway can be executed by two different sets of proteins it is important to consider that there is some degree of redundancy or overlap. For example, ATM kinase primarily activates CHEK2, but also phosphorylates CHEK1.

After broken end detection, BRCA1 is recruited to chromatin thus promoting end resection. Subsequently, RAD51, together with BRCA2 and PALB2 are involved in homology search and strand invasion, key steps of the HR pathway [2, 11]. After this, the sister chromatid is used as a template for new synthesis of the bro-

ken DNA; this final step will be completed by DNA ligases.

When double strand breaks are generated during the G1 phase of the cell cycle, cells will respond by activating the NHEJ pathway. Since NHEJ does not use the sister chromatid as a template, it is an error-prone mechanism. During G1 phase, cells try to rapidly ligate broken ends, resulting in inaccurate DNA repair. In the NHEJ cascade, the Ku70/80 heterodimer detects and binds to the broken ends. After recruitment, Ku70/80 activates the catalytic subunit of DNA-PKs, which auto-phosphorylates and phosphorylates other substrates. These kinases play an important role in the stabilization and protection of double strand breaks [12].

It is important to remark that in the NHEJ pathway, contrary to HR, there is no end resection step. A key protein of NHEJ is 53BP1, which hampers the end resection step. In the case of HR, BRCA1 inhibits 53BP1, thus enabling end resection during S/G2 phases [13]. After DNA-PK is activated, broken ends are processed by ARTEMIS. Finally, XRCC4 and LIG4 are recruited to the chromatin, and exert their functions on DNA ligation.

The Landscape of DNA Repair Gene Alterations in Prostate Cancer

Genomic Alterations in Prostate Cancer

Various alterations in the DNA sequence occur as a result of unrepaired damage e.g., mutations, small insertions or deletions (indels) and structural rearrangements (deletions, duplications, translocations and inversions; which can also result in big losses of DNA fragments), as well as deletions or amplifications of segments or whole chromosomes.

Mutations are alterations in the DNA sequence of bases. Sources of mutations include errors during DNA replication, exposure to radiation (UV or X-rays) and chemical agents among others. When only a single base pair is modified, it is called a point mutation. Depending on the effect that this point mutation has on protein coding, it can be classified as synonymous, which renders a change in the DNA but no effect on the encoded protein, or nonsynonymous (or missense) in which the amino acid sequence is altered, leading to the production of a different protein. Sometimes the altered amino acid sequence includes a stop codon, resulting in protein truncation (then the mutation is known as stop gain). A different type of mutation is the addition or deletion of a few bases in the DNA sequence, known commonly as indel (from the term insertion-deletion), which results in the DNA sequence of a gene being altered in size. Indels can cause change in the reading frame of the gene (frameshift mutations). Most of the loss-of-function mutations in DDR genes found in prostate cancer are frameshift mutations or point mutations resulting in a stop-gain.

DNA can be also altered by deletions or amplifications of a segment, which may result in an abnormal number of copies of one or more genes. These alterations, known as copy number variants (CNVs) affect genomic regions that can range in size from thousands to millions of bases, resulting in dosage changes of the genes within these regions. CNVs are of particular interest in prostate cancer, particularly amplifications of oncogenes, such as the androgen receptor (*AR*) gene or *MYC*, or by loss of critical tumor suppressor genes such as *PTEN*, *TP53*, *RB1* or *BRCA2*.

Structural chromosomal aberrations or gene fusions are the result of the inappropriate linkage of broken DNA ends. Gene fusions can result in recombinant proteins. In many tumors, including prostate cancer, recombinant proteins can act as cancer promoters. The canonical example in prostate cancer is the recurrent TMPRSS2-ERG fusion. Gene fusions can also result in the inactivation of tumor suppressor genes; for example, fusions with break points within *RB1*, *MSH2*, *MSH4* are examples of loss of protein function.

Besides direct changes to the DNA code, genes can also be affected by epigenetic modifications or gene transcription regulation. Broadly considered, epigenetics consists of chemical modifications, which do not affect the DNA

sequence itself, but result in either activation or inactivation of genes. The most common epigenetic changes are DNA methylation (generally associated with gene silencing) and histone acetylation (generally associated with gene activation). DNA hypermethylation alters expression of *GSTP1* and *MGMT*, two genes involved in prostate cancer DNA damage repair [14, 15]. Gene expression is also regulated by transcription factors. The canonical example of transcriptional regulation in prostate cancer is the role of the Androgen Receptor (AR). Prostate cancer is addicted to AR signaling, which drives a pro-oncogenic transcriptional program in cancer cells. Therapies targeting the AR are the mainstay of prostate cancer treatment. Interestingly, androgen deprivation therapy (which down-regulates AR signaling) enhances the cytotoxic effect of radiotherapy [16]. This observation might be explained by a role of AR in regulating DDR genes. As an hypothesis, radiotherapy induces DNA DSBs, resulting in DDR activation and also AR signaling induction. AR activation promotes the resolution of DNA DSBs, conferring resistance to radiotherapy. Based on this premise, the combination of radiotherapy and AR-inhibition (ADT) is synergistic by impairing double strand breaks repair, thus enhancing the effect of the former [17]. All of these observations suggest a role of AR as a regulator of the expression of DNA damage genes. For example, HR genes (*RAD54B*, *RAD51C*, *XRCC2* and *XCCR3*), NHEJ genes (*DNA-PKcs*, *Ku70*, *XRCC4*, *XRCC5* and *PRKDC*) and MMR-related genes (*MSH2* and *MSH6*) are regulated by AR. Moreover, PARP-1 can act as an AR cofactor.

Genomic Landscape of Localized Prostate Cancer

The Cancer Genome Atlas (TCGA) reported the genomic landscape of a cohort of 333 localized prostate cancer samples. Based on integration of exome DNA sequencing, RNA sequencing, miRNA sequencing, SNP arrays and DNA methylation arrays, seven distinct molecular subtypes were depicted. These seven subtypes displayed recurrent alterations such as ETS fusions (*ERG*, *ETV1*, *ETV4* and *FLI1*) or mutations in *SPOP*, *FOXA1*, or *IDH1*. Regarding DDR-related gene alterations; germline and somatic aberrations in *BRCA1/2*, *CDK12*, *ATM*, *FANCD2*, *RAD51C* were detected in 19% of the analyzed tumors. However, this 19% included some variants which may not have a significant impact in DDR function such as the common nonsense polymorphic variant *BRCA2* K3326∗, which truncates only a few amino acids at the end of the protein [18] and is currently classified as a benign variant, or *FANCD2* and *RAD51C* heterozygous, not biallelic losses [19].

Within cohorts of localized prostate cancer, DNA repair mutations, particularly in *BRCA2* and *ATM* genes, are enriched in those with Gleason grade group ≥3 and clinical stage ≥ cT3 disease [20]. Different genomic profiles can be found within one single tumor, what is commonly known as intra-tumor spatial heterogeneity. There are two main contributors to pre-treatment heterogeneity in prostate cancer. First, primary prostate tumors are commonly multi-focal, representing multiple tumors arising in parallel with different clonal origins; each having different grades of aggressiveness [21, 22]. One study using whole-genome sequencing compared different tumor foci in tumors resected from patients with clinically localized prostate cancer, and found a relatively low number of point mutations but significant heterogeneity between intra-tumor foci in the copy number profile and in terms of genomic rearrangements [23].

A second contributor to intra-patient spatial heterogeneity is the independent genomic evolution of different tumor areas, resulting in the emergence of different clones, but coming from a joint origin. Clonal evolution is highly driven by selective pressure from therapeutic interventions, such as androgen deprivation therapy. One study reconstructed the phylogenetic tree of 293 cases of localized prostate cancer, defining specific patterns of clonal evolution. Point mutations and deletions were characteristic of early stages (clonal), while amplifications and changes in trinucleotide mutational signatures were subclonal and occurred later. Early mutated genes included:

FOXP1, *ATM*, *RB1*, *NKX3-1*. Moreover, subclonal events affected genes such as *MTOR*, *TSC1*, *TSC2*, *BAD*, *BID*, and *BAK1*. Up to 59% of the analyzed tumors had multiple subclones, and clonal heterogeneity was clinically relevant as evidenced by patients with monoclonal tumors having a lower risk of relapse (7%) compared to polyclonal tumors (64%). Interestingly, patients with germline *BRCA2* mutations presented differentiated clonal evolution patterns, supporting these mutations as being key for genesis and progression of these tumors [24].

A study pursuing whole genome sequencing (WGS) analysis of multiple metastases from the same individual suggested that selective pressure from hypoxia or anticancer treatment (ADT) can result in a process of clonal convergence, by selecting the clones which are able to adapt [25].

Clonal and subclonal evolution of prostate tumors involves not only selection but also accumulation of events. Many times, these aberrations are accumulated sequentially. However, at other times, "catastrophic" genomic events can lead to massive genomic changes. These would include phenomena such as chromoplexy, which involves massive genomic structural restructuration [26], massive accumulation of DNA double-strand breaks (chromothripsis), or focal regions of DNA hypermutation (kataegis). Patterns compatible with chromothripsis and kataegis in 20% and 23% of cases respectively were observed in whole-genome sequencing data from a large number of localized prostate cancer [27].

Genomic Landscape of Advanced Prostate Cancer

Stand-Up-To-Cancer (SU2C)-Prostate Cancer Foundation (PCF) International Dream Team characterized the genomic landscape of metastatic prostate cancer by analyzing whole-exome and transcriptomic data from 150 samples of mCRPC [28]. Frequently mutated genes included *AR*, *TP53*, *RB1*, *SPOP*, and less frequently, *PIK3CA/B*, *RSPO*, *BRAF/RAF1*, *APC*, *β-catenin* and *ZBTB16*. Data integration highlighted molecular subtypes enriching the following pathways: AR signaling, PI3K, WNT, DNA repair, among others. Interestingly, 23% of the mCRPC samples harbored DNA repair pathway alterations. Aberrations in *BRCA1/2*, *CDK12*, *FANCA*, *RAD51B*, *RAD51C*, *ATM*, *MLH1* and *MSH2* were identified.

A small fraction of lethal prostate cancer exhibits a hypermutated phenotype. In many cases, this phenotype is related to pathogenic mutations or translocations in mismatch repair genes associated with microsatellite instability [29]. However, in general, prostate cancer is a disease with a lower burden of point mutations compared to other tumor types, and the relation of these hypermutated cases to the competence of the mismatch repair pathway and its role as potential predictive biomarker of response to immune checkpoint inhibitors (ICI) remains yet to be fully elucidated. While MMR deficient tumors seem to be more likely to responds to ICI, recent studies were not able to correlate MMR mutations with the MSI phenotype, suggesting further mechanisms may be involved [30, 31].

BRCA2 aberrations are the most common among DDR genes, accounting for 8–12% of metastatic prostate cancer. Approximately 90% of *BRCA2*-altered cases exhibit biallelic loss of the gene and hence complete loss of function. Mechanisms of *BRCA2* inactivation include germline or somatic mutations with loss of the second allele, homozygous deletions, or rearrangements that introduce breakpoints within *BRCA2*. BRCA1 gene alterations are very uncommon in prostate cancer (around 1% of metastatic prostate cancers).

The second most common event involves *ATM* mutations, including truncating mutations, which lead to an impaired protein lacking the kinase domain, albeit missense mutations within the kinase domain can also lead to dysfunctional protein. *ATM* can also be inactivated, although less commonly, by homozygous deletions of the gene. Taken together, events in the *ATM* gene predicted to cause loss of protein function occur in 6–8% of all metastatic prostate cancer.

Other HR genes are recurrently mutated or deleted in metastatic prostate cancer. These include *FANCA, CHEK1, CHEK2, PALB2, RAD51* and others. While the frequency of alterations in each individual gene is low (<2% each), taken together, these less common aberrations can account for 5–10% of metastatic prostate cancer.

A subset of prostate cancers is characterized by low AR-signaling and neuroendocrine-like features. While different studies have used slightly different definitions, this phenotype is generally referred to as neuroendocrine prostate cancer (NEPC). Enrichment of *TP53* loss has been identified in NEPC (>60%) compared to CRPC adenocarcinoma (30–40%). A recent study performed genomic characterization of 202 patients with metastatic CRPC who underwent metastatic biopsies [32]. The cohort was enriched for patients on androgen deprivation therapy (73% patients had their biopsies taken while on abiraterone acetate or enzalutamide). Seventeen percent of cases had phenotypic resemblances with NEPC, in what the authors termed "treatment-emergent NEPC", suggesting that the development of a basal-like phenotype, characterized by loss of tumor suppressor genes, could be related to resistance to AR targeting agents in some cases. Interestingly, mutations and deletions in HR and MMR genes were almost mutually exclusive with the treatment-emergent NEPC phenotype (8% of NEPC-like vs. 40% of non-NEPC-like cases harbored DDR gene defects, $p = 0.035$).

With easier access to improved sequencing technologies, discovery of additional drivers and determinants of the disease progression is likely. In one study using deep whole-genome (median coverage above 100×) and transcriptome sequencing of 101 prostate cancer metastatic biopsies, tumors with biallelic *BRCA2* aberrations were characterized by a higher frequency of microhomology deletions. Biallelic inactivation of *CDK12* instead associated with an increase of tandem duplication events [33]. Biallelic mutations in *CDK12* seem to define a distinct subset of prostate cancers, and are almost mutually exclusive with other DNA repair defects, ETS fusions, and *SPOP* mutations.

Integrative studies comparing localized and metastatic tumors revealed an enrichment in alterations of *TP53, RB1, AR, PTEN, FOXA1, APC* and *BRCA2* genes in the advanced setting. Moreover, pathway enrichment analysis using the frequently mutated genes showed that the DNA repair pathway is significantly more commonly altered in the setting of metastatic disease [34].

Further studies enabled the comparison across diseases states; not only between localized and metastatic castration-resistant but also including metastatic non-castrate prostate cancer. Somatic and germline alterations of *BRCA1/2, ATM, CHEK2, CDK12* account for 27% of the advanced prostate cancer tumor population. The analysis of a few matched samples (same-patient primary vs. metastatic) enabled the reconstruction of the mutations acquired during tumor evolution. *BRCA2* alterations were invariably present since the initial stage of the disease. In contrast, alterations in *AR* were mainly observed later during disease progression, suggesting that these alterations are acquired as a result of treatment and promote castration resistance in the metastatic clone [35].

Additional studies explored the impact of clonality at either stage of the disease. Results showed that localized prostate cancer is characterized by multifocality and multiclonality. On the other hand, the analysis of copy number alteration data from several metastases in the same individual (intraindividual), showed less marked heterogeneity in advanced stages of the disease, probably related to clonal selection when exposed to ADT. A different study integrated the mutational profiling and transcriptomic data of intraindividual metastatic lesions and reached a similar conclusion [36, 37].

Germline Mutations in DNA Repair Genes: Prevalence and Implications

Individuals who inherit certain gene mutations are at higher risk of developing prostate cancer. Recognition of those prostate cancer patients who carry germline mutations may be relevant at several levels, including identification of entire families whose members may be at risk of developing prostate cancer or other tumors, and may facilitate personalized therapeutic strategies for patients.

Prostate cancer is a tumor type where inherited genetic factors may have a higher impact in risk determination. The presence of other prostate cancers in a family is a well-known risk factor for cancer development [38]. Based on studies performed in twins, over 50% of risk of developing prostate cancer may be determined by inherited gene alterations [39].

Beyond identification of SNP profiles in genome-wide association studies related to increased risk, pathogenic mutations in certain genes is linked directly to prostate cancer risk. With regards to DNA repair, two different groups of inherited mutations in germline DNA are recognized as relevant: those occurring in HR genes (such as *BRCA1/2*, also involved in autosomal dominant patterns related to breast, ovarian and pancreatic cancer predisposition) and those occurring in mismatch repair genes, associated with Lynch Syndrome. Lynch syndrome, a multicancer syndrome cause by germline mutations in mismatch repair genes, is typically linked to gastrointestinal cancers and is also associated with increases prostate cancer risk [40].

Inherited mutations in the *BRCA2* gene determines the highest increase in risk for prostate cancer development. For instance, male *BRCA2* mutation carriers have an 8.6-fold higher risk than non-carriers to develop prostate cancer before the age of 65 [41]. Variants in *BRCA1* and *ATM* have also been linked to increased risk [42], whereas data for other genes such as *CHEK2* is less robust.

Identification of prostate cancer predisposition mutations, most commonly through cascade testing after a relative has been diagnosed with prostate, breast, ovarian or other cancers, could open opportunities for precision monitoring of these patients towards early diagnosis of prostate cancer. A number of studies are evaluating whether germline mutations can define a population that may benefit from targeted screening strategies. IMPACT (Identification of Men with a genetic predisposition to Prostate Cancer: Targeted screening in *BRCA1/2* mutation carriers and controls) is assessing the role of PSA screening and prompt prostate biopsies (triggered by PSA values of 3 or higher) in a population of over 3000 men carrying mutations in *BRCA1*, *BRCA2* or the Lynch Syndrome genes *MSH2*, *MSH6*, *MLH1*.

Germline studies of localized prostate cancer have traditionally focused in *BRCA1* and *BRCA2* mutations. Although such mutations are of low prevalence in this population (1–2% across studies), recent data suggests that these are clinically relevant, as will be discussed later in this chapter. The scenario is slightly different when considering population subsets with early age at diagnosis, or enriched for patients with significant family history, but rarely surpasses 2–2.5% in any series of localized prostate cancer patients [43]. In the TCGA landscape study for patients with primary prostate cancer, the prevalence of *BRCA1/2* mutations was 3%, but all six cases of germline *BRCA2* mutation were K3326*, which has unclear impact on protein function [19].

Prostate tumors associated with germline *BRCA2* mutations exhibit a characteristic genomic profile, supporting the concept that these mutations drive disease evolution. One study profiled in-depth 19 prostate tumors germline *BRCA2* mutation carriers with whole-genome sequencing and methylation profiling, and compared them to 200 sporadic prostate tumors [44]. *BRCA2*-mutated cases had a higher percent of the genome altered and were enriched for amplifications in *MYC* and chromosome 3q, suggesting higher genomic instability.

Due to the enrichment for DNA repair gene mutations in cohorts of advanced prostate tumors, including germline mutations, specific studies on the prevalence of germline mutations among patients with metastatic prostate cancer are needed. Germline DNA analysis of 20 DNA repair genes was conducted in a cohort of 692 patients with metastatic prostate cancers recruited into molecular characterization studies across several academic centers in the UK and US. Overall, 84 suspected pathogenic mutations in 82/692 (11.8%) patients were detected, including mutations in *BRCA2* (37 cases), *ATM* (11), *CHEK2* (10), *BRCA1* (6), *RAD51D* (3) and *PALB2* (3). This frequency was significantly higher than in the TCGA series, which consists of patients with localized or locoregional disease, even when considering only TCGA sub-groups with higher Gleason or higher risk of relapse based on clinic-pathological criteria. Although overall there was higher burden of cancer family history among relatives of germline mutation carriers than no carriers, the frequency of men with first-degree relatives affected by prostate cancer was the same in patients carrying or not carrying inherited DDR mutations (22% in both groups). The risk of carrying pathogenic germline mutations was also significantly higher than observed in the general population. The high prevalence of these mutations in late-stage prostate cancer, the lack of a complete association with the presence of prostate cancer cases in the family, and the potential clinical implications of such gene alterations has resulted in the NCCN clinical guidelines recommending to offer germline *BRCA1/2* testing to all men with metastatic prostate cancer.

The high prevalence and enrichment in late-stage disease population has been confirmed by other studies [45], and overall suggests that approximately 5% of all men with metastatic prostate cancer carry a germline pathogenic mutation in *BRCA2*. However, there may be differences when screening populations with different geographical or racial background are taken into account. For example, a comprehensive study of a Spanish population, mostly white Caucasians, showed that the prevalence of germline *BRCA2* mutations was lower (3%) but the prevalence of germline *ATM* mutations was slightly higher (2%) than in other cohorts [46].

A summary showing DDR alterations in different stages of the disease is presented in Table 1.

MMR Germline Mutations: Lynch Syndrome

Lynch syndrome (LS) is an inherited cancer predisposition syndrome characterized by mutations

Germline and Somatic Defects in DNA Repair Pathways in Prostate Cancer 287

Table 1 Summary of germline and somatic prevalence of mutations in key DDR genes

DDR mechanism	Gene	Localized		Metastatic	
		Germline [83]	Somatic [19]	Germline [83]	Somatic [28, 33]
HR	BRCA1	0.6%	1%	0.87%	0.7%
	BRCA2	0.2%	3%	5.35%	13.3%
	ATM	1%	4%	1.59%	7.3%
	CHEK2	0.4%	0%	1.87%	3.0%
	CDK12		2%		4.7%
	PALB2	0.4%	0%	0.43%	2%
	RAD51B		1%		
	RAD51C	0.4%	3%	0.14%	
	RAD51D	0.2%		0.43%	
	FANCD2		7%		
MMR	MLH1		0.3%		0.7%
	MSH2	0.2%	0.3%	0.14%	2%
	MSH6	0.2%	1.5%	0.14%	1%
NHEJ	PRKDC				8%
NER	ERCC2		0.6%		1.3%
	ERCC5		0.3%		1.3%

in genes involved in the MMR pathway genes. Specifically, germline mutations in *MLH1*, *MSH2*, *MSH6*, *PMS2* and *EPCAM* are associated with LS. Mismatches in repetitive sequences, such as microsatellite regions, are not efficiently repaired in LS associated tumors; resulting in high DNA microsatellite instability (MSI) phenotype [47, 48].

Lynch syndrome accounts for 2–5% of colorectal cancer [48]. However, other types of malignancies, such as ovarian, pancreatic, gastric and urinary cancer are associated with the disease. Men with Lynch syndrome have an increased risk for prostate cancer (2.13-fold) [49]; suggesting the need for prostate cancer screening in this population.

Impact of DNA Repair Defects in Clinical Outcome for Prostate Cancer

Most studies looking into the prognostic impact of DNA repair mutations in localized prostate cancer have focused on germline *BRCA1/2* mutations since these mutations are the most prevalent and it is recognized that they increase prostate cancer risk. We now know that a plethora of DNA repair mutations in different genes are found in

prostate cancer. However, we are still missing datasets that comprehensively assess the prognostic value of the different germline and somatic mutations.

A subset of localized prostate cancer patients are candidates for conservative management approaches including active surveillance. A recent study identified germline *BRCA2* mutations across a cohort of patients managed with active surveillance. Germline *BRCA2* mutation carriers are more likely to require stage reclassification during the follow-up period: tumor staging upgrade rate at 2-, 5- and 10-years was 27%, 50% and 78% in *BRCA2* mutation carriers compared to 10%, 22% and 40% in non-carriers (p = 0.001) [50]. While these data suggest that these patients may not be candidates for active surveillance, there is a lack of prospective data, and current guidelines do not include molecular testing among the criteria for advocating active surveillance.

Germline *BRCA2* mutations are a well-established poor prognosis factor for disease relapse after primary treatment for localized prostate cancer. In the largest study reported to date, outcome was retrospectively assessed in over 2000 patients treated with radical prostatectomy or radiotherapy for localized prostate cancer, stratifying the analysis based on the presence of germline *BRCA1* and *BRCA2* mutations. Mutation carriers (n = 79: 18 *BRCA1* and 61 *BRCA2*) had a shorter cancer-specific survival (8.6 years vs. 15.7 years) and higher risk of developing metastatic disease (23% of mutation carriers had developed metastatic relapse within 5 years, compared to 7% of non-carriers). In multivariate analysis, germline *BRCA2* mutations were identified as an independent poor prognosis factor. In a follow-up study, this worst-prognosis for germline *BRCA1/2* mutation carriers was confirmed for both cohorts of patients treated with surgery or radiotherapy [51]. While results were particularly poor for *BRCA1/2* mutation carriers undergoing radiation therapy, the results should not be read in terms of direct comparison between the impact of mutational status in radiotherapy vs. surgery outcome, as baseline characteristics of the cohorts were different and determined the selection of one or another treatment type.

The enrichment for DDR mutations in late-stage disease is in line with data suggesting that certain DDR mutations define a more aggressive course of prostate cancer. However, it remains to be explored if in the setting of metastatic prostate cancer, where there is a pre-selection of poor prognosis cases, DDR mutations are still informative of differential prognosis, and, more importantly, if such data can be used to define personalized cancer management studies.

The two pivotal studies in prostate cancer genomics (TCGA [19] and SU2C-PCF consortium [28] for localized and metastatic prostate cancer respectively did not report correlative clinical outcome data according to genomics, but data is expected once longer follow-up has been achieved. In the meantime, several studies have looked at the prognostic impact of DDR defects on mCRPC evolution. Some results may seem contradictory, but differences in study populations, the selected gene set, the use of tumor vs. plasma samples and the interrogation of germline vs. somatic defects, make the direct comparison of these studies challenging.

One study reported on a series of 319 mCRPC patients including 22 germline DDR mutation carriers retrospectively identified (16/22 carrying germline *BRCA2* mutations). It concluded that progression free survival (PFS) of mutation carriers on standard-of-care is shorter than that of non-carriers (3.3 months vs. 6.2 months, p = 0.01), although some of the carriers actually achieved long-lasting responses. This impact in PFS was not observed in the same patient population when receiving docetaxel [52]. In a later study, the same group pursued circulating tumor DNA genomic characterization from 202 treatment-naïve mCRPC patients who participated in a randomized trial of abiraterone plus prednisone (n = 101) or enzalutamide alone (n = 101). In 14/202 patients, germline or somatic alterations of *BRCA2* or *ATM* were detected, and these patients had a shorter time to progression (HR 5.27 in multivariate Cox regression models, p < 0.001). Detection of *TP53* mutations in circulating DNA was also associated with a worse outcome [53].

Other series have failed to identify association with prognosis. In a correlative analysis of data from patients in the UK, US and Australia (most of them were those included in the original genomic landscape study of the SU2C-PCF consortium), patients carrying germline mutations in a panel of DDR genes (namely, *ATM*, *BRCA1*, *BRCA2*, *CHEK2*, *MRE11A*, *MSH6*, *NBN*, *PALB2*, *RAD51C*, *RAD51D*, *GEN1*, *MSH2* and *ATR*), had a similar outcome on standard-of-care therapies (abiraterone, enzalutamide, taxane chemotherapy) than non-carriers [54]. Overall survival of those patients was also similar. However, the fact that this cohort was enriched for patients who participated in PARP inhibitor or platinum clinical trials makes it difficult to isolate the impact of the mutation on prognosis from the potential impact of the different therapeutic approaches. On the other hand, one series reported improved response rates for DDR mutation carriers on enzalutamide or abiraterone [55].

Differences in the study population and the methodology used may explain in part the different results. Nevertheless, with the envisioned use of personalized targeted therapeutics for patients with certain DDR defects in the near future, the prognostic impact of these mutations may be diluted by the predictive value to identify those patients who would benefit from specific therapeutics, similarly to what has happened with *HER2* amplifications in breast cancer. While these alterations determine a more aggressive phenotype, access to a specific therapeutic option targeting such alterations has resulted in a subset of patients achieving longer survival times when treated with anti-HER2 drugs [56]. Meanwhile, the only prospective study which has interrogated the impact of germline DDR mutations in patient outcome from mCRPC is the PROREPAIR study, conducted entirely in a Spanish population [46]. In this study, there were no significant differences in patient outcome for carriers of germline *BRCA1*, *BRCA2* and *ATM* mutations, albeit patients with germline *BRCA2* mutations presented shortened cause-specific survival than the rest of the study population (17 months vs. 33 months, p = 0.027). Response rates to abiraterone, enzalutamide or taxanes were similar for carriers and non-carriers, though there was a trend for a shorter time to progression on abiraterone and enzalutamide for germline *BRCA2* carriers.

In summary, all these studies suggest that some patients with DDR mutations, particularly within the germline *BRCA2* population, have aggressive cancers. However, none of the studies suggest that mutation carriers should not receive standard of care therapies for metastatic prostate cancer, as many of them derive significant benefit from these drugs. Further studies should refine the methods to identify these poor-prognosis patients, and plan personalized management strategies accounting for different disease behaviors. Moreover, since several of these therapies are currently used in the metastatic hormone-naïve setting, there is a need to determine if the classification of the disease based on DDR mutations could stratify patients for treatment selection.

Clinical Development of PARP Inhibitors in Prostate Cancer

The role of the DDR deficiency in DNA damage accumulation and generation of genome instability and cell death offers the opportunity for tar-

geting DDR signaling pathways as a therapeutic opportunity.

PARP inhibitors are a class of drugs that inhibit PARP1, a sensor of DNA SSB. PARP inhibitors are in advanced stages of clinical development in prostate cancer, after receiving approval in ovarian and breast cancer, where DDR is also a therapeutic target.

BRCA1/2 deficient tumors are sensitive to PARP inhibition primarily through a synthetic lethal interaction. Synthetic lethality is a biological principle by which two events, which are not lethal independently, become fatal for a cell when they occur simultaneously. In this case, synthetic lethal interactions in DDR involve a tumor harboring specific mutations in key DDR proteins and blockage of a backup component of the repair process on which cancer cells would have become dependent [57]. PARP-1 inhibition results in unsuccessful cell repair of DNA SSBs, which progress to DNA DSBs. In *BRCA*-deficient tumors, there is accumulation of DSB breaks since they are not efficiently repaired, leading to selective tumor cell death [58]. We now understand that PARP inhibitors also have direct cytotoxic effects, by trapping PARP1 when attached to the damaged DNA ends.

In the first-in-man clinical trials of the different PARP inhibitors in clinical development, a few prostate cancer patients were included, generating the very first dataset for this class of drugs in prostate cancer. These early-phase trials were primarily enriched for patients with germline *BRCA1/2* mutations, and led to subsequent trials that resulted in the approval of olaparib, niraparib and rucaparib for ovarian cancer (either as monotherapy for BRCA-mutated tumors or as maintenance therapy after response to platinum-based chemotherapy) and the approval of olaparib and talozaparib in breast cancer. Eight mCRPC patients were also enrolled in a tumor-agnostic basket trial of olaparib for patients with advanced cancers and *BRCA1/2* germline mutations, with half of them responding to therapy [59].

Several preclinical studies have characterized other alterations beyond *BRCA1/2* loss that can lead to PARP inhibitor sensitivity [60–62]. The term BRCAness is commonly used in the literature to refer to a phenotype characterized by HR deficiency and PARP inhibitor sensitivity despite these tumors not presenting BRCA1/2 defects [63]. The development of this drug class in ovarian cancer provided clinical evidence that there is an extended target patient population who can benefit beyond those carrying BRCA1/2 alterations.

Considering the above, and the high prevalence of germline and somatic mutations in HR genes in prostate cancer, the development of PARP inhibitors in advanced prostate cancer could result in the identification of a precision medicine strategy for a molecularly-defined subset of patients. The tolerability profile of this class of drugs is primarily characterized by bone marrow toxicity, namely anemia and thrombocytopenia, and mild gastrointestinal toxicities (nausea, diarrhea). Fatigue, mood changes and altered liver function can also appear. In the long term, the use of PARP inhibitors has raised concerns about the risk of secondary tumors, particularly myelodysplastic disorders, with a few cases of acute leukemias reported among patients receiving PARP inhibitor treatment. Further data is needed to conclude whether these myelodysplastic disorders are related to PARP inhibition or to the inherent risk of cancer in patients with genomic instability due to germline BRCA mutations.

The TOPARP-A trial, reported in 2015, provided proof-of-concept evidence for further development of PARP inhibitors in prostate cancer. Treatment of 50 patients with heavily pretreated mCRPC with the PARP inhibitor olaparib resulted in several durable responses being observed. Almost one third of treated patients experienced either a response by traditional criteria (radiological or PSA) and/or a significant decrease of circulating tumor cell counts (from >5 to <5 cells per 7.5 ml of blood sample), a response biomarker that has been associated with patient outcome in several prostate cancer trials [64, 65]. Retrospective next-generation sequencing of tumor biopsies collected from these patients at study entry showed that 14/16 patients responding to therapy by either criteria harbored

a somatic or germline alteration in HR or HR-related genes [66].

Matching the expected prevalence in the general mCRPC population, seven patients had biallelic loss of BRCA2 in their tumors, either by mutations with loss of the second allele or by a homozygous deletion. All seven patients responded to olaparib, with the longest response lasting for over 3 years in a patient who had already progressed on ADT, docetaxel and abiraterone. Some patients responding to therapy harbored germline or somatic defects in the ATM gene, although other ATM aberrations did not lead to a response, and not all the responses by CTC criteria resulted in significant PSA drops. Additionally, patients with BRCA1 or PALB2 mutations also responded to therapy. Sequencing of follow-up tumor or circulating tumor DNA samples from these patients also allowed for identification of PARPi secondary resistances mechanisms, which most commonly involve additional arising mutations in HR genes that compensate for the initial mutation, restoring the open reading frame of the gene and permitting translation of a probably functional form of the protein again [67, 68]. Interestingly, these secondary or "reversion" mutations seem to appear in a polyclonal manner and in specific regions of these genes defined by high density of microhomology areas, suggesting that the incapacity to successfully complete homologous recombination repair of these tumors would be favoring the emergence of these new events [69].

Data from the TOPARP trial emerged in parallel to several sets of genomics studies confirming the high prevalence of DNA repair defects in advanced prostate cancer, triggering the launch of numerous clinical trials of PARPi in this setting. A number of phase II and phase III clinical trials, trying to confirm the antitumor activity of this drug class in this disease and, more importantly, refine the predictive biomarker set for optimal patient stratification, are now underway. Interim analysis of a phase II trial of rucaparib seems to confirm the antitumor activity in BRCA-mutated tumors, although response rates in patients harboring other mutations were lower, and further data is needed to define the optimal selection strategy for PARP inhibitors in mCRPC.

Beyond the use of PARP inhibitors in HR deficient tumors, this class of drug is also being developed as a combination with AR-signaling directed therapies, which are the mainstay of prostate cancer treatment. Preclinical data demonstrate that anti-androgen therapies (i.e. enzalutamide) result in downregulation of HR genes [70, 71] generating a conditional HR deficiency. This represents a promising opportunity for the combination of antiandrogens and PARP1 inhibitors therapies in CRPC. Recent studies support the relationship between AR and PARP1; for example PARP1 has been shown to act as an AR cofactor [72], modulating its activity and function in DNA damage. Furthermore, ADT upregulates PARP1-mediated repair pathways [71]. Advancing this idea, a combination of ADT and PARP inhibitors could increase the sensitivity of current therapies. More specifically, it could be used in the advanced stages of the disease, as defects on the HR pathway in castration resistant prostate cancer (CRPC) patients sensitizes the tumors to PARP inhibition. A randomized phase II clinical trial found that the combination of olaparib and abiraterone prolonged patient's time to progression in mCRPC, even among those patients with no evidence of DDR defects in the tumor genomics analysis [73]. This is a field of ongoing research that aims to optimize the combined use of these two drug classes towards improving outcome for prostate cancer patients.

Targeting DNA Repair Beyond PARP1

Platinum-Based Chemotherapy

Platinum salts are part of the standard chemotherapy regimens used in many tumor types, but they are not commonly used in prostate cancer beyond infrequent small-cell prostate cancers. Platinum salts act primarily by inducing DNA adducts that result in damage. Tumors unable to correctly repair DNA damage may be more sensitive to platinum-based chemotherapy.

A phase II trial of an oral platinum-derivative, satraplatin, demonstrated tumor responses in a molecularly-unselected population of advanced prostate cancer. However, the phase III trial failed to show improved survival with satraplatin usage [74]. Now, with the identification of a subset of DDR-defective advanced prostate cancers, there is a growing interest for testing platinum-based therapy in a molecularly-defined subgroup of prostate cancer patients.

Retrospective studies of patients with prostate cancer receiving platinum-based chemotherapy (either as single agents or in combination with taxanes) have explored the genomics of those patients deriving the most benefit. The largest series reported so far, is a cohort of 141 men with mCRPC treated at Dana Farber Cancer Institute with carboplatin in combination with docetaxel [75]. A retrospective analysis of germline mutational status in this population revealed a higher PSA response rate for germline *BRCA2* mutation carriers (6/8 responses, 75%, compared to 23/133, 17%, in non-carriers; p = 0.001) and a longer survival (18.9 months vs. 9.5 months). Similarly, among 20 patients treated with carboplatin within a cohort of patients who underwent postmortem next-generation sequencing (NGS) of metastatic deposits, those with DNA repair mutations had prolonged times on therapy (p = 0.02) suggesting prolonged benefit [37].

Collectively, each of these studies represent a small series of patients with different DNA repair defects [76, 77]. Prospective trials will be required to elucidate the optimal role of platinum-based chemotherapy in patients with prostate cancer and DNA repair defects.

Other Inhibitors of DDR Proteins

Other components of DDR proteins are being tested as therapeutic targets, with some of these compounds having entered clinical development. One example are DNA-PKs, signaling kinase proteins mainly involved in NHEJ for repairing both exogenous and endogenous DNA DSBs. Inhibiting these proteins can generate defects in both NHEJ and indirectly on the HR pathway by impairing end resection. The key role of DNA-PKs in different DDR pathways highlight the potential of targeting these kinases. Indeed, DNA-PKs inhibition sensitizes cells and tumor xenografts to agents that generate DNA DSBs such as radiation and topoisomerase 2 inhibitors. Several inhibitors are currently being evaluated in early phases of clinical trials such as VX-984 and MSC2490484A [12].

ATR, the other main kinase that senses DNA DSB and starts the repair signaling cascade has been proposed as a suitable therapeutic target. In cells deficient for ATM-CHEK2-p53, the role of ATR-CHEK1 in response to DNA damage becomes more critical. ATR inhibition has also been proposed to have a synergy with an impaired ATM signaling [78]. Moreover, CHEK1 inhibition promotes cell death specifically in cells with ATM-CHEK2-p53 mutations [1].

One example is the ATR inhibitor VX-970, which has been demonstrated to chemo-sensitize cancer cells in combination with cisplatin. The combination of VX-970 with cisplatin has also shown an increased antitumor activity. Another example is AZD6738, an oral ATR inhibitor, which has been tested in combination with radiotherapy, olaparib, and carboplatin, among others. Clinical trials of ATR inhibitors are ongoing.

CHEK2 and CHEK1, the main substrates of ATM and ATR, are also being tested as therapeutic targets. Specific inhibitors MK8776 and CCT245737, which selectively inhibit CHEK1, are currently in phase I trials. Additionally, LY2606368 is a promising agent able to inhibit both CHEK1 and 2. Moreover, AZD7762, a novel checkpoint kinase inhibitor (CHEK1/2) enhances the cytotoxic effect of DNA inducing drugs in p53-deficient cells [1, 79].

In summary, several evidences support DDR proteins as potential therapeutic targets, with DDR defects being predictive of sensitivity to some of these drugs. Even though the main DDR pathways are widely described, it remains necessary to further understand the specific role of their components and characterize the effect of each cytotoxic agent. Clinical trials are now trying to determine the conditions and combinations of the different agents that will benefit each patient population.

Immunotherapy for Prostate Cancer and MMR Defects

Approximately 3–12% of advanced prostate cancer tumors harbor alterations in mismatch repair genes (*MSH2, MSH6* and *MLH1*). Inactivation of these genes triggers MSI coupled to a hypermutation phenotype [29]. Interestingly, immunotherapy-sensitive cancer types display a hypermutation phenotype associated with a high neoantigen burden. Mutation burden in mCRPC is commonly low (4.4 mutations/Mb); however, a small subset of patients with MMR gene alterations and a higher mutation rate (50 mutations/Mb) could benefit from current immune checkpoint inhibitors [28, 30]. The US FDA has approved one type of checkpoint immunotherapy for any cancer type with evidence of MMR deficiency, including prostate cancer.

The potential use of immunotherapy in prostate cancer has been highlighted by additional studies that characterize a subset of tumors with *CDK12* mutations. *CDK12* is a gene involved in genomic stability maintenance through the transcriptional regulation of DDR genes [80]. Loss of function mutations in *CDK12* are enriched in metastatic (6.9%) compared to primary (1.2%) prostate cancer [81]. *CDK12*-mutated tumors are enriched for focal tandem duplications [33], which results in increased rates of neoantigen formation, facilitating increased tumor T cell infiltration/clonal expansion [81, 82]. That has put *CDK12* mutations in the spotlight as putative predictive biomarkers of response to immune checkpoint inhibitors, and clinical studies are now interrogating this association.

Practical Aspects and Challenges to Stratify Patients Based on DNA Repair Defects

Sanger sequencing fueled genetic testing and became the 'gold standard' methodology for the identification of DNA mutations. For years, it was the preferred system for identification of *BRCA1/2* mutations in germline DNA, and is still used today in many centers for confirmation of clinically-relevant findings. However, the main limitation is that it only allows the analysis of one gene at a time.

The advent of NGS technologies, enabling multiplexed interrogations of the sequencing of large numbers of genes simultaneously, has transformed the field. Due to its cost reduction and high throughput, NGS is routinely applied in research and is making its way into the clinic, allowing for identification of DDR and other clinically-actionable defects in patient biopsies, opening the door for integrating genomics into prostate cancer management.

NGS encompass a broad range of applications that include genome sequencing, transcriptome analysis, and epigenome profiling, among many others. Below, we will focus on the genome sequencing applications: gene panels, whole exomes, and whole genomes.

Gene panels comprise a set of genes carefully selected based on certain criteria; for example, a particular tumor type, association with a phenotype or gene ontology. The sequencing of gene panels is considered a targeted approach that can be customized depending on the genes or regions of interest. Two variants of the targeted approach are: amplicon-based sequencing and capture hybridization-based sequencing. For capture hybridization, genomic DNA is sheared to small fragments. The fragments are then hybridized to biotinylated DNA or RNA single-stranded oligonucleotides, known as probes, which contain the sequences of the regions of interest. Captured fragments are then pulled down using streptavidin beads. The obtained DNA fragments are the starting material for library preparation. Variations of the capture hybridization-based sequencing include "in solution" or "in solid" phase hybridization and different probe chemistries. On the other hand, for amplicon-based sequencing, oligonucleotide capture probes are designed to flank the regions of interest. The designed probes are then hybridized to genomic DNA; followed by extension and ligation reactions to obtain the DNA amplicons containing the regions of interest. Sequencing primers are then added to each amplicon to further sequence. Both approaches enable the identification of single-

nucleotide variants (SNVs) and small insertions or deletions. Nevertheless, the selection of the sequencing approach depends on the type (formalin-fixed paraffin-embedded, FFPE and fresh frozen, FF) and amount of sample.

The development of gene panels for patient stratification has challenged the traditional regulatory approval for predictive biomarkers. Until recently, the approval of a biomarker or a medical test was based on the capacity of a technology to identify the true result within a range of pre-specified values (for example, HER2 protein expression by immunofluorescence test is classified in tiers, with a minimum and a maximum). Gene panels represent a breakthrough in biomarker development as (1) new mutations are discovered on a daily basis, hence the range of potential results varies and (2) raw sequencing output data needs to be curated through bioinformatics pipelines; hence the final results are dependent on the test and on the technology used to curate the results of the test.

Beyond targeted panels, it is now feasible to obtain the coding sequences of the entire exome or even the whole genome rather than only a set of genes. Whole exome sequencing (WES) can be used for the efficient identification of disease-causing variants. In addition to pinpointing mutations, whole exome analysis provides an opportunity for the detection of copy number variants. Although whole exome sequencing facilitates the discovery of novel DNA variants that could result in clinical phenotypes, it is limited to exonic regions (1% of the human genome). Variants in non-coding regions (untranslated regions, introns, promoters, regulatory elements, repetitive regions and non-coding functional RNA) can alter gene expression leading to disease. Novel variants will be missed with the whole exome approach. Whole genome sequencing (WGS) aids in the identification of all the existing variants at the base sequence level. Novel NGS technologies, the so called third generation, can yield longer reads. Longer sequencing reads provides an opportunity for better determination of CNVs, rearrangements, inversions, and translocations. Due to increased costs and lack of standardized pipelines for automated data analysis, whole exome and whole genome sequencing assays are still used predominantly in research, and targeted panels are used with more regularity in the clinical practice setting.

Analysis of blood samples offers a noninvasive alternative for molecular stratification of cancer patients. Tumor genomic material can be obtained from the several components of the blood, including circulating tumor cells (CTCs), and circulating, cell-free, tumor DNA (ctDNA). These approaches, collectively referred to as liquid biopsy, provide opportunities to monitor tumor evolution in response to cancer treatments and emergence of resistance mechanisms.

ctDNA consists of short fragments of tumor DNA shred into the circulation, mostly due to cell death processes (apoptosis and necrosis). Tumor genomics can be inferred from ctDNA analyses by NGS. Moreover, quantification of CTC and ctDNA correlates with tumor burden and has prognostic value in prostate cancer.

However, technical challenges to efficiently analyze ctDNA are manifold. For example, ctDNA cannot be directly extracted from blood since it co-occurs with non-tumor circulating DNA and its relative proportion is low. An additional challenge is the shorter length of DNA fragments in ctDNA, compared to genomic DNA extracted from tumor biopsies. These shorter DNA fragments make it challenging to assess copy number variants and structural rearrangements.

Conclusions

Alterations in DNA damage repair genes are common in prostate cancer, particularly in metastatic forms of prostate cancer. Approximately 50% of these defects are linked to an inherited mutation, which is relevant not only for the patient but potentially for his relatives, allowing for monitoring those individuals at risk of prostate or other cancers. Some of the DDR gene defects confer a more aggressive disease phenotype, showing worst prognosis, although further data is needed particularly for those genes with lower prevalence of mutations.

Importantly, the identification of DNA damage repair defects in prostate cancer offers the opportunity for precision medicine approaches, such as the use of PARP inhibitors or ATR inhibitors, currently in clinical development. Moreover, the identification of DNA damage repair defects may define a subset of prostate cancer patients who could benefit from platinum chemotherapy. There is an increased interest in implementing genomic stratification assays in clinical practice, which may transform significantly prostate cancer patient care.

References

1. S. Karanika, T. Karantanos, L. Li, P.G. Corn, T.C. Thompson, DNA damage response and prostate cancer: defects, regulation and therapeutic implications. Oncogene **34**(22), 2815–2822 (2015)
2. A. Ciccia, S.J. Elledge, The DNA damage response: making it safe to play with knives. Mol. Cell **40**(2), 179–204 (2010)
3. D. Hanahan, R.A. Weinberg, Hallmarks of cancer: the next generation. Cell **144**(5), 646–674 (2011)
4. G. Giglia-Mari, A. Zotter, W. Vermeulen, DNA damage response. Cold Spring Harb. Perspect. Biol. **3**(1), 1–19 (2011)
5. C.L. Peterson, J. Côté, Cellular machineries for chromosomal DNA repair. Genes Dev. **18**(6), 602–616 (2004)
6. K.W. Caldecott, Single-strand break repair and genetic disease. Nat. Rev. Genet. **9**(8), 619–631 (2008)
7. E. Sonoda, H. Hochegger, A. Saberi, Y. Taniguchi, S. Takeda, Differential usage of non-homologous end-joining and homologous recombination in double strand break repair. DNA Repair **5**(9–10), 1021–1029 (2006)
8. M.J. Schiewer, K.E. Knudsen, DNA Damage response in prostate cancer. Cold Spring Harb. Perspect. Med. **9**, a030486 (2019)
9. W.D. Wright, S.S. Shah, W.D. Heyer, Homologous recombination and the repair of DNA double-strand breaks. J. Biol. Chem. **293**(27), 10524–10535 (2018)
10. X. Li, W.-D. Heyer, NIH Public Access. Cell Res. **18**(1), 99–113 (2008)
11. W.K. Holloman, Unraveling the mechanism of BRCA2 in homologous recombination. Nat. Struct. Mol. Biol. **18**(7), 748–754 (2011)
12. J.S. Brown, B. O'Carrigan, S.P. Jackson, T.A. Yap, Targeting DNA repair in cancer: beyond PARP inhibitors. Cancer Discov. **7**(1), 20–37 (2017)
13. J.M. Daley, P. Sung, 53BP1, BRCA1, and the choice between recombination and end joining at DNA double-strand breaks. Mol. Cell. Biol. **34**(8), 1380–1388 (2014)

14. M. Nakayama, C.J. Bennett, J.L. Hicks, J.I. Epstein, E.A. Platz, W.G. Nelson, A.M. De Marzo, Hypermethylation of the human glutathione S-transferase-pi; gene (*GSTP1*) CpG island is present in a subset of proliferative inflammatory atrophy lesions but not in normal or hyperplastic epithelium of the prostate: a detailed study using laser-capture microdissection. Am. J. Pathol. **163**(3), 923–933 (2003)

15. G.H. Kang, S. Lee, H.J. Lee, K.S. Hwang, Aberrant CpG island hypermethylation of multiple genes in prostate cancer and prostatic intraepithelial neoplasia. J. Pathol. **202**(2), 233–240 (2004)

16. Y. Yin, R. Li, K. Xu, S. Ding, J. Li, G.H. Baek, S.G. Ramanand, S. Ding, Z. Liu, Y. Gao, M.S. Kanchwala, X. Li, R. Hutchinson, X. Liu, S.L. Woldu, C. Xing, N.B. Desai, F.Y. Feng, S. Burma, J.S. De Bono, S.M. Dehm, R.S. Mani, B.P.C. Chen, G.V. Raj, Androgen receptor variants mediate DNA repair after prostate cancer irradiation. Cancer Res. **77**(18), 4745–4754 (2017)

17. J.F. Goodwin, M.J. Schiewer, J.L. Dean, R.S. Schrecengost, R. de Leeuw, S. Han, T. Ma, R.B. Den, A.P. Dicker, F.Y. Feng, K.E. Knudsen, A hormone-DNA repair circuit governs the response to genotoxic insult. Cancer Discov. **3**(11), 1254–1271 (2013)

18. K. Wu, S.R. Hinson, A. Ohashi, D. Farrugia, P. Wendt, S.V. Tavtigian, A. Deffenbaugh, D. Goldgar, F.J. Couch, Functional evaluation and cancer risk assessment of BRCA2 unclassified variants. Cancer Res. **65**(2), 417–426 (2005)

19. A. Abeshouse, J. Ahn, R. Akbani, A. Ally, S. Amin, C.D. Andry, M. Annala, A. Aprikian, J. Armenia, A. Arora, J.T. Auman, M. Balasundaram, S. Balu, C.E. Barbieri, T. Bauer, C.C. Benz, A. Bergeron, R. Beroukhim, M. Berrios, A. Bivol, T. Bodenheimer, L. Boice, M.S. Bootwalla, R. Borges dos Reis, P.C. Boutros, J. Bowen, R. Bowlby, J. Boyd, R.K. Bradley, A. Breggia, F. Brimo, C.A. Bristow, D. Brooks, B.M. Broom, A.H. Bryce, G. Bubley, E. Burks, Y.S.N. Butterfield, M. Button, D. Canes, C.G. Carlotti, R. Carlsen, M. Carmel, P.R. Carroll, S.L. Carter, R. Cartun, B.S. Carver, J.M. Chan, M.T. Chang, Y. Chen, A.D. Cherniack, S. Chevalier, L. Chin, J. Cho, A. Chu, E. Chuah, S. Chudamani, K. Cibulskis, G. Ciriello, A. Clarke, M.R. Cooperberg, N.M. Corcoran, A.J. Costello, J. Cowan, D. Crain, E. Curley, K. David, J.A. Demchok, F. Demichelis, N. Dhalla, R. Dhir, A. Doueik, B. Drake, H. Dvinge, N. Dyakova, I. Felau, M.L. Ferguson, S. Frazer, S. Freedland, Y. Fu, S.B. Gabriel, J. Gao, J. Gardner, J.M. Gastier-Foster, N. Gehlenborg, M. Gerken, M.B. Gerstein, G. Getz, A.K. Godwin, A. Gopalan, M. Graefen, K. Graim, T. Gribbin, R. Guin, M. Gupta, A. Hadjipanayis, S. Haider, L. Hamel, D.N. Hayes, D.I. Heiman, J. Hess, K.A. Hoadley, A.H. Holbrook, R.A. Holt, A. Holway, C.M. Hovens, A.P. Hoyle, M. Huang, C.M. Hutter, M. Ittmann, L. Iype, S.R. Jefferys, C.D. Jones, S.J.M. Jones, H. Juhl, A. Kahles, C.J. Kane, K. Kasaian, M. Kerger, E. Khurana, J. Kim, R.J. Klein, R. Kucherlapati, L. Lacombe, M. Ladanyi, P.H. Lai, P.W. Laird, E.S. Lander, M. Latour, M.S. Lawrence, K. Lau, T. LeBien, D. Lee, S. Lee, K.-V. Lehmann, K.M. Leraas, I. Leshchiner, R. Leung, J.A. Libertino, T.M. Lichtenberg, P. Lin, W.M. Linehan, S. Ling, S.M. Lippman, J. Liu, W. Liu, L. Lochovsky, M. Loda, C. Logothetis, L. Lolla, T. Longacre, Y. Lu, J. Luo, Y. Ma, H.S. Mahadeshwar, D. Mallery, A. Mariamidze, M.A. Marra, M. Mayo, S. McCall, G. McKercher, S. Meng, A.-M. Mes-Masson, M.J. Merino, M. Meyerson, P.A. Mieczkowski, G.B. Mills, K.R.M. Shaw, S. Minner, A. Moinzadeh, R.A. Moore, S. Morris, C. Morrison, L.E. Mose, A.J. Mungall, B.A. Murray, J.B. Myers, R. Naresh, J. Nelson, M.A. Nelson, P.S. Nelson, Y. Newton, M.S. Noble, H. Noushmehr, M. Nykter, A. Pantazi, M. Parfenov, P.J. Park, J.S. Parker, J. Paulauskis, R. Penny, C.M. Perou, A. Piché, T. Pihl, P.A. Pinto, D. Prandi, A. Protopopov, N.C. Ramirez, A. Rao, W.K. Rathmell, G. Rätsch, X. Ren, V.E. Reuter, S.M. Reynolds, S.K. Rhie, K. Rieger-Christ, J. Roach, A.G. Robertson, B. Robinson, M.A. Rubin, F. Saad, S. Sadeghi, G. Saksena, C. Saller, A. Salner, F. Sanchez-Vega, C. Sander, G. Sandusky, G. Sauter, A. Sboner, P.T. Scardino, E. Scarlata, J.E. Schein, T. Schlomm, L.S. Schmidt, N. Schultz, S.E. Schumacher, J. Seidman, L. Neder, S. Seth, A. Sharp, C. Shelton, T. Shelton, H. Shen, R. Shen, M. Sherman, M. Sheth, Y. Shi, J. Shih, I. Shmulevich, J. Simko, R. Simon, J.V. Simons, P. Sipahimalani, T. Skelly, H.J. Sofia, M.G. Soloway, X. Song, A. Sorcini, C. Sougnez, S. Stepa, C. Stewart, J. Stewart, J.M. Stuart, T.B. Sullivan, C. Sun, H. Sun, A. Tam, D. Tan, J. Tang, R. Tarnuzzer, K. Tarvin, B.S. Taylor, P. Teebagy, I. Tenggara, B. Têtu, A. Tewari, N. Thiessen, T. Thompson, L.B. Thorne, D.P. Tirapelli, S.A. Tomlins, F.A. Trevisan, P. Troncoso, L.D. True, M.C. Tsourlakis, S. Tyekucheva, E. Van Allen, D.J. Van Den Berg, U. Veluvolu, R. Verhaak, C.D. Vocke, D. Voet, Y. Wan, Q. Wang, W. Wang, Z. Wang, N. Weinhold, J.N. Weinstein, D.J. Weisenberger, M.D. Wilkerson, L. Wise, J. Witte, C.-C. Wu, J. Wu, Y. Wu, A.W. Xu, S.S. Yadav, L. Yang, L. Yang, C. Yau, H. Ye, P. Yena, T. Zeng, J.C. Zenklusen, H. Zhang, J. Zhang, J. Zhang, W. Zhang, Y. Zhong, K. Zhu, E. Zmuda, The molecular taxonomy of primary prostate cancer. Cell **163**(4), 1011–1025 (2015)

20. C.H. Marshall, W. Fu, H. Wang, A.S. Baras, T.L. Lotan, E.S. Antonarakis, Prevalence of DNA repair gene mutations in localized prostate cancer according to clinical and pathologic features: association of Gleason score and tumor stage. Prostate Cancer Prostatic Dis. **22**(1), 59–65 (2019)

21. M. Løvf, S. Zhao, U. Axcrona, B. Johannessen, A.C. Bakken, K.T. Carm, A.M. Hoff, O. Myklebost, L.A. Meza-Zepeda, A.K. Lie, K. Axcrona, R.A. Lothe, R.I. Skotheim, Multifocal primary prostate cancer

exhibits high degree of genomic heterogeneity. Eur. Urol. **75**(3), 498–505 (2019)

22. C.S. Cooper, R. Eeles, D.C. Wedge, P. Van Loo, G. Gundem, L.B. Alexandrov, B. Kremeyer, A. Butler, A.G. Lynch, N. Camacho, C.E. Massie, J. Kay, H.J. Luxton, S. Edwards, Z. Kote-Jarai, N. Dennis, S. Merson, D. Leongamornlert, J. Zamora, C. Corbishley, S. Thomas, S. Nik-Zainal, M. Ramakrishna, S. O'Meara, L. Matthews, J. Clark, R. Hurst, R. Mithen, R.G. Bristow, P.C. Boutros, M. Fraser, S. Cooke, K. Raine, D. Jones, A. Menzies, L. Stebbings, J. Hinton, J. Teague, S. McLaren, L. Mudie, C. Hardy, E. Anderson, O. Joseph, V. Goody, B. Robinson, M. Maddison, S. Gamble, C. Greenman, D. Berney, S. Hazell, N. Livni, Group the IP, C. Fisher, C. Ogden, P. Kumar, A. Thompson, C. Woodhouse, D. Nicol, E. Mayer, T. Dudderidge, N.C. Shah, V. Gnanapragasam, T. Voet, P. Campbell, A. Futreal, D. Easton, A.Y. Warren, C.S. Foster, M.R. Stratton, H.C. Whitaker, U. McDermott, D.S. Brewer, D.E. Neal, Analysis of the genetic phylogeny of multifocal prostate cancer identifies multiple independent clonal expansions in neoplastic and morphologically normal prostate tissue. Nat. Genet. **47**, 367 (2015)

23. P.C. Boutros, M. Fraser, N.J. Harding, R. De Borja, D. Trudel, E. Lalonde, A. Meng, P.H. Hennings-Yeomans, A. McPherson, V.Y. Sabelnykova, A. Zia, N.S. Fox, J. Livingstone, Y.J. Shiah, J. Wang, T.A. Beck, C.L. Have, T. Chong, M. Sam, J. Johns, L. Timms, N. Buchner, A. Wong, J.D. Watson, T.T. Simmons, C. P'ng, G. Zafarana, F. Nguyen, X. Luo, K.C. Chu, S.D. Prokopec, J. Sykes, A.D. Pra, A. Berlin, A. Brown, M.A. Chan-Seng-Yue, F. Yousif, R.E. Denroche, L.C. Chong, G.M. Chen, E. Jung, C. Fung, M.H.W. Starmans, H. Chen, S.K. Govind, J. Hawley, A. D'Costa, M. Pintilie, D. Waggott, F. Hach, P. Lambin, L.B. Muthuswamy, C. Cooper, R. Eeles, D. Neal, B. Tetu, C. Sahinalp, L.D. Stein, N. Fleshner, S.P. Shah, C.C. Collins, T.J. Hudson, J.D. McPherson, T. Van Der Kwast, R.G. Bristow, Spatial genomic heterogeneity within localized, multifocal prostate cancer. Nat. Genet. **47**(7), 736–745 (2015)

24. S.M.G. Espiritu, L.Y. Liu, Y. Rubanova, V. Bhandari, E.M. Holgersen, L.M. Szyca, N.S. Fox, M.L.K. Chua, T.N. Yamaguchi, L.E. Heisler, J. Livingstone, J. Wintersinger, F. Yousif, E. Lalonde, A. Rouette, A. Salcedo, K.E. Houlahan, C.H. Li, V. Huang, M. Fraser, T. van der Kwast, Q.D. Morris, R.G. Bristow, P.C. Boutros, The evolutionary landscape of localized prostate cancers drives clinical aggression. Cell **173**(4), 1003–1013.e15 (2018)

25. G. Gundem, P. Van Loo, B. Kremeyer, L.B. Alexandrov, J.M.C. Tubio, E. Papaemmanuil, D.S. Brewer, H.M.L. Kallio, G. Högnäs, M. Annala, K. Kivinummi, V. Goody, C. Latimer, S. O'Meara, K.J. Dawson, W. Isaacs, M.R. Emmert-Buck, M. Nykter, C. Foster, Z. Kote-Jarai, D. Easton, H.C. Whitaker, ICGC Prostate Group, D.E. Neal, C.S. Cooper, R.A. Eeles, T. Visakorpi, P.J. Campbell, U. McDermott, D.C. Wedge, G.S. Bova, The evolutionary history of lethal metastatic prostate cancer. Nature **520**(7547), 353–357 (2015)

26. S.C. Baca, D. Prandi, M.S. Lawrence, J.M. Mosquera, A. Romanel, Y. Drier, K. Park, N. Kitabayashi, T.Y. MacDonald, M. Ghandi, E. Van Allen, G.V. Kryukov, A. Sboner, J.P. Theurillat, T.D. Soong, E. Nickerson, D. Auclair, A. Tewari, H. Beltran, R.C. Onofrio, G. Boysen, C. Guiducci, C.E. Barbieri, K. Cibulskis, A. Sivachenko, S.L. Carter, G. Saksena, D. Voet, A.H. Ramos, W. Winckler, M. Cipicchio, K. Ardlie, P.W. Kantoff, M.F. Berger, S.B. Gabriel, T.R. Golub, M. Meyerson, E.S. Lander, O. Elemento, G. Getz, F. Demichelis, M.A. Rubin, L.A. Garraway, Punctuated evolution of prostate cancer genomes. Cell **153**(3), 666–677 (2013)

27. M. Fraser, V.Y. Sabelnykova, T.N. Yamaguchi, L.E. Heisler, J. Livingstone, V. Huang, Y.J. Shiah, F. Yousif, X. Lin, A.P. Masella, N.S. Fox, M. Xie, S.D. Prokopec, A. Berlin, E. Lalonde, M. Ahmed, D. Trudel, X. Luo, T.A. Beck, A. Meng, J. Zhang, A. D'Costa, R.E. Denroche, H. Kong, S.M.G. Espiritu, M.L.K. Chua, A. Wong, T. Chong, M. Sam, J. Johns, L. Timms, N.B. Buchner, M. Orain, V. Picard, H. Hovington, A. Murison, K. Kron, N.J. Harding, C. P'ng, K.E. Houlahan, K.C. Chu, B. Lo, F. Nguyen, C.H. Li, R.X. Sun, R. De Borja, C.I. Cooper, J.F. Hopkins, S.K. Govind, C. Fung, D. Waggott, J. Green, S. Haider, M.A. Chan-Seng-Yue, E. Jung, Z. Wang, A. Bergeron, A. Dal Pra, L. Lacombe, C.C. Collins, C. Sahinalp, M. Lupien, N.E. Fleshner, H.H. He, Y. Fradet, B. Tetu, T. Van Der Kwast, J.D. McPherson, R.G. Bristow, P.C. Boutros, Genomic hallmarks of localized, non-indolent prostate cancer. Nature **541**(7637), 359–364 (2017)

28. D. Robinson, E.M. Van Allen, Y.M. Wu, N. Schultz, R.J. Lonigro, J.M. Mosquera, B. Montgomery, M.E. Taplin, C.C. Pritchard, G. Attard, H. Beltran, W. Abida, R.K. Bradley, J. Vinson, X. Cao, P. Vats, L.P. Kunju, M. Hussain, F.Y. Feng, S.A. Tomlins, K.A. Cooney, D.C. Smith, C. Brennan, J. Siddiqui, R. Mehra, Y. Chen, D.E. Rathkopf, M.J. Morris, S.B. Solomon, J.C. Durack, V.E. Reuter, A. Gopalan, J. Gao, M. Loda, R.T. Lis, M. Bowden, S.P. Balk, G. Gaviola, C. Sougnez, M. Gupta, E.Y. Yu, E.A. Mostaghel, H.H. Cheng, H. Mulcahy, L.D. True, S.R. Plymate, H. Dvinge, R. Ferraldeschi, P. Flohr, S. Miranda, Z. Zafeiriou, N. Tunariu, J. Mateo, R. Perez-Lopez, F. Demichelis, B.D. Robinson, M. Schiffman, D.M. Nanus, S.T. Tagawa, A. Sigaras, K.W. Eng, O. Elemento, A. Sboner, E.I. Heath, H.I. Scher, K.J. Pienta, P. Kantoff, J.S. De Bono, M.A. Rubin, P.S. Nelson, L.A. Garraway, C.L. Sawyers, A.M. Chinnaiyan, Integrative clinical genomics of advanced prostate cancer. Cell **161**(5), 1215–1228 (2015)

29. C.C. Pritchard, C. Morrissey, A. Kumar, X. Zhang, C. Smith, I. Coleman, S.J. Salipante, J. Milbank, M. Yu, W.M. Grady, J.F. Tait, E. Corey, R.L. Vessella,

T. Walsh, J. Shendure, P.S. Nelson, Complex MSH2 and MSH6 mutations in hypermutated microsatellite unstable advanced prostate cancer. Nat. Commun. **5**, 1–6 (2014)

30. D. Nava Rodrigues, P. Rescigno, D. Liu, W. Yuan, S. Carreira, M.B. Lambros, G. Seed, J. Mateo, R. Riisnaes, S. Mullane, C. Margolis, D. Miao, S. Miranda, D. Dolling, M. Clarke, C. Bertan, M. Crespo, G. Boysen, A. Ferreira, A. Sharp, I. Figueiredo, D. Keliher, S. Aldubayan, K.P. Burke, S. Sumanasuriya, M.S. Fontes, D. Bianchini, Z. Zafeiriou, L.S. Teixeira Mendes, K. Mouw, M.T. Schweizer, C.C. Pritchard, S. Salipante, M.-E. Taplin, H. Beltran, M.A. Rubin, M. Cieslik, D. Robinson, E. Heath, N. Schultz, J. Armenia, W. Abida, H. Scher, C. Lord, A. D'Andrea, C.L. Sawyers, A.M. Chinnaiyan, A. Alimonti, P.S. Nelson, C.G. Drake, E.M. Van Allen, J.S. de Bono, Immunogenomic analyses associate immuno-logical alterations with mismatch repair defects in prostate cancer. J. Clin. Invest. **128**, 4441 (2018)

31. W. Abida, M.L. Cheng, J. Armenia, et al., Analysis of the prevalence of microsatellite instability in prostate cancer and response to immune checkpoint blockade. JAMA Oncol. **5**, 471 (2018)

32. R. Aggarwal, J. Huang, J.J. Alumkal, L. Zhang, F.Y. Feng, G.V. Thomas, A.S. Weinstein, V. Friedl, C. Zhang, O.N. Witte, P. Lloyd, M. Gleave, C.P. Evans, J. Youngren, T.M. Beer, M. Rettig, C.K. Wong, L. True, A. Foye, D. Playdle, C.J. Ryan, P. Lara, K.N. Chi, V. Uzunangelov, A. Sokolov, Y. Newton, H. Beltran, F. Demichelis, M.A. Rubin, J.M. Stuart, E.J. Small, Clinical and genomic characterization of treatment-emergent small-cell neuroendocrine pros-tate cancer: a multi-institutional prospective study. J. Clin. Oncol. **36**(24), 2492–2503 (2018)

33. D.A. Quigley, H.X. Dang, S.G. Zhao, P. Lloyd, R. Aggarwal, J.J. Alumkal, A. Foye, V. Kothari, M.D. Perry, A.M. Bailey, D. Playdle, T.J. Barnard, L. Zhang, J. Zhang, J.F. Youngren, M.P. Cieslik, A. Parolia, T.M. Beer, G. Thomas, K.N. Chi, M. Gleave, N.A. Lack, A. Zoubeidi, R.E. Reiter, M.B. Rettig, O. Witte, C.J. Ryan, L. Fong, W. Kim, T. Friedlander, J. Chou, H. Li, R. Das, H. Li, R. Moussavi-Baygi, H. Goodarzi, L.A. Gilbert, P.N. Lara, C.P. Evans, T.C. Goldstein, J.M. Stuart, S.A. Tomlins, D.E. Spratt, R.K. Cheetham, D.T. Cheng, K. Farh, J.S. Gehring, J. Hakenberg, A. Liao, P.G. Febbo, J. Shon, B. Sickler, S. Batzoglou, K.E. Knudsen, H.H. He, J. Huang, A.W. Wyatt, S.M. Dehm, A. Ashworth, A.M. Chinnaiyan, C.A. Maher, E.J. Small, F.Y. Feng, Genomic hall-marks and structural variation in metastatic prostate cancer. Cell **174**(3), 758–769.e9 (2018)

34. J. Armenia, S.A.M. Wankowicz, D. Liu, J. Gao, R. Kundra, E. Reznik, W.K. Chatila, D. Chakravarty, G.C. Han, I. Coleman, B. Montgomery, C. Pritchard, C. Morrissey, C.E. Barbieri, H. Beltran, A. Sboner, Z. Zafeiriou, S. Miranda, C.M. Bielski, A.V. Penson, C. Tolonen, F.W. Huang, D. Robinson, Y.M. Wu, R. Lonigro, L.A. Garraway, F. Demichelis, P.W. Kantoff, M.-E. Taplin, W. Abida, B.S. Taylor, H.I. Scher, P.S. Nelson, J.S. de Bono, M.A. Rubin, C.L. Sawyers, A.M. Chinnaiyan, N. Schultz, E.M. Van Allen, Team PIPCD, The long tail of oncogenic drivers in prostate cancer. Nat. Genet. **50**(5), 645–651 (2018)

35. W. Abida, J. Armenia, A. Gopalan, R. Brennan, M. Walsh, D. Barron, D. Danila, D. Rathkopf, M. Morris, S. Slovin, B. McLaughlin, K. Curtis, D.M. Hyman, J.C. Durack, S.B. Solomon, M.E. Arcila, A. Zehir, A. Syed, J. Gao, D. Chakravarty, H.A. Vargas, M.E. Robson, V. Vijai, K. Offit, M.T.A. Donoghue, A.A. Abeshouse, R. Kundra, Z.J. Heins, A.V. Penson, C. Harris, B.S. Taylor, M. Ladanyi, D. Mandelker, L. Zhang, V.E. Reuter, P.W. Kantoff, D.B. Solit, M.F. Berger, C.L. Sawyers, N. Schultz, H.I. Scher, Prospective genomic profiling of prostate cancer across disease states reveals germline and somatic alterations that may affect clinical decision making. JCO Precis. Oncol. **1**(1), 1–16 (2017)

36. W. Liu, S. Laitinen, S. Khan, M. Vihinen, J. Kowalski, G. Yu, L. Chen, C.M. Ewing, M.A. Eisenberger, M.A. Carducci, W.G. Nelson, S. Yegnasubramanian, J. Luo, Y. Wang, J. Xu, W.B. Isaacs, T. Visakorpi, G.S. Bova, Copy number analysis indicates mono-clonal origin of lethal metastatic prostate cancer. Nat. Med. **15**, 559 (2009)

37. A. Kumar, I. Coleman, C. Morrissey, X. Zhang, L.D. True, R. Gulati, R. Etzioni, H. Bolouri, B. Montgomery, T. White, J.M. Lucas, L.G. Brown, R.F. Dumpit, N. DeSarkar, C. Higano, E.Y. Yu, R. Coleman, N. Schultz, M. Fang, P.H. Lange, J. Shendure, R.L. Vessella, P.S. Nelson, Substantial interindividual and limited intraindividual genomic diversity among tumors from men with metastatic prostate cancer. Nat. Med. **22**, 369 (2016)

38. K. Hemminki, J. Ji, A. Försti, J. Sundquist, P. Lenner, Concordance of survival in family members with pros-tate cancer. J. Clin. Oncol. **26**(10), 1705–1709 (2008)

39. L.A. Mucci, J.B. Hjelmborg, J.R. Harris, K. Czene, D.J. Havelick, T. Scheike, R.E. Graff, K. Holst, S. Möller, R.H. Unger, C. McIntosh, E. Nuttall, I. Brandt, K.L. Penney, M. Hartman, P. Kraft, G. Parmigiani, K. Christensen, M. Koskenvuo, N.V. Holm, K. Heikkilä, E. Pukkala, A. Skytthe, H.-O. Adami, J. Kaprio, Collaboration for the NTS of C (NorTwinCan), Familial risk and heritability of can-cer among twins in nordic countries. JAMA **315**(1), 68–76 (2016)

40. S. Angèle, A. Falconer, S.M. Edwards, T. Dörk, M. Bremer, N. Moullan, B. Chapot, K. Muir, R. Houlston, A.R. Norman, S. Bullock, Q. Hope, J. Meitz, D. Dearnaley, A. Dowe, C. Southgate, A. Ardern-Jones, The Cancer Research UK/British Prostate Group/Association of Urological Surgeons S

of OC, D.F. Easton, R.A. Eeles, J. Hall, ATM polymorphisms as risk factors for prostate cancer development. Br. J. Cancer **91**, 783 (2004)

41. C. Engel, M. Loeffler, V. Steinke, N. Rahner, E. Holinski-Feder, W. Dietmaier, H.K. Schackert, H. Goergens, M. von Knebel Doeberitz, T.O. Goecke, W. Schmiegel, R. Buettner, G. Moeslein, T.G.W. Letteboer, E.G. García, F.J. Hes, N. Hoogerbrugge, F.H. Menko, T.A.M. van Os, R.H. Sijmons, A. Wagner, I. Kluijt, P. Propping, H.F.A. Vasen, Risks of less common cancers in proven mutation carriers with lynch syndrome. J. Clin. Oncol. **30**(35), 4409–4415 (2012)

42. Z. Kote-Jarai, D. Leongamornlert, E. Saunders, M. Tymrakiewicz, E. Castro, N. Mahmud, M. Guy, S. Edwards, L. O'Brien, E. Sawyer, A. Hall, R. Wilkinson, T. Dadaev, C. Goh, D. Easton, T.U. Collaborators, D. Goldgar, R. Eeles, BRCA2 is a moderate penetrance gene contributing to young-onset prostate cancer: implications for genetic testing in prostate cancer patients. Br. J. Cancer **105**, 1230 (2011)

43. S.M. Edwards, Z. Kote-Jarai, J. Meitz, R. Hamoudi, Q. Hope, P. Osin, R. Jackson, C. Southgate, R. Singh, A. Falconer, D.P. Dearnaley, A. Ardern-Jones, A. Murkin, A. Dowe, J. Kelly, S. Williams, R. Oram, M. Stevens, D.M. Teare, A.J. Bruce Ponder, S.A. Gayther, D.F. Easton, R.A. Eeles, Two percent of men with early-onset prostate cancer harbor germline mutations in the *BRCA2* gene. Am. J. Hum. Genet. **72**(1), 1–12 (2003)

44. R.A. Taylor, M. Fraser, J. Livingstone, S.M.G. Espiritu, H. Thorne, V. Huang, W. Lo, Y.-J. Shiah, T.N. Yamaguchi, A. Sliwinski, S. Horsburgh, A. Meng, L.E. Heisler, N. Yu, F. Yousif, M. Papargiris, M.G. Lawrence, L. Timms, D.G. Murphy, M. Frydenberg, J.F. Hopkins, D. Bolton, D. Clouston, J.D. McPherson, T. van der Kwast, P.C. Boutros, G.P. Risbridger, R.G. Bristow, Germline BRCA2 mutations drive prostate cancers with distinct evolutionary trajectories. Nat. Commun. **8**, 13671 (2017)

45. R. Na, S.L. Zheng, M. Han, H. Yu, D. Jiang, S. Shah, C.M. Ewing, L. Zhang, K. Novakovic, J. Petkewicz, K. Gulukota, D.L. Helseth Jr., M. Quinn, E. Humphries, K.E. Wiley, S.D. Isaacs, Y. Wu, X. Liu, N. Zhang, C.-H. Wang, J. Khandekar, P.J. Hulick, D.H. Shevrin, K.A. Cooney, Z. Shen, A.W. Partin, H.B. Carter, M.A. Carducci, M.A. Eisenberger, S.R. Denmeade, M. McGuire, P.C. Walsh, B.T. Helfand, C.B. Brendler, Q. Ding, J. Xu, W.B. Isaacs, Germline mutations in *ATM* and *BRCA1/2* distinguish risk for lethal and indolent prostate cancer and are associated with early age at death. Eur. Urol. **71**(5), 740–747 (2017)

46. E. Castro, N. Romero-Laorden, A. del Pozo, R. Lozano, A. Medina, J. Puente, J.M. Piulats, D. Lorente, M.I. Saez, R. Morales-Barrera, E. Gonzalez-Billalabeitia, Y. Cendón, I. García-Carbonero, P. Borrega, M.J. Mendez Vidal, A. Montesa, P. Nombela, E. Fernández-Parra, A. Gonzalez del Alba, J.C. Villa-Guzmán,

K. Ibáñez, A. Rodriguez-Vida, L. Magraner-Pardo, B. Perez-Valderrama, E. Vallespín, E. Gallardo, S. Vazquez, C.C. Pritchard, P. Lapunzina, D. Olmos, PROREPAIR-B: a prospective cohort study of the impact of germline DNA repair mutations on the outcomes of patients with metastatic castration-resistant prostate cancer. J. Clin. Oncol. **18**, 00358 (2019)

47. V.M. Raymond, B. Mukherjee, F. Wang, S.-C. Huang, E.M. Stoffel, F. Kastrinos, S. Syngal, K.A. Cooney, S.B. Gruber, Elevated risk of prostate cancer among men with lynch syndrome. J. Clin. Oncol. **31**(14), 1713–1718 (2013)

48. H. Hampel, W.L. Frankel, E. Martin, M. Arnold, K. Khanduja, P. Kuebler, M. Clendenning, K. Sotamaa, T. Prior, J.A. Westman, J. Panescu, D. Fix, J. Lockman, J. LaJeunesse, I. Comeras, A. de la Chapelle, Feasibility of screening for lynch syndrome among patients with colorectal cancer. J. Clin. Oncol. **26**(35), 5783–5788 (2008)

49. S. Ryan, M.A. Jenkins, A.K. Win, Risk of prostate cancer in lynch syndrome: a systematic review and meta-analysis. Cancer Epidemiol. Biomarkers Prev. **23**(3), 437 LP–437449 (2014)

50. H.B. Carter, B. Helfand, M. Mamawala, Y. Wu, P. Landis, H. Yu, K. Wiley, R. Na, Z. Shi, J. Petkewicz, S. Shah, R.J. Fantus, K. Novakovic, C.B. Brendler, S.L. Zheng, W.B. Isaacs, J. Xu, Germline mutations in *ATM* and *BRCA1/2* are associated with grade reclassification in men on active surveillance for prostate cancer. Eur. Urol. **75**, 743 (2019)

51. E. Castro, C. Goh, D. Leongamornlert, E. Saunders, M. Tymrakiewicz, T. Dadaev, K. Govindasami, M. Guy, S. Ellis, D. Frost, E. Bancroft, T. Cole, M. Tischkowitz, M.J. Kennedy, J. Eason, C. Brewer, D.G. Evans, R. Davidson, D. Eccles, M.E. Porteous, F. Douglas, J. Adlard, A. Donaldson, A.C. Antoniou, Z. Kote-Jarai, D.F. Easton, D. Olmos, R. Eeles, Effect of *BRCA* mutations on metastatic relapse and cause-specific survival after radical treatment for localised prostate cancer. Eur. Urol. **68**(2), 186–193 (2015)

52. M. Annala, W.J. Struss, E.W. Warner, K. Beja, G. Vandekerkhove, A. Wong, D. Khalaf, I.-L. Seppälä, A. So, G. Lo, R. Aggarwal, E.J. Small, M. Nykter, M.E. Gleave, K.N. Chi, A.W. Wyatt, Treatment outcomes and tumor loss of heterozygosity in germline DNA repair-deficient prostate cancer. Eur. Urol. **72**(1), 34–42 (2017)

53. M. Annala, G. Vandekerkhove, D. Khalaf, S. Taavitsainen, K. Beja, E.W. Warner, K. Sunderland, C. Kollmannsberger, B.J. Eigl, D. Finch, C.D. Oja, J. Vergidis, M. Zulfiqar, A.A. Azad, M. Nykter, M.E. Gleave, A.W. Wyatt, K.N. Chi, Circulating tumor DNA genomics correlate with resistance to abiraterone and enzalutamide in prostate cancer. Cancer Discov. **8**(4), 444–457 (2018)

54. J. Mateo, H.H. Cheng, H. Beltran, D. Dolling, W. Xu, C.C. Pritchard, H. Mossop, P. Rescigno, R. Perez-Lopez, V. Sailer, M. Kolinsky, A. Balasopoulou, C. Bertan, D.M. Nanus, S.T. Tagawa, H. Thorne,

B. Montgomery, S. Carreira, S. Sandhu, M.A. Rubin, P.S. Nelson, J.S. de Bono, Clinical outcome of prostate cancer patients with germline DNA repair mutations: retrospective analysis from an international study. Eur. Urol. **73**(5), 687–693 (2018)

55. E.S. Antonarakis, F. Shaukat, P. Isaacsson Velho, H. Kaur, E. Shenderov, D.M. Pardoll, T.L. Lotan, Clinical features and therapeutic outcomes in men with advanced prostate cancer and DNA mismatch repair gene mutations. Eur. Urol. **75**, 378–382 (2019)

56. D.J. Slamon, B. Leyland-Jones, S. Shak, H. Fuchs, V. Paton, A. Bajamonde, T. Fleming, W. Eiermann, J. Wolter, M. Pegram, J. Baselga, L. Norton, Use of chemotherapy plus a monoclonal antibody against HER2 for metastatic breast cancer that overexpresses HER2. N. Engl. J. Med. **344**(11), 783–792 (2001)

57. N.J. O'Neil, M.L. Bailey, P. Hieter, Synthetic lethality and cancer. Nat. Rev. Genet. **18**(10), 613–623 (2017)

58. H.E. Bryant, N. Schultz, H.D. Thomas, K.M. Parker, D. Flower, E. Lopez, S. Kyle, M. Meuth, N.J. Curtin, T. Helleday, Specific killing of BRCA2-deficient tumours with inhibitors of poly(ADP-ribose) polymerase. Nature **434**(7035), 913–917 (2005)

59. B. Kaufman, R. Shapira-Frommer, R.K. Schmutzler, M.W. Audeh, M. Friedlander, J. Balmaña, G. Mitchell, G. Fried, S.M. Stemmer, A. Hubert, O. Rosengarten, M. Steiner, N. Loman, K. Bowen, A. Fielding, S.M. Domchek, Olaparib monotherapy in patients with advanced cancer and a germline BRCA1/2 mutation. J. Clin. Oncol. **33**(3), 244–250 (2014)

60. N. McCabe, N.C. Turner, C.J. Lord, K. Kluzek, A. Białkowska, S. Swift, S. Giavara, M.J. O'Connor, A.N. Tutt, M.Z. Zdzienicka, G.C.M. Smith, A. Ashworth, Deficiency in the repair of DNA damage by homologous recombination and sensitivity to poly(ADP-ribose) polymerase inhibition. Cancer Res. **66**(16), 8109–8115 (2006)

61. J. Murai, S.-Y.N. Huang, A. Renaud, Y. Zhang, J. Ji, S. Takeda, J. Morris, B. Teicher, J.H. Doroshow, Y. Pommier, Stereospecific PARP trapping by BMN 673 and comparison with olaparib and rucaparib. Mol. Cancer Ther. **13**(2), 433 LP–433443 (2014)

62. J. Murai, S.N. Huang, B.B. Das, A. Renaud, Y. Zhang, J.H. Doroshow, J. Ji, S. Takeda, Y. Pommier, Trapping of PARP1 and PARP2 by clinical PARP inhibitors. Cancer Res. **72**(21), 5588–5599 (2012)

63. C.J. Lord, A. Ashworth, BRCAness revisited. Nat. Rev. Cancer **16**(2), 110–120 (2016)

64. J.S. de Bono, H.I. Scher, R.B. Montgomery, C. Parker, M.C. Miller, H. Tissing, G.V. Doyle, L.W.W.M. Terstappen, K.J. Pienta, D. Raghavan, Circulating tumor cells predict survival benefit from treatment in metastatic castration-resistant prostate cancer. Clin. Cancer Res. **14**(19), 6302–6309 (2008)

65. H.I. Scher, G. Heller, A. Molina, G. Attard, D.C. Danila, X. Jia, W. Peng, S.K. Sandhu, D. Olmos, R. Riisnaes, R. McCormack, T. Burzykowski, T. Kheoh, M. Fleisher, M. Buyse, J.S. de Bono, Circulating tumor cell biomarker panel as an individual-level surrogate for survival in metastatic castration-resistant prostate cancer. J. Clin. Oncol. **33**(12), 1348–1355 (2015)

66. J. Mateo, S. Carreira, S. Sandhu, S. Miranda, H. Mossop, R. Perez-Lopez, D. Nava Rodrigues, D. Robinson, A. Omlin, N. Tunariu, G. Boysen, N. Porta, P. Flohr, A. Gillman, I. Figueiredo, C. Paulding, G. Seed, S. Jain, C. Ralph, A. Protheroe, S. Hussain, R. Jones, T. Elliott, U. McGovern, D. Bianchini, J. Goodall, Z. Zafeiriou, C.T. Williamson, R. Ferraldeschi, R. Riisnaes, B. Ebbs, G. Fowler, D. Roda, W. Yuan, Y.-M. Wu, X. Cao, R. Brough, H. Pemberton, R. A'Hern, A. Swain, L.P. Kunju, R. Eeles, G. Attard, C.J. Lord, A. Ashworth, M.A. Rubin, K.E. Knudsen, F.Y. Feng, A.M. Chinnaiyan, E. Hall, J.S. de Bono, DNA-repair defects and olaparib in metastatic prostate cancer. N. Engl. J. Med. **373**(18), 1697–1708 (2015)

67. J. Goodall, J. Mateo, W. Yuan, H. Mossop, N. Porta, S. Miranda, R. Perez-Lopez, D. Dolling, D.R. Robinson, S. Sandhu, G. Fowler, B. Ebbs, P. Flohr, G. Seed, D.N. Rodrigues, G. Boysen, C. Bertan, M. Atkin, M. Clarke, M. Crespo, I. Figueiredo, R. Riisnaes, S. Sumanasuriya, P. Rescigno, Z. Zafeiriou, A. Sharp, N. Tunariu, D. Bianchini, A. Gillman, C.J. Lord, E. Hall, A.M. Chinnaiyan, S. Carreira, J.S. De Bono, Circulating cell-free DNA to guide prostate cancer treatment with PARP inhibition. Cancer Discov. **7**(9), 1006–1017 (2017)

68. D. Quigley, J.J. Alumkal, A.W. Wyatt, V. Kothari, A. Foye, P. Lloyd, R. Aggarwal, W. Kim, E. Lu, J. Schwartzman, K. Beja, M. Annala, R. Das, M. Diolaiti, C. Pritchard, G. Thomas, S. Tomlins, K. Knudsen, C.J. Lord, C. Ryan, J. Youngren, T.M. Beer, A. Ashworth, E.J. Small, F.Y. Feng, Analysis of circulating cell-free DnA identifies multiclonal heterogeneity of BRCA2 reversion mutations associated with resistance to PARP inhibitors. Cancer Discov. **7**(9), 999–1005 (2017)

69. S.L. Edwards, R. Brough, C.J. Lord, R. Natrajan, R. Vatcheva, D.A. Levine, J. Boyd, J.S. Reis-Filho, A. Ashworth, Resistance to therapy caused by intragenic deletion in BRCA2. Nature **451**, 1111 (2008)

70. L. Li, S. Karanika, G. Yang, J. Wang, S. Park, B.M. Broom, G.C. Manyam, W. Wu, Y. Luo, S. Basourakos, J.H. Song, G.E. Gallick, T. Karantanos, D. Korentzelos, A.K. Azad, J. Kim, P.G. Corn, A.M. Aparicio, C.J. Logothetis, P. Troncoso, T. Heffernan, C. Toniatti, H.S. Lee, J.S. Lee, X. Zuo, W. Chang, J. Yin, T.C. Thompson, Androgen receptor inhibitor-induced "BRCAness" and PARP inhibition are synthetically lethal for castration-resistant prostate cancer. Sci. Signal. **10**(480), eaam7479 (2017)

71. M. Asim, F. Tarish, H.I. Zecchini, K. Sanjiv, E. Gelali, C.E. Massie, A. Baridi, A.Y. Warren, W. Zhao, C. Ogris, L.A. McDuffus, P. Mascalchi, G. Shaw, H. Dev, K. Wadhwa, P. Wijnhoven, J.V. Forment, S.R. Lyons, A.G. Lynch, C. O'Neill, V.R. Zecchini, P.S. Rennie, A. Baniahmad, S. Tavaré, I.G. Mills, Y. Galanty, N. Crosetto, N. Schultz, D. Neal,

T. Helleday, Synthetic lethality between androgen receptor signalling and the PARP pathway in prostate cancer. Nat. Commun. **8**(1), 374 (2017)

72. M.J. Schiewer, J.F. Goodwin, S. Han, J. Chad Brenner, M.A. Augello, J.L. Dean, F. Liu, J.L. Planck, P. Ravindranathan, A.M. Chinnaiyan, P. McCue, L.G. Gomella, G.V. Raj, A.P. Dicker, J.R. Brody, J.M. Pascal, M.M. Centenera, L.M. Butler, W.D. Tilley, F.Y. Feng, K.E. Knudsen, Dual roles of PARP-1 promote cancer growth and progression. Cancer Discov. **2**(12), 1134–1149 (2012)

73. N. Clarke, P. Wiechno, B. Alekseev, N. Sala, R. Jones, I. Kocak, V.E. Chiuri, J. Jassem, A. Fléchon, C. Redfern, C. Goessl, J. Burgents, R. Kozarski, D. Hodgson, M. Learoyd, F. Saad, Olaparib combined with abiraterone in patients with metastatic castration-resistant prostate cancer: a randomised, double-blind, placebo-controlled, phase 2 trial. Lancet Oncol. **19**(7), 975–986 (2018)

74. C.N. Sternberg, D.P. Petrylak, O. Sartor, J.A. Witjes, T. Demkow, J.-M. Ferrero, J.-C. Eymard, S. Falcon, F. Calabrò, N. James, I. Bodrogi, P. Harper, M. Wirth, W. Berry, M.E. Petrone, T.J. McKearn, M. Noursalehi, M. George, M. Rozencweig, Multinational, double-blind, phase III study of prednisone and either satraplatin or placebo in patients with castrate-refractory prostate cancer progressing after prior chemotherapy: the SPARC trial. J. Clin. Oncol. **27**(32), 5431–5438 (2009)

75. M.M. Pomerantz, S. Spisák, L. Jia, A.M. Cronin, I. Csabai, E. Ledet, A.O. Sartor, I. Rainville, E.P. O'Connor, Z.T. Herbert, Z. Szállási, W.K. Oh, P.W. Kantoff, J.E. Garber, D. Schrag, A.S. Kibel, M.L. Freedman, The association between germline BRCA2 variants and sensitivity to platinum-based chemotherapy among men with metastatic prostate cancer. Cancer **123**(18), 3532–3539 (2017)

76. H.H. Cheng, C.C. Pritchard, T. Boyd, P.S. Nelson, B. Montgomery, Biallelic inactivation of *BRCA2* in platinum-sensitive metastatic castration-resistant prostate cancer. Eur. Urol. **69**(6), 992–995 (2016)

77. Z. Zafeiriou, D. Bianchini, R. Chandler, P. Rescigno, W. Yuan, S. Carreira, M. Barrero, A. Petremolo, S. Miranda, R. Riisnaes, D.N. Rodrigues, B. Gurel, S. Sumanasuriya, A. Paschalis, A. Sharp, J. Mateo, N. Tunariu, A.M. Chinnaiyan, C.C. Pritchard, K. Kelly, J.S. de Bono, Genomic analysis of three metastatic prostate cancer patients with exceptional responses to carboplatin indicating different types of DNA repair deficiency. Eur. Urol. **75**(1), 184–192 (2019)

78. P.M. Reaper, M.R. Griffiths, J.M. Long, J.D. Charrier, S. MacCormick, P.A. Charlton, J.M.C. Golec, J.R. Pollard, Selective killing of ATM- or p53-deficient cancer cells through inhibition of ATR. Nat. Chem. Biol. **7**(7), 428–430 (2011)

79. H.J. Landau, S.C. McNeely, J.S. Nair, R.L. Comenzo, T. Asai, H. Friedman, S.C. Jhanwar, S.D. Nimer, G.K. Schwartz, The checkpoint kinase inhibitor AZD7762 potentiates chemotherapy-induced apoptosis of p53-mutated multiple myeloma cells. Mol. Cancer Ther. **11**(8), 1781–1788 (2012)

80. D. Blazek, J. Kohoutek, K. Bartholomeeusen, E. Johansen, P. Hulinkova, Z. Luo, P. Cimermancic, J. Ule, B.M. Peterlin, The cyclin K/Cdk12 complex maintains genomic stability via regulation of expression of DNA damage response genes. Genes Dev. **25**(20), 2158–2172 (2011)

81. Y.M. Wu, M. Cieślik, R.J. Lonigro, P. Vats, M.A. Reimers, X. Cao, Y. Ning, L. Wang, L.P. Kunju, N. de Sarkar, E.I. Heath, J. Chou, F.Y. Feng, P.S. Nelson, J.S. de Bono, W. Zou, B. Montgomery, A. Alva, D.R. Robinson, A.M. Chinnaiyan, Inactivation of CDK12 delineates a distinct immunogenic class of advanced prostate cancer. Cell **173**(7), 1770–1782.e14 (2018)

82. S.R. Viswanathan, G. Ha, A.M. Hoff, J.A. Wala, J. Carrot-Zhang, C.W. Whelan, N.J. Haradhvala, S.S. Freeman, S.C. Reed, J. Rhoades, P. Polak, M. Cipicchio, S.A. Wankowicz, A. Wong, T. Kamath, Z. Zhang, G.J. Gydush, D. Rotem, J.C. Love, G. Getz, S. Gabriel, C.Z. Zhang, S.M. Dehm, P.S. Nelson, E.M. Van Allen, A.D. Choudhury, V.A. Adalsteinsson, R. Beroukhim, M.E. Taplin, M. Meyerson, Structural alterations driving castration-resistant prostate cancer revealed by linked-read genome sequencing. Cell **174**(2), 433–447.e19 (2018)

83. C.C. Pritchard, J. Mateo, M.F. Walsh, N. De Sarkar, W. Abida, H. Beltran, A. Garofalo, R. Gulati, S. Carreira, R. Eeles, O. Elemento, M.A. Rubin, D. Robinson, R. Lonigro, M. Hussain, A. Chinnaiyan, J. Vinson, J. Filipenko, L. Garraway, M.-E. Taplin, S. AlDubayan, G.C. Han, M. Beightol, C. Morrissey, B. Nghiem, H.H. Cheng, B. Montgomery, T. Walsh, S. Casadei, M. Berger, L. Zhang, A. Zehir, J. Vijai, H.I. Scher, C. Sawyers, N. Schultz, P.W. Kantoff, D. Solit, M. Robson, E.M. Van Allen, K. Offit, J. de Bono, P.S. Nelson, Inherited DNA-repair gene mutations in men with metastatic prostate cancer. N. Engl. J. Med. **375**(5), 443–453 (2016)

The Role of RB in Prostate Cancer Progression

Deborah L. Burkhart, Katherine L. Morel, Anjali V. Sheahan, Zachary A. Richards, and Leigh Ellis

Introduction

Loss of chromosome arm 13q, which contains the RB tumor suppressor, was described in prostate carcinoma cells as early as 1990, and RB protein loss was confirmed shortly thereafter in a subset of primary prostate samples [1, 2]. More recent estimates suggest that RB is directly deleted or mutated in 1–2% of primary acinar disease [3, 4], although this may be an underestimate of the presence of RB deficient sub-clones within primary disease, with deep deletions detectable in microdissected foci [5]. RB-deficiency is highly correlated with recurrent, castration resistant disease [4, 6–8]. Specifically, RB loss has consistently been described in a subset of prostate tumors with phenotypically neuroendocrine features, known as small cell carcinoma. In this subset, RB loss has been observed in up to 90% [9]. In this chapter we review the pleiotropic functions

D. L. Burkhart · K. L. Morel · A. V. Sheahan
Z. A. Richards
Department of Oncologic Pathology, Dana-Farber
Cancer Institute, Boston, MA, USA

L. Ellis (✉)
Department of Oncologic Pathology, Dana-Farber
Cancer Institute, Boston, MA, USA

Department of Pathology, Brigham and Women's
Hospital, Boston, MA, USA

The Broad Institute, Cambridge, MA, USA
e-mail: leigh_ellis@dfci.harvard.edu

of RB and the consequences of RB loss specifically in prostate cancer development.

Structure and Classical Cell Cycle Functions of RB

As the first bona fide tumor suppressor to be identified, RB has been the subject of decades of studies into its structure and function. RB contains three major domains—the N- and C-terminal domains, and a central "pocket", which in turn is comprised of subdomains A and B connected by a linker [10]. Each domain includes several distinct protein-binding surfaces that are highly conserved [11, 12]. More than 200 cellular proteins physically associate with the RB protein through these regions, leading to the diverse plethora of functions associated with RB function. This extensive list of binding partners includes viral oncoproteins, transcription factors, and chromatin associated proteins (reviewed in [12, 13]), and, especially of note for prostate cancer, RB interacts directly with the androgen receptor (AR) [14], described in more detail in the sections below.

The RB-Pathway

Functional RB has a potent anti-proliferative effect [2, 15, 16]. Following the observation that both tumor-associated genomic deletions and viral

© Springer Nature Switzerland AG 2019
S. M. Dehm, D. J. Tindall (eds.), *Prostate Cancer*, Advances in Experimental Medicine
and Biology 1210, https://doi.org/10.1007/978-3-030-32656-2_13

oncoproteins disrupt an association between active, hypophosphorylated RB and the family of E2F transcription factors [17, 18], work focused on the role of RB as a transcriptional cofactor serving as a critical link between mitogenic signaling and commitment of the cell to cycle via E2F [19]. Mitogenic signaling results in expression and activation of cyclin dependent kinase activity, specifically Cyclin D/CDK4 and Cyclin D/CDK6 complexes that phosphorylate RB. RB is then sequentially phosphorylated by additional Cyclin/CDK complexes, such as Cyclin E/CDK2 during late G1 and S-phase. This hyperphosphorylation of RB results in release of the E2F transcription factors and activation of E2F target genes necessary for S-phase and cell cycle progression [20]. Genes in this so-called "RB-pathway" (Fig. 1) include the *CDKN2a* locus, which expresses the CDK4 inhibitor, p16^{INK4A}. Mutations in genes at any level of this pathway result in decreased RB function and consequently increased E2F-activity, and mutations in this pathway are observed frequently in human cancers in a mutually exclusive pattern.

The high frequency of cancer-associated alterations in the genes encoding p16, Cyclin D/CDK4, or RB resulted in the designation of alterations in this pathway as a hallmark of cancer [21–23]. Despite the linear description of this pathway, mutations at each level of the pathway are not equivalent, and alterations in individual components are more commonly observed in specific cellular contexts [24]. In the case of prostate cancer, both *RB* and *CDKN2A* deletions have been observed, as has amplification of *CCND1*; however, direct genomic loss of *RB* is highly overrepresented in most lethal, metastatic disease [6, 25]. Furthermore, increased expression of both E2F1 and E2F3 has been noted in prostate adenocarcinoma and castrate-resistant prostate cancer (CRPC) [6, 26]. Interestingly, loss of PTEN, one of the most commonly observed alterations in human prostate cancers, results in upregulation of Cyclin D and, therefore,

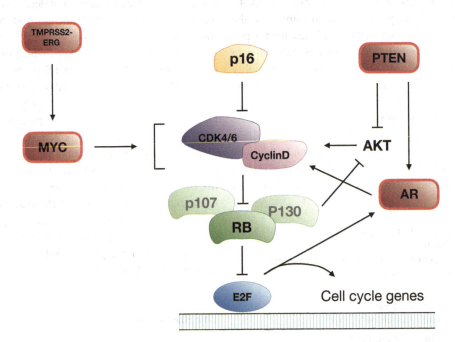

Fig. 1 RB-pathway downstream of major prostate cancer drivers. Simplified schematic of the RB-Pathway: RB and its family members, p107 and p130, negatively regulate the E2F family of transcription factors, thereby regulating the cell cycle through expression of genes required for the G1/S transition. The RB-family is functionally inactivated through hyperphosphorylation modifications via Cyclin D-dependent complexes. p16 is a negative regulator of the CDK/Cyclin complexes. In red are interactions between some of the genetic drivers most commonly observed in prostate cancer and their effect on the RB-pathway

hyperphosphorylation and inactivation of RB [27]. Similarly, overexpression of c-Myc, another of the most common prostate cancer driving oncogenes, is capable of immortalizing normal prostate cells without requiring additional RB mutations [28], suggesting that MYC-driven tumor initiation does not require RB loss. However, selection for direct loss of RB appears to still occur in more aggressive tumors, particularly metastatic CRPC, suggesting that indirect overexpression of cyclins, which results from other tumor-initiating mutations, alone is insufficient to remove all of tumor suppressive functions of RB.

The RB Family

RB is one member of a three-member family known as the "pocket proteins", so named because all three proteins contain an LXCXE binding pocket capable of binding to an overlapping set of protein binding partners, including the E2F family [29]. The activity of all three pocket proteins is regulated via cell cycle dependent phosphorylation as described for RB above (reviewed in [12, 30]), and activity of the entire family is therefore affected by altered expression of upstream regulators. In the mouse prostate, inactivation of the entire pocket protein family results in an increase of proliferation as well as Pten-dependent apoptosis [31], and direct and indirect inactivation of the entire RB-family by either loss of p18 or overexpression of CDK4 cooperates synergistically with loss of heterozygosity (LOH) of PTEN in order to activate Akt [32]. No known association exists between individual loss of either p107 or p130 expression in prostate cancer; however, some reports suggest that p107 and p130 mediate cell cycle arrest and apoptosis in prostate cells deficient for RB in response to therapeutic agents [33, 34] and radiation [35, 36].

RB and the Cell Cycle

As described above, the textbook description of the function of RB is to repress the activity of the E2F transcription factors; specifically, release of

the E2F transcription factors following inactivation of RB drives the G1/S transition of the cell cycle. Transcriptional targets of the E2F transcription factors regulate all stages of the cell cycle, from G1 (*CCND1, CCND3, MYC*), the G1/S transition (*CCNE1, CDK2, E2F1–3, MYB, MYBL2*), DNA synthesis (*CDC6, DHFR, MCM2–7, PCNA*), and the S/G2 transition (*AURKB, CCNA1, CDC2, PLK*) [37–39]. The progressive transition from hypophosphorylated RB in G0 to hyperphosphorylated RB and its release of E2F is thought to underpin the restriction point observed in mammalian cellular proliferation and is described in detail elsewhere [24, 39]. Active, hypophosphorylated RB is conversely an important mediator of cell cycle arrest.

Beyond direct regulation of genes required for proliferation, including DNA synthesis, RB-E2F complexes also regulate the cell cycle indirectly. For example, two components of the PRC2 Polycomb complex, *EZH2* and *EED* are direct E2F transcriptional targets [40, 41]. Unlike other E2F target genes that produce the machinery necessary for DNA synthesis and the cell cycle, the Polycomb complex methylates histones, generally for the purpose of long-term epigenetic silencing. Appropriate expression of these two genes is essential for cellular proliferation [41]. RB family members are also required for PRC2-dependent methylation of the p16 locus [42], which negatively regulates the cell cycle.

Furthermore, RB regulates cell cycle progression independently of its interaction with E2F. One of these functions is through the direct interaction between RB and SKP2. Through this interaction, RB physically blocks SKP2, thereby stabilizing the CDK inhibitor p27, and also mediates an interaction between SKP2 and the APC complex, resulting in degradation of SKP2 (reviewed in [12]). RB fulfills its function to prevent a cell from cycling through both of these interactions.

The Role of RB Regulation of the Cell Cycle in Prostate Cancer

The RB-pathway is a downstream mediator of the androgen receptor (AR); activation of AR through natural and synthetic androgens results in cellular proliferation, correlating with high levels of Cyclin/CDK activity and increased phosphorylation of RB [43–45]. Down-regulation of AR in both AR-dependent and independent prostate cancer cells increases hypophosphorylation of RB protein, which correlates with a G0/G1 cell cycle arrest [46–48]. Furthermore, overexpression of E2F1 alone can bypass androgen deprivation and enforce proliferation in otherwise androgen sensitive cells [49], confirming the concept that E2F is a critical mediator of RB in cell cycle progression downstream of AR (for further discussion of the RB-AR relationship, see below). Within prostate models, as in other tumor types, disruption of the RB-E2F complex by homozygous deletion of RB results in increased E2F activity and activation of E2F target genes *in vitro* and predisposes prostatic epithelium to hormone-driven carcinogenesis *in vivo* [50–52].

Loss of RB in many cell types and tissues leads to hyperproliferation, in some cases tied to an increase in apoptosis; whereas overexpression or re-introduction of RB reduces and arrests proliferation of cells. Similarly, in prostate cells, reintroduction of chromosome 13 reduces proliferation [53]. Broadly speaking, in many genetic systems, RB loss has only a mild effect on prostate-derived tumors and tissues. RB-deficient prostate epithelial cells retain the ability to undergo cell cycle arrest in the absence of serum, and the expression of many known cell cycle regulated genes is similar in wild-type and knockout cells [51]. Genetic loss of RB function induces hyperproliferation in the mouse prostate, correlating with an increase in expression of known E2F-regulated, cell cycle genes [50, 52, 54, 55]. Similarly, overexpression of the upstream regulator Cyclin D results in an increase in proliferation of LNCaP cells [56], and both overexpression of Cyclin D and inactivation of RB have minimal effects on an *ex vivo* stratified epithelial model of human primary prostate cells [57]. These proliferative alterations are generally mild, and loss of RB alone is insufficient to induce prostate tumor development in mice [55]. However, when loss of RB is combined with loss of the tumor suppressor p53 in genetically defined mouse models, loss of both genes is sufficient to induce metastatic carcinomas [58–63] (Table 1).

Although loss of RB alone has mild effects on the cellular proliferation of in the prostate, its absence is noticeable when prostate cells are

Table 1 Summary of Rb in mouse models of prostate cancer

Pb-Tag (TRAMP) [58–61] (Transgenic Adenocarcinoma of the Mouse Prostate)	The rat probasin promoter drives prostate-specific expression of the entire SV40 large T antigen, resulting in hyperplasia as early as 10 weeks, invasive adenocarcinoma as early as 18 weeks, and lymph node metastases in 31% of mice between 18 and 24 weeks, and pulmonary metastases in 36% by 24 weeks. Tumorigenesis is largely castration-resistant, with mice castrated at 12 weeks developing neuroendocrine carcinomas with metastases by 20–30 weeks of age. Large T antigen is known to bind to and inactivate both Rb and p53, as well as other binding partners.
Rb$^{-/-}$ reconstitution [52, 54]	PrE cells derived from Rb$^{-/-}$ fetal tissue have increased proliferation and DNA ploidy but retain differentiation potential when combined with wild-type UGSM *in vivo*, despite mild hyperplasia.
Rb$^{lox/lox}$;PbCre4 [55]	Prostate specific, conditional deletion of Rb1 in the mouse prostate results in multifocal hyperplasia that does not progress to advanced disease.
T$_{121}$ [61]	Prostate-specific expression of the Rb-family binding region of SV40 large T antigen (T$_{121}$) results in mPIN with increased proliferation and apoptosis. Microinvasive adenocarcinoma lesions positive for luminal markers developed. Pten heterozygosity, but not deletion of p53, reduced the apoptotic index; Pten heterozygosity accelerated the development of adenocarcinoma, without evidence of neuroendocrine tumors.
Rb$^{lox/lox}$;p53$^{lox/lox}$;PbCre4 [62, 63]	Combined deletion of Rb1 and Trp53 within the mouse prostate results in adenocarcinoma with neuroendocrine markers.
Rb$^{lox/lox}$;Pten$^{lox/lox}$;PbCre4 [64]	Loss of Rb1 in a Pten-initiated mouse model of prostate adenocarcinoma results in an increase of altered lineage markers, specifically with an increase in neuroendocrine markers. The expression pattern of these tumors reflects human neuroendocrine disease.
Rb$^{lox/lox}$;Pten$^{lox/lox}$;p53$^{lox/lox}$;PbCre4 [64]	Deletion of Rb1 and Trp53 in a Pten-initiated mouse model of prostate adenocarcinoma results in altered lineage markers, as well as castration resistance.

therapeutically challenged. A number of compounds which have an antiproliferative effect on human prostate cancer derived cells rely on the presence of a functional RB to exert their effects [65–67]. Furthermore, RB-status has an effect on sensitivity to irradiation and chemotherapeutics (reviewed in [68]). Most prominently, under conditions of androgen deprivation, LNCaP prostate cancer cells undergo G1-arrest, which is dependent on presence of functional RB [48] and can be overcome by overexpression of E2F1 [49]. Interestingly, overexpression of Cyclin D1 is unable to reverse this arrest in the absence of androgens, while transfection with viral oncoproteins, which bind to all three pocket proteins, is able to reverse the arrest [48]. Restoration of testosterone increases CDK/Cyclin activity and proliferation [69] (reviewed in [70]). These results suggest that the anti-tumor, antiproliferative effects of androgen deprivation therapy (ADT) rely on the presence of a functional RB protein, which is consistent with the observation that most recurrent tumors following ADT lack functional RB.

Complex Relationship Between RB and the Androgen Receptor

AR is a steroid hormone receptor that acts as a nuclear transcription factor following activation through binding to androgen. Following ligand-dependent activation, AR forms a dimer in order to activate transcriptional targets such as *TMPRSS2, NKX3.1 and KLK3* (PSA) and to repress other targets [71]. Androgen receptor signaling is required for development of a normal prostate gland, as well as for the maintenance of a mature adult gland. Therefore, castration or deprivation of androgens results in epithelial apoptosis (reviewed in [72]). In the normal prostate, active AR signaling promotes prostate epithelial differentiation while reducing proliferation [73], whereas in prostate cancer, AR signaling promotes proliferation [74]. In order to promote proliferation, AR induces expression of Cyclin D and enhances the formation of active CDK4/Cyclin D and CDK2-dependent complexes; these

changes culminate in the hyperphosphorylation and inactivation of RB (reviewed in [70]), as well as directly activating additional targets such as Cdc6 [75], potentially through interactions with E2F (see below).

As most prostate cancers are dependent on the AR signaling axis for growth and survival, ADT is the primary treatment option for patients with metastatic disease. However, the majority of metastatic tumors relapse following ADT, with growth resuming despite the presence of castration levels of hormones. Clinical progression to CRPC and mechanisms of ADT resistance have been reviewed elsewhere [76]. With recent, landmark clinical trials with second generation AR-pathway inhibitors in the metastatic, castrate-sensitive setting [77, 78], it is likely that an increased incidence of AR-indifferent forms of metastatic prostate cancer will continue to be observed as these more aggressive therapeutic strategies are adopted. A common mechanism of acquired ADT resistance is genetic alteration of AR itself, leading to persistent and sometimes ligand-independent AR signaling [79]. Several key AR target genes, including *KLK3* and *TMPRSS2*, are elevated in RB-deficient tumors, and loss of RB increases AR binding to the promoter and enhancer of the *KLK3* gene, both in the presence and absence of androgen [6]. RB loss increases recruitment of AR to key cell cycle AR-target genes, including *CDK1* and *CCNA2*, which are also targets of E2F [25]. Significantly, AR-mediated transcriptional upregulation of these mitotic genes is specific to CRPC, which frequently lacks functional RB, in contrast to AR-mediated suppression of proliferation in normal prostate epithelium [25]. However, the precise mechanisms regulating these changes require further elucidation (Fig. 2).

AR as a Transcriptional Target of RB/E2F

Chromatin immunoprecipitation experiments have confirmed direct binding of E2F1, E2F3, and RB, as well as the RB family members p107 and p130 [80], to the AR promoter [6], confirming that transcription of AR may be regulated by RB-E2F complexes. E2F1 is capable of

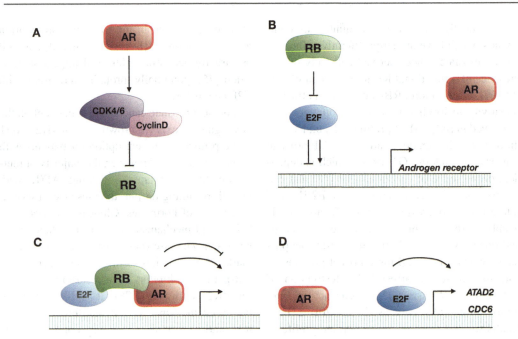

Fig. 2 RB-AR cross talk. (**a**) Activated AR induces expression of Cyclin D and enhances the formation of active CDK4/Cyclin D and CDK2-dependent complexes; these changes culminate in the hyperphosphorylation and inactivation of RB. (**b**) AR transcription is regulated by the E2F transcription factors. RB binds to E2F, altering target gene expression; as a result, AR may be up or down regulated through RB loss. (**c**) Direct interactions between RB and AR modulate the AR cistrome. RB-binding contributes to AR target gene activation and/or repression. (**d**) Functional collaboration between the AR and E2F proteins plays a role in regulation of *CDC6* and *ATAD2*, and possibly others

repressing AR mRNA and protein expression and inhibits AR promoter activity [80]. Moreover, the carboxyl-terminal transactivation domain is an essential component for E2F1 suppression of the AR promoter, and a reciprocal expression pattern of E2F1 and AR was observed in a human tissue microarray [80]. Also, AR levels decrease during cell cycle phases when E2F activity is high [81], suggestive of E2F-mediated repression of AR expression. However, AR expression and activity conversely increase when RB is depleted, and clinical CRPC samples show a correlation between RB loss, E2F1 overexpression, and AR overexpression [6], suggesting that RB-E2F regulation of the AR promoter is positive or negative depending on the cellular context.

RB as a Cofactor to AR

AR is co-activated by RB [14, 82]. This activity was first demonstrated by observations that RB can bind directly to AR in an androgen-independent manner to induce AR transcriptional activity [14]; the location of the RB interaction site lies within the N-terminal domain of the AR and suggests that RB potentiates AR activity due to the formation of AR-RB dimers [82]. Subsequent reports confirmed this physical interaction [83] and others showed that overexpression of RB leads to increased transcriptional activity of the AR, while loss of RB activity inhibits AR [82]. The observation that AR has little transcriptional activity in cells lacking RB held true regardless of whether RB activity was low due to loss of RB expression, mutation of RB, or expression of an RB-inactivating oncogene. The basis for this activity is not entirely clear but could possibly reflect sequestration of HDACs or other transcriptional corepressors. AR enhances RB binding to a series of E2F-regulated DNA replication genes [84], indicating that AR functions in concert with E2F to recruit RB and suppress these genes.

Importantly, there is a substantial overlap between RB and AR binding sites, and these overlapping sites correlate with AR transcriptional repression activity, suggesting that recruitment of RB to specific AR targets results in their AR-dependent repression [84]. Hyperphosphorylation and eventual loss of RB during prostate tumor development may therefore contribute to the altered AR-cistrome observed in tumors relative to normal prostate [85].

Interactions Between E2F and AR

AR has also been reported to associate with the canonical RB target, E2F. Co-immunoprecipitation experiments demonstrate AR can associate with both E2F1 and E2F3 [86], although the role of RB in this interaction is unclear. This association may enable AR-dependent activation of CDC6 expression following treatment with androgens [86].

Expression of ATAD2, a coactivator of AR and MYC, is directly regulated by AR via an AR binding sequence located in the distal enhancer of its regulatory region [87]. ATAD2 is also regulated by the E2F1 transcription factor [88] and the functional collaboration between AR and E2F1 results from a DNA looping over the ATAD2 promoter region between the AR binding sequence and E2F1 binding site in an androgen-dependent manner. Down-regulation of E2F1 further results in abrogation of the response of ATAD2 to androgens.

The Role of RB-E2F Non-cell Cycle Functions in Prostate Cancer

The E2F family of transcription factors comprises eight known members in mammalian cells, each capable of binding to classical E2F binding sites. Of these eight transcription factors, five (E2F1–5) contain RB-binding domains [89]. Thousands of genes have been shown to be bound by E2F family members and hundreds have been demonstrated to have altered expression following forced E2F expression or RB knockout. E2F target genes have known roles in a wide variety

of cellular functions beyond the G1/S transition of the cell cycle, including but not limited to mitosis, DNA damage repair, apoptosis, differentiation, senescence, and metabolism, such that inactivation of RB could have wide-ranging effects on these functions.

RB and Apoptosis

The role of RB in regulation of apoptosis is complex and context-dependent. Remarkably, in the earliest studies loss of RB was found to correlate with an induction of apoptosis, which is counterintuitive to its role as a tumor suppressor. Specifically, RB-deficient embryos displayed increased apoptosis in developing tissues in an E2F dependent fashion (reviewed in [90]). E2F1 specifically activates proapoptotic genes such as *p73* and *APAF-1* following release from RB (reviewed in [29, 91]). Other studies have shown the contrary—that intact RB is required for progression of apoptosis in other contexts [92]. The role of RB in induction of apoptosis in prostate cells is similarly complex.

In normal prostate, an increase in RB expression acutely follows castration in rats, which is followed by a wave of apoptosis [93]. Results from cell lines suggest that RB is required for apoptosis to occur in response to loss of epithelial-extracellular matrix contacts [93], potentially through loss of membrane E-cadherin [93]. In prostate cancer cells, overexpression of Cyclin D1 prevents regression of LNCaP xenograft tumors following castration [56]. Furthermore, reintroduction of RB and AR into DU145, an RB/AR-deficient metastatic prostate cancer cell line, resulted in apoptosis, rather than cell cycle arrest [83]. Together these results suggest that functional RB mediates apoptosis in these contexts. However, intact RB also generally blocks apoptosis in cell lines following DNA damage from either UV or gamma-irradiation [94, 95] or chemotherapeutics such as anti-microtubule agents or a topoisomerase inhibitor [96], whereas RB-deficient cells show increased sensitivity to these agents.

RB and Senescence

AR activity [97] as well as supraphysiological expression of androgen [98] have been associated with induction of senescence, which may serve an intrinsic tumor suppressive mechanism in prostate cancer. Cells lacking functional RB are prevented from exiting the cell cycle into the state of permanent arrest, which is known as senescence, consistent with several studies in prostate cells [99–101] and reinforcing a role for antiproliferative functions of RB in tumor suppression. Interestingly, mice engineered to express a mutant form of RB incapable of binding E2F transcription factors displayed wild-type patterns of senescence in response to loss of p53 and prostate tumor progression [102]. This result suggests that the RB-E2F interaction is dispensable for the ability of RB to induce senescence in developing prostate tumors.

RB and Metastasis

Loss of functional RB has been implicated in invasive, metastatic cancers of many tissue types, including hepatocellular carcinoma and breast cancers as well as prostate cancer. Recent work has for the first time identified a potential mechanism for RB regulation of metastasis in prostate cancer specifically. In this work, RB-deficient cells displayed increased metastatic potential in a tail vein transplant model, which was mediated by E2F-mediated transcription of RHAMM [103]. These mechanistic studies from cell lines suggest that the spread of prostate metastases harboring RB deletions are slowed through treatment with inhibitors of pathways downstream of RHAMM.

RB and Metabolism

Cancer cells adapt their metabolism to promote growth, proliferation, survival, and metastasis.

The metabolic profile of a tumor ultimately depends on the tissue of origin, the oncogenic alterations, the tumor stage, and the tumor microenvironment. Metabolic reprogramming is a hallmark of cancer, and selectively targeting tumor metabolism has been proposed in the recent years as a therapeutic strategy to treat cancer [104, 105].

RB is deeply implicated in the regulation of cellular metabolism [106, 107]. Human cancers without functional RB exhibit an increased glutamine-uptake; similarly, triple knockout of the RB family of genes in mouse embryonic fibroblasts (MEFs) increases glutamine consumption, due to upregulation of the glutamine transporter ASCT2 [108]. Loss of RB family members also results in higher glutamine utilization in the TCA cycle and glutathione accumulation [108]. Thus, RB family members may play a key role in re-wiring glutamine metabolism and glutathione synthesis in tumor cells. RB and E2F1 can also regulate oxidative metabolism by modulating the expression of several genes involved in mitochondrial biogenesis [109, 110]. The RB-E2F1 complex binds to the promoters of many of the genes implicated in oxidative metabolism in brown adipose tissue and muscle. When exposed to cold or fasting conditions, RB undergoes higher rates of phosphorylation, stimulating oxidative metabolism [109, 111].

Normal prostate epithelial cells have a relatively inefficient energy metabolism, using glucose to synthesize citrate that is secreted as part of the seminal fluid. However, prostate cancer cells modify their energy metabolism from inefficient to highly efficient, often taking advantage of the interaction with other cell types in the tumor microenvironment that are corrupted to produce and secrete metabolic intermediates used by cancer cells in catabolic and anabolic processes [112–114]. Increased glutamine utilization and consumption, as well as upregulation of RB-family regulated ASCT2 transporter, are markers of more aggressive prostate cancer and poorer oncological outcomes [115–118].

The Role of RB in Cellular Identity and Differentiation Status

As described earlier, RB is known to interact with hundreds of binding partners beyond the E2F transcription factors (reviewed in [12, 13]), and many studies have demonstrated that the tumor suppressor functions of RB extend well beyond cell cycle control (reviewed in [29]). In the classical model of the RB pathway, cyclin-dependent kinase activity results in the functional inactivation of RB through phosphorylation of key residues in a cell cycle-dependent manner. This model is consistent with the observation that some tumor types show strong mutual exclusivity between upregulation of Cyclin D and/or CDK4/6 and loss of RB. However, emerging work suggests that phosphorylated RB may not be completely nonfunctional. These functions may be particularly important in cancers with a strong association with deep deletion of the RB gene as opposed to mutations in other parts of the pathway, as is observed in CRPC; if functional inactivation of RB though phosphorylation was sufficient for progression, mutations at other levels of the RB pathway would be expected more frequently. For a recent perspective into these emerging non-canonical functions of RB, see [119], however of note is the role of RB on the chromatin and differentiation status of cancer cells. Specifically, the contribution of RB's regulation of the chromatin on lineage plasticity and therapeutic resistance is of particular interest.

RB-E2F Represses Transcription of Pluripotent Networks

Induced pluripotent stem cells and embryonic stem cells share some similarities to cancer cells, including the capacity to bypass senescence and form tumors upon transplantation [120]. Accordingly, some genes often associated with cancer, such as Myc [121, 122] and p53 [123] have been implicated in cellular reprogramming. Additionally, two factors necessary for reprogramming of differentiated into pluripotent cells, OCT4 and SOX2, can be oncogenic in some cellular contexts [124–126]. RB has been shown to repress the expression of both OCT4 and SOX2 through recruitment of histone modifying enzymes, including EZH2 [127, 128]. Loss of functional RB can therefore facilitate reprogramming of fibroblasts to a pluripotent state through de-repression of these pluripotency factors [127]. Indeed, samples of metastatic neuroendocrine prostate cancer (NEPC) show upregulation of SOX2 expression and gene targets, resulting in a stem cell-like gene signature [129]. *In vitro* and *in vivo* human prostate cancer models have been used to show that prostate tumors displaying lineage plastic characteristics can develop resistance to enzalutamide by a phenotypic shift from androgen AR-dependent luminal epithelial cells to AR-independent basal-like cells. This lineage plasticity, which is enabled by the co-loss of TP53 and RB or PTEN and RB function, is mediated by increased expression of the reprogramming transcription factor SOX2. This phenotype can be reversed by restoring gene function or by inhibiting SOX2 expression [64, 130].

RB and Chromatin Modifiers

The RB protein physically interacts with numerous chromatin modifiers. Classic examples of chromatin-associated proteins that RB is known to interact with include histone deacetylases [131–133], DNA methyl transferases [132], and histone methyl transferases [134, 135]. Increased telomere length is a key observation in RB deficient cells [136]. The RB and HDAC1/HDAC2 complex has also been shown to target long interspersed nuclear elements (LINE-1 elements). These associations generally result in epigenetic inactivation of target genes, including, but not limited to, classic E2F targets as well as tissue-specific genes needed for terminal differentiation (reviewed in [29, 137]). In the absence of RB, loss of repressive epigenetic marks results in activation of RB targets, including repetitive elements [138].

RB-PRC2

The polycomb repressive complex 2 (PRC2) methylates lysine 27 on histone 3, resulting in chromatin compaction and inactivation of targeted genes. PRC2 is vital for maintaining self-renewal capacity in embryonic and adult stem cells [139]. Two components of PRC2, *EZH2* and *EED*, are direct E2F transcriptional targets [40, 41, 140]. As expected of a classical E2F target, *EZH2* is repressed by androgens in the prostate, and its expression is subsequently upregulated following RB loss or RB-family inactivation [141]. Dysregulation of the E2F-RB pathway leads to overexpression of components of the PRC2 complex, including EZH2 and EED, and aberrant methylation patterns during development of small cell lung cancer [140] as well in mouse models of NEPC [64].

Beyond transcriptional regulation, RB also dimerizes with EZH2, contributing to the location and distribution of H3K27 tri-methylation (H3K27me3) sites, resulting in silencing of repeat elements such as endogenous retroviruses and LINE-1 elements [142]. These repetitive elements are usually silenced early in embryonic development and generally remain inactive [142].

Epigenetic Role of RB in Therapy Response

Prostate adenocarcinoma cells evade androgen deprivation therapies through a variety of mechanisms [143], including restoration of AR signaling, up regulation of glucocorticoid receptor (GR) activity to regulate AR target genes, or a change in differentiation status—referred to as lineage switching—to a cell type that no longer requires AR signaling [76]. The role of RB in modulating epigenetic changes is a major contributor to the evasion and relapse of CRPC in response to ADT through lineage switching from adenocarcinoma to a neuroendocrine phenotype [64].

Emerging Functions of RB

Although the RB gene was first discovered in the 1980s and was first knocked out in a mouse in 1992, novel functions of the RB protein continue to be discovered. The potential contribution of these more recently elucidated roles of RB to its tumor suppression function in prostate cancer are described below.

RB and Genome Stability

RB has a role in maintenance of genomic stability through a variety of direct and indirect mechanisms [29]. One example is through RB-E2F regulation of the mitotic checkpoint protein Mad2 [144]. Furthermore, interactions between RB and epigenetic modifiers are necessary for pericentric heterochromatin and chromosome stability [145]. In addition to its roles as a cis-acting transcriptional cofactor and epigenetic regulator, RB is also physically recruited to breaks in DNA and participates in the recruitment of chromatin regulators necessary for DNA repair [146]. RB also interacts with cohesin and condensin II at replicating DNA, promoting genomic stability [147, 148]. Broadly speaking, in prostate cancer genomic and chromosomal instability are increasingly frequent during prostate tumor progression—i.e., the most aggressive diseases display the most instability [149, 150]; this pattern bears a striking similarity to the pattern of inactivation of RB. However, the role of these potential RB functions in prostate cancer remains unknown.

Cancer Immunity

In some cases, RB functions as a protein adapter, bringing together two proteins independent of a DNA interaction, for example SKP2-APC/C (reviewed in [12]). Recent work has demonstrated that beyond the well-established role of RB as a transcriptional and chromatin modification cofactor, RB may act to suppress tumor development through a novel interaction with the NF-κB protein p65 [151]. Phosphorylated RB, which is thought to be inactive in the classical sense, is able to bind to p65 and thereby negatively regulate the NF-κB target PDL1, enhancing the ability of a tumor cell to evade the immune system. The implications of this recent finding have yet to be fully elucidated.

RB can repress expression of repetitive elements, including LINE-1s through an epigenetic mechanism [138], which particularly occurs in differentiated cells. Loss of RB therefore has the potential to increase expression of these repressed elements, thereby stimulating the defense mechanisms that combat exogenous viral infection via an interferon response pathway (reviewed in [152]). Furthermore, the upregulation of EZH2, which occurs in the absence of RB, may be a compensatory defense mechanism to re-repress repetitive elements that may exert deleterious effects following de-repression from RB loss. In this way, loss of RB may improve the response of a tumor to immune checkpoint therapy, either alone or in combination with inhibitors against the compensatory mechanisms, such as EZH2.

RB as a Biomarker

Loss of function mutations and deletion of the RB tumor suppressor provides a means for tumors to progress to metastatic and aggressive variants in the natural history of prostate cancer and suggests that RB is a plausible biomarker for diagnosis and predicting disease progression. However, the utility of RB as a biomarker remains challenging. Outgrowth of RB deleted cells occurs later in disease progression, and there is no difference in staining between benign and malignant glands [153]. Loss of heterozygosity at the RB locus associates with absence of immunohistochemical staining in only 33% of primary adenocarcinoma samples [154], suggesting that

retention of protein staining positivity is not necessarily indicative of protein function. Use of RB as a biomarker holds most promise in NEPC where absence of RB IHC staining has high concordance (93%) with tumor histology [9]. Conversely, RB staining is retained in the majority of high grade and metastatic adenocarcinoma, even with early presence of neuroendocrine morphologies. Despite this concordance, the translation of RB immunostaining to the clinic will be most useful in combination with positive markers of NEPC such as synaptophysin and chromogranin A, as absence of a single marker may produce additional false negative results. Overall, determination of RB status by IHC provides little utility, but absence of staining in combination with other methods may prove useful in validating neuroendocrine phenotypes.

Given these limitations, rather than detecting RB protein, several gene signatures have been developed to detect deficiency for RB-activity, rather than the protein itself [155, 156]. These signatures offer an improved method of identifying cells in which RB function has been compromised. However, as it would be most beneficial to identify the presence of a sub-clone of RB-deficient cells prior to androgen-deprivation therapy, these gene signatures may be lost within a bulk analysis from a biopsy or resected tumor. Recent work has demonstrated the feasibility of detecting circulating tumor cells in patients with prostate tumors [157, 158]. In the future it may be possible to apply single-cell RNA-seq to these circulating cells in order to identify an RB-loss signature and therefore to stratify patients for treatment based on the pre-existence of RB-deficient tumor subclones that may later lead to an outgrowth of aggressive, castration-resistant disease. Furthermore, detection of RB deletions may be possible from cell-free circulating DNA (cfDNA), obtained through minimally invasive blood or urine collection. Copy number variations (e.g. loss) of RB have been detected in cfDNA by array CGH; these results were consistent with the emergence of an RB-deficient clone following enzalutamide treatment [159].

Conclusions and Perspectives

The RB tumor suppressor plays an important cell cycle role in prostate cancer development. Although resulting hyperproliferation may be minor, RB is required in many cases for tumor cells to arrest in response to therapy. Specifically in response to androgen deprivation therapy, RB-deficiency may allow an escape mechanism to the reduction of androgen signaling, through multiple mechanisms. Loss of cell cycle arrest, an altered AR-cistrome, direct or indirect upregulation of AR expression, de-differentiation and lineage switching to reduced dependence on AR signaling, and increased mutation rate as a result of genetic instability, all may contribute to the survival of RB deficient clones within a primary prostate tumor, as well as the evolution of resistance mechanisms which arise in metastases (Fig. 3). Androgen deprivation therapy appears to transform a relatively neutral subclonal mutation into an incurable disease. It is further interesting to note that inactivation of RB through upstream mutations appears insufficient to remove all of the necessary functions of RB, as deletions of RB itself are especially common in CRPC, regardless of its canonical inactivation by overexpression of Cyclin D dependent complexes.

The ability to target RB-deficient cells, either small subclones within the primary disease or following the outgrowth of resistant disease, would therefore greatly benefit patients at risk for CRPC. RB-deficient cells are more sensitive to some types of chemotherapy as well as radiation [68], suggesting that use of these therapies prior to hormone-based therapies can enhance the therapeutic response by reducing the presence of pre-existing RB-depleted subclones, which may exist but escape detection using current screening methods. Increased sensitivity of techniques that would allow detection of RB-deletion in cfDNA would allow these combined therapies to be used only when necessary. More studies into therapies capable of specifically targeting RB-deficient cells, such as those which identified the Aurora kinases as synthetic lethal targets [160, 161], could refine the use of chemotherapy and radiation, and therefore reduce side-effects.

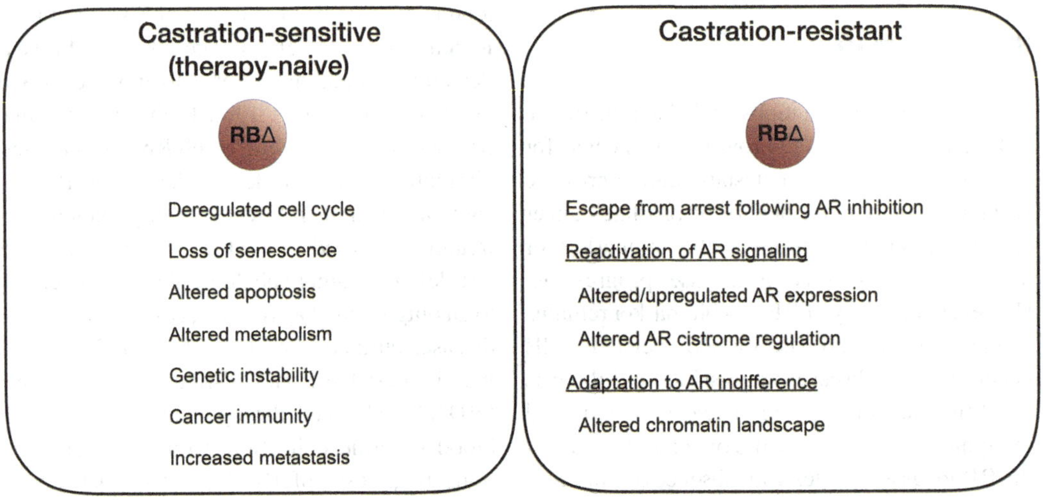

Castration-sensitive (therapy-naive)

RBΔ

Deregulated cell cycle
Loss of senescence
Altered apoptosis
Altered metabolism
Genetic instability
Cancer immunity
Increased metastasis

Castration-resistant

RBΔ

Escape from arrest following AR inhibition
Reactivation of AR signaling
Altered/upregulated AR expression
Altered AR cistrome regulation
Adaptation to AR indifference
Altered chromatin landscape

Fig. 3 Critical functions of RB loss; implications for ADT resistance. The pleiotropic functions of RB contribute to tumor suppression throughout tumor development through multiple mechanisms. The left box summarizes the RB functions which are generally cell-intrinsic following RB loss and may contribute to tumor evolution at any stage of tumor progression. The right box emphasizes RB loss-of-function mechanisms thought important following hormone deprivation therapy

Furthermore, recent work indicating a role for phosphorylated-RB in evasion of immunosurveillance, as well the potential for RB loss to upregulate novel antigens, together suggest that RB deficient tumor cells are poised for a response to immune checkpoint therapy, possibly in combination with additional epigenetic modifiers. In conclusion, therapies, particularly combination therapies, that take advantage of the phenotypic changes arising from the loss of RB merit further study.

References

1. B.S. Carter, C.M. Ewing, W.S. Ward, B.F. Treiger, T.W. Aalders, J.A. Schalken, et al., Allelic loss of chromosomes 16q and 10q in human prostate cancer. Proc. Natl. Acad. Sci. U. S. A. **87**(22), 8751–8755 (1990)
2. R. Bookstein, D.C. Allred, Recessive oncogenes. Cancer **71**(3 Suppl), 1179–1186 (1993)
3. W. Liu, C.C. Xie, C.Y. Thomas, S.T. Kim, J. Lindberg, L. Egevad, et al., Genetic markers associated with early cancer-specific mortality following prostatectomy. Cancer **119**(13), 2405–2412 (2013)
4. Cancer Genome Atlas Research Network, The molecular taxonomy of primary prostate cancer. Cell **163**(4), 1011–1025 (2015)
5. A.G. Sowalsky, H. Ye, M. Bhasin, E.M. Van Allen, M. Loda, R.T. Lis, et al., Neoadjuvant-intensive androgen deprivation therapy selects for prostate tumor foci with diverse subclonal oncogenic alterations. Cancer Res. **78**(16), 4716–4730 (2018)
6. A. Sharma, W.S. Yeow, A. Ertel, I. Coleman, N. Clegg, C. Thangavel, et al., The retinoblastoma tumor suppressor controls androgen signaling and human prostate cancer progression. J. Clin. Invest. **120**(12), 4478–4492 (2010)
7. J. Armenia, S.A.M. Wankowicz, D. Liu, J. Gao, R. Kundra, E. Reznik, et al., The long tail of oncogenic drivers in prostate cancer. Nat. Genet. **50**(5), 645–651 (2018)
8. A.A. Hamid, K.P. Gray, G. Shaw, L.E. MacConaill, C. Evan, B. Bernard, et al., Compound genomic alterations of TP53, PTEN, and RB1 tumor suppressors in localized and metastatic prostate cancer. Eur. Urol. **76**(1), 89–97 (2019)
9. H.L. Tan, A. Sood, H.A. Rahimi, W. Wang, N. Gupta, J. Hicks, et al., Rb loss is characteristic of prostatic small cell neuroendocrine carcinoma. Clin. Cancer Res. **20**(4), 890–903 (2014)
10. M. Hassler, S. Singh, W.W. Yue, M. Luczynski, R. Lakbir, F. Sanchez-Sanchez, et al., Crystal structure of the retinoblastoma protein N domain provides insight into tumor suppression, ligand interaction,

and holoprotein architecture. Mol. Cell. **28**(3), 371–385 (2007)
11. T.J. Liban, E.M. Medina, S. Tripathi, S. Sengupta, R.W. Henry, N.E. Buchler, et al., Conservation and divergence of C-terminal domain structure in the retinoblastoma protein family. Proc. Natl. Acad. Sci. U. S. A. **114**(19), 4942–4947 (2017)
12. F.A. Dick, S.M. Rubin, Molecular mechanisms underlying RB protein function. Nat. Rev. Mol. Cell Biol. **14**(5), 297–306 (2013)
13. E.J. Morris, N.J. Dyson, Retinoblastoma protein partners. Adv. Cancer Res. **82**, 1–54 (2001)
14. S. Yeh, H. Miyamoto, K. Nishimura, H. Kang, J. Ludlow, P. Hsiao, et al., Retinoblastoma, a tumor suppressor, is a coactivator for the androgen receptor in human prostate cancer DU145 cells. Biochem. Biophys. Res. Commun. **248**(2), 361–367 (1998)
15. H. Huang, J. Yee, J. Shew, P. Chen, R. Bookstein, T. Friedmann, et al., Suppression of the neoplastic phenotype by replacement of the RB gene in human cancer cells. Science **242**(4885), 1563–1566 (1988)
16. R. Takahashi, T. Hashimoto, H.J. Xu, S.X. Hu, T. Matsui, T. Miki, et al., The retinoblastoma gene functions as a growth and tumor suppressor in human bladder carcinoma cells. Proc. Natl. Acad. Sci. U. S. A. **88**(12), 5257–5261 (1991)
17. X.Q. Qin, T. Chittenden, D.M. Livingston, W.G. Kaelin Jr., Identification of a growth suppression domain within the retinoblastoma gene product. Genes Dev. **6**(6), 953–964 (1992)
18. J.R. Nevins, E2F: a link between the Rb tumor suppressor protein and viral oncoproteins. Science **258**(5081), 424–429 (1992)
19. R.A. Weinberg, The retinoblastoma protein and cell cycle control. Cell **81**(3), 323–330 (1995)
20. C.J. Sherr, D. Beach, G.I. Shapiro, Targeting CDK4 and CDK6: from discovery to therapy. Cancer Discov. **6**(4), 353–367 (2016)
21. D. Hanahan, R.A. Weinberg, The hallmarks of cancer. Cell **100**(1), 57–70 (2000)
22. D. Hanahan, R.A. Weinberg, Hallmarks of cancer: the next generation. Cell **144**(5), 646–674 (2011)
23. A. Deshpande, P. Sicinski, P.W. Hinds, Cyclins and cdks in development and cancer: a perspective. Oncogene **24**(17), 2909–2915 (2005)
24. N.J. Dyson, RB1: a prototype tumor suppressor and an enigma. Genes Dev. **30**(13), 1492–1502 (2016)
25. B.S. Taylor, N. Schultz, H. Hieronymus, A. Gopalan, Y. Xiao, B.S. Carver, et al., Integrative genomic profiling of human prostate cancer. Cancer Cell **18**(1), 11–22 (2010)
26. C.S. Foster, A. Falconer, A.R. Dodson, A.R. Norman, N. Dennis, A. Fletcher, et al., Transcription factor E2F3 overexpressed in prostate cancer independently predicts clinical outcome. Oncogene **23**(35), 5871–5879 (2004)
27. A. Radu, V. Neubauer, T. Akagi, H. Hanafusa, M.M. Georgescu, PTEN induces cell cycle arrest by decreasing the level and nuclear localization

28. J. Gil, P. Kerai, M. Lleonart, D. Bernard, J.C. Cigudosa, G. Peters, et al., Immortalization of primary human prostate epithelial cells by c-Myc. Cancer Res. **65**(6), 2179–2185 (2005)

29. D.L. Burkhart, J. Sage, Cellular mechanisms of tumour suppression by the retinoblastoma gene. Nat. Rev. Cancer **8**(9), 671–682 (2008)

30. C. Giacinti, A. Giordano, RB and cell cycle progression. Oncogene **25**(38), 5220–5227 (2006)

31. R. Hill, Y. Song, R.D. Cardiff, T. Van Dyke, Heterogeneous tumor evolution initiated by loss of pRb function in a preclinical prostate cancer model. Cancer Res. **65**(22), 10243–10254 (2005)

32. F. Bai, X.H. Pei, P.P. Pandolfi, Y. Xiong, p18 Ink4c and Pten constrain a positive regulatory loop between cell growth and cell cycle control. Mol. Cell. Biol. **26**(12), 4564–4576 (2006)

33. F. Bai, D.A. DeMason, Hormone interactions and regulation of Unifoliata, PsPK2, PsPIN1 and LE gene expression in pea (Pisum sativum) shoot tips. Plant Cell Physiol. **47**(7), 935–948 (2006)

34. A. Tyagi, C. Agarwal, R. Agarwal, Inhibition of retinoblastoma protein (Rb) phosphorylation at serine sites and an increase in Rb-E2F complex formation by silibinin in androgen-dependent human prostate carcinoma LNCaP cells: role in prostate cancer prevention. Mol. Cancer Ther. **1**(7), 525–532 (2002)

35. E.L. DuPree, S. Mazumder, A. Almasan, Genotoxic stress induces expression of E2F4, leading to its association with p130 in prostate carcinoma cells. Cancer Res. **64**(13), 4390–4393 (2004)

36. B.D. Lehmann, A.M. Brooks, M.S. Paine, W.H. Chappell, J.A. McCubrey, D.M. Terrian, Distinct roles for p107 and p130 in Rb-independent cellular senescence. Cell Cycle **7**(9), 1262–1268 (2008)

37. K. Helin, Regulation of cell proliferation by the E2F transcription factors. Curr. Opin. Genet. Dev. **8**(1), 28–35 (1998)

38. J.M. Trimarchi, J.A. Lees, Sibling rivalry in the E2F family. Nat. Rev. Mol. Cell Biol. **3**(1), 11–20 (2002)

39. A.P. Bracken, M. Ciro, A. Cocito, K. Helin, E2F target genes: unraveling the biology. Trends Biochem. Sci. **29**(8), 409–417 (2004)

40. H. Muller, A.P. Bracken, R. Vernell, M.C. Moroni, F. Christians, E. Grassilli, et al., E2Fs regulate the expression of genes involved in differentiation, development, proliferation, and apoptosis. Genes Dev. **15**(3), 267–285 (2001)

41. A.P. Bracken, D. Pasini, M. Capra, E. Prosperini, E. Colli, K. Helin, EZH2 is downstream of the pRB-E2F pathway, essential for proliferation and amplified in cancer. EMBO J. **22**(20), 5323–5335 (2003)

42. Y. Kotake, R. Cao, P. Viatour, J. Sage, Y. Zhang, Y. Xiong, pRB family proteins are required for H3K27 trimethylation and Polycomb repression complexes binding to and silencing p16INK4alpha tumor suppressor gene. Genes Dev. **21**(1), 49–54 (2007)

43. A.F. Fribourg, K.E. Knudsen, M.W. Strobeck, C.M. Lindhorst, E.S. Knudsen, Differential requirements for ras and the retinoblastoma tumor suppressor protein in the androgen dependence of prostatic adenocarcinoma cells. Cell Growth Differ. **11**(7), 361–372 (2000)

44. K. Hofman, J.V. Swinnen, G. Verhoeven, W. Heyns, E2F activity is biphasically regulated by androgens in LNCaP cells. Biochem. Biophys. Res. Commun. **283**(1), 97–101 (2001)

45. S.S. Taneja, S. Ha, M.J. Garabedian, Androgen stimulated cellular proliferation in the human prostate cancer cell line LNCaP is associated with reduced retinoblastoma protein expression. J. Cell. Biochem. **84**(1), 188–199 (2001)

46. X. Yuan, T. Li, H. Wang, T. Zhang, M. Barua, R.A. Borgesi, et al., Androgen receptor remains critical for cell-cycle progression in androgen-independent CWR22 prostate cancer cells. Am. J. Pathol. **169**(2), 682–696 (2006)

47. Y. Chen, A.I. Robles, L.A. Martinez, F. Liu, I.B. Gimenez-Conti, C.J. Conti, Expression of G1 cyclins, cyclin-dependent kinases, and cyclin-dependent kinase inhibitors in androgen-induced prostate proliferation in castrated rats. Cell Growth Differ. **7**(11), 1571–1578 (1996)

48. K.E. Knudsen, K.C. Arden, W.K. Cavenee, Multiple G1 regulatory elements control the androgen-dependent proliferation of prostatic carcinoma cells. J. Biol. Chem. **273**(32), 20213–20222 (1998)

49. S.J. Libertini, C.G. Tepper, M. Guadalupe, Y. Lu, D.M. Asmuth, M. Mudryj, E2F1 expression in LNCaP prostate cancer cells deregulates androgen dependent growth, suppresses differentiation, and enhances apoptosis. Prostate **66**(1), 70–81 (2006)

50. J.N. Davis, M.T. McCabe, S.W. Hayward, J.M. Park, M.L. Day, Disruption of Rb/E2F pathway results in increased cyclooxygenase-2 expression and activity in prostate epithelial cells. Cancer Res. **65**(9), 3633–3642 (2005)

51. M.T. McCabe, J.N. Davis, M.L. Day, Regulation of DNA methyltransferase 1 by the pRb/E2F1 pathway. Cancer Res. **65**(9), 3624–3632 (2005)

52. Y. Wang, S.W. Hayward, A.A. Donjacour, P. Young, T. Jacks, J. Sage, et al., Sex hormone-induced carcinogenesis in Rb-deficient prostate tissue. Cancer Res. **60**(21), 6008–6017 (2000)

53. M.S. Steiner, C.T. Anthony, Introduction of human chromosome 13 into retinoblastoma-negative metastatic human prostate cancer cells increases their sensitivity to growth inhibition by transforming growth factor-&be;1. Mol. Urol. **3**(3), 153–162 (1999)

54. K.C. Day, M.T. McCabe, X. Zhao, Y. Wang, J.N. Davis, J. Phillips, et al., Rescue of embryonic epithelium reveals that the homozygous deletion of the retinoblastoma gene confers growth factor independence and immortality but does not influence

epithelial differentiation or tissue morphogenesis. J. Biol. Chem. **277**(46), 44475–44484 (2002)

55. L.A. Maddison, B.W. Sutherland, R.J. Barrios, N.M. Greenberg, Conditional deletion of Rb causes early stage prostate cancer. Cancer Res. **64**(17), 6018–6025 (2004)

56. Y. Chen, L.A. Martinez, M. LaCava, L. Coghlan, C.J. Conti, Increased cell growth and tumorigenicity in human prostate LNCaP cells by overexpression to cyclin D1. Oncogene **16**(15), 1913–1920 (1998)

57. M.P. Gustafson, C. Xu, J.E. Grim, B.E. Clurman, B.S. Knudsen, Regulation of cell proliferation in a stratified culture system of epithelial cells from prostate tissue. Cell Tissue Res. **325**(2), 263–276 (2006)

58. N.M. Greenberg, F. DeMayo, M.J. Finegold, D. Medina, W.D. Tilley, J.O. Aspinall, et al., Prostate cancer in a transgenic mouse. Proc. Natl. Acad. Sci. U. S. A. **92**(8), 3439–3443 (1995)

59. J.R. Gingrich, R.J. Barrios, R.A. Morton, B.F. Boyce, F.J. DeMayo, M.J. Finegold, et al., Metastatic prostate cancer in a transgenic mouse. Cancer Res. **56**(18), 4096–4102 (1996)

60. J.R. Gingrich, R.J. Barrios, M.W. Kattan, H.S. Nahm, M.J. Finegold, N.M. Greenberg, Androgen-independent prostate cancer progression in the TRAMP model. Cancer Res. **57**(21), 4687–4691 (1997)

61. P.J. Kaplan-Lefko, T.M. Chen, M.M. Ittmann, R.J. Barrios, G.E. Ayala, W.J. Huss, et al., Pathobiology of autochthonous prostate cancer in a pre-clinical transgenic mouse model. Prostate **55**(3), 219–237 (2003)

62. Z. Zhou, A. Flesken-Nikitin, D.C. Corney, W. Wang, D.W. Goodrich, P. Roy-Burman, et al., Synergy of p53 and Rb deficiency in a conditional mouse model for metastatic prostate cancer. Cancer Res. **66**(16), 7889–7898 (2006)

63. Z. Zhou, A. Flesken-Nikitin, A.Y. Nikitin, Prostate cancer associated with p53 and Rb deficiency arises from the stem/progenitor cell-enriched proximal region of prostatic ducts. Cancer Res. **67**(12), 5683–5690 (2007)

64. S.Y. Ku, S. Rosario, Y. Wang, P. Mu, M. Seshadri, Z.W. Goodrich, et al., Rb1 and Trp53 cooperate to suppress prostate cancer lineage plasticity, metastasis, and antiandrogen resistance. Science **355**(6320), 78–83 (2017)

65. A.K. Tyagi, R.P. Singh, C. Agarwal, D.C. Chan, R. Agarwal, Silibinin strongly synergizes human prostate carcinoma DU145 cells to doxorubicin-induced growth Inhibition, G2-M arrest, and apoptosis. Clin. Cancer Res. **8**(11), 3512–3519 (2002)

66. R.P. Singh, C. Agarwal, R. Agarwal, Inositol hexaphosphate inhibits growth, and induces G1 arrest and apoptotic death of prostate carcinoma DU145 cells: modulation of CDKI-CDK-cyclin and pRb-related protein-E2F complexes. Carcinogenesis **24**(3), 555–563 (2003)

67. M.N. Washington, J.S. Kim, N.L. Weigel, 1alpha,25-dihydroxyvitamin D3 inhibits C4–2 prostate cancer cell growth via a retinoblastoma protein (Rb)-independent G1 arrest. Prostate **71**(1), 98–110 (2011)

68. A. Aparicio, R.B. Den, K.E. Knudsen, Time to stratify? The retinoblastoma protein in castrate-resistant prostate cancer. Nat. Rev. Urol. **8**(10), 562–568 (2011)

69. C.W. Gregory, R.T. Johnson Jr., S.C. Presnell, J.L. Mohler, F.S. French, Androgen receptor regulation of G1 cyclin and cyclin-dependent kinase function in the CWR22 human prostate cancer xenograft. J. Androl. **22**(4), 537–548 (2001)

70. S.P. Balk, K.E. Knudsen, AR, the cell cycle, and prostate cancer. Nucl. Recept. Signal. **6**, e001 (2008)

71. A. Grosse, S. Bartsch, A. Baniahmad, Androgen receptor-mediated gene repression. Mol. Cell. Endocrinol. **352**(1–2), 46–56 (2012)

72. G.R. Cunha, W. Ricke, A. Thomson, P.C. Marker, G. Risbridger, S.W. Hayward, et al., Hormonal, cellular, and molecular regulation of normal and neoplastic prostatic development. J. Steroid Biochem. Mol. Biol. **92**(4), 221–236 (2004)

73. L. Antony, F. van der Schoor, S.L. Dalrymple, J.T. Isaacs, Androgen receptor (AR) suppresses normal human prostate epithelial cell proliferation via AR/beta-catenin/TCF-4 complex inhibition of c-MYC transcription. Prostate **74**(11), 1118–1131 (2014)

74. J.T. Isaacs, W.B. Isaacs, Androgen receptor outwits prostate cancer drugs. Nat. Med. **10**(1), 26–27 (2004)

75. F. Jin, J.D. Fondell, A novel androgen receptor-binding element modulates Cdc6 transcription in prostate cancer cells during cell-cycle progression. Nucleic Acids Res. **37**(14), 4826–4838 (2009)

76. A.H. Davies, H. Beltran, A. Zoubeidi, Cellular plasticity and the neuroendocrine phenotype in prostate cancer. Nat. Rev. Urol. **15**(5), 271–286 (2018)

77. K. Fizazi, N. Tran, L. Fein, N. Matsubara, A. Rodriguez-Antolin, B.Y. Alekseev, et al., Abiraterone plus prednisone in metastatic, castration-sensitive prostate cancer. N. Engl. J. Med. **377**(4), 352–360 (2017)

78. N.D. James, J.S. de Bono, M.R. Spears, N.W. Clarke, M.D. Mason, D.P. Dearnaley, et al., Abiraterone for prostate cancer not previously treated with hormone therapy. N. Engl. J. Med. **377**(4), 338–351 (2017)

79. P.A. Watson, V.K. Arora, C.L. Sawyers, Emerging mechanisms of resistance to androgen receptor inhibitors in prostate cancer. Nat. Rev. Cancer **15**(12), 701–711 (2015)

80. J.N. Davis, K.J. Wojno, S. Daignault, M.D. Hofer, R. Kuefer, M.A. Rubin, et al., Elevated E2F1 inhibits transcription of the androgen receptor in metastatic hormone-resistant prostate cancer. Cancer Res. **66**(24), 11897–11906 (2006)

81. E.D. Martinez, M. Danielsen, Loss of androgen receptor transcriptional activity at the G(1)/S transition. J. Biol. Chem. **277**(33), 29719–29729 (2002)

82. J. Lu, M. Danielsen, Differential regulation of androgen and glucocorticoid receptors by retinoblastoma protein. J. Biol. Chem. **273**(47), 31528–31533 (1998)

83. X. Wang, H. Deng, I. Basu, L. Zhu, Induction of androgen receptor-dependent apoptosis in prostate cancer cells by the retinoblastoma protein. Cancer Res. **64**(4), 1377–1385 (2004)

84. S. Gao, Y. Gao, H.H. He, D. Han, W. Han, A. Avery, et al., Androgen receptor tumor suppressor function is mediated by recruitment of retinoblastoma protein. Cell Rep. **17**(4), 966–976 (2016)

85. F. Wang, H.K. Koul, Androgen receptor (AR) cistrome in prostate differentiation and cancer progression. Am. J. Clin. Exp. Urol. **5**(3), 18–24 (2017)

86. I. Mallik, M. Davila, T. Tapia, B. Schanen, R. Chakrabarti, Androgen regulates Cdc6 transcription through interactions between androgen receptor and E2F transcription factor in prostate cancer cells. Biochim. Biophys. Acta **1783**(10), 1737–1744 (2008)

87. D.M. Altintas, M.S. Shukla, D. Goutte-Gattat, D. Angelov, J.P. Rouault, S. Dimitrov, et al., Direct cooperation between androgen receptor and E2F1 reveals a common regulation mechanism for androgen-responsive genes in prostate cells. Mol. Endocrinol. **26**(9), 1531–1541 (2012)

88. M. Ciro, E. Prosperini, M. Quarto, U. Grazini, J. Walfridsson, F. McBlane, et al., ATAD2 is a novel cofactor for MYC, overexpressed and amplified in aggressive tumors. Cancer Res. **69**(21), 8491–8498 (2009)

89. J. DeGregori, D.G. Johnson, Distinct and overlapping roles for E2F family members in transcription, proliferation and apoptosis. Curr. Mol. Med. **6**(7), 739–748 (2006)

90. B.N. Chau, J.Y. Wang, Coordinated regulation of life and death by RB. Nat. Rev. Cancer **3**(2), 130–138 (2003)

91. P. Indovina, F. Pentimalli, N. Casini, I. Vocca, A. Giordano, RB1 dual role in proliferation and apoptosis: cell fate control and implications for cancer therapy. Oncotarget **6**(20), 17873–17890 (2015)

92. H.L. Borges, J. Bird, K. Wasson, R.D. Cardiff, N. Varki, L. Eckmann, et al., Tumor promotion by caspase-resistant retinoblastoma protein. Proc. Natl. Acad. Sci. U. S. A. **102**(43), 15587–15592 (2005)

93. M.L. Day, X. Zhao, C.J. Vallorosi, M. Putzi, C.T. Powell, C. Lin, et al., E-cadherin mediates aggregation-dependent survival of prostate and mammary epithelial cells through the retinoblastoma cell cycle control pathway. J. Biol. Chem. **274**(14), 9656–9664 (1999)

94. C.A. Carlson, S.P. Ethier, Lack of RB protein correlates with increased sensitivity to UV-radiation-induced apoptosis in human breast cancer cells. Radiat. Res. **154**(5), 590–599 (2000)

95. C. Bowen, M. Birrer, E.P. Gelmann, Retinoblastoma protein-mediated apoptosis after gamma-irradiation. J. Biol. Chem. **277**(47), 44969–44979 (2002)

96. A. Sharma, C.E. Comstock, E.S. Knudsen, K.H. Cao, J.K. Hess-Wilson, L.M. Morey, et al., Retinoblastoma tumor suppressor status is a critical determinant of therapeutic response in prostate cancer cells. Cancer Res. **67**(13), 6192–6203 (2007)

97. Y. Mirochnik, D. Veliceasa, L. Williams, K. Maxwell, A. Yemelyanov, I. Budunova, et al., Androgen receptor drives cellular senescence. PLoS One **7**(3), e31052 (2012)

98. J. Roediger, W. Hessenkemper, S. Bartsch, M. Manvelyan, S.S. Huettner, T. Liehr, et al., Supraphysiological androgen levels induce cellular senescence in human prostate cancer cells through the Src-Akt pathway. Mol. Cancer **13**, 214 (2014)

99. D.F. Jarrard, S. Sarkar, Y. Shi, T.R. Yeager, G. Magrane, H. Kinoshita, et al., p16/pRb pathway alterations are required for bypassing senescence in human prostate epithelial cells. Cancer Res. **59**(12), 2957–2964 (1999)

100. M.S. Steiner, Y. Wang, Y. Zhang, X. Zhang, Y. Lu, p16/MTS1/INK4A suppresses prostate cancer by both pRb dependent and independent pathways. Oncogene **19**(10), 1297–1306 (2000)

101. E.J. Noonan, R.F. Place, S. Basak, D. Pookot, L.C. Li, miR-449a causes Rb-dependent cell cycle arrest and senescence in prostate cancer cells. Oncotarget **1**(5), 349–358 (2010)

102. H. Sun, Y. Wang, M. Chinnam, X. Zhang, S.W. Hayward, B.A. Foster, et al., E2f binding-deficient Rb1 protein suppresses prostate tumor progression in vivo. Proc. Natl. Acad. Sci. U. S. A. **108**(2), 704–709 (2011)

103. C. Thangavel, E. Boopathi, Y. Liu, A. Haber, A. Ertel, A. Bhardwaj, et al., RB loss promotes prostate cancer metastasis. Cancer Res. **77**(4), 982–995 (2017)

104. N.N. Pavlova, C.B. Thompson, The emerging hallmarks of cancer metabolism. Cell Metab. **23**(1), 27–47 (2016)

105. J. Floter, I. Kaymak, A. Schulze, Regulation of metabolic activity by p53. Metabolites **7**(2), E21 (2017)

106. B.N. Nicolay, N.J. Dyson, The multiple connections between pRB and cell metabolism. Curr. Opin. Cell Biol. **25**(6), 735–740 (2013)

107. R. Iurlaro, C.L. Leon-Annicchiarico, C. Munoz-Pinedo, Regulation of cancer metabolism by oncogenes and tumor suppressors. Methods Enzymol. **542**, 59–80 (2014)

108. M.R. Reynolds, A.N. Lane, B. Robertson, S. Kemp, Y. Liu, B.G. Hill, et al., Control of glutamine metabolism by the tumor suppressor Rb. Oncogene **33**(5), 556–566 (2014)

109. E. Blanchet, J.S. Annicotte, S. Lagarrigue, V. Aguilar, C. Clape, C. Chavey, et al., E2F transcription factor-1 regulates oxidative metabolism. Nat. Cell Biol. **13**(9), 1146–1152 (2011)

110. V.G. Sankaran, S.H. Orkin, C.R. Walkley, Rb intrinsically promotes erythropoiesis by coupling cell cycle exit with mitochondrial biogenesis. Genes Dev. **22**(4), 463–475 (2008)

111. J.B. Hansen, C. Jorgensen, R.K. Petersen, P. Hallenborg, R. De Matteis, H.A. Boye, et al., Retinoblastoma protein functions as a molecular switch determining white versus brown adipocyte differentiation. Proc. Natl. Acad. Sci. U. S. A. **101**(12), 4112–4117 (2004)

112. F. Cutruzzola, G. Giardina, M. Marani, A. Macone, A. Paiardini, S. Rinaldo, et al., Glucose metabolism in the progression of prostate cancer. Front. Physiol. **8**, 97 (2017)

113. E. Eidelman, J. Twum-Ampofo, J. Ansari, M.M. Siddiqui, The metabolic phenotype of prostate cancer. Front. Oncol. **7**, 131 (2017)

114. N. Pertega-Gomes, S. Felisbino, C.E. Massie, J.R. Vizcaino, R. Coelho, C. Sandi, et al., A glycolytic phenotype is associated with prostate cancer progression and aggressiveness: a role for monocarboxylate transporters as metabolic targets for therapy. J. Pathol. **236**(4), 517–530 (2015)

115. N.M. Zacharias, C. McCullough, S. Shanmugavelandy, J. Lee, Y. Lee, P. Dutta, et al., Metabolic differences in glutamine utilization lead to metabolic vulnerabilities in prostate cancer. Sci. Rep. **7**(1), 16159 (2017)

116. M. Gacci, G.I. Russo, C. De Nunzio, A. Sebastianelli, M. Salvi, L. Vignozzi, et al., Meta-analysis of metabolic syndrome and prostate cancer. Prostate Cancer Prostatic Dis. **20**(2), 146–155 (2017)

117. K. Esposito, P. Chiodini, A. Capuano, G. Bellastella, M.I. Maiorino, E. Parretta, et al., Effect of metabolic syndrome and its components on prostate cancer risk: meta-analysis. J. Endocrinol. Investig. **36**(2), 132–139 (2013)

118. Q. Wang, R.A. Hardie, A.J. Hoy, M. van Geldermalsen, D. Gao, L. Fazli, et al., Targeting ASCT2-mediated glutamine uptake blocks prostate cancer growth and tumour development. J. Pathol. **236**(3), 278–289 (2015)

119. F.A. Dick, D.W. Goodrich, J. Sage, N.J. Dyson, Non-canonical functions of the RB protein in cancer. Nat. Rev. Cancer **18**(7), 442–451 (2018)

120. C.R. Goding, D. Pei, X. Lu, Cancer: pathological nuclear reprogramming? Nat. Rev. Cancer **14**(8), 568–573 (2014)

121. M. Nakagawa, M. Koyanagi, K. Tanabe, K. Takahashi, T. Ichisaka, T. Aoi, et al., Generation of induced pluripotent stem cells without Myc from mouse and human fibroblasts. Nat. Biotechnol. **26**(1), 101–106 (2008)

122. M. Wernig, A. Meissner, J.P. Cassady, R. Jaenisch, c-Myc is dispensable for direct reprogramming of mouse fibroblasts. Cell Stem Cell **2**(1), 10–12 (2008)

123. V. Krizhanovsky, S.W. Lowe, Stem cells: the promises and perils of p53. Nature **460**(7259), 1085–1086 (2009)

124. K. Hochedlinger, Y. Yamada, C. Beard, R. Jaenisch, Ectopic expression of Oct-4 blocks progenitor-cell differentiation and causes dysplasia in epithelial tissues. Cell **121**(3), 465–477 (2005)

125. C.M. Rudin, S. Durinck, E.W. Stawiski, J.T. Poirier, Z. Modrusan, D.S. Shames, et al., Comprehensive genomic analysis identifies SOX2 as a frequently amplified gene in small-cell lung cancer. Nat. Genet. **44**(10), 1111–1116 (2012)

126. A. Sarkar, K. Hochedlinger, The sox family of transcription factors: versatile regulators of stem and progenitor cell fate. Cell Stem Cell **12**(1), 15–30 (2013)

127. M.S. Kareta, L.L. Gorges, S. Hafeez, B.A. Benayoun, S. Marro, A.F. Zmoos, et al., Inhibition of pluripotency networks by the Rb tumor suppressor restricts reprogramming and tumorigenesis. Cell Stem Cell **16**(1), 39–50 (2015)

128. Y. Liu, B. Clem, E.K. Zuba-Surma, S. El-Naggar, S. Telang, A.B. Jenson, et al., Mouse fibroblasts lacking RB1 function form spheres and undergo reprogramming to a cancer stem cell phenotype. Cell Stem Cell **4**(4), 336–347 (2009)

129. B.A. Smith, A. Sokolov, V. Uzunangelov, R. Baertsch, Y. Newton, K. Graim, et al., A basal stem cell signature identifies aggressive prostate cancer phenotypes. Proc. Natl. Acad. Sci. U. S. A. **112**(47), E6544–E6552 (2015)

130. P. Mu, Z. Zhang, M. Benelli, W.R. Karthaus, E. Hoover, C.C. Chen, et al., SOX2 promotes lineage plasticity and antiandrogen resistance in TP53- and RB1-deficient prostate cancer. Science **355**(6320), 84–88 (2017)

131. L. Magnaghi-Jaulin, R. Groisman, I. Naguibneva, P. Robin, S. Lorain, J.P. Le Villain, et al., Retinoblastoma protein represses transcription by recruiting a histone deacetylase. Nature **391**(6667), 601–605 (1998)

132. K.D. Robertson, S. Ait-Si-Ali, T. Yokochi, P.A. Wade, P.L. Jones, A.P. Wolffe, DNMT1 forms a complex with Rb, E2F1 and HDAC1 and represses transcription from E2F-responsive promoters. Nat. Genet. **25**(3), 338–342 (2000)

133. A. Brehm, E.A. Miska, D.J. McCance, J.L. Reid, A.J. Bannister, T. Kouzarides, Retinoblastoma protein recruits histone deacetylase to repress transcription. Nature **391**(6667), 597–601 (1998)

134. S.J. Nielsen, R. Schneider, U.M. Bauer, A.J. Bannister, A. Morrison, D. O'Carroll, et al., Rb targets histone H3 methylation and HP1 to promoters. Nature **412**(6846), 561–565 (2001)

135. S. Gonzalo, M. Garcia-Cao, M.F. Fraga, G. Schotta, A.H. Peters, S.E. Cotter, et al., Role of the RB1 family in stabilizing histone methylation at constitutive heterochromatin. Nat. Cell Biol. **7**(4), 420–428 (2005)

136. M. Garcia-Cao, S. Gonzalo, D. Dean, M.A. Blasco, A role for the Rb family of proteins in controlling telomere length. Nat. Genet. **32**(3), 415–419 (2002)

137. S.X. Skapek, Y.R. Pan, E.Y. Lee, Regulation of cell lineage specification by the retinoblastoma tumor suppressor. Oncogene **25**(38), 5268–5276 (2006)

138. D.E. Montoya-Durango, Y. Liu, I. Teneng, T. Kalbfleisch, M.E. Lacy, M.C. Steffen, et al., Epigenetic control of mammalian LINE-1 retrotransposon by retinoblastoma proteins. Mutat. Res. **665**(1–2), 20–28 (2009)

139. A. Sparmann, M. van Lohuizen, Polycomb silencers control cell fate, development and cancer. Nat. Rev. Cancer. **6**(11), 846–856 (2006)

140. B.P. Coe, K.L. Thu, S. Aviel-Ronen, E.A. Vucic, A.F. Gazdar, S. Lam, et al., Genomic deregulation of the E2F/Rb pathway leads to activation of the oncogene EZH2 in small cell lung cancer. PLoS One **8**(8), e71670 (2013)

141. L.R. Bohrer, S. Chen, T.C. Hallstrom, H. Huang, Androgens suppress EZH2 expression via retinoblastoma (RB) and p130-dependent pathways: a potential mechanism of androgen-refractory progression of prostate cancer. Endocrinology **151**(11), 5136–5145 (2010)

142. C.A. Ishak, A.E. Marshall, D.T. Passos, C.R. White, S.J. Kim, M.J. Cecchini, et al., An RB-EZH2 complex mediates silencing of repetitive DNA sequences. Mol. Cell **64**(6), 1074–1087 (2016)

143. P.J. Vlachostergios, L. Puca, H. Beltran, Emerging variants of castration-resistant prostate cancer. Curr. Oncol. Rep. **19**(5), 32 (2017)

144. E. Hernando, Z. Nahle, G. Juan, E. Diaz-Rodriguez, M. Alaminos, M. Hemann, et al., Rb inactivation promotes genomic instability by uncoupling cell cycle progression from mitotic control. Nature **430**(7001), 797–802 (2004)

145. C.E. Isaac, S.M. Francis, A.L. Martens, L.M. Julian, L.A. Seifried, N. Erdmann, et al., The retinoblastoma protein regulates pericentric heterochromatin. Mol. Cell. Biol. **26**(9), 3659–3671 (2006)

146. R. Velez-Cruz, D.G. Johnson, The Retinoblastoma (RB) Tumor Suppressor: Pushing Back against Genome Instability on Multiple Fronts. Int J Mol Sci, **18**(8) (2017)

147. A. L. Manning et al. Suppression of genome instabilityin pRB- deficient cells by enhancement of chromosomecohesion. Mol. Cell **53**, 993–1004 (2014)

148. M. S. Longworth, A. Herr, J. Y. Ji, N. J Dyson. RBF1 promotes chromatin condensation through a conserved interaction with the Condensin II protein dCAP-D3. Genes Dev. **22**, 1011–1024 (2008)

149. G.A. Pihan, A. Purohit, J. Wallace, R. Malhotra, L. Liotta, S.J. Doxsey, Centrosome defects can account for cellular and genetic changes that characterize prostate cancer progression. Cancer Res. **61**(5), 2212–2219 (2001)

150. G.A. Pihan, J. Wallace, Y. Zhou, S.J. Doxsey, Centrosome abnormalities and chromosome instability occur together in pre-invasive carcinomas. Cancer Res. **63**(6), 1398–1404 (2003)

151. X. Jin, D. Ding, Y. Yan, H. Li, B. Wang, L. Ma, et al., Phosphorylated RB promotes cancer immunity by inhibiting NF-kappaB activation and PD-L1 expression. Mol. Cell **73**(1), 22–35.e6 (2019)

152. C.A. Ishak, M. Classon, D.D. De Carvalho, Deregulation of retroelements as an emerging therapeutic opportunity in cancer. Trends Cancer. **4**(8), 583–597 (2018)

153. J.M. Wolff, L.T. Brett, A.M. Lessells, F.K. Habib, Analysis of retinoblastoma gene expression in human prostate tissue. Urol. Oncol. **3**(5–6), 177–182 (1997)

154. M.M. Ittmann, R. Wieczorek, Alterations of the retinoblastoma gene in clinically localized, stage B prostate adenocarcinomas. Hum. Pathol. **27**(1), 28–34 (1996)

155. M.P. Markey, S.P. Angus, M.W. Strobeck, S.L. Williams, R.W. Gunawardena, B.J. Aronow, et al., Unbiased analysis of RB-mediated transcriptional repression identifies novel targets and distinctions from E2F action. Cancer Res. **62**(22), 6587–6597 (2002)

156. M.P. Markey, J. Bergseid, E.E. Bosco, K. Stengel, H. Xu, C.N. Mayhew, et al., Loss of the retinoblastoma tumor suppressor: differential action on transcriptional programs related to cell cycle control and immune function. Oncogene **26**(43), 6307–6318 (2007)

157. J. Lack, M. Gillard, M. Cam, G.P. Paner, D.J. Van der Weele, Circulating tumor cells capture disease evolution in advanced prostate cancer. J. Transl. Med. **15**(1), 44 (2017)

158. H. Beltran, A. Jendrisak, M. Landers, J.M. Mosquera, M. Kossai, J. Louw, et al., The initial detection and partial characterization of circulating tumor cells in neuroendocrine prostate cancer. Clin. Cancer Res. **22**(6), 1510–1519 (2016)

159. A.W. Wyatt, A.A. Azad, S.V. Volik, M. Annala, K. Beja, B. McConeghy, et al., Genomic alterations in cell-free DNA and enzalutamide resistance in castration-resistant prostate cancer. JAMA Oncol. **2**(12), 1598–1606 (2016)

160. M.G. Oser, R. Fonseca, A.A. Chakraborty, R. Brough, A. Spektor, R.B. Jennings, et al., Cells lacking the RB1 tumor suppressor gene are hyperdependent on Aurora B kinase for survival. Cancer Discov. **9**(2), 230–247 (2019)

161. X. Gong, J. Du, S.H. Parsons, F.F. Merzoug, Y. Webster, P.W. Iversen, et al., Aurora-A kinase inhibition is synthetic lethal with loss of the RB1 tumor suppressor gene. Cancer Discov. **9**(2), 248–263 (2019)

Interplay Among PI3K/AKT, PTEN/FOXO and AR Signaling in Prostate Cancer

Yuqian Yan and Haojie Huang

Introduction

Phosphatidylinositol-4,5-bisphosphate 3-kinase (PI3K) belongs to a family of lipid kinases involved in phosphorylating the 3-position hydroxyl group of the inositol ring of phosphatidylinositol (PtdIns) [1]. Products of PI3K activity, i.e., the lipid second messengers phosphatidylinositol (3,4,5) trisphosphate [$PI(3,4,5)P_3$ or PIP3] and $PI(3,4)P_2$ (PIP2), promote membrane association and activation of serine/threonine kinases such as AKT (or termed protein kinase B (PKB)). There are three highly-homologous AKT isoforms: AKT1/PKBα, AKT2/PKBβ, and AKT3/PKBγ [2]. These isoforms encoded by three different genes possess both common and isoform-specific functions.

AKT is activated by phosphorylation of two serine (Ser)/threonine (Thr) residues, one (Thr^{308} in AKT1) being phosphorylated by the phosphoinositide-dependent kinase 1 (PDK1) [3] and the other (Ser^{473} in AKT1) being phosphorylated by the mammalian target of rapamycin complex 2 (mTORC2) [4]. Therefore, this pathway is also known as the PI3K/AKT/mTOR signaling pathway. Appropriately 40% of primary and 70% of metastatic prostate cancers harbor genomic alterations leading to the activation of the PI3K signaling pathway [5, 6].

The PI3K signaling cascade transduces extracellular signals to intracellular targets. The extracellular signals include peptide hormones and growth factors, such as insulin [7], epidermal growth factor (EGF) [8], sonic hedgehog (shh) [9] and insulin-like growth factor 1 (IGF-1) [9]. Mechanistically, upon the stimulus of the extracellular signals, the signaling transduction cascade is activated by PI3K phosphorylation. AKT acts as an important mediator via recruitment to the membrane by interaction with phosphoinositide docking sites, where it becomes fully activated through its phosphorylation by PDK1 and mTORC2. Activated AKT phosphorylates Ser and Thr residues of its targets, primarily within a minimal consensus recognition motif of R-X-R-X-X-S/T-f (X: any amino acid; f: a preference for large hydrophobic residues) [10]. This pathway leads to CREB activation [11], p27 inhibition [12, 13], FOXO phosphorylation and cytoplasmic localization [14, 15], and activation of downstream effectors of mTORC1 such as

Y. Yan
Department of Biochemistry and Molecular Biology, Mayo Clinic College of Medicine and Science, Rochester, MN, USA

H. Huang (✉)
Department of Biochemistry and Molecular Biology, Mayo Clinic College of Medicine and Science, Rochester, MN, USA

Department of Urology, Mayo Clinic College of Medicine and Science, Rochester, MN, USA

Mayo Clinic Cancer Center, Mayo Clinic College of Medicine and Science, Rochester, MN, USA
e-mail: huang.haojie@mayo.edu

© Springer Nature Switzerland AG 2019
S. M. Dehm, D. J. Tindall (eds.), *Prostate Cancer*, Advances in Experimental Medicine and Biology 1210, https://doi.org/10.1007/978-3-030-32656-2_14

p70S6K and 4EBP1 [16]. There are myriad AKT targets with their phosphorylation sites listed in Fig. 1. Functionally, activation or inactivation of these downstream targets leads to nutrient metabolism, cell proliferation, survival, migration, and angiogenesis, which ensure prostate cancer cell survival and protection from apoptosis (Fig. 1).

Activation of PI3K Due to PTEN Genetic Alterations

PTEN Mutations Account for the Major Cause of PI3K Activation in Prostate Cancer

PI3K signaling in both primary and advanced prostate cancers is activated in a similar manner, mainly due to mutations in the tumor suppressor gene phosphatase and tensin homolog (*PTEN*). This gene encodes PTEN protein that acts as a PIP_3 phosphatase, therefore antagonizing the PI3K pathway [18]. Approximately 17% of primary prostate cancers in patients harbor *PTEN* mutations [19]. However, approximately 50% of metastatic castration-resistant prostate cancer (mCRPC) patients have somatic mutations in the PI3K pathway [20]. Among these mutations, *PTEN* mutations account for the highest frequency (approximately 40.7%), mainly biallelic inactivation of the phosphatase domain in the hotspots of this gene.

To date, PTEN deletion or mutations have been considered to be the major genetic alterations in PI3K/AKT signaling activation. Other genetic alterations, including amplifications and activating fusions in *PI3K3CA* and p.E17K activating mutations in *AKT1*, also contribute to the activation of PI3K signaling [20]. Mutations of another member of the PI3K catalytic subunit, *PI3K3CB*, were observed initially in a cohort of advanced prostate cancers [20]. In agreement with a previous study [21], mutations of *PI3K3CB* rather than *PI3K3CA* are most likely to occur in the context of *PTEN*-deficient cases, implying that some *PTEN* deficient cancers may depend on PIK3CB activation. The frequency of PI3K

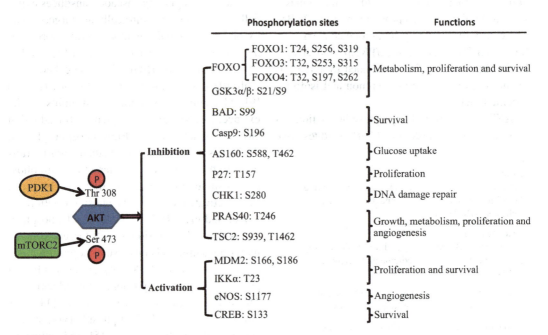

Fig. 1 Selected AKT-phosphorylated proteins (modified from a previous report [17]). AKT is phosphorylated by PDK1 and mTORC2, resulting in the phosphorylation at Thr308 and Ser473 respectively. The phosphorylated AKT subsequently phosphorylates a group of proteins through a recognition motif R-X-R-X-X-S/T-f, which leads to the inhibition or activation of AKT targets.

Table 1 The frequency of genetic alterations in PI3K signaling pathway genes in primary prostate cancer [22]

Gene	Altered frequency	Missense	Truncating	Frameshift	Other
PTEN	167/1013 (16%)	15	29	1	122 (Deep deletion)
PIK3CA	69/1013 (7%)	29	1	1	38 (Amplification)
PIK3CB	41/1013 (4%)	10	0	0	31 (Amplification)
AKT1	14/1013 (1%)	6	0	0	8 (Amplification)

Table 2 The frequency of genetic alterations in PI3K signaling pathway genes in advanced prostate cancer [20]

Gene	Altered frequency	Missense	Truncating	Frameshift	Other
PTEN	60/150 (40%)	1	10	0	39 (Deep deletion) 8 (Fusion) 2 (Deep deletion and fusion)
PIK3CA	8/150 (5%)	5	0	1	1 (Amplification) 1 (Amplification and fusion)
PIK3CB	10/150 (7%)	3	1	0	5 (Amplification) 1 (Amplification and fusion)
AKT1	5/150 (3%)	2	0	0	3 (Amplification)

pathway gene mutations in primary and advanced prostate cancer is listed in Tables 1 and 2, respectively.

Pten Deletion-Driven Prostate Cancer Mouse Models

Given that *PTEN* mutations comprise one of the most common genetic alterations in prostate cancer, it is important to assess its tumor suppressor function by generating *Pten* mutant mouse models. Conventional homozygous deletion of *Pten* causes embryonic lethality in mice [23]. Heterozygous loss of *Pten* in the mouse prostate results in a 100% penetrance of prostate intraepithelial neoplasia (PIN), a precursor of prostate cancer. However, on a Balb/c/129 genetic background, the latency of PIN is relatively long (approximately 10 months) and the PIN lesions rarely undergo metastasis [24, 25]. Thus, it is possible that mutations in other tumor suppressor genes or loss of another allele of *Pten* might be required for prostate tumorigenesis. Indeed, concomitant *Pten* heterozygous deletion and alterations in other genes, such as *p27* [26], *Nkx3.1* [27], *ERG* [28, 29], or *CREBBP* (*CBP*) [30] has given rise to prostate cancer in mice with various genetic backgrounds. Although these genetic alterations have accelerated formation of PIN

lesions and/or cancer, no metastatic prostate cancers have been observed in these models. Generation of a prostate-specific *Pten* homozygous deletion mouse model recapitulates the disease progression of human prostate cancer, mimicking the progression from PIN to invasive adenocarcinoma and, in very rare cases, metastasis [25]. This and other *Pten* deletion mouse models have been adapted to study the etiology of prostate cancer and the mechanisms of cancer progression [24, 25, 27, 30].

Activation of AKT/mTOR Signaling Pathway in SPOP Mutated Prostate Cancer

Mutations in other genes that appear to be irrelevant to the PI3K pathway can indirectly promote activation of AKT/mTOR signaling. The most striking example is mutation of the tumor suppressor gene, speckle-type POZ (*SPOP*). *SPOP* is mutated in approximately 10–15% of prostate cancer patients [19, 20, 22, 31]. Intriguingly, there is a mutually exclusive relationship between PTEN mutation and SPOP mutation in patients with primary prostate cancer (Fig. 2), implying that these two genetic alterations share a common downstream pathway during prostate cancer pathogenesis. In advanced prostate cancer this

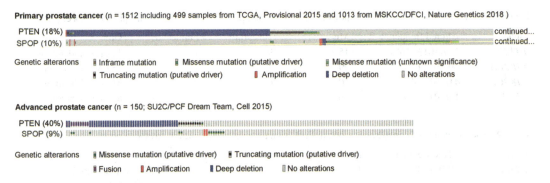

Fig. 2 *PTEN* mutations are mutually exclusive with *SPOP* mutations in primary prostate cancer. The frequency of *PTEN* and *SPOP* mutations was analyzed from primary and advanced prostate cancers. The primary prostate cancer data was combined from two studies, including 1512 samples, whereas the advanced prostate cancer data from one study consisting of 150 samples. Mutual exclusivity was observed between PTEN and SPOP mutations in primary prostate cancer (***$P < 0.001$) but not in advanced prostate cancer ($P = 0.484$).

mutual exclusivity is not evident in the small number of CRPC cases (150) examined to date (Fig. 2). Therefore, further investigation of the relationship between these two genetic alterations in large cohorts of advanced prostate cancers is warranted.

SPOP Mutations Induce AKT/mTORC1 Activation via Elevation of Bromodomain and Extra-Terminal (BET) Family Proteins

SPOP acts as an adaptor protein of the CULLIN3-based E3 ubiquitin ligase and promotes protein ubiquitylation and proteasome degradation. A number of prostate cancer relevant proteins such as bromodomain and extra-terminal motif (BET) proteins (BRD2, BRD3 and BRD4), SRC-3, TRIM24, ERG, and AR, are substrates of SPOP [32–40]. Almost all of these SPOP substrates are somehow involved in or associated with AKT signaling pathway and their relationships will be discussed throughout this chapter.

SPOP mutations result in an increased expression of BET family proteins including BRD2, BRD3, and BRD4 [40]. Subsequently, the stabilized BRD4 activates the transcriptional expression of the Rho GTPase family member RAC1 and cholesterol synthesis genes [40]. RAC1 is a canonical small GTPase that activates the AKT-mTORC1 pathway by binding directly to mTOR [40, 41]. Cholesterol-rich lipid rafts are linked to AKT activation and prostate cancer cell survival [42, 43].

SRC-3, which is also known as amplified in breast 1(*AIB1*), is encoded by the nuclear receptor coactivator 3 (*NCOA3*) gene. SRC-3 is a transcriptional coactivator that contains several nuclear receptor interacting domains and possesses an intrinsic histone acetyltransferase activity, which facilitates the accessibility of transcriptional factors to chromatin. IGF-1, which is a target of SRC-3 [44], is a potent upstream regulator of the AKT signaling pathway [10]. Moreover, SRC-3 can also contribute to the activation of AKT in SPOP-mutant prostate cancer cells by functioning as a transcriptional coactivator to facilitate expression of *RAC1* and cholesterol synthesis genes [40].

TRIM 24, a known AR coactivator, binds to the *PIK3CA* promoter to regulate the transcription of *PIK3CA* gene, leading to the upregualtion of PI3K-AKT signaling [45]. Intriguingly, TRIM24 is regulated by SPOP via proteasome pathway [37]. Therefore, in SPOP mutant prostate cancer cells, TRIM24 is stabilized at the protein level. Taken together, SPOP mutations augment AKT signaling through multiple mechanisms, and further investigation is warranted to fully elucidate the signaling pathways through which SPOP mutations lead to activation of AKT signaling.

SPOP Mutant Mouse Models

SPOP mutations represent a molecularly-distinct subtype of prostate cancer. Generation of mouse models that recapitulate the unique features of SPOP mutations is important for a full understanding of the etiology of these lesions and their role in prostate cancer. To mimic SPOP mutation-induced prostate cancer pathogenesis, the most-frequently occurring SPOP mutant, F133V, was knocked into the Rosa26 locus and specifically expressed in the mouse prostate through a lox-STOP-lox strategy [46]. Surprisingly, little or no histological or glandular architecture of the prostate was observed in this mouse model [46]. At the cellular level, proliferation was not significantly altered, and changes in AR expression were observed rarely, and rare cells exhibited cytological atypia with enlarged nuclei in a majority of all prostate lobes in mice at ≥ 12 months of age [46].

The above findings indicate that like many other known genetic alterations (*ERG*, *ETV1* and *TP53*) in human prostate cancer [28], SPOP mutations alone may not be sufficient to drive tumorigenesis, and other genetic alterations are required to promote or accelerate tumorigenesis and progression. As discussed above, *PTEN* heterozygous mutations alone results in minimal histologic changes in the prostate [25], even though *Pten* heterozygous mouse models have often been crossed with the other mutant mouse models to study the etiology of prostate cancer. Indeed, when the SPOP F133V mutation mouse was crossed with *Pten* heterozygous deletion mouse, high-grade PIN developed only in the compound mice [46]. However, it is worth noting that *SPOP* mutations and *PTEN* deletions are almost mutually exclusive in patients with primary prostate cancers (Fig. 2). Therefore, further development of clinically relevant *SPOP*-mutated prostate cancer mouse models is warranted to interrogate the molecular mechanisms underlying *SPOP* mutation-induced prostate tumorigenesis.

FOXO1 Dysregulation in Prostate Cancer

The forkhead box-O protein 1 (FOXO1) belongs to the FOXO family that includes three other members (FOXO3, FOXO4 and FOXO6). FOXO1 is a transcription factor that acts as a tumor suppressor by transcriptionally regulating expression of genes involved in apoptotic cell death, cell cycle, DNA damage repair, glucose metabolism, and carcinogenesis [47, 48]. Multiple mechanisms regulate FOXO1 functions, including, but not limited to, genomic deletion, transcriptional downregulation, and phosphorylation. *FOXO1* gene deletion as well as transcriptional downregulation have been found in a substantial proportion of prostate cancers from patients [19, 20, 49–52]. FOXO1 is a direct phosphorylation target of AKT. AKT-mediated phosphorylation of FOXO1 induces its translocation from the nucleus to the cytoplasm, resulting in inhibition of the transactivation of its target genes. We have summarized three major mechanisms that lead to FOXO1 inactivation and the effects on its downstream pathways in Fig. 3.

FOXO1 and AR

FOXO1 can bind directly to the AR in a ligand-independent manner, thereby inhibiting the transcriptional activity of both full-length AR and constitutively active splice variants of AR [53–56]. However, this inhibitory effect is dependent largely on FOXO1 phosphorylation status. Specifically, upon the stimulus of IGF-1 or insulin, AKT signaling can induce FOXO1 phosphorylation and subsequent translocation from the nucleus to the cytoplasm, thereby impairing FOXO1 inhibition of ligand-induced AR activation in the nucleus. Similarly, in PTEN-mutated prostate cancer cells, AKT signaling is activated and FOXO1 is transported from the nucleus to the cytoplasm, thereby favoring transactivation of AR [48, 55]. However, AR regulation by the

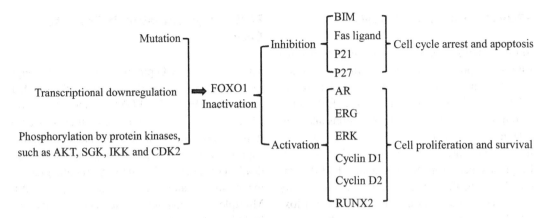

Fig. 3 Diagram depicting three major mechanisms leading to FOXO1 inactivation. There are at least three mechanisms leading to the inactivation of FOXO1, which result in either the inhibition or the activation of its downstream targets and signaling.

PI3K/AKT pathway is very complex, including a negative feedback between AR and AKT signaling [5, 57]. For instance, AKT phosphorylates AR at Ser210, thereby inhibiting AR transactivation [57].

Aberrant activation of AR is associated with the progression of CRPC. The downregulation of FOXO1 in PTEN-negative prostate cancer contributes to the hyperactivation of AR [55]. Interestingly, FOXO1 interacts physically with HDAC3 and acts as a corepressor that inhibits androgen-independent activation of AR. Thus, co-transfection of FOXO1 and HDAC3 in prostate cancer cells results in a greater inhibition of AR activity than transfection with FOXO1 or HDAC3 alone [55]. Specifically, a putative transcription repression domain in the NH2-terminus of FOXO1 appears to be responsible for FOXO1 inhibition of the AR. FOXO1 can bind to the transcription activation unit 5 (TAU5) motifs in the AR NH2-terminal domain (NTD) that is required for recruitment of p160 coactivators including SRC-1, subsequently inhibiting the ligand-independent activation of AR splice variants [53]. Moreover, PI3K-AKT-FOXO1 signaling regulates AR variant 7 [56], further indicating that PI3K is a potential therapeutic target in CRPC patients.

FOXO1 and ERG

The ETS-related gene (ERG) is a transcription factor belonging to the E-26 transformation-specific (ETS) family. It regulates a group of genes involved in vasculogenesis, angiogenesis, hematopoiesis, and bone development [58]. ERG is highly associated with prostate cancer development. Aberrant overexpression of ERG is found in approximately 50% of all human prostate cancer due to the fusion of the *ERG* gene body to androgen-regulated promoters and enhancers that normally regulate genes such as *TMPRSS2*, *SLC45A3*, and *NDRG1* [19, 20, 22, 59].

ERG genetic rearrangements and loss of PTEN often co-occur in human prostate cancers [19, 28], indicating an association between PI3K signaling and ERG. Indeed, the combination of transgenic expression of prostate cancer-associated *TMPRSS2-ERG* and heterozygous deletion of *Pten* induces high grade PIN and cancer in the mouse prostate [28, 29], although how the loss of PTEN works in concert with ERG overexpression to promote prostate tumorigenesis was unexplored in these studies. Also, FOXO1 binds directly to the DNA binding domain of ERG and inhibits ERG transcriptional activity in prostate cancer cells [60]. However, FOXO1 inhibition of ERG is abolished by AKT due to AKT mediated phosphorylation and exclusion of FOXO1 from the nucleus [60]. Importantly, homozygous deletion of *FOXO1* cooperates with overexpression of *TMPRSS2-ERG* to induce formation of HGPIN and cancerous phenotypes in the mouse prostate [60]. Therefore, functional loss or genetic deletion of FOXO1 results in an aberrant activation of ERG fusions and abnormal expression of ERG target genes, thereby contributing to prostate tumorigenesis.

FOXO1 and RUNX2

The Runt-related transcription factor 2 (RUNX2), also known as core-binding factor subunit alpha-1 (CBFA1), regulates many cellular proliferation genes, such as *c-Myc*, *C/EBP* [61], *TP53* [62], and the CDK inhibitor *p21^{cip1}* [63], at the transcription level. The DNA-binding affinity of RUNX2 is most likely dependent on its phosphorylation state [64], which is correlated with cellular proliferation. Thus, RUNX2 phosphorylation is related to RUNX2-mediated cellular proliferation and cell cycle control. In support of this concept, RUNX2 is phosphorylated at Ser451 by CDK1, which facilitates cell cycle progression through the regulation of G2 and M phases [65]. Phosphorylation of RUNX2 at Ser301 and Ser319 by MAPK-dependent activation also promotes RUNX2 transcriptional activity [64, 66].

Intriguingly, RUNX2 protein level fluctuates throughout the cell cycle, and this is most likely due to its regulation by both gene transcription and protein degradation. RUNX2 can also interact with several protein kinases that facilitate cell-cycle dependent dynamics. Thus far, most studies relating to RUNX2 have been focused on osteoblast proliferation and differentiation; however, the role of RUNX2 in prostate tumorigenesis is poorly understood.

RUNX2 forms a protein complex with AR in prostate cancer cells [67, 68]. AR can inhibit RUNX2 binding to DNA through protein-protein interaction [67]. AKT phosphorylates AR at Ser-210 and subsequently inhibits AR transactivation [57]. This supports the finding that PI3K signaling can stimulate the transcriptional activity of RUNX2. In contrast, FOXO1 acts as a repressor of RUNX2. Thus, loss of PTEN or FOXO1 leads to the upregulation of RUNX2 transcriptional activity and increased migration and invasion of prostate cancer cells [69]. FOXO1 inhibition of RUNX2 also occurs in osteoblasts [70]. Thus, FOXO1 is an important negative regulator of RUNX2. This concept is further supported by a recent study, which identified a signaling axis of AKT-FOXO1-RUNX2-OCN-GPRC6A-CREB, activation of which results in upregulation of cytochrome P450 (CYP) enzymes (CYP11A1, CYP17A1) and increased synthesis of testosterone in PTEN-null prostate cancer cells [71]. Abnormal RUNX2 activation plays a pivotal role in PTEN loss-induced intratumoral androgen synthesis and tumor microenvironment remodeling [71]. Deletion of *Runx2* in *Pten* homozygous knockout prostate decreased *Cyp11a1* and *Cyp17a1* gene expression, testosterone levels, and tumor growth in castrated mice [71]. Therefore, under AKT activation conditions, the cytoplasm exportation of FOXO1 results in the loss of its function and inhibition of the transcriptional activity of RUNX2. Moreover, aberrant activation of RUNX2 promotes intratumoral androgen biosynthesis through the RUNX2-OCN-GPRC6A-CREB signaling cascade.

Cross Talk Between PI3K Signaling and Other Pathways in Prostate Cancer

In prostate cancer, it is well accepted that a number of signaling pathways cross talk with each other either in an orchestrated fashion or in a negative feedback manner. An understanding of the cross talk among these pathways is critical for an effective treatment of prostate cancer.

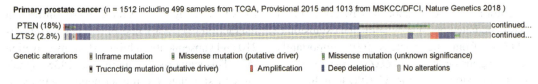

Fig. 4 *PTEN* mutations are concurrent with *LZTS2* mutations. The primary prostate cancer data was combined from two studies, including 1512 samples. The co-occurrence of *PTEN* and *LZTS2* mutations was analyzed and displayed significant co-occurrence (***$P < 0.001$).

AKT Signaling and AR

AKT and AR signaling are two major drivers of prostate cancer. AKT can phosphorylate AR at Ser-210 and subsequently inhibit AR transactivation [57]. On the other hand, AR inhibition activates AKT signaling by reducing expression of the AKT phosphatase PHLPP1 [5, 72]. Because inhibiting either of these two pathways often activates the other, the development of a dual inhibitor might be advantageous for therapy. Notably, HDAC3 can regulate both AKT signaling [39, 73] and AR transcriptional activity [74]. Moreover, a HDAC3-specific inhibitor (RGFP966) inhibits both AKT and AR pathways in prostate cancer in vitro and in vivo, including prostate cancer organoid and mouse xenograft models [39].

AKT Signaling and WNT/β-Catenin Signaling

WNT/β-catenin signaling is mediated by the extracellular signals of WNT proteins via cell membrane receptors, which converge on the transcription factor β-catenin. Genetic and epigenetic alterations have been identified in components of WNT signaling pathway in both primary and advanced prostate cancer [19, 20, 22], further suggesting that WNT signaling contributes prostate tumorigenesis. Notably, inhibition of WNT signaling in mice prevents prostate cancer progression [75].

There are at least three proposed mechanisms by which WNT signaling is activated in prostate cancer. Firstly, tumor stromal cells can secret WNT proteins to maintain the tumor microenvironment and support self-renewal or expansion of prostate cancer stem-like or progenitor cells and drug resistance via WNT/β-catenin signaling [75]. Secondly, AKT signaling can phosphorylate β-catenin and promote its transcriptional activity, which drives tumor cell invasion [76]. In the context of prostate cancer cells, the WNT co-receptor LRP6 increases aerobic glycolysis in a β-catenin-independent manner by directly activating AKT-mTORC1 signaling [77]. Thirdly, the leucine zipper tumor suppressor-2 (LZTS2), a β-catenin-binding protein, is a negative regulator of WNT signaling [78]. The *LZTS2* gene is approximately 15 Mb from the *PTEN* gene locus and is frequently deleted in a variety of human malignancies, including prostate cancer. Interestingly, *PTEN* deletions and *LZTS2* deletions frequently co-exist in primary prostate cancer (Fig. 4), implying a novel mechanism for the dysregulation of WNT/β-catenin signaling during prostate tumorigenesis [79].

AKT Signaling and MAPK/ERK Signaling

The mitogen-activated protein kinase (MAPK) (or called extracellular signal-regulated kinase (ERK)) pathway is also known as the Ras-Raf-MEK-ERK pathway. It transduces the extracellular signals from the cell surface to the DNA in the nucleus. In prostate cancer, AKT and MAPK signaling pathways are frequently activated in androgen-independent cancer types. The activation of either AKT or ERK signaling in an androgen-responsive prostate cancer cell line promotes hormone-independent

Table 3 PI3K/AKT inhibitors tested in prostate cancer [84]

Therapeutic regimen	Indication	Clinical trial status	Patient status	Reference or ClinicalTrials.gov identifier
BKM120	PI3K inhibitor	Phase II	mCRPC	NCT01385293
GSK2636771	PI3K inhibitor	Phase I, GSK2636771 ± Enzalutamide	mCRPC	NCT02215096
AZD8186	PI3K inhibitor	Phase I, AZD8186 ± Abiraterone or AZD2014	mCRPC	NCT01884285
LY3023414	PI3K + mTOR inhibitor	Phase II, Enzalutamide ± LY3023414	mCRPC	NCT02407054
AZD5363	AKT inhibitor	Phase I/phase II clinical trials	mCRPC	[85]
MK2206	AKT inhibitor	Phase II, Bicalutamide ± MK2206	mCRPC	NCT01251861

but AR-dependent growth in culture [80]. It is hypothesized that epithelial-stromal competition leads to androgen independence during prostate tumorigenesis, in which activation of AKT and ERK promotes AR activity in the prostate epithelium while counteracting the antagonistic effects in the stroma [80].

Inhibition of the PI3K/AKT signaling activates the MAPK pathway [81, 82], whereas activation of AKT by phosphorylation leads to FOXO1 exclusion from the nucleus and abolishment of its tumor suppressor functions in the nucleus. Intriguingly, AKT-phosphorylated FOXO1 can inhibit the MAPK pathway by binding to the scaffold protein IQGAP1 in the cytoplasm, thus impeding IQGAP1-dependent activation of ERK1/2 [83]. Thus, FOXO1 possesses tumor suppressor functions in both the nucleus and the cytoplasm.

Targeting PI3K/AKT Signaling for Prostate Cancer Treatment

Due to the critical role of PI3K/AKT in maintaining prostate cancer progression and cell survival, targeting the PI3K/AKT pathway represents a promising strategy for prostate cancer treatment. However, the inhibitors that are currently under testing often encounter issues such as low efficacy and acquired drug resistance. Thus, development of new single-targeting or dual inhibitors is urgently needed.

PI3K/AKT Inhibitors Tested in Prostate Cancer

A number of PI3K/AKT signaling pathway inhibitors have been tested, or are currently under investigation in clinical trials (Table 3). Mechanistically, most of them act on PI3K and AKT, with few on mTORC1 or mTORC2. Hereby, we have summarized a few of PI3K/AKT inhibitors which are currently under clinical trials to treat mCRPC patients either by administration alone or in combination with AR inhibitors as seen in Table 3. Specifically, BKM120 (NCT01385293), a PI3K inhibitor, is administrated alone and is currently under phase II clinical trial. GSK2636771 (NCT02215096) and AZD8186 (NCT01884285), two PI3K inhibitors, are treated in the combination with Enzalutamide or Abiraterone respectively and are under phase I clinical trials. LY3023414 (NCT02407054) is a dual inhibitor, which targets both PI3K and mTOR in an ATP-competitive manner [86]. AZD5365 [85] and MK2206 (NCT01251861) are two AKT inhibitors. Currently, AZD5363 is administrated alone and is under Phase I/phase II clinical trials. Interestingly, a preclinical study has found that AZD5363 significantly delayed Enzalutamide-resistant prostate cancer when it is combined with Enzalutamide [87]. MK2206 is under phase II clinical trial, and is administrated with or without Bicalutamide. Taken together, PI3K/AKT inhibitors appear to present a limited clinical outcome as single agents.

A Limitation of Monotherapy with the PI3K Inhibitors

In prostate cancer, PI3K and AR signaling pathways are the most frequently activated pathways in which a reciprocal feedback exists between these two pathways. The monotherapy with AR or AKT inhibitors often leads to activation of the other pathway to sustain the cell survival or drive acquired resistance in prostate cancer [5, 72]. In addition, both PI3K/AKT and AR signaling are extremely important for prostate pathogenesis, the combination treatment of PI3K/AKT inhibitors and AR signaling inhibitors is required for a better therapeutic perspective [72, 88]. Indeed, as described above (section "PI3K/AKT Inhibitors Tested in Prostate Cancer"), several clinical trials have been conducted by using the combination of PI3K inhibitors (such as GSK2636771 and LY3023414) and AR-inhibitory agents (such as Enzalutamide and Abiraterone).

Conclusions

The PI3K/AKT signaling pathway has been identified as a key driver of prostate tumorigenesis and drug resistance. Studies on the underlying mechanisms by which this pathway promotes prostate tumorigenesis have focused primarily on the phosphorylation of downstream target proteins. Cross talk between AKT signaling and the other parallel pathways has been underinvestigated. Thus, further exploration in this area could shed new light on our understanding of the molecular basis for prostate tumorigenesis and progression, and identify novel therapeutic targets for prostate cancer.

References

1. P. Liu, H. Cheng, T.M. Roberts, J.J. Zhao, Targeting the phosphoinositide 3-kinase pathway in cancer. Nat. Rev. Drug Discov. **8**, 627–644 (2009)
2. E. Gonzalez, T.E. McGraw, The Akt kinases: isoform specificity in metabolism and cancer. Cell Cycle **8**, 2502–2508 (2009)
3. D.R. Alessi, S.R. James, C.P. Downes, A.B. Holmes, P.R. Gaffney, C.B. Reese, P. Cohen, Characterization of a 3-phosphoinositide-dependent protein kinase which phosphorylates and activates protein kinase Balpha. Curr. Biol. **7**, 261–269 (1997)
4. D.D. Sarbassov, D.A. Guertin, S.M. Ali, D.M. Sabatini, Phosphorylation and regulation of Akt/PKB by the rictor-mTOR complex. Science **307**, 1098–1101 (2005)
5. B.S. Carver, C. Chapinski, J. Wongvipat, H. Hieronymus, Y. Chen, S. Chandarlapaty, V.K. Arora, C. Le, J. Koutcher, H. Scher, et al., Reciprocal feedback regulation of PI3K and androgen receptor signaling in PTEN-deficient prostate cancer. Cancer Cell **19**, 575–586 (2011)
6. B.S. Taylor, N. Schultz, H. Hieronymus, A. Gopalan, Y. Xiao, B.S. Carver, V.K. Arora, P. Kaushik, E. Cerami, B. Reva, et al., Integrative genomic profiling of human prostate cancer. Cancer Cell **18**, 11–22 (2010)
7. V.A. Rafalski, A. Brunet, Energy metabolism in adult neural stem cell fate. Prog. Neurobiol. **93**, 182–203 (2011)
8. L. Ojeda, J. Gao, K.G. Hooten, E. Wang, J.R. Thonhoff, T.J. Dunn, T. Gao, P. Wu, Critical role of PI3K/Akt/GSK3beta in motoneuron specification from human neural stem cells in response to FGF2 and EGF. PLoS One **6**, e23414 (2011)
9. J. Peltier, A. O'Neill, D.V. Schaffer, PI3K/Akt and CREB regulate adult neural hippocampal progenitor proliferation and differentiation. Dev. Neurobiol. **67**, 1348–1361 (2007)
10. B.D. Manning, A. Toker, AKT/PKB signaling: navigating the network. Cell **169**, 381–405 (2017)
11. K. Du, M. Montminy, CREB is a regulatory target for the protein kinase Akt/PKB. J. Biol. Chem. **273**, 32377–32379 (1998)
12. J.R. Graff, B.W. Konicek, A.M. McNulty, Z. Wang, K. Houck, S. Allen, J.D. Paul, A. Hbaiu, R.G. Goode, G.E. Sandusky, et al., Increased AKT activity contributes to prostate cancer progression by dramatically accelerating prostate tumor growth and diminishing p27Kip1 expression. J. Biol. Chem. **275**, 24500–24505 (2000)
13. I. Shin, F.M. Yakes, F. Rojo, N.Y. Shin, A.V. Bakin, J. Baselga, C.L. Arteaga, PKB/Akt mediates cell-cycle progression by phosphorylation of p27(Kip1) at threonine 157 and modulation of its cellular localization. Nat. Med. **8**, 1145–1152 (2002)
14. A. Brunet, A. Bonni, M.J. Zigmond, M.Z. Lin, P. Juo, L.S. Hu, M.J. Anderson, K.C. Arden, J. Blenis, M.E. Greenberg, Akt promotes cell survival by phosphorylating and inhibiting a Forkhead transcription factor. Cell **96**, 857–868 (1999)
15. X. Zhang, N. Tang, T.J. Hadden, A.K. Rishi, Akt, FoxO and regulation of apoptosis. Biochim. Biophys. Acta **1813**, 1978–1986 (2011b)
16. K. Hara, K. Yonezawa, M.T. Kozlowski, T. Sugimoto, K. Andrabi, Q.P. Weng, M. Kasuga, I. Nishimoto,

J. Avruch, Regulation of eIF-4E BP1 phosphorylation by mTOR. J. Biol. Chem. **272**, 26457–26463 (1997)

17. B.D. Manning, L.C. Cantley, AKT/PKB signaling: navigating downstream. Cell **129**, 1261–1274 (2007)

18. N. Chalhoub, S.J. Baker, PTEN and the PI3-kinase pathway in cancer. Annu. Rev. Pathol. **4**, 127–150 (2009)

19. Cancer Genome Atlas Research Network Network, The molecular taxonomy of primary prostate cancer. Cell **163**, 1011–1025 (2015)

20. D. Robinson, E.M. Van Allen, Y.M. Wu, N. Schultz, R.J. Lonigro, J.M. Mosquera, B. Montgomery, M.E. Taplin, C.C. Pritchard, G. Attard, et al., Integrative clinical genomics of advanced prostate cancer. Cell **162**, 454 (2015)

21. S. Wee, D. Wiederschain, S.M. Maira, A. Loo, C. Miller, R. deBeaumont, F. Stegmeier, Y.M. Yao, C. Lengauer, PTEN-deficient cancers depend on PIK3CB. Proc. Natl. Acad. Sci. U. S. A. **105**, 13057–13062 (2008)

22. J. Armenia, S.A.M. Wankowicz, D. Liu, J. Gao, R. Kundra, E. Reznik, W.K. Chatila, D. Chakravarty, G.C. Han, I. Coleman, et al., The long tail of oncogenic drivers in prostate cancer. Nat. Genet. **50**, 645–651 (2018)

23. A. Suzuki, J.L. de la Pompa, V. Stambolic, A.J. Elia, T. Sasaki, I. del Barco Barrantes, A. Ho, A. Wakeham, A. Itie, W. Khoo, et al., High cancer susceptibility and embryonic lethality associated with mutation of the PTEN tumor suppressor gene in mice. Curr. Biol. **8**, 1169–1178 (1998)

24. L.C. Trotman, M. Niki, Z.A. Dotan, J.A. Koutcher, A. Di Cristofano, A. Xiao, A.S. Khoo, P. Roy-Burman, N.M. Greenberg, T. Van Dyke, et al., Pten dose dictates cancer progression in the prostate. PLoS Biol. **1**, E59 (2003)

25. S. Wang, J. Gao, Q. Lei, N. Rozengurt, C. Pritchard, J. Jiao, G.V. Thomas, G. Li, P. Roy-Burman, P.S. Nelson, et al., Prostate-specific deletion of the murine Pten tumor suppressor gene leads to metastatic prostate cancer. Cancer Cell **4**, 209–221 (2003)

26. A. Di Cristofano, M. De Acetis, A. Koff, C. Cordon-Cardo, P.P. Pandolfi, Pten and p27KIP1 cooperate in prostate cancer tumor suppression in the mouse. Nat. Genet. **27**, 222–224 (2001)

27. M.J. Kim, R.D. Cardiff, N. Desai, W.A. Banach-Petrosky, R. Parsons, M.M. Shen, C. Abate-Shen, Cooperativity of Nkx3.1 and Pten loss of function in a mouse model of prostate carcinogenesis. Proc. Natl. Acad. Sci. U. S. A. **99**, 2884–2889 (2002)

28. B.S. Carver, J. Tran, A. Gopalan, Z. Chen, S. Shaikh, A. Carracedo, A. Alimonti, C. Nardella, S. Varmeh, P.T. Scardino, et al., Aberrant ERG expression cooperates with loss of PTEN to promote cancer progression in the prostate. Nat. Genet. **41**, 619–624 (2009)

29. J.C. King, J. Xu, J. Wongvipat, H. Hieronymus, B.S. Carver, D.H. Leung, B.S. Taylor, C. Sander, R.D. Cardiff, S.S. Couto, et al., Cooperativity of TMPRSS2-ERG with PI3-kinase pathway activation in prostate oncogenesis. Nat. Genet. **41**, 524–526 (2009)

30. L. Ding, S. Chen, P. Liu, Y. Pan, J. Zhong, K.M. Regan, L. Wang, C. Yu, A. Rizzardi, L. Cheng, et al., CBP loss cooperates with PTEN haploinsufficiency to drive prostate cancer: implications for epigenetic therapy. Cancer Res. **74**, 2050–2061 (2014)

31. C.E. Barbieri, S.C. Baca, M.S. Lawrence, F. Demichelis, M. Blattner, J.P. Theurillat, T.A. White, P. Stojanov, E. Van Allen, N. Stransky, et al., Exome sequencing identifies recurrent SPOP, FOXA1 and MED12 mutations in prostate cancer. Nat. Genet. **44**, 685–689 (2012)

32. J. An, S. Ren, S.J. Murphy, C. Dalangood, C. Chang, X. Pang, Y. Cui, L. Wang, Y. Pan, X. Zhang, et al., Truncated ERG oncoproteins from TMPRSS2-ERG fusions are resistant to SPOP-mediated proteasome degradation. Mol. Cell **59**, 904–916 (2015)

33. J. An, C. Wang, Y. Deng, L. Yu, H. Huang, Destruction of full-length androgen receptor by wild-type SPOP, but not prostate-cancer-associated mutants. Cell Rep. **6**, 657–669 (2014)

34. M. Blattner, D.J. Lee, C. O'Reilly, K. Park, T.Y. MacDonald, F. Khani, K.R. Turner, Y.L. Chiu, P.J. Wild, I. Dolgalev, et al., SPOP mutations in prostate cancer across demographically diverse patient cohorts. Neoplasia **16**, 14–20 (2014)

35. C. Geng, B. He, L. Xu, C.E. Barbieri, V.K. Eedunuri, S.A. Chew, M. Zimmermann, R. Bond, J. Shou, C. Li, et al., Prostate cancer-associated mutations in speckle-type POZ protein (SPOP) regulate steroid receptor coactivator 3 protein turnover. Proc. Natl. Acad. Sci. U. S. A. **110**, 6997–7002 (2013)

36. C. Geng, K. Rajapakshe, S.S. Shah, J. Shou, V.K. Eedunuri, C. Foley, W. Fiskus, M. Rajendran, S.A. Chew, M. Zimmermann, et al., Androgen receptor is the key transcriptional mediator of the tumor suppressor SPOP in prostate cancer. Cancer Res. **74**, 5631–5643 (2014)

37. A.C. Groner, L. Cato, J. de Tribolet-Hardy, T. Bernasocchi, H. Janouskova, D. Melchers, R. Houtman, A.C.B. Cato, P. Tschopp, L. Gu, et al., TRIM24 is an oncogenic transcriptional activator in prostate cancer. Cancer Cell **29**, 846–858 (2016)

38. C. Li, J. Ao, J. Fu, D.F. Lee, J. Xu, D. Lonard, B.W. O'Malley, Tumor-suppressor role for the SPOP ubiquitin ligase in signal-dependent proteolysis of the oncogenic co-activator SRC-3/AIB1. Oncogene **30**, 4350–4364 (2011)

39. Y. Yan, J. An, Y. Yang, D. Wu, Y. Bai, W. Cao, L. Ma, J. Chen, Z. Yu, Y. He, et al., Dual inhibition of AKT-mTOR and AR signaling by targeting HDAC3 in PTEN- or SPOP-mutated prostate cancer. EMBO Mol. Med. **10**, e8478 (2018)

40. P. Zhang, D. Wang, Y. Zhao, S. Ren, K. Gao, Z. Ye, S. Wang, C.W. Pan, Y. Zhu, Y. Yan, et al., Intrinsic BET inhibitor resistance in SPOP-mutated prostate cancer is mediated by BET protein stabilization and AKT-mTORC1 activation. Nat. Med. **23**, 1055–1062 (2017)

41. A. Saci, L.C. Cantley, C.L. Carpenter, Rac1 regulates the activity of mTORC1 and mTORC2 and controls cellular size. Mol. Cell **42**, 50–61 (2011)

42. R. Lasserre, X.J. Guo, F. Conchonaud, Y. Hamon, O. Hawchar, A.M. Bernard, S.M. Soudja, P.F. Lenne, H. Rigneault, D. Olive, et al., Raft nanodomains contribute to Akt/PKB plasma membrane recruitment and activation. Nat. Chem. Biol. **4**, 538–547 (2008)

43. L. Zhuang, J. Lin, M.L. Lu, K.R. Solomon, M.R. Freeman, Cholesterol-rich lipid rafts mediate akt-regulated survival in prostate cancer cells. Cancer Res. **62**, 2227–2231 (2002)

44. J. Xu, L. Liao, G. Ning, H. Yoshida-Komiya, C. Deng, B.W. O'Malley, The steroid receptor coactivator SRC-3 (p/CIP/RAC3/AIB1/ACTR/TRAM-1) is required for normal growth, puberty, female reproductive function, and mammary gland development. Proc. Natl. Acad. Sci. U. S. A. **97**, 6379–6384 (2000)

45. L.H. Zhang, A.A. Yin, J.X. Cheng, H.Y. Huang, X.M. Li, Y.Q. Zhang, N. Han, X. Zhang, TRIM24 promotes glioma progression and enhances chemoresistance through activation of the PI3K/Akt signaling pathway. Oncogene **34**, 600–610 (2015)

46. M. Blattner, D. Liu, B.D. Robinson, D. Huang, A. Poliakov, D. Gao, S. Nataraj, L.D. Deonarine, M.A. Augello, V. Sailer, et al., SPOP mutation drives prostate tumorigenesis in vivo through coordinate regulation of PI3K/mTOR and AR signaling. Cancer Cell **31**, 436–451 (2017)

47. H. Huang, D.J. Tindall, Dynamic FoxO transcription factors. J. Cell Sci. **120**, 2479–2487 (2007)

48. Y. Zhao, D.J. Tindall, H. Huang, Modulation of androgen receptor by FOXA1 and FOXO1 factors in prostate cancer. Int. J. Biol. Sci. **10**, 614–619 (2014)

49. X.Y. Dong, C. Chen, X. Sun, P. Guo, R.L. Vessella, R.X. Wang, L.W. Chung, W. Zhou, J.T. Dong, FOXO1A is a candidate for the 13q14 tumor suppressor gene inhibiting androgen receptor signaling in prostate cancer. Cancer Res. **66**, 6998–7006 (2006)

50. B.S. Haflidadottir, O. Larne, M. Martin, M. Persson, A. Edsjo, A. Bjartell, Y. Ceder, Upregulation of miR-96 enhances cellular proliferation of prostate cancer cells through FOXO1. PLoS One **8**, e72400 (2013)

51. V. Modur, R. Nagarajan, B.M. Evers, J. Milbrandt, FOXO proteins regulate tumor necrosis factor-related apoptosis inducing ligand expression. Implications for PTEN mutation in prostate cancer. J. Biol. Chem. **277**, 47928–47937 (2002)

52. Y. Yang, H. Hou, E.M. Haller, S.V. Nicosia, W. Bai, Suppression of FOXO1 activity by FHL2 through SIRT1-mediated deacetylation. EMBO J. **24**, 1021–1032 (2005)

53. L.R. Bohrer, P. Liu, J. Zhong, Y. Pan, J. Angstman, L.J. Brand, S.M. Dehm, H. Huang, FOXO1 binds to the TAU5 motif and inhibits constitutively active androgen receptor splice variants. Prostate **73**, 1017–1027 (2013)

54. W. Fan, T. Yanase, H. Morinaga, T. Okabe, M. Nomura, H. Daitoku, A. Fukamizu, S. Kato, R. Takayanagi, H. Nawata, Insulin-like growth factor 1/insulin signaling activates androgen signaling through direct interactions of Foxo1 with androgen receptor. J. Biol. Chem. **282**, 7329–7338 (2007)

55. P. Liu, S. Li, L. Gan, T.P. Kao, H. Huang, A transcription-independent function of FOXO1 in inhibition of androgen-independent activation of the androgen receptor in prostate cancer cells. Cancer Res. **68**, 10290–10299 (2008)

56. S.N. Mediwala, H. Sun, A.T. Szafran, S.M. Hartig, G. Sonpavde, T.G. Hayes, P. Thiagarajan, M.A. Mancini, M. Marcelli, The activity of the androgen receptor variant AR-V7 is regulated by FOXO1 in a PTEN-PI3K-AKT-dependent way. Prostate **73**, 267–277 (2013)

57. H.K. Lin, S. Yeh, H.Y. Kang, C. Chang, Akt suppresses androgen-induced apoptosis by phosphorylating and inhibiting androgen receptor. Proc. Natl. Acad. Sci. U. S. A. **98**, 7200–7205 (2001)

58. P. Adamo, M.R. Ladomery, The oncogene ERG: a key factor in prostate cancer. Oncogene **35**, 403–414 (2016)

59. S.A. Tomlins, D.R. Rhodes, S. Perner, S.M. Dhanasekaran, R. Mehra, X.W. Sun, S. Varambally, X. Cao, J. Tchinda, R. Kuefer, et al., Recurrent fusion of TMPRSS2 and ETS transcription factor genes in prostate cancer. Science **310**, 644–648 (2005)

60. Y. Yang, A.M. Blee, D. Wang, J. An, Y. Pan, Y. Yan, T. Ma, Y. He, J. Dugdale, X. Hou, et al., Loss of FOXO1 cooperates with TMPRSS2-ERG overexpression to promote prostate tumorigenesis and cell invasion. Cancer Res. **77**, 6524–6537 (2017)

61. I.A. San Martin, N. Varela, M. Gaete, K. Villegas, M. Osorio, J.C. Tapia, M. Antonelli, E.E. Mancilla, B.P. Pereira, S.S. Nathan, et al., Impaired cell cycle regulation of the osteoblast-related heterodimeric transcription factor Runx2-Cbfbeta in osteosarcoma cells. J. Cell. Physiol. **221**, 560–571 (2009)

62. D. Wysokinski, E. Pawlowska, J. Blasiak, RUNX2: a master bone growth regulator that may be involved in the DNA damage response. DNA Cell Biol. **34**, 305–315 (2015)

63. J.J. Westendorf, S.K. Zaidi, J.E. Cascino, R. Kahler, A.J. van Wijnen, J.B. Lian, M. Yoshida, G.S. Stein, X. Li, Runx2 (Cbfa1, AML-3) interacts with histone deacetylase 6 and represses the p21(CIP1/WAF1) promoter. Mol. Cell. Biol. **22**, 7982–7992 (2002)

64. C. Ge, G. Xiao, D. Jiang, Q. Yang, N.E. Hatch, H. Roca, R.T. Franceschi, Identification and functional characterization of ERK/MAPK phosphorylation sites in the Runx2 transcription factor. J. Biol. Chem. **284**, 32533–32543 (2009)

65. M. Qiao, P. Shapiro, M. Fosbrink, H. Rus, R. Kumar, A. Passaniti, Cell cycle-dependent phosphorylation of the RUNX2 transcription factor by cdc2 regulates endothelial cell proliferation. J. Biol. Chem. **281**, 7118–7128 (2006)

66. C. Ge, G. Zhao, Y. Li, H. Li, X. Zhao, G. Pannone, P. Bufo, A. Santoro, F. Sanguedolce, S. Tortorella, et al., Role of Runx2 phosphorylation in prostate can-

cer and association with metastatic disease. Oncogene **35**, 366–376 (2016)

67. S.K. Baniwal, O. Khalid, D. Sir, G. Buchanan, G.A. Coetzee, B. Frenkel, Repression of Runx2 by androgen receptor (AR) in osteoblasts and prostate cancer cells: AR binds Runx2 and abrogates its recruitment to DNA. Mol. Endocrinol. **23**, 1203–1214 (2009)

68. H. Kawate, Y. Wu, K. Ohnaka, R. Takayanagi, Mutual transactivational repression of Runx2 and the androgen receptor by an impairment of their normal compartmentalization. J. Steroid Biochem. Mol. Biol. **105**, 46–56 (2007)

69. H. Zhang, Y. Pan, L. Zheng, C. Choe, B. Lindgren, E.D. Jensen, J.J. Westendorf, L. Cheng, H. Huang, FOXO1 inhibits Runx2 transcriptional activity and prostate cancer cell migration and invasion. Cancer Res. **71**, 3257–3267 (2011a)

70. S. Yang, H. Xu, S. Yu, H. Cao, J. Fan, C. Ge, R.T. Fransceschi, H.H. Dong, G. Xiao, Foxo1 mediates insulin-like growth factor 1 (IGF1)/insulin regulation of osteocalcin expression by antagonizing Runx2 in osteoblasts. J. Biol. Chem. **286**, 19149–19158 (2011)

71. Y. Yang, Y. Bai, Y. He, Y. Zhao, J. Chen, L. Ma, Y. Pan, M. Hinten, J. Zhang, R.J. Karnes, et al., PTEN loss promotes intratumoral androgen synthesis and tumor microenvironment remodeling via aberrant activation of RUNX2 in castration-resistant prostate cancer. Clin. Cancer Res. **24**, 834–846 (2018)

72. D.J. Mulholland, L.M. Tran, Y. Li, H. Cai, A. Morim, S. Wang, S. Plaisier, I.P. Garraway, J. Huang, T.G. Graeber, H. Wu, Cell autonomous role of PTEN in regulating castration-resistant prostate cancer growth. Cancer Cell **19**, 792–804 (2011)

73. J. Long, W.Y. Fang, L. Chang, W.H. Gao, Y. Shen, M.Y. Jia, Y.X. Zhang, Y. Wang, H.B. Dou, W.J. Zhang, et al., Targeting HDAC3, a new partner protein of AKT in the reversal of chemoresistance in acute myeloid leukemia via DNA damage response. Leukemia **31**, 2761–2770 (2017)

74. D.S. Welsbie, J. Xu, Y. Chen, L. Borsu, H.I. Scher, N. Rosen, C.L. Sawyers, Histone deacetylases are required for androgen receptor function in hormone-sensitive and castrate-resistant prostate cancer. Cancer Res. **69**, 958–966 (2009)

75. V. Murillo-Garzon, R. Kypta, WNT signalling in prostate cancer. Nat. Rev. Urol. **14**, 683–696 (2017)

76. D. Fang, D. Hawke, Y. Zheng, Y. Xia, J. Meisenhelder, H. Nika, G.B. Mills, R. Kobayashi, T. Hunter, Z. Lu, Phosphorylation of beta-catenin by AKT promotes beta-catenin transcriptional activity. J. Biol. Chem. **282**, 11221–11229 (2007)

77. S.A. Tahir, G. Yang, A. Goltsov, K.D. Song, C. Ren, J. Wang, W. Chang, T.C. Thompson, Caveolin-1-LRP6 signaling module stimulates aerobic glycolysis in prostate cancer. Cancer Res. **73**, 1900–1911 (2013)

78. G. Thyssen, T.H. Li, L. Lehmann, M. Zhuo, M. Sharma, Z. Sun, LZTS2 is a novel beta-catenin-interacting protein and regulates the nuclear export of beta-catenin. Mol. Cell. Biol. **26**, 8857–8867 (2006)

79. E.J. Yu, E. Hooker, D.T. Johnson, M.K. Kwak, K. Zou, R. Luong, Y. He, Z. Sun, LZTS2 and PTEN collaboratively regulate ss-catenin in prostatic tumorigenesis. PLoS One **12**, e0174357 (2017)

80. H. Gao, X. Ouyang, W.A. Banach-Petrosky, W.L. Gerald, M.M. Shen, C. Abate-Shen, Combinatorial activities of Akt and B-Raf/Erk signaling in a mouse model of androgen-independent prostate cancer. Proc. Natl. Acad. Sci. U. S. A. **103**, 14477–14482 (2006)

81. S. Chandarlapaty, A. Sawai, M. Scaltriti, V. Rodrik-Outmezguine, O. Grbovic-Huezo, V. Serra, P.K. Majumder, J. Baselga, N. Rosen, AKT inhibition relieves feedback suppression of receptor tyrosine kinase expression and activity. Cancer Cell **19**, 58–71 (2011)

82. V. Serra, M. Scaltriti, L. Prudkin, P.J. Eichhorn, Y.H. Ibrahim, S. Chandarlapaty, B. Markman, O. Rodriguez, M. Guzman, S. Rodriguez, et al., PI3K inhibition results in enhanced HER signaling and acquired ERK dependency in HER2-overexpressing breast cancer. Oncogene **30**, 2547–2557 (2011)

83. C.W. Pan, X. Jin, Y. Zhao, Y. Pan, J. Yang, R.J. Karnes, J. Zhang, L. Wang, H. Huang, AKT-phosphorylated FOXO1 suppresses ERK activation and chemoresistance by disrupting IQGAP1-MAPK interaction. EMBO J. **36**, 995–1010 (2017)

84. M. Crumbaker, L. Khoja, A.M. Joshua, AR signaling and the PI3K pathway in prostate cancer. Cancers (Basel) **9**, E34 (2017)

85. T. McHardy, J.J. Caldwell, K.M. Cheung, L.J. Hunter, K. Taylor, M. Rowlands, R. Ruddle, A. Henley, A. de Haven Brandon, M. Valenti, et al., Discovery of 4-amino-1-(7H-pyrrolo[2,3-d]pyrimidin-4-yl) piperidine-4-carboxamides as selective, orally active inhibitors of protein kinase B (Akt). J. Med. Chem. **53**, 2239–2249 (2010)

86. J.C. Bendell, A.M. Varghese, D.M. Hyman, T.M. Bauer, S. Pant, S. Callies, J. Lin, R. Martinez, E. Wickremsinhe, A. Fink, et al., A first-in-human phase 1 study of LY3023414, an oral PI3K/mTOR dual inhibitor, in patients with advanced cancer. Clin. Cancer Res. **24**, 3253–3262 (2018)

87. P. Toren, S. Kim, T. Cordonnier, C. Crafter, B.R. Davies, L. Fazli, M.E. Gleave, A. Zoubeidi, Combination AZD5363 with enzalutamide significantly delays enzalutamide-resistant prostate cancer in preclinical models. Eur. Urol. **67**, 986–990 (2015)

88. A. Zoubeidi, M.E. Gleave, Co-targeting driver pathways in prostate cancer: two birds with one stone. EMBO Mol. Med. **10**, e8928 (2018)

Androgen Receptor Dependence

Aashi P. Chaturvedi and Scott M. Dehm

AR Structure and Function

AR is a member of the class I nuclear receptor transcription factor family, which includes the steroid receptors glucocorticoid receptor (GR), mineralocorticoid receptor (MR), estrogen receptor (ER), and progesterone receptor (PR). It is a 110 kDa phospho-protein encoded by the *AR* gene located on chromosome X at Xq11–12; hence XY males have 1 copy of *AR*. The *AR* gene comprises eight exons which encode four distinct functional domains of the full-length AR protein: (1) an intrinsically-disordered NH_2-terminal domain (NTD) encoded by exon 1; (2) a 2-zinc finger DNA-binding domain (DBD) encoded by exon 2 and the 5′ end of exon 3; (3) a short flexible hinge region harboring the nuclear localization signal (NLS) encoded by the 3′ end of exon 3 and 5′ end of exon 4; and (4) a ligand-binding domain (LBD) encoded by the 3′ end of exon 4 along with exons 5–8 (Fig. 1) [1, 2].

A. P. Chaturvedi
Masonic Cancer Center, University of Minnesota, Minneapolis, MN, USA

S. M. Dehm (✉)
Masonic Cancer Center, University of Minnesota, Minneapolis, MN, USA

Department of Laboratory Medicine and Pathology, University of Minnesota, Minneapolis, MN, USA

Department of Urology, University of Minnesota, Minneapolis, MN, USA
e-mail: dehm@umn.edu

The physiological ligands for AR include testosterone and dihydrotestosterone (DHT), which bind to the steroid binding site in the LBD. Like the other steroid receptors, there are two distinct transcriptional activation regions in AR: a strong activation function domain (AF-1) in the NTD and a weak activation function domain (AF-2) in the LBD, both of which can recruit various co-regulators of AR (Fig. 1). The relative roles of these two transcriptional activation domains have been studied extensively for AR as well as other steroid receptors. In the case of AR, it is AF-1 that appears to be necessary and sufficient for transcriptional activity [3–6]. This knowledge has generated considerable interest in dissecting the mechanisms of AF-1 function, and has led to the finding that AF-1 can be further sub-divided into two discrete transcriptional activation units, termed TAU-1 and TAU-5 [7–9]. TAU-1 (amino acids 101–360) contains two motifs: (1) an LKDIL motif, which is similar to the nuclear receptor box sequence found in nuclear receptor co-regulator proteins; and (2) an LX7LL motif, which is evolutionarily conserved in AR, ERα and PR (Fig. 1). Deletion of the LKDIL motif causes significant loss in transcriptional activity of AR, whereas the LX7LL motif is required for de-repression of a cohort of genes in response to inflammatory cytokine signaling [10, 11]. TAU-5 (amino acids 361–490) contains the WHTLF motif, which appears to play a selective transactivation role under conditions of no/low androgens

© Springer Nature Switzerland AG 2019
S. M. Dehm, D. J. Tindall (eds.), *Prostate Cancer*, Advances in Experimental Medicine and Biology 1210, https://doi.org/10.1007/978-3-030-32656-2_15

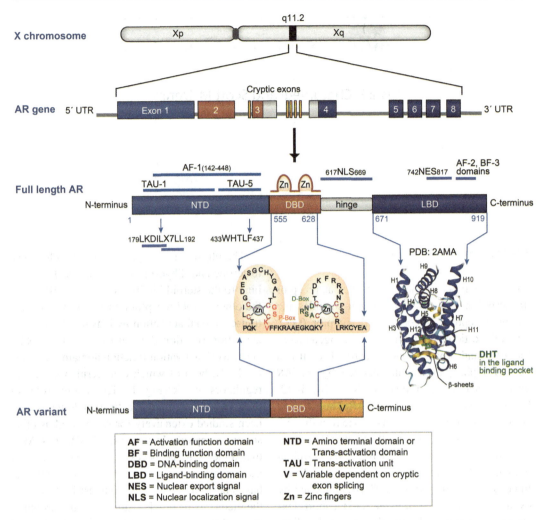

Fig. 1 AR gene and protein structure: AR is located on the X chromosome at position q11.2. The AR gene is encoded by eight exons that are color coded to represent the domains of the full-length AR protein they encode. The full-length AR is comprised of an amino terminal domain (NTD, in blue), DNA binding domain (DBD, in orange), a short hinge region (in grey) and a ligand binding domain (LBD, in purple). The amino acid sequence of the two zinc finger units containing the P-box and D-box of the AR DBD are shown. The structure of human AR LBD domain with a DHT bound in its ligand binding pocket is represented (PDB: 2AMA). AR variants contain the AR NTD and DBD but lack the LBD. The C-termini of AR variants have variable lengths (V, in yellow) and sequences based on the splicing of the cryptic exons in the AR gene

[12, 13]. Additionally, as elaborated below, this WHTLF motif mediates an intramolecular interaction between the AR amino and carboxyl termini by binding the AR AF-2 domain, indicating that accessibility of this transactivation motif is regulated, whether or not it is bound to AF-2 [14, 15].

The AR DBD is cysteine-rich and highly conserved among steroid receptors. There are two clusters of four cysteine residues, each of which coordinate a single zinc ion to make up the two zinc fingers of the DBD. As shown in Fig. 1, the first zinc finger contains the P or proximal box (amino acids 577–581), which specifically recognizes DNA androgen response elements (ARE). The second zinc finger contains the D or distal box (amino acids 596–600), which mediates dimerization between two AR monomers [16–18].

Like LBDs of other nuclear receptors, the structure of the AR LBD is arranged in a three-layer, antiparallel α-helical sandwich fold that surrounds an interior hydrophobic ligand binding pocket (Fig. 1). The AF-2 domain in the LBD is a shallow, hydrophobic groove formed by helices H12, H3 and H4 in the agonist-bound conformation. A domain proximal to AF-2, which is composed of a hydrophobic cleft made at the junction of H1 with the H3–H4 loop and H9 on the surface of the AR LBD, is referred to as binding function-3 domain (BF-3). BF-3 can allosterically regulate the binding of co-activators at AF-2 [19, 20]. The shallow AF-2 groove functions to bind LXXLL and LX7LL motifs found in nuclear receptor co-activator proteins, which are referred to as NR boxes [21]. As illustrated by AR LBD crystal structure 2AMA [22], deposited in The Protein Data Bank [23], agonists like testosterone or DHT, upon binding to the LBD re-position H12 to act as a lid and lock the agonist in the ligand-binding pocket. In contrast, when an antagonist binds the AR ligand binding pocket, it pushes H12 outwards to subsequently cause conformational changes in AF-2, thus rendering it incapable of binding co-activators [2, 19, 22]. In addition to binding NR boxes of co-activator proteins, the AF-2 domain also mediates interactions with the AR NTD, an intramolecular interaction referred to as the N/C interaction. The WHTLF motif of TAU-5 and the FXXLF motif both bind to the AF-2 domain of AR [14, 24].

Androgen Regulation of AR Nuclear Translocation and DNA Binding

AR shuttles between the cytoplasm and the nucleus in a manner that is regulated by binding to androgen ligand. In the un-liganded state, chaperones and co-chaperones, like members of the heat shock protein family, including Hsp23, Hsp40, Hsp56, Hsp70 and Hsp90, associate with the AR LBD and sequester AR in the cytoplasm in a conformation that is competent for ligand binding [1, 25, 26]. The principal androgen circu-

lating in the blood is testosterone, mostly produced by the Leydig cells in testes with a minor contribution from the adrenal cortex [27]. Synthesis of testosterone is regulated by the hypothalamus-pituitary-gonad and hypothalamus-pituitary-adrenal axes of the endocrine system. Several steroidogenic enzymes and isoenzymes are required to generate testicular and adrenal androgens from cholesterol in the canonical pathway. The hypothalamus secretes gonadotropin releasing hormone (GnRH) which acts upon the anterior pituitary to release the luteinizing hormone, subsequently signaling the release of testosterone from the testes [27, 28]. The anterior pituitary also releases the adrenocorticotropin hormone (ACTH) that acts on the adrenal cortex where the action of CYP17A1 and other enzymes produces dihydroepiandrosterone (DHEA), androstenedione and androstenediol. These weak adrenal androgens can then be converted to testosterone or DHT in peripheral tissues through various pathways such as the 5α-dione pathway or backdoor pathway [28, 29]. Although most of the testosterone in circulation is bound to sex hormone binding globulin (SHBG), ≤2% testosterone is free. When testosterone enters normal or cancerous prostate cells, it gets converted by 5α-reductase enzyme activity into DHT, which is a more potent androgen by virtue of it stabilizing the AR protein to a greater degree than testosterone and having a slower dissociation rate from the AR LBD. The binding of androgens to the AR LBD induces a conformational change in AR, thereby exposing the NLS and promoting translocation to the nucleus via direct interactions with the importin-α adapter protein and importin-β carrier protein, leading to transit through the nuclear pore complex [30–32]. In the nucleus, AR binds as a dimer via DBDs to androgen response elements. These AR dimers provide a platform for recruitment of a variety of co-regulators that govern the transcriptional program of AR. Androgen synthesis regulated by the hypothalamus-pituitary-gonad and hypothalamus-pituitary-adrenal axes and ultimate transmission of this hormonal signal via AR to the nucleus and genome of target cells is broadly referred to as the AR signaling axis

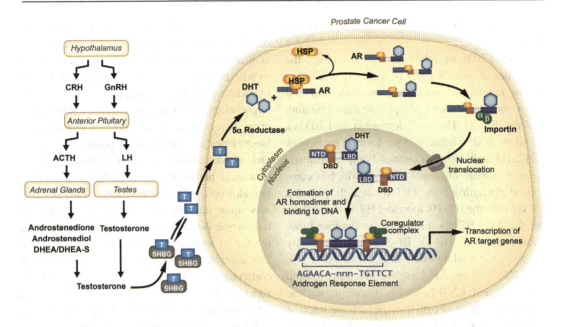

Fig. 2 The AR signaling axis: The production of androgens (e.g. testosterone) by the hypothalamus-pituitary-adrenal axis or hypothalamus-pituitary-gonadal axis is shown (left). In the bloodstream, testosterone is bound by sex-hormone binding globulin (SHBG), which releases free testosterone to enter cells where it is metabolized to DHT by 5α-reductase. AR bound to heat shock proteins (HSPs) in the cytoplasm binds DHT and translocates to the nucleus. In the nucleus, DHT-bound AR binds androgen response elements as dimers. The recruitment of various coactivators and corepressors determines the transcription profile of AR target genes

(Fig. 2). This AR signaling axis provides the foundation for the biological property of androgen-dependence of PCa cells.

AR Interactions with Chromatin

To understand the functional consequences of AR binding to AREs, researchers have focused their efforts on deciphering the AR transcriptome (the sets of mRNAs regulated by transcriptional activity of AR) and AR cistrome (the cis-regulatory elements in the genome to which AR binds). Genome-wide studies that have evaluated AR binding to AREs in various PCa models using ChIP-seq has provided fundamental information, although the exact number of AR binding events in PCa cells has not been clearly established. For instance, comparison of the number of AR-binding events in LNCaP cells (11,053) versus VCaP cells (51,811) demonstrated vastly different numbers. However, this is likely due to much higher expression of AR in VCaP cells due to *AR* gene amplification in this cell line. Nevertheless, despite this difference in number of AR binding events, the AR binding events observed in LNCaP cells displayed 90% overlap with the binding events observed in VCaP cells. In androgen-activated LNCaP cells, ChIP-seq studies have further revealed that recruitment of RNA Polymerase II to AR binding sites correlated with transcription of AR-upregulated genes. These AR-upregulated targets include genes involved in glucose uptake and glycolysis, biosynthetic pathways, regulators of cell cycle, and cellular metabolism [33].

Comparing and contrasting cistrome data from genome-wide ChIP-seq studies with structure/function studies of AR DNA binding has advanced the concept that there is flexibility with which AR binds ARE sites. For instance, global ChIP-seq studies have confirmed that the canonical ARE motif is a 15-mer sequence comprising of inverted repeats of a six base pair half-site (5′-AGAACA-3′) separated by three bases [34]. Structural studies have demonstrated that AR monomers engage with these ARE sequences as a homodimer arranged in a head-to-head symmetrical conformation. This leads to one AR monomer bound with high affinity to one ARE half-site, but the other AR monomer bound with lower affinity to the adjacent half-site. By reducing the stringency requirements for this adjacent half-site, AR can selectively bind its AREs [35]. This suboptimal binding of AR to its target DNA suggests an efficient way for AR to distinguish its various target genes and a mechanism to modulate transcription of ARE-driven AR target genes based on the strength of this binding interaction [17, 18]. Therefore, the way AR influences its target genes is non-uniform and heterogeneous, yet specific and strong.

Differential expression of AR target genes and variable occupancy of AR binding sites have been observed under different cellular contexts. For example, more than 50% of AR binding sites observed in CRPC tissue were not present in PCa cell lines, highlighting the divergence in AR signaling pathways under these conditions [36]. Further, comparative analysis of ChIP-seq data from 13 PCa tissue specimens versus 7 histologically normal prostate tissue specimens (6 of which were pair-matched from the same patient) revealed that prostate epithelial cells undergo reprogramming of the AR cistrome to achieve a neoplastic phenotype [37].

These genome wide studies reinforce the idea that under different cellular contexts and through different stages of PCa progression, AR displays alterations in the repertoire of transcriptional targets to which it binds and regulates. There are multiple mechanistic explanations for these alterations, including changes in *AR* gene expression levels, AR protein structure, changes in expression or activity of AR co-regulators, and global changes in the epigenome that affect the chromatin environment around AR binding sites [38]. Thus, global profiling of androgen-AR-ARE-co-regulator complexes in clinical specimens provides an important framework for understanding the role of the AR cistrome and transcriptome in disease progression and identifying new therapeutic avenues that could be exploited.

AR Interactions with Co-regulators and Other Transcription Factors

The co-regulators recruited as a result of AR-ARE interactions serve different roles in normal prostate function and PCa by fine-tuning AR transcriptional output. There is strong evidence that certain co-regulators display expression changes during PCa development and progression, and that these changes in expression re-direct or re-program AR chromatin binding and/or transcriptional output [39]. Therefore, there has been great interest in identifying the roles and regulatory mechanisms of AR co-regulators to better understand similarities and differences in regulation of AR action between normal and cancerous prostate tissue. This is an ambitious undertaking, since more than 200 co-activators (enhance transcription) and co-repressors (inhibit transcription) affect AR transcriptional activity and/or chromatin binding, and at least 50 have expression patterns that correlate with important clinical parameters in PCa specimens [40]. Mechanistically, co-regulators can affect stability and complex formation of AR, influence AR nuclear or cytoplasmic localization, DNA occupancy, chromatin remodeling, chromatin looping, interactions with other transcription factors and complexes, as well as priming and assembly of the overall transcription complex [41].

Some of the best-defined classes of AR co-regulators play important roles in regulating transcriptional output of many transcription factors. These co-regulators include molecular chaperones like FKBP1 (*FKBP1A*), FKBP2 (*FKBP2*), FKBP5 (*FKBP5*) and HSP90 (*HSP90AA1*), the p160 family of steroid receptor co-activators like SRC-1 (*NCOA1*), SRC-2/TIF-2/GRIP1 (*NCOA2*), SRC-3/AIB1 (*NCOA3*), p300 (*EP300*), CBP (*CREBBP*), ARA70 (*NCOA4*), ARA54 (*RNF14*) and ARA55 (*TGFB1I1*), as well as pioneer transcription factors like Oct1 (*POU2F1*) and GATA-2 (*GATA-2*) [42, 43]. AR-associated co-regulators are crucial to AR dependence in PCa.

Several such co-regulators affect AR binding to DNA and/or AR-gene regulation in genome-wide associated studies (GWAS) or AR-cistrome analysis. BAF57 (*SMARCE1*), an accessory subunit of the SWI/SNF chromatin-remodeling complex is one such cofactor, which is dramatically upregulated in metastatic PCa. Increased expression of BAF57 directed AR and the SWI/SNF complex to a distant intragenic region of the *ITGA2* gene, which encodes integrin alpha 2. *In vitro* studies confirmed that elevated levels of integrin alpha 2 protein results in an increased migratory and invasive phenotype in cells, supporting a prometastatic role for BAF57 [44].

FOXA1 and HOXB13 are key factors associated with growth and development of PCa through their binding interactions with AR [45, 46]. Physical interactions between AR-FOXA1 [47] and AR-HOXB13 [48] have been known for some time, but more recent global analyses have revealed that these interactions occur as a result of overlap with, and significant crosstalk between, the respective cistromes of AR, FOXA1 and HOXB13 to alter the transcriptional landscape of PCa cells. Furthermore, in a comparative analysis of FOXA1 and HOXB13 dependency across 102 cell lines from various tissue types, the PCa cell line LNCaP scored very high (second for HOXB13 and fifth for FOXA1), underscoring the relative importance of these factors in PCa cells [37]. For example, ectopic expression of FOXA1 and HOXB13 in immortalized LHSAR cells was sufficient to reprogram the AR cistrome to a state that was similar to that in a PCa cell line [37, 39, 49]. Additionally, FOXA1 is important for proliferation and cell cycle regulation in PCa, and knock down of FOXA1 expression in a PCa cell line led to an overall increase in other AR binding events. It is noteworthy that mutations in the coding sequence of FOXA1 occur in clinical PCa specimens, which are predicted to disrupt the forkhead DNA binding domain and thereby alter the affinity or specificity of FOXA1 for FOXA1 binding sites across the genome [50–52]. The role of these FOXA1 mutations in regulating the AR cistrome is an ongoing area of investigation.

In a recent study that used an unbiased proteomics technique termed RIME (rapid immunoprecipitation and mass spectrometry of endogenous proteins), Grainyhead-like 2 (GRHL2) was identified as a co-activator of AR with dichotomous roles in PCa development and progression. GRHL2 is pro-tumorigenic in early stages of PCa growth, but suppresses stromal invasion, intravasation of tumor cells, and survival of circulating tumor cells to reduce epithelial-mesenchymal transition and hence progression to metastatic PCa [53]. Another study used RIME and ChIP-seq to identify 66 known and novel interacting proteins of AR in LNCaP cells stimulated by a synthetic androgen R1881. These interaction partners were found to be members of the DNA repair machinery, chromatin remodeling factors, cell cycle regulators, cytoskeletal remodelers, and other transcriptional factors. These proteomics findings were subsequently followed by ChIP-seq studies to reveal that certain AR binding sites are co-occupied by AR and these interacting partners, including ARID1A, BRG1, FOXA1, HOXB13, TLE3, TRIM28 and WDHD1 [54].

Within PCa cells, co-regulators can modulate distinct sets of genes to affect AR regulated pathways. This is illustrated by a study wherein 18 clinically important AR co-regulators were selectively inhibited in a PCa cell line using siRNA knock-down. Inhibition of specific co-regulators was found to selectively activate or repress discrete sets of genes within a 452-AR-target gene panel. This demonstrated specific, context-dependent effects of individual AR co-regulators, providing a mechanistic basis for intracellular heterogeneity in AR gene regulation [55]. A precise definition of the mechanisms by which co-regulators affect AR target gene expression based on the availability of androgens, presence of different drugs, cell line under investigation, and other factors influencing PCa growth and progression, could ultimately enable a better assessment of this disease through various stages of PCa progression and enable the development of more effective therapeutics.

Therapeutic Targeting of the AR Signaling Axis

The concept of AR-dependence was first introduced by Charles Huggins and Clarence V. Hodges almost 75 years ago [56]. Since then, androgen depletion therapy (ADT) has remained the principal treatment strategy for locally advanced, metastatic, or relapsed PCa. ADT targets various points of the AR signaling axis, with the goal of inhibiting transcriptional activity of the AR, which is the most widely accepted driver of PCa development and progression [57]. The earliest implementation of ADT included orchiectomy to eliminate the testicular source of androgens, or treatment with the oral synthetic estrogen diethylstilbestrol. These castration-based ADT modalities, and benefit for advanced PCa patients, formed the basis for the 1966 Nobel Prize in Medicine being awarded to Charles Huggins. Gonadotropin-releasing hormone (GnRH) agonists and antagonists like leuprolide, goserelin, triptorelin, and histrelin have replaced diethylstilbestrol as the main castration-based therapies, due to increased risk of cardiovascular mortality with estrogen therapy. Additionally, AR antagonists including bicalutamide, flutamide, and nilutamide function as competitive antagonists by binding the testosterone binding site in the AR LBD [58]. These drugs, collectively referred to as "first-generation" ADT, lead to suppression of circulating testosterone levels and blockade of AR signaling. This is best exemplified by the ensuing reduction in serum levels of prostate specific antigen (PSA), an AR transcriptional gene target in PCa cells. The main limitation of ADT is that it is not curative, and the duration of the therapeutic response of patients varies from a few months to several years. This stage of the disease, where patients have stopped responding to ADT, is referred to as CRPC. This stage of the disease is lethal and often progresses quickly due to a lack of durable treatment options [58]. Progression to CRPC is usually indicated by rising serum PSA levels despite ADT, suggesting re-engagement of the AR signaling axis. This has driven efforts to understand the mechanisms by which AR signaling can resume under conditions of ADT, and develop new therapies that can counteract these mechanisms in patients with CRPC [58, 59].

AR Gene Amplification in CRPC

An early comparative genomic hybridization study with matched PCa tissues from patients collected pre-ADT and post-ADT demonstrated that 30% of patients displayed *AR* gene amplification, specifically in post-ADT tissues [60]. A follow-up study using fluorescence *in situ* hybridization confirmed these initial findings, and also demonstrated that *AR* mRNA expression was higher in tumors displaying *AR* gene amplification [61]. Comparing the global gene expression profiles of seven isogenic pairs of hormone sensitive and castration-resistant PCa xenografts revealed that the CRPC phenotype is consistently associated with increased expression of AR [62]. Mechanistically, this study further showed that higher expression of AR is sufficient for transition from hormone-sensitive PCa to a CRPC phenotype. For example, hormone sensitive LNCaP cells engineered to express a two- to threefold higher level of AR display increased growth under castrate conditions, as well as bicalutamide-stimulated growth. Consistent with these functional data, more contemporary DNA sequencing studies of localized PCa and CRPC-stage tumors demonstrated that *AR* gene amplification is the most frequent event in CRPC genomes, occurring in approximately 55–60% of CRPC cases but almost never in localized PCa [63]. Whole genome sequencing of multiple metastases from CRPC patients revealed that persistent selective pressure of ADT drives separate cancer cell clones within the same patient to undergo distinct *AR* amplification events in distinct metastatic lesions. This study reinforces the importance of AR amplification as a key mechanism of resistance to ADT in CRPC [64].

AR Somatic Mutations in CRPC

Primary PCa typically shows less mutational burden than other solid tumors, but upon progression of the disease, about 20% of patients progressing with CRPC show somatic mutations in the *AR* gene [65, 66]. Similar to *AR* gene amplification, AR point mutations are exceedingly rare in ADT-naïve PCa. The best-described AR mutations are T878A, H875Y, W742C and L702H in the AR LBD, which play a key role in promoting resistance to ADT. For example, T878A confers resistance to ADT by enabling AR activation in response to alternative ligands, including progesterone and the antiandrogen flutamide. Similarly, H875Y and W742C mutations enable AR activation in response to the antiandrogens bicalutamide and flutamide [52, 67, 68]. The L702H mutation, alone or in combination with T878A, also broadens the agonist repertoire of AR, enabling AR activation by glucocorticoids [69]. The frequency of these somatic AR point mutations appears to be enriched in CRPC patients treated with antiandrogens, indicating this is a major mechanism of resistance in patients under continuous selective pressure from AR antagonists.

Amplification of an Upstream *AR* Enhancer in CRPC

Three recent studies integrated whole genome sequencing datasets or copy number microarrays with epigenetic datasets to reveal an important enhancer region regulating expression of the AR in CRPC. One study analyzed genome-wide copy number alterations from 149 tumors and identified an amplification hotspot encompassing the *AR* gene body, and another amplification hotspot located 650 kb centromeric to the *AR* gene body [70]. This upstream genomic region coincides with a region of DNaseI hypersensitivity in LNCaP cells that is essential for LNCaP cell viability. Further analysis of H3K27ac ChIP-seq data in this study revealed that this upstream genomic region resembles a developmental enhancer that is selectively acetylated in CRPC,

indicating potential reactivation [70]. In a related study using linked read whole genome sequencing (WGS), 70–87% of metastatic CRPC patient samples showed tandem duplication events leading to amplification of this upstream *AR* enhancer region compared to only 2% of ADT naïve PCa cases [71]. Another study employed integrative deep WGS coupled with RNA-seq to find that 81% of 101 CRPC specimens displayed increased *AR* gene expression correlated with amplification of this enhancer region [72]. Collectively, these studies have demonstrated that amplification of an enhancer located ~650 kb upstream of the AR gene plays an important role in increasing the expression of *AR* mRNA in CRPC-stage tumors.

AR Variants in CRPC

Alternative splicing of *AR* mRNA to create AR variant (AR-V) proteins that lack the LBD represents a resistance mechanism where AR can function independent of androgen ligands to bypass ADT [73]. To date, several AR-Vs have been discovered and reported in PCa cell lines, xenograft tumors, primary tumors, metastatic lesions, and circulating tumor cells [74, 75]. However, the most widely-studied AR-V is termed AR-V7, composed of contiguously-spliced *AR* exons 1, 2, 3 and cryptic exon 3 (CE3). Development of antibodies specific to AR-V7 led to the finding that AR-V7 protein is rarely expressed (<1%) in primary PCa but detectable in >75% of CRPC cases. Expression of AR-V7 was homogenous within a tumor sample but was heterogeneous between different metastatic lesions from the same patient [76]. These studies aimed at evaluating the expression profiles of AR-V7 have suggested the potential to develop AR-Vs as biomarkers for resistance [77–79]. For example, detection of AR-V7 mRNA or protein in circulating tumor cells from patients with CRPC has been evaluated as a treatment selection biomarker that predicts poor treatment outcomes with second-generation AR targeted therapies abiraterone and enzalutamide, but better treatment outcomes with taxane chemotherapy [80–82]. Another AR-V expressed in clinical tissues that has been correlated with resistance to abiraterone acetate is AR-V9, composed of contiguously spliced AR exons 1/2/3/CE5 [79, 83]. Importantly, many AR-Vs are co-expressed in clinical CRPC [84, 85], raising the question of whether AR-Vs function alone, or cooperatively with other AR-Vs to promote resistance. More than 20 such variants have been reported in PCa models and clinical tissues in the last several years [86]. It also remains unresolved whether the functional effects of AR-V7 in CRPC cells requires the activity of full-length AR. For instance, knock-down of full-length AR in LNCaP cells engineered to overexpress AR-V7 inhibited androgen-independent growth [87]. Similarly, antisense oligonucleotides that blocked expression of full-length *AR* inhibited the growth of an AR-V7 positive LNCaP model of acquired resistance to enzalutamide [88]. Conversely, antisense oligonucleotides that blocked the expression of AR-V7 had no effect on growth of this enzalutamide-resistant LNCaP model. In light of these findings, it is important to note that AR-V7 is co-expressed with full-length AR, and the main mechanism underlying AR-V7 expression appears to be amplification of the *AR* gene [89]. These findings underscore the context-dependent roles of AR-Vs in PCa and point to a need to understand the interplay between full length AR and AR-Vs in disease staging, developing predictive biomarkers, and devising strategies for new therapies.

AR-V transcriptome and cistrome studies have provided important insights into the system-wide influence of these numerous AR isoforms in PCa. Gene expression profiling has shown that AR-Vs can activate many of the same transcriptional targets as full-length AR, while also displaying unique and distinct transcriptional targets. However, these differences may reflect different thresholds of activation between AR-Vs and full-length AR, and not absolute differences in transcriptional targets [87]. For example, AR-Vs were reported to uniquely activate genes involved in G2/M phase cell cycle progression like UBE2C and CCNA2 [90]. However, a subsequent study demonstrated that UBE2C and CCNA2 were also full-length AR targets that were induced depending on whether cells were maintained under conditions of low or high androgens [87]. In addition to differences in cell cycle regulation, differences in metabolic programs have been noted in cells expressing full-length AR vs. AR-V7 [91], with AR-V7-expressing cells displaying increased dependence on glutaminolysis and reductive carboxylation. One mechanism explaining differential regulation of transcriptional targets is differences in chromatin binding affinity, with AR-Vs having lower affinity for canonical AREs than full-length AR [92, 93].

AR Cross-Talk with Other Signaling Pathways

The AR signaling axis displays extensive cross-talk with other oncogenic pathways that are highly relevant in PCa. One such relevant pathway is the PI3K/AKT/PTEN pathway. About 20% of primary PCa samples display loss-of-function genomic alterations in PTEN, which increases to over 40% in CRPC. These PTEN alterations are in addition to somatic mutations or gene amplification of PIK3CA and PIK3CB in PCa [63, 68]. AR-mediated non-genomic activation of PI3K in the cytosol promotes cell survival and inhibits apoptosis in androgen-sensitive cells [94]. Mouse xenografts of LNCaP cells overexpressing AKT show accelerated tumor growth relative to control xenografts [95]. Mechanistically, AKT mediates direct phosphorylation of AR at Ser-213 and Ser-791, although the clinical relevance of these post-translational modifications has not yet been deciphered [96]. Collectively, these studies indicate that the PI3K signaling pathway positively regulates AR activity in PCa. However, PI3K signaling can negatively regulate AR and AR can negatively regulate PI3K. For example, FOXO3a binds to the AR promoter to upregulate AR expression, while FOXO1 recruits histone deacetylase 3 to decrease AR activity [97–99]. Further, PTEN loss results in suppression of androgen responsive transcription, while active expression of AR results in increased expression of FKBP5 and dephosphorylation of AKT, thereby suppressing AKT activity [100]. Using a PTEN-deficient murine PCa model, it was shown that this negative crosstalk between PI3K and AR is reciprocal, such that inhibition of one pathway leads to the activation of another to maintain tumor cell survival [101]. All these studies suggest that a combined therapeutic regimen targeting both AR and PI3K signaling would be more effective than targeting either pathway alone.

The role of AR in directing PCa cells towards distinct microenvironments like bone in advanced PCa provide an insight into the role of AR in tumor metastasis. Regulation of chemokine signaling via the Kruppel-Like Factor 5 (KLF5) transcription factor, chemokine receptor 4 (CXCR4), and the CXCR4 ligand CXCL12 is one such proposed mechanism in PCa cells [102]. The normal prostate gland expresses CXCR4, which becomes upregulated in response to androgens. Expression of CXCR4 is further elevated in bone metastatic lesions of PCa [103]. The ligand CXCL12 is a soluble chemoattractant highly enriched in bones. Upregulation of CXCR4 at the surface of LNCaP cells promotes cellular migration towards a CXCL12 gradient. Mechanistically, CXCR4 is indirectly regulated by AR via KLF5, which is an androgen-induced transcription factor necessary and sufficient for upregulation of CXCR4 and subsequent cellular functions in LNCaP cells [102]. The concept of increased androgen signaling, leading to increased CXCR4 expression to cause cellular migration to distant bony sites provides a foundation for future work to explore the roles and therapeutic vulnerabilities of chemokine signaling in aggressive metastatic PCa [46].

Recent studies have reported bidirectional cross-talk between AR and the nuclear receptor super family member peroxisome proliferator activated receptor gamma (PPAR-γ). PPAR-γ can either activate or repress the activity of AR, and AR can also repress the activity of PPAR-γ [104, 105]. These interactions between AR and PPAR-γ are mediated through PPAR coactivator 1 alpha (PGC1α) or through fatty acid binding proteins 4 and 5 (FABP4, FABP5), but also other, yet to be defined mechanisms [106–108]. Thus,

AR-dependent control of metabolic pathways appears to be central to PCa development and progression. PPAR-γ expression varies among PCa cell lines, with lower PPAR-γ expression in castration-sensitive cell lines like LNCaP and, higher PPAR-γ expression in castration resistant cell lines like C4-2 [109]. Although previous studies using PPAR-γ agonists suggested its role as a tumor suppressor in PCa [110], later transposon-based 'sleeping beauty' screen found that increased expression of PPAR-γ coupled with loss of PTEN promotes prostate tumorigenesis [111]. Further studies showed that PPAR-γ agonists increase AR signaling through an androgen-dependent and PPAR-γ-dependent mechanism [112, 113]. In larger studies using tissue microarray, RT-PCR and immunohistochemistry, PPAR-γ expression was found to be positively correlated with advanced PCa suggesting a more oncogenic role for PPAR-γ and its ligands [114, 115]. Gene set enrichment analysis of AR target genes regulating metabolism and biosynthetic pathways, showed enrichment for carbohydrate metabolism and PGC1α gene sets, further underscoring the relevance of this pathway in regulating AR and metabolic pathways in PCa cells [33]. As more ligands of AR and PPAR-γ enter clinical development, the intricacy of the bidirectional crosstalk between AR and PPAR-γ needs to be fully characterized in castration-sensitive PCa and CRPC.

Therapeutic Advances in AR Targeting for CRPC-Stage Disease

Studies of a cohort of CRPC tissues collected from PCa patients indicated that intra-tumoral levels of androgens were persistently high, despite castrate levels of androgens in the blood. This suggested that intracrine steroidogenesis in tumors could bypass the low levels of circulating androgens [116–118]. Understanding the mechanisms of AR re-activation in response to ADT in CRPC led to the development of second-generation AR-targeted therapies abiraterone acetate and enzalutamide [119–121]. Abiraterone acetate targets CYP17A1, an enzyme involved in conversion of cholesterol to the androgen precursor pregnenolone by blocking its 17,20 lyase and 17α-hydroxylase activities, thus inhibiting synthesis of DHT and hence reducing *de novo* production of androgens in the tumor tissue. Additionally, abiraterone acetate inhibits these CYP17A1 activities in the adrenal cortex, thus preventing the synthesis of adrenal androgens. More recently, AR antagonist activity was reported for a metabolite of abiraterone, Δ4-abiraterone, which provides further basis for its anti-tumor activity [122]. Enzalutamide (MDV-3100) acts as a competitive antagonist of the AR LBD, which reduces AR nuclear translocation and chromatin binding, and thereby blocks expression of AR target genes. As with first-generation ADT, development of resistance represents a major limitation of therapy with both abiraterone and enzalutamide. As discussed earlier, expression of AR-V7 and perhaps other AR-Vs is associated with resistance to both of these agents. Additionally, mechanisms like increased expression of steroidogenic enzyme AKR1C3 and activation of the 5α-dione pathway have been implicated in developing resistance to abiraterone [119, 123, 124]. Somatic mutations such as F876L in the AR LBD are associated with resistance to enzalutamide in models of CRPC progression, although the prevalence of F876L AR in clinical specimens appears to be low [125, 126].

Emerging Therapeutic Strategies to Target AR in CRPC

The ongoing durability of AR signaling in CRPC, which includes patients that have been treated with potent inhibitors such as enzalutamide and abiraterone, indicates an ongoing need to develop novel AR-targeted therapies. Broadly speaking, the current arsenal of AR-targeted therapies for PCa patients all exert their action by preventing androgen production, or by binding to the AR LBD. Given the importance of additional functional domains of the AR protein, one emerging strategy is to develop therapeutics that targets the AR NTD or the AR DBD. Additionally, there are

currently no approved PCa therapies that degrade or block expression of AR protein, which may be important for counteracting the widespread overexpression of AR observed in CRPC tumors harboring AR amplification. Below, we highlight experimental therapies that are being developed to target alternative domains on the AR protein, or block AR expression in PCa cells.

One strategy for targeted degradation of AR is using proteolysis targeting chimeric (PROTAC) technology. A PROTAC that has been developed to target AR is a bifunctional drug-like small molecule with one chemical moiety that binds the Von-Hippel-Lindau (VHL) E3 ubiquitin ligase complex and the other chemical moiety representing DHT, which binds the AR LBD [127, 128]. In treated PCa cells, these PROTACs bind to AR and recruited VHL E3 ligase, which induces AR polyubiquitination and degradation, leading to reduced levels of AR protein in cells and G1 growth arrest. Although these compounds are cell permeable and specific to AR, prolonged treatment with these PROTACs leads to cytotoxicity [128]. Recently, a more potent enzalutamide-based PROTAC called ARCC-4 has been developed and compared to enzalutamide under different cellular conditions. ARCC-4 selectively degraded about 95% of cellular AR in LNCaP cells. ARCC-4 was also very effective in LNCaP cells overexpressing AR point mutations F876L and T877A, as measured by reduced PSA levels in these cells upon treatment with ARCC-4. Unlike enzalutamide, ARCC-4 was able to block proliferation of VCaP cells under high androgen conditions, further demonstrating the advantage of this PROTAC over its parent compound [129]. The development of these AR degraders for therapeutic benefit in CRPC offers a new treatment strategy that can be tailored to create additional PROTACs targeting other proteins like bromodomain and extraterminal (BET) family proteins. Recently, the BET degrader ARV-771 was shown to indirectly target expression and activity of the AR-V7 splice variant [130].

A pressing challenge in the CRPC field is the development of agents that selectively target expression or activity of AR-Vs. Recently, selective AR degraders (SARDs) were developed that lead to efficient reduction in the activity of full length AR and AR-Vs even at sub-micromolar doses. SARDs UT-69 and UT-155 reduce AR expression and downstream transcription in LNCaP cells more effectively than enzalutamide. These SARDs are competitive antagonists of the AR LBD, but also bind the AR NTD domain at the AF-1 region. Further modification of UT-155 led to the development of R-UT-155, which could directly bind the AF-1 domain, but did not bind the AR LBD. Consistent with an AF-1-directed mechanism of action, R-UT-155 inhibited expression of AR and AR-Vs in the AR-V7-positive CRPC cell line 22Rv1. Moreover, R-UT-155 inhibits the growth of 22Rv1 xenografts in mice. These SARD compounds may provide a new avenue to inhibit AR by binding to and reducing expression of AR and AR-V proteins in CRPC cells [131].

In addition to the development of novel molecular entities for blocking AR expression, recent efforts have involved screening FDA-approved drugs for efficacy in CRPC cells. This led to the identification of niclosamide, an anti-helminthic drug, as a possible therapeutic that could be re-purposed for inhibition of AR in PCa [132]. Functional studies with niclosamide showed this drug could re-sensitize CRPC cells to treatment with both abiraterone and enzalu-tamide [133, 134]. Further, niclosamide was able to overcome the ability of AR-V7 to promote resistance to bicalutamide [135]. Based on these encouraging pre-clinical findings, niclosamide is being tested in combination with enzalutamide in a phase I clinical trial (NCT02532114).

Mutational hot spots that reside near the AR LBD, such as the binding function-3 (BF-3) pocket located near the AF-2 domain, have also been explored as targets in PCa cells resistant to enzalutamide. The BF-3 domain has functional significance in nuclear translocation of AR through interactions with cytoplasmic (like SGTA) and nuclear (e.g. FKBP52, BAG1L) co-chaperones [136–139]. VPC-13566 was developed as a potent and selective small molecule inhibitor of AR that binds specifically to the AR BF-3 domain [139]. In cells treated with VPC-13566, reduced AR transcriptional activity was observed. Mechanistically, this appears to be due in part to impaired translocation of AR to the nucleus. Because this compound inhibits AR BF-3 binding to cytoplasmic SGTA and nuclear BAG1L factors, it could be perceived to affect two separate pathways and therefore have less likelihood of promoting resistance. In xenograft studies, mice treated with VPC-13566 showed reduced tumor growth [139]. However, due to pharmacokinetic limitations, VPC13566 needs to be optimized for better *in vivo* stability and bio-availability before it can advance in clinical development [139].

Given that the AR NTD is responsible for the majority of *AR* transcriptional activity, the AR NTD represents an attractive therapeutic target to block activity of full length AR as well as AR-Vs. However, the AR NTD represents a challenging therapeutic target, because it is an intrinsically disordered domain of the AR [13]. Two classes of molecules, the EPI-series of bisphenol-like com-pounds, as well as Sintokamides, bind the AR NTD directly [140–142]. The compounds EPI-001 and EPI-002 engage and covalently bind to the AR NTD in treated cells, and thereby block the ability of the AR NTD to recruit co-activators such as CBP [140]. In NMR studies, EPI-001 was shown to bind to the AR TAU5 domain in the AR NTD, which is presumed to precede forma-tion of a covalent bond between TAU5 via a chlorhydrin moiety on EPI-001. However, the specificity of EPI-series compounds for binding the AR NTD is debatable, given that the highly-reactive chlorhydrin moiety of EPI-series com-pounds is required for the anti-AR action in cell models [140]. Indeed, EPI-001 was shown to have general non-specific alkylating activity in a pH-dependent manner, and also have PPAR-γ agonist activity, two properties which could also account for the anti-AR action of these com-pounds [143]. A pro-drug formulation of EPI-002, termed EPI-506, recently advanced to a Phase I/II clinical trial for metastatic PCa (NCT02606123) [99], but this trial was recently discontinued.

An additional domain of AR that could pro-vide a therapeutic targeting opportunity is the DBD. Small molecule inhibitors have been designed to target a small pocket exposed at the surface of the AR DBD and block the ability of the AR DBD to bind DNA. One such molecule termed VPC-14449 inhibits activity of full length AR and induced regression of LNCaP xenografts in mouse studies [144]. Mechanistically, VPC-14449 affects the chromatin binding interactions of wild-type and mutant forms of full length AR as well as AR-Vs. As a result, transcriptional programs mediated by full length AR, AR-Vs, or AR mutants such as F876L are all repressed. Interestingly, additive effects of VP-14449 and enzalutamide co-administered simultaneously suggest an attractive pre-clinical rationale for the development of combination therapies [145]. These studies led to the development of another lead compound termed VPC-17005, which binds selectively to the D-box of the AR DBD, thereby blocking AR dimerization. Consequently, this compound inhibits transcription of AR target genes [146].

Conclusions

AR is a master regulator in PCa that is crucial to disease development, progression and treatment. The presence of full-length AR along with generation of multiple AR-Vs creates intra-tumoral and intra-cellular heterogeneity of AR expression and activity in CRPC. There are myriad complexities to these heterogeneous transcriptomes and cistromes that are important for the field to decipher and understand. The biphasic nature of androgen signaling, escape from ADT, and rapid progression of aggressive CRPC present many variables that impact the androgen dependence and therapeutic responsiveness of PCa. The failure of several single-agent drug targets and pathway inhibitors in clinical trials that showed promising results in pre-clinical studies could be attributable to this vast heterogeneity. Efforts aimed at carefully selecting patients based on the presence of *AR* gene mutations, *AR* amplification, expression of AR-Vs, and status of related pathways including PTEN, could all impact the success of novel AR-targeted therapies in clinical trials. The myriad challenges also bring new and interesting solutions to target AR, AR-Vs, and AR target genes with potent and selective inhibitors that work alone or in combination with current anti-androgens.

References

1. C.A. Heinlein, C. Chang, Androgen receptor in prostate cancer. Endocr. Rev. **25**(2), 276–308 (2004)
2. M.H. Tan et al., Androgen receptor: structure, role in prostate cancer and drug discovery. Acta Pharmacol. Sin. **36**(1), 3–23 (2015)
3. L. Callewaert, N. Van Tilborgh, F. Claessens, Interplay between two hormone-independent activation domains in the androgen receptor. Cancer Res. **66**(1), 543–553 (2006)
4. C.L. Bevan et al., The AF1 and AF2 domains of the androgen receptor interact with distinct regions of SRC1. Mol. Cell. Biol. **19**(12), 8383–8392 (1999)
5. G. Jenster et al., Identification of two transcription activation units in the N-terminal domain of the human androgen receptor. J. Biol. Chem. **270**(13), 7341–7346 (1995)
6. V. Christiaens et al., Characterization of the two coactivator-interacting surfaces of the androgen receptor and their relative role in transcriptional control. J. Biol. Chem. **277**(51), 49230–49237 (2002)
7. S.M. Dehm, D.J. Tindall, Alternatively spliced androgen receptor variants. Endocr. Relat. Cancer **18**(5), R183–R196 (2011)
8. A. Warnmark et al., Activation functions 1 and 2 of nuclear receptors: molecular strategies for transcriptional activation. Mol. Endocrinol. **17**(10), 1901–1909 (2003)
9. I.J. McEwan, Molecular mechanisms of androgen receptor-mediated gene regulation: structure-function analysis of the AF-1 domain. Endocr. Relat. Cancer **11**(2), 281–293 (2004)
10. N.L. Chamberlain, D.C. Whitacre, R.L. Miesfeld, Delineation of two distinct type 1 activation functions in the androgen receptor amino-terminal domain. J. Biol. Chem. **271**(43), 26772–26778 (1996)
11. P. Zhu et al., Macrophage/cancer cell interactions mediate hormone resistance by a nuclear receptor derepression pathway. Cell **124**(3), 615–629 (2006)
12. S.M. Dehm et al., Selective role of an NH2-terminal WxxLF motif for aberrant androgen receptor activation in androgen depletion independent prostate cancer cells. Cancer Res. **67**(20), 10067–10077 (2007)
13. E. De Mol et al., Regulation of androgen receptor activity by transient interactions of its transactivation domain with general transcription regulators. Structure **26**(1), 145–152.e3 (2018)
14. B. He, J.A. Kemppainen, E.M. Wilson, FXXLF and WXXLF sequences mediate the NH2-terminal interaction with the ligand binding domain of the

androgen receptor. J. Biol. Chem. **275**(30), 22986–22994 (2000)

15. B. He et al., Dependence of selective gene activation on the androgen receptor NH2- and COOH-terminal interaction. J. Biol. Chem. **277**(28), 25631–25639 (2002)

16. D. Xu et al., Androgen receptor splice variants dimerize to transactivate target genes. Cancer Res. **75**(17), 3663–3671 (2015)

17. P.L. Shaffer et al., Structural basis of androgen receptor binding to selective androgen response elements. Proc. Natl. Acad. Sci. U. S. A. **101**(14), 4758–4763 (2004)

18. M. Nadal et al., Structure of the homodimeric androgen receptor ligand-binding domain. Nat. Commun. **8**, 14388 (2017)

19. E. Estebanez-Perpina et al., A surface on the androgen receptor that allosterically regulates coactivator binding. Proc. Natl. Acad. Sci. U. S. A. **104**(41), 16074–16079 (2007)

20. S. Grosdidier et al., Allosteric conversation in the androgen receptor ligand-binding domain surfaces. Mol. Endocrinol. **26**(7), 1078–1090 (2012)

21. W. Gao, C.E. Bohl, J.T. Dalton, Chemistry and structural biology of androgen receptor. Chem. Rev. **105**(9), 3352–3370 (2005)

22. K. Pereira de Jesus-Tran et al., Comparison of crystal structures of human androgen receptor ligand-binding domain complexed with various agonists reveals molecular determinants responsible for binding affinity. Protein Sci. **15**(5), 987–999 (2006)

23. H.M. Berman et al., The Protein Data Bank. Nucleic Acids Res. **28**(1), 235–242 (2000)

24. E.M. Wilson, Analysis of interdomain interactions of the androgen receptor. Methods Mol. Biol. **776**, 113–129 (2011)

25. A.J. Saporita et al., Identification and characterization of a ligand-regulated nuclear export signal in androgen receptor. J. Biol. Chem. **278**(43), 41998–42005 (2003)

26. A. Haelens et al., The hinge region regulates DNA binding, nuclear translocation, and transactivation of the androgen receptor. Cancer Res. **67**(9), 4514–4523 (2007)

27. T. Nishiyama, Serum testosterone levels after medical or surgical androgen deprivation: a comprehensive review of the literature. Urol. Oncol. **32**(1), 38.e17–38.e28 (2014)

28. C. Dai, H. Heemers, N. Sharifi, Androgen signaling in prostate cancer. Cold Spring Harb. Perspect. Med. **7**(9), a030452 (2017)

29. N. Sharifi, R.J. Auchus, Steroid biosynthesis and prostate cancer. Steroids **77**(7), 719–726 (2012)

30. N. Kaku et al., Characterization of nuclear import of the domain-specific androgen receptor in association with the importin alpha/beta and Ran-guanosine 5′-triphosphate systems. Endocrinology **149**(8), 3960–3969 (2008)

31. L. Ni et al., Androgen induces a switch from cytoplasmic retention to nuclear import of the androgen receptor. Mol. Cell. Biol. **33**(24), 4766–4778 (2013)

32. M.L. Cutress et al., Structural basis for the nuclear import of the human androgen receptor. J. Cell Sci. **121**(Pt 7), 957–968 (2008)

33. C.E. Massie et al., The androgen receptor fuels prostate cancer by regulating central metabolism and biosynthesis. EMBO J. **30**(13), 2719–2733 (2011)

34. F. Claessens, S. Joniau, C. Helsen, Comparing the rules of engagement of androgen and glucocorticoid receptors. Cell. Mol. Life Sci. **74**(12), 2217–2228 (2017)

35. B. Sahu et al., Androgen receptor uses relaxed response element stringency for selective chromatin binding and transcriptional regulation in vivo. Nucleic Acids Res. **42**(7), 4230–4240 (2014)

36. N.L. Sharma et al., The androgen receptor induces a distinct transcriptional program in castration-resistant prostate cancer in man. Cancer Cell **23**(1), 35–47 (2013)

37. M.M. Pomerantz et al., The androgen receptor cistrome is extensively reprogrammed in human prostate tumorigenesis. Nat. Genet. **47**(11), 1346–1351 (2015)

38. I.G. Mills, Maintaining and reprogramming genomic androgen receptor activity in prostate cancer. Nat. Rev. Cancer **14**(3), 187–198 (2014)

39. B.T. Copeland et al., The androgen receptor malignancy shift in prostate cancer. Prostate **78**(7), 521–531 (2018)

40. H.V. Heemers, D.J. Tindall, Unraveling the complexities of androgen receptor signaling in prostate cancer cells. Cancer Cell **15**(4), 245–247 (2009)

41. H.V. Heemers, D.J. Tindall, Androgen receptor (AR) coregulators: a diversity of functions converging on and regulating the AR transcriptional complex. Endocr. Rev. **28**(7), 778–808 (2007)

42. C. Foley, N. Mitsiades, Moving beyond the androgen receptor (AR): targeting AR-interacting proteins to treat prostate cancer. Horm. Cancer **7**(2), 84–103 (2016)

43. S.M. Dehm, D.J. Tindall, Ligand-independent androgen receptor activity is activation function-2-independent and resistant to antiandrogens in androgen refractory prostate cancer cells. J. Biol. Chem. **281**(38), 27882–27893 (2006)

44. S. Balasubramaniam et al., Aberrant BAF57 signaling facilitates prometastatic phenotypes. Clin. Cancer Res. **19**(10), 2657–2667 (2013)

45. T. Dadaev et al., Fine-mapping of prostate cancer susceptibility loci in a large meta-analysis identifies candidate causal variants. Nat. Commun. **9**(1), 2256 (2018)

46. M.A. Augello, R.B. Den, K.E. Knudsen, AR function in promoting metastatic prostate cancer. Cancer Metastasis Rev. **33**(2–3), 399–411 (2014)

47. N. Gao et al., The role of hepatocyte nuclear factor-3 alpha (Forkhead Box A1) and androgen receptor in transcriptional regulation of prostatic genes. Mol. Endocrinol. **17**(8), 1484–1507 (2003)

48. C. Jung et al., HOXB13 induces growth suppression of prostate cancer cells as a repressor of hormone-

activated androgen receptor signaling. Cancer Res. **64**(24), 9185–9192 (2004)

49. T. Whitington et al., Gene regulatory mechanisms underpinning prostate cancer susceptibility. Nat. Genet. **48**(4), 387–397 (2016)

50. J.L. Robinson, K.A. Holmes, J.S. Carroll, FOXA1 mutations in hormone-dependent cancers. Front. Oncol. **3**, 20 (2013)

51. C.E. Barbieri et al., Exome sequencing identifies recurrent SPOP, FOXA1 and MED12 mutations in prostate cancer. Nat. Genet. **44**(6), 685–689 (2012)

52. C.S. Grasso et al., The mutational landscape of lethal castration-resistant prostate cancer. Nature **487**(7406), 239–243 (2012)

53. S. Paltoglou et al., Novel androgen receptor coregulator GRHL2 exerts both oncogenic and antimetastatic functions in prostate cancer. Cancer Res. **77**(13), 3417–3430 (2017)

54. S. Stelloo et al., Endogenous androgen receptor proteomic profiling reveals genomic subcomplex involved in prostate tumorigenesis. Oncogene **37**(3), 313–322 (2018)

55. S. Liu et al., A comprehensive analysis of coregulator recruitment, androgen receptor function and gene expression in prostate cancer. Elife **6**, e28482 (2017)

56. C. Huggins, C.V. Hodges, Studies on prostatic cancer. I. The effect of castration, of estrogen and androgen injection on serum phosphatases in metastatic carcinoma of the prostate. CA Cancer J. Clin. **22**(4), 232–240 (1972)

57. L.J. Schmidt, D.J. Tindall, Androgen receptor: past, present and future. Curr. Drug Targets **14**(4), 401–407 (2013)

58. P.A. Watson, V.K. Arora, C.L. Sawyers, Emerging mechanisms of resistance to androgen receptor inhibitors in prostate cancer. Nat. Rev. Cancer **15**(12), 701–711 (2015)

59. K.E. Knudsen, W.K. Kelly, Outsmarting androgen receptor: creative approaches for targeting aberrant androgen signaling in advanced prostate cancer. Expert. Rev. Endocrinol. Metab. **6**(3), 483–493 (2011)

60. T. Visakorpi et al., In vivo amplification of the androgen receptor gene and progression of human prostate cancer. Nat. Genet. **9**(4), 401–406 (1995)

61. P. Koivisto et al., Androgen receptor gene amplification: a possible molecular mechanism for androgen deprivation therapy failure in prostate cancer. Cancer Res. **57**(2), 314–319 (1997)

62. C.D. Chen et al., Molecular determinants of resistance to antiandrogen therapy. Nat. Med. **10**(1), 33–39 (2004)

63. Cancer Genome Atlas Research Network, The molecular taxonomy of primary prostate cancer. Cell **163**(4), 1011–1025 (2015)

64. G. Gundem et al., The evolutionary history of lethal metastatic prostate cancer. Nature **520**(7547), 353–357 (2015)

65. A.A. Azad et al., Androgen receptor gene aberrations in circulating cell-free DNA: biomarkers of therapeutic resistance in castration-resistant prostate cancer. Clin. Cancer Res. **21**(10), 2315–2324 (2015)

66. S. Carreira et al., Tumor clone dynamics in lethal prostate cancer. Sci. Transl. Med. **6**(254), 254ra125 (2014)

67. M.E. Taplin et al., Selection for androgen receptor mutations in prostate cancers treated with androgen antagonist. Cancer Res. **59**(11), 2511–2515 (1999)

68. B.S. Taylor et al., Integrative genomic profiling of human prostate cancer. Cancer Cell **18**(1), 11–22 (2010)

69. X.Y. Zhao et al., Glucocorticoids can promote androgen-independent growth of prostate cancer cells through a mutated androgen receptor. Nat. Med. **6**(6), 703–706 (2000)

70. D.Y. Takeda et al., A somatically acquired enhancer of the androgen receptor is a noncoding driver in advanced prostate cancer. Cell **174**(2), 422–432.e13 (2018)

71. S.R. Viswanathan et al., Structural alterations driving castration-resistant prostate cancer revealed by linked-read genome sequencing. Cell **174**(2), 433–447.e19 (2018)

72. D.A. Quigley et al., Genomic hallmarks and structural variation in metastatic prostate cancer. Cell **175**(3), 889 (2018)

73. S.M. Dehm et al., Splicing of a novel androgen receptor exon generates a constitutively active androgen receptor that mediates prostate cancer therapy resistance. Cancer Res. **68**(13), 5469–5477 (2008)

74. K.M. Wadosky, S. Koochekpour, Androgen receptor splice variants and prostate cancer: from bench to bedside. Oncotarget **8**(11), 18550–18576 (2017)

75. A. Paschalis et al., Alternative splicing in prostate cancer. Nat. Rev. Clin. Oncol. **15**(11), 663–675 (2018)

76. A. Sharp et al., Androgen receptor splice variant-7 expression emerges with castration resistance in prostate cancer. J. Clin. Invest. **129**(1), 192–208 (2019)

77. E.S. Antonarakis et al., Androgen receptor variant-driven prostate cancer: clinical implications and therapeutic targeting. Prostate Cancer Prostatic Dis. **19**(3), 231–241 (2016)

78. H.I. Scher et al., Association of AR-V7 on circulating tumor cells as a treatment-specific biomarker with outcomes and survival in castration-resistant prostate cancer. JAMA Oncol. **2**(11), 1441–1449 (2016)

79. M. Kohli et al., Androgen receptor variant AR-V9 is coexpressed with AR-V7 in prostate cancer metastases and predicts abiraterone resistance. Clin. Cancer Res. **23**(16), 4704–4715 (2017)

80. E.S. Antonarakis et al., AR-V7 and resistance to enzalutamide and abiraterone in prostate cancer. N. Engl. J. Med. **371**(11), 1028–1038 (2014)

81. E.S. Antonarakis et al., Androgen receptor splice variant 7 and efficacy of taxane chemotherapy in patients with metastatic castration-resistant prostate cancer. JAMA Oncol. **1**(5), 582–591 (2015)

82. H.I. Scher et al., Assessment of the validity of nuclear-localized androgen receptor splice variant

7 in circulating tumor cells as a predictive biomarker for castration-resistant prostate cancer. JAMA Oncol. **4**(9), 1179–1186 (2018)

83. E.A. Mostaghel et al., Intraprostatic androgens and androgen-regulated gene expression persist after testosterone suppression: therapeutic implications for castration-resistant prostate cancer. Cancer Res. **67**(10), 5033–5041 (2007)

84. D. Robinson et al., Integrative clinical genomics of advanced prostate cancer. Cell **161**(5), 1215–1228 (2015)

85. D.T. Miyamoto et al., RNA-Seq of single prostate CTCs implicates noncanonical Wnt signaling in antiandrogen resistance. Science **349**(6254), 1351–1356 (2015)

86. T. van der Steen, D.J. Tindall, H. Huang, Posttranslational modification of the androgen receptor in prostate cancer. Int. J. Mol. Sci. **14**(7), 14833–14859 (2013)

87. P.A. Watson et al., Constitutively active androgen receptor splice variants expressed in castration-resistant prostate cancer require full-length androgen receptor. Proc. Natl. Acad. Sci. U. S. A. **107**(39), 16759–16765 (2010)

88. Y. Yamamoto et al., Generation 2.5 antisense oligonucleotides targeting the androgen receptor and its splice variants suppress enzalutamide-resistant prostate cancer cell growth. Clin. Cancer Res. **21**(7), 1675–1687 (2015)

89. C. Henzler et al., Truncation and constitutive activation of the androgen receptor by diverse genomic rearrangements in prostate cancer. Nat. Commun. **7**, 13668 (2016)

90. R. Hu et al., Distinct transcriptional programs mediated by the ligand-dependent full-length androgen receptor and its splice variants in castration-resistant prostate cancer. Cancer Res. **72**(14), 3457–3462 (2012)

91. A.A. Shafi et al., Differential regulation of metabolic pathways by androgen receptor (AR) and its constitutively active splice variant, AR-V7, in prostate cancer cells. Oncotarget **6**(31), 31997–32012 (2015)

92. S.C. Chan et al., Targeting chromatin binding regulation of constitutively active AR variants to overcome prostate cancer resistance to endocrine-based therapies. Nucleic Acids Res. **43**(12), 5880–5897 (2015)

93. S.C. Chan, Y. Li, S.M. Dehm, Androgen receptor splice variants activate androgen receptor target genes and support aberrant prostate cancer cell growth independent of canonical androgen receptor nuclear localization signal. J. Biol. Chem. **287**(23), 19736–19749 (2012)

94. S. Baron et al., Androgen receptor mediates non-genomic activation of phosphatidylinositol 3-OH kinase in androgen-sensitive epithelial cells. J. Biol. Chem. **279**(15), 14579–14586 (2004)

95. J.R. Graff et al., Increased AKT activity contributes to prostate cancer progression by dramatically accelerating prostate tumor growth and diminishing

p27Kip1 expression. J. Biol. Chem. **275**(32), 24500–24505 (2000)

96. L. Xin et al., Progression of prostate cancer by synergy of AKT with genotropic and nongenotropic actions of the androgen receptor. Proc. Natl. Acad. Sci. U. S. A. **103**(20), 7789–7794 (2006)

97. L. Yang et al., Induction of androgen receptor expression by phosphatidylinositol 3-kinase/Akt downstream substrate, FOXO3a, and their roles in apoptosis of LNCaP prostate cancer cells. J. Biol. Chem. **280**(39), 33558–33565 (2005)

98. P. Liu et al., A transcription-independent function of FOXO1 in inhibition of androgen-independent activation of the androgen receptor in prostate cancer cells. Cancer Res. **68**(24), 10290–10299 (2008)

99. J.K. Leung, M.D. Sadar, Non-genomic actions of the androgen receptor in prostate cancer. Front. Endocrinol. (Lausanne) **8**, 2 (2017)

100. D.J. Mulholland et al., Cell autonomous role of PTEN in regulating castration-resistant prostate cancer growth. Cancer Cell **19**(6), 792–804 (2011)

101. B.S. Carver et al., Reciprocal feedback regulation of PI3K and androgen receptor signaling in PTEN-deficient prostate cancer. Cancer Cell **19**(5), 575–586 (2011)

102. D.E. Frigo et al., Induction of Kruppel-like factor 5 expression by androgens results in increased CXCR4-dependent migration of prostate cancer cells in vitro. Mol. Endocrinol. **23**(9), 1385–1396 (2009)

103. Y.X. Sun et al., Expression of CXCR4 and CXCL12 (SDF-1) in human prostate cancers (PCa) in vivo. J. Cell. Biochem. **89**(3), 462–473 (2003)

104. E. Olokpa, P.E. Moss, L.V. Stewart, Crosstalk between the androgen receptor and PPAR gamma signaling pathways in the prostate. PPAR Res. **2017**, 9456020 (2017)

105. C. Elix, S.K. Pal, J.O. Jones, The role of peroxisome proliferator-activated receptor gamma in prostate cancer. Asian J. Androl. **20**(3), 238–243 (2018)

106. M. Shiota et al., Peroxisome proliferator-activated receptor gamma coactivator-1alpha interacts with the androgen receptor (AR) and promotes prostate cancer cell growth by activating the AR. Mol. Endocrinol. **24**(1), 114–127 (2010)

107. E. Olokpa, A. Bolden, L.V. Stewart, The androgen receptor regulates PPARgamma expression and activity in human prostate cancer cells. J. Cell. Physiol. **231**(12), 2664–2672 (2016)

108. Z. Bao et al., A novel cutaneous Fatty Acid-binding protein-related signaling pathway leading to malignant progression in prostate cancer cells. Genes Cancer **4**(7–8), 297–314 (2013)

109. E. Mueller et al., Effects of ligand activation of peroxisome proliferator-activated receptor gamma in human prostate cancer. Proc. Natl. Acad. Sci. U. S. A. **97**(20), 10990–10995 (2000)

110. J.I. Hisatake et al., Down-regulation of prostate-specific antigen expression by ligands for peroxi-

110. some proliferator-activated receptor gamma in human prostate cancer. Cancer Res. **60**(19), 5494–5498 (2000)

111. I. Ahmad et al., Sleeping Beauty screen reveals Pparg activation in metastatic prostate cancer. Proc. Natl. Acad. Sci. U. S. A. **113**(29), 8290–8295 (2016)

112. B.Y. Tew et al., Vitamin K epoxide reductase regulation of androgen receptor activity. Oncotarget **8**(8), 13818–13831 (2017)

113. P.E. Moss, B.E. Lyles, L.V. Stewart, The PPARgamma ligand ciglitazone regulates androgen receptor activation differently in androgen-dependent versus androgen-independent human prostate cancer cells. Exp. Cell Res. **316**(20), 3478–3488 (2010)

114. S. Rogenhofer et al., Enhanced expression of peroxisome proliferate-activated receptor gamma (PPAR-gamma) in advanced prostate cancer. Anticancer Res. **32**(8), 3479–3483 (2012)

115. Y. Segawa et al., Expression of peroxisome proliferator-activated receptor (PPAR) in human prostate cancer. Prostate **51**(2), 108–116 (2002)

116. T. Nishiyama, Y. Hashimoto, K. Takahashi, The influence of androgen deprivation therapy on dihydrotestosterone levels in the prostatic tissue of patients with prostate cancer. Clin. Cancer Res. **10**(21), 7121–7126 (2004)

117. R.B. Montgomery et al., Maintenance of intratumoral androgens in metastatic prostate cancer: a mechanism for castration-resistant tumor growth. Cancer Res. **68**(11), 4447–4454 (2008)

118. Y. Ho, S.M. Dehm, Androgen receptor rearrangement and splicing variants in resistance to endocrine therapies in prostate cancer. Endocrinology **158**(6), 1533–1542 (2017)

119. T. Chandrasekar et al., Mechanisms of resistance in castration-resistant prostate cancer (CRPC). Transl. Androl. Urol. **4**(3), 365–380 (2015)

120. H.I. Scher et al., Increased survival with enzalutamide in prostate cancer after chemotherapy. N. Engl. J. Med. **367**(13), 1187–1197 (2012)

121. J.S. de Bono et al., Abiraterone and increased survival in metastatic prostate cancer. N. Engl. J. Med. **364**(21), 1995–2005 (2011)

122. Z. Li et al., Conversion of abiraterone to D4A drives anti-tumour activity in prostate cancer. Nature **523**(7560), 347–351 (2015)

123. K.H. Chang et al., Dihydrotestosterone synthesis bypasses testosterone to drive castration-resistant prostate cancer. Proc. Natl. Acad. Sci. U. S. A. **108**(33), 13728–13733 (2011)

124. C. Liu et al., Inhibition of AKR1C3 activation overcomes resistance to abiraterone in advanced prostate cancer. Mol. Cancer Ther. **16**(1), 35–44 (2017)

125. H. Liu et al., Molecular dynamics studies on the enzalutamide resistance mechanisms induced by androgen receptor mutations. J. Cell. Biochem. **118**(9), 2792–2801 (2017)

126. Z. Culig, Molecular mechanisms of enzalutamide resistance in prostate cancer. Curr. Mol. Biol. Rep. **3**(4), 230–235 (2017)

127. J.S. Schneekloth Jr. et al., Chemical genetic control of protein levels: selective in vivo targeted degradation. J. Am. Chem. Soc. **126**(12), 3748–3754 (2004)

128. A. Rodriguez-Gonzalez et al., Targeting steroid hormone receptors for ubiquitination and degradation in breast and prostate cancer. Oncogene **27**(57), 7201–7211 (2008)

129. J. Salami et al., Androgen receptor degradation by the proteolysis-targeting chimera ARCC-4 outperforms enzalutamide in cellular models of prostate cancer drug resistance. Commun. Biol. **1**, 100 (2018)

130. K. Raina et al., PROTAC-induced BET protein degradation as a therapy for castration-resistant prostate cancer. Proc. Natl. Acad. Sci. U. S. A. **113**(26), 7124–7129 (2016)

131. S. Ponnusamy et al., Novel selective agents for the degradation of androgen receptor variants to treat castration-resistant prostate cancer. Cancer Res. **77**(22), 6282–6298 (2017)

132. C. Liu et al., Niclosamide inhibits androgen receptor variants expression and overcomes enzalutamide resistance in castration-resistant prostate cancer. Clin. Cancer Res. **20**(12), 3198–3210 (2014)

133. C. Liu et al., Niclosamide enhances abiraterone treatment via inhibition of androgen receptor variants in castration resistant prostate cancer. Oncotarget **7**(22), 32210–32220 (2016)

134. C. Liu et al., Niclosamide suppresses cell migration and invasion in enzalutamide resistant prostate cancer cells via Stat3-AR axis inhibition. Prostate **75**(13), 1341–1353 (2015)

135. C. Liu et al., Niclosamide and bicalutamide combination treatment overcomes enzalutamide- and bicalutamide-resistant prostate cancer. Mol. Cancer Ther. **16**(8), 1521–1530 (2017)

136. J.T. De Leon et al., Targeting the regulation of androgen receptor signaling by the heat shock protein 90 cochaperone FKBP52 in prostate cancer cells. Proc. Natl. Acad. Sci. U. S. A. **108**(29), 11878–11883 (2011)

137. G. Buchanan et al., Control of androgen receptor signaling in prostate cancer by the cochaperone small glutamine rich tetratricopeptide repeat containing protein alpha. Cancer Res. **67**(20), 10087–10096 (2007)

138. L.K. Philp et al., SGTA: a new player in the molecular co-chaperone game. Horm. Cancer **4**(6), 343–357 (2013)

139. N. Lallous et al., Targeting binding function-3 of the androgen receptor blocks its co-chaperone interactions, nuclear translocation, and activation. Mol. Cancer Ther. **15**(12), 2936–2945 (2016)

140. R.J. Andersen et al., Regression of castrate-recurrent prostate cancer by a small-molecule inhibitor of the amino-terminus domain of the androgen receptor. Cancer Cell **17**(6), 535–546 (2010)

141. J.K. Myung et al., An androgen receptor N-terminal domain antagonist for treating prostate cancer. J. Clin. Invest. **123**(7), 2948–2960 (2013)

142. M.D. Sadar et al., Sintokamides A to E, chlorinated peptides from the sponge Dysidea sp. that inhibit

transactivation of the N-terminus of the androgen receptor in prostate cancer cells. Org. Lett. **10**(21), 4947–4950 (2008)

143. L.J. Brand et al., EPI-001 is a selective peroxisome proliferator-activated receptor-gamma modulator with inhibitory effects on androgen receptor expression and activity in prostate cancer. Oncotarget **6**(6), 3811–3824 (2015)

144. K. Dalal et al., Selectively targeting the DNA-binding domain of the androgen receptor as a prospective therapy for prostate cancer. J. Biol. Chem. **289**(38), 26417–26429 (2014)

145. K. Dalal et al., Bypassing drug resistance mechanisms of prostate cancer with small molecules that target androgen receptor-chromatin interactions. Mol. Cancer Ther. **16**(10), 2281–2291 (2017)

146. K. Dalal et al., Selectively targeting the dimerization interface of human androgen receptor with small-molecules to treat castration-resistant prostate cancer. Cancer Lett. **437**, 35–43 (2018)

Wnt/Beta-Catenin Signaling and Prostate Cancer Therapy Resistance

Yunshin Yeh, Qiaozhi Guo, Zachary Connelly, Siyuan Cheng, Shu Yang, Nestor Prieto-Dominguez, and Xiuping Yu

Introduction to CRPCa and NEPCA

Prostate Cancer Progression

Prostate cancer (PCa) is the most common non-skin cancer and the second leading cause of cancer-related deaths among American men. Over 164,000 new cases and 29,000 deaths were reported in 2018. Large efforts have been made to study prostate carcinogenesis and to improve the strategies of treatment in the hopes of achieving complete remission of the disease. Currently, the localized disease, which is when there are no cancer cells identifiable in regional lymph nodes or distant organ metastasis, is treated with active surveillance, surgery, and radiation.

Y. Yeh
Department of Pathology, Overton Brooks Medical Center, Shreveport, LA, USA

Q. Guo
Merton College, University of Oxford, Oxford, UK

Z. Connelly · S. Cheng · S. Yang
N. Prieto-Dominguez
Department of Biochemistry & Molecular Biology, LSU Health Sciences Center, Shreveport, LA, USA

X. Yu (✉)
Department of Biochemistry & Molecular Biology, LSU Health Sciences Center, Shreveport, LA, USA

Department of Urology, LSU Health Sciences Center, Shreveport, LA, USA
e-mail: xyu@lsuhsc.edu

Metastatic or advanced PCa, on the other hand, is primarily treated with hormone therapy. In 1966, Charles B. Huggins received the Nobel Prize in Physiology for his discovery that PCa depends on the androgen hormone for tumor growth [1]. Since then, androgen deprivation therapy (ADT) has been the standard of care for treating metastatic or locally advanced PCa. This treatment can be achieved by administering either anti-androgen agents or androgen synthesis inhibitors and is largely based on the assumption that PCa cells will not survive in an androgen-depleted microenvironment. However, despite the initial effectiveness of this treatment, most advanced PCa patients develop resistance to ADT and the disease eventually progresses to castrate-resistant prostate cancer (CRPCa), which is defined by the following criteria [2]:

1. Castrate serum levels of testosterone <50 ng/dl or 1.7 nmol/l.
2. Three consecutive rises of PSA.
3. Anti-androgen withdrawal for at least 4 weeks.
4. Progression of pre-existing disease.
5. Two or more osseous or soft tissue lesions.

CRPCa is an incurable and highly lethal disease. Approximately 90% of the CRPCa patients develop distant organ metastases, and the mean survival period of CRPCa patients after diagnosis is only 16–18 months.

© Springer Nature Switzerland AG 2019
S. M. Dehm, D. J. Tindall (eds.), *Prostate Cancer*, Advances in Experimental Medicine and Biology 1210, https://doi.org/10.1007/978-3-030-32656-2_16

Mechanisms Driving CRPCa Progression

The prostate is an androgen-dependent organ. Androgens and androgen receptor (AR) signaling regulate the physiological development and growth of normal prostates. In the absence of androgens, cytoplasmic heat-shock proteins bind to and retain AR. Upon the binding of androgen, AR dissociates from the heat-shock proteins, translocates to the cell nucleus, and binds to androgen response elements in the DNA, activating genes important for cell survival, growth, and proliferation [3, 4]. These include genes that are involved in androgen biosynthesis, DNA synthesis and repair, cell cycle regulation, and proliferation.

Similar to normal prostate epithelial cells, prostate cancer cells are also dependent on androgens for cell proliferation, which is the basis for ADT. However, during the progression to castrate-resistance, PCa cells develop ways to survive, grow, and proliferate *despite* androgen ablation [5]. Several mechanisms, both AR-dependent and AR-independent, have been proposed to drive this [3, 6]. Studies have shown that many CRPCa cases retain AR signaling [7], and that AR alterations, including AR amplifications and mutations, play a significant role in the development of CRPCa [8]. A subset of therapy-resistant PCa, however, develop AR-independent mechanisms for survival after androgen deprivation, including cross talk with alternative signaling pathways and the acquisition of stem cell features and the neuroendocrine (NE) phenotype. Frequently, these different pathways are concurrently involved in the pathogenesis of CRPCa.

AR-Dependent Mechanisms

AR signaling is believed to remain active in most CRPCa cases [3, 4, 9]. AR-stimulated genes that are initially repressed during ADT subsequently rebound and promote cancer progression [10, 11]. Studies have identified several mechanisms that drive the persistence of AR signaling in CRPCa. It has been shown that despite androgen deprivation, intratumoral levels of androgen in patients with CRPCa are similar to those of patients that did not undergo hormone therapy

[12]. One reason is that enzymes which are involved in androgen biosynthesis were upregulated in the tumor microenvironment [12–14]. Also, studies have found significant heterogeneity in the expression of various steroidogenic enzymes and alternative androgen biosynthesis pathways in CRPCa patients [15, 16]. These mechanisms could continue to supply testosterone and sustain AR signaling in PCa cells following androgen deprivation.

Alterations of AR provide another way to compensate for androgen deprivation and promote cancer progression. Alterations occur either through mutations or amplifications, resulting in PCa cells being activated by other ligands or hypersensitive to low levels of androgens. For example, AR mutations allow for decreased ligand specificity, leading to AR activation by alternate steroids, such as estrogens, corticosteroids, and progesterone [17]; and AR amplifications sensitize PCa cells to castrate-levels of androgens [11]. Meanwhile, AR variants, which lack the C-terminus ligand binding domains, are constitutively active [18–25]. Moreover, a 5-amino-acid WHTLF motif located in the NH_2 terminal region of AR can mediate androgen-independent AR activation [26]. Although AR mutations or AR variants are not the principal driver of primary PCa, studies have found that treating PCa with AR antagonists increases the burden of AR mutations and the expression of AR variants [27–29]. This suggests that androgen deprivation acts as a selective pressure for mutations and splicing variants in the AR gene in advanced PCa.

AR signaling can also be activated in the absence of androgens by growth factors and cytokines, as well as the elevated expression of AR co-activators [3]. These mechanisms act on the transcriptional activity of AR, upregulating the transcription of downstream target genes that promote cell growth and proliferation. For example, chromatin immunoprecipitation analyses have revealed enhanced activities of the AR co-activators MED1, FOXA1 and GATA2 in CRPCa cells [30]. These results indicate that the distinctive pattern of AR transcriptional activity in castrate-resistant cells is determined to a large extent by coactivator stimulation and accompanying chromatin modifications.

Additionally, a recent study from our laboratory has shown that, Foxa2, a forkhead transcription factor downstream of Wnt/β-Catenin signaling, sustains the expression of AR target genes in PCa after castration [31]. This study was conducted using TRAMP SV40 T-antigen transgenic mice where the expression of T-antigen is driven by an androgen-responsive, prostate-specific Probasin promoter. It was found that T-antigen is expressed in prostatic tumors even after the castration of these mice and that Foxa2 is co-expressed in T-antigen positive cells. A subsequent functional study found that ectopic expression of Foxa2 is sufficient to drive the expression of T-antigen in prostate epithelial cells after androgen deprivation. Chromatin immunoprecipitation assays showed that Foxa2 binds to the promoter regions of AR target genes, regardless of the presence of androgens and concurrent with the occupancy of active transcription marks, H3K27Ac3. This indicates that Foxa2 provides an alternate mechanism for retaining AR signaling after ADT. Further identification and exploration of other alternative survival pathways could provide critical data on PCa biomarkers and combined therapies.

Because of the central role of AR signaling in CRPCa progression, many studies have focused on targeting AR signaling. This has led to the development of the second generation of the anti-androgen drug, enzalutamide, which inhibits nuclear localization and chromatin binding of AR [32], as well as abiraterone, which blocks the production of testosterone and other androgens in the testis, adrenal cortex, and tumor tissue [33]. Though these therapies are effective in treating symptoms of advanced PCa and prolonging life, they nevertheless remain ineffective in curing CRPCa and the tumors eventually progress.

AR-Independent Mechanisms

Although the newer and more potent treatments that block AR signaling in PCa have significantly improved patient survival, there remains a subpopulation of PCa that do not respond to the AR-targeted treatment. Even in patients who do respond initially, these drugs only extend life for several months and eventually fail. A possible reason for this is that some tumors, under selective pressure during ADT, develop mechanisms to bypass their dependency on AR signaling. Such mechanisms include the upregulation of glucocorticoid receptor expression, inactivation of tumor suppressor genes such as PTEN, and activation of oncogenic pathways such as PI3K/Akt and Wnt/β-Catenin signaling [34–36]. Also, late stage PCa cells display cellular plasticity, resulting from genetic and epigenetic alterations [37, 38]. The re-programmed transcriptome could cause the loss of expression of prostate differentiation genes such as AR, FOXA1, and SPDEF and the gain of expression of stemness genes in late stage PCa cells. These changes promote the PCa cells to undergo NE differentiation and/or acquire stem cell features, enabling the growth of PCa cells independent of AR signaling. The mechanisms that override AR signaling in PCa are the focus of this section and will be further discussed in the remainder of this chapter, beginning with an overview of NE differentiation.

Neuroendocrine Differentiation of Carcinoma Cells

An increasingly recognized mechanism of resistance to ADT in PCa involves epithelial plasticity, in which cancer cells lose prostate epithelial differentiation, express no or low levels of AR and AR targets, and display neuroendocrine (NE) features [39]. Although most human PCa cases are adenocarcinomas, NE differentiation in PCa is common after the failure of ADT [39–43]. With the use of new, more potent anti-androgens and drugs that block androgen synthesis in prostate, "therapy-induced" progression to NEPCa is seen in 25–30% of patients [39–43]. Similar to prostate small-cell carcinomas, which are typically castrate-resistant and highly aggressive, PCa exhibiting NE differentiation is associated with poor prognosis, as these tumors rapidly become resistant to hormone therapy, chemotherapy and radiotherapy [42]. Currently, there are no effective treatments for PCa with prominent NE differentiation.

Histologically, a normal adult prostate is comprised of epithelial components embedded in a background of fibromuscular stroma. The epithelial components consist of luminal secretory cells, basal cells, and NE cells, which account for less than 1% of the prostatic epithelial cells. In PCa, approximately 5% of early stage of PCa show NE differentiation. However, after long-term ADT, up to 40–100% of late stage prostatic tumors have an increase in NE cells [39–54]. Different from small cell carcinomas, which are sheets of poorly-differentiated cancer cells, PCa that have NE differentiation maintain glandular architecture with cells that express NE markers such as chromogranin A, synaptophysin, and/or enolase scattered among the adenocarcinoma cells.

NE cells, which express little or no AR, are more commonly observed in AR-independent than in AR-dependent tumors. It is likely that NE differentiation provides an escape mechanism for cancer cells to survive in a microenvironment of androgen deprivation [48]. Moreover, data from clinical studies suggest that androgen withdrawal induces a NE phenotype in human PCa [39, 40, 47, 48]. A similar phenomenon has also been observed in cell culture and animal models. Studies have shown that androgen depletion or knocking down AR induces NE differentiation in androgen-dependent LNCaP PCa cells [55–57]. In animal models, castration accelerates the emergence of NE tumors in TRAMP mice [58–60]. Additionally, androgen withdrawal through the castration of host mice induces NE differentiation in human PCa xenografts [61–63]. Taken together, these findings suggest that androgen deprivation selects for, or even accelerates, the process of NE differentiation.

In addition to ADT, a number of other mechanisms can induce NE differentiation in PCa, including exposure to db-cAMP [64] or IL-6 [65], over-expression of PCDH-PC [66] or MYCN [67, 68], inactivation of RB1 and TP53 [69, 70], hypoxia condition through the down-regulation of Notch signaling [71], and the cooperation of FOXA2 and HIF1a [72]. Increased NE differentiation is also observed in PCa after a

treatment with ionizing radiation or receptor tyrosine kinase inhibitors [44, 73]. It has been speculated that therapy-induced cellular stress reprograms the transcriptome in cancer cells, enabling them to develop alternative mechanisms, such as NE differentiation, for survival.

Not only are the NE cells resistant to therapies, studies suggest that they also promote the progression of non-NE PCa cells. For example, prostatic NE cells often preferentially reside in close proximity to non-NE proliferating cells, suggesting that NE cells may promote the proliferation of adjacent cells through the secretion of growth-modulating neuropeptides [74]. In line with this view, it has been shown that serotonin, a neuropeptide, promotes the proliferation and migration of PCa cells [75]. It has also been shown that co-culturing LNCaP cells with NEPCa cells promotes the androgen-independent growth of the former as well as their metastatic ability [75–79]. Taken together, results from these studies support that NEPCa cells release neuropeptides to promotes CRPCa progression.

The Cell of Origin of NEPCa

The cell of origin of NEPCa is not yet fully defined. During development, normal NE cells in prostates are derived from the neural crest [80]. However, NE cancer cells may arise via a different route. A recent lineage tracing study in mice found that NEPCa cells arise from the trans-differentiation of adenocarcinoma cells in Trp53 and Pten compound mutant PCa mouse models [81]. This suggests that normal NE cells and NE cancer cells have different origins. In humans, genetic data also support the concept that NEPCa cells arise from the trans-differentiation of prostate adenocarcinoma. For example, the same mutation of TP53 was found in both non-NE and NE areas on a prostatic tumor but not in the normal adjacent epithelia, indicating that this mutation is not a somatic mutation and that the two morphologically different types of cancer share the same cell of origin [82]. Moreover, it has been shown that the NEPCa and prostate adenocarcinoma cells which were micro-dissected from radical prostatectomy specimens shared all of the microsatellite markers examined, further supporting that NEPCa cells have common origin with non-NE PCa [83]. Finally, androgen-stimulated TMPRSS2-ERG fusions, which are frequently observed in prostate adenocarcinoma [84], are also detected in NEPCa [50, 85], suggesting that NEPCa arises from AR-positive prostate adenocarcinoma.

Similar indications of NE trans-differentiation have been observed in other types of cancer, after patients are heavily treated with chemo- or targeted therapies. For example, while small cell lung cancer (SCLC, a type of neuroendocrine cancer) predominantly arise from transformed pulmonary NE cells, a subset of SCLC tumors arise via trans-differentiation. One study found that 5/37 (14%) non-small cell lung cancer (NSCLC) cases progressed to SCLC after they developed resistance to EGFR-targeted therapies [86]. These SCLC tumors retained the original EGFR mutations, suggesting that the SCLC cells have the same origin as the NSCLC and that NE tumors can arise from epithelial cells via trans-differentiation.

Epigenetic Reprogramming Leads to NEPCa

Next-generation sequencing studies using DNA and RNA isolated from PCa patients indicate that NEPCa tumors exhibit a similar mutation burden but a different transcriptome profile from non-NE CRPCa [87]. This suggests that the acquisition of the NE phenotype is driven by epigenetic, transcriptional reprogramming but not genetic alterations. These epigenetically dysregulated pathways include the master regulators of neural lineage, cell-cell adhesion, epithelial-to-mesenchymal transition, and stem cell programming, all of which are thought to be involved in the development of NEPCa. For example, NEPCa tumors express neural markers Chromogranin A, Synaptophysin, and neuron specific enolase. BRN2, a master transcription factor that controls neuronal differentiation, is induced in NEPCa [7]. The expression of REST, a transcriptional

repressor that represses the expression of neural genes, is reduced/lost in NEPCa. SRRM4, an RNA splicing factor, is induced in NEPCa. Studies have shown that SRRM4 is involved in regulating the alternative splicing of REST and that ectopic expression of SRRM4 promotes NE differentiation [88].

Concurrent with the increased expression of neural markers, NEPCa tumors display reduced expression of genes that regulate prostate differentiation such as AR and its co-factor, FOXA1, as well as SPDEF, an ETS transcription factor that is highly expressed in the luminal epithelial cells of normal prostates [87, 89]. While the expression of prostate-differentiation genes is reduced, genes that are expressed during embryonic prostate development or in stem cells such as FOXA2 and SOX2 are up-regulated in NEPCa [54, 90, 91].

The altered transcriptome in NEPCa may result from epigenetic reprogramming by altered expression of key DNA and histone modification enzymes. This is supported by the observation that the DNA methylation profile of NEPCa is quite different from that of prostate adenocarcinoma [87]. In line with this, the expression of DNA modification enzymes including DNA methyltransferases (DNMT1 and DNMT3) and a putative DNA demethylase (TET1) is altered in NEPCa [87, 92]. Concordantly, promoter methylation is observed in 22% of the dysregulated genes, including SPDEF, in NEPCa. Additionally, NEPCa cells often exhibit altered expression of key histone modification enzymes such as the Polycomb Group (PcG) proteins and histone demethylases [92–94].

The most notable epigenetic regulator that is elevated in NEPCa is EZH2 [92–94]. EZH2 is the catalytic subunit of the Polycomb repressive complex 2 that regulates histone methylation and silences the expression of target genes. Polycomb proteins are highly expressed in embryonic stem cells and play critical roles in stem cell maintenance and cell lineage determination [95–101]. Aberrant expression of Polycomb proteins can cause misregulation of genes that are involved in controlling cell differentiation, preserving cell identity, and modulating the cell cycle [95–102]. A recent genomic profiling study identified EZH2, together with CBX2, another component of the Polycomb complex, as the most overexpressed genes in NEPCa [92]. In correlation with the elevated expression of EZH2, many target genes of EZH2, including DKK1, a Wnt inhibitor, and some key prostate differentiation genes such as HOXA13 and NKX3-1, are down regulated in NEPCa [87]. Together, these studies suggest that the elevated expression of EZH2 and other epigenetic regulators reprogram the transcriptome in therapy-resistant PCa cells, resulting in cellular plasticity and NE differentiation. In line with this concept, studies have shown that AR signaling suppresses the expression of EZH2 and inversely, androgen deprivation activates the CREB/EZH2 axis, promoting NE differentiation in PCa [103, 104]. Moreover, inhibition of EZH2 induces the re-expression AR and suppresses the growth of NEPCa [105], indicating that epigenetic reprogramming plays an important role in the induction and/or maintenance of the NE phenotype in PCa [106].

Finally, genetic alterations in the tumor suppressor genes RB1 and TP53 frequently occur in NEPCa [87, 107–111]. Loss of RB1, TP53, or concurrent loss of function of both genes, was detected with higher frequency in NEPCa compared with non-NE CRPCa [87]. The functional involvement of RB1 and TP53 in NEPCa has been well established. In genetically engineered mice, inactivation of Trp53 and Rb, either by the expression of SV40 T-antigen [112] or through genetic deletions of both Trp53 and Rb alleles [113], is sufficient to cause metastatic PCa with NE features. It is noteworthy that loss of RB1 and TP53 leads to the induction of EZH2 [114–117]. For example, studies have found that the expression of EZH2 is associated with inactivation of TP53/RB1 in SCLC and breast cancer [114, 117]. EZH2 is induced by altered E2F/Rb [116, 117], and TP53 binds directly to EZH2 promoter and suppresses its expression [115]. This suggests that the loss of function of RB1 and TP53 is mediated by epigenetic reprogramming in NEPCa.

Conclusion

In summary, ADT is the gold standard for treating advanced PCa. However, over time, these tumors inevitably become resistant to androgen deprivation and begin to regrow. After ADT fails, there is often an increase in the aggressive NE phenotype. It is suggested that epigenetic reprogramming plays a role in inducing cellular plasticity and the acquisition of NE features. During this process, stem cell signaling pathways can be re-activated and become involved in the induction of NE differentiation. In the following section, we will introduce a key stem cell signaling pathway, Wnt/β-Catenin signaling, and discuss how it contributes to PCa progression.

Wnt/Beta-Catenin Signaling Pathway Overview

One of the mechanisms that PCa cells use to bypass their dependency on AR signaling, resist chemotherapy, and progress to CRPCa/NEPCa is the activation of Wnt/β-Catenin signaling, which plays important roles in embryonic development and carcinogenesis of several types of cancer. Accumulating evidence indicates that Wnt/β-Catenin signaling is active in late stage PCa and promotes castrate-resistant growth [35]. This section will provide an overview of the components of the Wnt/β-Catenin pathway (Fig. 1).

Wnts are a family of secreted cysteine-rich glycoproteins that regulate cell fate determination, cell proliferation, and differentiation [118]. The binding of Wnt ligands to cell surface receptors activates signal transduction pathways. Transmembrane Frizzled (FZD) receptors and co-receptor Low-Density Lipoprotein Receptor-Related Proteins (LRPs) mediate Wnt signaling. Transmembrane E3 ligases, ZNRF3 and RNF43, promote the lysosomal degradation of the FZD receptors, negatively regulating the Wnt pathway. R-spondins, secreted proteins that are positive regulators of Wnt signaling, bind to Leucine Rich Repeat Containing G Protein-Coupled Receptors

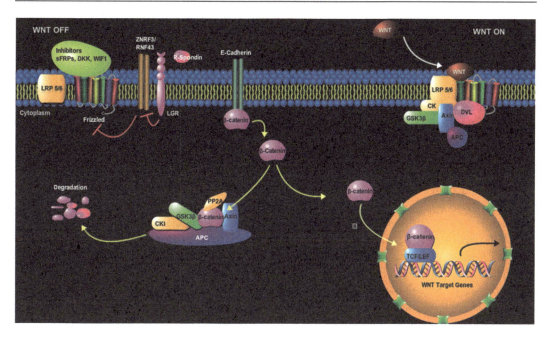

Fig. 1 Wnt/β-Catenin signaling. In the absence of Wnts, cytoplasmic β-Catenin is degraded through the proteasome. Wnt signaling prevents β-Catenin degradation and causes the cytoplasmic and nuclear accumulation of β-Catenin. In the nucleus, β-Catenin functions as a co-activator to activate the transcription of TCF/LEF targets, such as Myc and cyclin D1, as well as AR target genes

(LGRs) and inhibit the activities of the ZNRF3/RNF43 E3 ligases, leading to increased expression of FZD receptors on the cell surface and enhanced activation of Wnt signaling. The Wnt pathway is also modulated by endogenously secreted inhibitors including sFRPs, DKKs, and WIF [119]. These antagonists act on the Wnt ligands or the Wnt receptors/co-receptors to prevent the activation of the signaling pathway.

Activation of the Wnt pathway involves the canonical Wnt/β-Catenin pathway and the non-canonical Wnt/Ca^{2+} and Wnt/polarity pathways [118]. In the canonical pathway, β-Catenin is the main mediator of Wnt signaling in the nucleus. This section will focus on the involvement of canonical Wnt/β-Catenin pathway in PCa progression.

A hallmark of active Wnt/β-Catenin signaling is the stabilization of β-Catenin levels in the cell [120]. β-Catenin exists in two cellular pools: the membrane-bound pool and the cytoplasmic/nuclear pool. On the cell membrane, β-Catenin is a component of the E-Cadherin/α-Catenin/β-Catenin complex, which is an E-Cadherin-based adherens junction at the cell surface. E-Cadherin negatively regulates cytoplasmic/nuclear β-Catenin signaling by sequestering adherens junction-bound β-Catenin to the cell membrane. When the E-Cadherin complex is internalized or when its expression level is reduced, such as when the cells are induced to undergo EMT, β-Catenin is released to the cytoplasm. Tyrosine phosphorylation of β-Catenin can also release β-Catenin from the cell membrane by inducing the breakdown of the E-Cadherin/α-Catenin/β-Catenin complex [121, 122].

In the absence of Wnt signaling, cytoplasmic β-Catenin is rapidly degraded by a destruction complex that contains the adenomatous polyposis coli (APC) protein, AXIN, Casein Kinase, β-Transducin-Repeat-Containing Protein (β-TrCP), and Glycogen Synthase Kinase 3 (GSK3). YAP and TAZ, two transcriptional regulators of the Hippo pathway, are also involved in this destruction complex. In the absence of Wnts, β-Catenin is phosphorylated by GSK3 after first being primed through phosphorylation by Casein Kinase in the degradation complex. YAP/TAZ recruit β-TrCP to the destruction complex, causing the ubiquitination and subsequent degradation of β-Catenin by proteasomes. This constitutive proteasomal degradation of β-Catenin prevents its accumulation in the cytoplasm and maintains low cellular β-Catenin levels [118].

In the presence of the Wnt signal, the Wnt ligand forms a complex with cell-surface receptor FZD and co-receptor LRP5/6. LRP5/6 are phosphorylated by Casein Kinase and GSK3β. Dishevelled proteins, important regulators of Wnt/β-Catenin signaling, are then recruited to the cell membrane where they are hyperphosphorylated and bind to the FZD receptors. This results in the dissociation of the β-Catenin destruction complex, the displacement of YAP/TAZ and β-TrCP from the AXIN complex, and the inactivation of GSK3, all of which impairs the phosphorylation, ubiquitination, and proteasomal degradation of β-Catenin. The inactivation of the degradation complex causes cytoplasmic accumulation and nuclear translocation of β-Catenin [118].

In the nucleus, β-Catenin does not directly bind to DNA. Instead, it acts as a transcriptional co-activator by replacing co-repressor protein Groucho and forming a complex with transcription factors T-Cell Factors (TCFs) and Lymphoid Enhancer-Binding Factor 1 (LEF1). Many other co-factors are recruited to the TCF/β-Catenin complex such as Pygopus proteins and histone modification factors CBP/P300 to activate the transcription of downstream target genes, including ABCB1, AXIN2, CD44, CCND1, ISL1, LEF1, LGR5, MYC, MYCN, NEUROG1, NEUROD1, PLAU, SOX2, SUZ12, TWIST, and YAP [123]. Through these genes, many of which are oncogenes, aberrant Wnt/β-Catenin signaling contributes to the development of cancers. In the prostate, β-Catenin is also a co-activator of AR. It physically interacts with AR and enhances androgen-stimulated transcription of AR target genes [124]. In the next section, we will explore the role of β-Catenin in AR signaling and, more generally, its role in PCa progression.

Wnt/Beta-Catenin Signaling Is Involved in Prostate Carcinogenesis and Cancer Progression

Wnt/β-Catenin Signaling and Cancer

Wnts and their downstream effectors regulate multiple processes important for carcinogenesis including cancer initiation, tumor growth, and cell death. Active Wnt/β-Catenin signaling also promotes cancer progression through the induction of EMT, cancer metastasis, therapy resistance, and genome instability [118, 125].

Though Wnt genetic alterations have not been found to be directly involved, mutations of Wnt pathway *components* occur frequently in cancer. For example, APC, which binds AXIN and β-Catenin in the β-Catenin destruction complex, is mutated in more than 80% of sporadic colorectal adenomas and carcinomas [126]. In a mouse model, adenomas regressed to normal tissues once APC function was restored, indicating that continuous Wnt signaling is required for tumor maintenance [127]. In addition to APC, mutations in other Wnt/β-Catenin component genes such as RNF43 and R-spondin3 are often detected in colorectal cancer [128, 129].

Wnt/β-Catenin activity is also substantially increased in most leukemias and drives tumor development. In acute lymphoblastic leukemia (ALL), canonical Wnt signaling drives tumorigenesis in a subset of T-cell ALL. In acute myelogenous leukemia (AML) mouse models, active Wnt/β-Catenin signaling is essential for the development of cancer stem cells and their self-renewal [130, 131]. In chronic lymphocytic leukemia (CLL), canonical Wnt signaling is active in most CLL cells, often accompanied with somatic mutations in the Wnt pathway [132]. Knockdown of the mutated Wnt pathway proteins reduced the viability of those CLL cells [133], indicating that CLL cells depend on these mutations for survival.

Components of the Wnt/β-Catenin pathway are also frequently found to be altered in breast cancers and activation of this pathway is associated with reduced survival [134, 135]. For example, active Wnt/β-Catenin is enriched in and essential for the development of triple-negative breast cancers [136, 137]. In mice, overexpression of Wnt1 in mammary gland is oncogenic [138]. Also, overexpression of R-spondin2, a positive regulator of Wnt signaling, is sufficient to initiate mammary tumors [139]. Together, these studies show that canonical Wnt signaling contributes to breast cancer development and progression.

Wnt/β-Catenin Signaling in PCa

Accumulating evidence indicates that Wnt/β--Catenin signaling is active in late stage PCa and activation of this signaling pathway is oncogenic, enables castrate-resistant growth, induces EMT, and promotes NE differentiation in PCa cells [140–147]. For example, it has been shown that APC, the scaffold protein for the β-Catenin degradation complex, is among the top genes that are recurrently mutated in metastatic castrate-resistant PCa [28]. In line with this, increased nuclear β-Catenin levels are strongly correlated with metastatic CRPCa [145], indicating that this signaling pathway is active in these late stage tumors. Moreover, a recent study identified Wnt/β-Catenin signaling as a major mechanism for the development of enzalutamide resistance in PCa [148]. Inversely, antagonizing the Wnt pathway through small molecule inhibitors, knocking down β-Catenin, or ectopically expressing Wnt antagonists reverses enzalutamide resistance and inhibits the proliferation and invasion of PCa cells [148–152].

In mouse models, activation of the Wnt/β--Catenin signaling pathway either through the inactivation of APC or the deletion of exon 3 of β-Catenin, which prevents its phosphorylation by GSK3 and subsequent degradation, is oncogenic in prostates [35, 36, 153–155]. In both the APC-null and the β-Catenin exon 3-deleted mice, the tumors with active Wnt/β-Catenin signaling continued cell proliferation after castration, indicating that active Wnt/β-Catenin signaling enables the PCa cells to circumvent their dependence on AR signaling and progress to castrate-resistance. Furthermore, compound activation of the SV40 T-antigen and Wnt/β-Catenin pathways resulted in the development of invasive prostate adenocarcinoma with increased NE differentiation, linking this signaling pathway with the acquisition of the NE phenotype in PCa [156]. Taken together, these studies indicate that active Wnt/β-Catenin signaling provides a mechanism for PCa cells to survive androgen deprivation and progress to CRPCa.

Wnt/β-Catenin Signaling and PCa Metastasis

Metastasis is the major cause of cancer-related death. Cancer cells disseminate from the primary tumor site, invade into stroma, intravasate, travel to distant organs, extravasate, invade the parenchyma of secondary tissues, and establish colonies at distant sites. Wnt/β-Catenin signaling is involved in most steps of the cancer metastasis cascade. At the primary site, activation of Wnt/β--Catenin signaling induces EMT and enables cancer cells to invade. After the cancer cells disseminate, they often enter a dormant state before they develop ways to eventually grow into macroscopic lesions. It has been suggested that active Wnt/β-Catenin signaling provides a mechanism for dormant cancer cells to survive, be released from dormancy, and form macro-metastatic lesions [157, 158].

PCa cells preferentially metastasize to the bone. Wnt/β-Catenin signaling is involved in the establishment of PCa bone metastasis by mediating the reciprocal communication between PCa and bone cells, mainly osteoclasts and osteoblasts. PCa cells secrete factors that stimulate bone resorption mediated by osteoclasts. As osteoclasts resorb the bone matrix, they liberate growth factors, including Wnts, to support PCa growth and stimulate osteoblasts to form new bone. This process is important for creating a 'fertile' environment to support the metastatic PCa cell to escape apoptosis and survive, resulting in a vicious cycle of bone destruction, tumor growth, and new bone formation. High levels of Wnt-1 expression and nuclear β-Catenin were detected in 85% of PCa bone metastases [145]. Wnt7b, the expression of which is not detected in normal prostates, was present in 3/9 primary PCa and 16/38 PCa bone metastasis samples [159]. These data support the concept that Wnt/β--Catenin signaling is involved in PCa bone metastasis.

Wnt/β-Catenin Signaling and Therapy Resistance

Wnt/β-Catenin signaling has also been shown to be involved in the development of therapy resistance and likely endows PCa cells with a selective advantage during ADT. Although earlier studies found that activating mutations of the β-Catenin gene occur in only 5% of primary PCa [160, 161], data from patients who failed ADT indicate that mutations in the Wnt/β-Catenin pathway occur in roughly 18% of castrate-resistant PCa [28, 162]. RNA-seq analyses performed on paired pre- and post-ADT PCa samples have indicated that the Wnt signaling pathway is one of the top pathways that are significantly enriched in post-ADT PCa, containing the largest number of upregulated genes [162].

In a recent study, gene expression data collected from patients with hormone therapy-sensitive and enzalutamide-resistant PCa demonstrated that Wnt/β-Catenin signaling is active in enzalutamide-resistant tumors [148]. Further functional study indicates that activation of Wnt/β-Catenin conferred enzalutamide resistance and inhibition of Wnt/β-Catenin resensitized PCa cells to enzalutamide treatment [148]. In agreement with this, in another study that was conducted using samples from 101 mCRPC patients, Wnt/β-Catenin signaling was identified as the top pathway enriched in enzalutamide-resistant PCa [163]. Genomic DNA sequencing analysis revealed missense mutations in CTNNB1, the β-Catenin encoding gene, in four patients. Although the patient number is small, CTNNB1 mutations are significantly associated with enzalutamide-resistance [163].

Wnt/β-Catenin signaling pathway has also been implicated in PCa's resistance to abiraterone [164]. In a prospective clinical trial, mCRPCa patients were grouped into responders and non-responders based on their response to the treatment of abiraterone acetate/prednisone. Whole-exome sequencing and RNA sequencing were conducted on metastatic biopsies before the tumors developed resistance to the treatment. This study found that 18 out of 32 non-responders harbored mutations in the Wnt/β-Catenin network, compared to a lower mutation frequency (7/41) in responders. Functional annotation analysis revealed that genes which downregulate Wnt/β-Catenin signaling were lost more frequently in non-responders.

A similar phenomenon was also observed in PCa resistance to radiation. Wnt/β-Catenin signaling is one of the top pathways enriched in radio-resistant PCa progenitor cells. Activation of this signaling pathway increases the expression of ALDH1A1, a marker of PCa progenitor cells, and blocking this signaling pathway causes a reduction of the progenitor cell population and re-sensitizes PCa cells to radiation. This points to an essential role of active Wnt/β-Catenin signaling in the maintenance of radio-resistant PCa progenitor cells [165].

Finally, inducing the expression of ATP-binding cassette (ABC) transporters to enhance efflux is one of the major mechanisms of cancer drug resistance. It has been shown that the chemotherapy drug cisplatin induces the expression of canonical Wnt7b, leading to the up-regulation of ABCB1 and ABCG2 transporters [166]. Wnt/β-Catenin inhibition has been shown to decrease the expression of ABCB1 and ABCG2 [166], both of which have been involved in PCa therapy resistance [167, 168]. Taken together, these studies indicate that the activation of Wnt/β-Catenin signaling provides a mechanism for PCa cells to survive and develop resistance to cancer therapies.

Conclusion

Wnt/β-Catenin signaling has been implicated in essentially every stage of PCa progression including the development of CRPCa and therapy resistance. In agreement with data from other types of cancer, multiple studies have shown an upregulation of nuclear β-Catenin expression and Wnt/β-Catenin target genes in metastatic and therapy resistant PCa. With this overview of the importance of Wnt/β-Catenin signaling in PCa, we will next review how this pathway functionally contributes to PCa progression.

Wnt/Beta-Catenin Signaling Functionally Contributes to Prostate Cancer Progression

Animal studies indicate that active Wnt/β-Catenin signaling allows PCa to bypass its dependence on AR signaling. Studies have shown that activation of this signaling pathway promotes NE differentiation in PCa, induces cancer cells to undergo EMT, and contributes to the acquisition of stemness and therapy resistance [140–147]. Through these mechanisms, active Wnt/β-Catenin signaling promotes PCa progression.

Activation of Wnt/β-Catenin Signaling Bypasses PCa's Dependency on AR Signaling

As a central player in mediating CRPCa, AR signaling has been extensively examined to understand how active Wnt/β-Catenin signaling promotes castrate-resistant progression. β-Catenin is an AR co-activator. It physically interacts with AR and increases AR-dependent transactivation [169]. By enhancing the function of AR, β-Catenin makes PCa cells hypersensitive to androgens, resulting in the activation of AR signaling at low, post-castration levels of androgens [170]. Additionally, studies have shown that expression of LEF1, an effector of Wnt/β-Catenin signaling, increases 100-fold when LNCaP cells progress into an androgen-independent state. Moreover, ectopic expression of LEF1 induced AR-mediated cell proliferation and invasion, while knocking down LEF1 decreased AR expression and subsequent cell proliferation and invasion [171]. These studies indicate that active Wnt/β-Catenin signaling provides a mechanism to sustain AR signaling in CRPCa.

In mouse models, the activation of Wnt/β--Catenin signaling was found to increase the expression of AR target genes, as well as AR itself, in PIN lesions developed in 12-week-old dominant active β-Catenin mice [35]. However, as the tumors progressed to high-grade PIN and the tumor cells lost prostate differentiation, both AR and AR signaling were downregulated in the Wnt/β-Catenin active tumors [35]. This indicates that active Wnt/β-Catenin signaling negates the dependence of these tumor cells on AR signaling. Furthermore, it was found that in SV40 large T-antigen and Wnt/β-Catenin compound transgenic mice, there was increased NE differentiation in the invasive PCa that developed, but the expression of AR as well as large T-antigen, which was driven by an AR-responsive Probasin promoter, was reduced in these tumors compared to T-antigen alone tumors [156]. Taken together, these studies suggest that the activation of Wnt/β--Catenin in PCa has dual roles: enhancing AR signaling in PCa cells that still retain prostatic differentiation, but overriding AR signaling and providing a survival mechanism for the cells, such as NEPCa, that have lost prostate differentiation.

Wnt/β-Catenin Signaling Induces NE Differentiation

The Wnt/β-Catenin pathway is implicated in the acquisition of the NE phenotype in PCa. In prostates, active Wnt/β-Catenin signaling induces the expression of Foxa2 [35], a transcriptional factor that is expressed in NEPCa [90, 172, 173]. Additionally, the expression of other Wnt/β--Catenin target genes such as SOX2 [54, 174], CD44 [175–177], and MYCN [85, 178] is often increased in NEPCa, further supporting the involvement of Wnt/β-Catenin signaling in NEPCa. Furthermore, studies have shown that the elevated expression of these downstream targets of Wnt/β-Catenin signaling, such as FOXA2 [72], SOX2 [70], and MYCN [67, 68], could facilitate or lead to NEPCa progression, supporting the functional role of this signaling pathway in NEPCa.

In line with this concept, animal studies indicate that the activation of Wnt/β-Catenin signaling promotes the NE phenotype in PCa. When dominant active β-Catenin mice were bred with 12T-7S mice, which are SV40 large T-antigen transgenic mice that display slow tumor progression and develop PIN but not NEPCa in their lifetimes, the compound β-Catenin/T-antigen mice developed invasive PCa with increased NE differentiation [156]. This indicates a NE-promoting role of Wnt/β-Catenin signaling. A similar observation was reported in the human LNCaP PCa cell line, where expression of stabilized β-Catenin induced the expression of NE markers, NSE and chromogranin A [66, 179]. Inversely, blocking Wnt/β-Catenin by transfection of LNCaP cells with dominant negative TCF or with siRNA against β-Catenin attenuated NE differentiation induced by androgen deprivation or by the expression of Protocadherin-PC [66]. Together, these studies indicate that Wnt/β-Catenin signaling is functionally involved in NE differentiation in PCa.

Wnt/β-Catenin Signaling and Epithelial to Mesenchymal Transition

An important phenomenon in cancer metastasis is EMT. Studies have shown that EMT is associated with poor clinical outcomes in multiple types of cancers including PCa. During EMT, cancer cells lose epithelial characteristics, including cell polarity and the cell-cell adhesion complex that hold epithelial cells together, and acquire mesenchymal features, including spindle-shaped morphology, increased expression of mesenchymal genes such as N-Cadherin and Vimentin, and increased cell motility and invasiveness [180].

In addition to the increased mobility and invasive ability, cancer cells that undergo EMT display stem cell features and are more resistant to therapy [180–182]. For example, it has been shown that breast cancer cells that survive conventional therapies express many EMT-associated genes and exhibit stem cell features [183]. It has also been shown that EMT confers resistance to UV-induced apoptosis in mammary epithelial cells [182]. Additionally, lineage tracing experiments indicate that cells which undergo EMT are enriched in recurrent lung metastases after chemotherapy [184], suggesting that EMT cells have a selective advantage under the pressure of chemotherapy. In pancreatic cancer mouse models, suppression of EMT sensitizes tumors to gemcitabine treatment [185]. Moreover, inducing EMT generates stem-like cells, and these EMT-derived cells have the capacity to transdifferentiate into multiple mesodermal lineages [186, 187]. This suggests that the EMT-inducing signals confer a reprogramming of the transcriptome and activate the expression of stemness genes in these cancer cells, which could lead to their transdifferentiation into new lineages.

Wnt/β-Catenin signaling is implicated in EMT in several aspects. First, there is a multifaceted cross talk between Wnt/β-Catenin signaling and the loss of E-Cadherin, a critical step in EMT. β-Catenin is involved in both the adherens junction and Wnt signaling. The β-Catenin protein shuttles between the membrane and cytoplasm pools. During EMT, the loss of E-Cadherin releases β-Catenin from the membrane-bound adherens complex, resulting in an increase in the cytoplasmic and nuclear β-Catenin levels and the induction of Wnt/β-Catenin signaling [188]. Conversely, active Wnt/β-Catenin signaling induces the expression of transcription factors that promote EMT such as SNAIL, SLUG, and TWIST. These in turn downregulate the expression of E-Cadherin and activate the expression of N-Cadherin, resulting in the activation of EMT reprogramming [189–192].

Moreover, active Wnt/β-Catenin signaling induces the expression of proteases that degrade the extracellular matrix as well as adhesion molecules. For example, MMP7, a membrane type matrix metalloproteinase that is involved in inducing EMT and correlates with PCa pathological stage and PCa progression, is a direct down-stream target of Wnt/β-Catenin signaling [193]. Also, LEF-1, one of the downstream effectors of Wnt/β-Catenin signaling, induces EMT [194]. Together, these studies indicate that Wnt/β-Catenin signaling helps regulate and promote EMT.

Wnt/β-Catenin Signaling and Cancer Stem Cells

Late stage cancer cells resemble stem cells in several aspects including self-renewal ability, differentiation plasticity, and resistance to therapies. Master stem cell pathways such as Notch, Wnt, and Hedgehog that are involved in regulating the balance between stemness and differentiation in stem cells are frequently altered in late stage cancers, including PCa [195]. Although the cancer stem cell theory has not been fully tested, dysregulation of these stem cell pathway could induce a stem-like state in cancer cells and enable them to evade anti-cancer treatment.

There are many pieces of evidence pointing to the critical role of Wnt/β-Catenin signaling in stem cell self-renewal [196, 197] and the maintenance of adult stem cells [198]. This signaling pathway is also involved in somatic cell reprogramming [199, 200]. For example, it has been well established that the combined expression of Oct4, Sox2, c-Myc, and Klf4 induces adult tissue cells to become pluripotent stem cells. Wnt/β-Catenin signaling has been shown to cross talk with these reprogramming factors and enhance the reprogramming of somatic cells [199, 200]. One important feature of pluripotency is the telomere extension by telomerase. β-catenin has been shown to directly regulate the expression of TERT, the protein subunit of the telomerase complex [201]. Additionally, many other stem cell markers such as Sox2, Lgr5, CD133, and CD44 are targets of or regulated by Wnt/β-Catenin signaling [201–203]. All these studies point to the connection of Wnt/β-Catenin signaling with stem cell features.

In the prostate, Wnt/β-Catenin signaling targets are also associated with stem and progenitor cell features. For example, Lgr5, which is expressed in prostate stem/progenitor cells, is a positive regulator and downstream target of Wnt/β-Catenin [204]. Axin2, a direct target of Wnt/β-Catenin signaling, is co-expressed with the prostate progenitor cell marker, Sca-1 [205]. Taken together, these studies indicate an important role of Wnt/β-Catenin signaling in stem cell biology.

In many cancers, Wnt/β-Catenin is functionally involved in promoting the acquisition of stemness, which is believed to contribute to EMT and therapy resistance [206]. Active Wnt/β-Catenin signaling promotes the self-renewal of hematopoietic stem cells and leukemic stem and progenitor cells in acute myeloid leukemia [207, 208]. In PCa, Wnt3a treatment increases the population of cancer stem cells, spheroid formation, and the self-renewal ability of PCa cells [209]. Additionally, high TERT-expressing cells, which are enriched in the cancer stem cell population of PCa, display nuclear β-Catenin and elevated expression levels of Wnt/β-Catenin target genes, including Axin2, c-Myc, and cyclin D1 [210]. Moreover, this study found that the expression of β-Catenin is essential for TERT-mediated cancer stemness and therapy resistance [210], indicating that Wnt/β-Catenin signaling plays a role in promoting the stemness of PCa cells and that targeting Wnt/β-Catenin offers a way to overcome therapy resistance.

Importantly, accumulating evidence indicates that NEPCa tumors have stem cell features [211]. Genes that are expressed in stem cells (SOX2, OCT3/4, MYC, and CD44) or during embryonic prostate development (FOXA2 and SOX2) are often activated in NEPCa [54, 91, 176, 212, 213]. A similar connection of NE cells with stemness is also observed in small cell lung cancer (SCLC). SCLC cells are located in the lung stem cell niche and displayed stem cell features [214]. Given the discussed importance of the Wnt/β-Catenin pathway in NE differentiation, the stem cell features of NEPCa provide further indirect evidence of the involvement of Wnt/β-Catenin signaling in promoting stemness in PCa.

As for the mechanisms that drive the acquisition of stemness in cancer cells, a recent study has shown that chemotherapy-induced cellular senescence converts cancer cells into therapy-resistant stem-like cells [215]. This study also showed that the activation of canonical Wnt/β-Catenin signaling is involved in this process [215]. They found that cells undergoing senescence acquired stem cell features, expressed the signature genes of stem cells, and gained higher clonogenicity. Canonical Wnt/β-Catenin signaling was activated in these stem-like cells and blocking Wnt/β-Catenin diminished the enhanced tumor-initiating capacity, further indicating that the activation of Wnt/β-Catenin signaling is essential for maintaining the stemness in these therapy-resistant cells.

Taken together, these studies suggest that active Wnt/β-Catenin signaling, along with other key stem cell genes, enables cancer cells to transit into a stem-like state. This not only allows cancer cells to survive and accumulate additional genetic and epigenetic alterations that drive cancer progression, such as the loss of tumor suppressor genes RB1 and TP53, but also gives cancer cells the capacity to differentiate into different lineages and evade cancer therapies.

Mechanisms that Activate Wnt/Beta-Catenin Signaling in Prostate Cancer

Understanding the mechanisms through which the Wnt/β-Catenin signaling pathway is activated in PCa cells may help identify new targets for PCa therapy. In addition to mutations in the Wnt/β-Catenin pathway, such as mutations in APC, β-Catenin, and components of the β-Catenin destruction complex, other regulatory mechanisms appear to play important roles in its activation as well. These include androgen deprivation, reactive stroma, cross talk with other pathways, and abnormal expression of Wnt ligands, receptors, and inhibitors. In this section, we will review some of these mechanisms.

Activating Mutations of Wnt/β--Catenin Components in PCa

In PCa, mutations that activate the Wnt/β-Catenin signaling pathway are rare in primary tumors but enriched in late stage PCa. In a whole-exome and transcriptome sequencing study conducted on a cohort of 150 mCRPC specimens, 18% of CRPCa tumors carried mutations in the component genes of the Wnt/β-Catenin signaling pathway, including mutations in APC, β-Catenin, R-spondin, and RNF43 and ZNRF3, the ligases involved in the

ubiquitination and degradation of Wnt receptors [216]. In another study, whole exome sequencing was conducted on 1013 PCa specimens [217]. Because of the large patient number, this study was able to identify mutations of low frequency. The study found that mutations in the Wnt pathways occur in 10% of PCa specimens, including mutations in APC (5%), CTNNB1 (3%), RNF43 (1.3%), ZNRF3 (0.9%), AXIN1 (0.6%) and AXIN2 (0.5%), key components of the β-Catenin degradation complex. Loss-of-function mutations occurred in APC and RNF43, and missense mutations occurred in ZNRF3 and the N-terminal domain of CTNNB1. Overall, mutations in the Wnt/β-Catenin pathway were enriched in metastatic PCa (19%) compared to primary tumors (6%). Moreover, activating mutations in the β-Catenin gene were found to be enriched in enzalutamide-resistant PCa [163, 218]. Together, these studies suggest that activation of Wnt/β--Catenin signaling provides a mechanism that drives PCa progression.

Loss of Wnt/β-Catenin Signaling Inhibition

The activity of Wnt/β-Catenin signaling is modulated by secreted endogenous inhibitors, including sFRPs, DKKs, and WIF [119]. These inhibitors block the interactions between Wnt ligands and their receptors or co-receptors. Expression of these inhibitors is frequently downregulated in PCa [149, 219–221]. Promoter methylation is a major mechanism for silencing the expression of the genes that encode these inhibitors. For example, promoter methylation of sFRP1 and DKK3 was detected in 83% and 68% of PCa, respectively [149]. Concurrent with promoter methylation, sFRP1 protein expression was downregulated or lost in 29 of 39 PCa cases [149]. The promoter of sFRP2 was found to be hypermethylated in PCa versus benign tissues [220]. Inversely, ectopic expression of sFRPs, DKKs, or WIF1 in PCa cells suppresses cell proliferation, invasion and EMT [149–152].

Despite these studies, it is important to note that secreted Wnt inhibitors also sometimes function to *activate* Wnt/β-Catenin signaling. For example, sFRP1 has been shown to function as a pro-proliferative signal in prostate epithelial cells [222]. Moreover, although the expression of sFRP2 is decreased or lost in low grade PCa, moderate to strong sFRP2 expression has been detected in a subset of high grade PCa [223]. Nonetheless, decreased expression of these endogenous inhibitors could provide a mechanism to activate Wnt/β-Catenin signaling in PCa.

Reactive Stroma Activates Wnt/β–Catenin Signaling

Reactive stroma is associated with PCa in the tumor microenvironment. Recent studies have indicated that the growth factors and inflammatory factors secreted by fibroblasts and macrophages in the microenvironment activate Wnt/β-Catenin signaling. For example, carcinoma-associated fibroblasts (CAFs) produce Wnt ligands, growth factors, prostaglandins, and chemokines to directly or indirectly activate Wnt/β-Catenin signaling in epithelial cells [224, 225]. Immune cells such as macrophages in the tumor microenvironment can also secret growth factors and cytokines that activate Wnt/β-Catenin signaling and promote cancer progression [226, 227].

Additionally, therapy-induced inflammatory responses in the tumor microenvironment provide another mechanism to activate Wnt/β--Catenin signaling. One study identified a mechanism connecting treatment-induced tissue damage with therapy-resistance. It was found that cytotoxic therapy-induced tissue damage promotes the NF-κB-dependent expression of WNT16B in stromal cells, which in turn activated canonical Wnt signaling in PCa epithelial cells, promoted the survival of cancer cells, and conferred chemotherapy resistance.

Androgen Deprivation Activates Wnt/β-Catenin Signaling

Recent data from clinical studies support the hypothesis that inhibiting AR signaling promotes Wnt/β-Catenin signaling in PCa [146, 156]. For example, nuclear β-Catenin was detected in 10 of 27 metastatic castrate-resistant PCa cases and the nuclear expression of β-Catenin was inversely associated with AR expression, indicating that Wnt/β-Catenin signaling is active in these AR-negative tumors and that androgen deprivation possibly activates Wnt/β-catenin signaling in PCa [146]. This is further supported by animal studies that indicate androgen withdrawal induces the expression of canonical Wnts in the prostate [228]. Additionally, Axin2, a direct target of Wnt/β-Catenin signaling, is induced in the prostate by castration but repressed by androgen replacement [229]. Moreover, it has been shown that activation of Wnt/β-Catenin signaling is associated with resistance to enzalutamide. In a study that characterized the molecular features of PCa cells which survive enzalutamide treatment, AR and Wnt/β-Catenin were identified as the top pathways activated in the resistant cells [230]. Consistent with this, another study found that when LNCaP cells were grafted *in vivo*, castration of the host mice increased the mRNA levels of both AR and β-Catenin [231]. Similarly, the protein levels of cytoplasmic and nuclear β-Catenin were elevated in LNCaP tumors after castration and further increased in the tumors that developed resistance to castration. Concurrent with the nuclear localization of β-Catenin, the expression of multiple Wnt component genes was altered, including increased expression of canonical WNT2B as well as YES1 and LYN, tyrosine kinases that phosphorylate β-Catenin, and decreased expression of Wnt inhibitory genes [231]. Furthermore, it has been shown that blocking Wnt/β-Catenin signaling sensitizes therapy-resistant LNCaP-abl cells to enzalutamide [232]. Taken together, these studies indicate that Wnt/β-Catenin signaling is active in PCa cells which have survived androgen deprivation and targeting Wnt/β-Catenin offers a treatment option for CRPCa.

A possible mechanism underlying the activation of Wnt/β-Catenin signaling after androgen deprivation is the competition between AR and Wnt/β-Catenin signaling. In the presence of active AR signaling, AR competes with TCF for interaction with β-Catenin in the nucleus, sequestering β-Catenin and resulting in low Wnt/β-Catenin signaling activity in PCa cells [233–236]. After androgen deprivation, β-Catenin is released from the AR complex, resulting in its increased interaction with TCF and elevated expression of Wnt/β-Catenin target genes.

It must be noted that androgen deprivation has also been found to *suppress* Wnt/β-Catenin signaling by inducing the expression of Wnt inhibitory genes like sFRP1 [237]. Furthermore, it has been found that the Wnt/β-Catenin pathway is activated during androgen-induced prostate regeneration [238], suggesting a stimulatory role of AR on Wnt/β-Catenin signaling. Given the complexity of the Wnt/β-Catenin pathway, as well as the different gene expression and biological readouts that were used in these studies, it is possible that androgen deprivation modulates Wnt/β-Catenin signaling in a context-dependent manner. Further studies are needed to decipher whether androgen deprivation modulates Wnt/β--Catenin signaling differently at various stages of PCa progression.

Activation of Wnt/β-Catenin Signaling via Cross Talk with Other Signaling Pathways

In addition to the presence of activating mutations in genes that positively regulate the Wnt pathway and the downregulation of endogenous inhibitors, Wnt/β-Catenin signaling can also be activated via a number of other mechanisms, including cross talk with the PTEN/Akt, COX-2/PGE2, TGF-β, NF-κB, Hippo pathways, and SOX9 [226, 239, 240]. The following is a brief discussion of the PTEN/Akt pathway.

Loss of function mutations of phosphatase and tensin homolog (PTEN) can activate Wnt/β--Catenin signaling. PTEN is a tumor suppressor gene that encodes for a lipid phosphatase that counteracts the activity of PI3K and negatively regulates PI3K/Akt signaling. Loss of PTEN occurs frequently in late-stage PCa, and loss of PTEN or activation of PI3K signaling can activate the Wnt/β-Catenin pathway [241, 242]. Nuclear β-Catenin, a hallmark of active Wnt/β-Catenin, is associated with a PTEN-null phenotype in PCa cells, and restoration of PTEN in these cells suppresses Wnt/β-Catenin signaling [242].

Animal studies have shown that nuclear expression of β-Catenin is increased in intestinal tumors in PTEN null mice [243]. Similarly, mice that have PTEN knocked out in prostates develop PCa and display increased Wnt/β-Catenin signaling [242]. Moreover, overexpression of stabilized β-Catenin in combination with PTEN loss promotes the development of invasive PCa [155, 244]. However, deletion of β-Catenin in PTEN KO prostates did not prevent disease progression [155]. Thus, a detailed mechanism that explains how the interaction contributes to PCa progression remains unclear. Nevertheless, these data suggest that loss of PTEN promotes Wnt/β--Catenin signaling.

Loss of YAP/TAZ Expression Activates Wnt/β-Catenin Signaling

YAP/TAZ are transcriptional co-activators and the main effectors of the Hippo signaling pathway [245]. The Hippo signaling pathway is involved in regulating organ size by modulating multiple cellular functions, including cell proliferation and apoptosis. The Hippo pathway responds to a variety of signals, including cell-cell contact, mechano-transduction, cell polarity and growth factors. YAP/TAZ are usually inhibited by cell-cell contact in normal tissues. When the Hippo pathway is activated, canonical kinases MST1/2 and LATS1/2 mediate the phosphorylation and inactivation of YAP and TAZ. Upon phosphorylation by MST and LATS kinases, YAP and TAZ are sequestered in the cytoplasm, ubiquitylated by β-TrCP ubiquitin ligase, and marked for degradation by the proteasome (Fig. 2).

Over-activation of YAP/TAZ through aberrant regulation of the Hippo pathway has been noted in many solid tumors and associated with the acquisition of malignant traits, including resistance to anticancer therapies, maintenance of cancer stem cells, distant metastasis, and, in prostates, adenocarcinoma progression. When the Hippo core

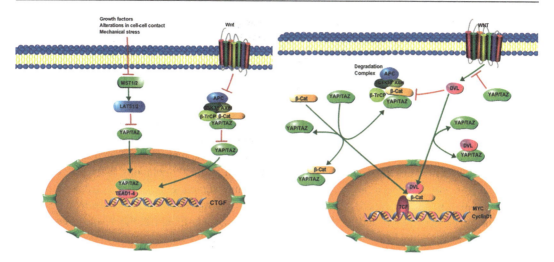

Fig. 2 Cross talk of the Hippo pathway with Wnt/β--Catenin signaling. (**a**) YAP and TAZ as effectors of Hippo signaling. In the Hippo-off condition, YAP and TAZ are stabilized and translocate to the nucleus. In the nucleus, YAP and TAZ interact with TEADs to activate the transcription of target genes such as CTGF and CYR61. (**b**) Mechanisms by which YAP and TAZ suppress Wnt/β--Catenin signaling. YAP/TAZ are essential for the recruitment of β-TrCP to the β-Catenin degradation complex. Also, YAP/TAZ inhibit Wnt-mediated phosphorylation of DVL and the subsequent dissociation of β-Catenin degradation complex. Furthermore, YAP/TAZ retain β-Catenin and DVL in the cytoplasm, inhibiting the activation of Wnt/β-Catenin signaling

kinases are "off," YAP/TAZ translocate into the nucleus, bind to TEAD1–4, which are four paralogous transcription factors, and activate the transcription of TEAD downstream target genes. YAP/TAZ also cross talk with Wnt, MAPK, Hedgehog, AR, and Notch pathways, leading to multiple oncogenic activities, including the loss of contact inhibition, increased cell proliferation, EMT, and resistance to apoptosis [245].

However, the role of YAP/TAZ in cancer progression is highly complex, as revealed through their many differing functions in different cancers [245]. While the pro-oncogenic function of YAP/TAZ has been well established in multiple types of cancer, YAP functions as a tumor suppressor in multiple myeloma and leukemias by regulating the apoptosis of cancer cells and is often found downregulated. In liver cancer, both activation and inactivation of YAP are oncogenic [246]. In non-small cell lung cancer, YAP is overexpressed, but in small-cell lung cancer, YAP expression is reduced [247].

In PCa, YAP is a binding partner of AR and promotes CRPCa [248]. YAP is up-regulated in LNCaP cells and, when expressed ectopically, activates AR signaling and confers castrate-resistance. Knocking down YAP greatly reduces the rates of migration and invasion of LNCaP cells and blocks cell division under androgen deprivation [249]. However, in NEPCa, YAP might play an inhibitory role. We recently found that, similar to the decreased levels of YAP expression in SCLC, the expression of YAP and TAZ is reduced in both human and mouse NEPCa. This suggests that YAP and TAZ function to suppress the progression of prostate adenocarcinoma to NEPCa and that loss of YAP/TAZ expression promotes the development of the NE phenotype in PCa.

The complexity of the role of YAP/TAZ in PCa may be explained by the multi-faceted cross talk between the Hippo pathway and Wnt/β--Catenin signaling. On the one hand, YAP and TAZ function as the positive downstream effectors of Wnt/β-Catenin signaling [250]. In the absence of Wnt signaling, cytoplasmic YAP/TAZ bind to the AXIN protein, a component of the β-Catenin destruction complex, and aid in recruiting β-TrCP ubiquitin ligase to induce β-Catenin degradation. The complex in turn serves as a cytoplasmic sink for YAP/TAZ and blocks their transcriptional activities. In the presence of Wnt signaling, YAP/TAZ are released from the β-Catenin degradation complex, which in turn blocks the recruitment of β-TrCP to the complex. This impairs the degradation of both β-Catenin and YAP/TAZ, leading to the activation of Wnt/β-Catenin signaling as well as the nuclear accumulation of YAP/TAZ and the activation of YAP/TAZ-dependent transcription, a branch of Wnt transcriptional effects that are independent of the Hippo signaling pathway [250]. In the nucleus, YAP can also interact with β-Catenin and function to activate the transcription of Wnt target genes [251, 252].

On the other hand, YAP and TAZ suppress Wnt/β-Catenin signaling via multiple mechanisms (Fig. 2). First, YAP and TAZ retain β-Catenin in the cytoplasm and inhibit the activation of Wnt/β-Catenin signaling [253]. Second, TAZ binds to and inhibits the activity of Dishevelled (DVL) proteins, crucial regulators of the Wnt/β-Catenin pathway. The binding of TAZ to DVL inhibits Wnt-induced phosphorylation of DVL, stabilizing the β-Catenin degradation complex and inhibiting Wnt/β-Catenin signaling [254]. Third, YAP and TAZ sequester DVL in the cytoplasm, resulting in the suppression of nuclear Wnt/β-Catenin signaling. Fourth, YAP and TAZ are essential components of the β-Catenin destruction complex and facilitate the degradation of β-Catenin. Loss of YAP and TAZ could enable β-Catenin to escape proteasomal degradation.

In line with the role of YAP and TAZ in repressing Wnt/β-Catenin signaling, a recent study found that YAP or TAZ knockout induces the expression of Lgr5, a positive regulator and downstream target of Wnt/β-Catenin signaling [255]. Additionally, loss of Yap results in hyperactive Wnt/β-Catenin signaling during intestinal regeneration, whereas ectopic Yap expression dampens Wnt/β-Catenin signaling [256]. Finally, knocking down YAP/TAZ activates Wnt/β--Catenin responsive promoters in HEK293 cells [253]. Taken together, these studies indicate that loss of YAP and TAZ can augment Wnt/β-Catenin signaling. Further research is needed to decipher the role of YAP and TAZ in NEPCa progression.

Targeting Wnt/β-Catenin for the Treatment of CRPCA

Conventional chemotherapy prolongs patient survival, but often only for a short time and accompanied by severe side effects. Targeted therapies have been shown to offer durable clinical responses and fewer side effects in patients selected based on molecular markers. The high frequency of Wnt/β-Catenin pathway mutations, activation of Wnt/β-Catenin signaling in different cancers, and the important role of this signaling pathway in cancer survival, metastasis, and the development of therapy resistance underscore the importance of Wnt-targeted treatment [120, 257, 258].

In PCa, patients can be stratified based on the mutations of Wnt/β-Catenin components, the expression of nuclear β-Catenin, and/or the elevated expression of Wnt/β-Catenin target genes such as FOXA2. Many agents have been developed and tested to block the various mechanisms that activate the Wnt/β-Catenin signaling. Small molecular inhibitors form a major category of drugs used in targeted therapies since they can penetrate the cell membrane and function inside cells. Many of these small-molecule inhibitors have been tested in clinical trials for cancer treatments. These inhibitors are shown to suppress different targets in Wnt/β-Catenin signaling, including Porcupine (PORCN), FZD, AXIN, GSK3, TCF/β-Catenin, and CBP/β-Catenin. In this section, we will briefly discuss some of these small-molecule inhibitors that may prove to be highly effective in targeting Wnt/β-Catenin signaling.

PRI-724: PRI-724 is a second-generation small molecule inhibitor that blocks the interaction of β-Catenin with its co-activator CBP [259, 260]. Clinical safety and efficacy studies on PRI-724 have shown promising results.

Niclosamide: Wnt signaling is activated when the Wnt ligands bind to the seven-transmembrane FZD receptors. It has been shown that niclosamide, an antihelminthic agent, blocks the signaling pathway by depleting upstream FZD and Dishevelled protein levels [261]. Niclosamide exerts its inhibitory action by promoting FZD1 endocytosis, downregulating Dishevelled-2 protein, and suppressing LEF/TCF transcriptional activity. Phase I clinical trials have included a study of best dosage and side effects of niclosamide plus enzalutamide in treating recurrent or metastatic castrate-resistant prostate cancer. A separate phase II trial has initiated a study of niclosamide in combination with abiraterone acetate and prednisone. One of the objectives of this study is to evaluate progression-free survival and the overall survival of castrate-resistant prostate cancer patients treated with the combined drugs.

NSC668036: The dishevelled protein modulates Wnt signaling by binding to the intracellular domain of FZD. NSC668036, an organic molecule, was designed to target the PDZ domain of dishevelled, an essential domain for interacting with FZD, and thereby inhibit the activation of the Wnt signaling pathway [262].

Quercetin: Quercetin (3,3',4',5,7-pentahydro xyflavone) interrupts Wnt signaling by interfering with the formation of the β-Catenin/TCF complex in the nucleus, thereby downregulating the expression of the downstream oncogenic genes [263]. It is a natural bioflavanoid and is abundant in onions, apples, and green tea. It is considered nontoxic and effectively inhibits growth of a variety of cancer cells including breast, colon, lung, and prostate cancers. Quercetin inhibits the proliferation of PCa cell lines by inducing cell cycle arrest in the G2 phase. It also arrests the cell cycle by downregulating cyclin D and E, CDK, and cdc25c, and upregulating p21, p53, p18, and p27. Quercetin also induces apoptosis of PC3 and LNCaP cells by increasing pro-apoptotic Bax and by decreasing anti-apoptotic Bcl-2 proteins. Additionally, quercetin exerts its anti-androgenic effect by inhibiting 5 alpha-reductase and suppressing the androgen receptor (AR) complex in prostate cancer cells. AR expression was found to be significantly decreased in LNCaP cells treated with quercetin in a dose-dependent fashion.

Tegavivint (BC2059): Tegavivint (BC2059), an anthraquinone oxime-analog, is a potent inhibitor of the Wnt/β-Catenin pathway [264]. When acting alone, Tegavivint decreases β-Catenin levels by enhancing its degradation process and attenuating LEF/TCF4 transcriptional activity. As a result, the expression levels of downstream targets such as cyclin D1, MYC are significantly decreased. Synergistic activities can be achieved when Tegavivint is administered in combination with the HDAC inhibitor panobinostat to treat primary acute myelogenous leukemia [264]. Co-treatment with BC2059 and JAK2-targeted tyrosine kinase inhibitor TG101209 induced synergistic apoptotic effect in myeloproliferative neoplasia. Synergism has also been shown in combined treatment of BC2059 with BCR-ABL-targeted tyrosine kinase inhibitor nilotinib in chronic myelogenous leukemia [264]. Phase I clinical trials are currently evaluating the safety of Tegavivint in treating patients with primary and recurrent unresectable desmoid tumors. Symptomatic and progressive desmid tumors are also included in the study.

Calphostin C (PKF115-584): Calphostin C (PKF115-584) inhibits many cancer cell lines by targeting β-Catenin/TCF (LEF) interaction [265] Calphostin disrupts the β-Catenin/TCF4 complex and inhibits binding of TCF proteins to protein kinase C in colon cancer.

Inhibitor of Wnt Production (IWP): Porcupine (PORCN), a membrane-bound O-acyltransferase family, produces functional and secretory Wnt proteins by transferring a palmitoyl group to Wnt ligands. IWP, including IWP-1 and IWP-2, inactivate PORCN function by binding to the critical functional determinant of the benzothiazole group of PORCN. Other IWP associated proteins (IWP-PEG-Biotin, IWP-PB) display similar mechanisms of inhibiting PORCN by competitive binding of benzothiazole group of PORCN to IWP-PB [266]. An alternate route of inhibiting the Wnt/β-Catenin signaling pathway by PORCN inhibitors has been hypothesized. In addition to binding to PORCN, IWPs intervene the signaling pathway by interacting with the isoforms of casein kinase (CK) family [267]. CK1δ/ε isoforms are involved in the Wnt signaling pathway and regulate cell growth by interacting with various proteins including DVL1-3, CTNNB1, AXIN1, and APC. It has been demonstrated that IWP-2 inhibits CK1δ/ε by binding to the selective ATP-binding site of the isoforms and exerting its anticancer activities in various cancer cell lines [267].

LGK974 (WNT974): LGK974 is a potent inhibitor of PORCN and blocks Wnt signaling *in vitro* and *in vivo* [268]. A phase I, dose escalation clinical trial has been initiated to test the efficacy of LGK974 in treating various cancers including pancreatic cancer, melanoma, breast cancer, and lung cancer. A phase II clinical trial of patients with squamous cell carcinoma of the head and neck is also underway.

Wnt-C59: Wnt-C59 targets PORCN and downregulates Wnt/β-Catenin signaling [269]. It is a potent inhibitor that exerts its biologic activity at nanomolar concentrations. Wnt-C59 prevents progression of mammary cancer in MMTV-WNT1 transgenic mice. No prominent cytotoxicity has been found, suggesting that it may be a safe molecule for treating cancer.

Inhibitor of Wnt Response (IWR): The targets of Inhibitor of Wnt Response (IWR) compounds are the tankyrase enzymes, intracytoplasmic poly-ADP-ribosylating enzymes that interact with Axin and regulate its ubiquitination and degradation. IWR compounds inhibit tankyrase enzymes, thereby promoting the accumulation of Axin. The excess Axin binds to the destruction complex and accelerates the degradation process of β-Catenin.

XAV939: XAV939 exerts its potent inhibitory activity on Wnt signaling through suppressing tankyrase 1 and tankyrase 2 [270]. Axin is then stabilized, stimulating the degradation of β-catenin and antagonizing Wnt signaling.

Conclusion

Dysregulation of the Wnt/β-Catenin signaling pathway is involved in the tumorigenesis and progression to therapy resistance of many types of cancer including PCa. Accumulating evidence indicates that epigenetic reprogramming induces cellular plasticity and a stem-like state in PCa cells, promoting EMT and/or the acquisition of NE phenotype. During this process, Wnt/β-- Catenin signaling is activated and enables PCa cells to survive cancer therapies. Small molecules are designed to target specific proteins and regulators of Wnt/β-Catenin signaling. So far,

they have shown promising results of inhibiting tumor cell growth. Clinical trials are currently being conducted in the hope of achieving clinical remission of advanced PCa.

References

1. C. Huggins, Effect of orchiectomy and irradiation on cancer of the prostate. Ann. Surg. **115**, 1192–1200 (1942)
2. M.S. Cookson et al., Castration-resistant prostate cancer: AUA Guideline. J. Urol. **190**, 429–438 (2013)
3. J.D. Debes, D.J. Tindall, Mechanisms of androgen-refractory prostate cancer. N. Engl. J. Med. **351**, 1488–1490 (2004)
4. S.M. Dehm, D.J. Tindall, Molecular regulation of androgen action in prostate cancer. J. Cell. Biochem. **99**, 333–344 (2006)
5. S.M. Dehm, D.J. Tindall, Androgen receptor structural and functional elements: role and regulation in prostate cancer. Mol. Endocrinol. **21**, 2855–2863 (2007)
6. T. Karantanos, P.G. Corn, T.C. Thompson, Prostate cancer progression after androgen deprivation therapy: mechanisms of castrate resistance and novel therapeutic approaches. Oncogene **32**, 5501–5511 (2013)
7. J.L. Bishop et al., The master neural transcription factor BRN2 is an androgen receptor-suppressed driver of neuroendocrine differentiation in prostate cancer. Cancer Discov. **7**, 54–71 (2017)
8. B.J. Feldman, D. Feldman, The development of androgen-independent prostate cancer. Nat. Rev. Cancer **1**, 34–45 (2001)
9. P.A. Watson, V.K. Arora, C.L. Sawyers, Emerging mechanisms of resistance to androgen receptor inhibitors in prostate cancer. Nat. Rev. Cancer **15**, 701–711 (2015)
10. M.J. Linja et al., Amplification and overexpression of androgen receptor gene in hormone-refractory prostate cancer. Cancer Res. **61**, 3550–3555 (2001)
11. C.D. Chen et al., Molecular determinants of resistance to antiandrogen therapy. Nat. Med. **10**, 33–39 (2004)
12. J.L. Mohler et al., The androgen axis in recurrent prostate cancer. Clin. Cancer Res. **10**, 440–448 (2004)
13. J.A. Locke et al., Androgen levels increase by intratumoral de novo steroidogenesis during progression of castration-resistant prostate cancer. Cancer Res. **68**, 6407–6415 (2008)
14. R.B. Montgomery et al., Maintenance of intratumoral androgens in metastatic prostate cancer: a mechanism for castration-resistant tumor growth. Cancer Res. **68**, 4447–4454 (2008)

15. T. Saloniemi, H. Jokela, L. Strauss, P. Pakarinen, M. Poutanen, The diversity of sex steroid action: novel functions of hydroxysteroid (17beta) dehydrogenases as revealed by genetically modified mouse models. J. Endocrinol. **212**, 27–40 (2012)

16. G.A. Potter, S.E. Barrie, M. Jarman, M.G. Rowlands, Novel steroidal inhibitors of human cytochrome P45017 alpha (17 alpha-hydroxylase-C17,20-lyase): potential agents for the treatment of prostatic cancer. J. Med. Chem. **38**, 2463–2471 (1995)

17. X.Y. Zhao et al., Glucocorticoids can promote androgen-independent growth of prostate cancer cells through a mutated androgen receptor. Nat. Med. **6**, 703–706 (2000)

18. R. Hu et al., Ligand-independent androgen receptor variants derived from splicing of cryptic exons signify hormone-refractory prostate cancer. Cancer Res. **69**, 16–22 (2009)

19. Z. Guo et al., A novel androgen receptor splice variant is up-regulated during prostate cancer progression and promotes androgen depletion-resistant growth. Cancer Res. **69**, 2305–2313 (2009)

20. S.C. Chan, Y. Li, S.M. Dehm, Androgen receptor splice variants activate androgen receptor target genes and support aberrant prostate cancer cell growth independent of canonical androgen receptor nuclear localization signal. J. Biol. Chem. **287**, 19736–19749 (2012)

21. S.M. Dehm, L.J. Schmidt, H.V. Heemers, R.L. Vessella, D.J. Tindall, Splicing of a novel androgen receptor exon generates a constitutively active androgen receptor that mediates prostate cancer therapy resistance. Cancer Res. **68**, 5469–5477 (2008)

22. S.M. Dehm, D.J. Tindall, Alternatively spliced androgen receptor variants. Endocr. Relat. Cancer **18**, R183–R196 (2011)

23. C. Henzler et al., Truncation and constitutive activation of the androgen receptor by diverse genomic rearrangements in prostate cancer. Nat. Commun. **7**, 13668 (2016)

24. Y. Li et al., Androgen receptor splice variants mediate enzalutamide resistance in castration-resistant prostate cancer cell lines. Cancer Res. **73**, 483–489 (2013)

25. Y. Ho, S.M. Dehm, Androgen receptor rearrangement and splicing variants in resistance to endocrine therapies in prostate cancer. Endocrinology **158**, 1533–1542 (2017)

26. S.M. Dehm, K.M. Regan, L.J. Schmidt, D.J. Tindall, Selective role of an NH2-terminal WxxLF motif for aberrant androgen receptor activation in androgen depletion independent prostate cancer cells. Cancer Res. **67**, 10067–10077 (2007)

27. M.E. Taplin et al., Mutation of the androgen-receptor gene in metastatic androgen-independent prostate cancer. N. Engl. J. Med. **332**, 1393–1398 (1995)

28. C.S. Grasso et al., The mutational landscape of lethal castration-resistant prostate cancer. Nature **487**, 239–243 (2012)

29. E.S. Antonarakis et al., AR-V7 and resistance to enzalutamide and abiraterone in prostate cancer. N. Engl. J. Med. **371**, 1028–1038 (2014)

30. Q. Wang et al., Androgen receptor regulates a distinct transcription program in androgen-independent prostate cancer. Cell **138**, 245–256 (2009)

31. Z.M. Connelly et al., Foxa2 activates the transcription of androgen receptor target genes in castrate resistant prostatic tumors. Am. J. Clin. Exp. Urol. **6**, 172–181 (2018)

32. C. Tran et al., Development of a second-generation antiandrogen for treatment of advanced prostate cancer. Science **324**, 787–790 (2009)

33. J.S. de Bono et al., Abiraterone and increased survival in metastatic prostate cancer. N. Engl. J. Med. **364**, 1995–2005 (2011)

34. S. Wang et al., Prostate-specific deletion of the murine Pten tumor suppressor gene leads to metastatic prostate cancer. Cancer Cell **4**, 209–221 (2003)

35. X. Yu et al., Activation of beta-Catenin in mouse prostate causes HGPIN and continuous prostate growth after castration. Prostate **69**, 249–262 (2009)

36. K.J. Bruxvoort et al., Inactivation of Apc in the mouse prostate causes prostate carcinoma. Cancer Res. **67**, 2490–2496 (2007)

37. A.H. Davies, H. Beltran, A. Zoubeidi, Cellular plasticity and the neuroendocrine phenotype in prostate cancer. Nat. Rev. Urol. **15**, 271–286 (2018)

38. V.K. Arora et al., Glucocorticoid receptor confers resistance to antiandrogens by bypassing androgen receptor blockade. Cell **155**, 1309–1322 (2013)

39. H. Beltran et al., Aggressive variants of castration-resistant prostate cancer. Clin. Cancer Res. **20**, 2846–2850 (2014)

40. D. Hirano, Y. Okada, S. Minei, Y. Takimoto, N. Nemoto, Neuroendocrine differentiation in hormone refractory prostate cancer following androgen deprivation therapy. Eur. Urol. **45**, 586–92; discussion 592 (2004)

41. R. Nadal, M. Schweizer, O.N. Kryvenko, J.I. Epstein, M.A. Eisenberger, Small cell carcinoma of the prostate. Nat. Rev. Urol. **11**, 213–219 (2014)

42. R. Aggarwal et al., Clinical and genomic characterization of treatment-emergent small-cell neuroendocrine prostate cancer: a multi-institutional prospective study. J. Clin. Oncol. **36**, 2492–2503 (2018)

43. J.I. Epstein et al., Proposed morphologic classification of prostate cancer with neuroendocrine differentiation. Am. J. Surg. Pathol. **38**, 756–767 (2014)

44. X. Deng et al., Ionizing radiation induces prostate cancer neuroendocrine differentiation through interplay of CREB and ATF2: implications for disease progression. Cancer Res. **68**, 9663–9670 (2008)

45. P.A. Abrahamsson, Neuroendocrine differentiation in prostatic carcinoma. Prostate **39**, 135–148 (1999)

46. G. Ahlgren et al., Regressive changes and neuroendocrine differentiation in prostate cancer after neoadjuvant hormonal treatment. Prostate **42**, 274–279 (2000)

47. T. Jiborn, A. Bjartell, P.A. Abrahamsson, Neuroendocrine differentiation in prostatic carcinoma during hormonal treatment. Urology **51**, 585–589 (1998)

48. S. Terry, H. Beltran, The many faces of neuroendocrine differentiation in prostate cancer progression. Front. Oncol. **4**, 60 (2014)

49. H. Beltran et al., Challenges in recognizing treatment-related neuroendocrine prostate cancer. J. Clin. Oncol. **30**, e386–e389 (2012)

50. J.M. Mosquera et al., Concurrent AURKA and MYCN gene amplifications are harbingers of lethal treatment-related neuroendocrine prostate cancer. Neoplasia **15**, 1–10 (2013)

51. J. Huang et al., Immunohistochemical characterization of neuroendocrine cells in prostate cancer. Prostate **66**, 1399–1406 (2006)

52. V. Tzelepi et al., Modeling a lethal prostate cancer variant with small-cell carcinoma features. Clin. Cancer Res. **18**, 666–677 (2012)

53. J.L. Yao et al., Small cell carcinoma of the prostate: an immunohistochemical study. Am. J. Surg. Pathol. **30**, 705–712 (2006)

54. X. Yu et al., SOX2 expression in the developing, adult, as well as, diseased prostate. Prostate Cancer Prostatic Dis. **17**, 301–309 (2014)

55. T.C. Yuan, S. Veeramani, M.F. Lin, Neuroendocrine-like prostate cancer cells: neuroendocrine transdifferentiation of prostate adenocarcinoma cells. Endocr. Relat. Cancer **14**, 531–547 (2007)

56. M.E. Wright, M.J. Tsai, R. Aebersold, Androgen receptor represses the neuroendocrine transdifferentiation process in prostate cancer cells. Mol. Endocrinol. **17**, 1726–1737 (2003)

57. R. Shen et al., Transdifferentiation of cultured human prostate cancer cells to a neuroendocrine cell phenotype in a hormone-depleted medium. Urol. Oncol. **3**, 67–75 (1997)

58. W.J. Huss et al., Origin of androgen-insensitive poorly differentiated tumors in the transgenic adenocarcinoma of mouse prostate model. Neoplasia **9**, 938–950 (2007)

59. J.R. Gingrich et al., Androgen-independent prostate cancer progression in the TRAMP model. Cancer Res. **57**, 4687–4691 (1997)

60. M.A. Johnson et al., Castration triggers growth of previously static androgen-independent lesions in the transgenic adenocarcinoma of the mouse prostate (TRAMP) model. Prostate **62**, 322–338 (2005)

61. D. Lin et al., High fidelity patient-derived xenografts for accelerating prostate cancer discovery and drug development. Cancer Res. **74**, 1272–1283 (2014)

62. M.A. Noordzij et al., Neuroendocrine differentiation in human prostatic tumor models. Am. J. Pathol. **149**, 859–871 (1996)

63. J. Jongsma et al., Kinetics of neuroendocrine differentiation in an androgen-dependent human prostate xenograft model. Am. J. Pathol. **154**, 543–551 (1999)

64. M.E. Cox, P.D. Deeble, E.A. Bissonette, S.J. Parsons, Activated 3′,5′-cyclic AMP-dependent protein kinase is sufficient to induce neuroendocrine-like differentiation of the LNCaP prostate tumor cell line. J. Biol. Chem. **275**, 13812–13818 (2000)

65. P.D. Deeble, D.J. Murphy, S.J. Parsons, M.E. Cox, Interleukin-6- and cyclic AMP-mediated signaling potentiates neuroendocrine differentiation of LNCaP prostate tumor cells. Mol. Cell. Biol. **21**, 8471–8482 (2001)

66. X. Yang et al., A human- and male-specific protocadherin that acts through the wnt signaling pathway to induce neuroendocrine transdifferentiation of prostate cancer cells. Cancer Res. **65**, 5263–5271 (2005)

67. E. Dardenne et al., N-Myc induces an EZH2-mediated transcriptional program driving neuroendocrine prostate cancer. Cancer Cell **30**, 563–577 (2016)

68. J.K. Lee et al., N-Myc drives neuroendocrine prostate cancer initiated from human prostate epithelial cells. Cancer Cell **29**, 536–547 (2016)

69. S.Y. Ku et al., Rb1 and Trp53 cooperate to suppress prostate cancer lineage plasticity, metastasis, and antiandrogen resistance. Science **355**, 78–83 (2017)

70. P. Mu et al., SOX2 promotes lineage plasticity and antiandrogen resistance in TP53- and RB1-deficient prostate cancer. Science **355**, 84–88 (2017)

71. G. Danza et al., Notch signaling modulates hypoxia-induced neuroendocrine differentiation of human prostate cancer cells. Mol. Cancer Res. **10**, 230–238 (2012)

72. J. Qi et al., Siah2-dependent concerted activity of HIF and FoxA2 regulates formation of neuroendocrine phenotype and neuroendocrine prostate tumors. Cancer Cell **18**, 23–38 (2010)

73. S.S. Yadav et al., Induction of neuroendocrine differentiation in prostate cancer cells by dovitinib (TKI-258) and its therapeutic implications. Transl. Oncol. **10**, 357–366 (2017)

74. H. Bonkhoff, N. Wernert, G. Dhom, K. Remberger, Relation of endocrine-paracrine cells to cell proliferation in normal, hyperplastic, and neoplastic human prostate. Prostate **19**, 91–98 (1991)

75. N. Dizeyi et al., Serotonin activates MAP kinase and PI3K/Akt signaling pathways in prostate cancer cell lines. Urol. Oncol. **29**, 436–445 (2011)

76. K. Uchida et al., Murine androgen-independent neuroendocrine carcinoma promotes metastasis of human prostate cancer cell line LNCaP. Prostate **66**, 536–545 (2006)

77. K. Hashimoto et al., The potential of neurotensin secreted from neuroendocrine tumor cells to promote gelsolin-mediated invasiveness of prostate adenocarcinoma cells. Lab. Investig. **95**, 283–295 (2015)

78. R. Grobholz et al., Influence of neuroendocrine tumor cells on proliferation in prostatic carcinoma. Hum. Pathol. **36**, 562–570 (2005)

79. R.J. Jin et al., NE-10 neuroendocrine cancer promotes the LNCaP xenograft growth in castrated mice. Cancer Res. **64**, 5489–5495 (2004)

80. J. Szczyrba et al., Neuroendocrine cells of the prostate derive from the neural crest. J. Biol. Chem. **292**, 2021–2031 (2017)

81. M. Zou et al., Transdifferentiation as a mechanism of treatment resistance in a mouse model of castration-resistant prostate cancer. Cancer Discov. **7**, 736–749 (2017)

82. D.E. Hansel et al., Shared TP53 gene mutation in morphologically and phenotypically distinct concurrent primary small cell neuroendocrine carcinoma and adenocarcinoma of the prostate. Prostate **69**, 603–609 (2009)

83. C.G. Sauer, A. Roemer, R. Grobholz, Genetic analysis of neuroendocrine tumor cells in prostatic carcinoma. Prostate **66**, 227–234 (2006)

84. N.C. Bastus et al., Androgen-induced TMPRSS2:ERG fusion in nonmalignant prostate epithelial cells. Cancer Res. **70**, 9544–9548 (2010)

85. H. Beltran et al., Molecular characterization of neuroendocrine prostate cancer and identification of new drug targets. Cancer Discov. **1**, 487–495 (2011)

86. M.G. Oser, M.J. Niederst, L.V. Sequist, J.A. Engelman, Transformation from non-small-cell lung cancer to small-cell lung cancer: molecular drivers and cells of origin. Lancet Oncol. **16**, e165–e172 (2015)

87. H. Beltran et al., Divergent clonal evolution of castration-resistant neuroendocrine prostate cancer. Nat. Med. **22**, 298–305 (2016)

88. Y. Li et al., SRRM4 drives neuroendocrine transdifferentiation of prostate adenocarcinoma under androgen receptor pathway inhibition. Eur. Urol. **71**, 68–78 (2017)

89. J. Kim et al., FOXA1 inhibits prostate cancer neuroendocrine differentiation. Oncogene **36**, 4072–4080 (2017)

90. J. Mirosevich et al., Expression and role of Foxa proteins in prostate cancer. Prostate **66**, 1013–1028 (2006)

91. J. Mirosevich, N. Gao, R.J. Matusik, Expression of Foxa transcription factors in the developing and adult murine prostate. Prostate **62**, 339–352 (2005)

92. P.L. Clermont et al., Polycomb-mediated silencing in neuroendocrine prostate cancer. Clin. Epigenetics **7**, 40 (2015)

93. T. Sato et al., PRC2 overexpression and PRC2-target gene repression relating to poorer prognosis in small cell lung cancer. Sci. Rep. **3**, 1911 (2013)

94. J.J. Findeis-Hosey et al., High-grade neuroendocrine carcinomas of the lung highly express enhancer of zeste homolog 2, but carcinoids do not. Hum. Pathol. **42**, 867–872 (2011)

95. A.P. Bracken, K. Helin, Polycomb group proteins: navigators of lineage pathways led astray in cancer. Nat. Rev. Cancer **9**, 773–784 (2009)

96. A. Kuzmichev et al., Composition and histone substrates of polycomb repressive group complexes change during cellular differentiation. Proc. Natl. Acad. Sci. U. S. A. **102**, 1859–1864 (2005)

97. H. Richly, L. Aloia, L. Di Croce, Roles of the Polycomb group proteins in stem cells and cancer. Cell Death Dis. **2**, e204 (2011)

98. A. Laugesen, K. Helin, Chromatin repressive complexes in stem cells, development, and cancer. Cell Stem Cell **14**, 735–751 (2014)

99. E. Conway, E. Healy, A.P. Bracken, PRC2 mediated H3K27 methylations in cellular identity and cancer. Curr. Opin. Cell Biol. **37**, 42–48 (2015)

100. P. Vizan, M. Beringer, C. Ballare, L. Di Croce, Role of PRC2-associated factors in stem cells and disease. FEBS J. **282**, 1723–1735 (2015)

101. A.P. Bracken, N. Dietrich, D. Pasini, K.H. Hansen, K. Helin, Genome-wide mapping of Polycomb target genes unravels their roles in cell fate transitions. Genes Dev. **20**, 1123–1136 (2006)

102. H. Chen et al., Polycomb protein Ezh2 regulates pancreatic beta-cell Ink4a/Arf expression and regeneration in diabetes mellitus. Genes Dev. **23**, 975–985 (2009)

103. L.R. Bohrer, S. Chen, T.C. Hallstrom, H. Huang, Androgens suppress EZH2 expression via retinoblastoma (RB) and p130-dependent pathways: a potential mechanism of androgen-refractory progression of prostate cancer. Endocrinology **151**, 5136–5145 (2010)

104. Y. Zhang et al., Androgen deprivation promotes neuroendocrine differentiation and angiogenesis through CREB-EZH2-TSP1 pathway in prostate cancers. Nat. Commun. **9**, 4080 (2018)

105. B. Kleb et al., Differentially methylated genes and androgen receptor re-expression in small cell prostate carcinomas. Epigenetics **11**, 184–193 (2016)

106. A.V. Sheahan, L. Ellis, Epigenetic reprogramming: a key mechanism driving therapeutic resistance. Urol. Oncol. **36**, 375–379 (2018)

107. H.L. Tan et al., Rb loss is characteristic of prostatic small cell neuroendocrine carcinoma. Clin. Cancer Res. **20**, 890–903 (2014)

108. H. Beltran et al., Targeted next-generation sequencing of advanced prostate cancer identifies potential therapeutic targets and disease heterogeneity. Eur. Urol. **63**, 920–926 (2013)

109. M. Peifer et al., Integrative genome analyses identify key somatic driver mutations of small-cell lung cancer. Nat. Genet. **44**, 1104–1110 (2012)

110. H. Chen et al., Pathogenesis of prostatic small cell carcinoma involves the inactivation of the P53 pathway. Endocr. Relat. Cancer **19**, 321–331 (2012)

111. A. Aparicio, R.B. Den, K.E. Knudsen, Time to stratify? The retinoblastoma protein in castrate-resistant prostate cancer. Nat. Rev. Urol. **8**, 562–568 (2011)

112. T. Chiaverotti et al., Dissociation of epithelial and neuroendocrine carcinoma lineages in the transgenic

adenocarcinoma of mouse prostate model of prostate cancer. Am. J. Pathol. **172**, 236–246 (2008)

113. Z. Zhou et al., Synergy of p53 and Rb deficiency in a conditional mouse model for metastatic prostate cancer. Cancer Res. **66**, 7889–7898 (2006)

114. A.M. Pietersen et al., EZH2 and BMI1 inversely correlate with prognosis and TP53 mutation in breast cancer. Breast Cancer Res. **10**, R109 (2008)

115. X. Tang et al., Activated p53 suppresses the histone methyltransferase EZH2 gene. Oncogene **23**, 5759–5769 (2004)

116. A.P. Bracken et al., EZH2 is downstream of the pRB-E2F pathway, essential for proliferation and amplified in cancer. EMBO J. **22**, 5323–5335 (2003)

117. B.P. Coe et al., Genomic deregulation of the E2F/Rb pathway leads to activation of the oncogene EZH2 in small cell lung cancer. PLoS One **8**, e71670 (2013)

118. J.R. Miller, The Wnts. Genome Biol. **3**, REVIEWS3001 (2002)

119. Y. Kawano, R. Kypta, Secreted antagonists of the Wnt signalling pathway. J. Cell Sci. **116**, 2627–2634 (2003)

120. J.N. Anastas, R.T. Moon, WNT signalling pathways as therapeutic targets in cancer. Nat. Rev. Cancer **13**, 11–26 (2013)

121. J. Lilien, J. Balsamo, The regulation of cadherin-mediated adhesion by tyrosine phosphorylation/dephosphorylation of beta-catenin. Curr. Opin. Cell Biol. **17**, 459–465 (2005)

122. T. Muller, A. Choidas, E. Reichmann, A. Ullrich, Phosphorylation and free pool of beta-catenin are regulated by tyrosine kinases and tyrosine phosphatases during epithelial cell migration. J. Biol. Chem. **274**, 10173–10183 (1999)

123. A. Herbst et al., Comprehensive analysis of beta-catenin target genes in colorectal carcinoma cell lines with deregulated Wnt/beta-catenin signaling. BMC Genomics **15**, 74 (2014)

124. F. Yang et al., Linking beta-catenin to androgen-signaling pathway. J. Biol. Chem. **277**, 11336–11344 (2002)

125. M.V. Hadjihannas et al., Aberrant Wnt/beta-catenin signaling can induce chromosomal instability in colon cancer. Proc. Natl. Acad. Sci. U. S. A. **103**, 10747–10752 (2006)

126. W. Giaretti et al., Chromosomal instability and APC gene mutations in human sporadic colorectal adenomas. J. Pathol. **204**, 193–199 (2004)

127. L.E. Dow et al., Apc restoration promotes cellular differentiation and reestablishes crypt homeostasis in colorectal cancer. Cell **161**, 1539–1552 (2015)

128. M. Giannakis et al., RNF43 is frequently mutated in colorectal and endometrial cancers. Nat. Genet. **46**, 1264–1266 (2014)

129. S. Seshagiri et al., Recurrent R-spondin fusions in colon cancer. Nature **488**, 660–664 (2012)

130. Y. Wang et al., The Wnt/beta-catenin pathway is required for the development of leukemia stem cells in AML. Science **327**, 1650–1653 (2010)

131. J. Yeung et al., beta-Catenin mediates the establishment and drug resistance of MLL leukemic stem cells. Cancer Cell **18**, 606–618 (2010)

132. D. Lu et al., Activation of the Wnt signaling pathway in chronic lymphocytic leukemia. Proc. Natl. Acad. Sci. U. S. A. **101**, 3118–3123 (2004)

133. L. Wang et al., Somatic mutation as a mechanism of Wnt/beta-catenin pathway activation in CLL. Blood **124**, 1089–1098 (2014)

134. C.C. Liu, J. Prior, D. Piwnica-Worms, G. Bu, LRP6 overexpression defines a class of breast cancer subtype and is a target for therapy. Proc. Natl. Acad. Sci. U. S. A. **107**, 5136–5141 (2010)

135. L.R. Howe, A.M. Brown, Wnt signaling and breast cancer. Cancer Biol. Ther. **3**, 36–41 (2004)

136. S.Y. Lin et al., Beta-catenin, a novel prognostic marker for breast cancer: its roles in cyclin D1 expression and cancer progression. Proc. Natl. Acad. Sci. U. S. A. **97**, 4262–4266 (2000)

137. A.I. Khramtsov et al., Wnt/beta-catenin pathway activation is enriched in basal-like breast cancers and predicts poor outcome. Am. J. Pathol. **176**, 2911–2920 (2010)

138. A.S. Tsukamoto, R. Grosschedl, R.C. Guzman, T. Parslow, H.E. Varmus, Expression of the int-1 gene in transgenic mice is associated with mammary gland hyperplasia and adenocarcinomas in male and female mice. Cell **55**, 619–625 (1988)

139. M. Klauzinska et al., Rspo2/Int7 regulates invasiveness and tumorigenic properties of mammary epithelial cells. J. Cell. Physiol. **227**, 1960–1971 (2012)

140. N.N. Yokoyama, S. Shao, B.H. Hoang, D. Mercola, X. Zi, Wnt signaling in castration-resistant prostate cancer: implications for therapy. Am. J. Clin. Exp. Urol. **2**, 27–44 (2014)

141. R.M. Kypta, J. Waxman, Wnt/beta-catenin signalling in prostate cancer. Nat. Rev. Urol. **9**, 418–428 (2012)

142. V. Murillo-Garzon, R. Kypta, WNT signalling in prostate cancer. Nat. Rev. Urol. **14**, 683–696 (2017)

143. S. Thiele et al., Expression profile of WNT molecules in prostate cancer and its regulation by aminobisphosphonates. J. Cell. Biochem. **112**, 1593–1600 (2011)

144. H. Zhu et al., Analysis of Wnt gene expression in prostate cancer: mutual inhibition by WNT11 and the androgen receptor. Cancer Res. **64**, 7918–7926 (2004)

145. G. Chen et al., Up-regulation of Wnt-1 and beta-catenin production in patients with advanced metastatic prostate carcinoma: potential pathogenetic and prognostic implications. Cancer **101**, 1345–1356 (2004)

146. X. Wan et al., Activation of beta-catenin signaling in androgen receptor-negative prostate cancer cells. Clin. Cancer Res. **18**, 726–736 (2012)

147. A. de la Taille et al., Beta-catenin-related anomalies in apoptosis-resistant and hormone-refractory prostate cancer cells. Clin. Cancer Res. **9**, 1801–1807 (2003)

148. Z. Zhang et al., Inhibition of the Wnt/beta-catenin pathway overcomes resistance to enzalutamide in castration-resistant prostate cancer. Cancer Res. **78**, 3147–3162 (2018)

149. D. Lodygin, A. Epanchintsev, A. Menssen, J. Diebold, H. Hermeking, Functional epigenomics identifies genes frequently silenced in prostate cancer. Cancer Res. **65**, 4218–4227 (2005)

150. L.G. Horvath et al., Secreted frizzled-related protein 4 inhibits proliferation and metastatic potential in prostate cancer. Prostate **67**, 1081–1090 (2007)

151. D.S. Yee et al., The Wnt inhibitory factor 1 restoration in prostate cancer cells was associated with reduced tumor growth, decreased capacity of cell migration and invasion and a reversal of epithelial to mesenchymal transition. Mol. Cancer **9**, 162 (2010)

152. X. Zi et al., Expression of Frzb/secreted Frizzled-related protein 3, a secreted Wnt antagonist, in human androgen-independent prostate cancer PC-3 cells suppresses tumor growth and cellular invasiveness. Cancer Res. **65**, 9762–9770 (2005)

153. B. Bierie et al., Activation of beta-catenin in prostate epithelium induces hyperplasias and squamous transdifferentiation. Oncogene **22**, 3875–3887 (2003)

154. F. Gounari et al., Stabilization of beta-catenin induces lesions reminiscent of prostatic intraepithelial neoplasia, but terminal squamous transdifferentiation of other secretory epithelia. Oncogene **21**, 4099–4107 (2002)

155. J.C. Francis, M.K. Thomsen, M.M. Taketo, A. Swain, beta-Catenin is required for prostate development and cooperates with Pten loss to drive invasive carcinoma. PLoS Genet. **9**, e1003180 (2013)

156. X. Yu, Y. Wang, D.J. DeGraff, M.L. Wills, R.J. Matusik, Wnt/beta-catenin activation promotes prostate tumor progression in a mouse model. Oncogene **30**, 1868–1879 (2011)

157. F.G. Giancotti, Mechanisms governing metastatic dormancy and reactivation. Cell **155**, 750–764 (2013)

158. P.I. Croucher, M.M. McDonald, T.J. Martin, Bone metastasis: the importance of the neighbourhood. Nat. Rev. Cancer **16**, 373–386 (2016)

159. Z.G. Li et al., Low-density lipoprotein receptor-related protein 5 (LRP5) mediates the prostate cancer-induced formation of new bone. Oncogene **27**, 596–603 (2008)

160. D.R. Chesire, C.M. Ewing, J. Sauvageot, G.S. Bova, W.B. Isaacs, Detection and analysis of beta-catenin mutations in prostate cancer. Prostate **45**, 323–334 (2000)

161. H.J. Voeller, C.I. Truica, E.P. Gelmann, Beta-catenin mutations in human prostate cancer. Cancer Res. **58**, 2520–2523 (1998)

162. P. Rajan et al., Next-generation sequencing of advanced prostate cancer treated with androgen-deprivation therapy. Eur. Urol. **66**, 32–39 (2014)

163. W.S. Chen et al., Genomic drivers of poor prognosis and enzalutamide resistance in metastatic castration-resistant prostate cancer. Eur. Urol. (2019). https://doi.org/10.1016/j.eururo.2019.03.020

164. L. Wang et al., A prospective genome-wide study of prostate cancer metastases reveals association of wnt pathway activation and increased cell cycle proliferation with primary resistance to abiraterone acetate-prednisone. Ann. Oncol. **29**, 352–360 (2018)

165. M. Cojoc et al., Aldehyde dehydrogenase is regulated by beta-catenin/TCF and promotes radioresistance in prostate cancer progenitor cells. Cancer Res. **75**, 1482–1494 (2015)

166. M. Vesel et al., ABCB1 and ABCG2 drug transporters are differentially expressed in non-small cell lung cancers (NSCLC) and expression is modified by cisplatin treatment via altered Wnt signaling. Respir. Res. **18**, 52 (2017)

167. M. Takeda et al., The establishment of two paclitaxel-resistant prostate cancer cell lines and the mechanisms of paclitaxel resistance with two cell lines. Prostate **67**, 955–967 (2007)

168. A.P. Lombard et al., ABCB1 mediates cabazitaxel-docetaxel cross-resistance in advanced prostate cancer. Mol. Cancer Ther. **16**, 2257–2266 (2017)

169. C.I. Truica, S. Byers, E.P. Gelmann, Beta-catenin affects androgen receptor transcriptional activity and ligand specificity. Cancer Res. **60**, 4709–4713 (2000)

170. L. Schweizer et al., The androgen receptor can signal through Wnt/beta-Catenin in prostate cancer cells as an adaptation mechanism to castration levels of androgens. BMC Cell Biol. **9**, 4 (2008)

171. Y. Li et al., LEF1 in androgen-independent prostate cancer: regulation of androgen receptor expression, prostate cancer growth, and invasion. Cancer Res. **69**, 3332–3338 (2009)

172. X. Yu et al., Foxa1 and Foxa2 interact with the androgen receptor to regulate prostate and epididymal genes differentially. Ann. N. Y. Acad. Sci. **1061**, 77–93 (2005)

173. J.W. Park, J.K. Lee, O.N. Witte, J. Huang, FOXA2 is a sensitive and specific marker for small cell neuroendocrine carcinoma of the prostate. Mod. Pathol. **30**(9), 1262–1272 (2017)

174. T.J. Van Raay et al., Frizzled 5 signaling governs the neural potential of progenitors in the developing Xenopus retina. Neuron **46**, 23–36 (2005)

175. V.J. Wielenga et al., Expression of CD44 in Apc and Tcf mutant mice implies regulation by the WNT pathway. Am. J. Pathol. **154**, 515–523 (1999)

176. R.A. Simon et al., CD44 expression is a feature of prostatic small cell carcinoma and distinguishes it from its mimickers. Hum. Pathol. **40**, 252–258 (2009)

177. G.S. Palapattu et al., Selective expression of CD44, a putative prostate cancer stem cell marker, in neuroendocrine tumor cells of human prostate cancer. Prostate **69**, 787–798 (2009)

178. D. ten Berge, S.A. Brugmann, J.A. Helms, R. Nusse, Wnt and FGF signals interact to coordinate growth with cell fate specification during limb development. Development **135**, 3247–3257 (2008)

179. M. Ciarlo et al., Regulation of neuroendocrine differentiation by AKT/hnRNPK/AR/beta-catenin signaling in prostate cancer cells. Int. J. Cancer **131**, 582–590 (2012)

180. R. Kalluri, R.A. Weinberg, The basics of epithelial-mesenchymal transition. J. Clin. Invest. **119**, 1420–1428 (2009)

181. K. Polyak, R.A. Weinberg, Transitions between epithelial and mesenchymal states: acquisition of malignant and stem cell traits. Nat. Rev. Cancer **9**, 265–273 (2009)

182. E.J. Robson, W.T. Khaled, K. Abell, C.J. Watson, Epithelial-to-mesenchymal transition confers resistance to apoptosis in three murine mammary epithelial cell lines. Differentiation **74**, 254–264 (2006)

183. C.J. Creighton et al., Residual breast cancers after conventional therapy display mesenchymal as well as tumor-initiating features. Proc. Natl. Acad. Sci. U. S. A. **106**, 13820–13825 (2009)

184. K.R. Fischer et al., Epithelial-to-mesenchymal transition is not required for lung metastasis but contributes to chemoresistance. Nature **527**, 472–476 (2015)

185. X. Zheng et al., Epithelial-to-mesenchymal transition is dispensable for metastasis but induces chemoresistance in pancreatic cancer. Nature **527**, 525–530 (2015)

186. S.A. Mani et al., The epithelial-mesenchymal transition generates cells with properties of stem cells. Cell **133**, 704–715 (2008)

187. V.L. Battula et al., Epithelial-mesenchymal transition-derived cells exhibit multilineage differentiation potential similar to mesenchymal stem cells. Stem Cells **28**, 1435–1445 (2010)

188. J. Heuberger, W. Birchmeier, Interplay of cadherin-mediated cell adhesion and canonical Wnt signaling. Cold Spring Harb. Perspect. Biol. **2**, a002915 (2010)

189. Z.Q. Wu et al., Canonical Wnt signaling regulates Slug activity and links epithelial-mesenchymal transition with epigenetic Breast Cancer 1, Early Onset (BRCA1) repression. Proc. Natl. Acad. Sci. U. S. A. **109**, 16654–16659 (2012)

190. M. Conacci-Sorrell et al., Autoregulation of E-cadherin expression by cadherin-cadherin interactions: the roles of beta-catenin signaling, Slug, and MAPK. J. Cell Biol. **163**, 847–857 (2003)

191. L.R. Howe, O. Watanabe, J. Leonard, A.M. Brown, Twist is up-regulated in response to Wnt1 and inhibits mouse mammary cell differentiation. Cancer Res. **63**, 1906–1913 (2003)

192. C. Jamora, R. DasGupta, P. Kocieniewski, E. Fuchs, Links between signal transduction, transcription and adhesion in epithelial bud development. Nature **422**, 317–322 (2003)

193. H.C. Crawford et al., The metalloproteinase matrilysin is a target of beta-catenin transactivation in intestinal tumors. Oncogene **18**, 2883–2891 (1999)

194. K. Kim, Z. Lu, E.D. Hay, Direct evidence for a role of beta-catenin/LEF-1 signaling pathway in induction of EMT. Cell Biol. Int. **26**, 463–476 (2002)

195. K. Yang et al., The evolving roles of canonical WNT signaling in stem cells and tumorigenesis: implications in targeted cancer therapies. Lab. Investig. **96**, 116–136 (2016)

196. B.J. Merrill, Wnt pathway regulation of embryonic stem cell self-renewal. Cold Spring Harb. Perspect. Biol. **4**, a007971 (2012)

197. J.D. Holland, A. Klaus, A.N. Garratt, W. Birchmeier, Wnt signaling in stem and cancer stem cells. Curr. Opin. Cell Biol. **25**, 254–264 (2013)

198. I. Malanchi et al., Cutaneous cancer stem cell maintenance is dependent on beta-catenin signalling. Nature **452**, 650–653 (2008)

199. N. Sato, L. Meijer, L. Skaltsounis, P. Greengard, A.H. Brivanlou, Maintenance of pluripotency in human and mouse embryonic stem cells through activation of Wnt signaling by a pharmacological GSK-3-specific inhibitor. Nat. Med. **10**, 55–63 (2004)

200. F. Lluis, E. Pedone, S. Pepe, M.P. Cosma, Periodic activation of Wnt/beta-catenin signaling enhances somatic cell reprogramming mediated by cell fusion. Cell Stem Cell **3**, 493–507 (2008)

201. K. Hoffmeyer et al., Wnt/beta-catenin signaling regulates telomerase in stem cells and cancer cells. Science **336**, 1549–1554 (2012)

202. J. Zeilstra et al., Stem cell CD44v isoforms promote intestinal cancer formation in Apc(min) mice downstream of Wnt signaling. Oncogene **33**, 665–670 (2014)

203. M. Katoh, Canonical and non-canonical WNT signaling in cancer stem cells and their niches: cellular heterogeneity, omics reprogramming, targeted therapy and tumor plasticity (Review). Int. J. Oncol. **51**, 1357–1369 (2017)

204. B.E. Wang et al., Castration-resistant Lgr5(+) cells are long-lived stem cells required for prostatic regeneration. Stem Cell Rep. **4**, 768–779 (2015)

205. C.S. Ontiveros, S.N. Salm, E.L. Wilson, Axin2 expression identifies progenitor cells in the murine prostate. Prostate **68**, 1263–1272 (2008)

206. E.J. Yun et al., Targeting cancer stem cells in castration-resistant prostate cancer. Clin. Cancer Res. **22**, 670–679 (2016)

207. T. Reya et al., A role for Wnt signalling in self-renewal of haematopoietic stem cells. Nature **423**, 409–414 (2003)

208. N. Kawaguchi-Ihara, I. Murohashi, N. Nara, S. Tohda, Promotion of the self-renewal capacity of human acute leukemia cells by Wnt3A. Anticancer Res. **28**, 2701–2704 (2008)

209. I. Bisson, D.M. Prowse, WNT signaling regulates self-renewal and differentiation of prostate cancer cells with stem cell characteristics. Cell Res. **19**, 683–697 (2009)

210. K. Zhang et al., WNT/beta-catenin directs self-renewal symmetric cell division of hTERT(high) prostate cancer stem cells. Cancer Res. **77**, 2534–2547 (2017)

211. B.A. Smith et al., A human adult stem cell signature marks aggressive variants across epithelial cancers. Cell Rep. **24**, 3353–3366.e5 (2018)

212. P. Sotomayor, A. Godoy, G.J. Smith, W.J. Huss, Oct4A is expressed by a subpopulation of prostate neuroendocrine cells. Prostate **69**, 401–410 (2009)

213. N. Monsef, M. Soller, M. Isaksson, P.A. Abrahamsson, I. Panagopoulos, The expression of pluripotency marker Oct 3/4 in prostate cancer and benign prostate hyperplasia. Prostate **69**, 909–916 (2009)

214. A. Cueto, F. Burigana, A. Nicolini, F. Lugnani, Neuroendocrine tumors of the lung: hystological classification, diagnosis, traditional and new therapeutic approaches. Curr. Med. Chem. **21**, 1107–1116 (2014)

215. M. Milanovic et al., Senescence-associated reprogramming promotes cancer stemness. Nature **553**, 96–100 (2018)

216. D. Robinson et al., Integrative clinical genomics of advanced prostate cancer. Cell **161**, 1215–1228 (2015)

217. J. Armenia et al., The long tail of oncogenic drivers in prostate cancer. Nat. Genet. **50**, 645–651 (2018)

218. A.W. Wyatt et al., Genomic alterations in cell-free DNA and enzalutamide resistance in castration-resistant prostate cancer. JAMA Oncol. **2**, 1598–1606 (2016)

219. C. Wissmann et al., WIF1, a component of the Wnt pathway, is down-regulated in prostate, breast, lung, and bladder cancer. J. Pathol. **201**, 204–212 (2003)

220. A.S. Perry et al., Gene expression and epigenetic discovery screen reveal methylation of SFRP2 in prostate cancer. Int. J. Cancer **132**, 1771–1780 (2013)

221. L.G. Horvath et al., Membranous expression of secreted frizzled-related protein 4 predicts for good prognosis in localized prostate cancer and inhibits PC3 cellular proliferation in vitro. Clin. Cancer Res. **10**, 615–625 (2004)

222. M.S. Joesting et al., Identification of SFRP1 as a candidate mediator of stromal-to-epithelial signaling in prostate cancer. Cancer Res. **65**, 10423–10430 (2005)

223. G. O'Hurley et al., The role of secreted frizzled-related protein 2 expression in prostate cancer. Histopathology **59**, 1240–1248 (2011)

224. Y. Zong et al., Stromal epigenetic dysregulation is sufficient to initiate mouse prostate cancer via paracrine Wnt signaling. Proc. Natl. Acad. Sci. U. S. A. **109**, E3395–E3404 (2012)

225. O. Dakhova, D. Rowley, M. Ittmann, Genes upregulated in prostate cancer reactive stroma promote prostate cancer progression in vivo. Clin. Cancer Res. **20**, 100–109 (2014)

226. N. Carayol, C.Y. Wang, IKKalpha stabilizes cytosolic beta-catenin by inhibiting both canonical and non-canonical degradation pathways. Cell. Signal. **18**, 1941–1946 (2006)

227. L. Yang, C. Lin, Z.R. Liu, P68 RNA helicase mediates PDGF-induced epithelial mesenchymal transition by displacing Axin from beta-catenin. Cell **127**, 139–155 (2006)

228. V.R. Placencio et al., Stromal transforming growth factor-beta signaling mediates prostatic response to androgen ablation by paracrine Wnt activity. Cancer Res. **68**, 4709–4718 (2008)

229. B.E. Wang, X.D. Wang, J.A. Ernst, P. Polakis, W.Q. Gao, Regulation of epithelial branching morphogenesis and cancer cell growth of the prostate by Wnt signaling. PLoS One **3**, e2186 (2008)

230. S. Kregel et al., Acquired resistance to the second-generation androgen receptor antagonist enzalutamide in castration-resistant prostate cancer. Oncotarget **7**, 26259–26274 (2016)

231. G. Wang, J. Wang, M.D. Sadar, Crosstalk between the androgen receptor and beta-catenin in castrate-resistant prostate cancer. Cancer Res. **68**, 9918–9927 (2008)

232. E. Lee et al., Inhibition of androgen receptor and beta-catenin activity in prostate cancer. Proc. Natl. Acad. Sci. U. S. A. **110**, 15710–15715 (2013)

233. J.A. Schneider, S.K. Logan, Revisiting the role of Wnt/beta-catenin signaling in prostate cancer. Mol. Cell. Endocrinol. **462**, 3–8 (2018)

234. L. Antony, F. van der Schoor, S.L. Dalrymple, J.T. Isaacs, Androgen receptor (AR) suppresses normal human prostate epithelial cell proliferation via AR/beta-catenin/TCF-4 complex inhibition of c-MYC transcription. Prostate **74**, 1118–1131 (2014)

235. D.J. Mulholland, J.T. Read, P.S. Rennie, M.E. Cox, C.C. Nelson, Functional localization and competition between the androgen receptor and T-cell factor for nuclear beta-catenin: a means for inhibition of the Tcf signaling axis. Oncogene **22**, 5602–5613 (2003)

236. D.R. Chesire, W.B. Isaacs, Ligand-dependent inhibition of beta-catenin/TCF signaling by androgen receptor. Oncogene **21**, 8453–8469 (2002)

237. X.D. Wang et al., Expression profiling of the mouse prostate after castration and hormone replacement: implication of H-cadherin in prostate tumorigenesis. Differentiation **75**, 219–234 (2007)

238. D.R. Chesire, C.M. Ewing, W.R. Gage, W.B. Isaacs, In vitro evidence for complex modes of nuclear beta-catenin signaling during prostate growth and tumorigenesis. Oncogene **21**, 2679–2694 (2002)

239. C. Lamberti et al., Regulation of beta-catenin function by the IkappaB kinases. J. Biol. Chem. **276**, 42276–42286 (2001)

240. F. Ma et al., SOX9 drives WNT pathway activation in prostate cancer. J. Clin. Invest. **126**, 1745–1758 (2016)

241. X. Wu et al., Rac1 activation controls nuclear localization of beta-catenin during canonical Wnt signaling. Cell **133**, 340–353 (2008)

242. S. Persad, A.A. Troussard, T.R. McPhee, D.J. Mulholland, S. Dedhar, Tumor suppressor PTEN inhibits nuclear accumulation of beta-catenin and T cell/lymphoid enhancer factor 1-mediated transcriptional activation. J. Cell Biol. **153**, 1161–1174 (2001)

243. D.S. Byun et al., Intestinal epithelial-specific PTEN inactivation results in tumor formation. Am. J. Physiol. Gastrointest. Liver Physiol. **301**, G856–G864 (2011)

244. M.T. Jefferies et al., PTEN loss and activation of K-RAS and beta-catenin cooperate to accelerate prostate tumourigenesis. J. Pathol. **243**, 442–456 (2017)

245. F.X. Yu, B. Zhao, K.L. Guan, Hippo pathway in organ size control, tissue homeostasis, and cancer. Cell **163**, 811–828 (2015)

246. U. Ehmer, J. Sage, Control of proliferation and cancer growth by the Hippo signaling pathway. Mol. Cancer Res. **14**, 127–140 (2016)

247. T. Ito et al., Loss of YAP1 defines neuroendocrine differentiation of lung tumors. Cancer Sci. **107**, 1527–1538 (2016)

248. G. Kuser-Abali, A. Alptekin, M. Lewis, I.P. Garraway, B. Cinar, YAP1 and AR interactions contribute to the switch from androgen-dependent to castration-resistant growth in prostate cancer. Nat. Commun. **6**, 8126 (2015)

249. L. Zhang et al., The Hippo pathway effector YAP regulates motility, invasion, and castration-resistant growth of prostate cancer cells. Mol. Cell. Biol. **35**, 1350–1362 (2015)

250. L. Azzolin et al., Role of TAZ as mediator of Wnt signaling. Cell **151**, 1443–1456 (2012)

251. T. Heallen et al., Hippo pathway inhibits Wnt signaling to restrain cardiomyocyte proliferation and heart size. Science **332**, 458–461 (2011)

252. J. Rosenbluh et al., beta-Catenin-driven cancers require a YAP1 transcriptional complex for survival and tumorigenesis. Cell **151**, 1457–1473 (2012)

253. L. Azzolin et al., YAP/TAZ incorporation in the beta-catenin destruction complex orchestrates the Wnt response. Cell **158**, 157–170 (2014)

254. X. Varelas et al., The Hippo pathway regulates Wnt/beta-catenin signaling. Dev. Cell **18**, 579–591 (2010)

255. S.W. Plouffe et al., The Hippo pathway effector proteins YAP and TAZ have both distinct and overlapping functions in the cell. J. Biol. Chem. **293**, 11230–11240 (2018)

256. E.R. Barry et al., Restriction of intestinal stem cell expansion and the regenerative response by YAP. Nature **493**, 106–110 (2013)

257. N. Takebe et al., Targeting Notch, Hedgehog, and Wnt pathways in cancer stem cells: clinical update. Nat. Rev. Clin. Oncol. **12**, 445–464 (2015)

258. R. Nusse, H. Clevers, Wnt/beta-catenin signaling, disease, and emerging therapeutic modalities. Cell **169**, 985–999 (2017)

259. K.H. Emami et al., A small molecule inhibitor of beta-catenin/CREB-binding protein transcription [corrected]. Proc. Natl. Acad. Sci. U. S. A. **101**, 12682–12687 (2004)

260. H.J. Lenz, M. Kahn, Safely targeting cancer stem cells via selective catenin coactivator antagonism. Cancer Sci. **105**, 1087–1092 (2014)

261. M. Chen et al., The anti-helminthic niclosamide inhibits Wnt/Frizzled1 signaling. Biochemistry **48**, 10267–10274 (2009)

262. J. Shan, D.L. Shi, J. Wang, J. Zheng, Identification of a specific inhibitor of the dishevelled PDZ domain. Biochemistry **44**, 15495–15503 (2005)

263. C.H. Park et al., Quercetin, a potent inhibitor against beta-catenin/Tcf signaling in SW480 colon cancer cells. Biochem. Biophys. Res. Commun. **328**, 227–234 (2005)

264. W. Fiskus et al., Pre-clinical efficacy of combined therapy with novel beta-catenin antagonist BC2059 and histone deacetylase inhibitor against AML cells. Leukemia **29**, 1267–1278 (2015)

265. K. Sukhdeo et al., Targeting the beta-catenin/TCF transcriptional complex in the treatment of multiple myeloma. Proc. Natl. Acad. Sci. U. S. A. **104**, 7516–7521 (2007)

266. B. Chen et al., Small molecule-mediated disruption of Wnt-dependent signaling in tissue regeneration and cancer. Nat. Chem. Biol. **5**, 100–107 (2009)

267. B. Garcia-Reyes et al., Discovery of inhibitor of Wnt production 2 (IWP-2) and related compounds as selective ATP-competitive inhibitors of casein kinase 1 (CK1) delta/epsilon. J. Med. Chem. **61**, 4087–4102 (2018)

268. J. Liu et al., Targeting Wnt-driven cancer through the inhibition of Porcupine by LGK974. Proc. Natl. Acad. Sci. U. S. A. **110**, 20224–20229 (2013)

269. K.D. Proffitt et al., Pharmacological inhibition of the Wnt acyltransferase PORCN prevents growth of WNT-driven mammary cancer. Cancer Res. **73**, 502–507 (2013)

270. S.M. Huang et al., Tankyrase inhibition stabilizes axin and antagonizes Wnt signalling. Nature **461**, 614–620 (2009)

Epigenetic Regulation of Chromatin in Prostate Cancer

Ramakrishnan Natesan, Shweta Aras, Samuel Sander Effron, and Irfan A. Asangani

Introduction

Prostate Cancer (PCa) is a disease characterized by genetic and epigenetic alterations. Accumulation of somatic genomic alterations such as mutations and chromosomal rearrangements and global changes to the chromatin landscape contribute to prostate cancer initiation, progression, and therapy resistance. Specifically, genomic alterations in genes that encode chromatin regulators and chromatin-remodeling factors are enriched in advanced, metastatic castration resistant prostate cancer (mCRPC). Epigenetic reprogramming of chromatin alters the accessible regions of the genome resulting in differential transcriptional output during prostate oncogenesis and progression.

Within the cell nucleus, the linear genomic DNA is organized into a highly compact form called chromatin, which helps fit the entire 2-m-long genomic DNA into a cell nucleus measuring only ~10 μm in diameter. This compaction is reversible and is mediated by interactions between the negatively charged DNA and a set of four proteins called histones. The fundamental unit of chromatin is a nucleosome that contains 145–147 bp of DNA wrapped about 1.65 times around a globular octamer complex formed from homo-dimers of the core histone proteins H2A, H2B, H3, and H4. This chromatin conformation resembles a "beads on a string" structure [1, 2]. A fifth histone protein H1 binds to the DNA between two adjacent nucleosomes and acts as a nucleosome-linker and further promotes compaction and stabilization of the chromatin into a 30 nm fiber representing a higher order chromatin structure [3, 4]. This higher order chromatin structure can be further classified into "heterochromatin" that represents a highly condensed "transcriptionally inactive" state not accessible to the transcriptional machinery, and "euchromatin" that represents an open "transcriptionally active" state containing most of the active genes and accessible to the transcription machinery.

Recent advances in chromosome conformation capture technologies (e.g., 4C, HiC, and its derivatives) [5, 6] to study the three dimensional (3D) organization of chromatin have shown that the chromatin architecture is complex and is organized in a less-random fashion, resulting in a higher order 3D organization of genome with heterochromatic and euchromatic compartments [7]. At the sub-chromosomal level, the gene regulatory regions are classified into *promoters* that are located near the Transcription Start Site (TSS)

R. Natesan · S. Aras · S. S. Effron
I. A. Asangani (✉)
Department of Cancer Biology, Abramson Family Cancer Research Institute, Epigenetics Institute, Perelman School of Medicine,
University of Pennsylvania, Philadelphia, PA, USA
e-mail: asangani@upenn.edu

© Springer Nature Switzerland AG 2019
S. M. Dehm, D. J. Tindall (eds.), *Prostate Cancer*, Advances in Experimental Medicine and Biology 1210, https://doi.org/10.1007/978-3-030-32656-2_17

of genes, and *enhancers* that are typically located at a considerable distance from the TSS [8]. Upon activation by tissue-specific transcription factors, these enhancers modulate gene expression by physically interacting with their target gene promoters through chromatin looping. This looping, which facilitates the physical interaction between distal enhancers and gene promoters, is further propagated by proteins that promote this contact, thereby constituting gene-modulatory landscape [9]. Based on such high levels of internal interactions, chromatin structure is sub-divided into many domains known as topologically associating domain (TAD). They are the chromatin domains with high levels of internal interactions and are separated from each other by regions of low interaction called boundary elements. TADs constitute fundamental units of the 3D organization of the genome, promoting enhancer-promoter interactions [10]. Characterization of these gene regulatory units like promoters, enhancers and their 3D interactions has opened avenues for understanding mechanisms of transcriptional control of genes. Chromosomal architecture is primarily mediated by the CCCTC-binding factor (CTCF)/cohesin complex. CTCF is an 11-zinc-finger transcription factor that is enriched at boundary elements of TADs and loop domains and is essential for the recruitment of cohesin to chromatin [11, 12]. This suggests that the microscopic structures of chromatin configurations are much more complex, and that the dynamic, yet controlled modulations of these chromatin conformations are essential for timely, tissue-specific, and coordinated gene expression. Efficient regulation of gene expression and cellular processes are mediated through modulating chromatin structure by three mechanisms involving the DNA and histones: *(1) Covalent modifications of DNA, (2) Post-translational modifications of histone tails by histone modifying enzymes, and (3) Disruption of histone-DNA contacts by ATP dependent chromatin remodeling proteins.* These epigenetic regulatory mechanisms work independently or in unison to modulate chromatin architecture, which in turn regulates gene expression, thereby governing cellular function.

DNA Methylation as an Epigenetic Code for Prostate Tumor Development

Methylation patterns of cytosine residues within CpG dinucleotide sequences play an important role in key cellular process such as DNA repair, recombination and replication, and regulation of gene expression [13, 14]. DNA methylation based regulatory mechanisms are highly dynamic where in nearly 60–80% of CpG sites in the mammalian genome display altered methylation patterns. Cytosine methylation is catalyzed by DNA methyltransferases DNMT1, DNMT3A, and DNMT3B that transfer a methyl group from *S*-adenosylmethionine to the 5′ carbon of the cytosine ring to form 5-methylcytosine (5-mC). CpG dinucleotide sequences are enriched at active gene promoters and hypermethylation of these regions leads to the preferential binding of methyl CpG binding domain (MBD) proteins, such as MBD1, MBD2, MBD3, MBD4, Kaiso, and MECP2 [15]. These MBD proteins contain transcriptional repression domains, which in association with histone deacetylases (HDACs), repress target gene transcription. In prostate cancer development, aberrant hypermethylation is commonly observed in the promoter regions of genes associated with tumor-suppressor activity (*e.g. APC, RARβ, RASSF1A, p16*), DNA-repair (*e.g. GSTP1, MGMT, GSTM1*), cell cycle control (*e.g. CCNA1, CDKN2A, CCND2, H1C1, SFN*), apoptosis (*e.g. PYCARD, DAPK, SLC5A8, SLC18A2, TNFRSF10C, RUNX3*), and maintenance of cell-cell contacts (*e.g. CDH1, CD44*), whose repression enables the growth and stabilization of neoplastic phenotypes [13, 16, 17]. Additionally, DNA hypermethylation represses the transcription of microRNAs leading to upregulation of their oncogenic targets that can drive tumorigenesis [18]. DNA demethylation proceeds primarily through oxidative reactions catalyzed by the TET family oxygenases that convert *5-methylcytosine → 5-hydroxymethylcytosine → 5-formylcytosine → 5-carboxylcytosine*, which is then converted to cytosine through the action of thymine DNA glycosylase [19]. DNA hypomethylation, which activates gene transcription primarily at the promoter regions of oncogenes, is observed more frequently in mCRPC compared to PCa [13]. Genes regulated by aberrant DNA hypomethylation in prostate cancer include *MYC, RAS, uPA, PLAU, HPSE, CYP1B1, WNT5A, S100P*, and *CRIP1*. More details on the role of DNA methylation in Pca and their potential for use as diagnostic and prognostic biomarkers may be found in several excellent reviews on this topic [13–18, 20].

Histone Post-translational Modifications

Amino acid residues in the unstructured N-terminal tails of histone proteins H3 and H4, and both N- and C-termini tails of H2A and H2B, display a variety of covalent, reversible, post-translational modifications (PTMs) that define the epigenetic code. PTMs are strongly associated with specific amino acid residues. Some of the well-studied PTMs include methylation (occurring on K and R residues), acetylation (K, S, T), phosphorylation (S, T, Y, H), ubiquitination (K), sumoylation (K), ADP ribosylation (K, E), succinylation (K), 2-hydroxyisobutyrylation (K) formylation (K), malonylation (K), propionylation (K), butyrylation (K), crotonylation (K), hydroxylation (Y), citrullination, a.k.a. deamination (R), O-GlcNAcylation (S, T), and proline isomerization [21–25]. The highly dynamic histone PTMs regulate cell function by altering DNA-histone interactions, nucleosomal assembly, and global/local higher order structures of chromatin, all of which in turn directly control the accessibility of genes to transcription factors. The major cellular processes regulated by the histone PTM epigenetic code include gene transcription, gene-repair, metabolism, replication, and chromatin condensation [23].

The vast number and combinations of histone modifications define highly complex epigenetic regulatory states of chromatin. Unlike the CpG specific DNA methylation discussed earlier, histones can be methylated at any of the lysine or arginine residues. Hence, the **position and state of methylations, and in effect that for all**

PTMs, represents integral part of the epigenetic code. For example, tri-methylations H3K9me3 and H3K27me3 are primarily linked to compact and closed chromatin conformations (*heterochromatin*) that repress gene transcription [26] (see Fig. 1a), while H3K4me1/me2/me3, and H3K36me3 are associated with relaxed chromatin conformations (*euchromatin*) that activate gene transcription [11] (see Fig. 1b).

In PCa, histone methylation and acetylation are often associated with progression and metastasis. Compared to non-malignant phenotypes, the levels of H3K4me1, H3K9me2, H3K9me3 are often reduced in primary prostate cancer and increased in mCRPC. Acetylation of H3K9, H318, and H4K12 also often display a similar trend [3, 17, 27]. Similarly, acetylation of the histone variant H2A.Z at active promoter sites is often associated with oncogene activation in PCa [28].

Factors governing epigenetic regulation can be broadly classified into three distinct groups namely **readers**—proteins that recognize a PTM, **writers**—enzymes that catalyze the addition of a PTM, and **erasers**—enzymes that catalyze the removal of a PTM. Each histone PTM has a specific set of reader, writer, and eraser proteins. In this chapter, we will specifically focus on the well-studied enzymatic machineries governing histone acetylation and methylation in PCa. Histone methylation and demethylation are mediated by histone methyltransferases (HMTs) and demethylases (HDMs), while acetylation and deacetylation are mediated by histone acetyltransferases (HATs), histone deacetylases (HDACs) and sirtuins. A significant number of these epigenetic readers, writers and erases are dysregulated and mutated in PCa.

Histone Methylation in Prostate Cancer

Histone methylation occurs primarily on the side chains of all basic amino acid residues, i.e. lysine (K), arginine (R) and histidine (H). Lysines can be mono (me1), di (me2), or tri (me3) methylated on their ε-amine group, Arginines can be mono

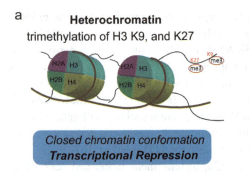 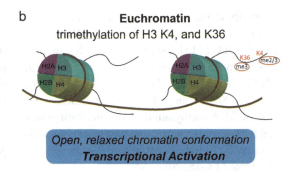

Fig. 1 Effect of state-dependent methylation on transcriptional activity. Tri-methylation (me3) of histone H3K9 or K27 leads to closed chromatin conformations, i.e., *heterochromatin*, resulting in transcriptional repression (**a**), while mono/di/tri-methylation of H3K4 and H3K36 leads to open and relaxed chromatin conformation, i.e., *euchromatin*, resulting in transcriptional activation (**b**)

(me1), symmetrically dimethylated (me2s), or asymmetrically dimethylated (me2a) on their guanidinyl group, whereas histidines have been reported to be monomethylated, although this is a rare form of methylation. Interestingly, unlike acetylation, histone methylation does not alter the charge of the histone protein [29]. Rather, the position and the state of methylation (me1/me2/me3) define different regulatory states. Histone methylation plays a central role in (1) transcriptional activation/silencing *via* chromosomal looping and chromatin remodeling, (2) recruitment of cell specific transcription factors *via* interactions with initiation and elongation factors, as well as (3) RNA splicing [30, 31]. Histone methylation dynamics regulates a variety of cell functions including cell-cycle regulation, DNA damage and stress response, development and differentiation. Thus, aberrations in histone methylation patterns and the enzymes regulating histone methylation play major roles in cancer development and growth.

Histone lysine methyltransferases (HMTs) catalyze the transfer of a methyl group from *S*-adenosylmethionine (SAM) to a lysine ε-amino group on the N-terminal tails of histones. Since the discovery of the first HMT, SUV39H1 that catalyzes H3K9me3, a variety of methyltransferases that target lysine on H3 and H4 histones have been identified [21]. A characteristic feature of all lysine methylating HMTs is that they contain a SET domain, which harbors the enzymatic activity, except for DOT1L, which lacks a SET domain. HMTs are further classified into the SUV39, EZH, SET1, SET2, PRDM, and SMYD sub-families based on the sequence homology in and around the SET domain [32]. Some of the most extensively studied histone methylation sites include H3K4, H3K9, H3K27, H3K36, H3K79, and H4K20. The enzymatic machinery that regulates the methylation of these residues is shown in Fig. 2. Generally, H3K4, H3K36, and H3K79 methylations mark sites of active transcription, while methylations of H3K9, H3K27, and H4K20 are associated with sites of silenced transcription [31]. Arginine methylation is catalyzed by arginine methyltransferases, which are divided into two subclasses—Class I and Class II enzymes. Together, these two types of arginine methyltransferases constitute a relatively large protein family with a total of 11 members that are referred to as PRMTs.

The highly stable methyl bonds on methylated lysines are removed through an oxidative mechanism catalyzed by Histone lysine demethylases (KDMs). Demethylases are classified into two major families: (1) those with an amine oxidase domain that use Flavin adenine dinucleotide (FAD) as a cofactor, and (2) those containing a Jumonji C (JmJC)-domain that use iron and α-ketoglutarate as cofactors to catalyze their oxidative reactions [31, 33, 35–37]. The first type belongs to the KDM1 family of KMTs, with two members LSD1/KDM1A and LSD2/KDM1B, that only catalyze demethylation of mono- and di-methyl groups. The JmJC family of KDMs

Epigenetic Regulation of Chromatin in Prostate Cancer

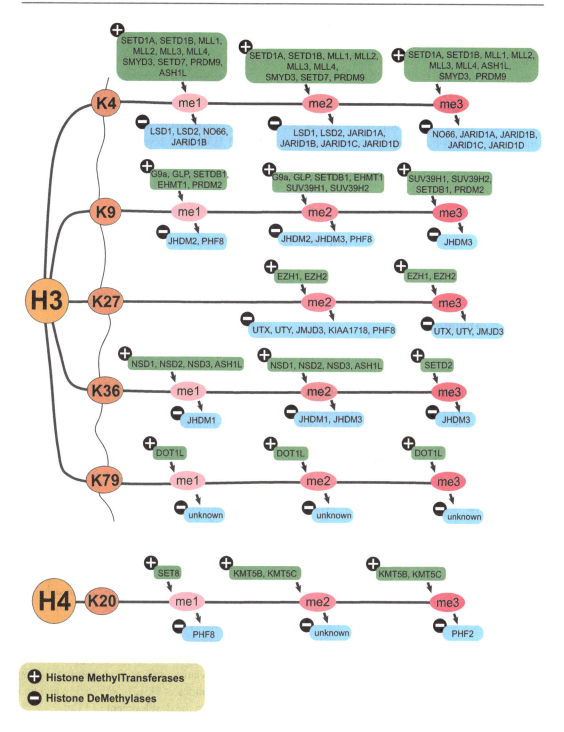

Fig. 2 Schematic showing the Histone methyltransferases and Histone demethylases involved in mono- (me1), di- (me2), and tri- (me3) methylation reactions in lysines 4, 9, 27, 36 and 79 of H3 and lysine 20 of H4. Data adapted from references [19, 21, 31, 33, 34]

contain 28 members that are further subdivided into the KDM2, KDM3, KDM4, KDM5, JARID2, KDM6, KDM7, and KDM8 subfamilies. Each subfamily is characterized by the similarities in their DNA binding, DNA recognition domains, and specific histone and non-histone substrates. The JmJC KDMs demethylate all mono-, di-, and tri-methylated histone lysines.

The following section will focus on the H3 and H4 lysine methylations/demethylations displayed in Fig. 2. It should be noted that in addition to these well studied marks, histone methylation has also been detected in all basic residues in all the four nucleosomal histones [29] and their specific roles in cell function remain to be elucidated.

H3K4 Methylation

A genome-wide study of H3K4 methylation at 44 loci in the human genome selected by the ENCODE consortium showed that H3K4me1 is a mark for active or poised enhancers, H3K4me3 is a mark for promoter regions of genes poised for or undergoing active transcription, while H3K4me2 marks both enhancer and promoter regions [38]. These findings were also verified in genome-wide studies of a panel of five human cell lines [39]. Though these marks are evolutionarily conserved, their exact role in active transcription of a specific gene depends on a number of factors such as the transcription frequency, elongation rate, and COMPASS activity [40]. H3K4me3 is highly enriched at TSSs, but totally depleted in the elongation regions [41] and CpG islands [42]. A majority of the highly transcribed genes have a gradient of H3K4me3 > H3K4me2 > H3K4me1 along the gene body from the 5' end to the 3' end, but the exact methylation patterns are gene-specific [40]. While observations on the function of a H3K4 mark holds true in most of the cases, its exact function is determined by the activity of the reader protein that recognizes the mark. For example, both H3K4me2 and H3K4me3 marks bound by plant homeodomain (PHD)-domain containing ING2 (inhibitor of growth family

member 2) proteins are associated with transcriptional repression. Also, H3K4 methylation levels correlate with DNA damage signaling [43].

A large body of evidence points to the importance of H3K4 methylation in PCa tumorigenesis and metastasis. Tissue microarray analysis of prostate tumor tissues revealed that patients with reduced H3K4me2 levels were at an increased risk for relapse [27, 44], and these results were verified through analysis of patient microarray data [45]. Studies using CRPC cell lines and tissues found that H3K4me1 and H3K4me2 marks were selectively enriched at androgen receptor (AR)-regulated enhancers of cell cycle genes such as UBE2C and CDK1 that promote CRPC growth [46]. Further, an increase in H3K4me3 marks correlated with activation of cell growth and survival genes FGFR1 and BCL2 [47]. A genome-wide ChIP-seq study using LNCaP cells showed that H3K4me2 precisely marked the nucleosomes flanking AR binding motifs at distal enhancers regulating the transcription of key AR target genes such as TMPRSS2, and PSA [48]. Alteration in H3K4 levels are also a result of dysregulation/mutations in the enzymatic machinery that regulates methylation. For example, the tumor suppressor gene PTEN is frequently deleted in primary PCa, and is lost to a greater extent in the mCRPC tumors [49], leading to deregulated PI3K signaling. This is turn affects the subcellular localization of the KDM5A demethylase [50] leading to genome-wide alteration of H3K4 levels. Reduced H3K4me2 levels in primary PCa was shown to be associated with increased risk for recurrence and metastasis [29]. Whole exome sequencing studies identified AR interactions with proteins of the KMT2 complex [51], specifically with Menin and ASH2L [52]. In the following section we will briefly describe the major methylation writers, and erasers that govern the dynamics of H3K4 methylation and their known role in the development of PCa.

H3K4 Methylation Writers

In mammalian cells, methylation of H3K4 is catalyzed by the KMT2/MLL/COMPASS family of methyltransferases; MLL1/KMT2A, MLL2/KMT2D, MLL3/KMT2C, MLL4/KMT2B,

MLL5/KMT2E, SETD1A/KMT2F, and SETD1B/KMT2G; SMYD family methyltransferases; SMYD1, SMYD2, and SMYD3; SETD7, and PRDM9 (see Fig. 2) [29, 31, 53].

KMT2/MLL/COMPASS family Methyltransferases: COMPASS (Complex proteins associated with Set1) proteins in mammalian cells are a family of six methyltransferases. All proteins in this family of methyltransferases contain a SET1 domain in complex with four common subunits—namely, WDR5, ASH2L, RbBP5, and DPY30—and other protein specific subunits chosen from CXXC1, WDR82, HCF1, HCF2, Menin, PTIP, PA1, and NCOA6 [31, 54]. The 130–140 amino acid long SET1 (Suppressor of variegation, Enhancer of Zeste, Trithorax) domain is conserved across all KMT2 family methyltransferases and is responsible for catalyzing lysine methylation activity [54–56]. An analysis of 12 different cancers in the cancer genome atlas (TCGA) dataset revealed that KMT2 family methyltransferases are frequently dysregulated and mutated in nearly all cancers, and prominently in bladder, lung, and endometrial cancers [57, 58].

KMT2A and KMT2B (MLL1 and MLL4) writers: MLL1/KMT2A is the founding member of the KMT2 family of methyltransferases, which was originally observed in the 11q23 chromosomal translocation known to be the key driver of acute lymphoblastic leukemia (ALL) and acute myeloid leukemia (AML). This family of methyltransferases carries the unique COMPASS protein subunits Menin, and HCF1 or HCF2. MLL1 and MLL4 both contain a N-terminal CXXC domain and a C-terminal SET domain. MLL1, which also contains an AT-hook, stably binds to mitotic chromatin at both enhancer and promoter regions through multivalent interactions of its AT-hook and CXCC domains with AT-rich and unmethylated CpG dinucleotides, respectively [58]. It remains bound to the chromatin during DNA replication and mitosis to ensure the propagation of a cell's transcription state to the daughter cells. MLL1 and MLL4 catalyze high levels of mono-, di-, and low levels of tri- methylation of H3K4, and MLL1 is localized to regions of RNA polymerase II activity, specifically at the 5′ end of actively transcribed genes [59]. MLL1 is also associated with microRNAs involved in cancer and hematopoiesis. Loss of MLL1 affects embryogenesis, transcriptional elongation and cancer development. Both MLL1 and MLL4 harbor cleavage sites for threonine aspartase, taspase1, hence their activity is also regulated by Taspase1. Furthermore, MLL1 is regulated by histone H2B ubiquitination. While the role of 11q23 translocations impacting MLL1 is well established in ALL and AML, analyses of the Catalogue of Somatic Mutations in Cancer (COSMIC) database showed that greater than 60% of lung, breast, bladder, endometrium, and large intestine cancer patients harbored one or more mutations in MLL1, but fewer in MLL4.

KMT2C and KMT2D (MLL3 and MLL2) writers: MLL2 and MLL3 primarily catalyze mono-methylation of H3K4 at enhancer regions [60, 61]. The unique COMPASS protein subunits in MLL2 and MLL3 are PTIP, PA1, NCOA6, and UTX. The stable binding of MLL2 and MLL3 to chromatin is mediated by their high mobility group I (HMG-I) and LXXLL binding motifs that are common in most transcription factors and coactivators [58]. Both MLL2 and MLL3 contain seven PHD domains via which they bind to arginine residues H3R3me0 and H3R3me2a within intergenic regions [62]. MLL2 is a known co-activator of the estrogen receptor (ER)-, and is required for ER-α transcriptional activity as well as proliferation of ER-α positive breast cancer cell lines, such as MCF7 [63]. MLL3 mono-methylation is essential for IgG class switching, adipogenesis, and nuclear receptor co-activators [54]. Loss of MLL3 results in developmental defects, reduction in white adipose tissue, embryogenesis and growth [58]. Analysis of the COSMIC database showed that MLL2 and MLL3 were mutated in >60% of lung, large intestine, breast, endometrium, and bladder cancers, and ~25% of non-Hodgkin lymphoma, medulloblastoma, and primitive neuroendodermal cancers. A recent clinical study on 46 Chinese PCa patients revealed that MLL2 is mutated in 63% of the samples and drives PCa progression by activating LIFR and KLF4 [64]. Whole exome sequencing studies of 50 mCRPC samples identified recurrent MLL2 mutations in 8.6% of patients [51].

KMT2F and KMT2G (SETD1A and SETD1B) writers: SETD1A and SETD1B catalyze mono-, di-, and tri-methylation of H3K4. In most cells, these enzymes catalyze the bulk of H3K4me3 methylations. They are the smallest subgroup of the KMT2 family and contain a SET domain that catalyzes methylation, an RNA recognition motif (RRM) at the N-terminal, a N-SET domain at the C-terminal, a WD-repeat 82 domain that interacts with RNA Pol II, and a CXXC finger domain that preferentially binds to CpG dinucleotides. Genome-wide ChIP-seq analysis of SETD1A and SETD1B binding patterns reveal that they preferentially bind to gene promoter regions justifying their strong preference for H3K4 tri-methylation [58, 65]. SETD1A is recruited in a transcription factor dependent manner and is essential for development, cell proliferation, and induced pluripotency [58]. Analysis of lung, large intestine, endometrium, liver and skin datasets in the COSMIC database found that ~62% of the chosen cancers harbored SETD1A mutations and ~60% harbored mutations in the SETD1B enzyme [58]. However, a similar analysis for the prostate adenocarcinoma dataset showed no evidence of SETD1A or SETD1B mutations (<0.5%) or dysregulation in PCa, consistent with other reports [54].

SMYD family methylation writers: SET and MYND domain-containing proteins (SMYD) are another family of five histone and non-histone substrate methyltransferases, of which only SMYD1, and SMYD3 are known to catalyze mono-, di-, and tri-methylation of H3K4 [66–68] (also see Fig. 1). SMYD2, on the other hand catalyzes H3K36 methylation [68]. SMYD3 monomethylates H4K5 [69] and trimethylates H4K20 [70]. SMYD family KMTs contains a bi-lobal architecture with the SET and MYND domains on the N-terminal lobe and a TPR-like domain on the C-terminal lobe. The interface between the N- and C-terminal lobes act as cavernous binding sites for protein substrates, which bind a wide variety of proteins including p53 (protein data bank ID: 3TG5), and ER-α (protein data bank ID: 406F) [67]. SMYD family proteins are key regulators of development and function of skeletal and cardiac muscles, and specifically SMYD1 is essential for thick filament organization of myosin [71].

SMYD3 is overexpressed in a broad range of cancers, including prostate, breast, and many colorectal and hepatocellular carcinomas. It regulates transcriptional activity of nearly 80 genes by forming a complex with RNA pol II and binding to their promoter regions. A well-studied example is the SMYD3 regulation of homeobox gene *NKX2.8* that is commonly upregulated in hepatocellular carcinoma. The role of SMYD3 in oncogenesis was demonstrated by its siRNA-mediated knockdown, which suppressed proliferation in HCC, while its introduction into 3T3 cells enhanced cell growth [72]. IHC staining in paired normal prostate and prostate tumor samples showed that SMYD3 was highly upregulated in tumors, and was primarily localized to the cytoplasm (in 92% of samples) compared to the nucleus (32% of samples) [73]. These findings suggest that the role of SMYD3 in cellular function extends beyond its methyltransferase activity. AR signaling was downregulated upon siRNA and shRNA mediated knockdown of SMYD3 leading to increased apoptosis and S phase accumulation, and decreased proliferation, colony formation, transwell cell migration, and invasion [73]. Interestingly, MMP-9, which is transcriptionally regulated by AR and associated with invasiveness in PCa cells, was upregulated by SMYD3 activity. This points to a cross-talk between the SMYD3 and AR regulatory networks [74, 75].

Currently, much research is focused on developing strategies to effectively inhibit SMYD3 [76]. For instance, one group synthesized a small molecule inhibitor BCl-121, that interrupts the SMYD3-substrate interaction at the histone peptide binding channel and thus prevents SMYD3 chromatin localization. Treatment with a high dose of 100 μM was able to induce S-phase arrest and thereby inhibit growth of cancer cells that specifically overexpress SMYD3. Noteworthy, this growth inhibitory effect of BCl-121 was observed in the LNCaP and DU145 prostate cancer cell lines, where proliferation was reduced by approximately 70% [77]. Epizyme has developed sulfonamides and sulfamides that inhibit the enzyme, including the orally bioavailable EPZ031686 [78]. GSK designed a small molecule, GSK2807, based on the crystal structure of SMYD3/MEKK2/SAH that can compete with SAM to link SMYD3 with its substrate [79]. H3K4 methylation is also catalyzed by other KMTs such as SETD7, and RMTs such as PRDM9. For detailed reviews on this topic see reference [67].

H3K4 Methylation Erasers

H3K4 demethylation is primarily catalyzed by KDM1A, KDM1B, KDM5A, KDM5B, KDM5C, KDM5D, and MAPJD KMTs [31, 33].

KDM1 family H3K4 methylation erasers: The KDM1 family consists of KDM1A/LSD1 and KDM1B/LSD2, both of which demethylate H3K4me1 and H3K4me2 residues (see Fig. 2) and is one of the well-studied lysine demethylases. LSD1 is a known activator of AR, and its overexpression is essential for the maintenance of malignant phenotypes in mCRPC, and also in other cancers including bladder, colorectal, AML, neuroblastoma, and estrogen-receptor-negative breast cancer [36, 66, 80]. LSD1 dysregulation is an oncogenic driver of PCa due to its ability to: (1) control phenotypic plasticity of PCa cells [81, 82], (2) promote AR-independent survival in mCRPC cells in a non-canonical, demethylase-independent manner [83], (3) regulate expression of the AR-V7 splice variant [84], (4) activate mCRPC gene networks independent of its demethylase activity [83], and (5) control metastasis of CRPC [85]. It also demethylates p53 and represses p53 function, inhibits differentiation of neuroblastoma and leukemia, and interacts with EWS/FLI1 in osteosarcoma [35]. LSD2 is a homolog of LSD1, but little is known about its role in cancer.

KDM5 family H3K4 methylation erasers: The KDM5/JARID1 family of lysine methyltransferases, KDM5A/JARID1A, KDM5B/JARID1B, KDM5C/JARID1C, and KDM5D/JARID1D, catalyze demethylation of di-, and trimethylated H3K4. The JARID1 family of proteins is unique among all JmJc family KDMs in that they contain a DNA binding ARID domain and a methylation binding PHD domain intersecting their JmJN and JmJC domains [86]. The KDM5 clusters are primarily associated with the removal of the active H3K4me2 and H3K36me2 marks that leads to gene silencing and are important for neuronal and hematopoietic development, as well as drug resistance [87]. KDM5A, KDM5B, and KDM5C function as oncogenic drivers [88] and are overexpressed in prostate cancer [33, 89], while KDM5D functions as a putative tumor suppressor and is down-regulated/deleted in prostate cancer [88]. KDM5A, originally identified as a retinoblastoma-binding protein (RBP) [90], is an integral part of the Notch/RBP-J repressor complex mediated gene silencing machinery that regulates Notch-signaling pathways. KDM5A plays an important role in homeostasis and carcinogenesis [35, 91] and is implicated in dysregulated Notch signaling with concomitant increase in Notch family proteins such as Jagged2, Notch3, and Hes6. KDM5A has been observed in highly invasive, high grade prostate cancer [92–94]. KDM5A overexpression has been linked to chemoresistance in both prostate and lung cancer cell lines [33, 80]. KDM5B, also known to demethylate H3K4me1 [29], is associated with normal development since it protects developmental genes from aberrant H3K4me3 modifications [36]. Overexpression of KDM5B attenuated transcription of genes related to melanoma progression [33] and promoted the aggressiveness of non-small cell lung cancer [95], while its knockdown inhibited tumor growth [33]. KDM5B is also regulated by AKT levels where it was recently shown that AKT inhibition in PTEN knockout mice reduced expression of KDM5B [96]. Analysis of Oncomine data showed that KDM5C expression was elevated in mCRPC and PCa, and could be used as a biomarker to predict the metastatic potential of primary tumors [88]. Furthermore, KDM5B has been shown to regulate a number of genes including BRCA1, CAV1, and HOXA5 and as a result is one of most altered KDMs in a number of cancers [31, 97]. The KDM5C gene, found on the X chromosome, is overexpressed in prostate cancer and is a prognostic marker for tumor relapse post radical prostatectomy [98]. Moreover, BRD4 mediated transcription of KDM5C led to mCRPC cell sensitivity to BET inhibitors [99]. Its inactivation triggered genomic instability in sporadic renal cancer [100]. KDM5C, independent of its enzymatic activity, functions as an oncogene by repressing the TGFβ-dependent transcription factor SMAD3 which in turn promotes PCa tumorigenesis [88]. KDM5D, a homolog of KDM5C found on the Y chromosome, is a putative tumor suppressor that is downregulated in prostate cancer. A recent analysis of the Oncomine datasets showed that prostate tumors with high Gleason score expressed lower levels of KDM5C, its loss lead to docetaxel resistance, and could be used as a marker for prostate cancer invasion and metastasis [101]. Analysis of whole exome sequencing data of metastatic prostate tumors showed that KDM5D was deleted in 25% of cases and perturbed nearly three times more frequently than other KDMs and KMTs [101]. An earlier study on prostate tumor samples obtained through radical prostatectomy found that 52% of prostate epithelial cells had a deletion of the genomic region containing KDM5D [102].

NO66 (nucleolar protein 66) is another JmJC-domain containing demethylase of H3K4me1 and H3K4me3. It is an oncogenic driver in prostate cancer that regulates genes associated with survival, invasion and metastasis [103].

Epigenetic Therapies Targeting H3K4 Methylation

Due to the high prevalence of mutations in many cancer types, there have been many efforts to target the MLL complex as a therapeutic target [104]. One group developed MM-401, the first small molecule inhibitor to block the KMT2A/WDR5 interaction, which was shown to ablate KMT2A HMTase activity while not affecting other KMT2 family members [105]. Recent studies have focused on peptidomimetic and small molecules that interrupt KMT2A/WDR5 binding, as well as small molecules that target the KMT2A/menin interaction [106–108]. Recently it was discovered that prostate cancer can specifically be targeted by using a small-molecule that inhibits this MLL1/menin interaction [52]. Additionally, a number of compounds have been developed to target the activity of KDM1A/LSD1 (see reference [109] for the complete list). In

Table 1 we provide a list of drugs in development and in clinical trials for epigenetic targeting of H3K4 methylation writers and erasers.

H3K9 Methylation

H3K9me1 is enriched at active promoters thus marking active gene transcription, while both H3K9me2, and H3K9me3 are significantly enriched within silent genes thus serving as a mark for heterochromatin and gene silencing [11]. Microarray analysis of PCa tumor samples showed that patients with increased levels of PSA also displayed elevated levels of H3K9me2 and H3K9me3 [3]. Methylated H3K9 preferentially recruits Heterochromatin protein 1 (HP1), which is known to repress transcription of euchromatic genes and promote the formation of heterochromatin [33].

H3K9 Methylation Writers

Methyltransferases known to be involved in the methylation of H3K9 belong to the KMT1 (SUV39H1/KMT1A, SUV39H2/KMT1B, G9a/EHMT2/KMT1C, GLP/EHMT1/KMT1D, SETDB1/KMT1E, and SETDB2/KMT1F), and the KMT8 (PRDM2) family of methyltransferases. Both of these families of methyltransferases contain a SET domain that mediates enzymatic activity. The SET domain in the KMT1/SUV39 (Suppressor of variegation 3–9) subfamily of KMTs are flanked by two cysteine-rich domains called pre-SET and post-SET. These domains bind to three zinc ions, a crucial step required for the enzymatic activity of the SET domain [110].

KMT1A/B (SUV39H1/2) writers: The SUV39 homologs SUV39H1 and SUV39H2 were the first SET domain containing methyltransferases identified. These enzymes primarily function within highly compact and transcriptionally silent chromatin regions called pericentric heterochromatin [111]. They are known to di- or trimethylate H3K9me1 substrates and their activity is critical for the establishment and maintenance of heterochromatin structure and genome stability [112–114]. SUV39H1/2 function as tumor suppressors that tightly control the repression of oncogenes within heterochromatin. They

Epigenetic Regulation of Chromatin in Prostate Cancer

Table 1 Epigenetic therapeutic drugs targeting H3K4 methylation writers and erasers

Drug class/name	Target protein	Clinical phase	Status in CRPC	Status in other cancers	Reference
KO-539	MLL/ menin	In development	–	–	Kura Oncology, Inc.
BCI-121	SMYD3	Pre-clinical	–	Colorectal, lung, pancreatic, prostate, ovarian cancer	[77]
EPZ031686	SMYD3	Pre-clinical	–	–	[78]
Seclidemstat (SP-2577)	LSD1	Phase I	–	Advanced solid tumors	NCT03895684
GSK2879552	LSD1	Clinical	–	Phase I AML, Phase I SCLC	NCT02034123, NCT02177812
INCB059872	LSD1	Phase I/II	–	Advanced cancer	NCT02712905

are aberrantly expressed in many cancers including prostate, AML, lung, liver and colorectal cancers. SUV39H1/2 both contain a chromatin modifier domain (chromodomain) that recognizes and binds to methylated lysines. SUV39H2 is a known co-activator of AR that is overexpressed in prostate cancer. It colocalizes with the AR coregulator melanoma antigen-11 (MAGE-A11) [99] that enhances AR transcriptional activity.

KMT1C and KMT1D (G9a and GLP) writers: G9a and GLP (G9a-like protein) are also members of the SUV39 KMT subfamily that catalyze mono- and di-methylation of H3K9 in euchromatin and facultative heterochromatin regions. Unlike SUV39H1/2, which are involved in stable repression, G9a and GLP proteins play central roles in dynamic transcriptional repression [110]. They form homo- and hetero-dimers and share 80% sequence identity in their catalytic domains [115]. G9a and GLP methylate both histone and non-histone substrates. For instance, G9a- and GLP-catalyzed methylation of LaminB1 plays an important role in regulating H3K9me2-marked heterochromatin anchorage to the nuclear periphery [116]. In addition to the SUV39 specific SET domain, G9a and GLP also contain seven ankyrin repeat domains that mediate protein–protein interactions [117, 118]. Interestingly, G9a can act both as a coactivator and a corepressor of transcription. For instance, G9a association with mediator complex MED1 leads to transcriptional activation whereas G9a association with the K3K4 demethylase KDM5A leads to transcriptional repression [119]. The activity of G9a is essential for prostate differen-

tiation. Its interactions with the homeobox gene *NKX3.1* form a highly regulated transcriptional network comprised of NKX3.1, G9a, and the H3K27me3 demethylase UTY, which regulates normal prostate differentiation. Dysregulation of this network results in predisposition to prostate cancer [120]. Similarly, G9a activity promotes coactivation of Runx2-induced expression of genes involved in processes such as epithelial to mesenchymal transition to induce prostate cancer progression and metastasis [121]. G9a is overexpressed in many types of cancers, including prostate cancer [122], esophageal squamous cell carcinoma, hepatocellular carcinoma, aggressive lung cancer, brain cancer, multiple myeloma, and aggressive ovarian carcinoma. Higher expression levels of G9a correlate with poor prognosis.

H3K9 Methylation Erasers

Enzymes demethylating mono-, di-, and tri-methylation marks on H3K9 belong to the KDM/ JHDM2, KDM4/JMJD2, and KDM7 subfamilies of lysine demethylases [34]. Additionally, the H3K4 demethylase, LSD1/KDM1A associates with AR and de-represses AR target genes by demethylating H3K9 without altering the status of the H3K4 methylations [123].

KDM3 methylation erasers: The KDM3 subfamily of demethylases only contain the catalytic JmJC domain and consist of three members, namely KDM3A/JHDM2A, KDM3B/JHDM2B, and KDM3C/JHDM3C. JHDM2A/B demethylate H3K9me1 and H3K9me2 [124]. However, there is no direct evidence of JHDM2C demethylating H3K9. JHDM2A is overexpressed in colorectal cancer and renal cell carcinoma [33]. In the former, higher expression of JHDM2C correlated with poor prognosis [33].

KDM4 methylation erasers: The KDM4 subfamily of KDMs consists of five members, KDM4A–E/JMJD2A–E, all of which contain the N-terminal JmJN and the catalytic JmJC domain. The KDM2A-C proteins also contain two PHD domains and two Tudor domains, that are essential for their function as histone methylation readers. These three proteins share higher than 50% sequence identity. KDM4A-D preferentially demethylate H3K9me2 and H3K9me3 while KDM4E only catalyzes the demethylation of H3K9me3 to H3K9me1 through the removal of two methyl groups [37]. Analysis of expression data for diseased and normal tissues showed that the KDM4 family is expressed in 6–9 diseased tissues, with the highest expression in the testes and spleen, making them highly relevant to prostate cancer [37]. Similarly, analysis of cancer mutation datasets from the CBio portal and TCGA showed that the KDM4 encoding genes, particularly KDM4C, harbor a variety of structural variations in prostate, lung, breast, and esophageal cancers, as well as lymphoma [37]. KDM4A interaction with activating protein transcription factors plays a major role in controlling proliferation, apoptosis and differentiation. KDM4B interacts with nuclear receptors and is particularly important in prostate and breast cancers. It is highly expressed in estrogen-positive breast cancer and regulates ER mediated transcription in an estrogen dependent manner [125]. In prostate cancer, KDM4B enhances AR stability through inhibition of AR ubiquitination [126]. Similarly, KDM4C strongly associates with AR and the H3K4 demethylase KDM1A/LSD1 and promotes H3K9me3 demethylation in an AR-ligand dependent manner, which in turn activates AR transcription [127].

Epigenetic Therapies Targeting H3K9 Methylation

There are currently no ongoing clinical trials targeting H3K9 methylation enzymes. However, several drugs are under development as shown in Table 2. Furthermore, compounds targeting LSD1 (GSK2879552, and INCB059872, see Table 1) are known to affect H3K9 levels in cells.

H3K27 Methylation

H3K27 methylation is a well-studied repressive histone PTM and is a hallmark of facultative chromatin [132]. H3K27 is either di- or tri-methylated by the enzymatic activity of the Polycomb group (PcG) protein complexes. Apart from its role as transcriptional repressor, H3K27 methylation is critical in regulating genes essential for cellular differentiation and proliferation. H3K27 methylation is dysregulated in a large number of cancers, including prostate cancer, particularly due to mutations and aberrations in its reader, writer, and eraser enzymes.

H3K27 Methylation Writers

KTM6A/B (EZH2/1) writers of H3K27: EZH1 (Enhancer of Zeste Homolog 1) and EZH2 (Enhancer of Zeste Homolog 2) are methyltransferases that catalyze the tri-methylation of H3K27. EZH1 or EZH2 constitute the catalytic subunit of the Polycomb repressor complex PRC2, that also contain the Suz12 zinc finger domain, and the EED domain that recognizes tri-methylated peptides [133]. EZH1- and EZH2-containing PRC2 complexes are found in 1:1 stoichiometry, and the latter effectively methylates H3K27. Thus, EZH2 plays a central role in governing H3K27 mediated gene repression. A number of studies have documented up-regulation of EZH2 in many tumor types, including prostate cancer, breast cancer, and lymphomas, where the expression level appears to correlate with disease progression [134, 135]. Overexpression of EZH2 in prostate cancer cell lines increases invasiveness, while EZH2 knockdown decreases proliferation, with the effect being more pronounced in AR-independent prostate cancer lines. Interestingly, EZH2 mRNA and protein levels are low in benign prostate and increase progressively from localized to metastatic tumors, suggesting that EZH2 could be a useful prognostic indicator as well as a potential therapeutic target [136].

Table 2 Epigenetic therapeutic drugs targeting H3K9 methylation writers and readers

Drug class/name	Target protein	Clinical phase	Status in CRPC	Status in other cancers	Reference
Chaetocin	SUV39H1	Preclinical	–	AML	[128]
BIX-01294	G9a	Preclinical	–	Neuroblastoma	[129]
UNC0638	G9a	Preclinical	–	Breast	[130]
BRD4770	G9a	Preclinical	–	Pancreas	[131]

EZH2 expression is also regulated by micro-RNA-101, which is encoded by a locus that is commonly deleted in prostate cancer. One or both alleles of the *miR-101* locus is deleted in 37.5% of clinically localized prostate cancer and 66.7% of metastatic prostate tumors, suggesting that loss of micro-RNA-101 leads to EZH2 overexpression and cancer progression mediated by deregulated epigenetic mechanisms [137]. Independent of its polycomb repressor function, EZH2 also has non-canonical oncogenic roles. For instance, in LNCaP cell lines, phosphorylation of EZH2 at Ser21 by the activity of AKT kinase lead to a massive increase in AR-regulated gene transcription [138]. In prostate cancer, EZH2 manifests its oncogenic activity primarily through repression of target genes including p16INK4alpha, DAB2IP, ADRB2, WNT pathway antagonists, VASH1 and CDH1, among others; a majority of these being tumor suppressors. EZH2 activity is also regulated by long noncoding RNAs (lncRNAs), which are implicated in PCa development and progression, e.g., Metastasis-associated lung adenocarcinoma transcript-1 (*MALAT1*). EZH2 binds to *MALAT1*, and knockdown of *MALAT1* impairs EZH2 recruitment to its target loci, thereby upregulating the expression of EZH2 repressed genes [139]. Furthermore, EZH2 activity affects the genome-wide three-dimensional structure of chromatin, thus making its role even more important in cancer development and progression. The gain of function mutation (EZH2 Y646X) in lymphoma completely inactivates selected topologically associated domains (TADs) that encode for tumor suppressor genes [140]. Together, these findings point to an all-encompassing role of EZH2 in cancer thus opening up new therapeutic avenues that target the canonical and non-canonical roles of EZH2 for efficacious therapeutic management and treatment of mCRPC.

H3K27 Methylation Erasers

KDM6 family of H3K27 erasers: The KDM6 family demethylases, KDM6A/UTX, KDM6B/JMJD3, and KDM6C/UTY catalyze demethylation of H3K27e2/3 substrates [31, 34, 141], as well as H3K27me1 substrates [141]. This subfamily contains the catalytic JmJC domain, and in addition KDM6A/C contains an eight TPR (tetratricopeptide) repeat (also see the uniport database) that mediates protein-protein interactions. KDM6A is dysregulated in a number of cancers including urothelial carcinoma, breast cancer and lymphoblastic leukemia [142]. Analysis of paired prostate cancer and high-grade prostate intraepithelial neoplasia found mutations in KDM6A that were specific to primary prostate cancer suggesting a role in early prostate cancer development [143]. Furthermore, KDM6A physically interacts with AR and functions as an oncogene that drives prostate tumor progression [51]. Similar functions have been observed for KDM6C, which is part of the NKX3.1-G9a-KDM6C transcriptional regulatory pathway that governs prostate differentiation [120] (also see section on "H3K9 Methylation Writer KMT1C").

Epigenetic Therapies Targeting H3K27 Methylation

Upregulation of EZH2 expression in wide variety of cancers has suggested opportunities to develop inhibitors targeting its oncogenic activity. Multiple early phase clinical trials of EZH2 inhibitors are currently ongoing in hematological malignancies as well as solid tumors, particularly in prostate cancer. One particular trial is a Phase 1b/2 study with oral administration of CPI-1205 (ClinicalTrials.gov Identifier: NCT03480646) in combination with either enzalutamide or abiraterone/prednisone in patients with CRPC. This study is designed to determine the maximum tolerated dose (MTD) and recommended Phase II dose (RP2D) based on safety, tolerability, pharmacokinetic, and efficacy profiles of CPI-1205 in combination with either enzalutamide or abiraterone/prednisone. A phase 1 dose escalation and expanded cohort study of PF-06821497 (ClinicalTrials.gov Identifier: NCT03460977) has been implemented for the treatment of adult patients with relapsed/refractory small cell lung cancer and CRPC. Another study testing GSK-J4, which inhibits KDM6A/B activity and thereby inhibits AR deletion mutants that lack LBD (ARΔLBD), significantly reduced proliferation of CRPC cells [144]. A list of pre-clinical and clinical trials targeting H3K27 is presented in Table 3.

H3K36 Methylation

H3K36 mono-, di-, and tri-methylation are well-established histone PTM marks for active transcription and euchromatin formation, and their removal triggers transcriptional repression. H3K36me2 levels are elevated in the promoter regions of actively transcribed genes and are localized to the 5' regions, while H3K36me3 is predominantly localized to the 3' regions [148, 149]. H3K36 methylation has been implicated in a variety of nuclear processes including transcriptional regulation, gene dosage compensation, pre-mRNA splicing, DNA replication, recombination, and DNA damage repair [150]. H3K36 methylation levels are highly dysregulated in prostate cancer mainly due to aberrations in the levels of its methylation enzymes and its upstream-binding partners. For example, we pre-

viously reported the interactions between EZH2 and H3K36 methyltransferase NSD2/MMSET in regulating levels of H3K36 [151].

H3K36 Methylation Writers

There are at least eight known H3K36 methyltransferases in mammalian cells, of which six belong to the KMT3 subfamily of methyltransferases (NSD1/KMT3B, NSD2/KMT3G, NSD3/KMT3F, SETD2/KMT3A, SETD3/KMT3E, SMYD2/KMT3C), with the other two being SETMAR, and ASH1L/KMT2H [33, 34]. Of these, only NSD1/2 and SETD2 and SETD3 have significant methyltransferase activity and display dysregulation in cancer. NSD2 (nuclear receptor binding SET domain containing protein 2, also known as KDM3G/MMSET/WHSC1), is the best-studied methyltransferase known to mono-, and di-methylate H3K36, and has an established role in prostate cancer progression. NSD2 is overexpressed in mCRPC compared to primary prostate tumors, and its expression is strongly correlated with disease stage and poor prognosis [152]. NSD2 functions in association with a number of other upstream proteins that regulate H3K36 methylation. We showed previously that the oncogenic activity of NSD2 is regulated by EZH2, where overexpression of EZH2 lead to oncogenic phenotypes that were characterized by increased proliferation, self-renewal, and invasion, only in cells expressing NSD2. Similarly, others have found NSD2 to be a strong coactivator of the NF-kB signaling pathway and NSD2 to be a transcriptional target highly enriched in components of the NF-kB network, including IL-6, IL-8, Birc5, and VEGFA. NSD2 has also been implicated in prostate cancer through its

Table 3 Epigenetic therapeutic drug targeting H3K27 methylation writers and readers

Drug class/ name	Target protein	Clinical phase	Status in CRPC	Status in other cancers	Reference
SGC0946	EZH2	Preclinical	–	Leukaemia	[145]
DZNeP	EZH2	Preclinical	Preclinical	Breast, colon, prostate	[146]
GSK126	EZH2	Phase I	–	Non-Hodgkin's lymphoma, solid tumours and multiple myeloma	[147]
CPI1205	EZH2	Clinical	Phase 1b/2	Prostate cancer, B-cell lymphomas	NCT03480646
Tazemetostat	EZH2	Phase II	–	Advanced solid tumors, non-Hodgkin lymphoma	NCT03213665

involvement in epithelia-mesenchymal-transition, which is known to drive metastasis in different cancer types. While NSD2 knockdown in benign prostate cells did not affect the levels of H3K36me2 and H3K27me3, knockdown of NSD2 in the CRPC cell lines, DU145 and PC3, altered both H3K36me2 and H3K27me3 levels and decreased cell proliferation, soft agar colony formation, migration, and invasion [153]. Further analysis revealed that NSD2 regulates the oncogenic phenotype through its interactions with TWIST1. These results establish NSD2-mediated epigenetic regulation as a major factor in cancer development.

H3K36 Methylation Erasers

H3K36 demethylation is catalyzed by the KDM2 subfamily of demethylases, KDM2A/JHDM1A and KDM2B/JHDM1B. Both KDM2A and its paralog KDM2B contain an N-terminal JmJC domain, followed by a CXXC zinc finger domain that binds to unmethylated CpG dinucleotides, a PHD domain that recognizes methylated lysine residues, and a Fbox domain [124]. Additionally, KDM2A contains six leucine-rich repeats at its C-terminal compared to two in KDM2B. They both catalyze demethylation of mono- and di-methylated H3K36. Interestingly, the KDM2A/B demethylase genes encode various spliced isoforms in addition to the full length proteins, but these isoforms are only expressed in mammals. Though the spliced short length proteins do not have any methyltransferase activity, they play important roles in biological processes [150]. A number of studies point to the role of KDM2B's oncogenic functions in cancer development and progression. KDM2B dysregulation is commonly seen in a majority of cancers, including, prostate, breast, pancreatic, gastric, lung, and bladder cancers, as well as ALL and AML. Its aberrant expression and activity leads to dysregulation of key cellular function such as apoptosis (inhibition of c-FLIP/c-Fos), proliferation (activation of PI3K/mTOR), inhibition of Wnt signaling (ubiquitination of β-Catenin), autophagy (inhibition of ERK) and cellular senescence (through inhibition of p53). KDM2B also interacts with EZH2 and promotes cell proliferation and metastasis, even-

tually leading to drug resistance [154]. Furthermore, a recent study on the cell line DU145 identified the role of KD2MB expression in modulating cell motility; KDM2B overexpression suppresses the expression of epithelial markers E-cadherin and the tight junction protein ZO-1 (thereby reducing cell-cell adhesion and increasing cell migration), and positively regulates the expression of the RhoA and RhoB GTPases (which leads to cytoskeletal rearrangement and increased cell motility) [155]. There have been significant challenges in the field to develop an inhibitor that exhibits specificity for NSD2, and several companies are actively pursuing this objective.

H3K79 Methylation

Lysine 79 on Histone H3 is located inside the globular domain, unlike K4, K9, K27, and K36 that are in the N-terminal unstructured tail. It is di- and tri-methylated by the non-SET domain containing methyltransferase KMT4/DOT1L and the H3K79me2 and H3K79me3 marks are associated with active transcription, cell cycle regulation, genome stability, and DNA damage response [156]. DOT1L methylation of H3K79 is an excellent example of crosstalk between histone PTMs. DOT1L interacts with ubiquitinated-H2BK120 and H4 to cooperatively reorient K79 buried in the H3 globular region to an accessible position [157, 158]. There are presently no known demethylases of H3K79.

H3K79 Methylation Writer

DOT1L: Disruptor of telomeric silencing 1 (DOT1) was first identified through a genetic screen for proteins whose over-expression lead to impaired telomeric silencing in yeast [159]. DOT1L transfers methyl groups from *S*-adenosyl-L-methionine (SAM) to mono-, di-, and tri-methylate H3K79. The structure of DOT1L is very similar to arginine methyltransferases but so far there has been no evidence for its involvement in arginine methylation. The seven-stranded beta sheets in the open α/β structure of DOT1L is characteristic of class I SAM-dependent methyltransferases [156]. DOT1L and H3K79 methylation have been implicated in diverse types of cancers including prostate, breast and lung cancer, as well as leukemia. Its role in the initiation and maintenance of mixed lineage leukemia (MLL)-rearranged leukemia has been widely studied. DOT1L interacts with MLL fusion proteins, leading to enhanced H3K79 methylation, maintenance of open chromatin, overexpression of downstream oncogenes and leukemogenesis. A DOT1L-HES6 gene fusion has been reported to drive AR-negative prostate cancer through overexpression of HES6 [160]. Loss of DOT1L has been shown to inhibit cell proliferation [161]. DOT1L was found to methylate AR and regulate its transcriptional activity through lncRNA-dependent mechanisms [162].

Epigenetic Therapies Targeting DOT1L

Therapeutic targeting of DOT1L holds significant promise due to its status as the sole H3K79 histone methyltransferase, its unique non-SET catalytic domain and its role in MLL leukaemogenesis. The first DOT1L specific small molecular inhibitor EPZ004777 displayed high specificity against DOT1L, and treatment with this inhibitor induced apoptosis in MLL-rearranged leukemia cells *in vitro* and also blocked leukemia progression in mice by suppressing the expression of HOXA cluster genes and Meis1 [163]. Due to its poor half-life and metabolic instability, modifications to EPZ00477 led to the synthesis of EPZ5676 [164] and SGC0946 [165] which showed improved binding affinity for DOT1L. EPZ5676 also displayed synergistic anticancer effects against MLL-rearrangement leukemia cells when used in combination with cytarabine and daunorubicin [164], and continuous intravenous treatment of rat xenografts with EPZ5676 led to dose-dependent leukemia regression [166]. Another novel DOT1L inhibitor SYC-522, synthesized by adding an additional urea group to the structure of SAH, showed increased specificity for DOT1L as well as increased anti-cancer efficacy. Treatment of MLL-rearranged leukemia cells with SYC-522 induced cell cycle arrest and cell differentiation, and treatment of primary MLL-rearranged AML

cells resulted in up to 50% decrease in colony formation and promotion of monocytic differentiation [167]. The role of DOT1L in prostate cancer has not been well characterized, and therapeutic targeting of DOT1L to treat PCa and CRCPC is still a work in progress [141].

Histone Lysine Methylation Readers

Methylation reader proteins display high specificity for methylated histone lysines. These proteins act upstream of the molecular machinery associated with specific methylation marks, and their function is indispensable for epigenetic regulation of cell function. They contain one of the following methyl-lysine recognition motifs/domains, namely ADD (ATRX-DNMT3-DNMT3L), ankyrin, bromo-adjacent homology (BAH), chromo-barrel, chromodomain, double chromodomain (DCD), MBT (malignant brain tumor), PHD, tandem PHD, PWWP (Pro-Trp-Trp-Pro), Tudor domin (TTD), tandem TTD, WD40/β-propeller, zinc finger CW (zf-CW). For specific details on these domains, see references [168, 169].

Histone Acetylation in Prostate Cancer

Histone acetylation is an important epigenetic modification in which an acetyl group from acetyl coenzyme A (acetyl-CoA) is transferred to the ε-amino group of lysine residues in histones and other proteins. Acetylation, in effect, abolishes the interaction of histones with the negatively charged DNA backbone leading to open chromatin structures that facilitate the binding of RNA Pol II. In prostate cancer, acetylation of histones at gene promoter and/or enhancer regions increases AR activity and cell survival. *In vitro* and *in vivo* studies of primary and metastatic prostate cancer cells have demonstrated a significant positive correlation between the levels of prostate specific antigen (PSA) mRNA and H3 acetylation at the PSA enhancers and proximal promoter [170]. Besides its role as a mark for active transcription, acetylation has also been found to be important for genome stability, protein stability, regulation of protein-protein interactions, and even chromatin compaction [171]. Tissue microarray analysis of primary prostate cancer patients showed that acetylation of H3K18, H3K9 and H4K12, in combination with demethylation of H3K4 and H4R3 could be used as a prognostic biomarker to estimate risk for prostate tumor recurrence in patients with low-grade tumors [27]. A genome wide study of acetylation patterns at 3286 gene promoter regions in CD4+ T cells detected the presence of 17 distinct acetylation patterns [172].

Histone acetylation is catalyzed by histone acetyltransferases (HATs) which are broadly divided into five subfamilies, namely GCN5/PCAF, MYST, TAFII250, CBP/p300, and SRC. Histone deacetylation is regulated by the activity of Histone deacetylases (HDACs) which are subdivided into four subfamilies, namely class I, II, III, and IV [173]. The histone acetyl marks in turn are recognized by bromodomain-containing proteins that act upstream of the molecular machinery associated with acetylation regulated cellular signaling pathways (see Fig. 3). Dysregulation and mutations in HATs and HDACs have been observed in a number of cancers and play a driver role in cancer initiation and growth [173]. We will next discuss the role of these enzymes in the onset and metastatic progression of prostate cancer.

Acetylation Writers: HATs in Prostate Cancer

CBP/p300: The CBP/p300 family of nuclear phosphoproteins contain the ubiquitously expressed and homologous (~61% homology) CBP (KAT3A) and p300 (KAT3B) proteins that participate in a number of physiological processes such as cell proliferation, differentiation, and apoptosis [174, 175]. Both CBP and p300 contain three CH and two ZZ zinc finger domains, a bromodomain that recognizes acetylated residues, and a catalytic acetyltransferase domain that catalyzes histone acetylation [175]. CBP/p300 is one of the well-studied histone acetyltransferases that acetylates H3 lysines 14 and 18, H4 lysines 5 and 8, H2A lysine 5, and H2B lysine 12 and 15 [21]. CBP/p300 HAT also acetylates other proteins; a recent study estimated that 411 proteins were part of the CBP/p300 interactome, and that 615 genes were part of the p300/CPB cistrome [175]. A majority of these proteins include components of the transcription complex and major signal transduction pathways, thus making CBP/p300 a central component of the cellular machinery that coordinates signal flow to regulate gene transcription [174].

CBP/p300 proteins are broadly linked to cancer as well as several human pathologies. The role of CBP/p300 in cancer as an oncogene or tumor suppressor is debatable and may be context dependent. In prostate cancer, CBP/p300 dis-

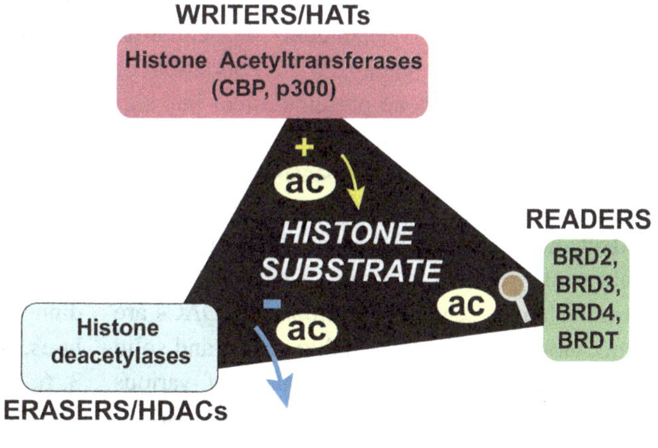

Fig. 3 Cartoon showing the function of histone acetyltransferases (HATs) and histone deacetylases (HDAC) and bromodomain containing histone reader proteins on a generic histone substrate

plays oncogenic properties by acting as a co-activator of the AR, which is the main oncogenic driver in prostate cancer. The formation of this AR–coactivator complex at AR-binding sites promotes chromatin opening and recruitment of the transcriptional machinery to target genes [176]. Expression of histone acetyltransferase p300 and AR also correlates positively in human prostate cancer specimens, especially those marked with PTEN loss. Mechanistically, PTEN loss induces AR phosphorylation at serine 81 (Ser81) to promote p300 binding and acetylation of AR, thereby precluding its polyubiquitination and degradation. Thus, p300 acetyltransferase regulates AR degradation and PTEN-deficient prostate tumorigenesis [177–180].

Acetylation Erasers: HDACs in Prostate Cancer

Histone Deacetylases (HDACs) belong to a family of enzymes that catalyze the removal of the acetyl group from ε-N-acetyl lysine amino acid. The 18 known HDACs are grouped into four major families: **class I** (reduced potassium dependency 3 (RPD3)-like proteins HDAC1, HDAC2, HDAC3, and HDAC8), **class II** (Histone deacetylase 1 (HDA1)-like proteins HDAC4, HDAC5, HDAC6, HDAC7, HDAC9, and HDAC10), **class III** (silent information regulator 2 (Sir2)-like proteins SIRT1, SIRT2, SIRT3, SIRT4, SIRT5, SIRT6, SIRT7), and **class IV** (protein HDAC11) [181].

The catalytic activity in class I, II and IV HDACs is driven by zinc (Zn^{2+})-dependent deacetylation reaction, while that in class III HDACs is driven by a nicotinamide adenine dinucleotide (NAD^+)-dependent deacetylation mechanism. HDAC catalyzed deacetylation of histone and non-histone proteins affect key signaling pathways including cell cycle, apoptosis, DNA damage response, metastasis, angiogenesis, and autophagy. Dysregulated HDAC functions affect one or more of the mentioned pathways that could serve as a driver of cancer development. Though genetic alterations in HDACs are rare, most hematological malignancies and solid tumors display aberrant expression of various HDACs (see reference [182] for the full list). In most cases, higher levels of HDAC expression are associated with advanced disease and poor prognosis. Studies on prostate cancer samples, cell lines, and mouse models have shown that HDAC1, 2, and 3 are overexpressed in prostate cancer and that increased HDAC2 expression correlates with shorter PSA relapse after radical prostatectomy [183].

The ubiquitous role of HDACs in cell function makes them attractive targets for therapeutic interventions in prostate cancer patients. So far, many compounds known as HDAC inhibitors (HDACi) have been developed to inhibit the activity of these HDAC complexes. Trichostatin A (TSA) was one of the first natural compounds found to be a potent HDACi. Some of the early studies tested the efficacy of TSA and sodium butyrate as HDAC inhibitors in the prostate cancer line PC3 and found them to induce differentiation and apoptosis [184]. Another compound, suberoylanilide hydroxamic acid (SAHA; Vorinostat) was the first HDACi approved by the US FDA for the treatment of cancer and was found to inhibit proliferation of the prostate cancer cell lines LNCaP, DU-145 and PC3 as well as prostate tumors in animal models of prostate cancer [185]. More recently, a compound called PAC-320 was shown to induce G2/M arrest and apoptosis in human prostate cancer cells 125178668 *in vitro*. After the initial success with these agents in other cancer types, there have been a number of Phase I and Phase II clinical trials on HDACi conducted in individuals with advanced prostate cancer where they have been tested as single agent or in combination with other anti-cancer agents [186–188]. These clinical trial results have been mixed with only marginal response and slightly improved outcomes. Therefore, further investigation is necessary to clarify the benefits and drawbacks of these medications.

Sirtuins: The Silent Information Regulator 2 (SIR2) proteins, Sirtuins (SIRT1–7), belong to the Class III family of Nicotinamide Adenine dinucleotide (NAD^+)-dependent histone deacetylases. Sirtuins deacetylate both histone (SIRT1–3, 6, 7) and nonhistone substrates (SIRT1–3, 5, 7). They differ greatly in their functions and localizations. SIRT1 is the best-characterized member of mammalian sirtuins and is involved in several cellular processes such as metabolism, DNA repair, recombination, aging, apoptosis and cellular senescence. Aberrant expression of sirtuin proteins has been reported in many diseases, including Bowen's disease, type I diabetic nephropathy, Alzheimer disease and amyotrophic lateral sclerosis, nonalcoholic fatty liver disease and cancer. SIRT1 and SIRT2 levels are upregulated in many cancers including prostate cancer, thus potentially functioning as oncogenes. One study [189] compared the expressions of SIRT1 and SIRT2 in a variety of CRPC cell lines (LNCaP, 22Rv1, PC-3 and DU145), with normal prostate epithelial PrEC cells, and normal prostate stromal PrSC cells. Their data demonstrated that SIRT1 and SIRT2 are significantly upregulated in all CRPC cell lines compared to normal prostate cells. Moreover, immunohistochemical analysis of human tissues showed that SIRT2 was significantly upregulated in prostate cancer compared to normal prostate. They also observed that chemical inhibition and/or genetic knockdown of Sirt1 caused a FoxO1-mediated inhibition in the growth and viability of human PCa cells [189]. Another study showed that SIRT1 levels are significantly elevated in mouse and human prostate cancer [190]. Overexpression of SIRT1 induces epithelial-mesenchymal transition (EMT) by inducing EMT transcription factors like ZEB1 to promote prostate cancer cell migration and metastasis [191]. Furthermore, SIRT1 associates with and deacetylates matrix metalloproteinase-2, and regulates its expression by controlling protein stability through the proteasomal pathway, and enhances tumor cell invasion in prostate cancer cells [192].

One study showed that SIRT1, by physically interacting and cooperating with MPP8, represses E-cadherin expression and promotes EMT in prostate cancer cells [193]. Nicotinamide N-methyltransferase (NNMT) is an important activator and stabilizer of SIRT1, and overexpression of NNMT in PC-3 prostate cancer cells upregulates SIRT1 expression, leading to enhanced cell invasion and migration [194].

Despite mounting evidences on the involvement of sirtuins in cancer development, their role as oncogenes or tumor suppressor genes is not well established. Evidence supporting the latter was provided in a study where mesenchymal stem cells overexpressing SIRT1 significantly suppressed prostate cancer growth by promoting the recruitment of Natural Killer cells and macrophages as anti-tumor effectors [195]. Another study showed that SIRT1 repressed androgen responsive gene expression and induced autophagy in the prostate, and that disruption of this SIRT1-dependent autophagy checkpoint in the prostate resulted in prostatic intraepithelial neoplasia (PIN) lesion formation [196]. Thus, these reports further highlight the role of SIRT1 as a tumor suppressor in prostate cancer, which is contradictory to previous reports suggesting its role in oncogenesis.

Acetylation Readers

BET bromodomain proteins: The bromodomain and extra-terminal (BET) family of proteins are an important class of epigenetic readers of acetylated histones regulating a vast network of protein expression across many different cancers. The BET subfamily is made up of the four members BRD2, BRD3, and BRD4, and BRDT all of which contain two bromodomains (BD1 and BD2) at the N terminal and an extra-terminal domain and a C-terminal domain [171, 197]. The bromodomain motifs are histone acetylation readers due to their ability to recognize and bind to acetylated lysines on histone tails (primarily on H4) and form a scaffold for the assembly of multi-protein complexes. They also recognize acetylated non-histone proteins. Well known examples of BET non-histone activity include its role in regulating the transcriptional activity of NF-kappaB, and of ERG in acute myeloid leukemia. BRD4 also binds to FLI1, MYB, SPI1, CEBPA, and p53 in a bromodomain independent manner [197].

BET proteins are usually part of large nuclear complexes and play decisive roles in cellular processes such as transcription, replication, chromatin remodeling, DNA damage and cell growth. Embryonic lethality upon knockdown of these proteins highlights their indispensable role in normal physiological processes. For example, BRD4 is associated with a coactivator complex of transcription factors [198] and promotes transcriptional elongation by increasing the processivity of RNA polymerase II, leading to expression of growth-promoting genes [199, 200]. The critical requirement of BET proteins in these basic cellular processes explains their dysregulation in terms of overexpression or recurrent translocations in many human cancers such as B-cell lymphoma and NUT midline carcinoma. BET proteins are essential components of the AR transcription machinery that drives AR signaling in both PCa and mCRPC [201] and are significantly over-expressed in mCRPC [202]. Overexpression of BET family proteins increase DNA accessibility, which could be used to identify advanced mCRPC from primary prostate tumors.

Therapeutic targeting of BET proteins with BET inhibitors (BETi) is an attractive area of clinical development [197]. BET inhibitors target bromodomains on BET proteins and abrogate oncogenic signaling commonly mediated by distal regulatory regions such as enhancers and superenhancers. The centrality of BET proteins in signaling mediated by AR, ETS fusions, and MYC, makes it an attractive candidate for BETi therapy. Treatment with the bromodomain inhibitor JQ1 suppressed c-Myc function and suppressed ligand-independent prostate cancer cell survival [203]. Our group recently demonstrated that JQ1 and I-BET762, two selective small-molecule inhibitors that target the dual N-terminal bromodomains of BRD4, exhibit antiproliferative effects in prostate cancer cells as well as xenograft mouse models [201]. We further showed that BET bromodomain inhibitors enhance efficacy and disrupt resistance to AR antagonists such as enzalutamide in the treatment of mCRPC as observed by enhanced prostate tumor growth inhibition when enzalutamide and JQ1 were combined together [204]. The next generation BET inhibitors, such as biBET, MT1, and AZD5153, aim to target both the BET bro-

modomains in a bivalent mode that could lead to stronger inhibitor activity. These compounds have shown promising result *in vitro*, and their clinical translation is underway. None-the-less, resistance to BETi therapy eventually develops, and mechanisms of BETi resistance have been documented [197]. For instance, BET resistance in prostate cancer patients carrying SPOP mutations has been shown to be due to increased or decreased degradation of BET proteins. In a recent study, we demonstrated that efficacy of BETi in prostate cancer might be limited due to acquired resistance mechanisms such as reactivation of AR signaling by CDK9 and PRC2 mediated silencing of DDR genes [205].

Targeted induced degradation is another effective approach to inhibit protein activity and function. Degradation of BET proteins using proteolysis targeting chimeras (PROTACs) is emerging as an effective strategy to inhibit their function [64, 206, 207]. The first BET PROTACs, including dBET1, MZ1 and ARV-825 [208–210], used JQ1 for the BET inhibitor and Thalidomide, VHL-ligand, and Thalidomide, respectively, as the small molecule targeting the E3-ubiquitin ligase. Of these, ARV-825 showed higher potency causing BET protein degradation in cell line models of Burkitt's lymphoma and led to MYC suppression and apoptosis induction [210]. Similarly, dBET1 also triggered apoptosis in primary human AML cells and tumor inhibition in xenograft studies [208]. Modifications to ARV-825, wherein the small molecule component was changed to a VHL-ligand, led to the development of ARV-771 with superior PK/PD and efficacy in 22Rv1 and VCaP xenograft models of castration-resistant prostate cancer [211].

Epigenetic Therapies Targeting Histone Acetylation

Histone acetylases and deacetylases have been a major target of epigenetics-based therapies. In Table 4 we provide a list of drugs presently in pre-clinical development or clinical trials for use as therapeutic agents in mCRPC and other cancers. The inherent dependency of AR activity on CBP/p300 coactivators makes them attractive tar-

Table 4 Epigenetic therapeutic drugs targeting BETs and HDACs in prostate clinical trails

Drug class/name	Target protein	Clinical phase	Status in mCRPC	Status in other cancers	Reference
OTX015/ MK-8628	BET BD	Phase I	Phase I	NUT midline carcinoma, triple negative breast, NSCL, pancreatic	NCT02259114
ZEN003694		Phase I	Phase I (mCRPC)		NCT02711956 (Recruiting)
		Phase I	Phase I (mCRPC)		NCT02705469 (Active)
GSK525762		Phase II	–	Solid tumors	NCT03266159
BMS-986158		Phase I/II	–	Advanced solid tumors	NCT02419417
ABBV-075		Phase I	Phase I	Breast, NCLS, AML, multiple myeloma, prostate, small-cell lung, non-Hodgkin's lymphoma	NCT02391480
GSK2820151		Phase I	–	Advanced or recurrent solid tumors	NCT02630251
GS-5829		Phase I/II	mCRPC + enzalutamide		NCT02607228
INCB057643		Phase I/II	–	Advanced solid tumors and hematologic malignancy	NCT02711137
Pracinostat (SB939)	HDAC	Phase II	mCRPC	–	NCT01075308
		Phase I	Prostate cancer	Head and neck cancer, esophageal cancer	NCT00670553
Panobinostat (LBH589)		Phase I/II	mCRPC (+bicalutamide)	–	NCT00878436
		Phase II	mCRPC	–	NCT00667862
		Phase I	mCRPC (+docetaxel and prednisone)	–	NCT00663832
		Phase I		–	NCT00493766
		Phase I		–	NCT00419536
Vorinostat (SAHA, MK0683)		Phase II	mCRPC	–	NCT00330161
		Phase II	Primary prostate cancer (+bicalutamide.)	–	NCT00589472

gets for therapeutic interventions to treat prostate cancer. Inhibitors such as MS2126, MS7972, and Ischemin were developed to target the CBP/p300 bromodomain, which in turn restores levels of the tumor suppressor protein p53. Second generation compounds such as SGC-CBP30, PF-CBP1 and I-CBP112 to target CBP/p300 bromodomains and restore p53 activity showed better selectivity at nanomolar concentrations. I-CBP112 confirmed potential involvement of CBP/p300 in self-renewal of leukemia cells, and more recently it has been reported to stimulate the catalytic

activity of CBP/p300 proteins with loss of function mutations in tumors with inherently low acetylation levels. A more advanced analogue, GNE-049 which showed low nanomolar potency and over 4000-fold selectivity for CBP/p300, demonstrated improved *in vitro* and *in vivo* activity in preclinical models of mCRPC [212]. Similarly, A-485, a potent, selective and drug-like catalytic inhibitor of p300 and CBP, selectively inhibited proliferation in lineage-specific tumor types, including AR-positive prostate cancer and several hematological malignancies.

A-485 inhibited the AR transcriptional program in both androgen-sensitive and castration-resistant prostate cancer and inhibited tumor growth in a castration-resistant xenograft model [213]. Taken together, these data strongly support CBP/p300 inhibition as a therapeutic strategy in mCRPC. To validate these findings in a clinical setting, CellCentric has initiated and is currently recruiting for a Phase I/IIa clinical trial (NCT03568656) for their lead CBP/p300 bromodomain inhibitor CCS1477 as a mono or combination therapy with Enzalutamide and/or Abiraterone in metastatic prostate cancer and other solid tumors.

Multiple early phase clinical trials of BET inhibitors are currently ongoing in hematological malignancies as well as solid tumors. These include a study of the novel BET inhibitor FT-1101 (ClinicalTrials.gov Identifier: NCT02543879) in patients with relapsed or refractory hematologic malignancy and a study of RO6870810 (ClinicalTrials.gov Identifier: NCT03068351) as single agent and combination therapy in advanced multiple myeloma. Another Phase I study of CPI0610 in patients with previously treated multiple myeloma and lymphoma, demonstrated changes in the expression of MYC and other genes in malignant tumor cells; changes in cellular proliferation and in the extent of apoptosis (ClinicalTrials.gov Identifier: NCT02157636). Many novel BET inhibitors are also being tested for safety, tolerability and efficacy for metastatic prostate cancer. For instance, a novel and highly potent small molecule BET inhibitor from Zenith Epigenetics called ZEN003694 is currently in Phase 1b/2a alone (ClinicalTrials.gov Identifier: NCT02705469) as well as in combination with enzalutamide in patients with abiraterone refractory but enzalutamide naïve mCRPC (ClinicalTrials.gov Identifier: NCT02711956).

Chromatin Remodeling Complexes

DNA accessibility to transcription factors can be modulated either by the deposition/removal of histone PTMs, as discussed earlier, or by the activity of chromatin remodeling enzymes/remodeler complexes. ATP-dependent chromatin remodeling enzymes include multi-subunit complexes of the Snf2 family, which are evolutionarily conserved from yeast to human. They are highly abundant in the cell with roughly one enzyme per ten nucleosomes [214]. Chromatin remodeling complexes utilize energy from ATP hydrolysis to disrupt DNA-histone contacts and slide, eject, or alter the position of histone octamers that allows for transcription of previously inaccessible DNA regions. They are recruited to specific sites through their interactions with cell specific transcriptional regulators. Remodeler complexes are broadly grouped into four subfamilies, namely: ISWI (imitation switch), CHD (chromodomain, helicase, DNA binding), INO80 (inositol requiring 80) and SWI/SNF (switching defective/sucrose nonfermenting). These proteins contain a conserved ATPase domain that facilitates nucleic acid binding and ATP hydrolysis, but differ in their flanking domains. For example, proteins of the SWI/SNF family contain a bromodomain that recognizes acetylated histone lysines, the CHD family contains two chromodomains that recognize methylated histone lysines, and the ISWI family contains HAND, SANT, and SLIDE domains that all recognizes nucleosomes and internucleosomal DNA. For more details on the mechanism of remodeler complexes, see references [214, 215]. Remodeler complexes have been implicated in a variety of cellular processes such as gene expression, DNA replication, repair, chromosomal recombination and mitosis, and their dysregulation, particularly SWI/SNF, has been observed in ~20% of human cancers. Even though mutations in SWI/SNF genes are not common in prostate cancer, accumulating evidence suggests its influence on AR signaling, ERG mediated transcription, cell cycle and DNA methylation. Understanding the role of these remodeler complexes and their interactions with other epigenetic marks and their enzymes would facilitate our understanding of how chromatin is regulated by non-genetic factors and their role in various pathologies including cancer.

Conclusions and Outlook

In this chapter, we have presented a brief overview of how histone methylation and acetylation based epigenetic modifications govern cell function through their ability to modulate DNA accessibility and recruit other effector proteins to regulate gene transcription. We have discussed the role of the major histone marks and their regulatory enzymes in prostate cancer development and metastasis. It is becoming increasingly clear that the epigenetic code, contained in the reversible, heritable, covalent modifications of DNA and histones, is as significant as the genetic code in diversity, complexity, and functionality. The advent of high throughput sequencing methods, particularly ChIP-seq, has greatly helped us gain a genome-wide view of the mechanisms and dynamics of histone post translational modifications. Interestingly, the role of epigenetic modifications has been reported in nearly all cellular functions including cell proliferation, differentiation, motility, cytoskeletal reorganization, apoptosis, chemoresistance and embryonic development.

Aberrant expression of PTM-regulating enzymes, specifically HMTs, HATs, HDACs and demethylases, has been reported in many types of blood and solid tumors. It is believed that the initial stages of tumor development are entirely regulated by epigenetic mechanisms until a stabilizing genomic mutation or structural variation appears and defines a tumor phenotype. Most histone PTMs play dual roles, both as tumor suppressors and oncogenes, and their interaction with nuclear receptors drive the progression of hormone-dependent cancers. In prostate cancer, AR interactions with acetylation and methylation enzymes regulate ligand-dependent and independent transcription of AR target genes. Epigenetic mechanisms play a central role in rewiring of the AR transcriptional network that drives a primary prostate tumor to mCRPC and also endows them with resistance to androgen deprivation therapies [216]. Given their importance in cancer, epigenetic regulatory enzymes have become major targets for therapeutic intervention. There is a heavy focus on the development of epigenetic drugs, and their number in clinical trials is steadily increasing. Presently there are nearly 250 monotherapy or combination therapies in phase I/II trials with epigenetic drugs (see ClinicalTrials. gov).

Acknowledgements We thank Reyaz Ur Rasool and Qu Deng for stimulating discussions. Research in the Asangani Laboratory is supported by the Department of Defense Idea Development Award (W81XWH17-1-0404 to I.A.A).

References

1. S.A. Grigoryev, C.L. Woodcock, Chromatin organization - the 30 nm fiber. Exp. Cell Res. **318**(12), 1448–1455 (2012)
2. D.V. Fyodorov et al., Emerging roles of linker histones in regulating chromatin structure and function. Nat. Rev. Mol. Cell Biol. **19**(3), 192–206 (2018)
3. J. Ellinger et al., Global levels of histone modifications predict prostate cancer recurrence. Prostate **70**(1), 61–69 (2010)
4. S. Venkatesh, J.L. Workman, Histone exchange, chromatin structure and the regulation of transcription. Nat. Rev. Mol. Cell Biol. **16**(3), 178–189 (2015)
5. J.M. Belton et al., Hi-C: a comprehensive technique to capture the conformation of genomes. Methods **58**(3), 268–276 (2012)
6. J. Dekker et al., The 4D nucleome project. Nature **549**(7671), 219–226 (2017)
7. G. Andrey, S. Mundlos, The three-dimensional genome: regulating gene expression during pluripotency and development. Development **144**(20), 3646–3658 (2017)
8. F. Spitz, E.E. Furlong, Transcription factors: from enhancer binding to developmental control. Nat. Rev. Genet. **13**(9), 613–626 (2012)
9. J.O. Carlsten, X. Zhu, C.M. Gustafsson, The multitalented Mediator complex. Trends Biochem. Sci. **38**(11), 531–537 (2013)
10. B. Bonev, G. Cavalli, Organization and function of the 3D genome. Nat. Rev. Genet. **17**(11), 661–678 (2016)
11. A. Barski et al., High-resolution profiling of histone methylations in the human genome. Cell **129**(4), 823–837 (2007)
12. S.-H. Song, T.-Y. Kim, CTCF, cohesin, and chromatin in human cancer. Genomics Inform. **15**(4), 114–122 (2017)
13. M. Nowacka-Zawisza, E. Wiśnik, DNA methylation and histone modifications as epigenetic regulation in prostate cancer (Review). Oncol. Rep. **38**(5), 2587–2596 (2017)

14. C.E. Massie, I.G. Mills, A.G. Lynch, The importance of DNA methylation in prostate cancer development. J. Steroid Biochem. Mol. Biol. **166**, 1–15 (2017)
15. Y. Wu, M. Sarkissyan, J.V. Vadgama, Epigenetics in breast and prostate cancer. Methods Mol. Biol. **1238**, 425–466 (2015)
16. R. Zelic et al., Global DNA hypomethylation in prostate cancer development and progression: a systematic review. Prostate Cancer Prostatic Dis. **18**(1), 1–12 (2015)
17. M. Ngollo et al., Epigenetic modifications in prostate cancer. Epigenomics **6**(4), 415–426 (2014)
18. S.B. Baylin, P.A. Jones, A decade of exploring the cancer epigenome - biological and translational implications. Nat. Rev. Cancer **11**(10), 726–734 (2011)
19. X. Wu, Y. Zhang, TET-mediated active DNA demethylation: mechanism, function and beyond. Nat. Rev. Genet. **18**(9), 517–534 (2017)
20. K. Ruggero et al., Epigenetic regulation in prostate cancer progression. Curr. Mol. Biol. Rep. **4**(2), 101–115 (2018)
21. T. Kouzarides, Chromatin modifications and their function. Cell **128**(4), 693–705 (2007)
22. P. Chi, C.D. Allis, G.G. Wang, Covalent histone modifications—miswritten, misinterpreted and miserased in human cancers. Nat. Rev. Cancer **10**(7), 457–469 (2010)
23. S.R. Bhaumik, E. Smith, A. Shilatifard, Covalent modifications of histones during development and disease pathogenesis. Nat. Struct. Mol. Biol. **14**(11), 1008–1016 (2007)
24. H. Huang et al., SnapShot: histone modifications. Cell **159**(2), 458–458.e1 (2014)
25. B.D. Strahl, C.D. Allis, The language of covalent histone modifications. Nature **403**(6765), 41–45 (2000)
26. B.A. Benayoun et al., H3K4me3 breadth is linked to cell identity and transcriptional consistency. Cell **158**(3), 673–688 (2014)
27. D.B. Seligson et al., Global histone modification patterns predict risk of prostate cancer recurrence. Nature **435**(7046), 1262–1266 (2005)
28. F. Valdés-Mora et al., Acetylated histone variant H2A.Z is involved in the activation of neo-enhancers in prostate cancer. Nat. Commun. **8**(1), 1346 (2017)
29. E.L. Greer, Y. Shi, Histone methylation: a dynamic mark in health, disease and inheritance. Nat. Rev. Genet. **13**(5), 343–357 (2012)
30. R.F. Luco et al., Regulation of alternative splicing by histone modifications. Science **327**(5968), 996–1000 (2010)
31. K. Hyun et al., Writing, erasing and reading histone lysine methylations. Exp. Mol. Med. **49**(4), e324 (2017)
32. V.K. Rao, A. Pal, R. Taneja, A drive in SUVs: from development to disease. Epigenetics **12**(3), 177–186 (2017)
33. I. Hoffmann et al., The role of histone demethylases in cancer therapy. Mol. Oncol. **6**(6), 683–703 (2012)
34. X. Zhang, H. Wen, X. Shi, Lysine methylation: beyond histones. Acta Biochim. Biophys. Sin. Shanghai **44**(1), 14–27 (2012)
35. A. D'Oto et al., Histone demethylases and their roles in cancer epigenetics. J. Med. Oncol. Therap. **1**(2), 34–40 (2016)
36. A. Janardhan et al., Prominent role of histone lysine demethylases in cancer epigenetics and therapy. Oncotarget **9**(76), 34429–34448 (2018)
37. R.M. Labbé, A. Holowatyj, Z.-Q. Yang, Histone lysine demethylase (kdm) subfamily 4: structures, functions and therapeutic potential. Am. J. Transl. Res. **6**(1), 1–15 (2014)
38. N.D. Heintzman et al., Distinct and predictive chromatin signatures of transcriptional promoters and enhancers in the human genome. Nat. Genet. **39**(3), 311–318 (2007)
39. N.D. Heintzman et al., Histone modifications at human enhancers reflect global cell-type-specific gene expression. Nature **459**(7243), 108–112 (2009)
40. L.M. Soares, Determinants of histone H3K4 methylation patterns Rpb4-Set1 fusion. Mol. Cell **68**, 773–785 (2017)
41. G. Liang et al., Distinct localization of histone H3 acetylation and H3-K4 methylation to the transcription start sites in the human genome. Proc. Natl. Acad. Sci. U. S. A. **101**(19), 7357–7362 (2004)
42. B. Jin, Y. Li, K.D. Robertson, DNA methylation: superior or subordinate in the epigenetic hierarchy? Genes Cancer **2**(6), 607–617 (2011)
43. D. Faucher, R.J. Wellinger, Methylated H3K4, a transcription-associated histone modification, is involved in the DNA damage response pathway. PLoS Genet. **6**(8), e1001082 (2010)
44. D.B. Seligson et al., Global levels of histone modifications predict prognosis in different cancers. Am. J. Pathol. **174**(5), 1619–1628 (2009)
45. T. Bianco-Miotto et al., Global levels of specific histone modifications and an epigenetic gene signature predict prostate cancer progression and development. Cancer Epidemiol. Biomark. Prev. **19**(10), 2611–2622 (2010)
46. Q. Wang et al., Androgen receptor regulates a distinct transcription program in androgen-independent prostate cancer. Cell **138**(2), 245–256 (2009)
47. X.S. Ke et al., Genome-wide profiling of histone h3 lysine 4 and lysine 27 trimethylation reveals an epigenetic signature in prostate carcinogenesis. PLoS One **4**(3), e4687 (2009)
48. H.H. He et al., Nucleosome dynamics define transcriptional enhancers. Nat. Genet. **42**(4), 343–347 (2010)
49. M.S. Geybels et al., PTEN loss is associated with prostate cancer recurrence and alterations in tumor DNA methylation profiles. Oncotarget **8**(48), 84338–84348 (2017)
50. J.M. Spangle et al., PI3K/AKT signaling regulates H3K4 methylation in breast cancer. Cell Rep. **15**(12), 2692–2704 (2016)

51. C.S. Grasso et al., The mutational landscape of lethal castration-resistant prostate cancer. Nature **487**(7406), 239–243 (2012)
52. R. Malik et al., Targeting the MLL complex in castration-resistant prostate cancer. Nat. Med. **21**(4), 344–352 (2015)
53. T. Kouzarides, SnapShot: histone-modifying enzymes. Cell **131**(4), 822 (2007)
54. A. Shilatifard, The COMPASS family of histone H3K4 methylases: mechanisms of regulation in development and disease pathogenesis. Annu. Rev. Biochem. **81**(1), 65–95 (2012)
55. T. Miller et al., COMPASS: a complex of proteins associated with a trithorax-related SET domain protein. Proc. Natl. Acad. Sci. U. S. A. **98**(23), 12902–12907 (2001)
56. J.J. Meeks, S. Ali, Multiple roles for the MLL/COMPASS family in the epigenetic regulation of gene expression and in cancer. Annu. Rev. Cancer Biol. **1**(1), 425–446 (2017)
57. C. Kandoth et al., Mutational landscape and significance across 12 major cancer types. Nature **502**(7471), 333–339 (2013)
58. R.C. Rao, Y. Dou, Hijacked in cancer: the KMT2 (MLL) family of methyltransferases. Nat. Rev. Cancer **15**(6), 334–346 (2015)
59. M.G. Guenther et al., Global and Hox-specific roles for the MLL1 methyltransferase. Proc. Natl. Acad. Sci. U. S. A. **102**(24), 8603–8608 (2005)
60. M. Wu et al., Molecular regulation of H3K4 trimethylation by Wdr82, a component of human Set1/COMPASS. Mol. Cell. Biol. **28**(24), 7337–7344 (2008)
61. L. Wu et al., ASH2L regulates ubiquitylation signaling to MLL: trans-regulation of H3 K4 methylation in higher eukaryotes. Mol. Cell **49**(6), 1108–1120 (2013)
62. S.S. Dhar et al., Trans-tail regulation of MLL4-catalyzed H3K4 methylation by H4R3 symmetric dimethylation is mediated by a tandem PHD of MLL4. Genes Dev. **26**(24), 2749–2762 (2012)
63. R. Mo, S.M. Rao, Y.-J. Zhu, Identification of the MLL2 complex as a coactivator for estrogen receptor alpha. J. Biol. Chem. **281**(23), 15714–15720 (2006)
64. S. Lv et al., Histone methyltransferase KMT2D sustains prostate carcinogenesis and metastasis via epigenetically activating LIFR and KLF4. Oncogene **37**(10), 1354–1368 (2018)
65. C. Deng et al., USF1 and hSET1A mediated epigenetic modifications regulate lineage differentiation and HoxB4 transcription. PLoS Genet. **9**(6), e1003524 (2013)
66. R.A. Varier, H.T. Timmers, Histone lysine methylation and demethylation pathways in cancer. Biochim. Biophys. Acta **1815**(1), 75–89 (2011)
67. P.A. Boriack-Sjodin, K.K. Swinger, Protein methyltransferases: a distinct, diverse, and dynamic family of enzymes. Biochemistry **55**(11), 1557–1569 (2016)
68. K. Leinhart, M. Brown, SET/MYND lysine methyltransferases regulate gene transcription and protein activity. Genes (Basel) **2**(1), 210–218 (2011)
69. G.S. Van Aller et al., Smyd3 regulates cancer cell phenotypes and catalyzes histone H4 lysine 5 methylation. Epigenetics **7**(4), 340–343 (2012)
70. F.Q. Vieira et al., SMYD3 contributes to a more aggressive phenotype of prostate cancer and targets Cyclin D2 through H4K20me3. Oncotarget **6**(15), 13644–13657 (2015)
71. S.J. Du, X. Tan, J. Zhang, SMYD proteins: key regulators in skeletal and cardiac muscle development and function. Anat. Rec. (Hoboken) **297**(9), 1650–1662 (2014)
72. R. Hamamoto et al., SMYD3 encodes a histone methyltransferase involved in the proliferation of cancer cells. Nat. Cell Biol. **6**(8), 731–740 (2004)
73. C. Liu et al., SMYD3 as an oncogenic driver in prostate cancer by stimulation of androgen receptor transcription. J. Natl. Cancer Inst. **105**(22), 1719–1728 (2013)
74. A.M. Cock-Rada et al., SMYD3 promotes cancer invasion by epigenetic upregulation of the metalloproteinase MMP-9. Cancer Res. **72**(3), 810–820 (2012)
75. T. Hara et al., Androgen receptor and invasion in prostate cancer. Cancer Res. **68**(4), 1128–1135 (2008)
76. G. Rajajeyabalachandran et al., Therapeutical potential of deregulated lysine methyltransferase SMYD3 as a safe target for novel anticancer agents. Expert Opin. Ther. Targets **21**(2), 145–157 (2017)
77. A. Peserico et al., A SMYD3 small-molecule inhibitor impairing cancer cell growth. J. Cell. Physiol. **230**(10), 2447–2460 (2015)
78. L.H. Mitchell et al., Novel oxindole sulfonamides and sulfamides: EPZ031686, the first orally bioavailable small molecule SMYD3 inhibitor. ACS Med. Chem. Lett. **7**(2), 134–138 (2016)
79. G.S. Van Aller et al., Structure-based design of a novel SMYD3 inhibitor that bridges the SAM-and MEKK2-binding pockets. Structure **24**(5), 774–781 (2016)
80. J. McGrath, P. Trojer, Targeting histone lysine methylation in cancer. Pharmacol. Ther. **150**, 1–22 (2015)
81. S. Hino, K. Kohrogi, M. Nakao, Histone demethylase LSD1 controls the phenotypic plasticity of cancer cells. Cancer Sci. **107**(9), 1187–1192 (2016)
82. L. Ellis, M. Loda, LSD1: a single target to combat lineage plasticity in lethal prostate cancer. Proc. Natl. Acad. Sci. U. S. A. **115**(18), 4530–4531 (2018)
83. A. Sehrawat et al., LSD1 activates a lethal prostate cancer gene network independently of its demethylase function. Proc. Natl. Acad. Sci. U. S. A. **115**(18), E4179–E4188 (2018)
84. S. Regufe da Mota et al., LSD1 inhibition attenuates androgen receptor V7 splice variant activation in castration resistant prostate cancer models. Cancer Cell Int. **18**, 71 (2018)

85. A. Ketscher et al., LSD1 controls metastasis of androgen-independent prostate cancer cells through PXN and LPAR6. Oncogene 3, e120 (2014)

86. P.A. Cloos et al., Erasing the methyl mark: histone demethylases at the center of cellular differentiation and disease. Genes Dev. **22**(9), 1115–1140 (2008)

87. J. Plch, J. Hrabeta, T. Eckschlager, KDM5 demethylases and their role in cancer cell chemoresistance. Int. J. Cancer **144**(2), 221–231 (2019)

88. F. Crea et al., The emerging role of histone lysine demethylases in prostate cancer. Mol. Cancer **11**, 52 (2012)

89. Y. Xiang et al., JARID1B is a histone H3 lysine 4 demethylase up-regulated in prostate cancer. Proc. Natl. Acad. Sci. U. S. A. **104**(49), 19226–19231 (2007)

90. J.R. Horton et al., Characterization of a linked Jumonji domain of the KDM5/JARID1 family of histone H3 lysine 4 demethylases. J. Biol. Chem. **291**(6), 2631–2646 (2016)

91. R. Liefke et al., Histone demethylase KDM5A is an integral part of the core Notch-RBP-J repressor complex. Genes Dev. **24**(6), 590–601 (2010)

92. F.L. Carvalho et al., Notch signaling in prostate cancer: a moving target. Prostate **74**(9), 933–945 (2014)

93. Q. Su, L. Xin, Notch signaling in prostate cancer: refining a therapeutic opportunity. Histol. Histopathol. **31**(2), 149–157 (2016)

94. L. Marignol et al., Hypoxia, notch signalling, and prostate cancer. Nat. Rev. Urol. **10**(7), 405–413 (2013)

95. K.T. Kuo et al., Histone demethylase JARID1B/KDM5B promotes aggressiveness of non-small cell lung cancer and serves as a good prognostic predictor. Clin. Epigenetics **10**(1), 107 (2018)

96. M.I. Khan et al., AKT inhibition modulates H3K4 demethylase levels in PTEN-null prostate cancer. Mol. Cancer Ther. **18**(2), 356–363 (2019)

97. J. Taylor-Papadimitriou, J. Burchell, JARID1/KDM5 demethylases as cancer targets? Expert Opin. Ther. Targets **21**(1), 5–7 (2017)

98. J. Stein et al., KDM5C is overexpressed in prostate cancer and is a prognostic marker for prostate-specific antigen-relapse following radical prostatectomy. Am. J. Pathol. **184**(9), 2430–2437 (2014)

99. Z. Hong et al., KDM5C is transcriptionally regulated by BRD4 and promotes castration-resistance prostate cancer cell proliferation by repressing PTEN. Biomed. Pharmacother. **114**, 108793 (2019)

100. B. Rondinelli et al., Histone demethylase JARID1C inactivation triggers genomic instability in sporadic renal cancer. J. Clin. Invest. **125**(12), 4625–4637 (2015)

101. N. Li et al., JARID1D is a suppressor and prognostic marker of prostate cancer invasion and metastasis. Cancer Res. **76**(4), 831–843 (2016)

102. G. Perinchery et al., Deletion of Y-chromosome specific genes in human prostate cancer. J. Urol. **163**(4), 1339–1342 (2000)

103. K.M. Sinha et al., Oncogenic and osteolytic functions of histone demethylase NO66 in castration-resistant prostate cancer. Oncogene **38**(25), 5038–5049 (2019)

104. M. Vedadi et al., Targeting human SET1/MLL family of proteins. Protein Sci. **26**(4), 662–676 (2017)

105. F. Cao et al., Targeting MLL1 H3K4 methyltransferase activity in mixed-lineage leukemia. Mol. Cell **53**(2), 247–261 (2014)

106. H. Karatas et al., High-affinity, small-molecule peptidomimetic inhibitors of MLL1/WDR5 protein–protein interaction. J. Am. Chem. Soc. **135**(2), 669–682 (2013)

107. G. Senisterra et al., Small-molecule inhibition of MLL activity by disruption of its interaction with WDR5. Biochem. J. **449**(1), 151–159 (2013)

108. J. Grembecka et al., Menin-MLL inhibitors reverse oncogenic activity of MLL fusion proteins in leukemia. Nat. Chem. Biol. **8**(3), 277–284 (2012)

109. A. Jambhekar, J.N. Anastas, Y. Shi, Histone lysine demethylase inhibitors. Cold Spring Harb. Perspect. Med. **7**(1), a026484 (2017)

110. C. Mozzetta et al., Sound of silence: the properties and functions of repressive Lys methyltransferases. Nat. Rev. Mol. Cell Biol. **16**(8), 499–513 (2015)

111. H. Wu et al., Structural biology of human H3K9 methyltransferases. PLoS One **5**(1), e8570 (2010)

112. D. Wang et al., Methylation of SUV39H1 by SET7/9 results in heterochromatin relaxation and genome instability. Proc. Natl. Acad. Sci. U. S. A. **110**(14), 5516–5521 (2013)

113. J.C. Rice et al., Histone methyltransferases direct different degrees of methylation to define distinct chromatin domains. Mol. Cell **12**(6), 1591–1598 (2003)

114. A.H. Peters et al., Partitioning and plasticity of repressive histone methylation states in mammalian chromatin. Mol. Cell **12**(6), 1577–1589 (2003)

115. M. Tachibana et al., Histone methyltransferases G9a and GLP form heteromeric complexes and are both crucial for methylation of euchromatin at H3-K9. Genes Dev. **19**(7), 815–826 (2005)

116. R.A. Rao et al., KMT1 family methyltransferases regulate heterochromatin-nuclear periphery tethering via histone and non-histone protein methylation. EMBO Rep. **20**(5), e43260 (2019)

117. S. Nakanishi et al., A comprehensive library of histone mutants identifies nucleosomal residues required for H3K4 methylation. Nat. Struct. Mol. Biol. **15**(8), 881–888 (2008)

118. C.M. Milner, R.D. Campbell, The G9a gene in the human major histocompatibility complex encodes a novel protein containing ankyrin-like repeats. Biochem. J. **290**(Pt 3), 811–818 (1993)

119. C. Chaturvedi et al., Maintenance of gene silencing by the coordinate action of the H3K9 methyltransferase G9a/KMT1C and the H3K4 demethylase Jarid1a/KDM5A. Proc. Natl. Acad. Sci. U. S. A. **109**(46), 18845–18850 (2012)

120. A. Dutta, Identification of an NKX3.1-G9a-UTY transcriptional regulatory network that controls prostate differentiation. Science **352**(6293), 1576–1580 (2016)

121. D.J. Purcell et al., Recruitment of coregulator G9a by Runx2 for selective enhancement or suppression of transcription. J. Cell. Biochem. **113**(7), 2406–2414 (2012)

122. Y. Kondo et al., Downregulation of histone H3 lysine 9 methyltransferase G9a induces centrosome disruption and chromosome instability in cancer cells. PLoS One **3**(4), e2037 (2008)

123. E. Metzger et al., LSD1 demethylates repressive histone marks to promote androgen-receptor-dependent transcription. Nature **437**(7057), 436–439 (2005)

124. N.R. Rose et al., Plant growth regulator daminozide is a selective inhibitor of human KDM2/7 histone demethylases. J. Med. Chem. **55**(14), 6639–6643 (2012)

125. J. Yang et al., The histone demethylase JMJD2B is regulated by estrogen receptor alpha and hypoxia, and is a key mediator of estrogen induced growth. Cancer Res. **70**(16), 6456–6466 (2010)

126. K. Coffey et al., The lysine demethylase, KDM4B, is a key molecule in androgen receptor signalling and turnover. Nucleic Acids Res. **41**(8), 4433–4446 (2013)

127. M. Wissmann et al., Cooperative demethylation by JMJD2C and LSD1 promotes androgen receptor-dependent gene expression. Nat. Cell Biol. **9**(3), 347–353 (2007)

128. T. Chiba et al., Histone lysine methyltransferase SUV39H1 is a potent target for epigenetic therapy of hepatocellular carcinoma. Int. J. Cancer **136**(2), 289–298 (2015)

129. Z. Lu et al., Histone-lysine methyltransferase EHMT2 is involved in proliferation, apoptosis, cell invasion, and DNA methylation of human neuroblastoma cells. Anti-Cancer Drugs **24**(5), 484–493 (2013)

130. M. Vedadi et al., A chemical probe selectively inhibits G9a and GLP methyltransferase activity in cells. Nat. Chem. Biol. **7**(8), 566–574 (2011)

131. Y. Yuan et al., A small-molecule probe of the histone methyltransferase G9a induces cellular senescence in pancreatic adenocarcinoma. ACS Chem. Biol. **7**(7), 1152–1157 (2012)

132. E.T. Wiles, E.U. Selker, H3K27 methylation: a promiscuous repressive chromatin mark. Curr. Opin. Genet. Dev. **43**, 31–37 (2017)

133. S. Aranda, G. Mas, L. Di Croce, Regulation of gene transcription by Polycomb proteins. Sci. Adv. **1**(11), e1500737 (2015)

134. E. Conway, E. Healy, A.P. Bracken, PRC2 mediated H3K27 methylations in cellular identity and cancer. Curr. Opin. Cell Biol. **37**, 42–48 (2015)

135. R. Margueron, D. Reinberg, The Polycomb complex PRC2 and its mark in life. Nature **469**(7330), 343–349 (2011)

136. S. Varambally et al., The polycomb group protein EZH2 is involved in progression of prostate cancer. Nature **419**(6907), 624–629 (2002)

137. S. Varambally et al., Genomic loss of microRNA-101 leads to overexpression of histone methyltransferase EZH2 in cancer. Science **322**(5908), 1695–1699 (2008)

138. K. Xu et al., EZH2 oncogenic activity in castration-resistant prostate cancer cells is Polycomb-independent. Science (New York, N.Y.) **338**(6113), 1465–1469 (2012)

139. D. Wang et al., LncRNA MALAT1 enhances oncogenic activities of EZH2 in castration-resistant prostate cancer. Oncotarget **6**(38), 41045–41055 (2015)

140. M.C. Donaldson-Collier et al., EZH2 oncogenic mutations drive epigenetic, transcriptional, and structural changes within chromatin domains. Nat. Genet. **51**(3), 517–528 (2019)

141. H. Kaniskan, M.L. Martini, J. Jin, Inhibitors of protein methyltransferases and demethylases. Chem. Rev. **118**(3), 989–1068 (2018)

142. W.A. Schulz et al., The histone demethylase UTX/KDM6A in cancer: progress and puzzles. Int. J. Cancer **145**(3), 614–620 (2019)

143. S.H. Jung et al., Genetic progression of high grade prostatic intraepithelial neoplasia to prostate cancer. Eur. Urol. **69**(5), 823–830 (2016)

144. V.M. Morozov et al., Inhibitor of H3K27 demethylase JMJD3/UTX GSK-J4 is a potential therapeutic option for castration resistant prostate cancer. Oncotarget **8**(37), 62131–62142 (2017)

145. W. Yu et al., Catalytic site remodelling of the DOT1L methyltransferase by selective inhibitors. Nat. Commun. **3**, 1288 (2012)

146. J. Tan et al., Pharmacologic disruption of Polycomb-repressive complex 2-mediated gene repression selectively induces apoptosis in cancer cells. Genes Dev. **21**(9), 1050–1063 (2007)

147. M.T. McCabe et al., EZH2 inhibition as a therapeutic strategy for lymphoma with EZH2-activating mutations. Nature **492**(7427), 108–112 (2012)

148. S. Schmähling et al., Regulation and function of H3K36 di-methylation by the trithorax-group protein complex AMC. Development **145**(7), dev163808 (2018)

149. D.K. Pokholok et al., Genome-wide map of nucleosome acetylation and methylation in yeast. Cell **122**(4), 517–527 (2005)

150. T. Vacík, D. Laďinović, I. Raška, KDM2A/B lysine demethylases and their alternative isoforms in development and disease. Nucleus **9**(1), 431–441 (2018)

151. I.A. Asangani et al., Characterization of the EZH2-MMSET histone methyltransferase regulatory axis in cancer. Mol. Cell **49**(1), 80–93 (2013)

152. N. Li et al., AKT-mediated stabilization of histone methyltransferase WHSC1 promotes prostate cancer metastasis. J. Clin. Invest. **127**(4), 1284–1302 (2017)

153. T. Ezponda et al., The histone methyltransferase MMSET/WHSC1 activates TWIST1 to promote an epithelial–mesenchymal transition and invasive properties of prostate cancer. Oncogene **32**(23), 2882–2890 (2013)

154. M. Yan et al., The critical role of histone lysine demethylase KDM2B in cancer. Am. J. Transl. Res. **10**(8), 2222–2233 (2018)

155. N. Zacharopoulou et al., The epigenetic factor KDM2B regulates cell adhesion, small rho GTPases, actin cytoskeleton and migration in prostate cancer cells. Biochim. Biophys. Acta, Mol. Cell Res. **1865**(4), 587–597 (2018)

156. A.T. Nguyen, Y. Zhang, The diverse functions of Dot1 and H3K79 methylation. Genes Dev. **25**(13), 1345–1358 (2011)

157. M.I. Valencia-Sanchez et al., Structural basis of Dot1L stimulation by histone H2B lysine 120 ubiquitination. Mol. Cell **74**(5), 1010–1019.e6 (2019)

158. E.J. Worden et al., Mechanism of cross-talk between H2B ubiquitination and H3 methylation by Dot1L. Cell **176**(6), 1490–1501.e12 (2019)

159. M.S. Singer et al., Identification of high-copy disruptors of telomeric silencing in Saccharomyces cerevisiae. Genetics **150**(2), 613–632 (1998)

160. M. Annala et al., DOT1L-HES6 fusion drives androgen independent growth in prostate cancer. EMBO Mol. Med. **6**(9), 1121–1123 (2014)

161. W. Kim et al., Deficiency of H3K79 histone methyltransferase Dot1-like protein (DOT1L) inhibits cell proliferation. J. Biol. Chem. **287**(8), 5588–5599 (2012)

162. L. Yang et al., LncRNA-dependent mechanisms of androgen-receptor-regulated gene activation programs. Nature **500**(7464), 598–602 (2013)

163. S.R. Daigle et al., Selective killing of mixed lineage leukemia cells by a potent small-molecule DOT1L inhibitor. Cancer Cell **20**(1), 53–65 (2011)

164. C.R. Klaus et al., DOT1L inhibitor EPZ-5676 displays synergistic antiproliferative activity in combination with standard of care drugs and hypomethylating agents in MLL-rearranged leukemia cells. J. Pharmacol. Exp. Ther. **350**(3), 646–656 (2014)

165. Y. Zhao et al., Prodrug strategy for PSMA-targeted delivery of TGX-221 to prostate cancer cells. Mol. Pharm. **9**(6), 1705–1716 (2012)

166. S.R. Daigle et al., Potent inhibition of DOT1L as treatment of MLL-fusion leukemia. Blood **122**(6), 1017–1025 (2013)

167. J.L. Anglin et al., Synthesis and structure-activity relationship investigation of adenosine-containing inhibitors of histone methyltransferase DOT1L. J. Med. Chem. **55**(18), 8066–8074 (2012)

168. C.A. Musselman et al., Perceiving the epigenetic landscape through histone readers. Nat. Struct. Mol. Biol. **19**(12), 1218–1227 (2012)

169. T.G. Kutateladze, SnapShot: histone readers. Cell **146**(5), 842–842.e1 (2011)

170. Z. Chen et al., Histone modifications and chromatin organization in prostate cancer. Epigenomics **2**(4), 551–560 (2010)

171. M. Pérez-Salvia, M. Esteller, Bromodomain inhibitors and cancer therapy: from structures to applications. Epigenetics **12**(5), 323–339 (2017)

172. Z. Wang et al., Combinatorial patterns of histone acetylations and methylations in the human genome. Nat. Genet. **40**(7), 897–903 (2008)

173. M. Han et al., Epigenetic enzyme mutations: role in tumorigenesis and molecular inhibitors. Front. Oncol. **9**, 194 (2019)

174. H.M. Chan, N.B. La Thangue, p300/CBP proteins: HATs for transcriptional bridges and scaffolds. J. Cell Sci. **114**(Pt 13), 2363–2373 (2001)

175. B.M. Dancy, P.A. Cole, Protein lysine acetylation by p300/CBP. Chem. Rev. **115**(6), 2419–2452 (2015)

176. M. Fu et al., Acetylation of androgen receptor enhances coactivator binding and promotes prostate cancer cell growth. Mol. Cell. Biol. **23**(23), 8563–8575 (2003)

177. M. Fu et al., p300 and p300/cAMP-response element-binding protein-associated factor acetylate the androgen receptor at sites governing hormone-dependent transactivation. J. Biol. Chem. **275**(27), 20853–20860 (2000)

178. R.M. Attar, C.H. Takimoto, M.M. Gottardis, Castration-resistant prostate cancer: locking up the molecular escape routes. Clin. Cancer Res. **15**(10), 3251–3255 (2009)

179. B. Comuzzi et al., The androgen receptor co-activator CBP is up-regulated following androgen withdrawal and is highly expressed in advanced prostate cancer. J. Pathol. **204**(2), 159–166 (2004)

180. J. Zhong et al., P300 acetyltransferase regulates androgen receptor degradation and pten-deficient prostate tumorigenesis. Cancer Res. **74**(6), 1870–1880 (2014)

181. E. Seto, M. Yoshida, Erasers of histone acetylation: the histone deacetylase enzymes. Cold Spring Harb. Perspect. Biol. **6**(4), a018713 (2014)

182. Y. Li, E. Seto, HDACs and HDAC inhibitors in cancer development and therapy. Cold Spring Harb. Perspect. Med. **6**(10), a026831 (2016)

183. W. Weichert et al., Histone deacetylases 1, 2 and 3 are highly expressed in prostate cancer and HDAC2 expression is associated with shorter PSA relapse time after radical prostatectomy. Br. J. Cancer **98**(3), 604–610 (2008)

184. H. Huang et al., Carboxypeptidase A3 (CPA3): a novel gene highly induced by histone deacetylase inhibitors during differentiation of prostate epithelial cancer cells. Cancer Res. **59**(12), 2981–2988 (1999)

185. L.K. Gediya et al., Improved synthesis of histone deacetylase inhibitors (HDIs) (MS-275 and CI-994) and inhibitory effects of HDIs alone or in combination with RAMBAs or retinoids on growth of human LNCaP prostate cancer cells and tumor xenografts. Bioorg. Med. Chem. **16**(6), 3352–3360 (2008)

186. P.N. Munster et al., Phase I trial of vorinostat and doxorubicin in solid tumours: histone deacetylase 2 expression as a predictive marker. Br. J. Cancer **101**(7), 1044–1050 (2009)

187. D. Rathkopf et al., A phase I study of oral panobinostat alone and in combination with docetaxel in patients with castration-resistant prostate cancer. Cancer Chemother. Pharmacol. **66**(1), 181–189 (2010)

188. L.R. Molife et al., Phase II, two-stage, single-arm trial of the histone deacetylase inhibitor (HDACi) romidepsin in metastatic castration-resistant prostate cancer (CRPC). Ann. Oncol. **21**(1), 109–113 (2009)

189. B. Jung-Hynes, N. Ahmad, Role of p53 in the antiproliferative effects of Sirt1 inhibition in prostate cancer cells. Cell Cycle **8**(10), 1478–1483 (2009)

190. D.M. Huffman et al., SIRT1 is significantly elevated in mouse and human prostate cancer. Cancer Res. **67**(14), 6612–6618 (2007)

191. V. Byles et al., SIRT1 induces EMT by cooperating with EMT transcription factors and enhances prostate cancer cell migration and metastasis. Oncogene **31**(43), 4619–4629 (2012)

192. J.D. Lovaas et al., SIRT1 enhances matrix metalloproteinase-2 expression and tumor cell invasion in prostate cancer cells. Prostate **73**(5), 522–530 (2013)

193. L. Sun et al., MPP8 and SIRT1 crosstalk in E-cadherin gene silencing and epithelial-mesenchymal transition. EMBO Rep. **16**(6), 689–699 (2015)

194. Z. You, Y. Liu, X. Liu, Nicotinamide N-methyltransferase enhances the progression of prostate cancer by stabilizing sirtuin 1. Oncol. Lett. **15**(6), 9195–9201 (2018)

195. Y. Yu et al., Mesenchymal stem cells overexpressing Sirt1 inhibit prostate cancer growth by recruiting natural killer cells and macrophages. Oncotarget **7**(44), 71112–71122 (2016)

196. M.J. Powell et al., Disruption of a Sirt1-dependent autophagy checkpoint in the prostate results in prostatic intraepithelial neoplasia lesion formation. Cancer Res. **71**(3), 964–975 (2011)

197. A. Stathis, F. Bertoni, BET proteins as targets for anticancer treatment. Cancer Discov. **8**(1), 24–36 (2018)

198. S. Malik, R.G. Roeder, The metazoan Mediator coactivator complex as an integrative hub for transcriptional regulation. Nat. Rev. Genet. **11**(11), 761–772 (2010)

199. K.J. Moon et al., The bromodomain protein Brd4 is a positive regulatory component of P-TEFb and stimulates RNA polymerase II-dependent transcription. Mol. Cell **19**(4), 523–534 (2005)

200. Z. Yang et al., Recruitment of P-TEFb for stimulation of transcriptional elongation by the bromodomain protein Brd4. Mol. Cell **19**(4), 535–545 (2005)

201. I.A. Asangani et al., Therapeutic targeting of BET bromodomain proteins in castration-resistant prostate cancer. Nature **510**(7504), 278–282 (2014)

202. A. Urbanucci et al., Androgen receptor deregulation drives bromodomain-mediated chromatin alterations in prostate cancer. Cell Rep. **19**(10), 2045–2059 (2017)

203. L. Gao et al., Androgen receptor promotes ligand-independent prostate cancer progression through c-Myc upregulation. PLoS One **8**(5), e63563 (2013)

204. I.A. Asangani et al., BET bromodomain inhibitors enhance efficacy and disrupt resistance to AR antagonists in the treatment of prostate cancer. Mol. Cancer Res. **14**(4), 324–331 (2016)

205. A. Pawar et al., Resistance to BET inhibitor leads to alternative therapeutic vulnerabilities in castration-resistant prostate cancer. Cell Rep. **22**(9), 2236–2245 (2018)

206. D.P. Bondeson, C.M. Crews, Targeted protein degradation by small molecules. Annu. Rev. Pharmacol. Toxicol. **57**, 107–123 (2016)

207. P.M. Cromm, C.M. Crews, Targeted protein degradation: from chemical biology to drug discovery. Cell Chem. Biol. **24**(9), 1181–1190 (2017)

208. G.E. Winter et al., Phthalimide conjugation as a strategy for in vivo target protein degradation. Science **348**(6241), 1376–1381 (2015)

209. M. Zengerle, K.H. Chan, A. Ciulli, Selective small molecule induced degradation of the BET bromodomain protein BRD4. ACS Chem. Biol. **10**(8), 1770–1777 (2015)

210. J. Lu et al., Hijacking the E3 ubiquitin ligase cereblon to efficiently target BRD4. Chem. Biol. **22**(6), 755–763 (2015)

211. K. Raina et al., PROTAC-induced BET protein degradation as a therapy for castration-resistant prostate cancer. Proc. Natl. Acad. Sci. U. S. A. **113**(26), 7124–7129 (2016)

212. L. Jin, J. Garcia, E. Chan, Therapeutic targeting of the CBP/p300 bromodomain blocks the growth of castration-resistant prostate cancer. Cancer Res. **77**(20), 5564–5575 (2017)

213. L.M. Lasko et al., Discovery of a selective catalytic p300/CBP inhibitor that targets lineage-specific tumours. Nature **550**(7674), 128–132 (2017)

214. G. Längst, L. Manelyte, Chromatin remodelers: from function to dysfunction. Genes (Basel) **6**(2), 299–324 (2015)

215. S.V. Saladi, I.L. de la Serna, ATP dependent chromatin remodeling enzymes in embryonic stem cells. Stem Cell Rev. **6**(1), 62–73 (2010)

216. D.P. Labbe, M. Brown, Transcriptional regulation in prostate cancer. Cold Spring Harb. Perspect. Med. **8**(11), a030437 (2018)

Oncogenic ETS Factors in Prostate Cancer

Taylor R. Nicholas, Brady G. Strittmatter, and Peter C. Hollenhorst

Introduction

Prostate cancer represents the only carcinoma containing a common chromosomal rearrangement, resulting in an oncogenic fusion gene: *TMRPSS2/ERG*. About one-half of prostate tumors have a rearrangement of chromosome 21 that fuses the promoter and 5′ untranslated region (5′ UTR) of *TMPRSS2* to the open reading frame of *ERG* [1]. *ERG* is normally silent in adult prostate epithelial cells, while *TMRPSS2* is androgen-responsive and highly expressed in the prostate. The *TMPRSS2/ERG* fusion gene usually results in prostate cells that express either a full-length, or N-terminally truncated ERG protein, depending on the breakpoint [2]. Transgenic mouse models indicate that *ERG* expression in prostate epithelial cells promotes carcinogenesis, but only in collaboration with a second oncogenic "hit" [3–6]. *ERG* encodes an ETS family transcription factor that has the potential to alter gene expression patterns. In addition to the *TMPRSS2/ERG*

T. R. Nicholas
Department of Biology, Indiana University, Bloomington, IN, USA

B. G. Strittmatter
Department of Molecular and Cellular Biochemistry, Indiana University, Bloomington, IN, USA

P. C. Hollenhorst (✉)
Medical Sciences, Indiana University School of Medicine, Bloomington, IN, USA
e-mail: pchollen@indiana.edu

rearrangement, prostate cancers have other recurrent rearrangements that similarly promote the expression of additional ETS factors such as *ETV1* and *ETV4* [2]. These findings indicate that ETS family transcription factors can play key roles in promoting prostate cancer. This chapter will discuss the role of ETS factors in prostate cancer with a focus on recently uncovered molecular mechanisms that could point the way to ETS-targeted therapeutics.

ETS Family Transcription Factors

What Is an ETS Factor?

The ETS family of transcription factors are encoded by 28 genes in the human genome [7]. ETS factors are defined by a conserved ETS DNA binding domain consisting of a winged-helix-turn-helix structure that can bind DNA monomerically. In vitro DNA sequence specificity of ETS domains have been extensively measured, and every ETS factor requires a GGA(A/T) sequence for high-affinity DNA binding. An extended consensus binding sequence of CCGGAAGT is common in the family, with a handful of members displaying slightly different preferences [8]. Due to this overlapping sequence preference it is difficult to predict which family member will bind an ETS binding sequence in vivo, and competition for binding sites between

© Springer Nature Switzerland AG 2019
S. M. Dehm, D. J. Tindall (eds.), *Prostate Cancer*, Advances in Experimental Medicine and Biology 1210, https://doi.org/10.1007/978-3-030-32656-2_18

family members is likely. For this reason, when a new ETS factor is expressed in a cell, such as when the *TMRPSS2/ERG* rearrangement occurs in prostate cells, it is possible that changes in gene expression could be the result of transcriptional changes driven by the newly expressed ETS factor, and/or changes caused by displacement of an endogenous ETS factor.

A phylogenetic comparison of the ETS domain divides the ETS family into subclasses of up to three members each, with ERG in the ERG subfamily and ETV1 and ETV4 in the PEA3 subfamily (Fig. 1). Within each of these subclasses there tends to be homology across the entire protein, however, a comparison between any two subclasses shows that the only homology is in the ETS DNA binding domain. This diversity of sequence outside of the DNA binding domain results in diverse molecular mechanisms, with some members acting as transcriptional activators, and others as transcriptional repressors, and with members responding in distinct ways to signaling pathways [7].

ETS transcription factors are extensively co-expressed in all cell types [9]. There are 8–10 ETS factors that are ubiquitously expressed, while others have varying levels of tissue specificity. Most cell types, normal prostate epithelia

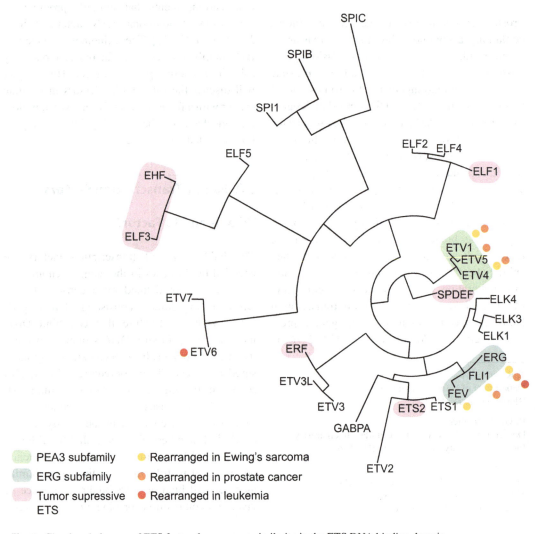

Fig. 1 Circular phylogeny of ETS factors by sequence similarity in the ETS DNA binding domain

included, express 15–20 ETS factors at the mRNA level [9]. SPDEF is the most highly expressed ETS mRNA in normal prostate, and one early name for this gene was prostate-derived ETS factor (PDEF). Likewise, the ETS factors ELF3 and EHF are highly expressed in multiple epithelial cell types including prostate epithelia and were historically named ESE1 and ESE3 for "Epithelial Specific ETS". Consistent with a role in maintaining normal prostate epithelial identity, SPDEF, ELF3 and EHF have all been reported to be potential tumor suppressive genes in the prostate [10–12]. Such tumor suppressive roles will be discussed in more detail later in this chapter.

ETS Factors Aberrantly Expressed in Prostate Cancer

Three members of the ETS transcription factor family, *ERG*, *ETV1*, and *ETV4* are not expressed in normal adult prostate cells [9], but are commonly expressed in prostate cancer cells due to chromosomal rearrangements [2]. Most common is *ERG* overexpression, which is due to *ERG* gene rearrangement in 40–50% of prostate tumors [13, 14]. *ETV1* rearrangement is the second most common and is found in 8–10% of tumors. ETV4 is rearranged in 2–5% of cases. These rearrangements tend to be mutually exclusive, indicating that they have similar roles as oncogenes (Fig. 2). Transgenic expression of *ERG*, *ETV1*, and *ETV4* in mouse prostates indicate that all three are oncogenic, although in all three cases additional "hits" are needed to promote robust tumorigenesis [3–5, 15–17]. These additional hits usually involve activation of oncogenic signaling pathways, and the necessity of these pathways will be discussed later in the chapter. Since ~85% of prostate tumors that harbor ETS rearrangements have the *TMPRSS2/ERG* rearrangement, most research has aimed to understand the molecular consequences of ectopic *ERG* expression in the prostate. Based on the incidence and these experimental findings, it is clear that *ERG*, *ETV1*, and *ETV4* are all prostate cancer oncogenes.

Recent paired-end sequencing of large numbers of prostate tumors has revealed that these tumors can have large numbers of gene rearrangements [14]. Due to this, thousands of genes have been found to be rearranged at low frequency (<2%) in prostate tumors. Inevitably, some of these genes are in the ETS family. But are these rearrangements passenger or driver mutations? In addition to the frequently rearranged *ERG*, *ETV1*, and *ETV4*, the ETS genes *ETV5*, *FLI1* and *ELK4* are rearranged at low frequency in prostate tumors [18–20]. Importantly, *ETV5* and *FLI1* are not expressed in normal prostate cells, however *ELK4* expression in normal prostate is relatively high [9], and it is possible that *ELK4* rearrangements discovered at the RNA level are due to trans-splicing, rather than DNA rearrangement. To date, no transgenic mouse models have been reported that can be used to determine if gene fusion of *FLI1*, *ETV5*, or *ELK4* in the prostate is oncogenic. We and others have used cell line models to test the role of overexpressing various ETS factors. In these studies, *ETV5* has similar roles to *ERG*, *ETV1*, and *ETV4* when expressed in normal prostate RWPE-1 cells [20, 21], suggesting that *ETV5* fusions are oncogenic. In all, we have tested 12 ETS genes, including *FLI1*, in in vitro assays, but only *ERG*, *ETV1*, *ETV4*, and *ETV5* promoted common oncogene-related functions such as cell migration [21, 22]. Intriguingly, these four ETS factors also share a common molecular mechanism, which will be discussed later in the chapter.

ETS Gene Fusions in Prostate Cancer

5′ Fusion Partners and Fusion Products

ETS gene fusions have been assayed in tumor samples by various techniques including reverse transcription polymerase chain reaction (RT-PCR), fluorescence *in situ* hybridization (FISH), and rapid amplification of cDNA ends (RACE). The frequency of each fusion type varies in the literature depending on cohort selection and detection technique, and these incongruities have been

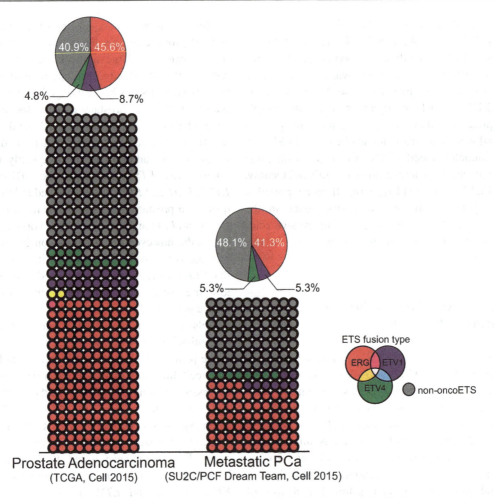

Fig. 2 Distribution of oncogenic ETS fusions across prostate cancer data sets. Three hundred and thirty-three primary prostate adenocarcinoma samples from the TCGA is depicted on the left. Hundred and fifty metastatic prostate cancer samples from the SU2C/PCF Dream Team is depicted on the right. Pie charts representing percentage of each data set with each feature type is displayed above the corresponding bar. Each circle represents a patient sample

reviewed elsewhere [23, 24]. However, recurrent ETS rearrangements exclusively involve fusion of the 3′ end of ETS genes, which include the ETS DNA binding domain. The 5′ partners most often donate promoters that drive high expression in prostate cells, allowing the aberrant expression of oncogenic ETS at high levels. The most common 5′ fusion partners are androgen responsive, yet androgen insensitive and androgen repressed fusion partners have been characterized [23]. *TMPRSS2*, an androgen responsive gene, is the most frequent 5′ partner [1]. While *TMPRSS2/ERG* accounts for the majority of ETS rearrangements, TMPRSS2 is also found rearranged with *ETV1*, *ETV4*, and *ETV5* at lower frequencies [20, 25, 26]. Rarer 5′ fusion partners of *ERG* are *SLC45A3*, *NDRG1*, and *HERPUD1* [23, 24, 27]. *ETV1* 5′ fusion partners include *TMPRSS2*, *SLC45A3*, *HERV-K*, *HERVK17*, *C15ORF21*, *HNRPA2B1*, *OR51E2*, *EST14*, *FLJ35294*, *FOXP1*, and *ACSLS* [1, 2, 24, 28]; and reported *ETV4* 5′ fusion partners are *TMPRSS2*, *SLC45A3*, *CANT1*, *KLK2*, *DDX5*, *HERVK17*, and *UBTF* [28–30]. While 5′ fusion partners are diverse, the congruency of 3′ fusion partners indicates an oncogenic ETS requirement for these rearrangement-driven prostate cancers. Rare cases in which patients with ERG rearrangements also have ETV1 or ETV4 fusions ([13, 19, 31]; Fig. 2) have been reported and sometimes arise from discrete foci; However, multiple ETS rearrangements have been reported in the same focus [32]. Regardless, these observations exemplify the heterogeneous nature of prostate tumors and tendency towards these genetic alterations.

Aside from diversity at the 5′ end, additional ETS rearrangement diversity can arise from breakpoint site variation. Many *TMPRSS2/ERG* fusions have been reported, and contain either the first, second, or third exon of *TMPRSS2* joined to exon 2, 3, 4, 5, or 6 of *ERG* [31, 33]. Briefly, the most common breakpoint in prostate cancer results in the fusion of the 5′ UTR exon 1 of TMPRSS2 to exon 3 of ERG, resulting in an N-terminally truncated ERG protein (ERGΔ32; named from NCBI ERG isoform 1). Other common fusion products are ERGΔ92 (exon 1 of *TMPRSS2* fused to exon 4 of ERG) and fusions that result in expression of full-length ERG (fusions prior to ERG exon 3). In more rare cases part of the open reading frame of *TMPRSS2* is fused to different breakpoints in *ERG* resulting in some *TMPRSS2* coded amino acids fused to the N-terminus of ERG. At the RNA level, alternative splicing of TMPRSS2/ERG can occur [31]. This potentiates the expression of multiple fusion protein products in the same tumor; however, the oncogenic contribution of different TMPRSS2/ERG variants is not fully understood. Transcripts encoding smaller ERG truncations can be found co-expressed in samples with larger fusion proteins [31]. These smaller truncations are likely alternatively spliced products of *TMPRSS2/ERG*, and it is unknown if they contribute to ERG-mediated oncogenesis.

Demographics of Patients with ETS-Positive Prostate Cancer

Occurrence of the *TMRPSS2/ERG* fusion in prostate cancer differs significantly among different demographic groups. Despite the higher incidence and mortality of prostate cancer in men of African descent [34], this group has a lower frequency of *TMPRSS2/ERG* rearrangement. In a recent meta-analysis, 49% of prostate tumors from men of European decent expressed ERG, while only 27% of tumors from men of Asian descent, and 25% of tumors from men of African descent expressed ERG [35]. ERG-positivity in North Indian patients resembles that of the incidence for Caucasians [36]. It is unknown if these discrepancies are due to common alternative mechanisms that drive prostate tumors in men of African and Asian descent, or if it is due to a lower likelihood of ERG gene rearrangement or repressed ERG function in these populations. The TMPRSS2/ERG fusion is more predominant in early onset prostate cancer, affecting men less than 50 years old [37, 38].

Molecular Stratification of ETS-Positive Prostate Cancers

Classification of prostate cancer into molecular subtypes is an important step in the development and use of precision medicines. ETS-positive prostate cancer has emerged as the largest molecular subtype, but the mutational landscape of ERG-positive (and ERG-negative) prostate cancers are still being investigated. Much of what we know about the molecular stratification of prostate cancer comes from the in depth molecular analysis of 333 primary prostate cancers by The Cancer Genome Atlas Research Network [13]. We will discuss prominent co-occurring and mutually exclusive genetic alterations with ETS fusions (profiling ERG expression in Fig. 3), but further reading should be directed to the above reference. The concurrence between the TMPRSS2/ERG rearrangement and PTEN silencing has been well characterized [36, 39]. The correlation is significant in prostatectomy samples and primary tumors but not in castration resistant prostate cancer (CRPC), where PTEN loss in ERG-positive cases typifies aggressive disease [40]. Other prostate cancer molecular subtypes, such as SPINK1 positive, SPOP mutant, and CHD1 mutant are mutually exclusive with ETS-rearrangements regardless of stage

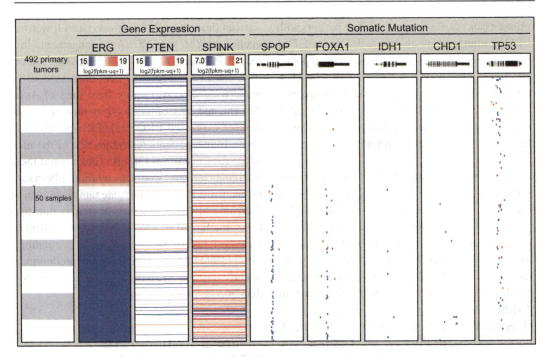

Fig. 3 ERG-centric molecular stratification of primary prostate cancers. The TCGA prostate adenocarcinoma (PRAD) data set was used to rank common molecular features by ERG gene expression and visualized by the UCSC Xena browser [200]. Gene expression profiles for ERG, PTEN, and SPINK are depicted as heat maps. Somatic mutations for the SPOP, FOXA1, IDH1, CHD1, and TP53 genes are shown in relation to gene structure and represented as pixels. Pixel color indicates mutation type, characterized by Xena browser. Samples with no data are excluded

(Fig. 3) [19, 39–44]. Interestingly, in some primary prostate cancers, ETV1 and ETV4 have been found overexpressed but *not* rearranged. Although the mechanism is undefined, these tumors are mutually exclusive with those harboring ETS rearrangements [13]. Tumor heterogeneity is an important factor in calling molecular subtypes; for example, mutational co-occurrence increases from single cell to single foci to whole tumor. The same tumor can be called ETS-positive or ETS-negative, depending on where the biopsy was taken [31]. Correct calling of the molecular subtype of a tumor underlies a critical obstacle in linking molecular subtypes to clinicopathological indicators and treatment course.

Clinicopathological Value of Oncogenic ETS

Because the TMPRSS2/ERG fusion gene is unique to prostate cancer cells and because ETS rearrangements result in much higher expression of oncogenic ETS in malignant tissue compared to adjacent normal tissue [1, 45, 46], the presence of the TMRPSS2/ERG fusion and/or expression of oncogenic ETS have emerged as clinical biomarkers of a cancerous prostate. Yet the connection between various clinicopathological indicators and ETS-positivity has not been fully determined. Cohort selection, evaluation, and mutational background differences potentially underlie the controversial prognostic value of oncogenic ETS. In a cohort of patients with localized prostate cancer undergoing "watchful waiting" therapy, a significant correlation was found with TMPRSS2/ERG-positive tumors and lethality [47]. Other groups have reported similar relationships between ERG positivity and active surveillance progression [48–50], suggesting the value of TMPRSS2/ERG as a prognosticator for risk of disease progression. However, in post-prostatectomy patients, ERG protein expression was found to correlate positively with longer progression free survival [40]. Similarly, ERG-positivity was found to be associated with longer recurrence free survival and longer overall survival in localized and castration resistant prostate cancer [51]. In contrast, no connection was found between ERG and prognosis in a separate radical prostatectomy cohort [52]. Additionally, patients with *TMPRSS2/ERG* and loss of PTEN were found to have shorter survival than ERG-/PTEN-patients [53]. ERG expression levels, rather than mere presence, correlates with aggression and disease stage and could be used for a more accurate prognosis [54]. It should be noted that because most ERG gene rearrangements result in androgen-driven ERG expression, the level of ERG in these prostate tumors may be indicative of high activity of the androgen receptor (AR), and it may be difficult to parse out whether phenotypes are due to high ERG or high AR activity. Clearly, a multivariate approach is necessary for understanding the risk associated with oncogenic ETS-positivity.

Recently, efforts have been made to develop screening assays to detect TMPRSS2/ERG by transcription-mediated amplification in the urine and serum to allow for early detection without biopsy. In this application TMPRSS2/ERG represents an ideal biomarker as it is an RNA that is completely tumor specific, and therefore the detection of even small amounts can be considered positive. The Mi-Prostate Score (MiPS) is a commercially available logistic regression model that uses detection of urine TMPRSS2/ERG as well as urine prostate cancer antigen 3 (PCA) normalized to PSA to predict prostate cancer and high-grade prostate cancer on biopsy [55]. MiPS has been shown to reduce biopsies by 51% [56]. PCA3 and TMPRSS2/ERG urine levels are significantly associated with high grade disease [53].

Since methods for detecting ETS fusions are abundant, relatively inexpensive, and precise, another hope is that variants of ETS fusions might contain more prognostic value. Several splice variants of TMPRSS2/ERG have been reported in clinical samples. Retention of both exons 10 and 11 of *ERG* is correlated with advanced disease [57]. Additionally, expression of *TMPRSS2/ERG*

fusions resulting from distinct breakpoints may have prognostic value. The type VI fusion, in which the *TMPRSS2* start codon in exon 2 is in frame with exon 4 of *ERG*, is associated with aggressive disease measured by early PSA recurrences and seminal vesicle invasion [33].

Generation of TMPRSS2/ERG Fusions

How do gene fusions occur in prostate epithelial cells? The prevalence of chromosomal rearrangements in leukemias and lymphomas could be due to the misapplication of the cellular machinery used to rearrange the immunoglobulin and T cell receptor loci. However, the discovery of common chromosomal rearrangements in prostate cancer was a surprise, as there is no such obvious mechanism to explain their occurrence. It is not clear why prostate cancer appears to be unique among carcinomas in this molecular cause. Prostate cancer is similar to many types of sarcoma, where oncogenic fusion genes are also common. Interestingly, the *TMPRSS2/ERG* fusion can be generated in rare cells in cell culture. Two prerequisites have been found necessary for the formation of the *TMPRSS2/ERG* fusion: (1) induced proximity of breakpoint regions and (2) DNA damage directed to breakpoint sites and subsequent aberrant repair. AR and topoisomerase II beta (TOP2B) bind to the TMPRSS2 and ERG breakpoints [58, 59], bringing these distal regions into close proximity in response to androgen stimulation in androgen responsive cells [58, 60]. TOP2B recruitment facilitates double stranded DNA breaks at the TMPRSS2 and ERG breakpoints, which upon recombination produce TMPRSS2/ERG [59]. Additionally, treatment with ionizing radiation [60] or TNFα [61] allows for *de novo* formation of TMPRSS2/ERG in androgen stimulated cells. Injecting a prostate cancer cell line into LPS treated air pouches of immunocompetent mice allowed formation of *TMPRSS2/ERG* [61]. Recently, the oncoprotein BRD4 has been shown to promote formation of TMPRSS2/ERG by facilitating DNA double stranded break repair via non-homologous end joining [62].

ETS Factors as Oncogenes and Tumor Suppressors

The Physiological Role of Oncogenic ETS

The normal function of oncogenic ETS in endogenous tissue should give clues to their functions when aberrantly expressed in prostatic tissue. ERG, ETV1 and ETV4 are tissue specific transcriptions factors and thus induce tissue specific gene programs important for development and maintenance. Oncogenic ETS factors belong to two subfamilies of ETS factors: the PEA3 and ERG subfamilies (Fig. 1). The PEA3 subfamily consists of ETV1, ETV4, and ETV5; and the ERG subfamily contains ERG, FLI1 and FEV. While homology between members of the same subfamily is high, members of different subfamilies only have homology in the ETS DNA binding domain, therefore ERG has little sequence similarity with ETV1 and ETV4. ETS factors are found throughout metozoan lineages. While some ETS factors are ubiquitous housekeepers, numerous developmental studies in mice and zebrafish show distinct spatio-temporal expression patterns of ERG, ETV1, and ETV4 orthologs. The tissue specific expression patterns of ERG, ETV1, and ETV4 suggest that these ETS factors control distinct gene programs that give rise to specialized organ function. ERG is predominantly expressed in hematopoietic stem cells and endothelial cells [63], regulating programs important for stem-cell self-renewal in hematopoietic cells and angiogenesis and migration in endothelial cells [64–67]. The ETVs, on the other hand, are mostly expressed in embryonic epithelial tissues including the developing lung, kidney, salivary gland, and mammary gland and promote branching morphogenesis during murine embryogenesis [68]. While the ETVs seem to function redundantly as fibroblast growth factor (FGF) signaling effectors [69–72], slightly divergent spatio-temporal expression throughout development and adulthood suggests tissue-specific functions [73, 74]. A common feature between ERG and the PEA3 factors may be functions related to the maintenance of tissue specific stem cells. Mice with ERG mutations have defects in hematopoietic stem cell function [75]. Similarly, mice with ETV5 knockouts are male-sterile due to loss of self-renewal in spermatogonial stem cells [76]. Interestingly, overexpression of ERG in mouse hematopoietic cells induces leukemia [77–79], suggesting that proper regulation of ERG levels is crucial for maintaining a normal cellular state.

Oncogenic ETS in Prostate Cancer Pathogenesis and Progression

Since the discovery of recurrent ETS rearrangements, a casual role for the oncogenic ETS in prostate carcinogenesis and disease progression has been under question. Oncogenic ETS gene rearrangements occur early, or even prior to disease onset, with evidence suggesting that ETS rearrangements drive prostatic neoplasia [31, 46]. Expression of oncogenic ETS factors alone in mouse models is not sufficient to drive formation of prostate tumors [6, 80]. Rather, a second hit, for example *PTEN* inactivation or *TP53* loss, is required for development of the disease [3, 4, 6, 16, 17, 81]. However, one mouse study reported that prostate specific expression of ERG at levels comparable to ERG-positive human prostate cancers allow 50% of mice over 2 years of age to develop tumors [82], suggesting an age-related component to ERG-mediated carcinogenesis.

Interestingly, the TMPRSS2/ERG fusion can drive expression from the wild-type ERG allele [83], creating a feed-forward loop. This allows continual, androgen-independent expression of ERG in CRPC, suggesting a requirement or oncogenic addiction to ERG. In fact, expression of oncogenic ETS is found in advanced metastatic disease, with ERG expression levels in CRPC comparable to expression in primary tumors [40, 84], with one study citing an increase in ERG expression from PIN to metastatic disease [51]; however, it is still unclear whether this expression actively contributes to an aggressive phenotype.

Cell line models have been used to address the causal role of oncogenic ETS in various disease stages. Introducing oncogenic ETS expression vectors into normal immortalized prostate cells results in increased oncogenic potential [21, 22]. This suggests normal prostate cells are susceptible to transformation when oncogenic ETS become expressed and recapitulates early-stage or indolent disease. Prostate cancer cell lines with ETS gene rearrangements include VCaP and NCI-H660 which have *TMPRSS2/ERG* [85]. Both LNCaP and MDA-PCa-2B cells harbor an ETV1 gene rearrangement [2]. PC3 and 22Rv1 both express high levels of ETV4 protein, but no rearrangement of this gene is apparent [86]. Knockdown of ERG in VCaP cells results in a loss of oncogenic functions [82, 87–90]; similar effects are observed when knocking down ETV1 and ETV4 in LNCaP and PC3 cell lines, respectively [15, 91, 92]. VCaP mouse xenograft studies suggest a requirement of ERG for tumor formation [88, 93]. Because the VCaP, LNCaP, and PC3 cell line models are all derived from advanced metastatic disease, these findings suggest that continued oncogenic ETS expression is important for this phenotype.

In an effort to reconcile molecular subtypes with cellular phenotypes, many groups are determining if certain cell types in the prostate are more susceptible to ETS rearrangements, and whether ETS rearrangements alter lineage outcomes during disease development. Multiple reports have found that ERG functions to block differentiation [94–96], reminiscent of ERG function in hematopoietic stem cells. Similarly, ERG repressed genes are involved in luminal and neuroendocrine differentiation in prostate-specific TMPRSS2/ERG transgenic mice. However recent work from multiple groups indicate that ERG expression actually promotes luminal cell fates in prostate cells. For example, ERG-positive patient specimens are classified as luminal [97]. Additionally, in a PTEN negative/TP53 mutant background, mice expressing transgenic ERG in the prostate grew tumors with a luminal epithelial phenotype [98]. These discrepancies suggest that ERG expression may function to define different cellular identities based on the mutational background, begging the need for additional mouse models that accurately represent co-occurring mutations in human prostate cancer.

Recurrent ETS Fusions in Other Cancers

Aside from prostate cancer, recurrent ETS fusions are found in the Ewing's family of tumors, and in leukemias (Fig. 1) [99–101]. However, these cases differ from prostate gene fusions in that Ewing's and leukemia gene fusions encode fusion proteins with emergent properties that depend on both fusion partners. In prostate cancer, ETS gene fusions usually only express full length or truncated ETS protein. Chromosomal translocations in Ewing's sarcoma create a fusion protein that includes the C-terminus and DNA binding domain of an ETS transcription factor and the N-terminus of another protein. The hallmark fusion, *EWS-FLI1*, occurs in ~85% of cases [102]. Rarer 3′ ETS fusion partners with EWS include: ERG, ETV1, ETV4, and FEV [100, 101, 103–106]. Additionally, ERG and FEV can be fused with the 5′ partner FUS, but these cases of Ewing's family tumors are rare [102]. In acute myeloid leukemia (AML), *ERG* can be fused to the 5′ partner *FUS* [107, 108]. Notably, EWS and FUS are paralogs, and therefore all of these fusions are likely to encode proteins with similar molecular properties. This similarity suggests a shared selective pressure that drives the formation of these events across distinct cell types. The ETS gene *ETV6* (TEL) is also commonly fused in lymphoid and myeloid leukemia [109], but the resulting fusion proteins do not include the ETS DNA binding domain, and therefore are likely to act through distinct mechanisms.

Tumor Suppressive ETS Factors

Of the six ETS genes with the highest mRNA expression in normal prostate (*SPDEF, EHF, ETS2, ELF3, ELF1,* and *ERF*), each exhibits tumor suppressive functions in prostate cells [10–12, 110–112]. In many cases these functions are due to binding site competition with oncogenic ETS. Although this likely plays a role, lower expression levels of these ETS genes do not always correlate with oncogenic ETS expression, so other mechanisms appear to be at work as well. ERF is mutated in a small portion of ETS tumors [113], and this mutation is mutually exclusive with oncogenic ETS rearrangements. ERF binds many of the same sites as oncogenic ETS factors, but acts as a repressor [110]; and loss of ERF results in gene signatures similar to expression of oncogenic ETS [113]. The role of ETS2 as a prostate tumor suppressor is particularly interesting, as the *ETS2* gene lies between *TMPRSS2* and *ERG* on chromosome 21, and one copy is lost in the interstitial deletion that is the most common cause of the *TMPRSS2/ERG* rearrangement. A mouse model indicates that this interstitial deletion results in more aggressive disease than expression of *ERG* without this deletion, and appears to result from the loss of *ETS2* [112]. Interestingly, *ETS2* can act as a tumor suppressor in other cell types, and the presence of *ETS2* on chromosome 21 has been attributed for the lower incidence of some tumor types in people with down syndrome [114]. *SPDEF*, *EHF* and *ELF3* all appear to promote epithelial differentiation, and deletion of these factors is implicated in epithelial to mesenchymal transition and carcinogenesis [10–12]. *ELF1* is frequently co-deleted with the tumor suppressor *RB1* in advanced prostate cancer. *RB1* deletions that co-occur with *TP53* loss of function are thought to promote resistance to hormonal therapies in castration resistant prostate cancer [115, 116], and the concomitant loss of *ELF1* may contribute to this resistant phenotype [111].

Molecular Mechanisms of Oncogenic Function

ERG, ETV1, and ETV4 all are able to activate similar transcriptional programs, which result in phenotypes related to oncogenesis including cell migration, invasion, and de-differentiation [22]. Understanding the mechanism by which oncogenic ETS initiate these gene expression programs is essential for guiding therapeutic efforts. Here we will detail what is known about DNA binding, chromatin accessibility, protein interacting partners, and post-translational modifications

that underlie mechanisms of oncogenic ETS in the prostate epithelium.

DNA Binding

All members of the ETS transcription factor family contain a highly conserved 84–90 amino acid ETS domain, which is sufficient to bind DNA as a monomer [117]. Structurally, the ETS DNA binding domain exists as a winged helix-turn-helix (wHTH) domain consisting of three alpha helices and four beta-strands [7]. To date 47 high resolution structures of the ETS domain have been solved by X-Ray crystallography or solution NMR and have been published and deposited into the Protein Data Bank (PDB), including every ETS factor rearranged in prostate tumors [118–130]. Many of these structures have been solved both in the presence and absence of DNA [121], as well as with an additional DNA binding partner [124, 128]. Taken together, these studies provide insight as to how the ETS domain interacts with DNA and how specific regions of the ETS domain or other binding partners alter this DNA binding specificity.

The core recognition motif of the ETS domain is a four base consensus 5′-GGA(A/T)-3′ in which two invariant arginine residues make contacts with the guanine bases [7]. Although no direct contacts are made with bases outside the GGAA core, the ETS domain binds DNA in a region spanning 12–15 base pairs, and binding is mediated by positioning of the phosphodiester backbone, water mediated hydrogen bonding, hydrophobic, and electrostatic interactions [7]. Although subtle differences exist in the primary amino acid sequences of various ETS factors, these differences do not dramatically alter the DNA-protein interface. The ETS transcription factor family can be grouped into four classes based on minor differences in their DNA sequence binding preference in vitro [8]. The first class contains half of the ETS factors including the PEA3, TCF, ETS, ERF, and ERG subfamilies and has highest affinity for 5′-ACCGGAAGT-3′ [8]. This class of ETS factors contains all of the oncogenic ETS factors. The second class includes all members of the TEL, ESE, and ELF subfamilies and differs from the first class only by a cytosine base rather than an adenine base at the beginning 5' nucleotide. The third class contains members of the SPI subfamily and prefers adenine rich sequences 5' to the core GGA sequence. The fourth class exhibits preferential binding for thymine at the final position in the GGA(A/T) core sequence and is comprised only of the ETS factor SPDEF. It is not clear to what degree these class differences influence binding site selectivity in vivo.

Autoinhibition of DNA binding plays an important regulatory role for many ETS factors and could be exploited as a therapeutic target. Autoinhibition is the process where regions of the protein outside of the DNA binding domain inhibit DNA binding, often in a regulated manner. All of the oncogenic ETS factors are subject to autoinhibition [121, 125]. Autoinhibition of DNA binding occurs by a similar mechanism in multiple members of the ETS family. In this mechanism, regions both N and C terminal of the ETS domain form an autoinhibitory module which is sustained by a hydrophobic pocket of amino acids [121, 124, 125, 131]. Full length ERG protein binds to a target DNA sequence in vitro with a $K_D = 120$ nM, whereas the ETS domain of ERG without its inhibitory modules binds to DNA with a $K_D = 37$ nM, indicating that the inhibitory modules on ERG play a modest (~3-fold) role in regulating DNA binding [125]. While autoinhibition of ERG plays a modest role, members of the PEA3 subfamily exhibit robust (~10–30-fold) autoinhibition when comparing minimal DNA binding domains to full length proteins [121]. The N- and C-terminal inhibitory regions of the ETS domain in the PEA3 family act independently to inhibit DNA binding, but can also act cooperatively to inhibit DNA binding at higher than additive levels. In the PEA3 family, the C-terminal inhibitory region forms an alpha helix, which packs against the ETS domain and distorts Helix 3 (H3), the helix responsible for direct DNA base contacts. The N-terminal region is predominantly disordered and also inhibits DNA binding through interactions with H3 [121]. Autoinhibition can be relived or enhanced through post-translational modifications and through interactions with other proteins [124, 129, 132]. The prototype for this mechanism comes from the ETS factor ETS1, where phosphorylation of multiple residues in the N terminal inhibitory domain by CAMKII reinforces autoinhibition of DNA binding by ~50-fold [132]. In terms of the oncogenic ETS factors, acetylation of members of the PEA3 subfamily in an N-terminal inhibitory domain relieves autoinhibition and increases transcriptional activation [121, 133]. ETV1 autoinhibition has also been demonstrated to be relieved by protein-protein contacts with USF1 [134].

Gene Regulation

Genome-wide mapping techniques such as chromatin immunoprecipitation sequencing (ChIP-seq) have been conducted on all of the oncogenic ETS factors in prostate cancer cell lines [6, 21, 22, 95], in mouse models of prostate cancer [89], and in normal prostate cells engineered to express exogenous oncogenic ETS [22]. To a lesser extent, oncogenic ETS factor DNA binding has been interrogated in other malignancies [135] and in the context of their normal physiological expression [136]. These ChIP-Seq experiments support and expand on basic biochemical studies of ETS factor DNA binding in vitro and provide critical insights into how ETS factors regulate their target genes in cells.

Studies of oncogenic ETS factor DNA binding coupled with whole transcriptome RNA-Sequencing (RNA-Seq) have further enhanced our understanding of the ETS regulome in prostate cancer and how it contributes to oncogenesis. These studies have also further interrogated ETS factor DNA binding and gene regulation in contexts relevant to prostate cancer such as PTEN deletions [6], with and without androgen treatment [95], and with knock-down/deletion of tumor suppressive ETS [95]. Taken together, these studies have provided insight into how ETS factor binding across the cistrome contributes to gene regulation in prostate cancer. This next section will summarize the aforementioned studies and will highlight recent key findings as to how oncogenic ETS factors regulate gene expression.

Members of the ETS family of transcription factors are extensively co-expressed in every cell type, and prostate epithelial cells are no exception [9]. As summarized in previous sections, the highly conserved ETS domain differs little in its sequence preference in vitro and accordingly, there is extensive competition between ETS factors to bind target genes in prostate cells [22]. Taking this information into account, measuring the contribution of a single ETS transcription factor's role in gene regulation has proved challenging. However, mis-expression of a single oncogenic ETS factor in the prostate results in dramatic changes in gene transcription. This information alongside studies of tumor suppressive ETS factor deletions in prostate cancer demonstrate the delicate balance of the ETS regulome.

Early studies mapping ETS family transcription factors discovered two distinct types of ETS binding site in the genome, "redundant" sites that can be bound by any ETS factor, and "specific" sites that tend to favor binding by one or a subset of ETS factors [7, 137]. Redundant ETS binding tends to occur at consensus ETS binding sequences common in the proximal promoters of CpG island-containing housekeeping genes. Specific binding tends to occur in tissue-specific enhancers and coincides with weaker match to consensus ETS binding sequences. Specificity for enhancer binding within the ETS family has been attributed to specific cooperative binding interactions with neighboring transcription factors such as ETS1 binding with RUNX1 or ELK1 binding with SRF [137, 138]. Like other ETS factors, oncogenic ETS factors expressed in prostate cells bind both housekeeping promoters and tissue-specific enhancers [22]. It is possible that oncogenic ETS could play a role in altering housekeeping gene expression, however this has not been described and it may be that redundancy of ETS function at these sites masks any role for this binding. Instead it has been proposed that it is the binding to tissue-specific enhancers that mediates oncogenic ETS function. As this specific binding is thought to be influenced by neighboring transcription factors, we will discuss several proposed interactions between oncogenic ETS and other transcription factors that may mediate oncogenic gene expression programs.

ETS/AP1

One class of enhancer bound by oncogenic ETS factors in prostate cancer cells contains an ETS binding sequence that is followed by a sequence recognized by the AP-1 class of transcription factors. AP-1 consists of a dimer of JUN and FOS family transcription factors. At an ETS/AP-1 sequence, a single ETS protein could bind next to a JUN homodimer or a JUN/FOS heterodimer [139]. In vitro biochemical experiments recently demonstrated that oncogenic ETS factors can bind with AP-1 cooperatively, where the affinity of the ETS factor increases when AP-1 is present. In contrast, several tumor-suppressive ETS factors showed anti-cooperative binding in the presence of AP-1 [140]. Amino acid substitutions in the ERG ETS domain interfere with the interaction with JUN and FOS, and the interaction mutants of ERG lose the ability to transcriptionally cooperate with AP-1 in luciferase assays [141]. Furthermore, ChIP-Seq data indicates that ERG exhibits preferential binding at ETS/AP1 sites in prostate cells compared to tumor suppressive ETS factor EHF [140]. ETS/AP1 enhancer elements regulated by oncogenic ETS factors in prostate cancer are found near genes involved in cell migration, cell morphogenesis, and cell development and include genes such as *PLAU*, *VIM*, and *ETS1* [22].

Both Ras/ERK signaling and differential binding of the JUN transcription factor family play important roles in regulation at ETS/AP1 enhancers [139]. ETS factors present in normal prostate epithelial cells, such as ETS1, can activate gene expression through binding ETS/AP-1 sequences, but this activation requires high levels of Ras/ERK signaling [142]. High Ras/ERK signaling can be caused by the activating KRAS mutations common in many types of carcinoma. However, prostate tumors rarely have activating KRAS mutations. One unique characteristic of the oncogenic ETS factors commonly found in prostate tumors is that they can activate ETS/AP1 enhancers in the presence of low levels of Ras/ERK activation [22]. Thus, one model suggests that oncogenic ETS factors essentially replace the role of activated KRAS in prostate cancer. Ras/ERK signaling is also involved in regulating Jun family transcription factors at ETS/AP1 enhancers. In the absence of Ras/ERK activation, c-Jun acts as a transcriptional activator and JunD as a transcriptional repressor of ETS/AP1 target genes, however, the converse is true in the presence of high Ras/ERK activation [139]. Thus, it has been proposed that c-Jun is the JUN family member likely to function with oncogenic ETS in prostate tumors.

GGAA Microsatellites

Within the preferred ETS binding sequence, the central GGA(A/T) nucleotides are the most important for high-affinity binding. Ewing's sarcoma is caused by a fusion oncogene that encodes a protein with an ETS DNA binding domain. This fusion protein most commonly consists of the N-terminus of the EWS protein fused to the ETS factor FLI1 [21]. Studies in Ewing's sarcoma indicate that this EWS/FLI1 fusion can bind an unusual regulatory sequence consisting of repeats of the sequence GGAA, which can extend over several hundred base pairs [143]. These GGAA microsatellite repeats regulate several oncogenes critical for tumorigenesis in Ewing's Sarcoma, including *NR0B1*, *NKX2-2*, *AND CCDN1* [144, 145]. The nature of these repeat elements suggests that EWS/FLI1 oligomerizes on DNA to form an activating transcriptional hub. Whether other ETS transcription factors also exhibit the ability to oligomerize and activate transcription in cells through GGAA repeats remains to be shown. However, the recent finding that the prostate cancer oncogenic ETS factors function with the EWS protein in transcription suggests common mechanisms between Ewing's sarcoma and prostate cancer [21]. In fact, preliminary studies have demonstrated that oncogenic ETS factors bind to GGAA microsatellite regions in prostate cells and that oncogenic ETS factors can activate a luciferase reporter regulated by a GGAA repeat element to a similar degree as EWS/FLI1 [21]. Thus, activation of genes regulated by GGAA repeats might be a common mechanism by which ETS factors promote both prostate cancer and Ewing's sarcoma.

Androgen Receptor

The function of oncogenic ETS factors in the prostate intersects in several ways with transcriptional regulation by the androgen receptor (AR). The most common oncogenic rearrangements, such as TMPRSS2/ERG put the oncogenic ETS factor under transcriptional control of AR. Further, oncogenic ETS proteins alter AR-dependent gene expression programs. Several studies report a physical interaction between oncogenic ETS, including ERG, with AR [95, 146]. ERG interacts with AR in normal endothelial cells, suggesting this interaction could play normal functions [89]. Importantly, non-oncogenic, and tumor-suppressive ETS factors have also been shown to bind AR, so the importance of the ETS/AR interaction in oncogenesis is still unclear [147]. Genomic interrogation shows co-enrichment of ERG and AR at many gene regulatory sequences [95]. Unlike ETS and AP-1, it is not clear whether the ERG and AR binding sequences at these enhancers or promoters are near each other, or that ERG and AR can influence each other's binding.

Contrasting work has been published as to whether ERG activates or attenuates AR transcriptional activity [6, 16]. In fact, one hypothesis is that this function is context dependent, and that ERG can promote oncogenic functions of AR, while inhibiting tumor suppressive AR functions. ERG attenuates AR transcription and prostate lineage specificity in VCaP cells, and epigenetic regulators EZH2 and HDAC1/2 have been implicated in this repression [148]. Conversely, ERG acts as an activator of androgen-independent genes that are responsible for prostate cancer cell invasion and growth [22, 89, 95, 149]. In the context of

PTEN loss, ERG can restore AR target gene expression and increases AR binding across the cistrome independent of AR protein levels and circulating testosterone [6, 16]. These results suggest that ERG can promote survival of cells that are dependent on AR transcriptional activity. Taken together these studies demonstrate that ERG regulation of AR transcription is context dependent and that PTEN status is a major determinant as to how ERG impacts transcription.

ETV1 is also intimately involved in AR transcriptional activity and is involved androgen metabolism in prostate cells [16]. ETV1 cooperates with AR signaling in LNCaP cells and in PTEN deficient mouse models of prostate cancer [16, 150]. ETV1 directly binds and upregulates genes associated with steroid hormone biosynthesis and androgen metabolism, which results in increased testosterone production and subsequent increased AR transcriptional activity [16]. To a lesser extent, ETV4 has been implicated in AR signaling, since survival and metastatic potential of mouse prostate cells is reduced upon ETV4 knockdown [15]. Interestingly, estrogen receptor (ER) signaling has been implicated in growth and migration of cells harboring TMPRSS2-ETV5 fusions suggesting that ETV5 is involved in steroid hormone signaling as well in prostate cancer [151].

Transcriptional Activation and Repression

Deciphering whether the role of a transcription factor in gene regulation is activating or repressive has proven to be a challenging task for multiple reasons: (1) transcription factors often regulate other transcriptional regulators, which themselves effect gene expression, (2) enhancer elements are often Kb-Mb in distance away from the genes that they regulate and may not always regulate the nearest gene, and (3) multiple transcription factors and enhancers regulate the transcription of any given gene. Families of transcription factors with similar consensus sequences and binding sites in vivo further complicate this problem as lower gene expression observed when a new factor is introduced may simply be due to less activation function than a factor that was displaced. Despite all of these complications, progress has been made to decipher the molecular mechanisms and contexts by which oncogenic ETS factors directly activate or repress transcription.

An activation function for a transcription factor can be predicted when there are direct interactions with known transcriptional co-activators such as CBP and p300. CBP/p300 are homologous acetyltransferases that directly acetylate histones and transcription factors at both promoters and enhancers [152]. All of the oncogenic ETS are acetylated by, and interact with, p300, however the mechanism of acetylation and interaction varies [153]. The N-terminus of ERG is acetylated by p300 at a KGGK motif, which results in recruitment of the transcriptional co-activator BRD4 that binds these acetylated lysines through its bromodomain [152, 154]. Initially, the ERG-BRD4 interaction was described in acute myeloid leukemia (AML), a cancer with ERG-dependencies [152], but now has also been shown in prostate cells. The interaction between ERG and BRD4 is of particular interest because of the potential therapeutic value of BRD4 inhibitors such as JQ1 and iBET, and because these inhibitors can decrease the ERG/BRD4 interaction [154]. ERG KGGK acetylmimetics can also increase invasion of normal prostate cell lines [154]. However, a highly prevalent ERG fusion product, ERG Δ92, does not contain the KGGK motif. Therefore, the relevance of the ERG-BRD4 interaction in prostate cancer, and whether it differs based on the fusion location is still unclear. Genome wide co-occupancy between oncogenic ETS factors and p300 in cells has not been investigated extensively in the context of prostate cancer and could provide further insights into specific sites of oncogenic ETS regulation.

While CBP/p300 is important for oncogenic ETS function, it also binds to and is important for the function of many non-oncogenic ETS factors. What then, is the mechanism that allows ERG, ETV1, and ETV4 to be oncogenic, when normal prostate cells express at least 15 other ETS factors? We recently reported that ERG, ETV1,

ETV4, and ETV5, and no other ETS factor interact with the Ewing Sarcoma Breakpoint Region 1 (EWS) protein [21]. In this context EWS acts as a co-activator, and the EWS interaction is essential for ETS-mediated cell migration, anchorage-independent growth, clonogenic survival, and tumor formation. EWS can also interact with ETV1 in developing limb buds of mouse and chick embryos, suggesting that EWS cooperates with oncogenic ETS in normal tissue [155]. The exact mechanism by which EWS regulates ERG transcriptional activity is not well understood. EWS has been implicated in a variety of co-activator and co-transcriptional activities including splicing, RNA Polymerase II CTD phosphorylation, and phase separation, all of which indicate that it is a multifunctional regulator of gene expression [21, 156]. Importantly, EWS is fused to the ETS DNA binding domain of various ETS factors in Ewing's Sarcoma, which suggests that there is a common oncogenic EWS/ETS function in both prostate cancer and Ewing's sarcoma [21].

A potential co-activator specific for the PEA3 subfamily is the mediator subunit MED25. ETV1, ETV4, and ETV5 interact with MED25 through a conserved N-terminal transactivation domain [157]. The ETV4 ETS domain also interacts with MED25, and both the transactivation domain and the ETS domain contact MED25 at multiple interfaces. MED25 has high affinity (K_d = 50 nM) for ETV1 and ETV4 but does not interact with ERG. The MED25 interaction with ETV4 relieves auto inhibition and promotes DNA binding. Genome-wide analysis of MED25 and ETV4 binding showed co-enrichment at ETS-motif containing enhancers [158]. While the specific interaction between the ETVs and MED25 suggests a functional link, studies have not been performed to explore the role on prostate cancer phenotypes.

In addition to transcriptional activators, ERG has been demonstrated to interact with epigenetic transcriptional repressors EZH2, HDAC1, and HDAC2 [148]. These repressors appear to be involved in ERG interplay with AR as addition of DHT alters binding of these repressors across AR and ERG binding sites and attenuates transcription

of specific ERG and AR co-bound genes. Transcription of the *VCL* gene in prostate cells is activated by AR and attenuated by ERG, however, the addition of an EZH2 enzymatic inhibitor relieves this attenuation [148]. FOXO1, a forkhead transcription factor frequently inactivated in prostate cancers, interacts directly with the N-terminus of ERG [89]. FOXO1 binding functions to inhibit ERG-mediated transcriptional activation. Unlike the ERG-AR interaction, FOXO1, which is also normally expressed in endothelial cells, does not interact with ERG in that cell type. This suggests that ERG function in prostate cells is specifically inhibited by FOXO1 whereas, in endogenous tissue, ERG function is unrestrained by this mechanism [89].

There is a dichotomy as to how ERG can act as a transcriptional activator in certain contexts and as a transcriptional repressor in others. It is still not clear if both activating and repression functions are critical for the oncogenic properties of ERG. It is possible that ERG acts as an activator at some regulatory elements and a repressor at others due to differing interactions with other proteins at these sites, or via displacement of endogenous ETS factors at a subset of sites. It is also possible that there is a dynamic switch between activating and repressing functions mediated by signaling pathways. ERG regulation by signaling will be summarized later in the chapter.

Oncogenic ETS and Chromatin

Expression of ERG in prostate cells can drastically influence chromosome topology and organization [159]. Upon ERG expression, novel intra- and inter-chromosomal interactions form with ERG binding sites enriched at these novel interacting loci. Dynamic nuclear co-localization of specific genes into transcriptional hubs contribute to gene activation or repression. To support this concept, the majority of ERG-regulated genes (65%) exhibit cis-interactions in cells and are enriched for genes involved in cell adhesion, skeletal system development, and migration [159]. However, the exact mechanism by which

ERG drives changes in chromosome structure and topology is not understood.

ERG interacts with various chromatin modifying enzymes including p300, HDAC1, HDAC2, EZH2, KDM4A, PRMT5, and SETDB1 [148, 152, 160, 161]. These interactions suggest that ERG modifies the local chromatin at its binding sites to alter gene transcription. ERG binding could facilitate an assortment of histone modifications as its interacting partners have the ability to methylate, demethylate, acetylate, and deacetylate various sites on histone tails. The influence of the PEA3 family on chromatin is less understood. ETV1/4/5 interact with P300, and ETV1 interacts with KAT2B [133, 153]. The enzymatic subunit of PRC2, EZH2, is upregulated in prostate cancers and is associated with aggressive disease [162]. Several groups have shown direct ERG binding and activation of the EZH2 locus [95, 163]. High levels of EZH2 results in down regulation of EZH2 targets through epigenetic silencing and dedifferentiation of prostate cancer cells [95]. ERG physically interacts with EZH2 in a manner that is regulated by post-translation modification. For example, phosphorylation of ERG Ser-96 by ERK disrupts EZH2 binding and allows ERG to function as a transcriptional activator [164]. However, modulation of ERG activity by the EZH2 interaction is not fully understood.

Signaling Pathways and Oncogenic ETS

Post translational modifications can alter the transcriptional output of transcription factors through multiple different mechanisms including effects on the rates of degradation, nuclear/cytoplasmic shuffling, altering interaction partners, and altering DNA binding. Multiple signaling pathways have been implicated in the activation of the oncogenic ETS transcription factors through a variety of post-translational modifications (Fig. 4). This section will summarize the various signaling pathways and post-translational modifications that regulate the transcriptional activities of oncogenic ETS factors in the prostate.

The Ras/MAPK pathway appears to play a role in activating the transcriptional activity of all of the prostate cancer oncogenic ETS factors. In vitro evidence demonstrates that all of the oncogenic ETS factors can be phosphorylated by multiple downstream kinases in the pathway (ERK2/JNK1/P38a) [165]. ERK2 can phosphorylate all of the oncogenic ETS factors in vitro and is critical for their transcriptional activation in vivo [166–168]. ERG can be phosphorylated by ERK2 at S96, S215, and S283 in various cell types. Phosphorylation at S283 in leukemic cells enhances stem cell features and cell proliferation and is associated with increased ERG binding at specific loci [169]. Phosphorylation of ERG at S96 and S215 occurs in prostate cells, and S96 phosphorylation is critical for ERG to activate transcription and promote cell migration [164]. Phosphorylation of S215 induces a conformational change in ERG necessary for subsequent S96 phosphorylation. ERG S96 phosphorylation decreases affinity for EZH2 and the polycomb repressive complex 2 (PRC2). This results in loss of the recruitment of EZH2 across the ERG cistrome and promotes activation of ERG target genes [164]. If Ras/MAPK signaling is important for ERG function, why do prostate tumors rarely have the Ras/MAPK pathway activating mutations that are common in other carcinomas? We have published findings suggesting that activating mutations in KRAS lead to phosphorylation of the ubiquitously expressed ETS protein ETS1, and subsequent activation of a gene expression program that promotes cell migration, invasion, and epithelial to mesenchymal transition (EMT) [142]. However, ERK2 has a higher affinity for ERG than for ETS1, and thus ERG requires lower levels of Ras/ERK signaling to be phosphorylated [165] and to activate the same cell migration/EMT gene expression program [22]. We hypothesize that these low levels of Ras/MAPK can be attributed to growth factors in the tumor microenvironment, rather than mutation. Less is understood about the mechanisms by which Ras/MAPK signaling effects the function of PEA3 subfamily members in prostate cells. The transactivation potential of ETV5 is greatly increased when Ras/MAPK signaling is activated above basal levels, and this increase in activation potential is

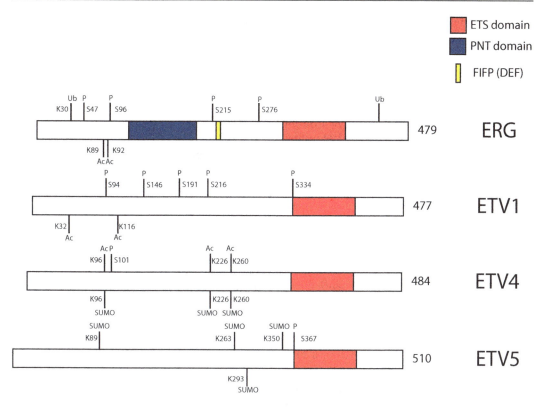

Fig. 4 Post-Translational Modifications of Oncogenic ETS Proteins. ERG (NP_001230358.1), ETV1 (NP-004947.2), ETV4 (NP_001073143.1), and ETV5 (NP_004445.1) proteins depicted with major structural domains: ETS DNA binding domain (Red), Pointed or PNT domain (Blue), and ERK docking sequence FIFP or DEF domain (Yellow). Post-translational modifications indicated by lines corresponding to respective amino acids on the protein. Acetylation (Ac), phosphorylation (P), ubiquitination (Ub), and sumoylation (SUMO) modifications have been curated from literature describing these modifications in cells. C-terminal ubiquitination site on ERG has not been assigned to a specific residue

dependent on the N-terminal activation domain [167]. Activation of the Ras/MAPK pathway has been implicated in increasing the metastatic potential of prostate cells, which express ETV4, and is important for ETV4 target gene activation in esophageal adenocarcinoma [170]. ETV1 transcriptional activity is also activated by ERK phosphorylation, and this phosphorylation occurs in the activation domain [166]. The exact mechanisms by which Ras/ERK signaling allows for PEA3 family activation remains to be investigated.

The PI3K/AKT signaling pathway is clearly important for ERG-mediated pathogenesis. ERG expression in prostate cells is not sufficient for transformation, but ERG can drive tumor growth when coupled with mutations that activate PI3K/AKT signaling [3–5]. In prostate tumors this PI3K/AKT activation is often associated with loss of the PI3K/AKT pathway inhibitor PTEN, and in fact, PTEN deletion is more common in ERG-positive tumors. While ERG and AKT could function in parallel pathways, one study suggests that AKT activation is necessary for ERG to activate target genes and cell migration in cell line models [86]. Interestingly, while this function required AKT, it did not require mTORC complexes, which are downstream of AKT. It has not been reported that AKT can directly phosphorylate the oncogenic ETS proteins, and so far, any molecular mechanism of synergy between ERG and AKT signaling is unclear [86].

Other kinases have been implicated in the activation of oncogenic ETS transcription factors. Protein Kinase A (PKA) activates ETV1 and

ETV5 through phosphorylation at the very N-terminal region of their respective ETS domains [167, 171]. Ribosomal S6 Kinase 1 (RSK1) also positively regulates ETV1 function through phosphorylation [171]. In vitro evidence demonstrates that P38a and JNK1 can phosphorylate all members of the PEA3 subfamily, however, only ETV1 has thus far been demonstrated to be phosphorylated by these kinases in cells [165]. Taken together, these data demonstrate that there is extensive cross talk between multiple signaling pathways and oncogenic ETS transcription factors. Phosphorylation by these different kinases, in general, results in increased activation of transcription by the oncogenic ETS factors.

Acetylation by CBP/p300 is another post-translational modification that effects transcriptional activation by oncogenic ETS transcription factors. ERG acetylation by CBP/p300 occurs at K89 and K92 and results in recruitment of the transcriptional regulator BRD4 [152]. Multiple hematopoietic transcription factors appear to utilize this mechanism to activate target gene transcription [152]. ETV1 and ETV4 are also targets of acetylation by CBP/p300, however, to date ETV5 has not been demonstrated to be acetylated [153]. Acetylation of ETV1 occurs at two residues at the N-terminus of the protein, which stabilizes the protein and increases its transcriptional activity [133]. Regulation of ETV4 by acetylation occurs through a dynamic mechanism, which involves cross talk of multiple post-translational modifications including phosphorylation, acetylation, sumoylation, and ubiquitination [172].

While phosphorylation and acetylation appear to regulate transcriptional activation of the oncogenic ETS proteins, both ubiquitination and sumoylation can repress their transcriptional activity. Although the exact SUMO-transferase has not been identified, ETV4 and ETV5 are sumoylated at multiple residues [172, 173]. The lysine residues sumolyated on ETV4 are also the same residues that are acetylated by CBP/P300, meaning that only one modification or the other can exist on the protein at any one time. Sumoylation of ETV4

opposes the acetylation-mediated stabilization of the protein and ultimately results in protein degradation and decreased transcriptional activation. Sumoylation of ETV5 also contributes to repression of target genes, however, this appears to occur by a mechanism other than altering protein stability [173]. The E3 ubiquitin ligase adaptor SPOP promotes ubiquitination of ERG at both the N- and C-terminal ends, however, the N-terminal ubiquitination seems to have a larger effect on protein stability [174, 175]. The E3 ubiquitin ligase adapter COP1 is involved in transcriptional activation and stabilization of the PEA3 subfamily [176]. ETV5 appears to have an N-terminal and C-terminal binding site for the complex, similar to ERG [177]. Different components of the ubiquitin ligase complex are also involved in regulation of ETV5 stability, as the presence of DET1 is necessary for its degradation.

ETS transcription factors are also able to modulate signaling pathways through the activation or repression of their downstream target genes. Ectopic expression of ERG in a prostate epithelial cell line can increase phospho-ERK and phospho-AKT levels indicating activation of both pathways [86]. Transcription of PTEN is repressed by ERG in prostate cancer cells [178]. However, in a mouse model ERG expression did not correlate with downregulation of the PTEN gene [3], indicating that regulation of these pathways is highly context dependent. The role of the PEA3 family in modulation of the Ras/ERK and PI3K/AKT pathways has not been extensively investigated in the prostate. ERG regulates Wnt/β-catenin/LEF signaling in both the context of vascularization and in prostate cancer [66, 179]. ERG is required for blood vessel development and sprouting angiogenesis in zebrafish, and this process is regulated through activation of Wnt/β-catenin signaling [66]. In the prostate, ERG binds to and increases transcription of various Wnt proteins and ultimately results in more active β-catenin. ERGs activation of Wnt/β-catenin signaling results in epithelial to mesenchymal transition of prostate cells and increases survival and invasive properties of prostate cancer cells [179].

Targeting Oncogenic ETS Factors

There are two major challenges to designing therapies that target oncogenic ETS factors. First, transcription factors such as ETS family members, do not have easily targetable binding pockets and have been described as "undruggable". However recent advances have been made in targeting transcriptional function by either developing small molecules that disrupt protein-protein interactions, or by indirectly targeting enzymes that work in transcriptional complexes. Second, since the oncogenic ETS factors belong to a large family of proteins, a challenge in therapeutic design apart from efficacy is specificity. To reduce off-target effects, oncogenic ETS targeting drugs should not cross react with non-oncogenic ETS and also should not affect the physiological function of these factors in normal tissue. For instance, inhibiting ERG requires that tumor suppressive ETS such as EHF remain active in the prostate and that ERG function is unaffected in endothelial cells. This section will describe efforts to target oncogenic ETS factors such as ERG (diagrammed in Fig. 5).

Understanding the differences between oncogenic ETS and tumor suppressive ETS may be key in developing targeted therapies. Therefore, identifying targetable regions of ETS factors that are specific for oncogenic mechanisms is crucial for future prostate cancer therapies. Two recent studies have used knowledge of the molecular mechanisms detailed above to construct a tumor suppressive ETS function more like oncogenic ETS. First, fusing the activation domain of an oncogenic ETS interacting partner, EWS, to the tumor suppressive ETS SPDEF and EHF results in the gain of oncogenic function [21]. This suggests that the oncogenic properties of ETS factors contribute to regions outside the ETS domain that allow for specific protein-protein interactions. Another key protein-protein interacting partner specific to the oncogenic ETS is the AP-1 transcription factor. Oncogenic ETS are able to activate ETS-AP1 sites by cooperatively binding with the AP-1 transcription factor. Tumor suppressive ETS instead bind anti-cooperatively with AP-1. Mutating positively charged residues

in the AP-1 interface of EHF to the corresponding residues in ERG disrupt this anti-cooperative binding [140]. This suggests that these residues contribute specifically to oncogenic ETS function.

ERG Targeting Peptide

An ERG inhibitory peptide was recently identified that forms high affinity interactions with the ETS DNA binding domain of ERG [180]. This peptide inhibits interactions between ERG and important protein binding partners AR and the catalytic subunit of DNA-PK (DNA-PKcs) in a dose-dependent fashion. Treatment of ERG expressing cells with the peptide caused a significant decrease in ERG-mediated cell migration by interfering with the DNA binding ability of ERG and triggering ERG degradation. Importantly, this peptide had little effect on normal ERG functions in endothelial cells, indicating that normal and oncogenic ERG functions are separable. Treatment of LNCaP, a cell line that overexpresses ETV1, with this peptide showed a similar effect. The mechanism that might allow this peptide to specifically target only oncogenic functions, or whether it targets only oncogenic ETS factors, remain unclear.

Small Molecule Inhibitors

A cell-based immuno-assay screen was used to identify ERGi-USU as a small molecule that altered ERG protein levels [181]. ERGi-USU was able to selectively inhibit ERG protein expression in the ERG-expressing prostate cancer cell line VCaP, but not ERG expressed in non-prostate cancers or in endothelial cells. ERGi-USU treatment inhibited VCaP xenograft growth in nude mice and was found to function by directly binding to and inhibiting RIOK2, a kinase important for ERG protein stability. One group targeted ERG-DNA complexes with the di-(phenyl-thiophene-amidine) compound DB1255 [182]. DB1255 interacts with ETS binding motifs in DNA and blocks ERG binding, impeding activation of reporter genes. An *in silico* approach was used to identify small mol-

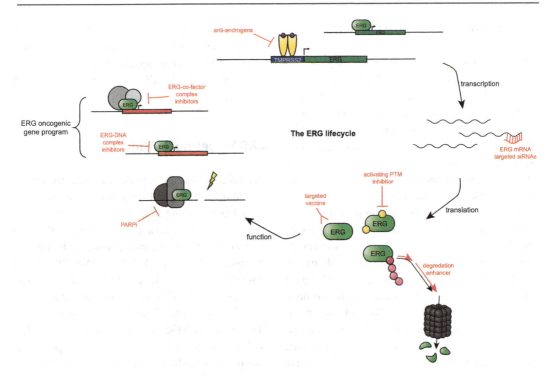

Fig. 5 A cartoon depiction of the ERG lifecycle. Diagramed is ERG expression from gene to protein and ERG function. Potential nodes of inhibition are depicted in red. Expression of TMPRSS2/ERG is driven by binding of ligand-bound AR (yellow pillars with purple circles) to the TMPRSS2 promoter. Additionally, ERG protein can bind the endogenous ERG promoter and drive ERG transcription. Transcription of the ERG gene produces ERG mRNA, which can be silenced by the RNAi pathway. ERG transcripts are translated into protein. Post translational modification of ERG can be activating (yellow circle), repressing (not shown), or target ERG protein for proteasomal degradation (pink circle). Potential ways of regulating ERG protein include ERG targeted vaccines, inhibition of activating post translational modifications, and promotion of degradation. To activate oncogenic gene programs, ERG binds gene regulatory elements with various co-factors and other transcription factors. Various inhibitors can target this function, such as ERG-DNA complex inhibitors and ERG-co-factor complex inhibitors. General transcription inhibitors may synergize with ERG inhibitors to yield a greater effect. When aberrantly expressed, ERG promotes DNA damage and thus PARP inhibitor treatment of ERG-rearranged cells promotes synthetic lethality

ecules that target the ETS domain of ERG. Compound VPC-18005 inhibited ERG-DNA complex formation and inhibited migration and invasion in cells expressing ERG. Additionally, VPC-18005 decreased metastasis in a zebrafish xenograft model [183]. The small molecule YK-4-279 was developed as an inhibitor of EWS-FLI1 for Ewing's Sarcomas [184]. Given that FLI1 is an ETS transcription factor with high homology to ERG, the effect of YK-4-279 has been tested for other oncogenic ETS factors. Treatment of the prostate cancer cell lines VCaP or LNCaP with YK-4-279 resulted in the inhibition of ERG or ETV1 transcriptional activity, respectively, as well as ETS-mediated cell invasion. However, YK-4-279 treatment of PC3 cells, which overexpress ETV4, had no effect on invasion [185]. In mouse xenograft tumor models YK-4-279 has been effective against prostate tumors expressing both ERG [186] and ETV1 [187]. BRD32048 was identified as a small molecule inhibitor of ETV1 via a microarray screen [188]. Treatment with BRD32048 caused a significant dose-responsive decrease in invasion of an ETV1 translocation-positive prostate cancer cell line but had no effect on prostate cancer cells that do not express an ETV1 fusion. BRD32048 functions by direct binding to ETV1, preventing p300 acetylation and promoting degradation.

PARP1 Inhibitors

ERG interacts with the DNA damage repair protein PARP1 and DNA-PKcs, a kinase in the PI3K pathway [189]. This complex is required for cell invasion, intravasation, and metastasis. Importantly, inhibiting PARP1 significantly decreased tumor growth in ETS-positive xenografts but had no effect on ETS-negative xenografts. It was hypothesized that aberrant ERG expression causes DNA damage, allowing for PARP inhibition to promote synthetic lethality specifically in ETS-positive cells. While this study suggested that ERG acts independently of XRCC4-mediated non-homologous end joining (NHEJ), another study found that ERG blocks XRCC1-mediated NHEJ to promote radiosensitization to PARP inhibition [190]. Treatment of ERG expressing prostate cancer cells with the PARP inhibitor olaparib significantly reduced clonogenic growth. Olaparib treatment increased the DNA damage marker gamma-H2AX in ERG-positive cells but not ERG-negative cells. Expression of ERG in a xenograft model was able to confer radiation resistance that was subject to reversal by the PARP inhibitor ABT-888 [191]. The PARP inhibitor rucaparib was shown to synergize with radiation in the ETS-positive cell lines VCaP and LNCaP [190]. Although more research is needed to pinpoint the mechanism, the PARP1-ERG axis provides an attractive prostate cancer target.

ERG Targeting Vaccine

The development of immune checkpoint inhibitors has renewed interest in the use of immunotherapy for many cancer types. However, prostate cancer tends to be resistant to such therapies. Since ERG expression is specific to prostate tumor cells, as compared to normal prostate, one effort has been to use this observation to promote an immune response. ERG peptides are processed and presented on the surface of prostate cancer cells that express the HLA-A2.1 type of MHC class 1 molecules [192]. HLA-A2.1+ cells are the most common human leukocyte antigen

in Caucasians, the demographic that most frequently has ERG rearrangements [193]. Mice expressing human HLA-A2.1 or both prostate-specific ERG and human HLA-A2.1, when immunized with an ERG derived peptide, were found to have significantly more cytotoxic T lymphocytes than the control mice. Additionally, co-culture of T-cells from ERG peptide immunized mice with ERG expressing prostate cancer cells resulted in an increased measures of T-cell activation compared to cells lacking ERG expression [192].

Chemoresistance

ERG-positive patients with CRPC are twice as likely to develop resistance to docetaxel treatment than ERG-negative CRPC patients. Immunoprecipitation experiments revealed that cytoplasmic ERG interacts with tubulin through the PNT domain. Deletion of the ERG PNT domain restores sensitivity to docetaxel treatment, suggesting that the ERG-tubulin interaction functionally contributes to taxane resistance in prostate cancer [194]. Additionally, detection of *TMPRSS2*/ERG in the blood of metastatic CRPC patients is predictive of docetaxel resistance [195]. ERG induces chemoresistance in leukemia cells [196]. These data may help clinicians determine a course of treatment by ERG subtyping.

Degradation of ETS Factors

It is clear that the protein stability of oncogenic ETS proteins is highly regulated, and it is possible that this process could be manipulated for therapeutic benefit. The tumor suppressor SPOP regulates ERG protein stability. Interestingly, SPOP loss-of-function mutations are present in 6–15% of prostate cancers but are mutually exclusive with ETS rearrangements [42, 175, 197, 198]. While wild type ERG protein is amenable to SPOP mediated degradation, most ERG fusion products are resistant through loss of the N-terminal degron [174, 175]. Similarly, the E3 ubiquitin ligase

COP1 targets the PEA3 factors for degradation through an N-terminal degron that is lost in most prostate cancer gene rearrangements [176]. It is possible that these mechanisms contribute to ERG and PEA3 factor upregulation, which is observed in some rearrangement-negative tumors. However, the E3 ubiquitin ligase TRIM25 is able to bind and ubiquitinate both full length and N-terminally truncated ERG, as it targets the C-terminal ERG degron [199]. The deubiquitase enzyme USP9X binds to and deubiquitinates ERG [93]. A small molecule inhibitor of USP9X, WP1130, destabilizes ERG protein and causes an increase in tumor growth in mice xenografted with ERG-expressing prostate cancer cells [93], suggesting that targeting ERG protein stability may be an effective prostate cancer treatment.

Conclusion

Recurrent gene rearrangements that involve ETS transcription factors typify the largest molecular subgroup of prostate cancer. Information on the molecular details of ETS factor function are essential to the movement towards precision therapies for prostate cancer patients. Genomic and biochemical studies have uncovered much about how ETS factors bind to and regulate chromatin to alter gene expression. However, the key to the development of effective and specific ETS therapies may be through teasing apart differences between oncogenic and tumor suppressive ETS function and understanding differences in physiological versus pathological roles.

References

1. S.A. Tomlins et al., Recurrent fusion of TMPRSS2 and ETS transcription factor genes in prostate cancer. Science **310**(5748), 644–648 (2005)
2. S.A. Tomlins et al., Distinct classes of chromosomal rearrangements create oncogenic ETS gene fusions in prostate cancer. Nature **448**(7153), 595–599 (2007)
3. B.S. Carver et al., Aberrant ERG expression cooperates with loss of PTEN to promote cancer progression in the prostate. Nat. Genet. **41**(5), 619–624 (2009)

4. J.C. King et al., Cooperativity of TMPRSS2-ERG with PI3-kinase pathway activation in prostate oncogenesis. Nat. Genet. **41**(5), 524–526 (2009)
5. Y. Zong et al., ETS family transcription factors collaborate with alternative signaling pathways to induce carcinoma from adult murine prostate cells. Proc. Natl. Acad. Sci. U. S. A. **106**(30), 12465–12470 (2009)
6. Y. Chen et al., ETS factors reprogram the androgen receptor cistrome and prime prostate tumorigenesis in response to PTEN loss. Nat. Med. **19**(8), 1023–1029 (2013)
7. P.C. Hollenhorst, L.P. McIntosh, B.J. Graves, Genomic and biochemical insights into the specificity of ETS transcription factors. Annu. Rev. Biochem. **80**, 437–471 (2011)
8. G.H. Wei et al., Genome-wide analysis of ETS-family DNA-binding in vitro and in vivo. EMBO J. **29**(13), 2147–2160 (2010)
9. P.C. Hollenhorst, D.A. Jones, B.J. Graves, Expression profiles frame the promoter specificity dilemma of the ETS family of transcription factors. Nucleic Acids Res. **32**(18), 5693–5702 (2004)
10. D. Albino et al., ESE3/EHF controls epithelial cell differentiation and its loss leads to prostate tumors with mesenchymal and stem-like features. Cancer Res. **72**(11), 2889–2900 (2012)
11. X. Gu et al., Reduced PDEF expression increases invasion and expression of mesenchymal genes in prostate cancer cells. Cancer Res. **67**(9), 4219–4226 (2007)
12. N. Longoni et al., ETS transcription factor ESE1/ELF3 orchestrates a positive feedback loop that constitutively activates NF-kappaB and drives prostate cancer progression. Cancer Res. **73**(14), 4533–4547 (2013)
13. Cancer Genome Atlas Research Network, The molecular taxonomy of primary prostate cancer. Cell **163**(4), 1011–1025 (2015)
14. D. Robinson et al., Integrative clinical genomics of advanced prostate cancer. Cell **161**(5), 1215–1228 (2015)
15. A. Aytes et al., ETV4 promotes metastasis in response to activation of PI3-kinase and Ras signaling in a mouse model of advanced prostate cancer. Proc. Natl. Acad. Sci. U. S. A. **110**(37), E3506–E3515 (2013)
16. E. Baena et al., ETV1 directs androgen metabolism and confers aggressive prostate cancer in targeted mice and patients. Genes Dev. **27**(6), 683–698 (2013)
17. J. Higgins et al., Interaction of the androgen receptor, ETV1, and PTEN pathways in mouse prostate varies with pathological stage and predicts cancer progression. Horm. Cancer **6**(2–3), 67–86 (2015)
18. D.S. Rickman et al., SLC45A3-ELK4 is a novel and frequent erythroblast transformation-specific fusion transcript in prostate cancer. Cancer Res. **69**(7), 2734–2738 (2009)

19. P. Paulo et al., FLI1 is a novel ETS transcription factor involved in gene fusions in prostate cancer. Genes Chromosomes Cancer **51**(3), 240–249 (2012)
20. B.E. Helgeson et al., Characterization of TMPRSS2:ETV5 and SLC45A3:ETV5 gene fusions in prostate cancer. Cancer Res. **68**(1), 73–80 (2008)
21. V. Kedage et al., An interaction with Ewing's sarcoma breakpoint protein EWS defines a specific oncogenic mechanism of ETS factors rearranged in prostate cancer. Cell Rep. **17**(5), 1289–1301 (2016)
22. P.C. Hollenhorst et al., Oncogenic ETS proteins mimic activated RAS/MAPK signaling in prostate cells. Genes Dev. **25**(20), 2147–2157 (2011)
23. C. Kumar-Sinha, S.A. Tomlins, A.M. Chinnaiyan, Recurrent gene fusions in prostate cancer. Nat. Rev. Cancer **8**(7), 497–511 (2008)
24. D. Hessels, J.A. Schalken, Recurrent gene fusions in prostate cancer: their clinical implications and uses. Curr. Urol. Rep. **14**(3), 214–222 (2013)
25. S.A. Tomlins et al., TMPRSS2:ETV4 gene fusions define a third molecular subtype of prostate cancer. Cancer Res. **66**(7), 3396–3400 (2006)
26. R. Mehra et al., Comprehensive assessment of TMPRSS2 and ETS family gene aberrations in clinically localized prostate cancer. Mod. Pathol. **20**(5), 538–544 (2007)
27. S. Minner et al., Marked heterogeneity of ERG expression in large primary prostate cancers. Mod. Pathol. **26**(1), 106–116 (2013)
28. J.D. Barros-Silva et al., Novel 5′ fusion partners of ETV1 and ETV4 in prostate cancer. Neoplasia **15**(7), 720–726 (2013)
29. K.G. Hermans et al., Two unique novel prostate-specific and androgen-regulated fusion partners of ETV4 in prostate cancer. Cancer Res. **68**(9), 3094–3098 (2008)
30. B. Han et al., A fluorescence in situ hybridization screen for E26 transformation-specific aberrations: identification of DDX5-ETV4 fusion protein in prostate cancer. Cancer Res. **68**(18), 7629–7637 (2008)
31. J. Clark et al., Diversity of TMPRSS2-ERG fusion transcripts in the human prostate. Oncogene **26**(18), 2667–2673 (2007)
32. M.A. Svensson et al., Testing mutual exclusivity of ETS rearranged prostate cancer. Lab. Investig. **91**(3), 404–412 (2011)
33. J. Wang et al., Expression of variant TMPRSS2/ERG fusion messenger RNAs is associated with aggressive prostate cancer. Cancer Res. **66**(17), 8347–8351 (2006)
34. M.C. Wong et al., Global incidence and mortality for prostate cancer: analysis of temporal patterns and trends in 36 countries. Eur. Urol. **70**(5), 862–874 (2016)
35. C.K. Zhou et al., TMPRSS2:ERG gene fusions in prostate cancer of West African men and a meta-analysis of racial differences. Am. J. Epidemiol. **186**(12), 1352–1361 (2017)
36. B. Ateeq et al., Molecular profiling of ETS and non-ETS aberrations in prostate cancer patients from northern India. Prostate **75**(10), 1051–1062 (2015)
37. J. Weischenfeldt et al., Integrative genomic analyses reveal an androgen-driven somatic alteration landscape in early-onset prostate cancer. Cancer Cell **23**(2), 159–170 (2013)
38. M.C. Tsourlakis et al., Heterogeneity of ERG expression in prostate cancer: a large section mapping study of entire prostatectomy specimens from 125 patients. BMC Cancer **16**, 641 (2016)
39. B.S. Taylor et al., Integrative genomic profiling of human prostate cancer. Cancer Cell **18**(1), 11–22 (2010)
40. K.A. Leinonen et al., Loss of PTEN is associated with aggressive behavior in ERG-positive prostate cancer. Cancer Epidemiol. Biomark. Prev. **22**(12), 2333–2344 (2013)
41. S.A. Tomlins et al., The role of SPINK1 in ETS rearrangement-negative prostate cancers. Cancer Cell **13**(6), 519–528 (2008)
42. C.E. Barbieri et al., Exome sequencing identifies recurrent SPOP, FOXA1 and MED12 mutations in prostate cancer. Nat. Genet. **44**(6), 685–689 (2012)
43. L. Burkhardt et al., CHD1 is a 5q21 tumor suppressor required for ERG rearrangement in prostate cancer. Cancer Res. **73**(9), 2795–2805 (2013)
44. J. Shoag et al., SPOP mutation drives prostate neoplasia without stabilizing oncogenic transcription factor ERG. J. Clin. Invest. **128**(1), 381–386 (2018)
45. J. Clark et al., Complex patterns of ETS gene alteration arise during cancer development in the human prostate. Oncogene **27**(14), 1993–2003 (2008)
46. M.M. Shen, C. Abate-Shen, Molecular genetics of prostate cancer: new prospects for old challenges. Genes Dev. **24**(18), 1967–2000 (2010)
47. F. Demichelis et al., TMPRSS2:ERG gene fusion associated with lethal prostate cancer in a watchful waiting cohort. Oncogene **26**(31), 4596–4599 (2007)
48. K.D. Berg et al., ERG protein expression in diagnostic specimens is associated with increased risk of progression during active surveillance for prostate cancer. Eur. Urol. **66**(5), 851–860 (2014)
49. C. Hagglof et al., TMPRSS2-ERG expression predicts prostate cancer survival and associates with stromal biomarkers. PLoS One **9**(2), e86824 (2014)
50. U. Lokman et al., PTEN loss but not ERG expression in diagnostic biopsies is associated with increased risk of progression and adverse surgical findings in men with prostate cancer on active surveillance. Eur. Urol. Focus **4**(6), 867–873 (2018)
51. M. Taris et al., ERG expression in prostate cancer: the prognostic paradox. Prostate **74**(15), 1481–1487 (2014)

52. S. Terry et al., Clinical value of ERG, TFF3, and SPINK1 for molecular subtyping of prostate cancer. Cancer 121(9), 1422–1430 (2015)
53. D.W. Lin et al., Urinary TMPRSS2:ERG and PCA3 in an active surveillance cohort: results from a baseline analysis in the Canary Prostate Active Surveillance Study. Clin. Cancer Res. 19(9), 2442–2450 (2013)
54. A. Font-Tello et al., Association of ERG and TMPRSS2-ERG with grade, stage, and prognosis of prostate cancer is dependent on their expression levels. Prostate 75(11), 1216–1226 (2015)
55. S.A. Tomlins et al., Urine TMPRSS2:ERG plus PCA3 for individualized prostate cancer risk assessment. Eur. Urol. 70(1), 45–53 (2016)
56. H. Amir, C.M.R. Lebastchi, A.M. Helfand, T. Osawa, J. Siddiqui, R. Siddiqui, A.M. Chinnaiyan, P. Kunju, R. Mehra, D. Snyder, S.A. Tomlins, J.T. Wei, T.M. Morgan, Michigan Prostate Score (MIPS): an analysis of a novel urinary biomarker panel for the prediction of prostate cancer and its impact on biopsy rates. J. Urol. 197(4), e128 (2007)
57. R.M. Hagen et al., Quantitative analysis of ERG expression and its splice isoforms in formalin-fixed, paraffin-embedded prostate cancer samples: association with seminal vesicle invasion and biochemical recurrence. Am. J. Clin. Pathol. 142(4), 533–540 (2014)
58. C. Lin et al., Nuclear receptor-induced chromosomal proximity and DNA breaks underlie specific translocations in cancer. Cell 139(6), 1069–1083 (2009)
59. M.C. Haffner et al., Androgen-induced TOP2B-mediated double-strand breaks and prostate cancer gene rearrangements. Nat. Genet. 42(8), 668–675 (2010)
60. R.S. Mani et al., Induced chromosomal proximity and gene fusions in prostate cancer. Science 326(5957), 1230 (2009)
61. R.S. Mani et al., Inflammation-induced oxidative stress mediates gene fusion formation in prostate cancer. Cell Rep. 17(10), 2620–2631 (2016)
62. X. Li et al., BRD4 promotes DNA repair and mediates the formation of TMPRSS2-ERG gene rearrangements in prostate cancer. Cell Rep. 22(3), 796–808 (2018)
63. V. Vlaeminck-Guillem et al., The Ets family member Erg gene is expressed in mesodermal tissues and neural crests at fundamental steps during mouse embryogenesis. Mech. Dev. 91(1–2), 331–335 (2000)
64. F. Ellett, B.T. Kile, G.J. Lieschke, The role of the ETS factor erg in zebrafish vasculogenesis. Mech. Dev. 126(3–4), 220–229 (2009)
65. F. McLaughlin et al., Combined genomic and antisense analysis reveals that the transcription factor Erg is implicated in endothelial cell differentiation. Blood 98(12), 3332–3339 (2001)
66. G.M. Birdsey et al., The endothelial transcription factor ERG promotes vascular stability and growth through Wnt/beta-catenin signaling. Dev. Cell 32(1), 82–96 (2015)
67. G.M. Birdsey et al., Transcription factor Erg regulates angiogenesis and endothelial apoptosis through VE-cadherin. Blood 111(7), 3498–3506 (2008)
68. A. Chotteau-Lelievre et al., PEA3 transcription factors are expressed in tissues undergoing branching morphogenesis and promote formation of duct-like structures by mammary epithelial cells in vitro. Dev. Biol. 259(2), 241–257 (2003)
69. A. Garg et al., FGF-induced Pea3 transcription factors program the genetic landscape for cell fate determination. PLoS Genet. 14(9), e1007660 (2018)
70. W.A. Znosko et al., Overlapping functions of Pea3 ETS transcription factors in FGF signaling during zebrafish development. Dev. Biol. 342(1), 11–25 (2010)
71. J.C. Herriges et al., FGF-regulated ETV transcription factors control FGF-SHH feedback loop in lung branching. Dev. Cell 35(3), 322–332 (2015)
72. Z. Zhang et al., FGF-regulated Etv genes are essential for repressing Shh expression in mouse limb buds. Dev. Cell 16(4), 607–613 (2009)
73. A. Chotteau-Lelievre et al., Differential expression patterns of the PEA3 group transcription factors through murine embryonic development. Oncogene 15(8), 937–952 (1997)
74. A. Chotteau-Lelievre et al., Expression patterns of the Ets transcription factors from the PEA3 group during early stages of mouse development. Mech. Dev. 108(1–2), 191–195 (2001)
75. S.J. Loughran et al., The transcription factor Erg is essential for definitive hematopoiesis and the function of adult hematopoietic stem cells. Nat. Immunol. 9(7), 810–819 (2008)
76. G. Tyagi et al., Loss of Etv5 decreases proliferation and RET levels in neonatal mouse testicular germ cells and causes an abnormal first wave of spermatogenesis. Biol. Reprod. 81(2), 258–266 (2009)
77. C.L. Carmichael et al., Hematopoietic overexpression of the transcription factor Erg induces lymphoid and erythro-megakaryocytic leukemia. Proc. Natl. Acad. Sci. U. S. A. 109(38), 15437–15442 (2012)
78. S. Tsuzuki, O. Taguchi, M. Seto, Promotion and maintenance of leukemia by ERG. Blood 117(14), 3858–3868 (2011)
79. J.A. Thoms et al., ERG promotes T-acute lymphoblastic leukemia and is transcriptionally regulated in leukemic cells by a stem cell enhancer. Blood 117(26), 7079–7089 (2011)
80. O.M. Casey et al., TMPRSS2-driven ERG expression in vivo increases self-renewal and maintains expression in a castration resistant subpopulation. PLoS One 7(7), e41668 (2012)
81. A. Srivastava, D.K. Price, W.D. Figg, Prostate tumor development and androgen receptor function alterations in a new mouse model with ERG overexpression and PTEN inactivation. Cancer Biol. Ther. 15(10), 1293–1295 (2014)

82. L.T. Nguyen et al., ERG activates the YAP1 transcriptional program and induces the development of age-related prostate tumors. Cancer Cell **27**(6), 797–808 (2015)

83. R.S. Mani et al., TMPRSS2-ERG-mediated feedforward regulation of wild-type ERG in human prostate cancers. Cancer Res. **71**(16), 5387–5392 (2011)

84. C. Cai et al., Reactivation of androgen receptor-regulated TMPRSS2:ERG gene expression in castration-resistant prostate cancer. Cancer Res. **69**(15), 6027–6032 (2009)

85. K.D. Mertz et al., Molecular characterization of TMPRSS2-ERG gene fusion in the NCI-H660 prostate cancer cell line: a new perspective for an old model. Neoplasia **9**(3), 200–206 (2007)

86. N. Selvaraj et al., Prostate cancer ETS rearrangements switch a cell migration gene expression program from RAS/ERK to PI3K/AKT regulation. Mol. Cancer **13**, 61 (2014)

87. S.A. Tomlins et al., Role of the TMPRSS2-ERG gene fusion in prostate cancer. Neoplasia **10**(2), 177–188 (2008)

88. L. Shao et al., Highly specific targeting of the TMPRSS2/ERG fusion gene using liposomal nanovectors. Clin. Cancer Res. **18**(24), 6648–6657 (2012)

89. Y. Yang et al., Loss of FOXO1 cooperates with TMPRSS2-ERG overexpression to promote prostate tumorigenesis and cell invasion. Cancer Res. **77**(23), 6524–6537 (2017)

90. S. Gupta et al., FZD4 as a mediator of ERG oncogene-induced WNT signaling and epithelial-to-mesenchymal transition in human prostate cancer cells. Cancer Res. **70**(17), 6735–6745 (2010)

91. P.C. Hollenhorst et al., The ETS gene ETV4 is required for anchorage-independent growth and a cell proliferation gene expression program in PC3 prostate cells. Genes Cancer **1**(10), 1044–1052 (2011)

92. D. Mesquita et al., Specific and redundant activities of ETV1 and ETV4 in prostate cancer aggressiveness revealed by co-overexpression cellular contexts. Oncotarget **6**(7), 5217–5236 (2015)

93. S. Wang et al., Ablation of the oncogenic transcription factor ERG by deubiquitinase inhibition in prostate cancer. Proc. Natl. Acad. Sci. U. S. A. **111**(11), 4251–4256 (2014)

94. Z. Mounir et al., TMPRSS2:ERG blocks neuroendocrine and luminal cell differentiation to maintain prostate cancer proliferation. Oncogene **34**(29), 3815–3825 (2015)

95. J. Yu et al., An integrated network of androgen receptor, polycomb, and TMPRSS2-ERG gene fusions in prostate cancer progression. Cancer Cell **17**(5), 443–454 (2010)

96. C. Sun et al., TMPRSS2-ERG fusion, a common genomic alteration in prostate cancer activates C-MYC and abrogates prostate epithelial differentiation. Oncogene **27**(40), 5348–5353 (2008)

97. S. You et al., Integrated classification of prostate cancer reveals a novel luminal subtype with poor outcome. Cancer Res. **76**(17), 4948–4958 (2016)

98. A.M. Blee et al., TMPRSS2-ERG controls luminal epithelial lineage and antiandrogen sensitivity in PTEN and TP53-mutated prostate cancer. Clin. Cancer Res. **24**(18), 4551–4565 (2018)

99. K. Shimizu et al., An ets-related gene, ERG, is rearranged in human myeloid leukemia with t(16;21) chromosomal translocation. Proc. Natl. Acad. Sci. U. S. A. **90**(21), 10280–10284 (1993)

100. M. Giovannini et al., EWS-erg and EWS-Fli1 fusion transcripts in Ewing's sarcoma and primitive neuroectodermal tumors with variant translocations. J. Clin. Invest. **94**(2), 489–496 (1994)

101. T. Dunn et al., ERG gene is translocated in an Ewing's sarcoma cell line. Cancer Genet. Cytogenet. **76**(1), 19–22 (1994)

102. T.G.P. Grunewald et al., Ewing sarcoma. Nat. Rev. Dis. Primers **4**(1), 5 (2018)

103. I.S. Jeon et al., A variant Ewing's sarcoma translocation (7;22) fuses the EWS gene to the ETS gene ETV1. Oncogene **10**(6), 1229–1234 (1995)

104. M. Peter et al., A new member of the ETS family fused to EWS in Ewing tumors. Oncogene **14**(10), 1159–1164 (1997)

105. F. Urano et al., Molecular analysis of Ewing's sarcoma: another fusion gene, EWS-E1AF, available for diagnosis. Jpn. J. Cancer Res. **89**(7), 703–711 (1998)

106. P.H. Sorensen et al., A second Ewing's sarcoma translocation, t(21;22), fuses the EWS gene to another ETS-family transcription factor, ERG. Nat. Genet. **6**(2), 146–151 (1994)

107. D.D. Prasad et al., TLS/FUS fusion domain of TLS/FUS-erg chimeric protein resulting from the t(16;21) chromosomal translocation in human myeloid leukemia functions as a transcriptional activation domain. Oncogene **9**(12), 3717–3729 (1994)

108. I. Panagopoulos et al., Fusion of the FUS gene with ERG in acute myeloid leukemia with t(16;21) (p11;q22). Genes Chromosomes Cancer **11**(4), 256–262 (1994)

109. P. Peeters et al., Fusion of TEL, the ETS-variant gene 6 (ETV6), to the receptor-associated kinase JAK2 as a result of t(9;12) in a lymphoid and t(9;15;12) in a myeloid leukemia. Blood **90**(7), 2535–2540 (1997)

110. R. Bose et al., ERF mutations reveal a balance of ETS factors controlling prostate oncogenesis. Nature **546**(7660), 671–675 (2017)

111. J.A. Budka et al., Common ELF1 deletion in prostate cancer bolsters oncogenic ETS function, inhibits senescence and promotes docetaxel resistance. Genes Cancer **9**(5–6), 198–214 (2018)

112. D.E. Linn et al., Deletion of interstitial genes between TMPRSS2 and ERG promotes prostate cancer progression. Cancer Res. **76**(7), 1869–1881 (2016)

113. F.W. Huang et al., Exome sequencing of African-American prostate cancer reveals loss-of-function ERF mutations. Cancer Discov. **7**(9), 973–983 (2017)
114. T.E. Sussan et al., Trisomy represses Apc(Min)-mediated tumours in mouse models of Down's syndrome. Nature **451**(7174), 73–75 (2008)
115. S.Y. Ku et al., Rb1 and Trp53 cooperate to suppress prostate cancer lineage plasticity, metastasis, and antiandrogen resistance. Science **355**(6320), 78–83 (2017)
116. P. Mu et al., SOX2 promotes lineage plasticity and antiandrogen resistance in TP53- and RB1-deficient prostate cancer. Science **355**(6320), 84–88 (2017)
117. B.J. Graves, J.M. Petersen, Specificity within the ets family of transcription factors. Adv. Cancer Res. **75**, 1–55 (1998)
118. S. De et al., Steric mechanism of auto-inhibitory regulation of specific and non-specific DNA binding by the ETS transcriptional repressor ETV6. J. Mol. Biol. **426**(7), 1390–1406 (2014)
119. X. Xu et al., Structural basis for reactivating the mutant TERT promoter by cooperative binding of p52 and ETS1. Nat. Commun. **9**(1), 3183 (2018)
120. R. Sharma, S.P. Gangwar, A.K. Saxena, Comparative structure analysis of the ETSi domain of ERG3 and its complex with the E74 promoter DNA sequence. Acta Crystallogr. F Struct. Biol. Commun. **74**(Pt 10), 656–663 (2018)
121. S.L. Currie et al., Structured and disordered regions cooperatively mediate DNA-binding autoinhibition of ETS factors ETV1, ETV4 and ETV5. Nucleic Acids Res. **45**(5), 2223–2241 (2017)
122. M. Shiina et al., A novel allosteric mechanism on protein-DNA interactions underlying the phosphorylation-dependent regulation of Ets1 target gene expressions. J. Mol. Biol. **427**(8), 1655–1669 (2015)
123. C.D. Cooper et al., Structures of the Ets protein DNA-binding domains of transcription factors Etv1, Etv4, Etv5, and Fev: determinants of DNA binding and redox regulation by disulfide bond formation. J. Biol. Chem. **290**(22), 13692–13709 (2015)
124. T. Shrivastava et al., Structural basis of Ets1 activation by Runx1. Leukemia **28**(10), 2040–2048 (2014)
125. M.C. Regan et al., Structural and dynamic studies of the transcription factor ERG reveal DNA binding is allosterically autoinhibited. Proc. Natl. Acad. Sci. U. S. A. **110**(33), 13374–13379 (2013)
126. N.D. Babayeva, O.I. Baranovskaya, T.H. Tahirov, Structural basis of Ets1 cooperative binding to widely separated sites on promoter DNA. PLoS One **7**(3), e33698 (2012)
127. N.D. Babayeva et al., Structural basis of Ets1 cooperative binding to palindromic sequences on stromelysin-1 promoter DNA. Cell Cycle **9**(15), 3054–3062 (2010)
128. M. Hassler, T.J. Richmond, The B-box dominates SAP-1-SRF interactions in the structure of the ternary complex. EMBO J. **20**(12), 3018–3028 (2001)
129. Y. Mo et al., Structures of SAP-1 bound to DNA targets from the E74 and c-fos promoters: insights into DNA sequence discrimination by Ets proteins. Mol. Cell **2**(2), 201–212 (1998)
130. Y. Mo et al., Structure of the elk-1-DNA complex reveals how DNA-distal residues affect ETS domain recognition of DNA. Nat. Struct. Biol. **7**(4), 292–297 (2000)
131. M.D. Jonsen et al., Characterization of the cooperative function of inhibitory sequences in Ets-1. Mol. Cell. Biol. **16**(5), 2065–2073 (1996)
132. D.O. Cowley, B.J. Graves, Phosphorylation represses Ets-1 DNA binding by reinforcing autoinhibition. Genes Dev. **14**(3), 366–376 (2000)
133. A. Goel, R. Janknecht, Acetylation-mediated transcriptional activation of the ETS protein ER81 by p300, P/CAF, and HER2/Neu. Mol. Cell. Biol. **23**(17), 6243–6254 (2003)
134. A. Greenall et al., DNA binding by the ETS-domain transcription factor PEA3 is regulated by intramolecular and intermolecular protein.protein interactions. J. Biol. Chem. **276**(19), 16207–16215 (2001)
135. P. Chi et al., ETV1 is a lineage survival factor that cooperates with KIT in gastrointestinal stromal tumours. Nature **467**(7317), 849–853 (2010)
136. J.E. Fish et al., Dynamic regulation of VEGF-inducible genes by an ERK/ERG/p300 transcriptional network. Development **144**(13), 2428–2444 (2017)
137. P.C. Hollenhorst et al., Genome-wide analyses reveal properties of redundant and specific promoter occupancy within the ETS gene family. Genes Dev. **21**(15), 1882–1894 (2007)
138. J. Boros et al., Elucidation of the ELK1 target gene network reveals a role in the coordinate regulation of core components of the gene regulation machinery. Genome Res. **19**(11), 1963–1973 (2009)
139. N. Selvaraj et al., Extracellular signal-regulated kinase signaling regulates the opposing roles of JUN family transcription factors at ETS/AP-1 sites and in cell migration. Mol. Cell. Biol. **35**(1), 88–100 (2015)
140. B.J. Madison et al., Electrostatic repulsion causes anticooperative DNA binding between tumor suppressor ETS transcription factors and JUN-FOS at composite DNA sites. J. Biol. Chem. **293**(48), 18624–18635 (2018)
141. A. Verger et al., Identification of amino acid residues in the ETS transcription factor Erg that mediate Erg-Jun/Fos-DNA ternary complex formation. J. Biol. Chem. **276**(20), 17181–17189 (2001)
142. J.P. Plotnik et al., ETS1 is a genome-wide effector of RAS/ERK signaling in epithelial cells. Nucleic Acids Res. **42**(19), 11928–11940 (2014)
143. K. Gangwal et al., Emergent properties of EWS/FLI regulation via GGAA microsatellites in Ewing's sarcoma. Genes Cancer **1**(2), 177–187 (2010)
144. K. Gangwal et al., Microsatellites as EWS/FLI response elements in Ewing's sarcoma. Proc. Natl. Acad. Sci. U. S. A. **105**(29), 10149–10154 (2008)
145. A.L. Kennedy et al., Functional, chemical genomic, and super-enhancer screening identify sensitivity to cyclin D1/CDK4 pathway inhibition in Ewing sarcoma. Oncotarget **6**(30), 30178–30193 (2015)

146. T.L. Sreenath et al., ETS related gene mediated androgen receptor aggregation and endoplasmic reticulum stress in prostate cancer development. Sci. Rep. **7**(1), 1109 (2017)
147. N.L. Sharma et al., The ETS family member GABPalpha modulates androgen receptor signalling and mediates an aggressive phenotype in prostate cancer. Nucleic Acids Res. **42**(10), 6256–6269 (2014)
148. K.R. Chng et al., A transcriptional repressor co-regulatory network governing androgen response in prostate cancers. EMBO J. **31**(12), 2810–2823 (2012)
149. J. Wang et al., Pleiotropic biological activities of alternatively spliced TMPRSS2/ERG fusion gene transcripts. Cancer Res. **68**(20), 8516–8524 (2008)
150. C. Cai et al., ETV1 is a novel androgen receptor-regulated gene that mediates prostate cancer cell invasion. Mol. Endocrinol. **21**(8), 1835–1846 (2007)
151. H. Kim et al., Estradiol-ERbeta2 signaling axis confers growth and migration of CRPC cells through TMPRSS2-ETV5 gene fusion. Oncotarget **8**(38), 62820–62833 (2017)
152. J.S. Roe et al., BET bromodomain inhibition suppresses the function of hematopoietic transcription factors in acute myeloid leukemia. Mol. Cell **58**(6), 1028–1039 (2015)
153. S. Oh, S. Shin, R. Janknecht, ETV1, 4 and 5: an oncogenic subfamily of ETS transcription factors. Biochim. Biophys. Acta **1826**(1), 1–12 (2012)
154. A.M. Blee et al., BET bromodomain-mediated interaction between ERG and BRD4 promotes prostate cancer cell invasion. Oncotarget **7**(25), 38319–38332 (2016)
155. Y. Yamamoto-Shiraishi et al., Etv1 and Ewsr1 cooperatively regulate limb mesenchymal Fgf10 expression in response to apical ectodermal ridge-derived fibroblast growth factor signal. Dev. Biol. **394**(1), 181–190 (2014)
156. A. Gorthi et al., EWS-FLI1 increases transcription to cause R-loops and block BRCA1 repair in Ewing sarcoma. Nature **555**(7696), 387–391 (2018)
157. A. Verger et al., The Mediator complex subunit MED25 is targeted by the N-terminal transactivation domain of the PEA3 group members. Nucleic Acids Res. **41**(9), 4847–4859 (2013)
158. S.L. Currie et al., ETV4 and AP1 transcription factors form multivalent interactions with three sites on the MED25 activator-interacting domain. J. Mol. Biol. **429**(20), 2975–2995 (2017)
159. D.S. Rickman et al., Oncogene-mediated alterations in chromatin conformation. Proc. Natl. Acad. Sci. U. S. A. **109**(23), 9083–9088 (2012)
160. T.D. Kim, S. Shin, R. Janknecht, ETS transcription factor ERG cooperates with histone demethylase KDM4A. Oncol. Rep. **35**(6), 3679–3688 (2016)
161. Z. Mounir et al., ERG signaling in prostate cancer is driven through PRMT5-dependent methylation of the Androgen Receptor. Elife **5**, e13964 (2016)
162. N. Melling et al., Overexpression of enhancer of zeste homolog 2 (EZH2) characterizes an aggressive subset of prostate cancers and predicts patient prognosis independently from pre- and postoperatively assessed clinicopathological parameters. Carcinogenesis **36**(11), 1333–1340 (2015)
163. P. Kunderfranco et al., ETS transcription factors control transcription of EZH2 and epigenetic silencing of the tumor suppressor gene Nkx3.1 in prostate cancer. PLoS One **5**(5), e10547 (2010)
164. V. Kedage et al., Phosphorylation of the oncogenic transcription factor ERG in prostate cells dissociates polycomb repressive complex 2, allowing target gene activation. J. Biol. Chem. **292**(42), 17225–17235 (2017)
165. N. Selvaraj, V. Kedage, P.C. Hollenhorst, Comparison of MAPK specificity across the ETS transcription factor family identifies a high-affinity ERK interaction required for ERG function in prostate cells. Cell Commun. Signal **13**(1), 12 (2015)
166. R. Janknecht, Analysis of the ERK-stimulated ETS transcription factor ER81. Mol. Cell. Biol. **16**(4), 1550–1556 (1996)
167. R. Janknecht et al., The ETS-related transcription factor ERM is a nuclear target of signaling cascades involving MAPK and PKA. Oncogene **13**(8), 1745–1754 (1996)
168. R.C. O'Hagan et al., The activity of the Ets transcription factor PEA3 is regulated by two distinct MAPK cascades. Oncogene **13**(6), 1323–1333 (1996)
169. Y. Huang et al., MAPK/ERK2 phosphorylates ERG at serine 283 in leukemic cells and promotes stem cell signatures and cell proliferation. Leukemia **30**(7), 1552–1561 (2016)
170. R. Keld et al., The ERK MAP kinase-PEA3/ETV4-MMP-1 axis is operative in oesophageal adenocarcinoma. Mol. Cancer **9**, 313 (2010)
171. J. Wu, R. Janknecht, Regulation of the ETS transcription factor ER81 by the 90-kDa ribosomal S6 kinase 1 and protein kinase A. J. Biol. Chem. **277**(45), 42669–42679 (2002)
172. B. Guo et al., Dynamic modification of the ETS transcription factor PEA3 by sumoylation and p300-mediated acetylation. Nucleic Acids Res. **39**(15), 6403–6413 (2011)
173. C. Degerny et al., SUMO modification of the Ets-related transcription factor ERM inhibits its transcriptional activity. J. Biol. Chem. **280**(26), 24330–24338 (2005)
174. W. Gan et al., SPOP promotes ubiquitination and degradation of the ERG oncoprotein to suppress prostate cancer progression. Mol. Cell **59**(6), 917–930 (2015)
175. J. An et al., Truncated ERG oncoproteins from TMPRSS2-ERG fusions are resistant to SPOP-mediated proteasome degradation. Mol. Cell **59**(6), 904–916 (2015)
176. A.C. Vitari et al., COP1 is a tumour suppressor that causes degradation of ETS transcription factors. Nature **474**(7351), 403–406 (2011)
177. J.L. Baert et al., The E3 ubiquitin ligase complex component COP1 regulates PEA3 group member stability and transcriptional activity. Oncogene **29**(12), 1810–1820 (2010)

178. P. Adamo et al., The oncogenic transcription factor ERG represses the transcription of the tumour suppressor gene PTEN in prostate cancer cells. Oncol. Lett. **14**(5), 5605–5610 (2017)

179. L. Wu et al., ERG is a critical regulator of Wnt/LEF1 signaling in prostate cancer. Cancer Res. **73**(19), 6068–6079 (2013)

180. X. Wang et al., Development of peptidomimetic inhibitors of the ERG gene fusion product in prostate cancer. Cancer Cell **31**(6), 844–847 (2017)

181. A.A. Mohamed et al., Identification of a small molecule that selectively inhibits ERG-positive cancer cell growth. Cancer Res. **78**(13), 3659–3671 (2018)

182. R. Nhili et al., Targeting the DNA-binding activity of the human ERG transcription factor using new heterocyclic dithiophene diamidines. Nucleic Acids Res. **41**(1), 125–138 (2013)

183. M.S. Butler et al., Discovery and characterization of small molecules targeting the DNA-binding ETS domain of ERG in prostate cancer. Oncotarget **8**(26), 42438–42454 (2017)

184. H.V. Erkizan et al., A small molecule blocking oncogenic protein EWS-FLI1 interaction with RNA helicase A inhibits growth of Ewing's sarcoma. Nat. Med. **15**(7), 750–756 (2009)

185. S. Rahim et al., YK-4-279 inhibits ERG and ETV1 mediated prostate cancer cell invasion. PLoS One **6**(4), e19343 (2011)

186. B. Winters et al., Inhibition of ERG activity in patient-derived prostate cancer xenografts by YK-4-279. Anticancer Res. **37**(7), 3385–3396 (2017)

187. S. Rahim et al., A small molecule inhibitor of ETV1, YK-4-279, prevents prostate cancer growth and metastasis in a mouse xenograft model. PLoS One **9**(12), e114260 (2014)

188. M.S. Pop et al., A small molecule that binds and inhibits the ETV1 transcription factor oncoprotein. Mol. Cancer Ther. **13**(6), 1492–1502 (2014)

189. J.C. Brenner et al., Mechanistic rationale for inhibition of poly(ADP-ribose) polymerase in ETS gene fusion-positive prostate cancer. Cancer Cell **19**(5), 664–678 (2011)

190. P. Chatterjee et al., PARP inhibition sensitizes to low dose-rate radiation TMPRSS2-ERG fusion gene-expressing and PTEN-deficient prostate cancer cells. PLoS One **8**(4), e60408 (2013)

191. S. Han et al., Targeted radiosensitization of ETS fusion-positive prostate cancer through PARP1 inhibition. Neoplasia **15**(10), 1207–1217 (2013)

192. H.T. Kissick et al., Development of a peptide-based vaccine targeting TMPRSS2:ERG fusion-positive prostate cancer. Cancer Immunol. Immunother. **62**(12), 1831–1840 (2013)

193. C. Magi-Galluzzi et al., TMPRSS2-ERG gene fusion prevalence and class are significantly different in prostate cancer of Caucasian, African-American and Japanese patients. Prostate **71**(5), 489–497 (2011)

194. G. Galletti et al., ERG induces taxane resistance in castration-resistant prostate cancer. Nat. Commun. **5**, 5548 (2014)

195. O. Reig et al., TMPRSS2-ERG in blood and docetaxel resistance in metastatic castration-resistant prostate cancer. Eur. Urol. **70**(5), 709–713 (2016)

196. L.H. Mochmann et al., ERG induces a mesenchymal-like state associated with chemoresistance in leukemia cells. Oncotarget **5**(2), 351–362 (2014)

197. C.S. Grasso et al., The mutational landscape of lethal castration-resistant prostate cancer. Nature **487**(7406), 239–243 (2012)

198. M.F. Berger et al., The genomic complexity of primary human prostate cancer. Nature **470**(7333), 214–220 (2011)

199. S. Wang et al., The ubiquitin ligase TRIM25 targets ERG for degradation in prostate cancer. Oncotarget **7**(40), 64921–64931 (2016)

200. M. Goldman, B. Craft, A. Kamath, A. Brooks, J. Zhu, D. Haussler, The UCSC Xena Platform for cancer genomics data visualization and interpretation. bioRxiv (2018). https://doi.org/10.1101/326470

Neural Transcription Factors in Disease Progression

Daksh Thaper, Sepideh Vahid,
and Amina Zoubeidi

Prostate Cancer Progression

Prostate Cancer Cell Plasticity and Disease Progression

All somatic cells in the human body (apart from red blood cells and B-cells) contain two sets of 23 chromosome pairs and therefore have equivalent genetic potential. Under perfect conditions, it should theoretically be possible for a terminally differentiated cell to de-differentiate, and then re-differentiate into a completely different cell type. The term cell plasticity succinctly defines the phenomenon of cells ascending the differentiation ladder and adopting an alternate cell fate. In the context of cancer biology, it refers to the ability of cancer cells to "bend like plastic" into any shape the environment requires in order to survive [1]. A considerable body of research has focused on the concept of cellular plasticity. For cancer research, this topic is of particular interest since the factors that direct plasticity are often anti-cancer therapies. Supporting evidence for treatment-induced phenotypic plasticity can

be found in studies of cell dormancy and cancer stem cells (CSC).

Tumor dormancy is a well-documented phenomenon that leads to the process of tumor relapse after years of remission. In order to survive through treatments, cancer cells often adopt a low proliferative quiescent phenotype. This small cadre of tumor cells likely exists in an equilibrium of cell renewal and death as there is little or no overall growth of the tumor [2]. This growth equilibrium is also held in check by two major factors, immune evasion and localized angiogenesis [2]. The hypotheses of tumor dormancy and cancer stem cells (CSCs) have considerable similarities since the low-proliferating, immune-evasive dormant cells exhibit stem cell properties [3]. The CSC theory has two possible interpretations. One hypothesizes that the stem-like cancer cells always exist as a small sub-population similar to normal stem cell populations, while the other hypothesizes that cancer cells gain self-renewal properties to become CSCs. Nonetheless, CSCs are inherently drug-resistant and survive therapies in order to re-populate a tumor [4, 5]. A recent study in support of the second hypothesis showed that some cancer cells gain enhanced self-renewal and migratory properties in response to external stressors (anti-cancer treatments) allowing for their continued survival and disease progression [5]. Parallel to the CSC phenotype, in response to therapy, cancer cells utilize another differentiation-based escape mechanism known

D. Thaper · A. Zoubeidi (✉)
Department of Urologic Sciences, Faculty of Medicine, University of British Columbia, Vancouver, BC, Canada

Vancouver Prostate Centre, Vancouver, BC, Canada
e-mail: azoubeidi@prostatecentre.com

S. Vahid
Vancouver Prostate Centre, Vancouver, BC, Canada

© Springer Nature Switzerland AG 2019
S. M. Dehm, D. J. Tindall (eds.), *Prostate Cancer*, Advances in Experimental Medicine and Biology 1210, https://doi.org/10.1007/978-3-030-32656-2_19

as epithelial to mesenchymal plasticity (EMP), loosely defined as a loss of epithelial markers and cell polarity combined with a gain of cell migration and invasion. Moreover, mechanisms of disease progression are not limited to phenotypic changes in cancer cells; i.e., auto/endo- or paracrine signaling can enhance angiogenesis [6] and neurogenesis [7, 8] in the tumor micro-environment, thus promoting tumor growth.

The arsenal of PCa therapies has been relatively effective as evidenced from the decline of deaths over the past 25 years. Apart from surgery, patients have a number of treatments available, including radiotherapies, chemotherapies and hormone therapies. The importance of cell plasticity in PCa progression following hormonal therapy should be emphasized as there is a relatively long delay (10–13 years) between first generation hormone therapy and emergence of castration resistant prostate cancer (CRPC) [9, 10]. More importantly, about 84% of CRPC patients are diagnosed with metastatic disease and 33% of those who don't have metastases at the time of diagnosis, are expected to develop metastases within 2 years [11]. These statistics highlight the significance of cell plasticity processes in the progression of PCa as EMP is the primary suspect for the mechanisms behind dissemination of cancer cells and metastasis. Therefore studying and understanding these various plasticity processes is important to improve patient survival.

Although an intermediate plasticity exists prior to CRPC [12], studies capturing this state are rare as approximately 90% of CRPC-stage patients present with enhanced AR activity. Therefore, AR remains the cornerstone of targeted therapy in CRPC.

Treatment-Induced Cell Plasticity: Emergent Phenotypes

In 2011 and 2012, two breakthrough trials, CUO-AA-301 [13] and AFFIRM [14] lead to the approval of the anti-androgens abiraterone (ABI) and enzalutamide (ENZ) respectively. Resistance to these second generation anti-androgens created a shift in the disease landscape as the ratio of AR-driven disease went from >90% at CRPC to ~60% at resistance to ABI and ENZ [13]. Resistant patients present with a mixture of adenocarcinoma, squamous and small cell carcinoma pathologies that might be either positive or negative for AR, AR activity (PSA) or neuroendocrine (NE) markers. The disease phase captured during biopsy/autopsy can be highly variable, exemplifying this plasticity process [13].

Examples of these emergent phenotypes are the lethal small cell/neuroendocrine prostate cancer and a newly characterized Intermediate Atypical Carcinoma (IAC) observed in a striking 17% and 28% of patients respectively [15, 16]. The IAC subtype carries a worse prognosis in comparison to resistant patients with adenocarcinoma [15].

Relative to the new IAC subtype, NEPC is better characterized and understood. Aside from the unique small cell morphology and positive staining for NE markers, namely chromogranin A (CHGA), neuron specific enolase (NSE), and synaptophysin (SYP), it is often distinguished from prostatic adenocarcinoma by reduced AR activity [17]. Poorly differentiated NE tumors can arise from a variety of different mechanisms. In epithelial cancers, genomic and epigenomic alterations can drive this phenotype as the cancer progresses. Thus, the acquisition of this phenotype as a response to targeted therapy in epithelial cancers has generated a renewed interest [18]. Specifically, the mechanisms behind the increased incidence of NEPC have become a major topic of research in PCa. In an effort to understand the biology of NEPC, the cell of origin for this form of PCa is a focus of intense debate as there are two competing hypotheses. The "lurker cell" hypothesis [19] states that NEPC cells arise from the sparse NE cell population already present in the prostate. Until recently, this was a reasonable hypothesis as the prevalence of *de novo* NEPC is approximately 1% [20]. Furthermore, the presence of tumor cells with NEPC characteristics within well-differentiated adenocarcinomas is well documented and can be interpreted to support this hypothesis [21, 22]. The other hypothesis states that these tumors arise from the plasticity of PCa cells as an adaptive response to the more potent therapies ENZ and ABI. Combined with the detection of typical adenocarcinoma mutations like TMPRSS-ERG fusions and a general lack of gross genomic differences in terminal NEPC tumors, a model of divergent evolution from a common precursor PCa cell is conceivable [23].

Another example of treatment-induced cell plasticity was demonstrated in ~25% of treated patients that developed Double Negative Prostate Cancer (DNPC), a CRPC subtype that was negative for AR/PSA and NE markers, but displayed increased FGFR signaling [24]. It remains to be seen if there are any commonalities between the IAC subtype and the DNPC subtype and how they relate to the NEPC stage. Therefore, it is important to study the genetic and signaling mechanisms within cancer cells that change in response to anti-androgen therapies in order to develop better therapeutics.

The dramatic changes in shape a cancer cell would not be possible without an extreme shift in transcriptional activity. Dysregulation of transcription factors under the pressure of internal and external stimuli plays an important role in cellular plasticity [25]. Also, progression to a malignant state is fundamentally dependent on the quality and quantity of transcription in cells, as cancer cells require an optimum abundance of different proteins to proliferate, metastasize or resist treatment [25, 26].

Figure 1 summarizes prostate cancer cellular plasticity in the context of androgen deprivation therapies. We suggest that cells undergo de-differentiation under the pressure from AR pathway inhibitors and can manifest metastatic, stem cell-like or NE characteristics that can overlap. Differential transcriptional activity provides momentum in these plastic states towards a final stage of the disease presented in patients.

Transcription Factors

Transcription factors (TF) are proteins that have the ability to bind to specific DNA sequences either as monomers, homo-dimers or hetero-dimers and regulate gene expression as a direct result of this process. While their overall number is still undefined, the most recent catalog with this definition identified 1639 human TFs [27]. Secondary protein structures like α-helix, β-sheet, loops and the ability of proline to create kinks/turns promote the assembly of small domains such as helix-turn-helix, helix-loop-sheet and helix-loop-helix, which have evolved into domains with efficient DNA-binding capabilities. A majority of these DNA binding domains bind to the major groove of the DNA, although there are examples like the HMG-box-containing proteins that bind the minor groove [28]. In eukaryotes, some of the major TF families are zinc finger (ZF), Homeodomain (HD), basic helix-loop-helix (bHLH), basic leucine zipper (bZIP), Forkhead (FOX), nuclear receptors, E-twenty-Six (ETS) and HMG/SOX transcription factors [27]. Moreover, a number of human TFs have more than one DNA binding domain which can be similar or different (hetero-domain TFs) in structure. With a total of 33 defined families and more than 100 TFs with undefined DNA binding domains, understanding the role of TFs in gene regulation remains an exciting opportunity and a daunting challenge.

In the context of PCa, AR is still considered the key transcription factor. However, as disease progresses to a treatment-resistant state, the classic AR activity fades slowly and transforms to a new role to coordinate with a complex network of transcription factors and epigenetic elements to support a highly malignant undifferentiated disease state.

Deciphering the progression of prostate cancer to the neuroendocrine state has generated immense interest as the prevalence of NEPC has increased in recent years. The natural suspects for this process are transcription factors involved in neural differentiation. During development, complex transcriptional programs work towards establishing diverse neural cell types. High expression levels of strong neural cell-fate-determining transcription factors can trigger a cascade of events leading to neuronal lineage commitment. Positive or negative feedback loops and activity of transcriptional regulators can lead to epigenetic changes that reinforce the expression

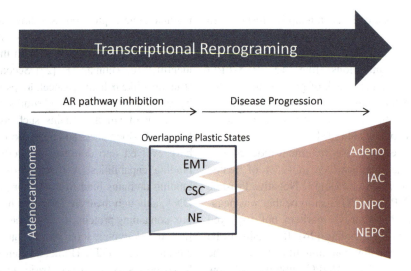

Fig. 1 Cell plasticity and different stages of prostate cancer. Transdifferentiation of adenocarcinoma of prostate under the pressure of AR pathway inhibitors is a dynamic process where epithelial cells gain a plastic phenotype first and then progress to different stages of disease based on the activity of transcription factors. AR, androgen receptor; EMT, epithelial-mesenchymal transition; CSC, cancer stem cell; NE, neuroendocrine; CRPC, castration-resistant prostate cancer; IAC, intermediate atypical carcinoma; DNPC, double negative prostate cancer; NEPC, neuroendocrine prostate cancer

of downstream target genes and facilitate the neuronal phenotype. During development, multipotent neural progenitor cells give rise to neurons and supporting cells in the central nervous system. Exogenous expression of some of these transcription factors can also endow non-neural cells with neuronal properties [29]. Indeed, a combination of only three factors (BRN2, ASCL1 AND MYT1L—BAM cocktail) is sufficient to convert mouse embryonic fibroblasts and human pluripotent stem cells into functional neurons *in vitro*. Addition of a fourth TF, NEUROD1 to the BAM cocktail increased the efficiency of neuronal reprogramming of human fetal fibroblasts from ~20% to ~65% [30]. Conversely, human embryonic fibroblasts can be converted into neurons with just the addition of BRN2 and ASCL1 [31]. A recent study screened 598 pairs of transcription factors and identified 76 pairs of transcription factors that induced neuronal differentiation in mouse embryonic fibroblasts [32].

Terminally differentiated cells can also be converted into neurons. Adult human fibroblasts can be transformed into functional neurons (>20% efficiency) with overexpression of BRN2 and microRNA (miR) 124 and can be enhanced to >50% with the addition of MYT1L [33]. Analogously in PCa, terminally-differentiated prostate cancer cells traverse up the differentiation ladder gaining expression of stem cell and neuronal genes as they gain resistance to treatments and progress to NEPC. For example, LNCaP cells stably overexpressing MYCN exhibit a neuronal phenotype compared to control [17]. Moreover, BRN2 expression appears to be driving NEPC in a unique treatment-resistant model of trans-differentiation [12].

Table 1 is a list of major TF families and their members with roles in neurogenesis or neuronal development that contribute to prostate cancer progression. These key transcription factors are capable of driving different plastic stages in prostate cancer progression including, but not limited to, NE differentiation and EMP.

Zinc Finger (ZF) Transcription Factors

Proteins with zinc finger domains comprise a group of 1723 proteins that is highly diverse and contains approximately 750 transcription factors [27, 34]. The most common ZF domain consists

Table 1 Neural transcription factor in prostate cancer disease progression

Transcription factor family	Members discussed in this chapter
Zinc Finger (ZF) transcription factors	Steroid hormone receptors, SNAI1, SNAI2, GATA family, GLI family, ZBTB46, INSM1
Helix-loop-Helix (bHLH) transcription factors	ASCL1, NGNs, NEUROD1, HIF1α, HES6, TWIST, IDs
Homeodomain transcription factors	LHX family, NKX2.1, NKX2.2, NKX3.1, NANOG, HOX family
Forkhead domain transcription factors	FOXA(1/2), FOXB1, FOXC(1/2), FOXG1 and FOXP(1/2), FOXM1
Leucine Zipper (bZIP)	CREB, cJUN, cFOS
Hetero-domain transcription factors	POU2F1, POU3F2, POU4F1, POU5F1, ZEB1/2, MYC, MYCL, MYCN

of an α-helix, a linker sequence and a β-sheet held together by a zinc ion through interactions with two cysteine and two histidine residues (Cys2 His2 or C2H2) or four cysteine (Cys4) or six cysteine (Cys6) residues [35]. The residues in the α-helix and the linker sequence, tertiary structures and tandem repeats of ZF domains determine specificity for DNA sequences in the promoters of target genes. ZF domains recognize not only DNA, but also RNA, other proteins and even lipids [34]. Of utmost importance to PCa is the steroid hormone receptor class of the ZF family, including AR, glucocorticoid receptor (GR), estrogen receptor (ER) and progesterone receptor (PR). The DNA-binding domain of the steroid receptors contains two zinc finger domains that bind to DNA upon interaction with the ligand (e.g. dihydrotestosterone with AR) [36]. Interestingly, steroid hormones and their receptors play crucial roles in neurogenesis [37] and their role in PCa cell plasticity is continuously being studied. Throughout this chapter, we show that a large number of neural transcription factors can cooperate with/against AR to advance this disease to a more malignant state.

Cell plasticity in response to positive and/or negative AR pathway stimulation is full of interesting and contradictory research. Loss of

AR signaling through androgen deprivation consistently results in downregulation of E-cadherin, one of the hallmarks of EMP [38]. Furthermore, AR pathway inhibition also causes a shift from CDH1 (E-cadherin) to CDH2 (N-cadherin) [39] and a simultaneous upregulation in both CSC and NE markers [40]. Conversely, positive AR signaling also promotes metastatic progression. PCa metastatic tumor sites, specifically in the bone, are generally AR positive [41]. In ERG-fusion positive PCa, ERG activity through AR activates SOX9 expression (discussed later), which can enhance metastasis [42]. The bivalent relationship between AR and cell plasticity also extends to its downstream regulation of another class of ZF transcription factors SNAIL (SNAI1) and SLUG (SNAI2), which have well-defined roles in driving EMP [43]. SNAIL expression is directly repressed by AR [44] and interestingly, SNAIL is upregulated during NEPC transdifferentiation and co-operates with another key NEPC protein PEG10 to promote the aggressiveness of NEPC [45]. Conversely, SLUG (SNAI2) expression is positively regulated by AR, which can drive EMP and metastatic progression of prostate cancer [46]. Altogether, AR activity is a double-edge sword, whereby activating or inhibiting AR transcriptional activity can promote EMP and metastatic disease. In addition to driving EMP, exogenous expression of SNAIL expression can also drive NE differentiation [47, 48]. Over-expression of SNAIL in LNCaP cells triggers neurite formation and upregulation of NEPC markers such as CHGA and NSE [48]. While not nearly as prominent, ER, PR and GR all have roles in PCa disease progression. ERβ plays a tumor suppressive role, whereas a switch to ERα can promote PCa tumor growth [49, 50]. Similarly, the role of PR in prostate cancer is controversial as both tumor promoting and tumor suppressing roles have been reported [51]. GR is an important mediator of resistance to AR inhibitors like ENZ in xenograft models [52, 53]. Clinical data have validated GR as a target in PCa, as GR is upregulated in patients treated with ENZ and/or ABI, and GR can transcriptionally regulate many of the same target genes that are regulated by active AR [54].

While the AR is paramount for PCa progression, many other ZF transcription factors play important roles in neural development as well as in PCa and often function in concert with the AR. For example, the GATA family member GATA2 not only regulates AR expression, but also co-operates with AR and FOXA1 to promote specific AR signaling and tumor proliferation [55]. One the other hand, GATA3 plays a tumor suppressive role where it regulates epithelial progenitor cell division via atypical protein kinase C to control lineage commitment during prostate development [56, 57]. The GLI-class (named after Glioma) TFs also play a crucial role in cell fate determination and PCa disease progression as all three family members (GLI1, GLI2, GLI3) interact with AR [58–60]. The GLI family of proteins mediates the hedgehog (Hh) signaling pathway [61]. Binding of the Hh ligand to receptor Patched (PTCH1) initiates a signaling cascade through Smoothened (SMO), ultimately activating the GLI transcription factors. The interaction between GLI family members and AR occurs at the Tau5 domain of the AR, which is important for ligand-independent AR transcriptional activity [59]. This interplay is particularly important for CRPC because inhibition of AR increases Hh dependent activation of GLI as a compensatory mechanism for ligand-independent AR activity. Treatment with an Hh inhibitor is effective in treating CRPC xenograft tumors [62, 63]. Thus, GLI proteins cooperate with AR in both CSCs [64] and prostate development [65], via acquisition of self-renewal properties and subsequent tumor growth. With a new focus on AR pathway inhibitor driven de-differentiation, many groups have examined the role of ZF family members' in progression of PCa to NEPC. For example, in response to AR inhibition, the Zinc finger and BTB domain containing protein (ZBTB46) drives NE-differentiation in multiple PCa cell line models [66]. Moreover, another ZF protein, Insulinoma-associated protein 1 (INSM1) is expressed in 93% of NEPC patient samples [67].

ZF family transcription factors can also play a role in suppressing PCa. For example, multiple members of the Krüppel-like family (KLFs) as well as REST can negatively affect PCa proliferation. These proteins are discussed later in the chapter in more detail.

Helix-Loop-Helix (bHLH) Transcription Factors

The basic helix-loop-helix domain is comprised of two α-helices connected together by a loop region. Within a typical bHLH domain, one helix is smaller than the other helix. Protein dimers use the longer helix for DNA binding activities while flexibility of the smaller helix creates space for dimerization [68]. bHLH transcription factors are expressed in cells from many tissues throughout the body, but often in a cell- or organ-specific manner. For example, Achaete-scute homolog 1 (ASCL1), Neurogenin (NGN) and NeuroD/G sub-classes of the bHLH family are pro-neuronal transcription factors that induce the differentiation of fibroblasts to neurons [68], and may play a role in neuroendocrine differentiation. ASCL1 or its mouse counterpart mASH1, is expressed in the SV40 T-antigen-driven TRAMP mouse model [69], and its role in human NEPC is currently under investigation [70]. The bHLH family member HES6 and cell cycle proteins BUB and CDKN2D are direct targets of ASCL1 in cell lines derived from TRAMP tumors [71]. Similarly, NeuroD1 drives cell proliferation and migration in several PCa cell lines [72]. In addition to proliferation, NeuroD1 appears to be linked to NE-differentiation as its expression is elevated in NEPC patients and induced sharply upon cyclic-adenosine monophosphate (cAMP)-treated AR-independent PC3 and DU145 cells [73]. NGN3 is another neural TF from the Neurogenin sub-class that is highly expressed in invasive NE tumors in the 12T-10 prostate model [69]. The expression pattern of NGN3 is particularly interesting since it was observed only in lung metastases of these NE tumors but not the liver metastases, suggesting that it regulates organ site-specific metastasis. NGN3 is also upregulated in response to hypoxia in PCa, together with suppression of another bHLH protein HES1 [74]. Hypoxia inducible factor 1 (HIF1) appears to be a driver of NE-differentiation under hypoxic conditions. Hypoxia is a common pathology in solid tumors and the resultant micro-environmental stress is a potent inducer of cell plasticity [75]. Upregulation of HIF1α is an early event in PCa tumorigenesis [76] and therefore, its expression and activity in EMP, CSC and/or NE-differentiation is an important area of research in PCa. The HIF1 complex is comprised of the bHLH transcription factor HIF1α and the β-subunit of aryl hydrocarbon receptor nuclear translocator (ARNT). In addition to NEPC, HIF1α has a near pan-cancer role in promoting aggressive tumor types through EMP or CSCs as it can promote key players like vascular endothelial growth factor (VEGF) and SOX9 (A HMG domain TF discussed later in detail) [77]. Stabilization of HIF1α by hypoxia prolongs its cooperation with forkhead box protein A2 (FOXA2), thus driving progression of TRAMP tumors to NEPC [78]. Interestingly, activated HIF1α promotes transcription of the bHLH family member TWIST1 [79], enhancing the metastatic nature of NEPC. TWIST is an E-box suppressor that reduces E-cadherin expression and promotes metastasis in multiple cancer types including prostate [80]. TWIST1 also regulates cell fate decisions in neural crest cells [81].

Another sub-class of bHLH repressor proteins is the inhibitor of DNA binding/differentiation (ID) proteins [82]. In contrast to other bHLH TFs, the ID proteins lack a DNA binding domain but instead, associate with other members of the family and prevent them from binding DNA or forming active heterodimers. The ID family consists of four proteins (ID1 to ID4); among them, ID1 has been thoroughly studied for its role in hormone therapy resistance and promoting androgen-independent disease. ID1 is upregulated in prostate cancer [83], and exogenous expression of ID1 makes the AR-dependent cell line LNCaP insensitive to hormone depravation by suppressing AR signaling [84], increasing epithelial growth factor (EGF) [85] and mitogen activated protein kinase (MAPK) signaling [86]. ID1 and ID3 are associated with disease progression [87], and appear to promote EMP through transforming growth factor (TGFβ) signaling

[88, 89]. Moreover, ID1 downregulation is associated with cellular differentiation, whereas ID1 overexpression inhibits differentiation [90]. ID1 can transcriptionally inhibit the expression of cyclin-dependent kinase inhibitors p21 (CDKN1A), and p27 (CDKN1B), thereby enabling cells to progress through the cell cycle faster, without being able to differentiate [91]. ID1 is also linked to chemoresistance, as its inactivation sensitizing PC3, DU145, LNCaP, and C4-2 prostate cancer cells to taxane chemotherapy [92, 93]. ID1 is also associated with DNPC, a prime example of inhibition of NE-differentiation during PCa disease progression [24]. Characterized by a loss of AR signaling as well as a lack of NEPC marker expression, DNPC tumors contain higher ID1 levels and ID1 is upregulated in DNPC metastases. All four members of the ID family (ID1–4) are significantly upregulated in DNPC patient-derived xenografts and appear to promote androgen-independent disease progression downstream of fibroblast growth factor (FGF) signaling [24].

While the pattern of ID2 expression in PCa is similar to ID1 and ID3, ID4 appears to have a more growth suppressing function [94] as it promotes senescence [95] and functions as an inhibitor of ID1, ID2 and ID3 activity [96].

HES and HEY family of proteins also belong to the bHLH family of TFs; however, they have an inhibitory role in neuronal differentiation. Their role in progression of PCa is discussed later in this chapter. Altogether, the bHLH transcriptional activators and repressors interact with each other as well as other proteins to manage the equilibrium between undifferentiated and terminally differentiated populations of cells.

Homeodomain Transcription Factors

The homeodomain (HD) is approximately 60–61 amino acids long and contains three α-helices with the second and third helix forming a helix-turn-helix motif with DNA binding capabilities [97]. While HD-containing TFs make-up the second largest family of TFs, other structural properties sub-divide them into many different classes such as the HOX-class, CUT-class and HNF-class and many more [97]. Mammalian HD-containing TFs have fundamental roles in embryonic development and organogenesis [95], and some are linked to prostate cancer. For example, The LIM sub-class of HD transcription factors was recently linked to NEPC, as the LIM sub-class motifs (LIM Homeobox—LHX) are highly enriched in regions of open chromatin identified using the ATAC-seq technique with prostate cells transformed to NEPC [98]. Functioning as transcriptional activators and/or repressors, members of the NK-like homeobox (NKX) transcription factors comprise a large sub-class of homeodomain TFs with a total of 48 genes and 19 pseudogenes [99]. Several other members of the NKX sub-class participate in patterning of the spinal cord [100]. Specifically, both NKX2.1, also known as thyroid transcription factor 1 (TTF1), and NKX2.2 are expressed in NEPC [69, 101]. Moreover, NKX2.2 regulates a network of microRNAs that increases PCa metastasis, and whose expression in patient tumors correlates with poor overall survival [102]. As an AR driven tumor suppressor, NKX3.1 is an important mediator of growth suppression in normal prostate epithelium in response to active AR signaling, and its loss is an important variable in PCa tumorigenesis [103, 104]. Interestingly, PCa patients have a loss of NKX3.1 heterozygosity in ~75% of tumors [105]. Another family member, NANOG, promotes pluripotency and is often upregulated in undifferentiated cancers with poor prognosis [106]. NANOG reprograms PCa cells to resist castration-based therapy, thus promoting CSC properties in prostate cancer [107, 108].

Within the HD family, the role of HOX-class TFs has been investigated in PCa due to the interplay of the family member HOXB13 with AR [109]. In vertebrates, 39 HOX proteins control the anterior-posterior vertebrate body plan during development. Interestingly, they exist in clusters on chromosome 7 (HOXA), chromosome 17 (HOXB), chromosome 12 (HOXC) and chromosome 2 (HOXD) [110]. A large majority of these HOX genes are expressed within the central nervous system and play a crucial role in its development [111]. Outside of the urogenital sinus,

HOXB13 is expressed only in the spinal cord and the vertebrate tail bud [112]. It is highly upregulated in CRPC [113], and HOXB13 mutations increase the risk of developing PCa [114]. HOXB13's governance of the AR cistrome is complex as it activates and/or represses different sets of genes in the presence or absence of androgens [109] and therefore could play an interesting yet undefined role in anti-androgen induced cell plasticity. In addition to pro-proliferative roles through direct suppression of p21 [115] and indirect suppression of Survivin (BIRC5) [116], HOXB13 also promotes NFκB activity leading to increased cell invasion and metastasis [117]. Several other HOX-class TFs promote EMP, endowing PCa cells with enhanced migration and invasion properties. Coincidently four different HOX proteins promote expression of the extracellular matrix (ECM) modifiers, matrix metalloproteases (MMP). Degradation of the ECM by MMPs is crucial for cancer cells to escape through the basement membrane and metastasize to different sites. Moreover, this alteration of the microenvironment itself disrupts cell polarity and promotes EMP within the cancer cells [118]. HOXA1 regulates expression of SNAI1 and MMP-3 [119], while HOXA3 regulates MMP-14 [120]. In breast cancer cells, HOXB7 regulates MMP-9 expression [121], and HOXC11 governs both MMP-2 and MMP-8 in vascular smooth muscle cells [122]. Though sparse, some research has linked development of NEPC to the HOX-class of TFs. The entire HOXD cluster, including surrounding genes like CREB and NEUROD1, is activated in response to cAMP [123]. Additionally, an evolutionary cousin to the HOX-class, the para-HOX gene PDX1, was recently implicated in treatment-induced NEPC (t-NEPC) [16], further demonstrating the importance of the HOX and para-HOX sub-class as regulators of PCa disease progression.

HMG Domain Transcription Factors

The high mobility domain (HMG) was first identified in the protein sex determining region of the Y chromosome (SRY). Therefore, the SOX family derives its name from SRY related HMG box [124]. The HMG domain TFs contain the SOX and the LEF/TCF sub-class of proteins. The SOX sub-class is comprised of ~20 TFs [125] that play crucial roles in development of various tissues. They are sub-divided into groups A–J due to differences and similarities in protein domains and other structural properties [125]. Based on expression patterns, members of SOXB1 (SOX1, SOX2 and SOX3), SOXB2 (SOX14, SOX21), SOXC (SOX4, SOX11, SOX12), SOXD (SOX5, SOX6 and SOX13) and SOXE (SOX8, SOX9, SOX10) are expressed in the nervous system and may be involved with neural differentiation [125]. Structurally, the HMG domain contains both nuclear export [126] and import [127] signals and modulates nucleocytolasmic localization of SOX family proteins as well as binding to the consensus DNA sequence (A/T)(A/T)CAA(A/T) [128]. In contrast to most TFs, SOX family proteins bind to the minor groove of DNA across both sense and anti-sense strands (reverse complement sequence: TTGTT) while bending the helix about $60°$–$70°$ [129]. This property has interesting implications for prostate cancer as it resembles the 5′ AR half-site AGAACA or the 3′ AR half-site of TGTTC [130], creating opportunities for agonistic and/or antagonistic behavior in PCa cells. Indeed, AR co-operates positively with SOX9 [131] in a complex with SOX4 [132], and can be negatively regulated by SRY itself [133].

Due to their roles in development, members of the SOX family are linked to mechanisms of cell plasticity in prostate cancer, like CSCs, NE differentiation and EMP. In particular, SOX4 and SOX5 are important regulators if EMP. For instance, SOX5 promotes cancer cell invasion and metastatic potential through TGFβ signaling and upregulation of the EMP transcription factor TWIST1 [134]. Interestingly, SOX4 is an AR suppressed gene [135] that promotes plasticity by upregulating EMP pathways through collaboration with TGFβ signaling [136] and through interactions with ERG [137]. SOX4 was also identified in a TF network associated with prostate cancer subtype 1 (PCS1) and progression to

metastatic disease [138]. While data is limited to single publications, both SOX6 and SOX7 appear to have tumor suppressive functions in PCa [139, 140]. In comparison, there have been considerably more studies investigating the role of SOX9 in prostate cancer. SOX9, which is downstream of ERG and AR signaling, promotes PCa tumorigenesis and metastasis in TMPRSS-ERG fusion positive PCa [42]. Moreover, SOX9 regulates multiple components of the WNT pathway [141], including but not limited to receptors like LRP5/6 and FZD5/7 as well as other HMG domain TFs like LEF1 and TCF4/TCF7L2 [142]. Also, SOX9 appears to play a role in prostate cancer initiation and progression [143, 144].

Perhaps the most well-studied SOX family protein in PCa (or any other cancer) is the SOXB1 member, SOX2. SOX2 is one of four Yamanaka TFs (OCT3/4, SOX2, MYC, KLF4) that induces pluripotency [145]. SOX2 is an AR-suppressed gene that plays a crucial role in development of both CRPC [146, 147] as well as the more undifferentiated phenotypes of prostate cancer discussed earlier [148]. SOX2 promotes all three types of prostate cancer cell plasticity (CSC, NE, EMP), and regulates EMP through WNT/β-catenin signaling in both breast and prostate cancer cells [149]. Inhibiting SOX2 reduces EMP marker expression and reverses the loss of the cell adhesion molecule E-cadherin in AR independent PCa cell lines [150]. A majority of the SOX2 research in PCa focuses on its role in promoting CSC and NE phenotypes. SOX2 is upregulated in NEPC patients as a consequence of ENZ-resistance, and SOX2 is abnormally expressed in the context of TP53 and RB1 loss. PCa cells undergo a lineage switch through upregulation of SOX2 and differentiation to a NE phenotype. Moreover, suppressing SOX2 expression in cells with TP53 and RB1 loss greatly reduces their ability to switch lineages, which is a mechanism of treatment resistance [148]. In addition, SOX2 is upregulated in DNPC patient tumors in the context of AR deletion [24]. Thus, SOX2 is a context-dependent promoter, if not a driver, of the different aspects of cell plasticity observed in prostate cancer.

Forkhead Domain Transcription Factors

A combination of helix-turn-helix and β-sheets forms the forkhead box DNA-binding domain (FOX domain). The FOX family of proteins is highly variable (44 members) and the ~110 amino acids in the forkhead domain can be comprised of anywhere between two to four α-helices and β-sheets [151]. The FOX proteins are divided into sub-classes from A to S based on sequence homology and have a wide range of biological functions [152]. Within the family, FOXA(1/2), FOXB1, FOXC(1/2), FOXG1 and FOXP(1/2) promote neural development or participate in the emergence of cell plasticity in PCa cells [152, 153].

FOXA1 and FOXA2, also known as HNF3A and HNF3B are pioneer factors with the unique ability to interact with and open up closed regions of chromatin, thus facilitating other TF complexes to promote specific gene transcription. This ability to bind DNA within closed chromatin is attributable to the structure of the FOX domains in FOXA1 and FOXA2, which bear striking structural similarities with linker histone H1 [154]. FOXA1 and FOXA2, respectively, can interact with AR and are integral to the NE differentiation process in PCa cells. For example, in LNCaP cells, more than half of the AR-binding events overlap with FOXA1 binding sites [154]. Interestingly, the loss of FOXA1 promotes NE differentiation [155] and FOXA2 is almost universally expressed in NEPC tumors [156]. Considering the role of FOXA1 as a pioneering factor essential for AR activity [157–159], a switch from FOXA1 to FOXA2 could be an important event in NE differentiation. FOXA1 also functions as a pioneering factor for ER, creating an open and easily accessible chromatin conformation for ER to bind in breast cancer [160]; however, its correlation with ER in prostate cancer has not been confirmed [161]. Thus, FOXA1/2, through their interactions with hormone receptors, play important functions in treatment resistance and cell plasticity, and may alter the course of PCa progression.

Additionally, FOXM1 is one of the most commonly investigated FOX family proteins in PCa. As a pro-mitotic TF, FOXM1 regulates transcription of genes important for G2/M transition of the cell cycle [162]. FOXM1 expression is associated with biochemical recurrence (BCR) and more rapid progression to metastasis [163]. Loss of microRNA (miR) 101 and 27a in metastatic PCa results in enhanced FOXM1 signaling and is associated with ENZ resistance [164]. Interestingly, FOXM1 is also upregulated in prostate cancer subtype 1 (PCS1) characterized by low AR signaling and lack of response to AR pathway inhibitors like ENZ [165]. FOXM1 activity is enhanced in an LNCaP-derived model of ENZ-resistance, and inhibiting FOXM1 reduces both proliferative and tumor initiating capacity [166]. FOXM1 is also upregulated in NEPC and remains an attractive target for future investigation.

Several other FOX family members appear to have roles in NEPC. For example, FOXC2 is a convergent signaling node downstream of ZEB1 and SNAI1 that regulates EMP, CSC and NE genes in PCa [167]. FOXC2 expression is upregulated in response to AR pathway inhibition and also appears to play a role in treatment-induced NEPC (t-NEPC) [44, 168]. Expression of FOXP2 is linked to BCR in ERG fusion negative patients [169] and promotes PCa cell migration and invasion [170]. FOXP1 nuclear expression, but not cytoplasmic expression, correlates with expression of AR, HIF1α and VEGF [171]. FOXP1 also appears to play a tumor suppressor role by inhibiting proliferation and migration of the AR positive prostate cancer cell line LNCaP [172]. Interestingly, FOXP1 binding regions overlap with AR binding sites in an androgen-dependent manner [173].

Leucine Zipper (bZIP)

Structurally, the bZIP DNA binding domain typically contains 30 amino acids, forming an α-helix that fits into the major groove of the DNA. The α-helix domain contains a leucine residue every seven amino acids, creating an opportunity for interaction with bZIP domains of other TFs as either homo- or hetero-dimers. The interaction "zips" up the arms of the helices, squeezing them together, allowing the other end of each helix to bind DNA, similar to tongs. In humans, 53 proteins contain the bZIP domain [174], and their importance in PCa cell plasticity can be traced back to one of the first papers characterizing NE-differentiation [176]. PCa cells undergo NE-differentiation in response to cAMP signaling [175], which results in downstream activation of the bZIP transcription factor cAMP response element-binding protein (CREB) [176], thereby upregulating several neuropeptides that maintain neural plasticity and long term memory [177]. CREB activates G-protein coupled receptor kinase (GRK3), which promotes NE-differentiation through its kinase activity [178]. AR pathway inhibition in itself can activate CREB signaling and promote NE-differentiation and angiogenesis through a key epigenetic player, Enhancer of Zeste homolog 2 (EZH2) [179]. The positioning of the Leucine residues along the helix also creates highly specific hetero-dimerization opportunities within members of the bZIP family. Activator protein 1 (AP1) is a homo- or hetero-dimer of the bZIP transcription factors cFOS and cJUN, which play key roles in the central nervous system during propagation of the action potential [180]. The AP1 complex governs the AR cistrome in prostatic fibroblasts [181], and although AP1 activity is enhanced by androgen [182], it also inhibits AR signaling in prostate adenocarcinoma cells [183]. It would be interesting to determine whether suppression of AR signaling via AP1 expression promotes NE-differentiation.

Members of the MYC family, which are some of the most potent oncogenes in cancer, contain a bZIP domain along with a bHLH domain. These hetero-typic transcription factors will be discussed in the next section.

Hetero-Domain Transcription Factors

Of the approximately 1640 defined TFs, only 47 (~3%) contain two different DNA binding domains [27]. These rare heterotypic TFs, which are among the most potent inducers of cell plasticity, include Yamanaka factors OCT3/4 and

MYC as well as pan-cancer inducers of EMP, ZEB1 and ZEB2 [184].

Over half of these heterotypic TFs contain homeodomains, and the Pituitary Octamer UNC86 (POU) domain sub-class is the largest among them. All 16 POU family transcription factors contain a POU_S (POU specific—four α-helices) and a POU_H (homeodomain—three α-helices) connected together by a linker sequence. As a family, these proteins are ubiquitously expressed, and different family members play important roles in guiding organ development [185]. While the POU domains are highly conserved between the family members, the variability and flexibility of the linker region combined with co-factor activity alters the DNA binding capabilities of the POU family members. OCT4 is a principal mediator of embryonic pluripotency; however, it is also important for the transition from embryonic to neural stem cells [186]. Several other POU family members are important for neurogenesis and neural crest formation, specifically POU3F(1–4), POU4F1 and POU6F1 [185].

OCT4 promotes PCa initiation [187], resistance to chemotherapy [188], and biochemical recurrence [189]. Computational modeling of gene expression data sets identified OCT4 along with OCT8 (POU3F3) and OCT1 (POU2F1) as major transcriptional nodes in metastatic CRPC. Previous research had already identified OCT1 is a co-factor that modulates AR cistrome and activity downstream of GATA2 and FOXA1 [190]. Interestingly, PCa cells preferentially express the short isoform of BRN3a (POU4F1), and exogenous expression of BRN3a promotes PCa cell proliferation and expression of sodium channel SCN9A [191], suggesting a role in neuronal differentiation. BRN2 (POU3F2), also known as OCT7, is a key mediator of NE-differentiation and may play a role in acquired resistance to ENZ as evidenced in an *in vivo* derived model of ENZ-resistance [12]. BRN2 expression is significantly higher in NEPC samples, and introduction of BRN2 in CRPC cell lines is sufficient to induce NE-differentiation by upregulating NEPC drivers like SOX2 and markers like CHGA and NCAM1. Importantly, BRN2 and AR suppress each other.

In addition to the POU family, the heterotypic Zinc finger E-box-binding homeobox (ZEB) sub-class of TFs contains a homeodomain and C2H2 ZF domains; specifically ZEB1 has seven ZF domains, whereas its paralog ZEB2 contains eight. While their expression patterns vary among tissues, both play important roles in neural development as they are crucial for migration of the neural crest cells and formation of derivative structures during embryonic development [192]. Both ZEB1 and ZEB2 activate the EMP and CSC signaling network in multiple cancers [193]. These functions extend to PCa where both proteins promote EMP and CSC properties [194–198] as well as chemotherapy resistance [199, 200]. Their relationship with AR however, is likely contextual since AR regulates ZEB1 and ZEB2 either positively or negatively, depending on the model [201–203]. Interestingly, RNA-seq data showed that AR activity goes down, while ZEB1 expression increases and ZEB2 expression decreases during NEPC trans-differentiation [45]. Importantly, ZEB1 promotes treatment-induced NE-differentiation [204], and targeting ZEB1 reduces expression of EMP, CSC and NEPC markers in PCa cells [168]. Furthermore, ZEB1 likely promotes progression of CRPC as its expression is detected in over 50% of visceral and bone metastasis [205]. Also, ZEB1 expression correlates with more rapid time to metastasis and biochemical recurrence [206]. Lastly, there are three other heterotypic sub-classes of homeodomain TFs, PAX (paired domain + homeodomain), CUT (cut domain + homeodomain) and PROX (prospero-homeodomain). Interestingly, almost all of these proteins play important roles in neurogenesis and neuronal development [207–209]. ONECUT2 from the CUT-class promotes both PCa progression and neuroendocrine transdifferentiation by suppressing AR expression [210] and by activating HIF1α signaling [211]. PROX1 appears to activate HIF1α and promote EMP [212]. There are nine members of the PAX sub-class, however, only four of them contain a complete homeodomain (PAX3, PAX4, PAX6 and PAX7). While research of these proteins is limited, two studies examining PAX6 showed that it interacts with AR and suppresses its activity [213, 214]. Interestingly, a majority of these heterotypic homeodomain TFs reduce canonical AR activity by either altering AR expression or the AR cistrome. Combined with the fact that PCa tumors with low AR activity are more aggressive and have higher Gleason grade due to lack of differentiation, further investigation of these heterotypic neural TFs is warranted as they may promote this phenotype.

The last major heterotypic TFs belong to the MYC and MAX sub-class containing a bHLH and bZIP domain (bHLHZ). Acquired from the viral oncogene v-MYC, the cellular protein MYC and its family members (MYCN, MYCL) are a major research topic in cancer biology [215]. Similar to other bZIP proteins, the MYC TFs often function as heterodimers with MAX or MAD [215]. As one of the Yamanaka factors, MYC is integral for maintenance of embryonic stem cells [145], and interestingly, MYCN plays an interchangeable role in regulating the pluripotency of stem cells [216, 217]. Both proteins are also important for maintenance and expansion of the neural progenitor population of cells [218–220]. In normal prostate epithelium, AR activity suppresses MYC expression, while in cancer, AR up-regulates MYC (likely through ERG upregulation) [221]. Considering the prevalence of the TMPRSS-ERG fusion in PCa (~60%), the dysregulation of MYC signaling could be an early event in PCa development [222]. MYC promotes cell proliferation by enhancing the efficiency of ribosomal RNA (rRNA) synthesis, consequently increasing ribosome number and protein production capacity of cancer cells [223]. Moreover, amplification of the MYC locus at 8q24 is frequent in PCa, correlating with increased metastasis and poor prognosis [224]. MYC also promotes EMP [225, 226] and CSC [227] phenotypes in PCa and, together with a cocktail of proteins, can transform prostate epithelium to NEPC [98].

Within the MYC family, expression of MYCL is limited to low Gleason grade PCa patients [228], while MYCN is almost exclusively expressed in patients with aggressive disease [17, 23]. MYCN is amplified in 40% of NEPC patients, and disrupting its protein stability by targeting Aurora kinase A (AURKA) reduces growth of NEPC xenografts [17]. In 2016, three landmark studies consolidated the importance of MYCN in NEPC. One study demonstrated that MYCN, over-expressed with a constitutively active myristoylated AKT, converted prostate epithelial cells to castration resistant tumors with focal NEPC histology [229]. Another study showed that NEPC tumors exhibit enhanced MYCN and EZH2 messages [23]. These findings were complemented with patient data showing that NEPC is an epigenetically driven disease with increases in expression of genes responsible for both histone and DNA methylation. Mechanistically, MYCN was shown to cooperate with EZH2 in order to drive NEPC specific histone methylation patterns [230]. On the bases of these studies, a clinical trial was established to evaluate the efficacy of MYCN inhibition by targeting AURKA (NCT01799278) in NEPC. Even though the trial was unsuccessful [231], a small subset of patients responded positively to AURKA inhibition warranting further exploration with better patient selection. Interestingly, MYCN also regulates PARP1 expression, and inhibition of either protein (PARP inhibitor Olaparib and AURKA inhibitor PHA1.5) reduces tumor growth *in vivo* [232]. Several strategies for therapeutic inhibition of MYC and MYCN are under active investigation. For example, BET bromodomain family proteins regulate expression of both MYC and MYCN, and treatment with BET inhibitors drastically reduces their expression across multiple cancers. Several BET inhibitors are currently under clinical investigation and may prove useful for targeting this potently oncogenic family in PCa.

Inhibition of Neural Transcription to Attenuate Disease Progression

The expression of terminal differentiation genes in each cell type depends on both positive and negative transcriptional controls [233, 234]. In other words, prostate cancer cells not only employ neural transcription factors to promote a more malignant state, but they also silence the natural inhibitory pathways designed to repress the transcription of neural genes. Some of these inhibitors of neural differentiation are discussed below (Table 2).

REST

One of the most important repressors of transcription is RE1-silencing transcription factor (REST), also known as neuron-restrictive silencer factor (NRSF) which was first identified in 1995 as a master repressor of neurogenesis, silencing multiple neuron-specific promoters [235]. Expression of REST is high in stem cells but drops rapidly in neural progenitors and is maintained at very low levels after differentiation, thus ensuring precise development [236]. REST is therefore fundamental during early lineage commitment as it prevents premature expression of terminal differentiation genes in non-neuronal cells [233]. Moreover, REST maintains transcriptional silencing of a range of neuronal genes in differentiated cells and mediates transcriptional responses associated with neural plasticity [236].

REST is a 116 kDa transcription factor, containing a zinc factor DNA-binding domain flanked by two independent repressor domains located at the amino and carboxy termini. The DNA binding domain recognizes a 21–23 base pair sequence in the regulatory region of target genes known as RE1 elements or neuron-restrictive silencer elements. The repressor domains interact with co-repressors such as Sin3A and coREST to recruit DNA modifying agents, namely histone deacetylases (HDACs) and methyl CpG binding protein 2 (MeCP2) to alter the conformation of chromatin from transcriptionally active euchromatin to transcriptionally silent heterochromatin [237].

Table 2 Neural transcription factors with tumor suppressor activity in prostate cancer

Transcription factor family	Members discussed in this section
Zinc Finger (ZF) transcription factors	REST, KLF family
Helix-loop-Helix (bHLH) transcription factors	HES family, HEY family
P53-Specific DNA Binding Domain	TP53

In prostate cancer, downregulation of REST appears to be associated with development of NE phenotype in patients and prostate cancer cell lines [238–240]. One interesting protein that has been shown to downregulate REST is Ser/Arg repetitive matrix 4 (SRRM4), a neural-specific mRNA splicing factor that is upregulated in NEPC [240, 241]. SRRM4 facilitates the alternative splicing of REST into a truncated splice variant called REST4. REST4 binds weakly to RE1 elements and therefore has diminished repressor function. REST4 isoforms can also bind directly to REST and inhibit its function [237, 242].

HES and HEY Families of Transcription Factors

The mammalian bHLH-Orange proteins belong to the repressor family of bHLH TFs and are divided into four subfamilies: Hes, Hey, Helt and Stra13/Dec. They have a distinctive motif called the "Orange domain", in addition to the bHLH domain that mediates DNA-binding and protein dimerization [243]. The "Orange domain" is a ~35 amino acid motif at the end of the bHLH c terminal region and provides a platform for protein-protein interactions.

The Hes (Hairy/enhancer of Split) and the Hey (Hairy/Enhancer of Split related with YRPW motif) subfamilies are downstream targets of the NOTCH signaling pathway and regulate various cell fates during development by repressing the transcription of neural genes and inhibiting premature neuronal differentiation [244, 245]. Hes1 represses the expression of human ASCL1, an important developmental protein in neuronal and endocrine cells, which is upregulated in NEPC [246]. Similarly, Hes1 inhibits neurogenin3 expression in a model of hypoxia-induced NE differentiation [74]. Moreover, a dominant-negative form of Hes1 induces increased levels of NE markers under normoxic conditions [74]. Altogether, the suppression of HES1 and upregulation of ASCL1 and NGN3 may play a crucial role in the lineage commitment of PCa cells to a NEPC phenotype. All three members of the HEY family (HEY1, HEY2, HEYL) function as repressors for AR transcriptional activity [247, 248]. Interestingly Hey1/AR interaction is strong in patients with benign prostatic hyperplasia (BPH), but lost in patients with PCa [247].

KLFs

Krüppel-like factors are highly conserved ZF transcription factors with roles ranging from proliferation to differentiation, migration, pluripotency and axon regeneration [13, 249, 250]. Many KLFs are involved in reprogramming somatic cells into inducible embryonic stem cells [251]. Some KLF proteins function predominantly as activators of transcription (KLFs 1, 2, 4, 5, 6, and 7), while others repress transcription. For example, KLFs 9, 10, 11, 13, 14, and 16 have repressor activity through their interaction with the common transcriptional corepressor Sin3A. Sin3A recruits HDACs to attenuate transcription [251]. The Yamanaka factor KLF4, an inhibitor of neurite growth [252], is a tumor suppressor in PCa and has been suggested to be a predictor of good prognosis in this malignancy [14]. KLF4 regulates several different pathways, including Hedgehog and the non-canonical WNT/Ca$^+$ pathway, involved in the development and progression of aggressive prostate cancer and therapy-resistant states. KLF4 suppresses EMP in PCa by binding directly to the SLUG promoter and inhibiting its transcription [253]. Moreover, KLF4 expression correlates negatively with Gleason score, with highest the expression found in indolent disease [14]. Another family member, KLF6, also upregulates CDKN1A (in a p53-independent manner) and acts as a tumor suppressor in PCa [254]. KLF6 is frequently deleted or mutated in human PCa [255]. Interestingly, splice variants of KLF6 such as KLF6-SV1 demonstrate an opposite role, i.e., decreasing CDKN1A expression and antagonizing the tumor suppressive function of KLF6. Increased KLF6-SV1 expression accelerates PCa progression and metastasis and is associated with poor prognosis in patients [256]. Both KLF9, which is important in late-phase neuronal maturation [257], and KLF13 [258] inhibit PCa proliferation by suppressing AKT signaling.

TP53

Aberrations of the tumor suppressor TP53 are more common in CRPC compared to adenocarcinoma and in NEPC compared to CRPC [259, 260]. Most TP53 mutations found in human malignancies are missense mutations (80%) within the DNA-binding domain (DBD), which negatively affect 3D folding and chromatin binding. Wild-type tumor suppressor protein 53 contains two N-terminal transcriptional activation domains, a proline-rich domain, a DBD, a tetramerization domain and a carboxy-terminal region with basic residues and is able to orchestrate a wide array of cellular functions, including apoptosis, transient cell cycle arrest and senescence in response to different stressors [261]. p53 also controls differentiation as it can negatively regulate the transcription of genes like WNT7A, NANOG and OCT-4, which are elevated in embryonic stem cells and are required for maintaining an undifferentiated state [262]. Accordingly, activation of p53 enhances differentiation of embryonic stem cells and decreases OCT-4 and NANOG [263]. Interestingly, mutant p53 not only exerts a dominant-negative effect on the wild-type protein but also displays enhanced malignant behaviors such as migration/invasion, increased proliferation and survival [261].

The regulatory processes that prevent neural differentiation and control transcription of neural genes are not limited to transcriptional repressors. Other proteins (lacking a DBD) also have the ability to influence disease progression towards a more malignant state by either acting as co-activators or co-repressors for TFs or by controlling the epigenetic reprogramming via post-translational modifications. However, discussing these players is out of the scope of this review, and readers are encouraged to consult other resources [264–266].

Conclusion

There is an increase in the emergence of treatment-induced resistant NEPC upon APIs. A review of key players among the diverse classes of TFs driving such differentiation reveals a common theme that these key TFs are often protagonists of multiple cellular plasticity phenotypes. For example, E-cadherin-suppressing TF, SNAIL, regulates both EMP and CSCs [267] and NE master regulator, BRN2, modulates metastasis [268, 269] as well as maintenance of CSCs [270, 271]. MYCN overexpressing model retains remarkable tumor initiating capacity and most importantly, have the ability to give rise to both adenocarcinoma and NEPC tumors [229]. More importantly, upregulation of CSC genes in NEPC models is common (as evident from stem cell signatures enriched in NEPC tumors) [272].

In summary, it would be tempting to speculate that upon inhibition of an identity-defining protein like the androgen receptor, PCa cells exist in a gradient-like state with adenocarcinoma at one end and a meta-stable cell type sharing EMT, CSC and NE properties at the other (Fig. 1). Subsequently, trigger events in specific combinations and/or sequence may dictate whether resistant CRPC tumors become adenocarcinoma, IAC, DNPC or NEPC tumors observed in patients. Therefore, understanding the transcriptional processes that lead to IAC/DNPC/NEPC pathologies is fundamentally important as it might alter survival of PCa patients.

Acknowledgement Authors declare no conflict of interest.

References

1. J. Varga, F.R. Greten, Cell plasticity in epithelial homeostasis and tumorigenesis. Nat. Cell Biol. **19**(10), 1133–1141 (2017). Epub 2017/09/26
2. A.C. Yeh, S. Ramaswamy, Mechanisms of cancer cell dormancy—another hallmark of cancer? Cancer Res. **75**(23), 5014–5022 (2015). Epub 2015/09/12
3. S. Kleffel, T. Schatton, Tumor dormancy and cancer stem cells: two sides of the same coin? Adv. Exp. Med. Biol. **734**, 145–179 (2013). Epub 2012/11/13
4. A. Fabian, G. Vereb, J. Szollosi, The hitchhikers guide to cancer stem cell theory: markers, pathways and therapy. Cytometry A **83**(1), 62–71 (2013). Epub 2012/09/22
5. E. Batlle, H. Clevers, Cancer stem cells revisited. Nat. Med. **23**(10), 1124–1134 (2017). Epub 2017/10/07

6. A.H. Zahalka, A. Arnal-Estape, M. Maryanovich, F. Nakahara, C.D. Cruz, L.W.S. Finley, et al., Adrenergic nerves activate an angio-metabolic switch in prostate cancer. Science **358**(6361), 321–326 (2017). Epub 2017/10/21. eng
7. G.E. Ayala, H. Dai, M. Powell, R. Li, Y. Ding, T.M. Wheeler, et al., Cancer-related axonogenesis and neurogenesis in prostate cancer. Clin. Cancer Res. **14**(23), 7593–7603 (2008). Epub 2008/12/03. eng
8. S.W. Cole, A.S. Nagaraja, S.K. Lutgendorf, P.A. Green, A.K. Sood, Sympathetic nervous system regulation of the tumour microenvironment. Nat. Rev. Cancer **15**, 563 (2015)
9. F. Crea, N.R. Nur Saidy, C.C. Collins, Y. Wang, The epigenetic/noncoding origin of tumor dormancy. Trends Mol. Med. **21**(4), 206–211 (2015). Epub 2015/03/17
10. N.S. Ruppender, C. Morrissey, P.H. Lange, R.L. Vessella, Dormancy in solid tumors: implications for prostate cancer. Cancer Metastasis Rev. **32**(3–4), 501–509 (2013). Epub 2013/04/25
11. M. Kirby, C. Hirst, E.D. Crawford, Characterising the castration-resistant prostate cancer population: a systematic review. Int. J. Clin. Pract. **65**(11), 1180–1192 (2011). Epub 2011/10/15. eng
12. J.L. Bishop, D. Thaper, S. Vahid, A. Davies, K. Ketola, H. Kuruma, et al., The master neural transcription factor BRN2 is an androgen receptor-suppressed driver of neuroendocrine differentiation in prostate cancer. Cancer Discov. **7**(1), 54–71 (2017). Epub 2016/10/28. eng
13. A. Apara, J.L. Goldberg, Molecular mechanisms of the suppression of axon regeneration by KLF transcription factors. Neural Regen. Res. **9**(15), 1418–1421 (2014). Epub 2014/10/16
14. X. Xiong, M. Schober, E. Tassone, A. Khodadadi-Jamayran, A. Sastre-Perona, H. Zhou, et al., KLF4, a gene regulating prostate stem cell homeostasis, is a barrier to malignant progression and predictor of good prognosis in prostate cancer. Cell Rep. **25**(11), 3006–20.e7 (2018). Epub 2018/12/13
15. E.J. Small, R.R. Aggarwal, J. Huang, A. Sokolov, L. Zhang, J.J. Alumkal, et al., Clinical and genomic characterization of metastatic small cell/neuroendocrine prostate cancer (SCNC) and intermediate atypical prostate cancer (IAC): results from the SU2C/PCF/AACRWest Coast Prostate Cancer Dream Team (WCDT). J. Clin. Oncol. **34**(15_suppl), 5019 (2016)
16. R. Aggarwal, J. Huang, J.J. Alumkal, L. Zhang, F.Y. Feng, G.V. Thomas, et al., Clinical and genomic characterization of treatment-emergent small-cell neuroendocrine prostate cancer: a multi-institutional prospective study. J. Clin. Oncol. **36**(24), 2492–2503 (2018). Epub 2018/07/10
17. H. Beltran, D.S. Rickman, K. Park, S.S. Chae, A. Sboner, T.Y. MacDonald, et al., Molecular characterization of neuroendocrine prostate cancer and identification of new drug targets. Cancer Discov. **1**(6), 487–495 (2011)
18. A.H. Davies, H. Beltran, A. Zoubeidi, Cellular plasticity and the neuroendocrine phenotype in prostate cancer. Nat. Rev. Urol. **15**(5), 271–286 (2018). Epub 2018/02/21. eng
19. B.J. Feldman, D. Feldman, The development of androgen-independent prostate cancer. Nat. Rev. Cancer **1**(1), 34–45 (2001). Epub 2002/03/20. eng
20. R. Aggarwal, T. Zhang, E.J. Small, A.J. Armstrong, Neuroendocrine prostate cancer: subtypes, biology, and clinical outcomes. J. Natl. Compr. Cancer Netw. **12**(5), 719–726 (2014). Epub 2014/05/09. eng
21. A. Berruti, A. Mosca, F. Porpiglia, E. Bollito, M. Tucci, F. Vana, et al., Chromogranin A expression in patients with hormone naive prostate cancer predicts the development of hormone refractory disease. J. Urol. **178**(3 Pt 1), 838–43; quiz 1129 (2007). Epub 2007/07/17. eng
22. D. Hirano, Y. Okada, S. Minei, Y. Takimoto, N. Nemoto, Neuroendocrine differentiation in hormone refractory prostate cancer following androgen deprivation therapy. Eur. Urol. **45**(5), 586–92; discussion 92 (2004). Epub 2004/04/15. eng
23. H. Beltran, D. Prandi, J.M. Mosquera, M. Benelli, L. Puca, J. Cyrta, et al., Divergent clonal evolution of castration-resistant neuroendocrine prostate cancer. Nat. Med. **22**(3), 298–305 (2016)
24. E.G. Bluemn, I.M. Coleman, J.M. Lucas, R.T. Coleman, S. Hernandez-Lopez, R. Tharakan, et al., Androgen receptor pathway-independent prostate cancer is sustained through FGF signaling. Cancer Cell **32**(4), 474–89.e6 (2017). Epub 2017/10/11. eng
25. B. Ell, Y. Kang, Transcriptional control of cancer metastasis. Trends Cell Biol. **23**(12), 603–611 (2013). Epub 2013/07/11
26. D.P. Labbe, M. Brown, Transcriptional regulation in prostate cancer. Cold Spring Harb. Perspect. Med. **8**(11), a030437 (2018)
27. S.A. Lambert, A. Jolma, L.F. Campitelli, P.K. Das, Y. Yin, M. Albu, et al., The human transcription factors. Cell **172**(4), 650–665 (2018). Epub 2018/02/10. eng
28. C.A. Bewley, A.M. Gronenborn, G.M. Clore, Minor groove-binding architectural proteins: structure, function, and DNA recognition. Annu. Rev. Biophys. Biomol. Struct. **27**, 105–131 (1998)
29. T. Vierbuchen, A. Ostermeier, Z.P. Pang, Y. Kokubu, T.C. Sudhof, M. Wernig, Direct conversion of fibroblasts to functional neurons by defined factors. Nature **463**(7284), 1035–1041 (2010). Epub 2010/01/29. eng
30. Z.P. Pang, N. Yang, T. Vierbuchen, A. Ostermeier, D.R. Fuentes, T.Q. Yang, et al., Induction of human neuronal cells by defined transcription factors. Nature **476**(7359), 220–223 (2011). Epub 2011/05/28. eng
31. U. Pfisterer, A. Kirkeby, O. Torper, J. Wood, J. Nelander, A. Dufour, et al., Direct conversion of

human fibroblasts to dopaminergic neurons. Proc. Natl. Acad. Sci. U. S. A. **108**(25), 10343–10348 (2011). Epub 2011/06/08. eng

32. R. Tsunemoto, S. Lee, A. Szucs, P. Chubukov, I. Sokolova, J.W. Blanchard, et al., Diverse reprogramming codes for neuronal identity. Nature **557**(7705), 375–380 (2018). Epub 2018/05/11. eng

33. R. Ambasudhan, M. Talantova, R. Coleman, X. Yuan, S. Zhu, S.A. Lipton, et al., Direct reprogramming of adult human fibroblasts to functional neurons under defined conditions. Cell Stem Cell **9**(2), 113–118 (2011). Epub 2011/08/02. eng

34. M. Cassandri, A. Smirnov, F. Novelli, C. Pitolli, M. Agostini, M. Malewicz, et al., Zinc-finger proteins in health and disease. Cell Death Discov. **3**, 17071 (2017). Epub 2017/11/21. eng

35. A. Klug, The discovery of zinc fingers and their development for practical applications in gene regulation and genome manipulation. Q. Rev. Biophys. **43**(1), 1–21 (2010)

36. J.W. Schwabe, D. Rhodes, Beyond zinc fingers: steroid hormone receptors have a novel structural motif for DNA recognition. Trends Biochem. Sci. **16**(8), 291–296 (1991)

37. C. Heberden, Sex steroids and neurogenesis. Biochem. Pharmacol. **141**, 56–62 (2017)

38. Y.N. Liu, Y. Liu, H.J. Lee, Y.H. Hsu, J.H. Chen, Activated androgen receptor downregulates E-cadherin gene expression and promotes tumor metastasis. Mol. Cell. Biol. **28**(23), 7096–7108 (2008)

39. K. Jennbacken, T. Tesan, W. Wang, H. Gustavsson, J.E. Damber, K. Welen, N-cadherin increases after androgen deprivation and is associated with metastasis in prostate cancer. Endocr. Relat. Cancer **17**(2), 469–479 (2010)

40. H. Tanaka, E. Kono, C.P. Tran, H. Miyazaki, J. Yamashiro, T. Shimomura, et al., Monoclonal antibody targeting of N-cadherin inhibits prostate cancer growth, metastasis and castration resistance. Nat. Med. **16**(12), 1414–1420 (2010)

41. E. Hornberg, E.B. Ylitalo, S. Crnalic, H. Antti, P. Stattin, A. Widmark, et al., Expression of androgen receptor splice variants in prostate cancer bone metastases is associated with castration-resistance and short survival. PLoS One **6**(4), e19059 (2011)

42. C. Cai, H. Wang, H.H. He, S. Chen, L. He, F. Ma, et al., ERG induces androgen receptor-mediated regulation of SOX9 in prostate cancer. J. Clin. Invest. **123**(3), 1109–1122 (2013)

43. Y. Wang, J. Shi, K. Chai, X. Ying, B.P. Zhou, The role of Snail in EMT and tumorigenesis. Curr. Cancer Drug Targets **13**(9), 963–972 (2013)

44. L. Miao, L. Yang, R. Li, D.N. Rodrigues, M. Crespo, J.T. Hsieh, et al., Disrupting androgen receptor signaling induces Snail-mediated epithelial-mesenchymal plasticity in prostate cancer. Cancer Res. **77**(11), 3101–3112 (2017)

45. S. Akamatsu, A.W. Wyatt, D. Lin, S. Lysakowski, F. Zhang, S. Kim, et al., The placental gene PEG10 promotes progression of neuroendocrine prostate cancer. Cell Rep. **12**(6), 922–936 (2015)

46. K. Wu, C. Gore, L. Yang, L. Fazli, M. Gleave, R.C. Pong, et al., Slug, a unique androgen-regulated transcription factor, coordinates androgen receptor to facilitate castration resistance in prostate cancer. Mol. Endocrinol. **26**(9), 1496–1507 (2012)

47. K.E. Ware, J.A. Somarelli, D. Schaeffer, J. Li, T. Zhang, S. Park, et al., Snail promotes resistance to enzalutamide through regulation of androgen receptor activity in prostate cancer. Oncotarget **7**(31), 50507–50521 (2016)

48. D. McKeithen, T. Graham, L.W. Chung, V. Odero-Marah, Snail transcription factor regulates neuroendocrine differentiation in LNCaP prostate cancer cells. Prostate **70**(9), 982–992 (2010)

49. T. Fujimura, S. Takahashi, T. Urano, J. Kumagai, T. Ogushi, K. Horie-Inoue, et al., Increased expression of estrogen-related receptor alpha (ERRalpha) is a negative prognostic predictor in human prostate cancer. Int. J. Cancer **120**(11), 2325–2330 (2007)

50. T. Fujimura, K. Takayama, S. Takahashi, I.S. Estrogen, Androgen blockade for advanced prostate cancer in the era of precision medicine. Cancers **10**(2), E29 (2018)

51. T. Grindstad, E. Richardsen, S. Andersen, K. Skjefstad, M. Rakaee Khanehkenari, T. Donnem, et al., Progesterone receptors in prostate cancer: progesterone receptor B is the isoform associated with disease progression. Sci. Rep. **8**(1), 11358 (2018)

52. V.K. Arora, E. Schenkein, R. Murali, S.K. Subudhi, J. Wongvipat, M.D. Balbas, et al., Glucocorticoid receptor confers resistance to antiandrogens by bypassing androgen receptor blockade. Cell **155**(6), 1309–1322 (2013)

53. M. Isikbay, K. Otto, S. Kregel, J. Kach, Y. Cai, D.J. Vander Griend, et al., Glucocorticoid receptor activity contributes to resistance to androgen-targeted therapy in prostate cancer. Horm. Cancer. **5**(2), 72–89 (2014)

54. M. Puhr, J. Hoefer, A. Eigentler, C. Ploner, F. Handle, G. Schaefer, et al., The glucocorticoid receptor is a key player for prostate cancer cell survival and a target for improved antiandrogen therapy. Clin. Cancer Res. **24**(4), 927–938 (2018)

55. V. Rodriguez-Bravo, M. Carceles-Cordon, Y. Hoshida, C. Cordon-Cardo, M.D. Galsky, J. Domingo-Domenech, The role of GATA2 in lethal prostate cancer aggressiveness. Nat. Rev. Urol. **14**(1), 38–48 (2017)

56. A.H. Nguyen, M. Tremblay, K. Haigh, I.H. Koumakpayi, M. Paquet, P.P. Pandolfi, et al., Gata3 antagonizes cancer progression in Pten-deficient prostates. Hum. Mol. Genet. **22**(12), 2400–2410 (2013)

57. M.E.R. Shafer, A.H.T. Nguyen, M. Tremblay, S. Viala, M. Beland, N.R. Bertos, et al., Lineage

specification from prostate progenitor cells requires Gata3-dependent mitotic spindle orientation. Stem Cell Rep. **8**(4), 1018–1031 (2017)

58. N. Li, M. Chen, S. Truong, C. Yan, R. Buttyan, Determinants of Gli2 co-activation of wildtype and naturally truncated androgen receptors. Prostate **74**(14), 1400–1410 (2014)

59. N. Li, S. Truong, M. Nouri, J. Moore, N. Al Nakouzi, A.A. Lubik, et al., Non-canonical activation of hedgehog in prostate cancer cells mediated by the interaction of transcriptionally active androgen receptor proteins with Gli3. Oncogene **37**(17), 2313–2325 (2018)

60. G. Chen, Y. Goto, R. Sakamoto, K. Tanaka, E. Matsubara, M. Nakamura, et al., GLI1, a crucial mediator of sonic hedgehog signaling in prostate cancer, functions as a negative modulator for androgen receptor. Biochem. Biophys. Res. Commun. **404**(3), 809–815 (2011)

61. A. Gonnissen, S. Isebaert, K. Haustermans, Hedgehog signaling in prostate cancer and its therapeutic implication. Int. J. Mol. Sci. **14**(7), 13979–14007 (2013)

62. D.L. Suzman, E.S. Antonarakis, Clinical implications of hedgehog pathway signaling in prostate cancer. Cancers **7**(4), 1983–1993 (2015)

63. N. Ibuki, M. Ghaffari, M. Pandey, I. Iu, L. Fazli, M. Kashiwagi, et al., TAK-441, a novel investigational smoothened antagonist, delays castration-resistant progression in prostate cancer by disrupting paracrine hedgehog signaling. Int. J. Cancer **133**(8), 1955–1966 (2013)

64. C.R. Cochrane, A. Szczepny, D.N. Watkins, J.E. Cain, Hedgehog signaling in the maintenance of cancer stem cells. Cancers **7**(3), 1554–1585 (2015)

65. W. Bushman, Hedgehog signaling in prostate development, regeneration and cancer. J. Dev. Biol. **4**(4), E30 (2016)

66. W.Y. Chen, T. Zeng, Y.C. Wen, H.L. Yeh, K.C. Jiang, W.H. Chen, et al., Androgen deprivation-induced ZBTB46-PTGS1 signaling promotes neuroendocrine differentiation of prostate cancer. Cancer Lett. **440–441**, 35–46 (2019)

67. Z. Xin, Y. Zhang, Z. Jiang, L. Zhao, L. Fan, Y. Wang, et al., Insulinoma-associated protein 1 is a novel sensitive and specific marker for small cell carcinoma of the prostate. Hum. Pathol. **79**, 151–159 (2018)

68. S. Jones, An overview of the basic helix-loop-helix proteins. Genome Biol. **5**(6), 226 (2004)

69. A. Gupta, Y. Wang, C. Browne, S. Kim, T. Case, M. Paul, et al., Neuroendocrine differentiation in the 12T-10 transgenic prostate mouse model mimics endocrine differentiation of pancreatic beta cells. Prostate **68**(1), 50–60 (2008)

70. S. Sinha, M.D. Nyquist, A. Corella, I. Coleman, P.S. Nelson, Abstract LB-199: Role of ASCL1 in neuroendocrine prostate cancer progression. Cancer Res. **78**(13 Suppl), LB-199 (2018)

71. Y. Hu, T. Wang, G.D. Stormo, J.I. Gordon, RNA interference of achaete-scute homolog 1 in mouse prostate neuroendocrine cells reveals its gene targets and DNA binding sites. Proc. Natl. Acad. Sci. U. S. A. **101**(15), 5559–5564 (2004)

72. J.K. Osborne, J.E. Larsen, M.D. Shields, J.X. Gonzales, D.S. Shames, M. Sato, et al., NeuroD1 regulates survival and migration of neuroendocrine lung carcinomas via signaling molecules TrkB and NCAM. Proc. Natl. Acad. Sci. U. S. A. **110**(16), 6524–6529 (2013)

73. L. Cindolo, R. Franco, M. Cantile, G. Schiavo, G. Liguori, P. Chiodini, et al., NeuroD1 expression in human prostate cancer: can it contribute to neuroendocrine differentiation comprehension? Eur. Urol. **52**(5), 1365–1373 (2007)

74. G. Danza, C. Di Serio, F. Rosati, G. Lonetto, N. Sturli, D. Kacer, et al., Notch signaling modulates hypoxia-induced neuroendocrine differentiation of human prostate cancer cells. Mol. Cancer Res. **10**(2), 230–238 (2012)

75. G.L. Semenza, Hypoxia-inducible factor 1 and cancer pathogenesis. IUBMB Life **60**(9), 591–597 (2008)

76. H. Zhong, G.L. Semenza, J.W. Simons, A.M. De Marzo, Up-regulation of hypoxia-inducible factor 1alpha is an early event in prostate carcinogenesis. Cancer Detect. Prev. **28**(2), 88–93 (2004)

77. T.S. Eisinger-Mathason, M.C. Simon, HIF-1alpha partners with FoxA2, a neuroendocrine-specific transcription factor, to promote tumorigenesis. Cancer Cell. **18**(1), 3–4 (2010)

78. J. Qi, K. Nakayama, R.D. Cardiff, A.D. Borowsky, K. Kaul, R. Williams, et al., Siah2-dependent concerted activity of HIF and FoxA2 regulates formation of neuroendocrine phenotype and neuroendocrine prostate tumors. Cancer Cell **18**(1), 23–38 (2010)

79. M.H. Yang, M.Z. Wu, S.H. Chiou, P.M. Chen, S.Y. Chang, C.J. Liu, et al., Direct regulation of TWIST by HIF-1alpha promotes metastasis. Nat. Cell Biol. **10**(3), 295–305 (2008)

80. R.P. Gajula, S.T. Chettiar, R.D. Williams, S. Thiyagarajan, Y. Kato, K. Aziz, et al., The twist box domain is required for Twist1-induced prostate cancer metastasis. Mol. Cancer Res. **11**(11), 1387–1400 (2013)

81. J.W. Vincentz, B.A. Firulli, A. Lin, D.B. Spicer, M.J. Howard, A.B. Firulli, Twist1 controls a cell-specification switch governing cell fate decisions within the cardiac neural crest. PLoS Genet. **9**(3), e1003405 (2013)

82. A. Lasorella, R. Benezra, A. Iavarone, The ID proteins: master regulators of cancer stem cells and tumour aggressiveness. Nat. Rev. Cancer **14**(2), 77–91 (2014)

83. X.S. Ouyang, X. Wang, D.T. Lee, S.W. Tsao, Y.C. Wong, Over expression of ID-1 in prostate cancer. J. Urol. **167**(6), 2598–2602 (2002)

84. A.J. Zielinski, S. Fong, J. Allison, M. Kawahara, J.P. Coppe, H. Feiler, et al., The helix-loop-helix Id-1 inhibits PSA expression in prostate cancer cells. Int. J. Cancer **126**(10), 2490–2496 (2010)

85. M.T. Ling, X. Wang, D.T. Lee, P.C. Tam, S.W. Tsao, Y.C. Wong, Id-1 expression induces androgen-independent prostate cancer cell growth through activation of epidermal growth factor receptor (EGF-R). Carcinogenesis **25**(4), 517–525 (2004)
86. M.T. Ling, X. Wang, X.S. Ouyang, T.K. Lee, T.Y. Fan, K. Xu, et al., Activation of MAPK signaling pathway is essential for Id-1 induced serum independent prostate cancer cell growth. Oncogene **21**(55), 8498–8505 (2002)
87. P. Sharma, D. Patel, J. Chaudhary, Id1 and Id3 expression is associated with increasing grade of prostate cancer: Id3 preferentially regulates CDKN1B. Cancer Med. **1**(2), 187–197 (2012)
88. N. Strong, A.C. Millena, L. Walker, J. Chaudhary, S.A. Khan, Inhibitor of differentiation 1 (Id1) and Id3 proteins play different roles in TGFbeta effects on cell proliferation and migration in prostate cancer cells. Prostate **73**(6), 624–633 (2013)
89. W. Zhao, Q. Zhu, P. Tan, A. Ajibade, T. Long, W. Long, et al., Tgfbr2 inactivation facilitates cellular plasticity and development of Pten-null prostate cancer. J. Mol. Cell Biol. **10**(4), 316–330 (2018)
90. J. Perk, A. Iavarone, R. Benezra, Id family of helix-loop-helix proteins in cancer. Nat. Rev. Cancer **5**(8), 603–614 (2005)
91. A. Lasorella, T. Uo, A. Iavarone, Id proteins at the cross-road of development and cancer. Oncogene **20**(58), 8326–8333 (2001)
92. X. Zhang, M.T. Ling, X. Wang, Y.C. Wong, Inactivation of Id-1 in prostate cancer cells: a potential therapeutic target in inducing chemosensitization to taxol through activation of JNK pathway. Int. J. Cancer **118**(8), 2072–2081 (2006)
93. H. Geng, B.L. Rademacher, J. Pittsenbarger, C.Y. Huang, C.T. Harvey, M.C. Lafortune, et al., ID1 enhances docetaxel cytotoxicity in prostate cancer cells through inhibition of p21. Cancer Res. **70**(8), 3239–3248 (2010)
94. J.P. Carey, A.J. Asirvatham, O. Galm, T.A. Ghogomu, J. Chaudhary, Inhibitor of differentiation 4 (Id4) is a potential tumor suppressor in prostate cancer. BMC Cancer **9**, 173 (2009)
95. J.P. Carey, A.E. Knowell, S. Chinaranagari, J. Chaudhary, Id4 promotes senescence and sensitivity to doxorubicin-induced apoptosis in DU145 prostate cancer cells. Anticancer Res. **33**(10), 4271–4278 (2013)
96. P. Sharma, S. Chinaranagari, J. Chaudhary, Inhibitor of differentiation 4 (ID4) acts as an inhibitor of ID-1, -2 and -3 and promotes basic helix loop helix (bHLH) E47 DNA binding and transcriptional activity. Biochimie **112**, 139–150 (2015)
97. S. Samuel, H. Naora, Homeobox gene expression in cancer: insights from developmental regulation and deregulation. Eur. J. Cancer **41**(16), 2428–2437 (2005)
98. J.W. Park, J.K. Lee, K.M. Sheu, L. Wang, N.G. Balanis, K. Nguyen, et al., Reprogramming normal human epithelial tissues to a common, lethal neuroendocrine cancer lineage. Science **362**(6410), 91–95 (2018)
99. I. Homminga, R. Pieters, J.P. Meijerink, NKL homeobox genes in leukemia. Leukemia **26**(4), 572–581 (2012)
100. A.P. McMahon, Neural patterning: the role of Nkx genes in the ventral spinal cord. Genes Dev. **14**(18), 2261–2264 (2000)
101. E. Rodriguez-Zarco, A. Vallejo-Benitez, S. Umbria-Jimenez, S. Pereira-Gallardo, S. Pabon-Carrasco, A. Azueta, et al., Immunohistochemical study of the neural development transcription factors (TTF1, ASCL1 and BRN2) in neuroendocrine prostate tumours. [Estudio inmunohistoquimico de los factores de trancripcion de desarrollo neural (TTF1, ASCL1 y BRN2) en los tumores neuroendocrinos de prostata]. Actas Urol. Esp. **41**(8), 529–534 (2017)
102. M. Xue, H. Liu, L. Zhang, H. Chang, Y. Liu, S. Du, et al., Computational identification of mutually exclusive transcriptional drivers dysregulating metastatic microRNAs in prostate cancer. Nat. Commun. **8**, 14917 (2017)
103. C. Abate-Shen, M.M. Shen, E. Gelmann, Integrating differentiation and cancer: the Nkx3.1 homeobox gene in prostate organogenesis and carcinogenesis. Differentiation **76**(6), 717–727 (2008)
104. P.Y. Tan, C.W. Chang, K.R. Chng, K.D. Wansa, W.K. Sung, E. Cheung, Integration of regulatory networks by NKX3-1 promotes androgen-dependent prostate cancer survival. Mol. Cell. Biol. **32**(2), 399–414 (2012)
105. C.D. Vocke, R.O. Pozzatti, D.G. Bostwick, C.D. Florence, S.B. Jennings, S.E. Strup, et al., Analysis of 99 microdissected prostate carcinomas reveals a high frequency of allelic loss on chromosome 8p12-21. Cancer Res. **56**(10), 2411–2416 (1996)
106. L.E. Iv Santaliz-Ruiz, X. Xie, M. Old, T.N. Teknos, Q. Pan, Emerging role of nanog in tumorigenesis and cancer stem cells. Int. J. Cancer **135**(12), 2741–2748 (2014)
107. C.R. Jeter, B. Liu, Y. Lu, H.P. Chao, D. Zhang, X. Liu, et al., NANOG reprograms prostate cancer cells to castration resistance via dynamically repressing and engaging the AR/FOXA1 signaling axis. Cell Discov. **2**, 16041 (2016)
108. C.R. Jeter, B. Liu, X. Liu, X. Chen, C. Liu, T. Calhoun-Davis, et al., NANOG promotes cancer stem cell characteristics and prostate cancer resistance to androgen deprivation. Oncogene **30**(36), 3833–3845 (2011)
109. J.D. Norris, C.Y. Chang, B.M. Wittmann, R.S. Kunder, H. Cui, D. Fan, et al., The homeodomain protein HOXB13 regulates the cellular response to androgens. Mol. Cell **36**(3), 405–416 (2009)
110. H. Brechka, R.R. Bhanvadia, C. VanOpstall, D.J. Vander Griend, HOXB13 mutations and binding partners in prostate development and cancer:

function, clinical significance, and future directions. Genes Dis. **4**(2), 75–87 (2017)

111. E.M. Carpenter, Hox genes and spinal cord development. Dev. Neurosci. **24**(1), 24–34 (2002)

112. X. Warot, C. Fromental-Ramain, V. Fraulob, P. Chambon, P. Dolle, Gene dosage-dependent effects of the Hoxa-13 and Hoxd-13 mutations on morphogenesis of the terminal parts of the digestive and urogenital tracts. Development **124**(23), 4781–4791 (1997). Epub 1998/01/15

113. Y.R. Kim, K.J. Oh, R.Y. Park, N.T. Xuan, T.W. Kang, D.D. Kwon, et al., HOXB13 promotes androgen independent growth of LNCaP prostate cancer cells by the activation of E2F signaling. Mol. Cancer **9**, 124 (2010). Epub 2010/05/28

114. C.M. Ewing, A.M. Ray, E.M. Lange, K.A. Zuhlke, C.M. Robbins, W.D. Tembe, et al., Germline mutations in HOXB13 and prostate-cancer risk. N. Engl. J. Med. **366**(2), 141–149 (2012). Epub 2012/01/13

115. Y.R. Kim, T.W. Kang, P.K. To, N.T. Xuan Nguyen, Y.S. Cho, C. Jung, et al., HOXB13-mediated suppression of p21WAF1/CIP1 regulates JNK/c-Jun signaling in prostate cancer cells. Oncol. Rep. **35**(4), 2011–2016 (2016). Epub 2016/01/20

116. I.J. Kim, T.W. Kang, T. Jeong, Y.R. Kim, C. Jung, HOXB13 regulates the prostate-derived Ets factor: implications for prostate cancer cell invasion. Int. J. Oncol. **45**(2), 869–876 (2014). Epub 2014/06/06

117. Y.R. Kim, I.J. Kim, T.W. Kang, C. Choi, K.K. Kim, M.S. Kim, et al., HOXB13 downregulates intracellular zinc and increases NF-kappaB signaling to promote prostate cancer metastasis. Oncogene **33**(37), 4558–4567 (2014). Epub 2013/10/08

118. S. Kumar, A. Das, S. Sen, Extracellular matrix density promotes EMT by weakening cell-cell adhesions. Mol. BioSyst. **10**(4), 838–850 (2014). Epub 2014/02/01

119. H. Wang, G. Liu, D. Shen, H. Ye, J. Huang, L. Jiao, et al., HOXA1 enhances the cell proliferation, invasion and metastasis of prostate cancer cells. Oncol. Rep. **34**(3), 1203–1210 (2015). Epub 2015/07/03

120. K.A. Mace, S.L. Hansen, C. Myers, D.M. Young, N. Boudreau, HOXA3 induces cell migration in endothelial and epithelial cells promoting angiogenesis and wound repair. J. Cell Sci. **118**(Pt 12), 2567–2577 (2005). Epub 2005/05/26

121. A. Care, F. Felicetti, E. Meccia, L. Bottero, M. Parenza, A. Stoppacciaro, et al., HOXB7: a key factor for tumor-associated angiogenic switch. Cancer Res. **61**(17), 6532–6539 (2001). Epub 2001/08/28

122. N.D. Pruett, Z. Hajdu, J. Zhang, R.P. Visconti, M.J. Kern, D.M. Wellik, et al., Changing topographic Hox expression in blood vessels results in regionally distinct vessel wall remodeling. Biol. Open **1**(5), 430–435 (2012). Epub 2012/12/06

123. M. Cantile, A. Kisslinger, L. Cindolo, G. Schiavo, V. D'Anto, R. Franco, et al., cAMP induced modifications of HOX D gene expression in prostate cells allow the identification of a chromosomal area involved in vivo with neuroendocrine differentiation of human advanced prostate cancers. J. Cell Physiol. **205**(2), 202–210 (2005). Epub 2005/05/17

124. F. Poulat, F. Girard, M.P. Chevron, C. Goze, X. Rebillard, B. Calas, et al., Nuclear localization of the testis determining gene product SRY. J. Cell Biol. **128**(5), 737–748 (1995). Epub 1995/03/01

125. Z.Y. She, W.X. Yang, SOX family transcription factors involved in diverse cellular events during development. Eur. J. Cell Biol. **94**(12), 547–563 (2015). Epub 2015/09/06

126. S. Gasca, J. Canizares, P. De Santa Barbara, C. Mejean, F. Poulat, P. Berta, et al., A nuclear export signal within the high mobility group domain regulates the nucleocytoplasmic translocation of SOX9 during sexual determination. Proc. Natl. Acad. Sci. U. S. A. **99**(17), 11199–11204 (2002). Epub 2002/08/10

127. P. Sudbeck, G. Scherer, Two independent nuclear localization signals are present in the DNA-binding high-mobility group domains of SRY and SOX9. J. Biol. Chem. **272**(44), 27848–27852 (1997). Epub 1997/11/05

128. V.R. Harley, R. Lovell Badge, P.N. Goodfellow, Definition of a consensus DNA binding site for SRY. Nucleic Acids Res. **22**(8), 1500–1501 (1994). Epub 1994/04/25

129. L. Hou, Y. Srivastava, R. Jauch, Molecular basis for the genome engagement by Sox proteins. Semin. Cell Dev. Biol. **63**, 2–12 (2017). Epub 2016/08/16

130. S. Wilson, J. Qi, F.V. Filipp, Refinement of the androgen response element based on ChIP-Seq in androgen-insensitive and androgen-responsive prostate cancer cell lines. Sci. Rep. **6**, 32611 (2016). Epub 2016/09/15

131. H. Wang, N.C. McKnight, T. Zhang, M.L. Lu, S.P. Balk, X. Yuan, SOX9 is expressed in normal prostate basal cells and regulates androgen receptor expression in prostate cancer cells. Cancer Res. **67**(2), 528–536 (2007). Epub 2007/01/20

132. B. Bilir, A.O. Osunkoya, W.G. Wiles IV, S. Sannigrahi, V. Lefebvre, D. Metzger, et al., SOX4 is essential for prostate tumorigenesis initiated by PTEN ablation. Cancer Res. **76**(5), 1112–1121 (2016). Epub 2015/12/25

133. X. Yuan, M.L. Lu, T. Li, S.P. Balk, SRY interacts with and negatively regulates androgen receptor transcriptional activity. J. Biol. Chem. **276**(49), 46647–46654 (2001). Epub 2001/10/05

134. J. Hu, J. Tian, S. Zhu, L. Sun, J. Yu, H. Tian, et al., Sox5 contributes to prostate cancer metastasis and is a master regulator of TGF-beta-induced epithelial mesenchymal transition through controlling Twist1 expression. Br J Cancer **118**(1), 88–97 (2018). Epub 2017/11/11

135. M. Yang, J. Wang, L. Wang, C. Shen, B. Su, M. Qi, et al., Estrogen induces androgen-repressed SOX4 expression to promote progression of prostate cancer cells. Prostate **75**(13), 1363–1375 (2015). Epub 2015/05/28

136. W. Fu, T. Tao, M. Qi, L. Wang, J. Hu, X. Li, et al., MicroRNA-132/212 upregulation inhibits TGF-beta-mediated epithelial-mesenchymal transition of prostate cancer cells by targeting SOX4. Prostate **76**(16), 1560–1570 (2016). Epub 2016/10/19

137. L. Wang, Y. Li, X. Yang, H. Yuan, X. Li, M. Qi, et al., ERG-SOX4 interaction promotes epithelial-mesenchymal transition in prostate cancer cells. Prostate **74**(6), 647–658 (2014). Epub 2014/01/18

138. N.V. Sharma, K.L. Pellegrini, V. Ouellet, F.O. Giuste, S. Ramalingam, K. Watanabe, et al., Identification of the transcription factor relationships associated with androgen deprivation therapy response and metastatic progression in prostate cancer. Cancers **10**(10), E379 (2018). Epub 2018/10/14

139. Y. Yu, Z. Wang, D. Sun, X. Zhou, X. Wei, W. Hou, et al., miR-671 promotes prostate cancer cell proliferation by targeting tumor suppressor SOX6. Eur. J. Pharmacol. **823**, 65–71 (2018). Epub 2018/01/23

140. L. Guo, D. Zhong, S. Lau, X. Liu, X.Y. Dong, X. Sun, et al., Sox7 is an independent checkpoint for beta-catenin function in prostate and colon epithelial cells. Mol. Cancer Res. **6**(9), 1421–1430 (2008). Epub 2008/09/30

141. V. Murillo-Garzon, R. Kypta, WNT signalling in prostate cancer. Nat. Rev. Urol. **14**(11), 683–696 (2017). Epub 2017/09/13

142. F. Ma, H. Ye, H.H. He, S.J. Gerrin, S. Chen, B.A. Tanenbaum, et al., SOX9 drives WNT pathway activation in prostate cancer. J. Clin. Invest. **126**(5), 1745–1758 (2016). Epub 2016/04/05

143. J.C. Francis, A. Capper, J. Ning, E. Knight, J. de Bono, A. Swain, SOX9 is a driver of aggressive prostate cancer by promoting invasion, cell fate and cytoskeleton alterations and epithelial to mesenchymal transition. Oncotarget **9**(7), 7604–7615 (2018). Epub 2018/02/28

144. Z. Huang, P.J. Hurley, B.W. Simons, L. Marchionni, D.M. Berman, A.E. Ross, et al., Sox9 is required for prostate development and prostate cancer initiation. Oncotarget **3**(6), 651–663 (2012). Epub 2012/07/05

145. K. Takahashi, S. Yamanaka, Induction of pluripotent stem cells from mouse embryonic and adult fibroblast cultures by defined factors. Cell **126**(4), 663–676 (2006)

146. S. Kregel, K.J. Kiriluk, A.M. Rosen, Y. Cai, E.E. Reyes, K.B. Otto, et al., Sox2 is an androgen receptor-repressed gene that promotes castration-resistant prostate cancer. PLoS One **8**(1), e53701 (2013). Epub 2013/01/18

147. L. de Wet, A. Williams, M. Gillard, S. Kregel, T. Garcia, E. McAuley, et al., Abstract 3348: The role of SOX2 in promoting resistance to AR-targeted therapies in prostate cancer. Cancer Res. **78**(13 Suppl), 3348 (2018)

148. P. Mu, Z. Zhang, M. Benelli, W.R. Karthaus, E. Hoover, C.C. Chen, et al., SOX2 promotes lineage plasticity and antiandrogen resistance in TP53- and RB1-deficient prostate cancer. Science **355**(6320), 84–88 (2017)

149. X. Li, Y. Xu, Y. Chen, S. Chen, X. Jia, T. Sun, et al., SOX2 promotes tumor metastasis by stimulating epithelial-to-mesenchymal transition via regulation of WNT/beta-catenin signal network. Cancer Lett. **336**(2), 379–389 (2013)

150. S. Kar, D. Sengupta, M. Deb, N. Pradhan, S.K. Patra, SOX2 function and Hedgehog signaling pathway are co-conspirators in promoting androgen independent prostate cancer. Biochim. Biophys. Acta Mol. basis Dis. **1863**(1), 253–265 (2017)

151. M.L. Golson, K.H. Kaestner, Fox transcription factors: from development to disease. Development **143**(24), 4558–4570 (2016)

152. G. Tuteja, K.H. Kaestner, SnapShot: forkhead transcription factors I. Cell **130**(6), 1160 (2007)

153. G. Tuteja, K.H. Kaestner, Forkhead transcription factors II. Cell **131**(1), 192 (2007)

154. K.S. Zaret, J.S. Carroll, Pioneer transcription factors: establishing competence for gene expression. Genes Dev. **25**(21), 2227–2241 (2011)

155. J. Kim, H. Jin, J.C. Zhao, Y.A. Yang, Y. Li, X. Yang, et al., FOXA1 inhibits prostate cancer neuroendocrine differentiation. Oncogene **36**(28), 4072–4080 (2017)

156. J.W. Park, J.K. Lee, O.N. Witte, J. Huang, FOXA2 is a sensitive and specific marker for small cell neuroendocrine carcinoma of the prostate. Mod. Pathol. **30**(9), 1262–1272 (2017)

157. D. Jones, M. Wade, S. Nakjang, L. Chaytor, J. Grey, C.N. Robson, et al., FOXA1 regulates androgen receptor variant activity in models of castrate-resistant prostate cancer. Oncotarget **6**(30), 29782–29794 (2015)

158. J.L. Robinson, T.E. Hickey, A.Y. Warren, S.L. Vowler, T. Carroll, A.D. Lamb, et al., Elevated levels of FOXA1 facilitate androgen receptor chromatin binding resulting in a CRPC-like phenotype. Oncogene **33**(50), 5666–5674 (2014)

159. Y.A. Yang, J. Yu, Current perspectives on FOXA1 regulation of androgen receptor signaling and prostate cancer. Genes Dis. **2**(2), 144–151 (2015)

160. J.S. Carroll, X.S. Liu, A.S. Brodsky, W. Li, C.A. Meyer, A.J. Szary, et al., Chromosome-wide mapping of estrogen receptor binding reveals long-range regulation requiring the forkhead protein FoxA1. Cell **122**(1), 33–43 (2005)

161. J. Gerhardt, M. Montani, P. Wild, M. Beer, F. Huber, T. Hermanns, et al., FOXA1 promotes tumor progression in prostate cancer and represents a novel hallmark of castration-resistant prostate cancer. Am. J. Pathol. **180**(2), 848–861 (2012)

162. J. Laoukili, M.R. Kooistra, A. Bras, J. Kauw, R.M. Kerkhoven, A. Morrison, et al., FoxM1 is required for execution of the mitotic programme and chromosome stability. Nat. Cell Biol. **7**(2), 126–136 (2005)

163. A. Aytes, A. Mitrofanova, C. Lefebvre, M.J. Alvarez, M. Castillo-Martin, T. Zheng, et al., Cross-species regulatory network analysis identifies a synergistic interaction between FOXM1 and CENPF that drives

prostate cancer malignancy. Cancer Cell **25**(5), 638–651 (2014)

164. S.C. Lin, C.Y. Kao, H.J. Lee, C.J. Creighton, M.M. Ittmann, S.J. Tsai, et al., Dysregulation of miRNAs-COUP-TFII-FOXM1-CENPF axis contributes to the metastasis of prostate cancer. Nat. Commun. **7**, 11418 (2016)

165. S. You, B.S. Knudsen, N. Erho, M. Alshalalfa, M. Takhar, H. Al-Deen Ashab, et al., Integrated classification of prostate cancer reveals a novel luminal subtype with poor outcome. Cancer Res. **76**(17), 4948–4958 (2016)

166. K. Ketola, R.S.N. Munuganti, A. Davies, K.M. Nip, J.L. Bishop, A. Zoubeidi, Targeting prostate cancer subtype 1 by forkhead box M1 pathway inhibition. Clin. Cancer Res. **23**(22), 6923–6933 (2017)

167. T. Wang, L. Zheng, Q. Wang, Y.W. Hu, Emerging roles and mechanisms of FOXC2 in cancer. Clin. Chim. Acta **479**, 84–93 (2018)

168. A.N. Paranjape, R. Soundararajan, S.J. Werden, R. Joseph, J.H. Taube, H. Liu, et al., Inhibition of FOXC2 restores epithelial phenotype and drug sensitivity in prostate cancer cells with stem-cell properties. Oncogene **35**(46), 5963–5976 (2016). Findings described in this manuscript. The other authors declare no conflict of interest

169. L. Stumm, L. Burkhardt, S. Steurer, R. Simon, M. Adam, A. Becker, et al., Strong expression of the neuronal transcription factor FOXP2 is linked to an increased risk of early PSA recurrence in ERG fusion-negative cancers. J. Clin. Pathol. **66**(7), 563–568 (2013)

170. X.L. Song, Y. Tang, X.H. Lei, S.C. Zhao, Z.Q. Wu, miR-618 inhibits prostate cancer migration and invasion by targeting FOXP2. J. Cancer **8**(13), 2501–2510 (2017)

171. A.H. Banham, J. Boddy, R. Launchbury, C. Han, H. Turley, P.R. Malone, et al., Expression of the forkhead transcription factor FOXP1 is associated both with hypoxia inducible factors (HIFs) and the androgen receptor in prostate cancer but is not directly regulated by androgens or hypoxia. Prostate **67**(10), 1091–1098 (2007)

172. K. Takayama, K. Horie-Inoue, K. Ikeda, T. Urano, K. Murakami, Y. Hayashizaki, et al., FOXP1 is an androgen-responsive transcription factor that negatively regulates androgen receptor signaling in prostate cancer cells. Biochem. Biophys. Res. Commun. **374**(2), 388–393 (2008)

173. K. Takayama, T. Suzuki, S. Tsutsumi, T. Fujimura, S. Takahashi, Y. Homma, et al., Integrative analysis of FOXP1 function reveals a tumor-suppressive effect in prostate cancer. Mol. Endocrinol. **28**(12), 2012–2024 (2014)

174. J.A. Rodriguez-Martinez, A.W. Reinke, D. Bhimsaria, A.E. Keating, A.Z. Ansari, Combinatorial bZIP dimers display complex DNA-binding specificity landscapes. Elife **6**, e19272 (2017). Epub 2017/02/12

175. M.E. Cox, P.D. Deeble, S. Lakhani, S.J. Parsons, Acquisition of neuroendocrine characteristics by prostate tumor cells is reversible: implications for prostate cancer progression. Cancer Res. **59**(15), 3821–3830 (1999)

176. P. Sassone-Corsi, The cyclic AMP pathway. Cold Spring Harb. Perspect. Biol. **4**(12), a011148 (2012). Epub 2012/12/05

177. C.A. Saura, J.R. Cardinaux, Emerging roles of CREB-regulated transcription coactivators in brain physiology and pathology. Trends Neurosci. **40**(12), 720–733 (2017). Epub 2017/11/04

178. M. Sang, M. Hulsurkar, X. Zhang, H. Song, D. Zheng, Y. Zhang, et al., GRK3 is a direct target of CREB activation and regulates neuroendocrine differentiation of prostate cancer cells. Oncotarget **7**(29), 45171–45185 (2016). Epub 2016/05/19

179. Y. Zhang, D. Zheng, T. Zhou, H. Song, M. Hulsurkar, N. Su, et al., Androgen deprivation promotes neuroendocrine differentiation and angiogenesis through CREB-EZH2-TSP1 pathway in prostate cancers. Nat. Commun. **9**(1), 4080 (2018). Epub 2018/10/06

180. A.H. Ahmad, Z. Ismail, c-fos and its consequences in pain. Malays. J. Med. Sci. **9**(1), 3–8 (2002). Epub 2002/01/01

181. D.A. Leach, V. Panagopoulos, C. Nash, C. Bevan, A.A. Thomson, L.A. Selth, et al., Cell-lineage specificity and role of AP-1 in the prostate fibroblast androgen receptor cistrome. Mol. Cell Endocrinol. **439**, 261–272 (2017). Epub 2016/09/17

182. D.R. Church, E. Lee, T.A. Thompson, H.S. Basu, M.O. Ripple, E.A. Ariazi, et al., Induction of AP-1 activity by androgen activation of the androgen receptor in LNCaP human prostate carcinoma cells. Prostate **63**(2), 155–168 (2005). Epub 2004/10/16

183. N. Sato, M.D. Sadar, N. Bruchovsky, F. Saatcioglu, P.S. Rennie, S. Sato, et al., Androgenic induction of prostate-specific antigen gene is repressed by protein-protein interaction between the androgen receptor and AP-1/c-Jun in the human prostate cancer cell line LNCaP. J Biol Chem **272**(28), 17485–17494 (1997). Epub 1997/07/11

184. M. Korpal, E.S. Lee, G. Hu, Y. Kang, The miR-200 family inhibits epithelial-mesenchymal transition and cancer cell migration by direct targeting of E-cadherin transcriptional repressors ZEB1 and ZEB2. J. Biol. Chem. **283**(22), 14910–14914 (2008). Epub 2008/04/16

185. V. Malik, D. Zimmer, R. Jauch, Diversity among POU transcription factors in chromatin recognition and cell fate reprogramming. Cell. Mol. Life Sci. **75**(9), 1587–1612 (2018)

186. S.H. Lee, J.N. Jeyapalan, V. Appleby, D.A. Mohamed Noor, V. Sottile, P.J. Scotting, Dynamic methylation and expression of Oct4 in early neural stem cells. J. Anat. **217**(3), 203–213 (2010)

187. K.M. Bae, Z. Su, C. Frye, S. McClellan, R.W. Allan, J.T. Andrejewski, et al., Expression of pluripotent stem cell reprogramming factors by prostate tumor

initiating cells. J. Urol. **183**(5), 2045–2053 (2010). Epub 2010/03/23

188. D.E. Linn, X. Yang, F. Sun, Y. Xie, H. Chen, R. Jiang, et al., A role for OCT4 in tumor initiation of drug-resistant prostate cancer cells. Genes Cancer **1**(9), 908–916 (2010). Epub 2011/07/23

189. T. Kosaka, S. Mikami, S. Yoshimine, Y. Miyazaki, T. Daimon, E. Kikuchi, et al., The prognostic significance of OCT4 expression in patients with prostate cancer. Human Pathol. **51**, 1–8 (2016). Epub 2016/04/14

190. D. Obinata, K. Takayama, K. Fujiwara, T. Suzuki, S. Tsutsumi, N. Fukuda, et al., Targeting Oct1 genomic function inhibits androgen receptor signaling and castration-resistant prostate cancer growth. Oncogene **35**(49), 6350–6358 (2016). Epub 2016/06/09

191. J.K. Diss, D.J. Faulkes, M.M. Walker, A. Patel, C.S. Foster, V. Budhram-Mahadeo, et al., Brn-3a neuronal transcription factor functional expression in human prostate cancer. Prostate Cancer Prostatic Dis. **9**(1), 83–91 (2006). Epub 2005/11/09

192. A. Gheldof, P. Hulpiau, F. van Roy, B. De Craene, G. Berx, Evolutionary functional analysis and molecular regulation of the ZEB transcription factors. Cell. Mol. Life Sci. **69**(15), 2527–2541 (2012)

193. E. Sanchez-Tillo, L. Siles, O. de Barrios, M. Cuatrecasas, E.C. Vaquero, A. Castells, et al., Expanding roles of ZEB factors in tumorigenesis and tumor progression. Am. J. Cancer Res. **1**(7), 897–912 (2011). Epub 2011/10/22

194. O. Leshem, S. Madar, I. Kogan-Sakin, I. Kamer, I. Goldstein, R. Brosh, et al., TMPRSS2/ERG promotes epithelial to mesenchymal transition through the ZEB1/ZEB2 axis in a prostate cancer model. PLoS One **6**(7), e21650 (2011). Epub 2011/07/13

195. D. Ren, M. Wang, W. Guo, S. Huang, Z. Wang, X. Zhao, et al., Double-negative feedback loop between ZEB2 and miR-145 regulates epithelial-mesenchymal transition and stem cell properties in prostate cancer cells. Cell Tissue Res. **358**(3), 763–778 (2014). Epub 2014/10/10

196. G.C. Chu, H.E. Zhau, R. Wang, A. Rogatko, X. Feng, M. Zayzafoon, et al., RANK- and c-Met-mediated signal network promotes prostate cancer metastatic colonization. Endocr. Relat. Cancer **21**(2), 311–326 (2014). Epub 2014/01/31

197. L.A. Selth, R. Das, S.L. Townley, I. Coutinho, A.R. Hanson, M.M. Centenera, et al., A ZEB1-miR-375-YAP1 pathway regulates epithelial plasticity in prostate cancer. Oncogene **36**(1), 24–34 (2017). Epub 2016/06/09

198. T.R. Graham, H.E. Zhau, V.A. Odero-Marah, A.O. Osunkoya, K.S. Kimbro, M. Tighiouart, et al., Insulin-like growth factor-I-dependent up-regulation of ZEB1 drives epithelial-to-mesenchymal transition in human prostate cancer cells. Cancer Res. **68**(7), 2479–2488 (2008). Epub 2008/04/03

199. K. Hanrahan, A. O'Neill, M. Prencipe, J. Bugler, L. Murphy, A. Fabre, et al., The role of epithelial-mesenchymal transition drivers ZEB1 and ZEB2 in mediating docetaxel-resistant prostate cancer. Mol. Oncol. **11**(3), 251–265 (2017). Epub 2017/01/31

200. M. Marin-Aguilera, J. Codony-Servat, O. Reig, J.J. Lozano, P.L. Fernandez, M.V. Pereira, et al., Epithelial-to-mesenchymal transition mediates docetaxel resistance and high risk of relapse in prostate cancer. Mol. Cancer Ther. **13**(5), 1270–1284 (2014). Epub 2014/03/25

201. T.R. Graham, R. Yacoub, L. Taliaferro-Smith, A.O. Osunkoya, V.A. Odero-Marah, T. Liu, et al., Reciprocal regulation of ZEB1 and AR in triple negative breast cancer cells. Breast Cancer Res. Treat. **123**(1), 139–147 (2010). Epub 2009/11/19

202. B.M. Anose, M.M. Sanders, Androgen receptor regulates transcription of the ZEB1 transcription factor. Int. J. Endocrinol. **2011**, 903918 (2011). Epub 2011/12/23

203. S. Jacob, S. Nayak, G. Fernandes, R.S. Barai, S. Menon, U.K. Chaudhari, et al., Androgen receptor as a regulator of ZEB2 expression and its implications in epithelial-to-mesenchymal transition in prostate cancer. Endocr. Relat. Cancer **21**(3), 473–486 (2014). Epub 2014/05/09

204. V.S. Viswanathan, M.J. Ryan, H.D. Dhruv, S. Gill, O.M. Eichhoff, B. Seashore-Ludlow, et al., Dependency of a therapy-resistant state of cancer cells on a lipid peroxidase pathway. Nature **547**(7664), 453–457 (2017)

205. M. Haider, X. Zhang, I. Coleman, N. Ericson, L.D. True, H.M. Lam, et al., Epithelial mesenchymal-like transition occurs in a subset of cells in castration resistant prostate cancer bone metastases. Clin. Exp. Metastasis **33**(3), 239–248 (2016)

206. K.G. Sweeney, N. Stylianou, G. Tevz, A. Taherianfard, K. Pirlo, A. Upadhyaya, et al., Abstract 4909: Androgen targeted therapy induces ZEB1 expression and is associated with suppression of androgen signalling and therapy resistance. Cancer Res. **77**(13 Suppl), 4909 (2017)

207. N. Chi, J.A. Epstein, Getting your Pax straight: Pax proteins in development and disease. Trends Genet. **18**(1), 41–47 (2002). Epub 2001/12/26

208. R. Madelaine, P. Mourrain, Endogenous retinal neural stem cell reprogramming for neuronal regeneration. Neural Regen. Res. **12**(11), 1765–1767 (2017). Epub 2017/12/15

209. D.X. Yu, F.P. Di Giorgio, J. Yao, M.C. Marchetto, K. Brennand, R. Wright, et al., Modeling hippocampal neurogenesis using human pluripotent stem cells. Stem Cell Rep. **2**(3), 295–310 (2014). Epub 2014/03/29

210. M. Rotinen, S. You, J. Yang, S.G. Coetzee, M. Reis-Sobreiro, W.C. Huang, et al., ONECUT2 is a targetable master regulator of lethal prostate cancer that suppresses the androgen axis. Nat. Med. **24**(12), 1887–1898 (2018). Epub 2018/11/28

211. H. Guo, X. Ci, M. Ahmed, J.T. Hua, F. Soares, D. Lin, et al., ONECUT2 is a driver of neuroendocrine prostate cancer. Nat. Commun. **10**(1), 278 (2019). Epub 2019/01/19

212. B. Wang, J. Huang, J. Zhou, K. Hui, S. Xu, J. Fan, et al., DAB2IP regulates EMT and metastasis of prostate cancer through targeting PROX1 transcription and destabilizing HIF1alpha protein. Cell. Signal. **28**(11), 1623–1630 (2016)

213. C.R. Shyr, M.Y. Tsai, S. Yeh, H.Y. Kang, Y.C. Chang, P.L. Wong, et al., Tumor suppressor PAX6 functions as androgen receptor co-repressor to inhibit prostate cancer growth. Prostate **70**(2), 190–199 (2010)

214. J. Elvenes, E.I. Thomassen, S.S. Johnsen, K. Kaino, E. Sjottem, T. Johansen, Pax6 represses androgen receptor-mediated transactivation by inhibiting recruitment of the coactivator SPBP. PLoS One **6**(9), e24659 (2011)

215. C.V. Dang, MYC on the path to cancer. Cell **149**(1), 22–35 (2012)

216. N.V. Varlakhanova, R.F. Cotterman, W.N. de Vries, J. Morgan, L.R. Donahue, S. Murray, et al., myc maintains embryonic stem cell pluripotency and self-renewal. Differentiation **80**(1), 9–19 (2010)

217. T.S. Cliff, T. Wu, B.R. Boward, A. Yin, H. Yin, J.N. Glushka, et al., MYC controls human pluripotent stem cell fate decisions through regulation of metabolic flux. Cell Stem Cell **21**(4), 502–16.e9 (2017)

218. L. Kerosuo, K. Piltti, H. Fox, A. Angers-Loustau, V. Hayry, M. Eilers, et al., Myc increases self-renewal in neural progenitor cells through Miz-1. J. Cell Sci. **121**(Pt 23), 3941–3950 (2008)

219. J.T. Zhang, Z.H. Weng, K.S. Tsang, L.L. Tsang, H.C. Chan, X.H. Jiang, MycN is critical for the maintenance of human embryonic stem cell-derived neural crest stem cells. PLoS One **11**(1), e0148062 (2016)

220. P.S. Knoepfler, P.F. Cheng, R.N. Eisenman, N-myc is essential during neurogenesis for the rapid expansion of progenitor cell populations and the inhibition of neuronal differentiation. Genes Dev. **16**(20), 2699–2712 (2002)

221. C. Sun, A. Dobi, A. Mohamed, H. Li, R.L. Thangapazham, B. Furusato, et al., TMPRSS2-ERG fusion, a common genomic alteration in prostate cancer activates C-MYC and abrogates prostate epithelial differentiation. Oncogene **27**(40), 5348–5353 (2008)

222. D.J. Vander Griend, I.V. Litvinov, J.T. Isaacs, Conversion of androgen receptor signaling from a growth suppressor in normal prostate epithelial cells to an oncogene in prostate cancer cells involves a gain of function in c-Myc regulation. Int. J. Biol. Sci. **10**(6), 627–642 (2014)

223. C. Grandori, N. Gomez-Roman, Z.A. Felton-Edkins, C. Ngouenet, D.A. Galloway, R.N. Eisenman, et al., c-Myc binds to human ribosomal DNA and stimulates transcription of rRNA genes by RNA polymerase I. Nat. Cell Biol. **7**(3), 311–318 (2005)

224. G. Fromont, J. Godet, A. Peyret, J. Irani, O. Celhay, F. Rozet, et al., 8q24 amplification is associated with Myc expression and prostate cancer progression and is an independent predictor of recurrence after radical prostatectomy. Hum. Pathol. **44**(8), 1617–1623 (2013)

225. K.B. Cho, M.K. Cho, W.Y. Lee, K.W. Kang, Overexpression of c-myc induces epithelial mesenchymal transition in mammary epithelial cells. Cancer Lett. **293**(2), 230–239 (2010)

226. M.D. Amatangelo, S. Goodyear, D. Varma, M.E. Stearns, c-Myc expression and MEK1-induced Erk2 nuclear localization are required for TGF-beta induced epithelial-mesenchymal transition and invasion in prostate cancer. Carcinogenesis **33**(10), 1965–1975 (2012)

227. C.M. Koh, C.J. Bieberich, C.V. Dang, W.G. Nelson, S. Yegnasubramanian, A.M. De Marzo, MYC and prostate cancer. Genes Cancer **1**(6), 617–628 (2010)

228. P.C. Boutros, M. Fraser, N.J. Harding, R. de Borja, D. Trudel, E. Lalonde, et al., Spatial genomic heterogeneity within localized, multifocal prostate cancer. Nat. Genet. **47**(7), 736–745 (2015)

229. J.K. Lee, J.W. Phillips, B.A. Smith, J.W. Park, T. Stoyanova, E.F. McCaffrey, et al., N-Myc drives neuroendocrine prostate cancer initiated from human prostate epithelial cells. Cancer Cell **29**(4), 536–547 (2016)

230. E. Dardenne, H. Beltran, M. Benelli, K. Gayvert, A. Berger, L. Puca, et al., N-Myc induces an EZH2-mediated transcriptional program driving neuroendocrine prostate cancer. Cancer Cell **30**(4), 563–577 (2016)

231. H. Beltran, C. Oromendia, D.C. Danila, B. Montgomery, C. Hoimes, R.Z. Szmulewitz, et al., A phase II trial of the aurora kinase A inhibitor alisertib for patients with castration-resistant and neuroendocrine prostate cancer: efficacy and biomarkers. Clin. Cancer Res. **25**(1), 43–51 (2019)

232. W. Zhang, B. Liu, W. Wu, L. Li, B.M. Broom, S.P. Basourakos, et al., Targeting the MYCN-PARP-DNA damage response pathway in neuroendocrine prostate cancer. Clin. Cancer Res. **24**(3), 696–707 (2018)

233. Z.F. Chen, A.J. Paquette, D.J. Anderson, NRSF/REST is required in vivo for repression of multiple neuronal target genes during embryogenesis. Nat. Genet. **20**(2), 136–142 (1998). Epub 1998/10/15

234. S.E. McGrath, A. Michael, H. Pandha, R. Morgan, Engrailed homeobox transcription factors as potential markers and targets in cancer. FEBS Lett. **587**(6), 549–554 (2013)

235. C.J. Schoenherr, D.J. Anderson, The neuron-restrictive silencer factor (NRSF): a coordinate repressor of multiple neuron-specific genes. Science **267**(5202), 1360–1363 (1995). Epub 1995/03/03

236. J.M. Coulson, Transcriptional regulation: cancer, neurons and the REST. Curr. Biol. **15**(17), R665–R668 (2005). Epub 2005/09/06

237. M.D. Binder, N. Hirokawa, U. Windhorst, M.C. Hirsch, *Encyclopedia of neuroscience* (Springer, Berlin, 2009)

238. A.V. Lapuk, C. Wu, A.W. Wyatt, A. McPherson, B.J. McConeghy, S. Brahmbhatt, et al., From sequence to molecular pathology, and a mechanism driving the neuroendocrine phenotype in prostate cancer. J. Pathol. **227**(3), 286–297 (2012). Epub 2012/05/04

239. C. Svensson, J. Ceder, D. Iglesias-Gato, Y.C. Chuan, S.T. Pang, A. Bjartell, et al., REST mediates androgen receptor actions on gene repression and predicts early recurrence of prostate cancer. Nucleic Acids Res. **42**(2), 999–1015 (2014). Epub 2013/10/29

240. Y. Li, N. Donmez, C. Sahinalp, N. Xie, Y. Wang, H. Xue, et al., SRRM4 drives neuroendocrine transdifferentiation of prostate adenocarcinoma under androgen receptor pathway inhibition. Eur. Urol. **71**(1), 68–78 (2017). Epub 2016/05/18

241. X. Zhang, I.M. Coleman, L.G. Brown, L.D. True, L. Kollath, J.M. Lucas, et al., SRRM4 expression and the loss of REST activity may promote the emergence of the neuroendocrine phenotype in castration-resistant prostate cancer. Clin. Cancer Res. **21**(20), 4698–4708 (2015)

242. A.R. Lee, N. Che, J.M. Lovnicki, X. Dong, Development of neuroendocrine prostate cancers by the Ser/Arg repetitive matrix 4-mediated RNA splicing network. Front. Oncol. **8**, 93 (2018). Epub 2018/04/19

243. H. Sun, S. Ghaffari, R. Taneja, bHLH-orange transcription factors in development and cancer. Transl. Oncogenomics **2**, 107–120 (2007). Epub 2007/01/01

244. R. Kageyama, T. Ohtsuka, The Notch-Hes pathway in mammalian neural development. Cell Res. **9**(3), 179–188 (1999). Epub 1999/10/16

245. M. Sakamoto, H. Hirata, T. Ohtsuka, Y. Bessho, R. Kageyama, The basic helix-loop-helix genes Hesr1/Hey1 and Hesr2/Hey2 regulate maintenance of neural precursor cells in the brain. J. Biol. Chem. **278**(45), 44808–44815 (2003). Epub 2003/08/30

246. I. Rapa, P. Ceppi, E. Bollito, R. Rosas, S. Cappia, E. Bacillo, et al., Human ASH1 expression in prostate cancer with neuroendocrine differentiation. Mod. Pathol. **21**(6), 700–707 (2008). Epub 2008/03/04

247. B. Belandia, S.M. Powell, J.M. Garcia-Pedrero, M.M. Walker, C.L. Bevan, M.G. Parker, Hey1, a mediator of notch signaling, is an androgen receptor corepressor. Mol. Cell. Biol. **25**(4), 1425–1436 (2005). Epub 2005/02/03

248. D.N. Lavery, M.A. Villaronga, M.M. Walker, A. Patel, B. Belandia, C.L. Bevan, Repression of androgen receptor activity by HEYL, a third member of the Hairy/Enhancer-of-split-related family of Notch effectors. J. Biol. Chem. **286**(20), 17796–17808 (2011). Epub 2011/04/02

249. R. Limame, K. Op de Beeck, F. Lardon, O. De Wever, P. Pauwels, Kruppel-like factors in cancer progression: three fingers on the steering wheel. Oncotarget **5**(1), 29–48 (2014). Epub 2014/01/17

250. M.P. Tetreault, Y. Yang, J.P. Katz, Kruppel-like factors in cancer. Nat. Rev. Cancer **13**(10), 701–713 (2013). Epub 2013/09/26

251. B.B. McConnell, V.W. Yang, Mammalian Kruppel-like factors in health and diseases. Physiol. Rev. **90**(4), 1337–1381 (2010). Epub 2010/10/21

252. D.L. Moore, M.G. Blackmore, Y. Hu, K.H. Kaestner, J.L. Bixby, V.P. Lemmon, et al., KLF family members regulate intrinsic axon regeneration ability. Science **326**(5950), 298–301 (2009). Epub 2009/10/10

253. Y.N. Liu, W. Abou-Kheir, J.J. Yin, L. Fang, P. Hynes, O. Casey, et al., Critical and reciprocal regulation of KLF4 and SLUG in transforming growth factor beta-initiated prostate cancer epithelial-mesenchymal transition. Mol. Cell. Biol. **32**(5), 941–953 (2012). Epub 2011/12/29

254. D. Li, S. Yea, G. Dolios, J.A. Martignetti, G. Narla, R. Wang, et al., Regulation of Kruppel-like factor 6 tumor suppressor activity by acetylation. Cancer Res. **65**(20), 9216–9225 (2005). Epub 2005/10/19

255. C. Chen, E.R. Hyytinen, X. Sun, H.J. Helin, P.A. Koivisto, H.F. Frierson Jr., et al., Deletion, mutation, and loss of expression of KLF6 in human prostate cancer. Am. J. Pathol. **162**(4), 1349–1354 (2003). Epub 2003/03/26

256. G. Narla, A. DiFeo, Y. Fernandez, S. Dhanasekaran, F. Huang, J. Sangodkar, et al., KLF6-SV1 overexpression accelerates human and mouse prostate cancer progression and metastasis. J. Clin. Invest. **118**(8), 2711–2721 (2008). Epub 2008/07/04

257. P. Shen, J. Sun, G. Xu, L. Zhang, Z. Yang, S. Xia, et al., KLF9, a transcription factor induced in flutamide-caused cell apoptosis, inhibits AKT activation and suppresses tumor growth of prostate cancer cells. Prostate **74**(9), 946–958 (2014). Epub 2014/04/17

258. Q. Wang, R. Peng, B. Wang, J. Wang, W. Yu, Y. Liu, et al., Transcription factor KLF13 inhibits AKT activation and suppresses the growth of prostate carcinoma cells. Cancer Biomark. **22**(3), 533–541 (2018). Epub 2018/05/31

259. D.S. Rickman, H. Beltran, F. Demichelis, M.A. Rubin, Biology and evolution of poorly differentiated neuroendocrine tumors. Nat. Med. **23**(6), 1–10 (2017)

260. D. Robinson, E.M. Van Allen, Y.M. Wu, N. Schultz, R.J. Lonigro, J.M. Mosquera, et al., Integrative clinical genomics of advanced prostate cancer. Cell **161**(5), 1215–1228 (2015)

261. K.T. Bieging, S.S. Mello, L.D. Attardi, Unravelling mechanisms of p53-mediated tumour suppression. Nat. Rev. Cancer **14**(5), 359–370 (2014)

262. M.V. Glazova, The role of p53 protein in the regulation of neuronal differentiation. Neurosci. Behav. Physiol. **46**(9), 984–991 (2016)

263. T. Maimets, I. Neganova, L. Armstrong, M. Lako, Activation of p53 by nutlin leads to rapid differentiation of human embryonic stem cells. Oncogene **27**(40), 5277–5287 (2008)

264. K. Gupta, S. Gupta, Neuroendocrine differentiation in prostate cancer: key epigenetic players. Transl. Cancer Res. **6**(Suppl 1), S104–S1S8 (2017). Epub 2017/02/01

265. K. Ruggero, S. Farran-Matas, A. Martinez-Tebar, A. Aytes, Epigenetic regulation in prostate cancer progression. Curr. Mol. Biol. Rep. **4**(2), 101–115 (2018). Epub 2018/06/12

266. S. Terry, H. Beltran, The many faces of neuroendocrine differentiation in prostate cancer progression. Front. Oncol. **4**, 60 (2014). Epub 2014/04/12

267. I. Ota, T. Masui, M. Kurihara, J.I. Yook, S. Mikami, T. Kimura, et al., Snail-induced EMT promotes cancer stem cell-like properties in head and neck cancer cells. Oncol. Rep. **35**(1), 261–266 (2016)

268. M.E. Fane, Y. Chhabra, D.E.J. Hollingsworth, J.L. Simmons, L. Spoerri, T.G. Oh, et al., NFIB mediates BRN2 driven melanoma cell migration and invasion through regulation of EZH2 and MITF. EBioMedicine **16**, 63–75 (2017)

269. H. Zeng, A. Jorapur, A.H. Shain, U.E. Lang, R. Torres, Y. Zhang, et al., Bi-allelic loss of CDKN2A initiates melanoma invasion via BRN2 activation. Cancer Cell **34**(1), 56–68.e9 (2018)

270. M.L. Suva, E. Rheinbay, S.M. Gillespie, A.P. Patel, H. Wakimoto, S.D. Rabkin, et al., Reconstructing and reprogramming the tumor-propagating potential of glioblastoma stem-like cells. Cell **157**(3), 580–594 (2014)

271. D.H. Naoko Kobayashi, C. Wang, J. Yamashiro, J. Guan, R.E. Reiter, The role of POU3F2 in the aggressive behavior of neuroendocrine prostate cancer. Cancer Res. **78**(13), 5901 (2018)

272. B.A. Smith, A. Sokolov, V. Uzunangelov, R. Baertsch, Y. Newton, K. Graim, et al., A basal stem cell signature identifies aggressive prostate cancer phenotypes. Proc. Natl. Acad. Sci. U. S. A. **112**(47), E6544–E6552 (2015)

Index

A

Abiraterone, 151
  and AR inhibitors, 265
  clinical effect, 258
  DHEA-S, 259, 261
  DHT, 241
  dose-escalation, 265
  effectiveness, 262
  efficacy, 258
  and enzalutamide, 239, 259, 261, 262
  mechanisms, 258
  metabolism, 259–260
  neoadjuvant trial, 259
  precision medicine, 263–264
  pregnenolone derivative, 258
  progression, 261
  resistant LAPC4 cells, 258
  tumor progression, 259
Acetyl coenzyme A (acetyl-CoA), 394
Acinar adenocarcinoma, 75–77, 79
ACLY activation, 193
Activating enhancer binding protein 2 alpha (AP-2 alpha), 187
Activator protein 1 (AP1), 447
Acute lymphoblastic leukemia (ALL), 358, 385
Acute myeloid leukemia (AML), 385, 417, 422
Acylation, 193
Adenocarcinoma, 302
Adenomatous polyposis coli (APC), 357
Adipocytes, 206
Adiposity, 17
Adrenal androgens, 239–242, 246, 248–250, 252, 254, 256–259, 261, 263, 266
Adrenal gland, 242, 244
Adrenal steroid, 239
Adrenalectomy, 241
Adrenocorticotropin hormone (ACTH), 335
Aggressive PCa, 151
Aggressive prostate cancer, 79–80
AKR1C3, 262, 263
AKT-phosphorylated proteins, 320
AKT signaling
  AR, 326

isoforms, 319
MAPK/ERK, 326, 327
phosphorylation, 319
Wnt/β-catenin signaling, 326
Alcohol consumption, 34, 35
Aldo-keto reductase 1C1 (AKR1C1), 245
Alpha-linolenic acid (ALA), 6
Alpha-tocopherol, 12
Alpha-Tocopherol, Beta-Carotene Cancer Prevention (ATBC), 3, 13
Alternative non-homologous end joining (alt-NHEJ) pathway, 280
Amino acid metabolism, 194–196, 282
Aminoimidazoarenes (AIAs), 37
Amplicon-based sequencing, 293
Amplification, 94, 96, 99, 101, 103
Amplified in breast 1(AIB1), 322
Anaerobic metabolism, 190
Androgen biosynthesis
  adrenal gland, 242, 244
  backdoor pathway, 245, 246
  candidate enzymes, 246
  classical pathway, 245, 246
  de novo, 242
  5α-dione pathway, 246
  peripheral tissues, 242, 244
  prostate and pre-receptor control of DHT metabolism, 244, 245
Androgen deprivation therapy (ADT), 88, 91, 101, 151, 210, 239, 240, 247, 263, 282–285, 290, 291, 312, 339, 351
  antiproliferative effects, 305
  mechanisms, 305
  metastatic disease, 305
  prostate biopsies, 247
  resistance, 312
Androgen receptor (AR), 115, 178–180, 352, 415, 421, 422
  activation, 252
  AKT, 326
  AR-positive and negative tumors, 208
  chromatin, 336, 337
  coactivator, 248

© Springer Nature Switzerland AG 2019
S. M. Dehm, D. J. Tindall (eds.), *Prostate Cancer*, Advances in Experimental Medicine and Biology 1210, https://doi.org/10.1007/978-3-030-32656-2

Index

Androgen receptor (AR) (cont.)
- coactivators/corepressors, 241
- co-regulators, 337, 338
- crosstalk, 341, 342
- cross-talk and regulation, 251
- diverse metabolic pathways, 200
- DNA binding, 335, 336
- down-regulation, 304
- enzalutamide, 217, 239
- FASN, 193
- FOXO1, 323, 324
- gene and protein structure, 334
- inhibitors, 241
- in LNCaP cells, 253
- mechanisms, 241
- nuclear translocation, 335, 336
- ODC, 196
- PCa therapy, 240
- and RB
  - castration/deprivation, 305
  - Cdc6, 305
  - clinical trials, 305
  - cofactors, 306, 307
  - ligand-dependent activation, 305
  - signaling, 305
  - target genes, 305
  - transcriptional target of RB/E2F, 305, 306
- restoration, 241
- SCAP transcription, 192
- signaling axis, 336, 339
- signaling in CRPC, 254
- structure and function, 333–335
- transcription factors, 337, 338
- upregulation, AR signaling, 192
Androgen-dependent LNCaP cells, 253
Androgen-responsive elements (AREs), 90, 334
Androgens, 352
- and steroidogenic enzymes, 240
Androstenediol, 335
Androstenedione, 335
Anemia, 290
Angiogenesis, 155
Annexin2 (ANXA2), 172
Anti-androgens, 345
Antigen presenting cells (APCs), 122–130, 135
Antiproliferative effect, 305
Apoptosis, 307
Apurinic-apyrimidinic (AP) sites, 280
AR signaling, 213
AR variant (AR-V), 340, 341
AR variant 7 (AR-V7), 249
Arachidonic acid (AA), 30, 31
Arginine methylation, 382
Arginines, 381
Argininosuccinate synthetase (ASS), 195
AR-mediated gene expression, 239
ATP-binding cassette (ABC) transporters, 360
ATP citrate lyase (ACLY), 193
Autophagy, 198, 199
Auto-phosphorylation, 281

B

Backdoor pathway, 245, 246, 251
Basal cell heterogeneity, 73
Basal epithelial cells, 245
Basal stem/progenitor cell theory, 70–71
Base excision repair (BER), 280
Benign prostatic hyperplasia (BPH), 240
BET inhibitors (BETi), 398
Beta oxidation, 194
Biallelic inactivation, 284
Binding function-3 (BF-3), 344
Biomarkers, 311, 312
Body mass index (BMI), 16
Bone marrow microenvironment
- cell-autonomous features, 171
- CTC to DTC, 171
- HSCs (see Hematopoietic stem cells (HSCs))
- implication, resident cells, 171
- malignant phenotypes, 171
Bone microenvironment, 252–254
Bone-marrow stromal cells, 253
BRCA1, 281, 284–286, 288–290
BRCA2, 101–102, 281–291
BRCAness, 289
Bromodomain and extra-terminal (BET), 322, 397

C

Calcium
- dietary, 9
- homeostasis, 9
- and prostate cancer risk, 9
- and vitamin D, 8
cAMP response element-binding protein (CREB), 447
Cancer associated fibroblasts (CAFs), 205
Cancer Genome Atlas Research Network, 413
Cancer immunity, 310, 311
Cancer Prevention Study II Nutrition Study (CPS II), 3
Cancer stem cells (CSCs), 363, 364, 437
Candidate enzymes, 245, 246
Capsule, 151
Capture genomic subtypes
- ERG, 112
- ETS, 112
- genomic data, 112
- PTEN loss signature, 114
- SPINK1, 112
- SPOP mutant signature, 114
Capture hybridization, 293
Carbonic anhydrase IX (CAIX), 205
Carcinogenesis
- HAAs, 39
Carcinoma-associated fibroblasts (CAFs), 365
Carnitine palmitoyltransferase 1 (CPT1), 194
Castrate-resistant prostate cancer (CRPC), 302, 413
- abiraterone, 343
- amplification, 340
- androgens, 352
- AR DBD, 345
- AR NTD, 344

ARCC-4, 343
AR-dependent mechanisms, 352, 353
AR-independent mechanisms, 353
AR-V, 340, 341
BET, 343
BF-3, 344
calphostin C (PKF115-584), 369
enzalutamide, 343
gene amplification, 339
IWP, 369, 370
IWR, 370
niclosamide, 344, 368
NSC668036, 369
PRI-724, 368
PROTAC, 343
quercetin, 369
SARDs, 343
small molecular inhibitors, 368
somatic mutations, 340
Tegavivint (BC2059), 369
VHL E3 ligase, 343
Castration resistant disease, 241
Castration resistant prostate cancer (CRPC), 284, 291, 438
characterization, 239
development, 239
low androgen environment, 239
progression, 240
residual prostate tumor androgens, 240–241
steroidogenesis (see Steroidogenesis)
Castration therapy, 241
Castration-resistant Nkx3.1-expressing (CARN) cells, 72
Castration-resistant PCa (CRPC), 151
Castration-resistant prostate cancer (CRPC), 57, 175, 178
Catalogue of Somatic Mutations in Cancer (COSMIC), 385
Catastrophic genomic events, 283
CBP/p300, 395, 422, 426
CD8+ T cells, 121
cell exhaustion
LCMV model, 128
stem-like model, 129–130
and DCs, 128
Cell cycle and RB
AR, 303
cellular proliferation, 304
components, 303
Cyclin D, 304
direct regulation, 303
E2F, 303, 304
genetic loss, 304
mouse models, 304
PRC2-dependent methylation, 303
prostate epithelial cells, 304
regulated genes, 304
SKP2, 303
testosterone, 305
Cell Cycle Progression (CCP), 117
Cell line models, 240
Cell plasticity, 437, 439–441, 451
Cells of origin
aggressive prostate cancer, 79–80

cell-of-mutation, 67
cellular entities, 67
epithelial cells, 68–69
prostate epithelial cells, 68
rodent and human prostates, 68
Cellular dormancy, 172
Cellular reprogramming, 309
Cellular stressors
genotoxic stressors, 58, 59
oxidative stress, 59
topological stress, 59–60
Checkpoint blockade, 121
Chemokine receptor 4 (CXCR4), 342
Chemotherapy, 312
Childhood obesity, 17
Chimeric antigen receptor T cells (CAR-T), 133, 134, 140
Chromatin, 336, 337, 379
Chromatin immunoprecipitation, 305
Chromatin immunoprecipitation assays, 353
Chromatin immunoprecipitation sequencing (ChIP-seq), 419
Chromatin modifiers, 309, 310
Chromatin remodeling, 400
Chromoplexy, 63, 64, 93, 99, 283
Chromothripsis, 63, 64, 93, 94, 96, 97
Chronic lymphocytic leukemia (CLL), 358
Cigarette smoking, 15
Circulating tumor cells (CTCs), 171, 290, 294
Classical pathway, 246, 251
androgen biosynthesis, 245
Clinico-genomics
heterogeneity, 104–105
molecular biomarkers, 103–104
prognostic biomarkers
CNAs and genome altered, 102–103
multi-modal profiling, 103
Clonal evolution, 283
Clonality, 285
c-Myc models, 137–138
Cockayne syndrome factors B (CSB), 280
Cohort studies, 1–3
Colonization, 172, 175, 180
Combined androgen blockade (CAB), 241
Computational modeling, 63
Copy number aberrations (CNAs), 91–93, 96, 97, 100, 103, 104
Copy number variants (CNVs), 282
Core-binding factor subunit alpha-1 (CBFA1), 325
CRPC-stage disease, 342, 343
Cryptic exon 3 (CE3), 340
ctDNA, 294
Cyclin/CDK complexes, 302
Cyclin-dependent kinase activity, 309
CYP11B1, 261, 262
CYP17A1, 241, 243–246, 248, 250–252, 254, 258–260, 262–265
Cytokeratins, 69, 70, 76
Cytokines, 251
Cytosine methylation, 380
Cytotoxic T lymphocyte antigen-4 (CTLA-4), 128, 135, 136, 139

D

Dairy products, 8, 9, 32
Danger-associated molecular patterns (DAMPs), 124
D-bifunctional protein (DBP), 194, 247
dCTP pyrophosphatase 1 (DCTPP1), 199
De novo androgen biosynthesis, 251
De novo NE tumors, 76
De novo steroidogenesis, 248
    CRPC, 250, 251
Decipher, 116
Dehydroepiandrosterone (DHEA), 239
Dehydroepiandrosterone sulfate (DHEA-S), 239
Dendritic cells (DCs), 122
    and CD8+ T cells, 123
    CD141+ DCs expressing CCR7, 125
    CD8+ T cell activation, 125
    cDC2, 125
    classification, 125
    cross-presenting (cDC1), 125
    infiltration in tumors, 125
    necrotic cells, 124
    pDCs, 125
    PLAMPs, 122
    pro-inflammatory cytokines, 122
    secondary and tertiary lymphoid organs, 122
    and T cell, 124
    TME, 126
Diagnostic imaging, 206
Diet as a risk factor, human PC
    alcohol, 34, 35
    dairy products, 32
    fatty acids, 30
    heme iron, 36
    high-calcium intake, 32
    high-fat diet, 29
    IGF-1, 33, 34
    inflammation, 30, 31
    NOCs, 36
    oxidative stress, 31
    PAHs, 36, 37
    red and processed meat, 35
    saturated fat intake, 32
    vitamin D, 32, 33
Dietary fat intake, 5, 6
Dietary phytoestrogens, 11
Dihydroepiandrosterone (DHEA), 335
5α-Dihydrotestosterone (DHT), 239, 333
    castrate tumors, 241
    in CRPC, 246
    in mouse testis, 246
    in PCa cells, 250
    inhibiting conversion, 246
    pre-receptor control, 240
    pre-receptor metabolism, 246
    prostate and pre-receptor control, 244, 245
    RIA, 240
Dihydroxyvitamin D [1,25(OH)$_2$D], 10
Disease-causing variants, 293
Disruptor of telomeric silencing 1 (DOT1), 393, 394

Disseminated tumor cells (DTCs), 171–173, 175, 176
DNA adducts, 39
DNA binding domain, 441, 449
DNA damage, 59, 60, 62–64
DNA damage repair (DDR)
    amplicon-based sequencing, 293
    APE1, 280
    ATM and ATR kinases, 281
    blood samples, 294
    BRCA1, 281
    capture hybridization, 293
    cellular processes, 280
    in clinical outcome, 287–289
    ctDNA, 294
    deficient, 279
    deleterious effects, 279
    DSBs, 280, 281
    endogenous, 279
    exogenous, 279
    gene panels, 293
    genomic alterations, 281–282
    genomic damage, 280
    genomic landscape, 282–285
    germline mutations, 285–287
    HR pathway, 281
    immunotherapy, 292
    MSH and MLH family, 281
    NER, 280
    NGS technologies, 293
    NHEJ pathway, 281
    PARP inhibitors, 289–291
    PARP1, 280
    restore genome integrity, 280
    RPA, 281
    Sanger sequencing fueled genetic testing, 292
    signaling pathways, 279
    single-DNA strand, 280
    SNVs, 293
    SSBs, 280
    targeting, 291–292
    tumor genomic material, 294
    WES, 293
    WGS, 293
DNA damage response (DDR), 58, 97, 101
DNA double strand breaks (DSBs)
    cellular stressors
        genotoxic stress, 58, 59
        IR-induced ROS formation, 60
        oxidative stress, 59
        topological stress, 59, 60
    DNA sequence and epigenetic features, 60–61
    gene rearrangements, 58
    human cells, 61
DNA methylation, 282, 380, 381
DNA sequence, 60
DNA-PKs, 291
DNase I hypersensitivity sites (DHSs), 61
Dormancy, 172, 173, 180
Double Negative Prostate Cancer (DNPC), 439

Index
467

Double-strand breaks (DSBs), 280–282, 289, 291, 292
Ductal adenocarcinoma, 76
Dunning R3327 prostate carcinoma model, 249
Dynamic nuclear polarization (DNP), 207
Dynamic reciprocity, 151

E
E-26 transformation-specific (ETS), 324
E-cadherin, 155, 357
11-Oxygenated androgens
    activation by AKR1C3, 256, 257
    activation of wild-type AR, 256
    adrenal vein, 254
    AKR1C3, 255
    androgen pool in CRPC, 256
    CYP11B1, 261, 262
    HSD11B2, 261, 262
    HSD17B3, 255
    inactivation by UGT2B, 256, 257
    reference levels, 255
Embryonic epithelial tissues, 416
Emergent phenotypes, 438
Engulfed necrotic tissue, 199–200
Enzalutamide, 151, 239, 241, 259–262, 265
Enzalutamide (MDV-3100), 343
Enzyme-linked immunosorbent assay (ELISA), 212
Epidermal growth factor (EGF), 319
Epigenetic modifications/gene transcription regulation, 282
Epigenetic reprogramming, 356
Epigenetic role, RB, 310
Epigenetics
    chromatin, 379
    DNA methylation, 380, 381
    heterochromatin, 379
    mechanisms, 380
    methylation (*see* Histone methylation)
    nucleosomes, 379
    PTMs, 381
    TAD, 380
Epithelial and stromal cell protein, 240
Epithelial to mesenchymal plasticity (EMP), 438
Epithelial to mesenchymal transition (EMT), 155, 156, 362, 397, 424
ERG lifecycle, 428
ERG-centric molecular stratification, 414
ERGi-USU, 427
Error-prone mechanism, 281
Estrogen pathway-related genes, 253
Estrogen receptor (ER), 333, 422
Ethanol, 35
ETS family transcription factors
    cell migration, 411
    cell types, 410
    chromosomal rearrangements, 411
    DNA binding domain, 410
    in vitro DNA sequence, 409
    prostate tumors, 411

ETS fusions, 283
ETS gene fusions
    clinicopathological value, 414, 415
    distribution of, 412
    positive (*see* ETS-positive prostate cancer)
    TMPRSS2/ERG fusions, 415
ETS-positive prostate cancer
    demographics, 413
    molecular stratification, 413, 414
ETS-related gene (ERG), 324, 325
European Prospective Investigation into Cancer and Nutrition (EPIC), 2
Ewing's sarcoma, 417, 421
Exome DNA sequencing, 282
Extracellular signal-regulated kinase (ERK), 326
Extracellular vesicle (EV), 151
    cell-to-cell communication, 160
    cisplatin, 160
    derivative names, 159
    in vitro, 160
    MR imaging, 160
    MVs, 159
    natural role, 160
    routes and mechanisms, uptake, 160

F
Family history, 101
Fatty acid synthase (FASN), 193, 215
Fibroblast growth factor (FGF), 416
Fish intake, 7, 8
5α-androstanedione (5α-dione) pathway, 246
5α-reduced 11-oxygenated androgens, 256
$5'$ untranslated region ($5'$ UTR), 409
5-Androstenediol (A5-diol), 241
Flavin adenine dinucleotide (FAD), 382
Fluciclovine, 212
Fluorescence *in situ* hybridization (FISH), 339, 411
Forkhead box-O protein 1 (FOXO1)
    AKT, 323
    AR, 323, 324
    ERG, 324, 325
    gene deletion, 323
    mechanisms, 324
    RUNX2, 325
    tumor suppressor, 323
Forkhead domain transcription factors
    FOX family, 446
    FOXC2, 446
    FOXM1, 446
    FOXP1, 446
    HNF3A and HNF3B, 446
Functional annotation analysis, 360

G
Gamma-tocopherol, 12
Gene expression, 282
Gene fusions, 282

Gene panels, 293
Genome
   chromoplexy, 93
   CNAs, 93–96
   DDR, 97
   exome sequencing, 88
   PSA testing, 88
   SNVs, 91
   structural variation, 91
   systemic therapies, 88
Genome sequencing, 88
Genome-wide associated studies (GWAS), 101, 338
Genome-wide mapping techniques, 419
Genomic alterations
   in prostate cancer, 281–282
Genomic damage, 280
Genomic heterogeneity, 87, 97, 98, 104
Genomic landscape
   prostate cancer
      advanced, 283–285
      localized, 282–283
Genomic Prostate Score (GPS), 117
Genomic rearrangements (GR), 57, 89, 91, 92
   DNA DSBs, formation
      cellular stressors, 58–60
      DNA sequence and epigenetic features, 60–61
   TMPRSS2-ERG gene fusion, 57
   triggers, 58
Genomics
   catastrophic, 283
   characterization, 284
   DNA, 293
   instability, 279, 286, 290
   prostate cancer, 288
   rearrangements, 283
   stability, 292
   structural restructuration, 283
   tumor material, 294
Genomic stability, 310
Genotoxic stressors, 58, 59
Germline mutations
   DDR, 285–287
   MMR, 286, 287
Germline-somatic interaction, 101–102
Germline variation
   HSD3B1, 263, 264
   SLCO transport genes, 263
GGAA microsatellites, 421
Glandular escape, 152
Gleason grading, 240
Gleason score (GS), 149, 208
Global genome NER pathways (GG-NER), 280
Glucocorticoid receptor (GR), 250, 310, 333
Glucosamine-phosphate N-acetyltransferase 1
   (GNPNAT1), 197
Glucose metabolism
   AP-2 alpha, 187
   Bax-associated mitochondrial pore formation, 187
   FDG-PET scans, 188
   GLUTs, 188

   HKs, 188, 189
   LDH, 190
   MCTs, 190, 191
   OXPHOS, 188
   PDC, 189
   PK, 189, 190
   PPP, 191
   TCA cycle, 186
   ZIP expression, 187
Glucose transporters (GLUTs), 188
Glucose-6-phosphate isomerase (GPI), 204
Glucuronidating enzymes, 245
Glutaminolysis, 209
Glycine-N-methyl transferase (GNMT), 196
Glycogen synthase kinase 3 (GSK3), 357
Glycolysis, 185, 188–191, 197, 200–202, 204, 206, 207,
   213
Gonadotropin-releasing hormone (GnRH), 335, 339
G-protein coupled receptor family C group 6 member A
   (GPRC6A), 253, 254
Grainyhead-like 2 (GRHL2), 338
GRB2-Like Endophilin B2 (SH3GLB2) peptide, 135
Growth factors, 251
Growth-arrest specific factor (GAS6), 173

H
H3K27 methylation
   epigenetic therapeutic targeting, 391, 392
   KDM6 family erasers, 391
   KTM6A/B (EZH2/1) writers, 390, 391
H3K27 tri-methylation (H3K27me3) sites, 310
H3K36 methylation
   epigenetic therapeutic targeting, 392
   erasers, 393
   writers, 392, 393
H3K4 methylation
   cell cycle genes, 384
   DNA damage signaling, 384
   epigenetic therapeutic targeting, 388
   genome-wide study, 384
   KDM1 family erasers, 387
   KDM5 family erasers, 387, 388
   KMT2/MLL/COMPASS family methyltransferases,
      385
   KMT2A and KMT2B (MLL1 and MLL4) writers, 385
   KMT2C and KMT2D (MLL3 and MLL2) writers,
      385
   KMT2F and KMT2G (SETD1A and SETD1B)
      writers, 386
   LNCaP cells, 384
   PCa tumorigenesis and metastasis, 384
   SMYD family methylation writers, 386, 387
   tissue microarray analysis, 384
   whole exome sequencing studies, 384
H3K79 methylation
   DOT1L, 394
   writers, 393
H3K9 methylation
   epigenetic therapeutic targeting, 390, 391

Index

KDM3 erasers, 389
KDM4 erasers, 390
KMT1A/B (SUV39H1/2) writers, 388
KMT1C and KMT1D (G9a and GLP) writers, 389
HDAC inhibitors (HDACi), 396
Health Professionals Follow-up Study (HPFS), 2
Hedgehog (Hh) signaling pathway, 253
Helix-loop-Helix (bHLH) transcription factors
cells, 442
DNA binding/differentiation (ID), 443
HES and HEY, 443
HIF1, 443
NGN3, 442
pro-neuronal, 442
TWIST, 443
Hematopoietic stem cells (HSCs)
ANXA2, 172
CXCR4/CXCL12 signaling, 172
function, HSC niche, 172
homing, CXCL12 role, 172
M-CSF, 174
niche, 172
and prostate cancer cells, 172
transplantation, 172
Heme iron, 36
Heterochromatin protein 1 (HP1), 388
Heterocyclic amines (HCA), 7
Heterocyclic aromatic amines (HAAs), 35
AIAs, 37
carcinogenesis, 39
classification, 37
DNA adducts, 39
formation, 38
N-hydroxy-HAAs, 38
pyrolytic, 37
sources of exposure, 37
Hetero-domain transcription factors
homeodomains, 447
MYC, 448
MYCL, 448
MYCN, 449
NEPC, 449
OCT4, 447
OCT7, 447
POU, 447
ZEB, 448
ZEB1 expression, 448
Heterogeneity, 104, 105
Hexokinases (HKs), 188, 189
Hexosamine biosynthetic pathway (HBP), 197
High mobility domain (HMG) transcription factors
SOX sub-class, 445
SRY, 445
High-calcium intake, 32
High mobility group I (HMG-I), 385
Histidines, 382
Histone acetylation, 282
erasers
HDACs, 396

PSA, 394
readers
BET bromodomain proteins, 397, 398
epigenetic therapeutic targeting, 398–400
tissue microarray analysis, 395
writers
HATs, 395
Histone acetyltransferases (HATs), 381, 395
Histone deacetylases (HDACs), 380, 381, 395, 396
Histone demethylases (HDMs), 381
Histone lysine methyltransferases, 382
Histone methylation
acetylation (*see* Histone acetylation)
amino acid residues, 381
chromatin remodeling, 400
demethylases, 382
H3K27 (*see* H3K27 methylation)
H3K36 (*see* H3K36 methylation)
H3K4 (*see* H3K4 methylation)
H3K79 (*see* H3K79 methylation)
H3K9 (*see* H3K9 methylation)
lysine readers, 394
RNA splicing, 382
transcription factors, 382
transcriptional activation/silencing, 382
Histone methyltransferases (HMTs), 381
HMG-CoA synthase, 248
Homeodomain (HD) transcription factors
embryonic development and organogenesis, 444
HOX-class, 444
NKK, 444
Homologous recombination (HR), 61, 280–282, 284,
288, 290, 291
HSD17B2, 249
HSD3B1, 263, 264
Human Genome Project, 87
Human prostate stromal cells (PrSCs), 253
Human prostate tissue, 249–250
Human tumors, 240
Hypercholesterolemia, 250
Hyperphosphorylation, 302
Hyperpolarization, 207
Hyperpolarized ^{13}C Magnetic Resonance (HP-MR), 207,
210, 211
Hypophosphorylated RB, 302
Hypophysectomy, 241
Hypoxia, 204
angiogenesis, 155
CSH imaging method, 154
degradation, ECM, 156
EMT, 155–156
genes, 154
local tumor and systemic, 159
low oxygen tension, 154
markers, 154
and PCa invasion, 154
prostate carcinogenesis, 154
remodeling, ECM, 157
Hypoxia inducible factor 1 (HIF1), 442–443

I

IGF system, 33
IGF-binding proteins (IGFBPs), 34
Imaging
    biofluid and tissue metabolite biomarkers, 212–213
    clinical hyperpolarized metabolic imaging, 210
    clinical translation, hyperpolarized metabolic imaging, 211
    diagnostic imaging, 206
    drug development, 213–216
    hyperpolarized imaging, 207–208 (*see also* Hyperpolarized ^{13}C Magnetic Resonance (HP-MR))
    PET imaging, 211, 212
    preclinical hyperpolarized metabolic imaging, 208–210
Immature mouse, 246
Immune checkpoint inhibitors (ICI), 284
Immune response
    to cancer, 124
        antigen presentation, 125
        danger sensing, 124
        DCs, 125–126
        lymphoid organization, 126–127
        macrophages, 127–128
    to viruses
        antigen presentation, 122
        CD8$^+$ T cell effector function, 123–124
        danger sensing, 122
        lymphoid organization, 122, 123
Immune system, 122
Immunogenicity, 135
Immunoscore, 122
Immunostaining, 69–71, 73
Immunotherapy
    bi-specific antibodies, 140
    CAR-T cells, 140
    CD8$^+$ T cell exhaustion, 129
    checkpoint blockade, 121
    CTLA-4 blockade, 139
    DCs, 122
    future, 130
    MMR, 292
    PD-1/PD-L1 blockade, 139–140
    preclinical models, 131
    and NY-ESO-1 vaccine therapy, 135
    vaccines, 138–139
In vitro and *in vivo* human prostate cancer models, 309
In vivo MRS, 212
Indels, 282
Independent lineage theory, 72
Inflammation, 30, 31
Inhibition, neural transcription
    HES and HEY families, 450
    KLFs, 450
    REST, 449, 450
    TP53, 451
    tumor suppressor activity, 449
Inhibitor of Wnt production (IWP)
    LGK974 (WNT974), 370
    Wnt-C59, 370

Inhibitor of Wnt response (IWR)
    Axin, 370
    XAV939, 370
Insulin, 251, 319
Insulin/IGF signaling, prostate carcinogenesis, 34
Insulin-like growth factor 1 (IGF-1), 33, 34, 319
Integrins
    α7 integrin, 152
    description, 152
    LBIs, 152
Intermediate Atypical Carcinoma (IAC), 438
Intracellular fat, 192
Intracrine
    androgen biosynthesis, 239
    biosynthesis, 257
    CRPC, 256
    generation and activity, 257, 266
    utilization of adrenal androgens, 248
Intracrine steroidogenesis
    bone microenvironment, 252–254
    de novo, 250, 251
    in vitro and *in vivo* models, 249
    intratumoral androgen biosynthesis, 251, 252
    normal prostate and prostate cancer tissue, 249–250
    stromal cells, 252–254
Intracrinology, 244
Intraductal carcinoma of the prostate (IDC-P), 102
Intra-lineage heterogeneity, 73, 80
Intra-patient spatial heterogeneity, 283
Intra-tumor spatial heterogeneity, 283
Intratumoral androgens, 239, 241, 250–252
Intrinsically disordered regions (IDRs), 61
Ipilimumab, 139, 140
Isoflavonoids, 11

K

Knockout models, 254
Kruppel-like factor 5 (KLF5), 342, 450
Ku70/80 heterodimer, 281

L

Lactate dehydrogenase (LDH), 190, 209, 212
LADY model, 137
Laminin-binding integrin (LBIs), 152, 153
Laminin extracellular matrix proteins, 152
Lanosterol synthase, 248
Leucine Zipper (bZIP), 446, 447
Leucine zipper tumor suppressor-2 (LZTS2), 326
Leydig cells, 244, 245
LH releasing hormone (LHRH) agonist therapy, 251
Ligand binding, 241
Ligand binding domain (LBD), 249, 333
Ligand-dependent activation, 305
Lignans, 11
Lineage plasticity, 75, 309
Lineage tracing study, 68, 69, 71–74, 78
Lipid metabolism
    CPT1, 194

Index

fatty acid synthesis, 193
intracellular fat, 192
LXRs, 194
metabolic pathways, 191
mevalonate, 193
PI3K/AKT signaling, 193
SREBP, 192
TCA cycle, 193
Liquid chromatography/mass spectrometry (LC/MS), 213
Liver X receptors (LXR), 194
LNCaP cells, 251–253, 256
Localized PCa, 149, 154
Localized prostate cancer
ETS-family oncogene, 89
molecular interrogation, 88
sub-clonal reconstruction, 99–101
T2E, 90
Long-chain acyl-coenzyme A (CoA) synthetase 3 (ACSL3), 252
Long non-coding RNAs (lncRNAs), 391
Low AR-signaling, 284
Lox-STOP-lox strategy, 323
LRH-1, 252
Luminal cell heterogeneity, 74
Luminal epithelial cells, 245
Luminal stem/progenitor cell theory, 71–72
Luteinizing hormone (LH), 251
Lycopene intake, 11
Lymphocytic choriomeningitis virus (LCMV), 128
Lymphoid enhancer-binding factor 1 (LEF1), 358
Lynch syndrome (LS), 285–287
Lysines, 381

M
Macrophage colony-stimulating factor (M-CSF), 174
Macrophages, 127, 128
Macropinocytosis, 199
Magnetic resonance spectroscopy (MRS), 212
Malate dehydrogenase 2 (MDH2), 202
Mammalian target of rapamycin complex 2 (mTORC2), 319
Marine fatty acids, 7, 8
Mass spectrometry, 213
Matrix metalloprotease 9 (MMP9), 174
Matrix metalloproteases (MMP), 444
Maximum tolerated dose (MTD), 391–392
Meat intake, 6, 7
Messenger RNA (mRNA), 111
Metabolic reprogramming
amino acid metabolism, 194–197
glucose metabolism (see Glucose metabolism)
HBP (see Hexosamine biosynthetic pathway (HBP))
lipid metabolism, 191–194
metabolic scavenging, 198–200
regulation
miRNAs (see MicroRNAs (miRNAs))
signal transduction, 200–201
tumor microenvironment, 204–206

Metabolic scavenging, 198, 199
Metabolism, 209
Metastasis, 149, 152, 154–159
bone metastasis specimens, 179
experimental animal model, 175
mouse model, 172
osteoclasts role, 173
p38 expression, 173
randomized controlled trails, 176
Metastatic castration resistant prostate cancer (mCRPC), 115, 131–134, 136, 139, 140, 283, 285, 288–292, 379
Metastatic non-castrate prostate cancer, 285
Metformin, 213
Methyl CpG binding domain (MBD), 380
Mevalonate, 193
Microenvironments, 150
dynamic reciprocity, 151
laminin-binding integrins, 153
lethal PCa, 150
normal prostate gland, 150
PIN, 150
smooth muscle and neural invasion, 151
Micrometastasis, 172
MicroRNAs (miRNAs), 201, 202
Microsatellite instability (MSI), 287
Microvesicles (MVs), 159
Mineralocorticoid receptor (MR), 333
Mi-Prostate Score (MiPS), 415
miRNA sequencing, 282
Mismatch repair (MMR)
deficient tumors, 284
germline mutations, 286, 287
and HR, 284
immunotherapy, 292
mechanism, 280
pathogenic mutations/translocations, 284
UV radiation, 280
Mis-repair, DNA DBSs, 57, 58
chromothripsis and chromoplexy, 63–64
error-prone end-joining, 62
genomic rearrangements, 62
mechanisms, 61
NHEJ, 62
novel mediators, ETS gene rearrangements, 62–63
spatial proximity, 61
Mitogen-activated protein kinase (MAPK), 326
Mitogenic signaling, 302
Mixed lineage leukemia (MLL), 394
Modifiable risk factors
obesity, 1
smoking, 1, 15, 16
Molecular biomarkers, 104
Molecular imaging, 206
Molecular subtype, see Genomic subtypes
Monocarboxylate transporters (MCTs), 190, 191
Mouse embryonic fibroblasts (MEFs), 308
Mucin-1 (MUC1), 134
Multiclonality, 285
Multi-ethnic Cohort Study (MEC), 3

472   Index

Multifocality, 98, 285
Murine prostate cancer models
   c-Myc, 137–138
   LADY, 137
   PTEN knockout, 138
   TRAMP, 136–137
Muscle stroma, 151–153
Mutations, 281, 282

N

Neuroendocrine (NE) cells, 68, 69, 352–354
   heterogeneity, 74
   synaptophysin and neuropeptides, 68
Neuroendocrine prostate cancer (NEPC), 116, 284, 309,
     438–451
   cell of origin, 354, 355
   epigenetic reprogramming, 355, 356
Neurogenin (NGN), 442
Neuron-restrictive silencer factor (NRSF), 449
Next generation sequencing (NGS) technologies, 293
NH2-terminal domain (NTD), 324, 333
Niclosamide, 344
Nicotinamide N-methyltransferase (NNMT), 397
NIH-AARP Diet and Health Study, 2
Nivolumab, 139, 140
NK-like homeobox (NKX) transcription factors, 444
Nkx3.1-CreER^{T2} model, 72, 80
NMR spectroscopy, 212
N-nitroso compounds (NOCs), 36
Non-acinar carcinoma, 76
Noncanonical androgens
   in CRPC, 254
   steroid metabolizing enzymes, 264–265
Non-homologous end joining (NHEJ), 61, 62, 64,
     280–282, 291, 429
Non-small cell lung cancer (NSCLC), 355
Nuclear localization signal (NLS), 333
Nuclear magnetic resonance (NMR), 208, 212
Nuclear receptor (NR), 251
Nuclear receptor coactivator 3 (NCOA3), 322
Nuclear transcription factors, 251
Nucleotide excision repair (NER), 280
NY-ESO-1, 134, 135

O

Obesity, 16–18
   adiposity, 17
   BMI, 16
   childhood, 17
   weight change, 18
O-linked β-N-acetylglucosamine transferase (OGT), 197
Oncogenic ETS factors
   endogenous tissue, 416
   ERG subfamily, 416
   ETVs, 416
   PEA3 subfamily, 416
   prostate carcinogenesis and disease progression,
     416, 417

   recurrent ETS fusions, 417
   spermatogonial stem cells, 416
   targeting (*see* Targeting oncogenic ETS factors)
Oncogenic function
   AR, 421, 422
   chromatin, 423, 424
   DNA binding, 418, 419
   ETS/AP1, 420, 421
   gene regulation, 419, 420
   GGAA microsatellites, 421
   post-translational modifications, 424, 425
   signaling pathways, 424–426
   transcriptional activation and repression, 422, 423
Oncogenic fusion gene, 409
Oncotype Dx, 117
One-carbon metabolism, 196
Open-chromatin, 63
Orange domain, 450
Organoids, 72, 75, 79
Ornithine carbamoyl transferase (OCT), 195
Ornithine decarboxylase (ODC), 196
Osteoblast conditioned media, 253
Osteoblasts promote CRPC, 253
Osteoblast-secreted osteocalcin (OCN), 253, 254
Osteocalcin, 174
Osteoclasts
   cytokines, 173
   ineffectiveness, 175
   life span, 174
   matrix deposition, 174
   normal bone turnover, 174
   osteoclastogenesis, 174
   pre-osteoclasts, 174
   and prostate cancer cells, 175
   RANKL inhibitor, 175
   skeletal colonization, 175
   in skeletal metastases, 173, 176
   spatial relationship, with tumor cells, 177
   TRAP, 176
Osteolytic PC3, 253
Oxidative phosphorylation (OXPHOS), 188, 202, 215,
     216
Oxidative stress, 31, 32, 58–61, 204

P

PAC-320, 396
PAM50, 119
Paracrine cellular interaction, 251
Paracrine role, 253
Paracrine-stimulated steroidogenesis, 254
Parathyroid-hormone related peptide (PTHrP), 174
PARP1, 280, 291–292
Pathogen-associated molecular patterns (PAMPs), 122
Pattern recognition receptors (PRRs), 122
PD-1 blockade, 129, 130, 134
Pembrolizumab, 139, 140
Pentose phosphate pathway (PPP), 191
Pericentric heterochromatin, 388
Perineural invasion (PNI), 157, 158

Index

Peripheral nerves, 158
Peroxisome proliferator activated receptor gamma (PPAR-γ), 342
Phenylimidazo[4,5-b]pyridine (PhIP), 36, 38, 39
  animal toxicity, 39
  description, 37
  and dG-C8-PhIP formation, 42
  extensive inflammation, 39
  formation in tobacco smoke, 37
  human PC patients, 40–42
  MeIQx, 38
  metabolism, 39
  rodent models, 39, 40
Phosphatase and tensin homolog (PTEN), 138
Phosphatidylinositol-4,5-bisphosphate 3-kinase (PI3K)
  extracellular signals, 319
  genetic alterations, 321
  mTORC1, 319
  serine/threonine, 319
Phosphoinositide-dependent kinase 1 (PDK1), 319
Phosphorus, 9
Physical activity, 18, 19
Physicians' Health Study I and II (PHS), 2
Phytoestrogens, 11, 12, 19
PI3K, 114
PI3K/AKT signaling, 193
  drug resistance, 328
  inhibitors, 327
  monotherapy, 328
  tumorigenesis, 328
PIM kinases, 155
Pituitary Octamer UNC86 (POU), 447
Plant homeodomain (PHD)-domain, 384
Platelet derived growth factor receptor (PDGFR), 209
Platinum-based chemotherapy, 291
Pluripotent networks, 309
Pocket proteins, 303
Point mutations, 281, 283
Poly(ADP) ribose polymerase (PARP)
  and ATR inhibitors, 294
  clinical development, 289–291
  and platinum clinical trials, 288
Polycomb Group (PcG), 355
Polycomb repressive complex 2 (PRC2), 310, 424
Polycyclic aromatic hydrocarbons (PAHs), 36, 37
Polyethylene glycol 20 (PEG20), 195
Positron emission tomography (PET), 211, 212
Post translational modifications (PTMs), 381
Post-Operative Radiation Therapy Outcomes Score (PORTOS), 118
PPAR coactivator 1 alpha (PGC1α), 342
Precision medicine, 290, 294
  germline variation
    HSD3B1, 263, 264
    SLCO transport genes, 263
  prognostic and predictive biomarkers, 263
Primary PCa, 247, 248
Primary prostate tumors, 240, 283
Progesterone receptor (PR), 333
Programmed cell death 1 (PD-1), 127–130, 139

Programmed cell death ligands 1 (PD-L1), 139
Progression
  cell plasticity, 437
  CSCs, 437
  EMP, 438
  PCa therapies, 438
  treatment-induced cell plasticity, 438, 439
  tumor dormancy, 437
Progression free survival (PFS), 288
PRO-IMPACT trial, 116
Pro-inflammatory cytokines, 31, 33
Prolaris, 117
Prostate acid phosphatase (PAP), 131, 132
Prostate cancer (PC/PCa)
  adaptation, 151
  aggressive, 151
  cause of deaths, 149
  CD8+ T cells, 121–122
  challenges, genome analysis, 87, 88
  challenges, metabolism, 216, 217
  chronic inflammation, 59
  cohesive invasion, 151
  diet and lifestyle factors, 1
  ETS family, 57
  ETS gene fusions, 59
  genomic-based classification, 149
  genomics, 88
  germline variation and risk, 101
  Gleason score (GS), 149
  incidence, 1
  invasion, human PCa, 152
  lymphomas and leukemias, 57
  metastasis, 359, 360
  microenvironment contextual changes, 150
  non-ETS fusions, 57
  occurrence, 29
  peripheral zone, gland, 151
  progression, 351
  risk factors, 4, 29
  standard of care, 151
  TMPRSS2 and *ERG* genes, 61
  Wnt/β-catenin signaling, 359
Prostate Cancer Prevention Trial (PCPT), 2, 7
Prostate epithelial cells, 69
Prostate intraepithelial neoplasia (PIN), 321
Prostate specific antigen (PSA), 68, 79, 150, 151, 154, 174, 178, 241, 339, 394
Prostate specific homeobox gene, *NKX3.1*, 59
Prostate stem cell antigen (PSCA), 75, 133, 134
Prostate-derived ETS factor (PDEF), 411
Prostate-specific antigen (PSA), 4, 131, 134, 135, 138, 140, 206
Prostate-specific membrane antigen (PSMA), 133
Prostatic intraepithelial neoplasia (PIN), 77–79, 150, 397
PROSTVAC, 131, 138, 139
Protein Data Bank (PDB), 418
Protein Kinase A (PKA), 425
Protein kinase B (PKB), 319
Protein sex determining region of the Y chromosome (SRY), 444

474   Index

Protein-protein interactions, 427
Proteolysis targeting chimeras (PROTACs), 343, 398
Protocadherin (PC), 362
PSA screening
  on epidemiological studies, 4
PTEN mutations
  deletion-driven prostate cancer, 321
  LZTS2, 326
  PI3K activation, 320, 321
Pyruvate, 208
Pyruvate dehydrogenase complex (PDC), 189
Pyruvate kinase (PK), 189, 190

R
Radiation Sensitivity Index (RSI), 117
Radioimmunoassay (RIA), 240
Radiotherapy, 282
Rapid amplification of cDNA ends (RACE), 411
Rapid immunoprecipitation and mass spectrometry of
  endogenous proteins (RIME), 338
Ras/MAPK pathway, 424, 425
RB
  and AR, 305–307
  biomarker, 311, 312
  cancer immunity, 310, 311
  CDK4/6, 309
  and cell cycle, 303–305
  and chromatin modifiers, 309, 310
  cyclin D, 309
  deficient cells, 312
  deficient sub-clones, 301
  epigenetic role, 310
  family, 303
  function, 301
  N-and C-terminal domains, 301
  pathway, 301–303
  phosphorylation, 309
  pleiotropic functions, 301
  structure, 301
  tumor suppressor, 301, 309
RB/E2F
  non-cell cycle functions
    apoptosis, 307
    mammalian cells, 307
    metabolism, 308
    metastasis, 308
    senescence, 308
  represses transcription of pluripotent networks, 309
  transcriptional target, 305, 306
RB-PRC2, 310
RE1-silencing transcription factor (REST), 449, 450
Reactive oxygen species (ROS), 31, 32, 59, 60, 127, 204,
  280
Receptor activator of nuclear factor κ-B ligand
  (RANKL), 174, 175, 177
Receptor tyrosine kinases, 192
Recommended Phase II dose (RP2D), 392
Recurrent amplification, 96
Red wine, 35
Replication protein A (RPA), 281

Repressor element-1 silencing transcription factor
  (REST), 199
Residual prostate tumor androgens, 240–241
Residual tissue androgens, 241
Restore genome integrity, 280
Retinoblastoma-binding protein (RBP), 387
Reverse transcription polymerase chain reaction
  (RT-PCR), 411
Ribosomal S6 Kinase 1 (RSK1), 426
RNA polymerase II (RNA Pol II), 60
RNA recognition motif (RRM), 386
RNA sequencing, 282, 360
RNA-seq technologies
  gene expression, 111
  subtyping
    ADT-RS, 118
    capture genomic subtypes, 112, 114
    commercial classifiers, 116, 117
    definition, 112
    non-commercial classifiers, 117
    PAM50, 119
    PORTOS, 118
    RSI, 117
    supervised clustering, 115, 116
    unsupervised hierarchical clustering, 114, 115
Runt-related transcription factor 2 (RUNX2), 325

S
S-adenosyl-homocysteine (SAH), 196
S-adenosylmethionine (SAM), 196, 380, 382, 393
Sanger sequencing fueled genetic testing, 292
Sanger-based approaches, 87
Saturated fat, 32
Schwann cells, 158
Selective AR degraders (SARDs), 343
Selenium
  dietary intake, 14
  and prostate cancer risk, 14
  SELECT trial, 14, 15
  and vitamin E, 13, 14
Selenium and Vitamin E Cancer Prevention Trial
  (SELECT), 2, 7
17α-Hydroxypregnenolone, 246
17α-Hydroxyprogesterone, 245, 246
17β-Hydroxysteroid dehydrogenase type 3 (HSD17B3),
  244
Sex hormone binding globulin (SHBG), 335
Signal transduction, 200, 201
Single nucleotide variants (SNVs), 91, 293
Single-strand breaks (SSBs), 59, 280
Sirtuins, 381, 396, 397
Six transmembrane epithelial antigen of the prostate 1
  (STEAP1), 136
Small cell carcinoma, 301
Small cell lung cancer (SCLC), 355, 363
Smooth muscle
  cohesive invasion, 153
  glandular escape, 152
  in vitro and one *in vivo* model, 152
  mouse xenograft model, 153

Index 475

and neural invasion, 151
 prostatic capsule, 151
 stroma, 151, 152
 time-lapse microscopy, 152
SNP arrays, 282
SNP profiles, 285
Solute carrier organic anion (SLCO) genes, 263
Somatic defects, 288, 290
Somatic genome heterogeneity
 multifocality, 97–98
Sonic hedgehog (shh), 319
Soy intake, 11, 12
SPARTAN trial, 117
Sperm, 185
Splice variants, 249, 265
SPOP mutations
 BET proteins, 322
 mouse models, 323
 primary prostate cancer, 322
 PTEN, 321
Squalene epoxidase (SQLE), 248
Squalene synthetase, 248
SRp40, 249
SRSF1, 249
SRSF5, 249
Stand-Up-To-Cancer (SU2C)-Prostate Cancer Foundation
 (PCF) International Dream Team, 283
Stem-like model, CD8+ T cell exhaustion, 129–130
Steroid 5α-reductase type 2 (SRD5A2), 245
Steroid hormone receptor, 305
Steroid sulfatase (STS), 245, 250, 260, 261
Steroidogenesis
 abiraterone, 259–260
 AKR1C3, 262, 263
 androgen metabolizing enzymes, 248, 249
 CRPC, 248
 CYP11B1, 261, 262
 CYP17A1, 258–259
 HSD11B2, 261, 262
 intracrine (see Intracrine steroidogenesis)
 primary PCa, 247, 248
 steroid sulphatase, 260, 261
Steroidogenic acute regulatory protein (StAR), 242
Steroidogenic factor 1 (SF1, NR5A1), 252
Sterol regulatory element-binding protein (SREBP), 192
Stop gain, 282
Stromal cells, 252–254
Structural chromosomal aberrations, 282
Suberoylanilide hydroxamic acid (SAHA), 396
Sulfation, 247
Supervised clustering
 AR-activity, 115
 NEPC signature, 116
 prostate cancer subtypes 1-3, 115

T
Tammar wallaby, 246
Targeting DNA repair
 inhibitors of DDR proteins, 291–292
 platinum-based chemotherapy, 291

Targeting oncogenic ETS factors
 chemoresistance, 429
 degradation, 429, 430
 ERG inhibitory peptide, 427
 ERG targeting vaccine, 429
 PARP1 inhibitors, 429
 prostate cancer therapies, 427
 small molecule inhibitors, 427, 428
 tumor suppressive, 427
 undruggable, 427
Tartrate-resistant acid phosphatase (TRAP), 176
T-cell factors (TCFs), 358
T-cell receptor alternate reading frame protein (TARP),
 135, 140
Terminally differentiated cells, 440
Tertiary lymphoid structures (TLS), 126, 127
Testosterone (T), 240, 305, 333
Thrombocytopenia, 290
Tissue-based alterations, 239
TMPRSS2/ERG fusions, 90–93, 282, 409, 415
Tobacco, 15
Tomatoes, 10
TOPARP-A trial, 290
Topoisomerase II beta (TOP2B), 415
Topological stress, 58, 60
Topologically associated domains (TADs), 60, 380, 391
TRAMP murine model, 136–137
Transcription activation unit 5 (TAU5), 324
Transcription factors (TF)
 AR, 439
 bHLH, 442–443
 bZIP, 446, 447
 DNA binding domains, 439
 exogenous expression, 440
 families and members, 440
 forkhead domain, 446
 HD, 443–444
 hetero-domain, 447–449
 HMG, 444–445
 human embryonic fibroblasts, 440
 inhibition (see Inhibition, neural transcription)
 neural differentiation, 439
 progression, 441
 proteins, 439
 secondary protein structures, 439
 terminally differentiated cells, 440
 ZF, 440–442
Transcription Start Site (TSS), 379
Transcription-coupled NER (TC-NER), 280
Transcriptomics, 111, 112
Transforming growth factor beta-1 (TGF-β1), 173
Transketolase-like protein 1 (TKTL1), 191
Treatment-emergent NEPC, 284
Tricarboxylic acid (TCA) cycle, 185–190, 193, 194, 200,
 202, 208, 213, 215, 216
Trichostatin A (TSA), 396
Tumor dormancy, 437
Tumor genomic material, 294
Tumor heterogeneity, 414
Tumor microenvironment (TME), 124–128, 130
Tumor suppressive ETS factors, 417, 418

Tumor-associated antigens (TAA)
  MUC-1, 134
  NY-ESO-1, 134, 135
  PAP, 131, 132
  PSA, 131
  PSCA, 133, 134
  PSMA, 133
  SH3GLB2 peptide, 135
  STEAP1, 136
  TARP, 135
Tumorigenesis, 216

U
UDP-glucose 6-dehydrogenase (UGDH), 245
Unsupervised hierarchical clustering
  TCGA- 3 clusters, 114, 115
  tissue microarray, 115
Urokinase receptor (uPAR), 156, 158

V
Vaccines, 138
VCaP cells, 252
Vicious cycle, 173, 175
Vitamin D
  diet as a risk factor, human PC, 32, 33
  genetic variants, 10
  and risk, prostate cancer, 10
  source, 10
Vitamin E
  anti-carcinogenic actions, 13
  ATBC Study, 13
  fat-soluble compounds, 12
  PHS II, 13
  and prostate cancer risk, 13
  PSA screening, 13
  SELECT, 13
  VITAL study, 13
VITamins And Lifestyle (VITAL), 2, 3, 13
Von-Hippel-Lindau (VHL) E3 ubiquitin ligase, 343

W
Warburg effect, 188–190, 205, 207
Western dietary patterns, 5, 7
  dairy products and calcium, 8–9
  fat intake, 5
    ALA, 6
    and survival, 6
    long-chain n-3 fatty acids, 6
  fish intake and marine fatty acids, 7–8
  height, 5
  IGF-1, 5
  meat intake
    and survival, 7
    HCAs, 7
    meta-analysis, 7
    red and processed, 6

selenium, 14–15
soy and phytoestrogens, 11–12
tomatoes and lycopene, 10–11
vitamin D, 9–10
vitamin E and alpha-tocopherol, 12–14
Whole exome sequencing (WES), 293
Whole genome sequencing (WGS), 63, 283, 293
  multiple metastases, 283
Whole-exome sequencing, 360
Wine, 35
Winged helix-turn-helix (wHTH), 418
Wnt/β-catenin signaling, 426
  AKT, 326
  androgen deprivation, 365, 366
  AR signaling, 361
  cancer, 358, 359
  cancer stem cells, 363, 364
  carcinogenesis, 356
  cellular pools, 357
  dysregulation, 370
  embryonic development, 356
  EMT, 362
  FZD receptors, 357
  Hippo pathway, 367
  inhibition, 364, 365
  LRP5/6, 358
  mutations, 364
  NE differentiation, 353, 354, 361, 362
  PCa, 351, 359, 360
  PTEN, 366
  reactive stroma, 365
  target genes, 358
  therapy resistance, 360
  YAP/TAZ, 357, 366–368
Wound repair, 127

X
Xenograft LNCaP model, 251
Xenograft model, 153
Xenografts, 240, 250
X-ray crystallography, 418

Z
ZEN003694, 400
Zinc Finger (ZF) transcription factors
  AR, 442
  AR activity, 441
  AR pathway inhibition, 441, 442
  AR vs. cell plasticity, 441
  cell plasticity, 441
  DNA-binding domain, 441
  PCa, 442
  PR, 441
Zinc finger E-box-binding homeobox (ZEB), 448
Zinc transporters (ZIPs), 185–187
Zona fasciculata, 244
Zona reticularis, 244